International Scientific-Practical Conference "Agriculture and Food Security: Technology, Innovation, Markets, Human Resources" (FIES 2022)

BIO Web of Conferences Volume 52 (2022)

Kazan, Russia
26-28 May 2022

Editors:

Ayrat Valiev
Bulat Ziganshin
Farida Nezhmetdinova
Aksu Taylan
Rustam Nizamov

ISBN: 978-1-7138-6265-9

Printed from e-media with permission by:

Curran Associates, Inc.
57 Morehouse Lane
Red Hook, NY 12571

Some format issues inherent in the e-media version may also appear in this print version.

For additional information, please contact EDP Sciences – Web of Conferences
at the address below.

EDP Sciences – Web of Conferences
17, Avenue du Hoggar
Parc d'Activité de Courtabœuf
BP 112
F-91944 Les Ulis Cedex A
France

Phone: +33 (0) 1 69 18 75 75
Fax: +33 (0) 1 69 28 84 91

contact-edps@webofconferences.org

Additional copies of this publication are available from:

Curran Associates, Inc.
57 Morehouse Lane
Red Hook, NY 12571 USA
Phone: 845-758-0400
Fax: 845-758-2633
Email: curran@proceedings.com
Web: www.proceedings.com

TABLE OF CONTENTS

IV International scientific-practical conference
"Agriculture and Food Security:
Technology, Innovation, Markets, Human Resources" dedicated to the 100th anniversary of Kazan State Agrarian University

May 26–28, 2022, Kazan, Samara, and Ulyanovsk, Russian Federation

From 26 to 28 May, IV International scientific and practical conference "Agriculture and food security: technologies, innovations, markets, human resources", dedicated to the 100th anniversary of Kazan State Agrarian University, was held at Kazan State Agrarian University, Kazan, Russian Federation and at Samara State Agrarian University, Samara, Russian Federation and at Ulyanovsk State Agrarian University, Ulyanovsk, Russian Federation.

Organizing Institutions

- Kazan State Agrarian University
- Kazan State Academy of Veterinary Medicine named after N. E. Bauman
- Kazan State Power Engineering University
- Samara State Agrarian University
- Ulyanovsk State Agrarian University named after P. A. Stolypin

In Cooperation with

- Ministry of Agriculture of the Russian Federation
- Ministry of Agriculture and Food of the Republic of Tatarstan
- TatNIISH FRC KazSC RAS
- Ministry of Agriculture and Food of the Samara Region
- Ministry of Agro-Industrial Complex and Rural Development of the Ulyanovsk Region
- Ministry of Education and Science of the Samara Region
- Ministry of Education and Science of the Republic of Tatarstan

Associate International Partners:

- FAO/Food and Agriculture Organization of the United Nations
- Yüzüncü Yıl Universty (Turkey)
- LOGO (Germany)
- AMAZONEN-WERKE H (Germany)

Programme Committee

Chairman

Rector of the Kazan State Agrarian University, Doctor of Technical Sciences, Associate Professorб
Valiyev Ayrat Rasimowich

Viktor ***Buksman*** *– PhD, regional representative for the CIS, adviser to DreyerGmbH&Co. KG, Germany*

Kurbanov Eldar Alikramovich *– Doctor of Agricultural Sciences, Professor of the Volga State Technological University, Yoshkar-Ola, Russian Federation*

Mazitov Nazib Kayumovich *– Doctor of Agricultural Sciences, Professor, Corresponding Member of the Russian Academy of Sciences, Russian Federation*

Tino Hochmuth *– Germany*

Shogenov Yury Khasanovich *– Doctor of Technical Sciences, Professor, Head of the Sector of Electrification and Automation of the Department of Mechanization, Electrification and Automation of the Federal State Budgetary Institution of the Russian Academy of Sciences, Corresponding Member of RAS*

Tsoi Yuri Alekseevich *– Doctor of Technical Sciences, Professor, Corresponding Member of the Russian Academy of Sciences, Gen. Director of LLC NPO Femax*

Sakhapov Rustem Lukmanovich *– Doctor of Technical Sciences, Professor Kazan State University of Architecture and Civil Engineering, Corresponding Member of AS RT*

Vyshtakalyuk Alexandra Borisovna *– Doctor of Biological Sciences, Deputy Head of the Laboratory of Chemical and Biological Research, IOPC named after A.E. Arbuzov – a separate subdivision of the FRC KazSC RAS, Russian Federation*

Rukol Vasily Mikhailovich *– Doctor of Veterinary Sciences, Professor of the Department of General, Private and Operative Surgery, EE "Vitebsk Order of the Badge of Honor" State Academy of Veterinary Medicine, Republic of Belarus*

Dnekeshev Amanzhol Kusainovich *– Associate Professor of the Department of Non-communicable Diseases and Morphology, West Kazakhstan Agrarian Technical University. Zhangir Khana, Republic of Kazakhstan*

Lugovtsev Vladimir Yuryevich *– Candidate of Veterinary Sciences, Researcher at the Center for the Evaluation and Study of Biologicals of the Food and Drug Administration (USA)*

Akhmetova Irina Gareevna *– Doctor of Technical Sciences, Associate Professor, Vice-Rector for Development and Innovation of Kazan State Power Engineering University*

Conference Secretariat

Executive Secretary – ***Nezhmetdinova Farida Tansykovna*** *– Candidate of Philosophy, Head of the Department of Philosophy and Law, Kazan State Agrarian University, Kazan*

Scientific secretary of the conference – ***Kalimullin Marat Nazipovich*** *– Doctor of Technical Sciences, Professor of the Kazan State Agrarian University, Kazan*

Scientific secretary of the Section "Actual issues of veterinary medicine and zootechnics, biotechnology of production and processing of agricultural products" – ***Shagivaliev Lenar Rifatovich*** *– Candidate of Economic Sciences, Head of the Scientific Information Department of the Kazan State Aviation Administration, Kazan*

Statement of Peer review

In submitting conference proceedings to *Web of Conferences*, the editors of the proceedings certify to the Publisher that

1. They adhere to its **Policy on Publishing Integrity** in order to safeguard good scientific practice in publishing.
2. All articles have been subjected to peer review administered by the proceedings editors.
3. Reviews have been conducted by expert referees, who have been requested to provide unbiased and constructive comments aimed, whenever possible, at improving the work.
4. Proceedings editors have taken all reasonable steps to ensure the quality of the materials they publish and their decision to accept or reject a paper for publication has been based only on the merits of the work and the relevance to the journal.

Title, date and place of the conference

International scientific-practical conference «Agriculture and food security: technology, innovation, markets, human resources» (FIES 2022)

May 26-28, 2022 Kazan, Russia

Proceedings editor(s):

Ayrat Valiev
Bulat Ziganshin
Farida Nezhmetdinova
Aksu Taylan
Rustam Nizamov

Date and editor's signature

September 02, 2022

Use of fat-soluble vitamins and feed additives in cow diets

V. V. Golovey*, *G. N.* Vyayzenen, and *A. V.* Golovey

Novgorod State University named after Yaroslav the Wise, Institute of Agriculture and Natural Resources, 7, Sovetskoi Armii Ave., Veliky Novgorod, 173000, Russian Federation

Abstract. The authors have created modern effective methods for controlling the vitamin content of highly productive lactating cows during the first period of lactation. The results were obtained on the content of vitamins in the diet, milk and blood when feeding with supplements Vitaminol and phytocomplex at 10, 15 and 20 g per animal unit daily separately and with mixed feed. The authors monitored the levels of vitamins A, carotene, D and E in cow's milk and serum. Scientific research was carried out for the agricultural enterprise "Novgorodsky Becon" of the Novgorod region. The cows ration was balanced by 50 nutrients and biologically active substances in accordance with the recommendations for animal feeding (Moscow, Russian Academy of Agricultural Sciences, 2003). Studies of the content of vitamins were carried out using the silage-concentration type of ration fed to lactating black-and-white cows in conditions without grazing. We conducted 2 scientific experiments and 2 physiological experiments with cows during the first 100 days of their lactation. The increase in the level of vitamin A in milk and blood was due to the use of feed additives in the diets. The use of feed additives increased the vitamin D levels in milk and serum. The balance of calcium and phosphorus metabolism in cows is positive. The use of "Vitaminol" together with compound feeds increased the level of calcium by 5.6–6.3 and phosphorus by 4.4–4.9 grams per 1 kg of milk with the natural fat content. The diets were admixed with the "Vitaminol" and phytocomplex ratio of Ca: P (1.3: 1) and Na: K (0.2: 1). With the same doses of feed additives, the ratio of acid and alkaline elements in the diets was the same: Vitaminol was 0.743 and phytocomplex was 0.709. Due to the use of "Vitaminol", the concentration of vitamin E in milk is 4.6-6 times higher than that in the control group, in blood and blood serum.

1 Introduction

An increase in the production of vitamin feeds, feed additives (preparations), an increase in their quality, stabilization of vitamins in plant-based feed are a means of increasing the usefulness of feeding highly productive lactating cows [1, 2, 3 ,4].

It is known that the assimilation of vitamins depends on many factors: the ratio of vitamins to each other in diets, milk, the balance of diets in terms of minerals and amino acids [5, 6].

In the available literature there are very few modern experimental data on the problem of increasing the vitamin value of feed, rations, milk and blood of cows. The studies related to solving the problem of increasing the vitamin nutritional value of specific zonal typical rations and types of feeding of dairy cows in a year-round stable method of keeping are rare in the zootechnical literature, veterinary medicine and biology. When searching a solution to the problem of increasing the milk productivity of cows, many researchers have deviated from studying one of the main tasks of modern domestic and foreign science about animal feeding. In particular, they did not consider the influence of typical diets and types of feeding of dairy cattle on the degree of vitamin sufficiency or deficiency, vitamins in their body and milk. The quality of life of people largely depends on the vitamin value of milk [7, 8].

2 Materials and methods

The composition and structure of diets affect the absorption of vitamins by the body of highly productive lactating cows during the milking period; therefore, it is advisable to show the interaction of vitamins with crude protein, sugars, crude fiber, crude fat, metabolic energy and a number of minerals and amino acids.

In this regard, it was necessary to study the influence of some factors on the dynamics of the content of vitamins in the milk of high-yielding lactating black-and-white cows during the milking period.

The studies were carried out in Novgorodsky Becon LLC, Novgorod Region, the milk production of cows for the previous lactation was 6000 kg. The rations for feeding cows are balanced by 50 indicators of nutrients and biologically active substances in accordance with the norms of feeding farm animals (M., RAAS, 2003).

Two scientific, economic and two physiological (balance) experiments were carried out. The diets of the cows of the first scientific and economic experiment included an additional feed additive (preparation)

* Corresponding author: valentin-golovej@yandex.ru

 https://doi.org/10.1051/bioconf/20225200001

"Vitaminol", and the second scientific and economic experiment – a feed additive in the form of a phytocomplex.

For each scientific and economic experiment, 4 groups of cows, 10 animal units in each, were formed. The animals of the first experimental group, in addition to the main diet (RR), were given 10 g of "Vitaminol", the second experimental had 15 g, the third experimental group had 20 g per animal unit per day. For the cows of the control group, the drug was not included in the diets.

Similar doses of feeding the phytocomplex were made for cows of the experimental groups in the second scientific and economic experiment. For carrying out physiological experiments, 3 animal units with cows were selected according to the principle of paired analogs, taking into account the origin, breed, productivity and live weight.

The research methodology is based on scientific principles set forth in the works of domestic and foreign researchers on vitamin nutrition of animals. Zootechnical, physiological, biochemical and biometric research methods were used.

The purpose of the research is to study the effect of feeding cows during the milking period with feed additives "Vitaminol" and phytocomplex on the vitamin nutritional value of milk (A, D, E).

3 Results and discussion

The reasoning is as follows. During the milking period of cows, silage – 21.7%, haylage – 28.9% and compound feed – 36.2% were in the structure of rations in terms of nutritional value with a daily milk yield of 26-29 kg when using "Vitaminol" (Table 1). And when adding different doses of the phytocomplex to the diet, its structure consisted of silage – 20.0%, haylage – 32.8% and compound feed – 31.3%. The provision of the body of cows and milk with basic nutrients and biologically active substances, in particular, vitamins, was achieved through the inclusion of hay, silage, haylage, compound feed and feed additives in the diets [9].

Table 1. The composition of the diets of dairy cows in winter, % by nutritional value.

Kind of feed	Daily milk yield with natural content of fat during first part of lactation, during experimental period, kg	
	26 -29	26-29
	Main ration (MR) plus "Vitaminol"	Main ration (MR) plus phytocomplex
Hay from perennial grasses	9.6	11.1
Silage from mixed grasses	21.7	20.0
Wilted silage from mixed grasses	28.9	32.8
Concentrates (mixed feeds)	36.2	31.3
Sugar beet molasses	3.6	4.8
Total	100	100

The diets of cows using different doses of "Vitaminol" and the phyto-complex (separately) contained the following nutrients (Table 2).

Table 2. Some data on the composition of diets.

Data	Name of feed additive	
	"Vitaminol"	phytocomplex
Crude protein, % of DM	15.6	15.5
Fat, % of DM	3.3	3.3
Crude fibre, % of DM	27.4	28.7
Sugar, % of DM	8.3	9.0
Sugar-protein relation	0.82:1	0.90:1
Relation of acids elements to alkaline elements	0.743	0.709
Digestible protein in 1 feeding unit	111.2	115.9
Feeding unit in 1 kg of DM	0.86	0.84
Content of ME in 1 kg of DM	9.02	8.67
Carotene, mg	1838	1613
Vitamin E, mg	6470	6011

The existing structure of rations corresponds to the type of feeding and fodder production, depending on the prevailing conditions of the farm, its applicability and reliability.

According to experimental studies [10, 11], it was found that when including in the rations of feeds rich in protein, fiber, sugar, carotene and minerals, there was an unequal content of vitamins in cow's milk. The content of fat-soluble vitamins in cow's milk largely depends on: breed, lactation phase, diet structure, type of feeding, physiological state, season of the year, keeping conditions, quality and chemical composition of voluminous plant feed, shelf life, harvesting technology and their processing [12].

The content of fat-soluble vitamins in the milk of cows was revealed with the silage-haylage-concentrate type of feeding under the conditions of a year-round stall-keeping method.

3.1. Methods for monitoring vitamin A supply

When adding different doses of the Vitaminol feed additive to the diets of cows, the vitamin A content ranged from 0.09 ± 0.002 to 0.24 ± 0.07 mg per 100 g of milk versus 0.06 ± 0.01 mg in cows of the control group (table 3). When feeding it to cows in the ratio of 10 g per animal units per day compared with the control, it was possible to note an increase in the content of this vitamin in milk by 50%. P <0.001, and 15 g/animal units/day was 3 times, while when using the maximum doses of the drug there was 20 g/animal units/day, this indicator increased 4 times.

When using different doses of the phytocomplex in the diets (out of 13 medicinal feed crops), the vitamin A content in cow's milk ranged from 0.08 ± 0.003 to 0.18 ± 0.03 mg (P <0.001) versus 0.07 ± 0.001 mg in peers of the control group (Table 3).

3.2. Methods for monitoring vitamin D supply

Vitamin D is involved in the regulation of calcium-phosphorus metabolism. Direct control methods include: accounting for the content of calcium, phosphorus in the diets, their ratio, the ratio of acid gram-equivalents to basic gram-equivalents, the structure of the diet, type of diet, type of feeding, conditions of detention, control of the content of basic nutrients and biologically active substances in the diet.

The inclusion of different doses of "Vitaminol" in mixed foods with feeds increased the content of vitamin D in milk from 0.29 ± 0.09 to 0.41 ± 0.03 mg versus 0.18 ± 0.01 mg in the control (see table 3). Its maximum content in milk was found in cows of the experimental group when this preparation was included in the diet of 20 g/animal units/day, which is 2.3 times higher than that of peers in the control.

The existing structure of rations for feeding highly productive cows with the use of domestic feed additives corresponded to an improvement in the absorption of vitamin D in the body and its accumulation in milk, which is confirmed by the results of experimental studies.

When a phytocomplex of 10 g/ animal units/day was included in the diet, milk contained 0.28 ± 0.06 mg of vitamin D, which is 1.5 times higher than the control level, and 15 g/animal units/day – 0.31 ± 0.09 mg (1.6 times higher) and, finally, 20 g/animal units/day – 0.40 ± 0.09 mg (2.1 times higher).

It should be kept in mind that in the diets of cows with the addition of "vitaminol" and a phytocomplex, the Ca: P ratio was 1.3: 1 and the Na: K ratio was 0.2: 1. With the same doses of additives, but not the same in the structure, chemical composition and properties, close relationships between these mineral macronutrients in diets are obtained. And the ratio of acid gram equivalent to the main gram equivalents in the diets of cows during milking when using different doses of Vitaminol was 0.743, and the phytocomplex was 0.709.

As a result, in the first case (Vitaminol) with milk, 35.14 ± 0.35 (P <0.001) to 39.03 ± 0.18 g of calcium (P <0.001) were excreted versus 33.67 ± 0, 20 g in the peers of the control group. Due to the increase in the milk production level of cows, the degree of calcium deposition in the body decreased from 37.02 to 29.5 g against 37.9 g in animals in the control.

In the second case (phytocomplex), a similar trend was observed in the use of calcium by cows.

As for the phosphorus metabolism in the body of cows, when using Vitaminol with milk, from 21.52 ± 0.32 (P <0.001) to 26.51 ± 0.39 g (P <0.001) were excreted, while peers of the control group were 19.35 ± 0.24 g. The phosphorus retention in the body is positive; however, it decreased from 36.40 to 23.7 g, depending on the increase in milk production of cows.

When various doses of the phytocomplex are included in the diet, phosphorus is excreted per day from 24.66 ± 0.20 (P <0.001) to 25.63 ± 0.12 g (P <0.001), while in animals of the control group this indicator is 22.56 ± 0.18 g. In cows of the control and experimental groups, the phosphorus balance is positive.

When feeding cows with Vitaminol (with mixed feed) per 1 kg of milk, the fat content accounted for 5.6 - 6.3 g of calcium and 4.4 - 4.9 g of phosphorus, and the phytocomplex amounted to 4.9 - 5 for calcium, 6 g for phosphorus and others to 3.8 - 4.4 g.

3.3. Methods for monitoring vitamin E supply

Most often, the lack of vitamin E in the body is manifested, especially in high-yielding cows, during the dry period of pregnancy, as well as during the period of milking in conditions of year-round stall keeping. Vitamin E (tocopherol) has antioxidant properties (reproduction vitamin). In animals provided with vitamin E, vitamin A is better absorbed and accumulates in milk and blood. It is believed that cows on diets low in vitamin E have an increased need for vitamin A and ascorbic acid [13]. In this regard, it is advisable to control the activity of these vitamins in finished food products.

When adding different doses of "Vitaminol" to the diets, an increase in the content of vitamin E in cow's milk was established during milking (see table 3). So, the use of this drug in a mixture with animal feed contributed to an increase in the activity of vitamin E in milk from 4.6 to 6.0 times in comparison with the control.

The ratio of vitamin E to vitamin A in milk can serve as a sensitive, effective test of high availability, vitamin

E sufficiency in the body of cows. When using the Vitaminol preparation at a dose of 10 g/animal units/day, the ratio of vitamin E: vitamin A in cow's milk was 3.5: 1, and at a dose of 15 g/animal units/day, it was 1.9: 1 and at a dose of 20 g/animal units/day was 1.7: 1 versus 1.2: 1 in the control. As for this indicator in blood serum, it had the following form: at a dose of 10 g/animal units/day there is 31.5: 1, a dose of 15 g/animal units/day gives 22.8: 1 and a dose of 20 g/animal units/day is 20.6: 1.

When different doses of the vitamin E phytocomplex were included in diets, milk contained it from 0.30 (P <0.001) to 0.39 ± 0.004 mg versus 0.08 ± 0.001 mg in the control. In the case of applying the minimum amount of additive in the milk of cows of the experimental group, vitamin E is noted 3.7 times more than that in milk of the peers of the control group, and the maximum amount is 4.9 times.

Table 3. Content of vitamins in cow's milk in the first part of lactation, in 100 ml.

Experiment	Cows group	Vitamin A	Vitamin D	Vitamin E
1	Control group - MR	0.06 ± 0.001	0.18 ± 0.01	0.07 ± 0.002
	1-st experimental group (MR + "Vitaminol", 10 g per animal unit daily	***	0.29 ± 0.09	*
		0.09 ± 0.002		0.32 ± 0.11
	2-nd experimental group (MR + "Vitaminol", 15 g per animal unit daily	*	0.34 ± 0.17	*
		0.18 ± 0.05		0.35 ± 0.12
	3-rd experimental group (MR + "Vitaminol", 20 g per animal unit daily	*	***	0.42 ± 0.14
		0.24 ± 0.07	0.41 ± 0.03	
2	Control group - MR	0.07 ± 0.001	0.19 ± 0.25	0.08 ± 0.001
	1-st experimental group (MR + phytocomplex, 10 g per animal unit daily	0.08± 0.003	0.28 ± 0.06	***
				0.30 ± 0.001
	2-nd experimental group (MR + phytocomplex, 15 g per animal unit daily	***	0.31 ± 0.09	***
		0.15 ± 0.02		0.32 ± 0.001
	3-rd experimental group (MR + phytocomplex, 20 g per animal unit daily	***	0.40 ± 0.09	***
		0.18 ± 0.03		0.3 9 ± 0.004

*P<0,05; *** P<0,001

4 Conclusion

On the basis of experimental data, modern, effective methods of monitoring the supply of fat-soluble vitamins in highly productive lactating cows during milk production with a silage-hay-concentrate type of feeding in a year-round stall method of keeping are presented.

The activity of the vitamins under study was noted not only in their content in diets, but also in milk. At the same doses (10, 15 and 20 g/animal unit/day), using different types of feed additives (Vitaminol and phytocomplex), but not identical in structure, chemical composition and properties, close ratios of Ca:P and Na:K were obtained in diets. The dosed inclusion of feed additives in the composition of rations contributed to an increase not only in the average daily milk yield, but also in the content of fat-soluble vitamins (A, D, and E) in milk.

To improve the vitamin value of milk, metabolic processes in the body, it is necessary to use such feed additives as "Vitaminol" and phytocomplex in a mixture with compound feed at a dose of 20 g per animal unit per day.

References

1. N. Emelina, V. Krilova, E. Petukhova, N Bromley, *Vitamins in the feeding agricultural animals and birds* (Moscow: Kolos, 1970).
2. G. Tverdokhleb, G. Sazhinov, R. Ramanauskas, *Milk and milk products technology* (Moscow: DeLi print, 2006).
3. L. Romanenko, V. Volgin, *Agricultural biology*, **4,** 20–27, (2007).
4. L. Romanenko, V. Volgin, Main zooingeneer **2**, 22–25, (2008).

5. L. Romanenko, Gerald of Russian academy of agricultural sciences **5**, 88–90, (2008).
6. V. Kuznetsov, G. Shiler *Pocket-book about milk technology. Technology and content* (Cheeses – SPb: GIORD, 2005)
7. G. Vyayzenen, N. Ivanova, *Saving and proceeding of agricultural raw material* **10**, 62–66 (2011)
8. P. J. Gorden, M. D. Kleinhenz, R. Warner, P. K. Sidhu, *J. Dairy Sci.*, **102**, 11465–11469 (2019)
9. R. Rauch, J. Martín-Tereso, J.-B. Daniel, J. Dijkstra, *J. Dairy Sci.*, **104**, 20211–20219 (2021)
10. M. K. Connelly, L. L. Hernandez, *JDS Communications*, **2**, 0108 (2021)
11. J. Karlsson, M. Lindberg, M. Åkerlind, K. Holtenius, *Livestock Science* **242**, 8922–8937 (2020).
12. K. L. Swanson, H. M. Bill, J. Asmus, J. M. Heguy, E. J. DePeters, *J. Dairy Sci.* **104**, 8846–8856 (2020)
13. C. B. M. Müller, B. Kuhla, *Science of the Total Environment* **788**, 147813 (2021)

Water-soluble vitamins in the feed of high-yield lactating cows during the milking period

V. V. Golovey, *G. N.* Vyayzenen, and *A. V.* Golovey

Novgorod State University named after Yaroslav the Wise, Institute of Agriculture and Natural Resources, 7, Sovetskoi Armii Ave., Veliky Novgorod, 173000, Russian Federation

Abstract. The aim of the work is to show how to increase the productivity of cows, to improve the feeding system, to introduce intensive, promising, effective, resource-saving milk production technologies, including the implementation of measures for the production and use of domestic vitamin feed products (preparations). The studies were carried out by the example of highly productive dairy cows with an annual productivity of over 6000 kg of milk per lactation. The type of feeding is the silage – hayage silage concentrate, the rations are balanced in terms of nutrients, biologically active substances and metabolic energy. It corresponded to the generally accepted norms of animal feeding in the country (M., RAAS, 2003). The diets of cows during the milking period included Vitaminol and a phytocomplex (of 13 medicinal forage herbs), 10 g, 15 g and 20 g per head per day. When milking cows (100 days), fed with Vitaminol, more milk was produced ranging from 2.7 to 9.9 %, the phytocomplex increased milk yield from 8.4 to 16.3 %. The consumption of metabolic energy of diets per 1 kg of milk is lower in cows of the experimental groups when using different doses of Vitaminol ranging from 2.8 to 9.9 %, and the phytocomplex allowed ranging it from 8.3 to 16.1 % compared to the control variant. To increase metabolic processes in the body, productivity and supply of water-soluble vitamins and mineral macronutrients in cows under the conditions of a year-round stable method of keeping, the rations included the optimal dose of Vitaminol and the phytocomplex (separately) amounting to 20 g per head a day. Vitaminol serves to replenish the body of highly productive lactating cows during the period of milk production with missing biologically active substances, vitamin nutritional value of milk, increase the digestibility of organic substances in rations, and improve metabolic processes.

1 Introduction

Over the past 15–20 years, more and more attention has been paid to the distribution of a number of vitamin preparations in dairy farming. It allows not only preventing vitamin deficiency diseases, but also increasing the productivity and vitamin value of cow milk [1, 2, 3], in particular with a year-round stall method of keeping [3, 4, 5].

At present, the possibility of using domestic, more effective, natural plant feed additives with the inclusion of wild crops in the diets of cows during the period of their milking has not been sufficiently substantiated [6, 7].

The main role in improving the vitamin nutrition of cows is played by natural sources of water-soluble vitamins (in particular, groups B, C): green fodder (feeding green mass) of pastures, silage, haylage, hay, grass flour [8, 9, 10]. However, in many regions of the country and abroad, dairy cattle are not grazed [11, 12]. Green fodder is delivered to the animals directly to the feeding trough [13, 14].

In this regard, it is necessary to make wider use of domestic feed additives Vitaminol and phytocomplex. Vitaminol contains a complex of water-soluble B vitamins, vitamin C, vitamin K and a number of dispersible forms of fat-soluble vitamins A, D_3 and E. Vitaminol is a white to cream powder made in accordance with GOST 35/095 - 97. The purpose of the phytocomplex is to significantly improve the taste and smell of freshly milked fresh milk due to the content in its composition of a range of beneficial essential oils and many healing (fodder) substances in the diet [15].

The phytocomplex included the dried medicinal fodder crops at the rate of 100 kg (flour). Milfoil was 2 kg (2.0 %), perennial timothy was 2 kg (2.0 %), orach was 3 kg (3.0 %), tansy was 4 kg (4.0 %), blueberries without berries were 4 kg (4.0 %), white sweet clover was 4 kg (4.0 %). White clover was 5 kg (5.0 %), pink clover was 5 kg (5.0 %), red clover, meadow were 5 kg (5.0%), goatweed was 6 kg (6.0%), big-sting nettle was 15 kg (15.0 %), fireweed was 15 kg (15.0 %) and alfalfa purple was 30 kg (30.0 % by mass). In total there was 100 %.

An increase in the production of vitamin complexes, medicinal herbs and supplements, an increase in their assortment, quality and safety in different seasons of the calendar year are a means of increasing the usefulness of feeding highly productive cows.

The aim of the study is to determine the effect of feeding Vitaminol and the phytocomplex to cows during the period of milk production on the productivity and content of water-soluble vitamins in milk. The task was to study the use of various doses of the domestic vitamin complex Vitaminol and phytocomplex (of 13 medicinal herbs) on the productivity and vitamin value of milk.

2 Materials and methods

The studies were carried out by the example of lactating black-and-white cows on the basis of an experimental dairy farm of OOO Novgorodsky Bacon. The groups of cows were formed by the method of analogue groups, taking into account the age, live weight, productivity for the previous lactation of 6.5 thousand kg of milk, origin, pedigree and timing of insemination. To solve the set tasks, two scientific, economic and physiological experiments were carried out in accordance with generally accepted methods. Scientific and economic experiments consisted of periods: preparatory – 10 days, accounting – 100 days.

The accounting of actually consumed feed and feed additives was carried out daily. The rations were formulated taking into account the chemical composition and nutritional value of the feed based on detailed norms recommended by the Russian Academy of Agricultural Sciences and the Russian Academy of Sciences (2003). The dairy productivity of cows was monitored on a daily basis. 10 days before the end of the scientific and economic experiment, a physiological experiment was carried out to determine the mass of excretory products (faeces, urine and milk) by weighing them on a scale, the digestibility of nutrients in rations, the use and assimilation of nitrogen, calcium and phosphorus by the body.

The number of cows in scientific and economic experiments was 10 heads in each of the groups, physiological experiments were 3 heads. The cows were fed on a leash in stalls three times a day in individual feeders. The faeces were stored in desiccators with lids, and the urine was stored in bottles containing 2 - 4 thymol crystals. Waste products were weighed once a day. A 10 % average sample was taken from the daily amount for preservation with a 10 % HCl solution. The content of nutrients in biological samples was determined in accordance with generally accepted methods, minerals - by the method of atomic absorption spectrophotometry (Perkin Elmer USA, Analyst 400) in the central chemical laboratory of PJSC Akron, Novgorod region.

The content of water-soluble vitamins in bioassays was determined using a Kapel device at the Epidemiological Center of the Russian Federal Service for Supervision of Consumer Rights Protection and Human Welfare in the Novgorod Region.

3 Results and discussion

3.1. Use of feed additives

When feeding lactating cows during the milking period, the rations included bulky plant foods: hay produced from cereal perennial grasses, silage from cereal and leguminous grasses, haylage from cereal and leguminous perennial grasses. The feed additive Vitaminol was added to the diets of cows (in the first scientific and economic experiment), and in the second scientific and economic experiment, the phytocomplex, respectively, 10 g, 15 g and 20 g per head per day (depending on each experimental group). Animals of the control and all experimental groups were given a mineral supplement in the form of sodium chloride and feed precipitate. It contained 23.980 kg of the dry substance, 15.6 % of crude protein, 3.3 % of crude fat and 27.4 % of crude fiber from dry substance, 111.2 g of digestible protein per 1 ECU, 8.3 % of sugar from dry substance, 21.63 ECU and 216.3 MJ of metabolizable energy.

The main diet of cows with the addition of a phytocomplex in doses of 10 g, 15 g and 20 g per head a day in a mixture with a concentrate feed also included hay, silage, molasses (molasses), table salt, and fodder precipitate. The diets of the cows of the first experimental group, the second experimental and the third experimental group contained 21.368 kg of dry matter, 15.5 % of crude protein, 28.7 % of crude fiber, 3.3 % of fat, 9.0% of sugar from dry matter, 115.9 g of digestible protein per 1 ECU, 18.53 ECU and 185.3 MJ of metabolic energy. The cows of the control group were not fed the fodder supplement in the form of a phytocomplex.

Medicinal fodder crops were collected in the tracts of the Novgorod district of the Novgorod region.

When added to the diet, for example, 10 g/head/day of Vitaminol, the average daily milk yield of natural fat cows was 27.24 ± 1.85 kg, which is 2.7 % higher than that of the control group. And in the case of an increase in the dose to 15 g/head/day, the average daily milk yield of cows in the second experimental group increased to 28.27 ± 1.85 kg, which is 6.6 % higher than that in the animals of the control group. The inclusion in the diet of cows of the third experimental group of 20 g of this feed additive per head per day contributed to an increase in milk yield of natural fat content to 29.13 ± 1.15 kg, which is 9.9 % ($P>0.05$) higher than the control level.

The use of a number of nutrients and biologically active substances of medicinal fodder and other crops as part of the phytocomplex against the background of the main diet contributed to an increase in the average daily milk yield of natural fat content in cows of the first experimental group. Thus, it was 28.11 ± 1.16 kg (8.4 %), the second experimental group had 28.67 ± 1.24 kg (10.6 %) and the third experimental group had 30.15 ± 1.82 kg (16.2 % of the control). This indicator in cows of the control group was 25.93 ± 1.80 kg.

Dosed use of Vitaminol as a source of water-soluble and a number of fat-soluble vitamins (A, D and E) in the composition of diets positively influenced not only the level of productivity, but also corrected the ratio of Ca: P in the diet, milk and during their retention (accumulation) in the body due to the improvement of this indicator among the main chemical minerals in the optimal nutrition of cows, the quality, composition and

 https://doi.org/10.1051/bioconf/20225200002

taste of milk in the conditions of a year-round stall method of keeping (Table 1).

Table 1. Influence of feeding various feed additives to dairy cows during the milking period on the dynamics of the Ca:P ratio in the diet, milk and during retention (accumulation) in the body.

Group of cows	When using Vitaminol, the ratio of CA:P			When using the phytocomplex, the CA:P ratio		
	in the diet	in the milk	their retention in the body	in the diet	in the milk	their retention in the body
Control (OR)	1.04:1	1.74:1	0.87:1	1.09:1	1.47:1	0.98:1
Experimental I (OR + 10 g/head/day)	1.04:1	1.63:1	1.01:1	1.09:1	1.46:1	1.23:1
Experimental II (OR + 15 g/head/day)	1.04:1	1.49:1	1.19:1	1.09:1	1.56:1	0.81:1
Experimental III (OR + 20 g/head/day)	1.04:1	1.47:1	1.25:1	1.09:1	1.73:1	0.93:1

The use of Vitaminol and phytocomplex with high antioxidant and sorption properties in the diets simultaneously allowed more fully realizing the level of productivity. It was manifested in an increase in the average daily milk yield of natural fat milk in the first case from 2.7 to 9.9 %, in the second – from 8.4 to 16.2 % in comparison with the control variant. There was also the absence of salt poisoning and gastrointestinal tract disorders, impairment of the functional and excretory systems of the kidneys. The use of nutrients and biologically active substances in the herbs of the phytocomplex in the composition of the rations for feeding cows is given in Table 2.

A relatively high concentration of iodine was noted in the composition of stinging nettle (0.16 mg/kg), fireweed (willow weed, 0.16 mg/kg), pink clover (flowers, 0.20 mg/kg), meadow clover (0.14 mg/kg) and orach (0.25 mg/kg). As for the content of ferrous iron, its concentration increased with the use of the following medicinal herbs: St. John's wort (248 mg/kg) > meadow timothy (333 mg/kg) > blue lucerne (367 mg/kg) > stinging nettle (375 mg/kg).

A high concentration of copper was found in the following herbs: orach (7.95 mg/kg), meadow clover (8.40 mg/kg), goat weed (8.51 mg/kg), pink clover (8.82 mg/kg), milfoil (9.14 mg/kg) and tansy (10.4 mg/kg).

Natural plant resources at the regional level are suitable for enriching the body of highly productive lactating cows with mineral microelements (J, Fe, Cu, Mn, Zn). The leaves and stems of blueberries contain 559 mg / kg of manganese, which is 3-15 times more than that in other studied green herbs. Also, the types of medicinal herbs in the phytocomplex of antioxidant and sorption action had a positive effect not only on the level of productivity, but also increased the body's resistance to stressful situations, defences for the functioning of the nervous system, resistance to overload, the condition of the skin, skeleton, blood vessels, ligaments and joints.

3.2. The influence of the use of Vitaminol and the phytocomplex in the diets of cows during the milking period on the content of water-soluble vitamins in milk.

To increase the productivity and vitamin value of milk in lactating cows during the milking period, especially in the conditions of a year-round stable method of keeping, it is necessary to take into account all possible sources on which the supply of the body with vitamins depends.

Improvement of metabolic processes in the body of cows throughout the year-round stable method of keeping is closely related to the balance of daily rations for nutrients and biologically active substances and metabolic and energy, and with the use of domestic highly effective feed additives in animal feeding. In view of the fact that these domestic feed additives are currently insufficiently studied, especially in feeding highly productive lactating cows to the discord phase, with a year-round stall (tethered) method of keeping, solving the problem of increasing milk production is of extremely important economic importance.

3.3. The influence of the use of Vitaminol and the photo complex in the diets of cows during the milking period on the content of water-soluble vitamins in milk.

To increase the productivity and vitamin value of milk in lactating cows during the milking period, especially in the conditions of a year-round stable method of keeping, it is necessary to take into account all possible sources on which the supply of the body with vitamins depends.

Improvement of metabolic processes in the body of cows throughout the year-round stable method of keeping is closely related to the balance of daily rations for nutrients and biologically active substances and metabolic and energy, and with the use of domestic highly effective feed additives in animal feeding.

Table 2. The content of nutrients and biologically active substances in the herbs of the phytocomplex.

The name of the green mass of different herbs of the phytocomplex	Organic substance, %				Mineral macroelements, g/kg						Mineral microelements, g/kg					Heavy metals, mg/kg		
	dry substance	protein	fibre	nitrogen-free extractive substances (NFES)	Mg	S	Na	P	K	Ca	Fe^{2+}	J	Mn	Cu	Zn	Pb	Cd	Ni
Goatweed	90.4	11.87	26.4	52.1	0.15	0.09	1.79	17.3	10.46	7.07	248	0.07	159	8.51	41.28	0.230	0.186	1.15
Orach	90.9	22.06	14.2	54.6	0.31	0.48	2.48	4.19	47.75	9.13	158	0.25	138	7.95	52.04	0.304	0.047	0.41
Big-sting nettle	89.7	23.42	15.1	51.2	0.39	0.82	2.82	5.93	25.03	18.80	375	0.16	122	7.65	38.75	0.499	0.054	0.40
Fireweed	90.0	13.31	25.6	51.1	0.17	0.20	2.58	3.98	15.11	8.78	132	0.16	65.7	5.54	33.26	0.279	0.031	0.76
Blueberry (leaves and stems)	91.3	8.98	32.4	49.9	0.11	0.10	2.01	1.40	5.40	9.81	106	0.10	559	5.73	17.07	0.125	0.041	0.42
Tansy	90.1	12.30	23.5	54.3	0.15	0.28	2.34	5.67	25.87	9.81	167	0.10	60.2	10.4	37.11	0.162	0.121	0.91
Milfoil	89.5	9.95	31.5	48.1	0.13	0.84	3.80	3.08	13.94	7.60	141	0.08	95.2	9.14	50.11	0.309	0.362	1.54
Perennial timothy	91.4	9.97	32.8	48.6	0.14	0.17	1.91	2.88	12.10	4.55	333	0.08	68.2	4.96	37.92	0.197	0.011	0.69
White sweet clover	90.1	16.50	36.7	36.9	0.16	0.20	2.08	2.45	10.29	8.21	128	0.08	29.0	6.57	19.33	0.239	0.047	2.16
White clover	88.7	15.91	26.9	45.9	0.11	0.10	2.54	2.59	16.05	8.73	88	0.02	57.7	6.04	26.97	0.256	0.018	2.26
Alsike clover	90.0	18.78	21.9	49.3	0.24	0.20	2.42	2.70	23.93	13.30	176	0.20	73.7	8.82	35.75	0.252	0.026	1.22
Red clover (meadow)	90.4	16.56	23.2	50.6	0.21	0.10	2.71	2.79	17.55	11.50	167	0.14	37.7	8.40	26.26	0.247	0.018	1.25
Alfalfa purple	90.7	15.86	28.0	46.8	0.14	0.19	1.94	2.41	17.72	9.45	367	0.10	92.2	6.60	26.44	0.164	0.027	0.77
Herbal mixture of phytocomplex	83.8	16.07	24.8	42.9	0.28	0.18	2.88	2.90	17.69	6.30	193	0.16	102	5.00	27.66	0.200	0.041	1.09

In view of the fact that these domestic feed additives are currently insufficiently studied, especially in feeding highly productive lactating cows to the discord phase, with a year-round stall (tethered) method of keeping, solving the problem of increasing milk production is of extremely important economic importance.

Some regional features of the properties of medicinal fodder crops and their phytocide activity were used to improve the vitamin value of fodder, rations, body resistance, health status and quality of milk of cows.

The effect of the dosed use of the studied feed additives in the diets of cows on the content of water-soluble vitamins in milk is given in Table 3.

The biological activity of B vitamins is closely related to protein metabolism, the provision of cows with protein, carbohydrates, and the lack – to a decrease in glycogen in the liver, a delay in the oxidation of lactic and pyruvic acids. And they are needed for protein optimization [16].

The need for highly productive cows for water-soluble vitamins is not well understood. They take part in the exchange of mineral trace elements. Both the antioxidant and sorption effects of feed additives, as well as the biologically active substances introduced into the composition, determine when feeding them to cows, along with an increase in the activity of enzyme systems, an increase in the anti-toxic function of the liver and the body as a whole. At the same time, the degree of localization of heavy metals (Pb, Cd, Ni) in the phytocomplex itself significantly decreased during the formation of certain medicinal herbs for their inclusion in the diet, and the functional activation of the digestive system of the cow's body increased. Therefore, in the phytocomplex itself, the total content of lead (0.200 mg/kg) and cadmium (0.041 mg/kg) is significantly lower than that in the rest of the studied laconic herbs.

With increasing doses of Vitaminol and phytocomplex from 10 to 20 g per head per day, there was a tendency to increase the content of all water-soluble vitamins in cow's milk during milk production, which is explained by an increase in the activity of the functional activity of enzyme systems, blood circulation and mammary gland. And the value of a particular water-soluble vitamin in the body of lactating cows is increased in combination with many minerals, amino acids, protein, fat and carbohydrates in the diet.

Vitamin B_{12} (cyanocobalamin) participates in the metabolism of protein and carbohydrates. Iron protects vitamin B_{12} from interaction with vitamin B_5, and copper accelerates the loss of activity of this vitamin. For the synthesis of vitamin B_{12} by microorganisms, the presence of carbon, nitrogen and cobalt is necessary. This vitamin is associated with the biological activity of B vitamins, accumulation (retention) and assimilation of nitrogen and a number of amino acids in the body. It promotes the purification of blood corpuscles and is rich in animal feed.

The composition of the diet influences the intestinal synthesis of vitamin B_{12}. The presence of a feed additive in the form of a phytocomplex in the diets of the silage-hay-concentrate type of feeding improves B_{12}, the vitamin supply of the organism of highly productive lactating cows during the milking period and increases its content in milk.

4 Conclusion

To normalize metabolic processes, increase milk productivity and vitamin nutritional value of cows' milk during the milking period, it is advisable to use domestic, environmentally friendly feed additives in a mixture with compound feed.

According to the results of the experimental studies, the parameters of the content of water-soluble vitamins in the milk of black-and-white cows during the milking period (in 100 g) are given in Table 4.

Table 3. Vitamin content in milk of cows during the milking period (per 100g).

Experiment	Group of cows	Water soluble								
		B_1 (thiamine), mg	B_2 (riboflavin), mg	B_3 (pantothenic acid), mg	B_5 (PP, nicotinic acid), mg	B_6 (pyridoxine), mg	B_7 (biotin), µg	B9 (folic acid), µg	B_{12} (cyanocobalamin), µg	C (ascorbic acid), mg
1	Control (OR)	0.02±0.001	0.09±0.001	0.27±0.011	0.04±0.001	0.02±0.001	1.82±0.05	0.10±0.01	0.18±0.06	0.64±0.09
	Experimental I (OR + 10g/head/day "Vitaminol")	0.03±0.002	0.11±0.011	*** 0.35±0.02	*** 0.09±0.001	*** 0.08±0.002	*** 2.21±0.07	*** 0.25±0.03	0.28±0.07	*** 1.10±0.04
	Experimental II (OR + 15g/head/day "Vitaminol")	*** 0.04±0.002	** 0.18±0.013	0.44±0.12	*** 0.10±0.001	*** 0.16±0.01	*** 2.86±0.09	*** 0.40±0.02	* 0.31±0.09	*** 1.89±0.07
	Experimental III (OR + 20g/head/day "Vitaminol")	*** 0.06±0.003	*** 0.23±0.011	* 0.57±0.11	*** 0.20±0.02	*** 0.18±0.12	*** 3.14±0.08	*** 0.54±0.01	*** 0.44±0.04	*** 2.04±0.07
2	Control group (OR)	0.03±0.001	0.09±0.002	0.29±0.04	0.04±0.004	0.03±0.001	1.86±0.06	0.12±0.009	0.22±0.07	0.58±0.08
	Experimental I (RR + 10g/head/day of phytocomplex)	0.04±0.002	*** 0.14±0.001	** 0.42±0.03	*** 0.11±0.001	*** 0.10±0.001	*** 2.17±0.05	0.37±0.16	0.27±0.05	*** 1.17±0.12
	Experimental II (OR + 15g/head/day of the phytocomplex)	*** 0.05±0.003	*** 0.15±0.002	*** 0.49±0.03	*** 0.16±0.001	*** 0.12±0.001	*** 2.74±0.06	*** 0.49±0.02	** 0.49±0.02	*** 1.79±0.06
	Experimental III (OR + 20g/head/day of phytocomplex)	*** 0.05±0.001	*** 0.22±0.003	*** 0.57±0.04	*** 0.26±0.02	*** 0.17±0.002	*** 2.96±0.02	*** 0.63±0.07	*** 0.52±0.01	*** 2.09±0.09

*P < 0.05; **P < 0.01; ***P < 0.001

Table 4. Parameters of the content of water-soluble vitamins in the milk of black-and-white cows during the milking period (in 100 g).

Name of the water-soluble vitamin	The name of the feed additive	
	Vitaminol	Phytocomplex

B_1 (thiamine), mg	0.03 – 0.06 (control 0.02)	0.04 – 0.05 (control 0.03)
B_2 (riboflavin), mg	0.11 – 0.23 (control 0.09)	0.14 – 0.22 (control 0.09)
B_3 (pantothenic acid), mg	0.35 – 0.57 (control 0.27)	0.42 – 0.57 (control 0.29)
B_5 (PP, nicotinic acid), mg	0.09 – 0.20 (control 0.04)	0.11 – 0.26 (control 0.04)
B_6 (pyridoxine), µg	0.08 – 0.18 (control 0.02)	0.10 – 0.17 (control 0.03)
B_7 (biotin), µg	2.21 – 3.14 (control 1.82)	2.17 – 2.96 (control 1.86)
B9 (folic acid), µg	0.25 – 0.54 (control 0.10)	0.37 – 0.63 (control 0.12)
B_{12} (cyanocobalamin), µg	0.28 – 0.44 (control 0.18)	0.27 – 0.52 (control 0.22)
C (ascorbic acid), mg	1.10 - 2.04 (control 0.64)	1.17 – 2.09 (control 0.58)

To increase the efficiency of the use of water-soluble vitamins in the diets, it is necessary to include feed additives Vitaminol and phytocomplex (separately) in a dose of 20 g per head per day.

When improving the methods of providing water-soluble vitamins in cows with a silage-hay-concentrate type of feeding in conditions of year-round stall keeping, it is recommended to optimize the parameters of their content in milk per 100 g as compared with peers in the control group.

References

1. S. A. Aleksashina, N. V. Makarova, *Storage and processing of agricultural raw materials* **6,** 29–31 (2016)
2. N.V. Babiy, *Storage and processing of agricultural raw materials,* **6,** 38–42 (2016)
3. N. V. Ermakova, Science and education: a new times **3 (20),** 25–29 (2017)
4. N. V. Makarova, Samara State Technical University, 7–101 (2014)
5. A. I. Ovsyannikov, *Basics of experimental work* (Moscow: Kolos, 1976)
6. N. Razumovsky, D. Sobolev, Animal husbandry of Russia **5,** 61–64 (2017)
7. B. V. Usha, *Storage and processing of agricultural raw materials* **4,** 42–44 (2017)
8. S. Agnew, *The facts about grain aeration* (PAMI) (2017)
9. M. Cenariu, E. Pall, C. Cernea, I. Groza, *J.* Biomedicine and Biotechnology, 541384 (2012)
10. R. C. Hansen, H. M. Keener, R. J. Gustafson, *Ohio State University* AEX–202.2006 (2006)
11. E. D. Hilborn, A. Bloom, *Centers for Disease Control MM WR* **63 (1),** 11–15 (2014)
12. L. F. Tranel, *Iowa State University Animal Industry Report* (2013) doi:10.31274/ans_air-180814-606
13. V. R. Lopes, V. M. Vasconcelos, *Mar. Drugs,* **9,** 790–802 (2011)
14. Queensland Goverment Department of Primary Industries, *Rewiew of hydroponic fodder for beef cattle* (Meat & Livestock Australia Limited) (2003)
15. E. M. Serba, M. B. Overchenko, V. E. Davydkina, N. V. Shelekhova, L. V. Rimareva, V. A. Polyakov, *Storage and processing of agricultural raw materials* **2,** 44–50 (2015)
16. R. V. De Sousa, C. R. da Silva Cardoso, G. Butzke, M. A. Dode Nunes, R. Rumpf, M. M. Franco, Theriogenology **90,** 25–31 (2017)

https://doi.org/10.1051/bioconf/20225200003

Cultivation efficiency of new varieties of winter soft wheat in the north-eastern zone of the Krasnodar Territory

A. S. Golovko and *E. K.* Kuvshinova*

Azov-Black Sea Engineering Institute of Don State Agrarian University, 21, Lenin St., Zernograd, 347740, Russian Federation

Abstract. The paper presents a comparative assessment of new varieties of winter soft wheat cultivated in the north-eastern zone of the Krasnodar Territory. It provides the results on yield, grain quality and cost efficiency of their cultivation. On average over the years of study the Alekseich and Bezostaya 100 varieties have consistently provided high actual grain yields for the predecessor of peas – 6.67 and 6.55 t/ha, respectively. The highest grain quality was formed by the Bezostaya 100 variety. The average protein content over three years was 13.3%, gluten content 2 – 4.6%, grain unit – 806 g/l. Bezostaya 100 and Alekseich varieties ensured a high profit of 38,625-39,885 rub/ha and a level of profitability of 128-132% when cultivating the predecessor of peas. The study resulted in conclusions and proposals to the production to increase the efficiency of growing winter soft wheat varieties in Kuban.

1 Introduction

The main industry of domestic agriculture is grain farming, which provides the industry with high-quality raw materials, livestock with feed, and the population with food. The Krasnodar Territory is one of the leading regions of the South of Russia in terms of gross winter soft wheat harvest, which receives about 10 million tons of grain annually. The average yield of winter soft wheat in the region varies over the years and averages about 6.0 t/ha [1].

One of the factors that increases the stability of grain production, and its gross yield is the cultivation of new high-yielding varieties with good technological qualities [2–4].

Russian breeders have created a number of varieties, but only a few have played an outstanding role in increasing grain production. A critical review of the achievements of the world's breeders shows that those who use the richest and most genetically diverse source material, as well as apply the most progressive, scientifically based methods of work at all stages of the breeding process, achieve great success. The wheat plant during the growing season is greatly influenced by a large number of environmental factors. Some of them contribute to the realization of the potential of high productivity, while others hinder it. The most important task is to make the most of the former and neutralize the latter [5].

The importance of the grain variety is now especially increasing due to the great demand in the consumer market for high-quality grain suitable for baking bread. However, modern varieties have not only positive but also negative properties, which manifest themselves in different ways in individual years. In particular, they form a high yield only in favorable weather and agrotechnological conditions of cultivation and sharply reduce it when they deteriorate [6, 7].

Consequently, grain producers must move faster from old to new varieties of winter wheat and, above all, to their domestic ones, which are better adapted to local conditions. In addition, the shortage of winter wheat grain from the use of old varieties is 900–1000 thousand tons [8, 9].

The studies covered new varieties of winter soft wheat – Alekseich and Bezostaya 100 – and compared it with the Grom variety approved for cultivation since 2010 ordinary chernozem of the Prikubanskaya plain. The issue of cultivating new varieties according to the most common predecessors in the region for agricultural producers is always relevant, because currently the variety has become the main factor in the economic growth of agricultural production.

The choice of a predecessor is one of the most important factors in winter wheat cultivation technology, because among the variety of agronomic issues, the observance of crop rotation and appropriate soil preparation are considered a strong measure in increasing the agriculture crop and improving the yield of agricultural plants, in particular winter soft wheat [10].

In this regard, the purpose of this study was to assess the efficiency of cultivation of new varieties of winter soft wheat for various predecessors in the north-eastern zone of the Krasnodar Territory. This requires a comparative assessment of new varieties in terms of yield, grain quality and cost efficiency of cultivation according to different predecessors.

* Corresponding author: kuv.ek61@yandex.ru

2 Materials and methods

The studies were carried out in 2018–2020 in the north-eastern zone of the Krasnodar Territory on the territory of the Gorbach S.E. Farm located in the Pavlovsky District. The farm is located in the zone of risky agriculture, so the yield is largely determined not only by agrotechnical, but also by agroclimatic conditions. The limiting factor in the cultivation of crops in this zone is the soil moisture. The average annual rainfall in the district is 579.8-611.0 mm and their total amount is determined as sufficient, but they are distributed unevenly throughout the year. In summer there are frequent droughts and destructive effects of dry winds with the predominance of easterly winds. Such conditions pose the need for strict adherence to a science-based farming system in the cultivation of crops, and winter wheat in particular.

The soil of the experimental site is represented by ordinary chernozem. The grain-size distribution is heavy loamy. The soil cover is heterogeneous, mosaic. The humus layer thickness makes 120–140 cm. The humus content – 3.3–3.6%, however, the total supply of organic matter in the entire humus horizon is quite large and amounts to 400-500 t/ha. The reaction of the soil medium is neutral, pH – 6.7–7.0, which is a favorable factor for the cultivation of crops.

Field and laboratory tests were conducted in accordance with GOSTs and generally accepted methods.

The material for the study included promising and new varieties of winter soft wheat: Grom, Alekseich and Bezostaya 100 selected in the P.P. Lukyanenko National Grain Center (Krasnodar).

The varieties studied were sown according to three predecessors:

- winter wheat,
- grain maize,
- pea.

Diagrams of field experiments will be given in the tables below.

The main cultivation and preparation of soil, crop sowing and tending were carried out according to the farming system of the Krasnodar Territory [11].

3 Results and discussion

The most important economic and valuable sign of the variety is its yield. In 2018, the yield of varieties ranged from 5.18 to 7.09 t/ha. The minimum yield was obtained from Grom variety in terms of its predecessor winter wheat – 5.18 t/ha. According to this predecessor for Alekseich and Bezostaya 100 it was also minimal and amounted to 5.91 and 6.07 t/ha, respectively. The maximum yield in 2018 was obtained from Alekseich variety according to the predecessor of peas 7.09 t/ha.

The most favorable conditions in the 2018-2019 agricultural years contributed to higher grain yield of the studied varieties compared to the previous year. The yield in 2019 was 5.64-7.71 t/ha. The maximum yield, similar to 2018, was obtained from the Alekseich variety according to the predecessor of peas – 7.71 t/ha. The yield for this predecessor was quite high for the Grom variety – 7.58 t/ha and for the Bezostaya 100 – 7.62 t/ha. 2020 was not quite favorable for winter soft wheat, when the yield ranged from 4.11 to 5.20 t/ha. The minimum yield in 2020 was obtained from the Grom variety according to its predecessor winter wheat – 4.11 t/ha, and the maximum – from Alekseich according to its predecessor peas – 5.20 t/ha (Table 1).

Table 1. Impact of predecessors on winter wheat yields (2018–2020).

Variety	Predecessor	Yield, t/ha			
		2018	2019	2020	$\bar{x}$
Grom	winter wheat	5.18	5.64	4.11	4.98
	grain maize	5.96	5.75	4.42	5.38
	pea	6.27	7.58	5.00	6.28
Alekseich	winter wheat	5.91	5.97	4.24	5.37
	grain maize	6.24	6.02	4.61	5.62
	pea	7.09	7.71	5.20	6.67
Bezostaya 100	winter wheat	6.07	6.26	4.13	5.49
	grain maize	6.62	6.83	4.59	6.01
	pea	6.90	7.62	5.14	6.55
	HCP_{05}	0.58	0.41	0.34	–

On average, over three years, all varieties formed the highest yield according to the predecessor of peas: Grom – 6.28 t/ha, Bezostaya 100 – 6.55 t/ha, and Alekseich – the maximum yield of 6.67 t/ha.

In 2018 a reliable increase was formed by the Alekseich variety according to the predecessor of peas and the Bezostaya 100 variety according to the predecessor of winter wheat compared to the Grom zoned variety. In 2019 the reliability of the increase was proven only by the Bezostaya 100 variety according to the predecessor of grain maize. In 2020 the reliability of increase has not been proven.

Thus, on average, over three years of study, the responsiveness of winter wheat varieties to the predecessor of peas was positive. Grain maize can also serve an acceptable predecessor for the Bezostaya 100 variety.

The successful solution of the grain problem involves not only an increase in gross grain yield, but also an increase in its quality. Grain quality indicators are important in assessing certain methods of cultivation, since the yield and quality of grain processing products strongly depend on them. This evaluation makes it possible to make mix compositions of grain with satisfactory baking properties. Let us consider the grain quality for protein content, gluten, gluten quality and grain unit as the average of all assessments made for each grade after storage.

In 2018 the grain unit of the varieties was quite high and amounted to 808-814 g/l. The highest grain unit was obtained from the Bezostaya variety – 100-814 g/l. In 2019 the grain unit was also high, but slightly lower than in the previous year and amounted to 800-804 g/l, while for the Bezostaya 100 variety it was also the largest. The lowest grain unit was noted in 2020 – 780-799 g/l, however, according to the requirements for food wheat in all varieties, it met the first-grade quality.

The protein content in the grain of the studied varieties in 2018 ranged from 11.9 for Grom to 12.6% for Bezostaya 100. The gluten content in the grain also changed in the same way: from 22.3% for Grom to 23.9% for Bezostaya 100. Such grain met the requirements of the third and fourth grades for food wheat. In 2019 the protein and gluten content of the grain was higher than in 2018. The maximum content of protein (15.0%) and gluten (26.0%) was noted in Bezostaya 100. In 2019 all varieties formed a grade 3 food grain that meets the requirements for "valuable" wheat. In 2020 Bezostaya 100, as in previous years, formed the highest quality grain with a protein content of 12.3% and a gluten content of 24.0% compared to other varieties. Such indicators also characterized grain of this grade as "valuable" wheat (Table 2).

Table 2. Winter wheat grain quality (2018–2020).

Variety	Grain unit, g/l	Protein content, %	Gluten		Grade*
			%	FDM, c.u.	
2018					
Grom	808	11.9	22.3	79	4
Alekseich	812	12.1	23.6	85	3
Bezostaya 100	814	12.6	23.9	81	3
2019					
Grom	801	13.1	23.5	80	3
Alekseich	800	14.5	24.6	79	3
Bezostaya 100	804	15.0	26.0	83	3
2020					
Grom	781	11.5	21.4	78	4
Alekseich	780	11.9	22.1	80	4
Bezostaya 100	799	12.3	24.0	81	3

* Grain grade as per GOST 9353-2016.

There was a strong positive correlation between protein and gluten content in all years of studies: $r = 0.83 \pm 0.16$; $r = 0.94 \pm 0.18$; $r = 0.78 \pm 0.14$ according to years.

The issue of economic feasibility is always relevant in any sector of agriculture. The general methods formed the basis for the calculation of the economic efficiency indicators of cultivation of new varieties of winter soft wheat. The cost of gross production per hectare was calculated at the average purchase prices for wheat grain over 3 years, which amounted to 10.500 rub/t.

Winter wheat grain production costs amounted to 31.628 rub/ha for winter wheat; 30.956 rub/ha for grain maize and 30.150 rub/ha for peas. Higher predecessor costs of winter wheat are associated with the use of more expensive insecticides to control crop pests after the spiked predecessor (Table 3).

Table 3. Cost efficiency of winter wheat cultivation (2018–2020).

Variety	Predecessor	Yield, t/ha	Cost, rub/ha	Production costs, rub/ha	Gross product cost, rub/ha	Profit, rub/ha	Cost efficiency , %
Grom	winter wheat	4.98	6351	31628	52290	20662	65
	maize	5.38	5754	30956	56490	25534	82

	pea	6.28	4801	30150	65940	35790	118
Alekseich	winter wheat	5.37	5890	31628	56385	24757	78
	maize	5.62	5508	30956	59010	28054	90
	pea	6.67	4520	30150	70035	39885	132
Bezostaya 100	winter wheat	5.49	5761	31628	57645	26017	82
	maize	6.01	5151	30956	63105	32149	104
	pea	6.55	4603	30150	68775	38625	128

A comparative evaluation of efficiency indicators for different predecessors showed that from an economic point of view the most efficient for all varieties due to grain yield is the predecessor of peas. Thus, the Alekseich variety in crops for this predecessor had the maximum profit of 39.885 rub/ha and the highest production profitability of 132%. A slightly lower profit was received for Grom – 35.790 rub/ha and Bezostaya 100 – 38.625 rub/ha with a profitability level of 118 and 128%, respectively. It is worth noting the economic efficiency of Bezostaya 100 according to the predecessor of grain maize, where the profit amounted to 32.149 rub/ha, and the profitability level – to 104%.

The calculations found that the cultivation of all the winter soft wheat varieties under study is cost effective, especially for the predecessor of peas. This is confirmed by the positive profit and high profitability of production. The cultivation of the studied varieties according to other predecessors will also generate income, which in absolute terms will be slightly less.

4 Conclusion

On average, over the years of study the winter soft wheat variety Alekseich and Bezostaya 100 formed a consistently high actual grain yield for the predecessor of peas – 6.67 and 6.55 t/ha, respectively.

The highest grain quality was formed by the Bezostaya 100 variety. On average, over three years, the protein content was 13.3%, gluten – 24.6%, grain unit – 806 g/l. Such grain met the requirements of the third food grade and was characterized as "valuable" wheat. There was a strong positive correlation between protein and gluten content in all years ($r = 0.83 \pm 0.16$; $r = 0.94 \pm 0.18$ and $r = 0.78 \pm 0.14$).

The highest profit of 38.625–39.885 rub/ha and the profitability level of 128–132% were obtained from the Bezostaya 100 and Alekseich varieties according to the predecessor of peas.

References

1. N. N. Grey, S. V Dyrda, *Innovative Economy: Prospects for Development and Improvement* **8 (34),** 315–319 (2018)
2. L. A. Bespalova, I. N. Kudryashov, I. B. Ablova, F. A. Kolesnikov, G. D. Nabokov, V. A. Filobok, O. Yu. Puzyrnaya, *Proceedings of the All-Russian Scientific and Practical Conference with International Participation* (June 26–29, 2017, Derbent; Makhachkala) pp 72–78 (2017)
3. A. S. Golovko, E. V. Khronyuk, *Scientific Foundations for Improving the Efficiency of Agricultural Production IV: Proceedings of the International Scientific and Practical Conference* pp 143–146 (2020)
4. V. I. Nechaev, N. A. Glechikova, A. A Seregin, Lecture notes in networks and dystems **205,** 395–401 (2021)
5. V. I. Kovtun, *Selection of Highly Adaptive Varieties of Winter Soft Wheat and Non-Traditional Technologies for Their Cultivation in Arid Conditions in the South of Russia* (Rostov-on-Don: Kniga) (2002)
6. A. V. Alabushev Adaptive potential of varieties of grain crops, *Leguminous and cereal crops* **2 (6),** 47–51 (2013)
7. V. L. Gaze, E. V. Ionova, D. M. Marchenko, V. A. Likhovidova, Variety change of winter soft wheat as a mechanism for increasing productivity and resistance to abiotic environmental factors, *Grain Farming in Russia* **6 (60)** 16–20 (2018)
8. A. Bespalova, I. B. Ablova, "Mosaic" of varieties as a method of managing and monitoring the phytopathological situation in wheat agrocenoses in *Protection of plants from harmful organisms: Materials of the IX International Scientific and Practical Conference*, pp 28–31 (2019)
9. Ministry of Agriculture of Russia, *National report "On the progress and results of the implementation in 2016 of the State Program for the Development of Agriculture and Regulation of Agricultural Products, Raw Materials and Food Markets for 2013–2020"* (Moscow) (2017)
10. L. P. Beltyukov, V. B Khronyuk., E. K. Kuvshinova, R. G. Bershansky, I. S. Tatyanchenko, V. V. Denisenko, Bulletin of Agrarian Science of Don State Agrarian University **1 (37.1),** 52–64 (2017)
11. *The agriculture system of the krasnodar territory on an agro-landscape basis* (Krasnodar) (2015)

On the designing method for express assessment of the state of winter crops in the process of their overwintering

A. Feklistov[1], *I.* Yudaev[2], *S.* Volobuev[1], *Yu.* Daus[2*], and *V.* Panchenko[3]

[1] FSBEI HE "Volgograd State Agrarian University", 26, Universitetsky ave., Volgograd, 400002, Russian Federation
[2] FSBEI HE "Kuban State Agrarian University", 13, Kalinina, Krasnodar, 350044, Russian Federation
[3] Russian University of Transport, 2, Minaevsky per., Moscow, 127055, Russian Federation

Abstract. Winter crops, having high yields in comparison with spring crops, determine the overall grain harvest, thereby providing significant contribution to strengthening the country's food security. But at the same time, climatic metamorphoses of recent years seriously affect the preliminary forecast of the amount of collected products and its final result. Therefore, the problem of current control over the safety of cereals cultivated using the technology of winter sowing is very acute today. The review of existing control methods in the autumn-winter-spring period shows that almost all methods are laborious, time-consuming, many are the result of a visual subjective assessment, there is no possibility of express assessment of the state of plants in the place where cultivated crops grow. Analysis of the mechanism of intracellular processes under plant tissue exposure to negative temperature allows saying that intracellular changes occur in it, leading to a change in the electrical conductivity of the "intercellular substance and cell contents" solution. Based on the already used methods for assessing the physiological state of plants by measuring the resistance of plant tissue, it was decided to use the same approach to create algorithm for the functioning and technical means for monitoring the winter hardiness of cultivated crops. Conducted studies have shown that the electrical resistance of plant tissue has a systematically repeating nonlinear character during freezing and thawing during measurement. Resistance measurements must be carried out at different frequencies, but due to the limited capabilities of measuring instruments to measure the high resistance of plant tissue, preference should be given to frequencies from 0.5 kHz and higher.

1 Introduction

Today, in addition to the pandemic spread on the planet and the manifestation of negative climatic factors, as well as due to the overpopulation of certain regions, the food crisis has become an acute topic. According to statistics provided by the Food and Agriculture Organization of the United Nations (FAO), over 1 billion people suffer from various forms and stages of malnutrition in the world, and about 240 million people are doomed to illness and death due to hunger. The production of crop products and raw materials in the required quantity and the required quality is just that vital necessity that can help stabilize the situation at the acceptable level for existence and create the groundwork for its improvement. At present, the agricultural sector is everywhere concerned about the preservation of plantings and sowing of plants, their protection from pests and diseases, seeks to reduce crop losses during harvesting and storage.

In many countries of the world, large percentage of arable land is allocated for winter crops, since it is well known that the latter are more productive than spring crops are. In accordance with their agricultural techniques, winter crops must be regularly monitored, their condition evaluated, and a variety of measures for their care, protection and health improvement should be designed. To take measures, it is necessary to conduct systematic surveys, which are carried out in the autumn and spring periods of the year, with mandatory control in the winter for winter crops. The features of the implementation of such surveys and methods of their implementation are to be considered.

Winter control of the state of winter crops is carried out at specified periods, as well as after each severe frost. To do this, soil samples with plants for subsequent assessment of their viability, are taken using the appropriate methods [1].

The method of monoliths or direct growing of plants in soil is the most reliable among the existing ones. In the field, monoliths with cultivated plants are cut from the frozen ground, which are transported to a room, where they are thawed at a slight positive temperature, covering the boxes with wet bags. After complete thawing, the boxes are moved to the dry, bright room with an air temperature of +16°C and after 2-3 days, when the plants grow back, all old and dead leaves are cut off. Two weeks later, the number of living and dead plants is counted, referring to living plants only those in which new leaves have grown [2].

The method of growing plants on water allows quickly assessing the viability of winter crops. As with the previous method, soil samples with plants are cut down, which are covered with burlap and delivered to the

* Corresponding author: zirochka2505@gmail.com

premises. After separating the plants from the soil after thawing, they are washed with water, the roots and leaves are cut off, and placed in the growth chamber with water, which is changed once a day. Through the holes in the lid, the plants grow in good light and at temperature of about +15°C. Viability is determined after 7 days, taking into account the plants that form new leaves and roots [2].

The sugar method practically repeats the water method, but with the difference that the washed plants are placed instead of water for 14 hours in a 2-5% sugar solution, and only after that they are transferred to clean water at room temperature. In living plants, regrowth of roots can be seen on the second day, and leaves - after five days, but the assessment of viability is also determined after a week [2].

The tillering node growing method is an accelerated method based on the ability of tillering nodes to grow back under favorable conditions. The sampling and thawing process is the same as for growing on water. After thawing and washing the plants, the shoots of the latter are cut off at some distance from the tillering node, and the roots are cut off at the tillering node. The nodes themselves are placed in Petri dishes or in glass jars with filter paper at the bottom. The containers are closed and kept for 16-24 hours at the positive temperature of +24-28°C. Viability is determined by the fact that sprouts appear from the meristem tissue in living plants, but not in those that have not wintered [2].

There are other methods that make it possible to determine the viability of the plant at any stage of its development, these include, for example, the electrometric method of Nizenkov based on measuring the electromotive force (EMF) taken from a pair of electrodes introduced into the plant and evaluating, according to the deflection of the arrow of the galvanometer, which is previously calibrated, the degree of vitality: the greater the deflection of the arrow, the higher the frost resistance [3].

A survey of crops in the spring is carried out before harrowing, two weeks after the beginning of the spring growing season carried out visually with the quantitative assessment. Visual control of overwintered and healthy plants is carried out based on such signs as: the plants must be green; they should actively and amicably resume vegetation, but some may have frozen tops of leaves; the tillering node is white in appearance, dense - when growing, new (white) nodal roots appear from it. In plants that have not overwintered, the underground part is brown, the tillering node is yellow-brown, watery, new roots are not formed. Dead plants usually appear from under the snow in a green state, but with the onset of warmth they quickly turn brown. The distribution of dead plants in crops is diffuse, in which the dead are more or less evenly distributed among living plants throughout the field, or spotty - in the form of bald patches with dead plants [4].

Assessment of the condition of winter plants in winter and winter-spring time is of very great practical importance for planning and carrying out field work to care for winter crops. At the same time, all widely applicable methods of winter control are very laborious and time-consuming to assess, because of which the results obtained cannot be operational. It is obvious that some new methods and approaches are needed for a rapid assessment of the state of winter crops.

The purpose of the article is to substantiate and experimentally test the possibility of implementing the method of operational assessment (express assessment) of the state of winter crops in the winter and spring before the start of the active phase of field work.

2 Materials and methods

To design technical tool and substantiate the algorithm for its functioning, it is necessary to establish a physical phenomenon or law of natural science, based on which the device can work properly and provide reliable results for analysis [5]. Let us consider the essence of the processes occurring in the plant organism when exposed to negative temperatures.

It is well known from the course of plant physiology and morphology that ice formed during slow freezing in the intercellular spaces and cell walls of plant tissue draws water from the cells; the cell sap becomes concentrated, and as a result, the pH of the medium changes. At the same time, there is a loss of a very large amount of calcium ions by the membranes, above the threshold value. As a result, the number of calcium ions in the cytoplasm increases and the normal functioning of membrane structures is disrupted - "proton pumps" in membrane walls are inactivated, and a group of enzymes that catalyze the hydrolysis of phospholipids, on the contrary, are activated, which causes ion leakage, thereby stimulating the degradation of membrane lipids. The main reasons for the death of plants exposed to negative temperatures are an irreversible increase in membrane permeability, damage to cell metabolism and the accumulation of toxic substances [6, 7].

Having analyzed the processes occurring in the intracellular and intercellular spaces during freezing and thawing of plant tissue, they will be correlated with the theory of electrical conductivity.

It is well known from the courses of biophysics and electrophysiology of plants that the electric substitution circuit of the cell is a mixed (series-parallel) connection of active and capacitive elements, as well as sources of EMF. The sources of EMF characterize the fact established by research that there is an ion exchange between the intra- and extracellular environment, creating a potential difference of 60-80 mV on the membrane surface. Capacitive properties in the cell are possessed by the cell membrane, the membrane of the vacuole and the nucleus, and the active resistance - by the electrolytic components: the intracellular and intercellular environment [8].

As already noted, organic intracellular and intercellular spaces are filled with liquid media, which in their electrical conductive properties are close to electrolyte solutions or to conductors of the second kind, and which, depending on the amount of dissolved salts, sugars or other minerals, changes in the pH value of the medium, the presence of water change their characteristics, thereby affecting the electrical conductivity not only of the individual cell, but also of

https://doi.org/10.1051/bioconf/20225200004

the plant tissue as a whole.

In the practice of electrical technology, methods for determining damage to plant tissue, for example, by electric or thermal effects, are known and are successfully used, which are based on fixing the change in the impedance of tissue resistance (or its components) by the frequency of the measuring current, as well as assessing the depth of damage determined by the ratio of tissue resistance measured at high and low frequencies of the measuring current [9]. In addition, the measurement of electrical conductivity or electrical impedance and its components is widely used to determine the physiological state of plant tissues, for example, to detect damage to apples [10]; assessment of the ripening of various fruits [11, 12]; determination of the state of water in tomatoes [13]; determination of the objective dependence of the resistance of plant tissue on temperature [14], etc.

Thus, having considered the mechanism of intracellular reactions and studied the behavior of the cellular structures of plant tissues exposed to low, negative temperatures, the possibility of designing the algorithm and creating technical means that fixes changes in electrical conductivity characteristic of "intracellular and intercellular electrolytic solutions" depending on temperature can be disscussed.

To implement the above provisions, series of experiments was carried out to study the effect of negative and positive temperatures on the electrical resistance of plant tissue of cultures during freezing and thawing, as well as the behavior of experimental dependences as a function of the measuring current frequency.

The dependences were obtained experimentally on winter wheat plants in winter and spring using a SHIVAKI SFR-81W freezer. To control the freezer, or rather to set, regulate and maintain the required temperature, a special control unit was made. The resistance of plant tissue to an alternating measuring current with a variable frequency during freezing and thawing was monitored using a GW Instek LCR-821 RLC-meter, and the air and soil temperature was monitored with the DS1820 digital sensor [15]. The block diagram and appearance of the experimental setup are presented in Figure 1.

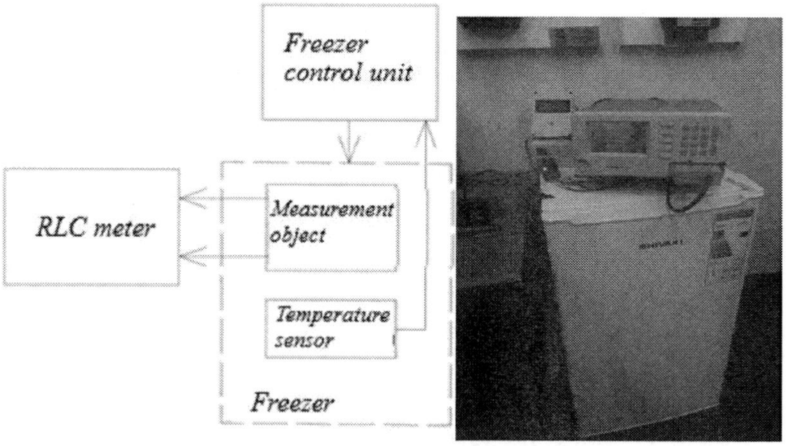

(a) (b)

Fig. 1. Block diagram (a) and external view (b) of the experimental setup.

In the educational farm "Gornaya Polyana" of Volgograd State Agrarian University, monolith (40x40x15cm) was cut down on the field of winter wheat crops, which was warmed up under room conditions and divided into seedling pots. In turn, the pots were placed in the SHIVAKI SFR-81W freezer and frozen at temperatures: -5; -10; -15; -17.5 and -33°C. Then it thawed at room temperature. Both during freezing and thawing, the resistance of plant tissue to alternating current was measured at frequencies: 0.012; 0.02; 0.1; 0.5; 1 and 2 kHz. Preliminary studies have releaved that at the temperature of -10°C all winter wheat plants were alive, at the same time at -17.5°C they perished. The temperature range from -12 to -13 ° C was taken as the survival limit.

3 Results and discussion

Basing on the results of the conducted studies, the dependences of electrical resistance on temperature were built.

As an example, the graphs of changes in the resistance of plant tissue at frequencies of the measuring current of 12 Hz and 1 kHz, obtained during freezing to -10 and -17.5 ° C, were considered (Fig. 2).

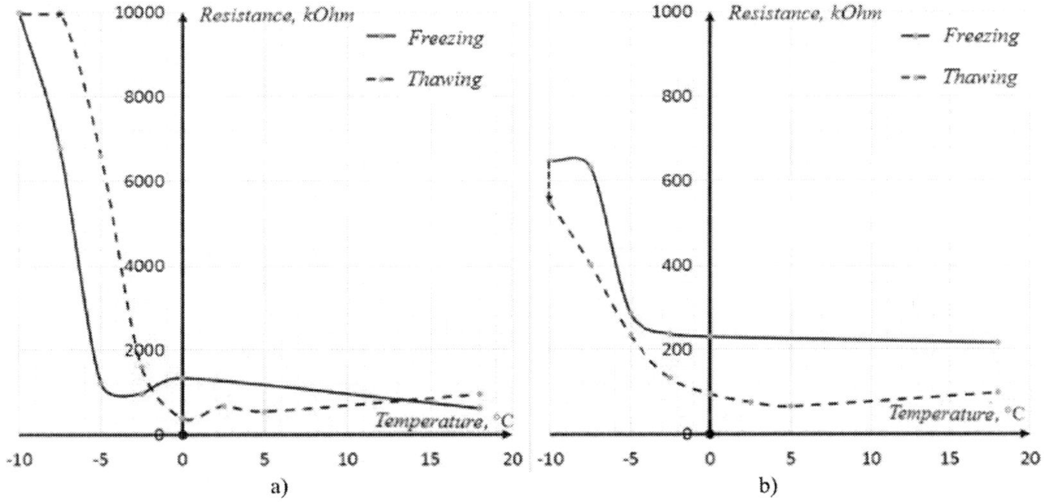

Fig. 2. Change in the plant tissue resistance during freezing at up to -10°C and subsequent thawing at frequencies of 12 Hz (a) and 1 kHz (b).

From the analysis of the first two graphs (Fig. 2), it can be seen that changes in the electrical resistance of plant tissue with the decrease in temperature at first appear insignificant, and only after the temperature in the freezer reaches -5°C we observed a sharp increase. The behavior of the dependences of changes in the plant tissue resistance at both frequencies is practically the same, but with the increase in the measuring current frequency, the difference in the numerical values of tissue resistances during freezing and thawing increases. At the same time, the plant remained alive.

In the next pair of graphs (Fig. 3), this resistance difference has increased significantly. The nature of the change in resistance is almost the same as in Figure 1, but in numerical values, the difference in resistances before and after freezing changed almost tenfold. In this case, the plant perished.

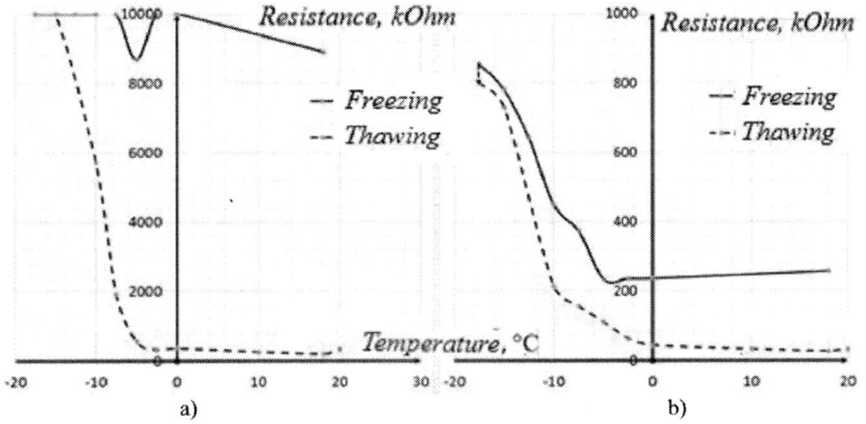

Fig. 3. Change in the plant tissue resistance during freezing at up to -17.5°C and subsequent thawing at frequencies of 12 Hz (a) and 1 kHz (b).

The summary graph of all the taken experimental data is presented in the form of changes in the electrical plant tissue resistance after thawing relatively the resistance before freezing (Fig. 4). Experimental studies were carried out at 6 frequencies of the measuring current in the temperature range from 0 to -34°C (the minimum is taken from the possibility of the available freezer).

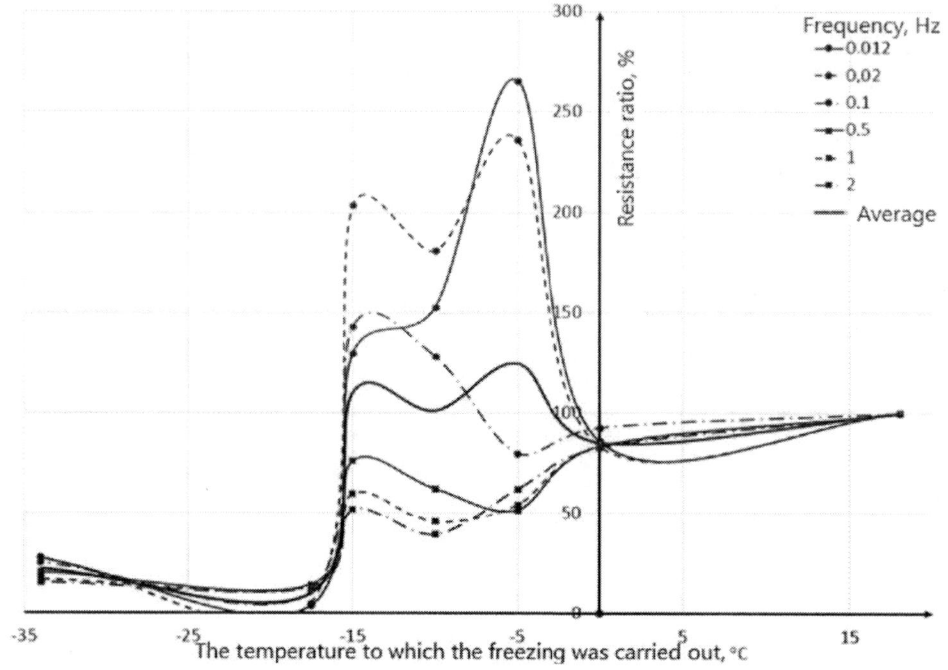

Fig. 4. Change in the plant tissue resistance after exposure to negative temperatures at different frequencies.

From the analysis of the behavior of the dependences presented in Figure 4, it can be seen that all graphs in the temperature range from 0 to -15 ° C are sharply nonlinear. It is this nonlinearity that should be used to find an objective way to assess the plant tissue state of winter crops. And although the nature of the nonlinearity is different for various plots, significant numerical differences in the relative resistances at the frequencies adopted for measurement allow one to continue research, collection and processing of experimental results for the final substantiation of the method for determining the plant viability by measuring the electrical resistance of its tissue and creating a portable technical means for its implementation.

4 Conclusion

Basing on the results of the carried out research and the analysis of the obtained results, the following can be stated:

1) Plant tissue resistances during freezing and thawing under laboratory conditions at different frequencies are nonlinear.

2) Resistance measurements must be carried out at different frequencies, but due to the limited capabilities of measuring instruments to measure the high resistance of plant tissue, preference should be given to frequencies from 0.5 kHz and higher.

3) In further studies, special attention should be paid to the range of negative temperatures of -17.5...-10°C.

4) The obtained results of experimental studies make it possible to substantiate the measurement of the electrical conductivity of plant tissue as a way to determine the state of winter crops and assess their viability.

References

1. P. V. Datsyuk, O. A. Antoshina, V. I. Petrakova, V. Z. Venevtsev, *Assessment of the state of winter wheat crops by the phases of vegetation in the conditions of the Central region of the Non-Chernozem zone (methodology)* (Ryazan: Ryazan NIPTI APK, 2007).
2. V. F. Moiseichenko, T. F. Trifonova, A. X. Zaveryukha, V. E. Eshchenko, *Fundamentals of scientific research in agronomy* (Moscow: Kolos, 1996).
3. N. P. Nizenkov, *Pat. 69841, Method for determining frost resistance, heat tolerance and drought resistance of plants and their seeds* (USSR).
4. V. M. Lichikaki, *Wintering of winter crops* (Moscow: Kolos, 1974).
5. A. S. Feklistov, Strategic development of the agro-industrial complex and rural territories of the Russian Federation in modern international conditions, *Materials of an International Scientific-Practical Conference* **2**, 377–382 (2016)
6. E. I. Koshkin, *Physiology of agricultural crop stability* (Moscow: Bustard, 2010).
7. Yu. P. Fedulov, *Methods for determining plant resistance* (Krasnodar: KubSAU, 2015).
8. I. V. Yudaev, *Surface Engineering and Applied Electrochemistry* **55 (2)**, 219–224 (2019).
9. V. I. Baev, I. V. Yudaev, V. A. Petrukhin et al. *Handbook of Research on Renewable Energy and*

Electric Resources for Sustainable Rural Development, 149–175 (2018).
10. F. Euring, W. Russ, W. Wilke, U. Grupa, *Procedia – Food Science* **01,** 1188–1194 (2011).
11. J Juansah, I. W. Budiastra, K. Dahlan, K. K. B. Seminar, *International Journal of Engineering & Technology* **12 (04)** 1-8 (2012)
12. W. L. Bertemes, R. Cavalieri, A. T. Paré, Ju Spessatto, D. Savi, P. Bertemes-Filho, *International Journal of Biosensors & Bioelectronics* **11 (4),** 99–101 (2020).
13. J.-X. He, Z.-Y. Wang, Y.-L. Shi, Y. Qin, D.-J. Zhao, L. Huang, *Sens. Lett.* **9,** 1151–1156 (2011).
14. I. N. Vorotnikov, A. V. Atanov, I. K. Sharipov, D. G. Ushkur, *Rural mechanic* **4,** 10–11 (2018).
15. V. I. Baev, A. S. Feklistov, *Ecological and reclamation aspects of rational environmental management,* 167–171 (2017).

Persistence factors and antibiotic susceptibility of enterobacteria isolated from various animal species

Vladimir Ermakov* and *Nikolai* Titov

Department of Epizootiology, Pathology, and Pharmacology, Samara State Agrarian University, town Ust-Kinelsky, 2, Uchebnaya St., Samara, 446442, Russian Federation

Abstract. During the study, we isolated 575 strains of enterobacteria from various animal species belonging to the genus Escherichia, Shigella, Salmonella, Klebsiella, Proteus, Providencia, Hafnia, Morganella, Enterobacter, Citrobacter, Serratia, Erwinia, Kluyvera, Yersinia. In farm animals, the amount of transient pathogenic enterobacteria Salmonella enteritidis and Yersinia enterocolitica did not exceed 0.22-0.12%. Among wild animals, the proportion of pathogenic Salmonella enteritidis and Yersinia enterocolitica did not exceed 0.24-0.11%, while Shigella dysenteriae and Shigella flexneri were isolated at a low concentration of 0.01%. In the feces of zoo animals, the amount of Salmonella enteritidis and Yersinia enterocolitica did not exceed 0.21-0.10%, while Shigella dysenteriae and Shigella flexneri were detected at the level of 0.01% of the total concentration of enterobacteria. The share of Salmonella enteritidis and Yersinia enterocolitica in domestic animals did not exceed 0.012-0.04%. Persistence factors in Salmonella enteritidis, Shigella dysenteriae and Shigella flexneri, Klebsiella oxytoca, and Yersinia enterocolitica were observed to be the highest among all isolated enterobacteria. Enterobacteria showed high resistance to benzylpenicillin from the group of natural penicillins to streptomycin, cephalothin from the group of cephalosporins of the first generation, polymyxin B, ofloxacin (tarivid), and metronidazole. Carbenicillin from the group of carboxypenicillins and piperacillin from the group of ureidopenicillins, kanamycin, amikacin, and gentamicin, cefepime from the group of IV generation cephalosporins, tetracycline, doxycycline and chloramphenicol, nalidixic acid, trimethoprim showed the highest antimicrobial activity against all cultures of enterobacteria isolated in this study.

1 Introduction

Recently, the role of the etiological diagnosis of infectious diseases has increased significantly, which requires the study of new and improvement of existing microbiological methods. Interest in biotechnological research has increased in terms of the etiological laboratory diagnosis of various diseases of a non-contagious and infectious nature. The incidence of so-called opportunistic infections caused by opportunistic microorganisms remains quite high in humans and animals [1, 9, 10, 4].

Leading microbiologists note that everywhere there is an activation of opportunistic bacteria and fungi, which are characterized by the absence of nosological specificity and localization of the infectious process. Opportunistic enterobacteria of the Enterobacteriaceae family are one of the main causative agents of opportunistic infections that can cause pathology of the gastrointestinal tract, meningitis, encephalitis, multiple neuritides, pyelitis, pyelonephritis, cystitis, cholecystitis, peritonitis, appendicitis, pancreatitis, pneumonia, nasopharyngitis, otitis, conjunctivitis, ophthalmitis, toxic-septic complications. This pathology is characterized by a polymorphism of clinical manifestations, associated not so much with the epidemic or epizootic situation, but with the age and state of the patient's defenses. Opportunistic enterobacteria, under certain conditions, also play an active role in the etiology of acute intestinal infections in humans and animals [5, 6, 7, 8, 12].

Currently, several new groups (genera) of enterobacteria have been introduced into the Enterobacteriaceae family. These are the bacteria Cedecea, Ewingella, Kluyvera, Lecrecia, Moellerella, Pantoea, Pragia, Rahnella, Tatumella, Xenorhabdus, and Jokenella. The clinical significance of these new members of the Enterobacteriaceae family is under investigation. The biological properties of these enterobacteria are being studied, and methods for their isolation and identification are being developed [2, 3, 11].

2 Materials and methods

During the study, 575 enterobacteria isolates obtained from the intestinal contents of various animal species were experimental material. Among the farm animals involved cows, sheep, goats, pigs, horses, and birds (chickens, guinea fowl, ducks, and geese). Among domestic animals, the material was obtained from cats

*Corresponding author: Vladimir_21_2010@mail.ru

and cats, dogs, ferrets, raccoons, and chinchillas. The material was also obtained from wild and zoo animals such as wild boars, elks, foxes, ponies, and camels.

The suspension of the material was sown on elective and differential diagnostic nutrient media. The grown cultures of enterobacteria were identified by specific biological properties. Enterobacteria were also identified using Escherichiosis, Shigella, Klebsiella, Proteus phage, and a set of Salmonella phages.

Determination of pathogenicity factors of enterobacteria was carried out according to generally accepted methods. Hemolytic and gelatinase, and catalase activity of cultures of enterobacteria were detected during the cultivation of microorganisms on enriched media and by setting up biochemical tests. The protease activity of enterobacteria cultures was determined by the decrease in albumin after co-incubation with the studied microorganisms by the biuret method. Among the persistence factors of enterobacteria, antilysozyme and anticarnosine activity were determined by the photometric method. The ability of enterobacteria to form biofilms was determined by the degree of binding of crystal violet by microorganisms in polystyrene plates.

Serological properties of enterobacteria were studied in reactions with specific diagnostic sera in reactions, agglutination, complement fixation, and precipitation. The determination of the sensitivity to antimicrobial drugs isolated by us from various animal isolates of enterobacteria was carried out by the disk diffusion method on AGV medium and Muller-Hinton agar, as well as by the method of serial dilutions in MPB broth and on Endo agar. The results obtained during the study were processed statistically according to the generally accepted method using a PC Pentium using the Microsoft Office Excel application.

3 Results and discussion

During the research, 215 intestinal isolates of enterobacteria were isolated from farm animals - cows, sheep, goats, pigs, horses, and poultry (chickens, guinea fowl, ducks, and geese). We found that the total number of identified enterobacteria in 1 g of feces in farm animals was 1.159510×106±1.32 (Table 1).

Table 1. Cultures of enterobacteria isolated from farm animals.

Enterobacteria, pure culture	Number of enterobacteria, 10^n	Correlation with the total number of enterobacteria, %
Escherichia coli	$5.43 \times 10^5 \pm 0.82$	46.83
Shigella dysenteriae	not identified	not identified
Shigella flexneri	not identified	not identified
Salmonella enteritidis	$2.54 \times 10^3 \pm 0.26$	022
Klebsiella oxytoca	$3.47 \times 10^4 \pm 0.68$	2.99
Proteus vulgaris	$3.59 \times 10^3 \pm 0.61$	0.31
Providencia alcalifaciens	$3.74 \times 10^4 \pm 0.13$	3.22
Hafnia alvei	$4.78 \times 10^4 \pm 0.47$	4.12
Morganella morganii	$4.72 \times 10^3 \pm 0.23$	0.41
Enterobacter cloacae	$4.54 \times 10^4 \pm 0.26$	3.92
Citrobacter freundii	$2.37 \times 10^4 \pm 0.53$	2.04
Serratia marcescens	$3.55 \times 10^5 \pm 0.13$	30.62
Erwinia amylovora	$3.28 \times 10^4 \pm 0.16$	2.83
Kluyvera cryocrescens	$2.75 \times 10^4 \pm 0.15$	2.37
Yersinia enterocolitica	$1.36 \times 10^3 \pm 0.12$	0.12

The species composition of enterobacteria isolated from agricultural ones was dominated by representatives of the genus Escherichia coli 46.83% and Serratia marcescens 30.62%. During the research, 105 intestinal isolates of enterobacteria were isolated from wild animals, namely wild boars, elks, and foxes. We found that the total number of identified enterobacteria in 1 g of feces in wild animals was 1.155454×106±1.18 (Table 2).

Table 2. Enterobacteria cultures isolated from wild animals.

Enterobacteria, pure culture	Number of enterobacteria, 10^n	Correlation with the total number of enterobacteria, %
Escherichia coli	$4.77 \times 10^5 \pm 0.38$	41.28
Shigella dysenteriae	$1.48 \times 10^2 \pm 0.72$	0.01
Shigella flexneri	$1.06 \times 10^2 \pm 0.38$	0.01
Salmonella enteritidis	$2.83 \times 10^3 \pm 0.33$	0.24
Klebsiella oxytoca	$2.84 \times 10^4 \pm 0.37$	2.46
Proteus vulgaris	$4.37 \times 10^3 \pm 0.53$	0.38
Providencia alcalifaciens	$3.82 \times 10^4 \pm 0.35$	3.32
Hafnia alvei	$4.19 \times 10^4 \pm 0.67$	3.62
Morganella morganii	$4.62 \times 10^3 \pm 0.24$	0.40
Enterobacter cloacae	$5.12 \times 10^4 \pm 0.32$	4.43
Citrobacter freundii	$3.16 \times 10^4 \pm 0.42$	2.73
Serratia marcescens	$4.06 \times 10^5 \pm 0.16$	35.14
Erwinia amylovora	$3.66 \times 10^4 \pm 0.64$	3.17
Kluyvera cryocrescens	$3.12 \times 10^4 \pm 0.36$	2.70
Yersinia enterocolitica	$1.28 \times 10^3 \pm 0.32$	0.11

The species composition of enterobacteria isolated from wild animals was dominated by representatives of the genus Escherichia coli 41.28% and Serratia marcescens 35.14%. During the research, 85 intestinal isolates of enterobacteria were isolated from zoo animals, namely, ponies, and camels. We found that the total number of identified enterobacteria in 1 g of feces in zoo animals was 1.586852x106±1.46 (Table 3).

Table 3. Cultures of enterobacteria isolated from zoo animals.

Enterobacteria, pure culture	Number of enterobacteria, 10^n	Correlation with the total number of enterobacteria, %
Escherichia coli	$5.24x10^5±1.26$	33.03
Shigella dysenteriae	1.28 $x10^2±0.94$	0.01
Shigella flexneri	1.64 $x10^2±0.76$	0.01
Salmonella enteritidis	$3.36x10^3±0.58$	0.21
Klebsiella oxytoca	$3.88x10^4±0.82$	2.44
Proteus vulgaris	$5.94x10^3±1.46$	0.37
Providencia alcalifaciens	$4.16x10^4±0.92$	2.62
Hafnia alvei	$6.32x10^4±1.56$	3.98
Morganella morganii	$6.12x10^3±0.88$	0.39
Enterobacter cloacae	$7.48x10^4±1.58$	4.71
Citrobacter freundii	$4.86x10^4±1.26$	3.06
Serratia marcescens	$6.74x10^5±1.62$	42.48
Erwinia amylovora	$4.78x10^4±1.38$	3.01
Kluyvera cryocrescens	$5.68x10^4±1.24$	3.58
Yersinia enterocolitica	$1.54x10^3±0.72$	0.1

The species composition of enterobacteria isolated from zoo animals was dominated by representatives of the genus Escherichia coli 33.03% and Serratia marcescens 42.48%. At the same time, the proportion of serrations was higher compared to Escherichia. During the research, we isolated 180 intestinal isolates of enterobacteria from domestic animals, namely, cats and cats, dogs, ferrets, raccoons, and chinchillas. We found that the total number of identified enterobacteria in 1 g of feces in domestic animals was 1.529452x106±1.32 (Table 4).

Table 4. Cultures of enterobacteria isolated from domestic animals

Enterobacteria, pure culture	Number of enterobacteria, 10^n	Correlation with the total number of enterobacteria, %
Escherichia coli	$6.72x10^5±1.16$	43.94
Shigella dysenteriae	$0.62x10^2±0.44$	0.003
Shigella flexneri	$0.86x10^2±0.68$	0.005
Salmonella enteritidis	$1.84x10^3±0.48$	0.012
Klebsiella oxytoca	$4.78x10^4±0.82$	3.12
Proteus vulgaris	$4.62x10^3±1.18$	0.30
Providencia alcalifaciens	$4.28x10^4±0.52$	2.80
Hafnia alvei	$5.68x10^4±0.85$	3.71
Morganella morganii	$4.12x10^3±0.46$	0.27
Enterobacter cloacae	$5.74x10^4±0.82$	3.75
Citrobacter freundii	$4.22x10^4±0.66$	2.76
Serratia marcescens	$5.18x10^5±1.48$	33.87
Erwinia amylovora	$4.86x10^4±1.08$	3.18
Kluyvera cryocrescens	$3.42x10^4±0.55$	2.24
Yersinia enterocolitica	$0.58x10^3±0.26$	0.04

The species composition of enterobacteria isolated from domestic animals was dominated by representatives of the genus Escherichia coli 43.94% and Serratia marcescens 33.87%. Enterobacteria isolated by us from various animal species were identified by specific biological properties. In the process of identifying persistence factors in enterobacteria isolated from various animals, we determined the indicators of the manifestation of antilysozyme, anticarnosine activity, and the ability of enterobacteria to biofilm formation (Table 5).

Table 5. Persistence factors of enterobacteria isolated from farm animals.

Enterobacteria	Persistence Factors		
	Antilysozyme activity µg/ml	Anticarnosine activity mg/ml	Ability of biofilm formation, %
Escherichia coli	3.28±0.12	3.06±0.08	72.4±5.2
Serratia marcescens	2.86±0.14	2.94±0.12	58.7±3.6
Kluyvera cryocrescens	2.74±0.16	2.88±0.12	49.6±2.8
Providencia alcallifaciens	2.12±0.06	2.04±0.04	58.2±3.4
Morganella morganii	2.04±0.04	1.74±0.08	52.8±2.6
Hafnia alvei	2.44±0.32	2.86±0.46	58.6±2.8
Erwinia amylovora	2.06±0.24	1.34±0.22	64.6±3.2
Klebsiella oxytoca	3.68±0.37	2.52±0.18	54.6±3.4
Proteus vulgaris	2.34±0.14	2.78±0.12	62.8±2.4
Enterobacter cloacae	2.78±0.18	2.94±0.24	66.4±4.8
Citrobacter freundii	2.54±0.12	3.08±0.16	48.2±2.8
Salmonella enteritidis	3.86±0.86	2.18±0.72	60.8±5.6
Yersinia enterocolitica	1.46±0.26	3.38±0.48	56.2±3.8

Enterobacteria in farm animals Salmonella Enteritidis and Klebsiella oxytoca showed the highest rates of anti-lysozyme activity. And representatives of Yersinia enterocolitica had higher anticarnosine activity. At the same time, representatives of Escherichia coli showed the highest ability to biofilm formation. Yersinia enterocolitica, on the contrary, showed minimal anti-lysozyme activity compared to other enterobacteria. Erwinias and morganellas showed minimal anticarnosine activity, while citrobacters and kluivers had the lowest ability to biofilm formation. Table 6 shows the persistence factors of enterobacteria isolated from wild animals.

Table 6. Persistence factors for enterobacteria isolated from wild animals.

Enterobacteria	Persistence Factors		
	Antilysozyme activity µg/ml	Anticarnosine activity mg/ml	Ability of biofilm formation, %
Escherichia coli	2.54±0.05	2.48±0.04	68.4±4.8
Shigella dysenteriae	4.38±0.68	3.76±0.88	74.8±6.4
Shigella flexneri	3.16±0.48	3.54±0.36	78.6±5.2
Serratia marcescens	2.12±0.04	2.26±0.02	52.3±4.2
Kluyvera cryocrescens	2.22±0.06	2.56±0.08	45.7±1.5
Providencia alcallifaciens	2.37±0.08	2.28±0.07	42.8±2.3
Morganella morganii	1.14±0.02	1.25±0.03	33.5±1.6
Hafnia alvei	1.08±0.03	1.06±0.04	38.7±1.3
Erwinia amylovora	not identified	not identified	43.2±1.4
Klebsiella oxytoca	2.63±0.07	3.24±0.25	74.3±3.8
Proteus vulgaris	2.98±0.08	2.43±0.16	69.8±3.5
Enterobacter cloacae	2.05±0.05	1.54±0.05	52.6±1.9
Citrobacter freundii	1.32±0.04	1.04±0.02	40.6±1.2
Salmonella enteritidis	3.07±0.23	2.77±0.26	86.4±4.2
Yersinia enterocolitica	2.64±0.18	2.34±0.17	61.7±2.7

The following Shigella showed the highest levels of antilysozyme activity and anticarnosine activity in wild animals: Shigella dysenteriae and Shigella flexneri. In these species of Shigella, high ability for biofilm formation was also observed. Salmonella enteritidis also had high anti-lysozyme and anticarnosine activity and surpassed pathogenic Shigella representatives Shigella dysenteriae and Shigella flexneri in biofilm formation ability.

Hafnia showed minimal antilysozyme and anticarnosine activity compared to other enterobacteria. Morganella also had insignificant antilysozyme activity, while citrobactors also showed low anticarnosine activity. Hafnia and morganella had a low ability to biofilm formation, compared with other enterobacteria. At the same time, the anti-lysozyme and anti-carnosine activity in enterobacteria Erwinia amylovora could not be revealed, and their ability to biofilm formation was also insignificant. Table 7 shows the persistence factors of enterobacteria isolated from zoo animals.

Table 7. Persistence factors for enterobacteria isolated from zoo animals.

Enterobacteria	Persistence Factors		
	Antilysozyme activity µg/ml	Anticarnosine activity mg/ml	Ability of biofilm formation, %
Escherichia coli	2.94±0.38	2.52±0.12	66.8±3.4
Shigella dysenteriae	1.36±0.44	2.24±0.86	62.6±2.6
Shigella flexneri	1.84±0.52	1.32±0.88	70.8±2.2
Serratia marcescens	1.52±0.12	2.58±0.56	62.6±2.4
Kluyvera cryocrescens	1.32±0.22	1.56±0.34	54.2±3.6
Providencia alcallifaciens	1.74±0.18	1.14±0.22	63.8±2.8
Morganella morganii	1.88±0.26	2.38±0.44	64.4±3.8
Hafnia alvei	1.46±0.12	1.54±0.10	60.2±3.4
Erwinia amylovora	2.72±0.56	1.78±0.68	74.8±5.8
Klebsiella oxytoca	4.56±3,84	3.98±1.36	80.8±7.2
Proteus vulgaris	4.82±1.72	3.12±1.84	76.6±3.2
Enterobacter cloacae	1.73±0.26	1.52±0.10	70.4±2.4
Citrobacter freundii	1.18±0.08	2.34±0.82	58.4±3.6
Salmonella enteritidis	5.18±2.32	3.72±0.98	52.6±2.4
Yersinia enterocolitica	6.48±2.84	4.76±2.18	66.8±1.6

The indicators of antilysozyme and anticarnosine activity were the highest in pathogenic Yersinia enterocolitica and Salmonella enteritidis. At the same time, in Klebsiella oxytoca, the anticarnosine activity was slightly higher than in Salmonella enteritidis. Klebsiella oxytoca and Proteus Vulgaris were the most capable of biofilm formation. The ability for biofilm formation was also high in representatives of Erwinia amylovora and Enterobacter cloacae.

Indicators of anti-lysozyme activity in enterobacteria Citrobacter freundii, anticarnosine activity in bacteria Shigella flexneri were the lowest compared to other enterobacteria. Salmonella (Salmonella enteritidis) and Kluyvera (Kluyvera cryocrescens) had a slight ability to biofilm formation, compared with other enterobacteria. Table 8 shows the persistence factors of enterobacteria isolated from domestic animals.

Enterobacteria of the genus Klebsiella oxytoca and Salmonella enteritidis showed the highest antilysozyme and anticarnosine activity compared to other members of the Enterobacteriaceae family. Escherichia coli and Klebsiella Klebsiella oxytoca showed the highest ability for biofilm formation. The lowest values of antilysozyme activity were registered by us in enterobacter Enterobacter cloacae, and anticarnosine activity in Morganella morganii. At the same time, pathogenic Salmonella enteritidis showed the lowest ability to biofilm formation compared to other enterobacteria.

Table 8. Persistence factors for Enterobacteriaceae isolated from domestic animals.

Enterobacteria	Persistence Factors		
	Antilysozyme activity µg/ml	Anticarnosine activity mg/ml	Ability of biofilm formation, %
Escherichia coli	2.04±0.08	2.34±0.12	80.6±6.8
Shigella dysenteriae	3.36±1.18	2.44±1.22	74.6±3.4
Shigella flexneri	3.48±1.22	3.78±1.46	72.8±4.8
Serratia marcescens	1.58±0.10	1.32±0.18	60.7±4.2
Kluyvera cryocrescens	2.06±0.34	1.66±0.88	56.8±3.4
Providencia alcallifaciens	1.54±0.34	1.84±0.08	64.8±2.6
Morganella morganii	1.34±0.12	1.26±0.18	58.4±5.8
Hafnia alvei	1,38±0.14	1.44±0.36	52.8±3.2
Erwinia amylovora	1,44±0.52	1.92±0.80	62.2±4.6
Klebsiella oxytoca	6,36±2.78	4.18±1.74	80.2±5.4
Proteus vulgaris	3,76±0.33	3.24±0.48	78.2±3.2
Enterobacter cloacae	1,18±0.12	3.14±0.88	74.4±6.4
Citrobacter freundii	1,76±0.34	2.18±0.94	56.2±3.2
Salmonella enteritidis	6,18±6,74	5,74±3,96	48,2±3,4
Yersinia enterocolitica	4,73±2,16	3,06±1,26	62,8±5,6

Pathogenic Yersinia enterocolitica, Shigella dysenteriae, and Shigella flexneri also showed relatively high antilosizyme, anticarnosine activity, and ability to biofilm formation against the background of persistence indicators of other enterobacteria cultures. However, the biofilm-forming ability of Yersinia enterocolitica was lower than that of Proteus Vulgaris. Trimethoprim showed the highest antimicrobial activity against all cultures of enterobacteria isolated by us. Most enterobacteria were resistant to metronidazole. At the same time, pathogenic Shigella dysenteriae, S. flexneri, Salmonella Enteritidis, and Yersinia enterocolitica were also more sensitive to trimethoprim (Table 9).

As a result, we found that Escherichia coli are highly sensitive to carbenicillin and piperacillin, kanamycin, amikacin, cefepime, cefotaxime, cefozopran, polymyxin B, chloramphenicol, nalidixic acid, ciprofloxacin, trimethoprim. Escherichia coli are relatively resistant to benzylpenicillin, gentamicin, cephalothin, doxycycline, ofloxacin, metronidazole.

Shigella dysenteriae and S. flexneri are sensitive to carbenicillin and piperacillin, kanamycin, amikacin, cefepime, cefozopran, cefquin, chloramphenicol, ciprofloxacin, trimethoprim. Shigella dysenteriae and S. flexneri are relatively resistant to benzylpenicillin, streptomycin, cephalothin, polymyxin B, ofloxacin, metronidazole.

Salmonella enteritidis (S. enterica subsp. enterica enteritidis) are sensitive to carbenicillin and piperacillin, kanamycin, amikacin, cefepime, chloramphenicol, nalidixic acid, co-trimoxazole, and trimethoprim. Salmonella enteritidis relatively resistant to benzylpenicillin, streptomycin, cephalothin, polymyxin B, ciprofloxacin, metronidazole.

Klebsiella oxytoca is sensitive to carbenicillin and piperacillin, gentamicin and kanamycin, cefepime, chloramphenicol, nalidixic acid, co-trimoxazole, and trimethoprim. Klebsiella oxytoca is relatively resistant to benzylpenicillin, amikacin, cephalothin, dioxycycline, ofloxacin, furadonin, and metronidazole.

Proteus Vulgaris is sensitive to carbenicillin and piperacillin, gentamicin and kanamycin, cefepime, tetracycline, chloramphenicol, nalidixic acid, ciprofloxacin, cotrimoxazole, and trimethoprim. Proteus Vulgaris is relatively resistant to benzylpenicillin, streptomycin, amikacin, cephalothin, polymyxin B, ofloxacin, metronidazole, furadonin.

Table 9. Antibiotic susceptibility of enterobacteria to metronidazole, furadonin, cotrimoxazole, and trimethoprim (growth inhibition zone value (mm).

Enterobacteria, pure culture	Antimicrobials			
	metronidazole	furadonin (nitrofurantoin)	co-trimoxazole	trimethoprim
Escherichia coli	8±0.8	18±1.2	17±0.5	26±2.3
Shigella dysenteriae	6±0.5	22±1.9	20±1.2	28±2.8
Shigella flexneri	5±0.4	26±2.2	25±1.5	30±2.6
Salmonella enteritidis	8±1.4	15±0.8	20±1.6	22±1.8
Klebsiella oxytoca	10±1.6	8±1.8	25±3.2	24±1.5
Proteus vulgaris	12±1.8	12±1.3	25±3.8	28±2.5
Providencia alcalifaciens	8±0.8	20±1.4	25±2.6	25±1.8
Hafnia alvei	10±1.2	18±2.2	23±2.5	23±2.0
Morganella morganii	12±1.5	10±3.5	27±4.5	26±1.7
Enterobacter cloacae	10±1.4	8±1.4	26±1.9	27±2.2
Citrobacter freundii	8±0.5	20±1.6	25±1.5	30±2.4
Serratia marcescens	8±1.2	19±0.6	22±0.7	35±3.4
Erwinia amylovora	10±2.2	12±1.3	37±6.2	34±2.6
Kluyvera cryocrescens	12±2.6	18±1.2	21±0.9	32±3.6
Yersinia enterocolitica	10±1.5	28±0.6	21±0.7	30±2.7

Providencia alcalifaciens are sensitive to carbenicillin and piperacillin, amikacin, cefepime, tetracycline, cmprofloxacin, cotrimoxazole, and trimethoprim. Providencia alcalifaciens is relatively resistant to benzylpenicillin, streptomycin, kanamycin, cephalothin, polymyxin B, and metronidazole.

Hafnia alvei are sensitive to carbenicillin and piperacillin, gentamicin, amikacin, cefepime, chloramphenicol, nalidixic acid, co-trimoxazole, and trimethoprim. Hafnia alvei are relatively resistant to benzylpenicillin, streptomycin, kanamycin, cephalothin, polymyxin B, ciprofloxacin, metronidazole.

Morganella morganii is susceptible to carbenicillin and piperacillin, gentamicin, kanamycin, cefepime, cefquin, cefosoprane, doxycycline, chloramphenicol, ciprofloxacin, co-trimoxazole, and trimethoprim. Morganella morganii is relatively resistant to benzylpenicillin, streptomycin, amikacin, cephalothin, polymyxin B, ofloxacin, metronidazole, furadonin.

Enterobacter cloacae are sensitive to carbenicillin and piperacillin, kanamycin, amikacin, cefepime, doxycycline, nalidixic acid, co-trimoxazole, and trimethoprim. Enterobacter cloacae are relatively resistant to benzylpenicillin, streptomycin, cephalothin, polymyxin B, ciprofloxacin, furadonin, metronidazole.

Citrobacter freundii are sensitive to piperacillin, amikacin, cefepime, doxycycline, ofloxacin, and trimethoprim. Citrobacter freundii is relatively resistant to benzylpenicillin, streptomycin, cephalothin, polymyxin B, ciprofloxacin, and metronidazole.

Serratia marcescens are sensitive to piperacillin, gentamicin, cefepime, tetracycline, nalidixic acid, and trimethoprim. Serratia marcescens is relatively resistant to benzylpenicillin, amikacin, cephalothin, polymyxin B, ofloxacin, and metronidazole.

Erwinia amylovora is sensitive to piperacillin, kanamycin, cefepime, cefquin, chloramphenicol, nalidixic acid, co-trimoxazole, and trimethoprim. Erwinia amylovora relatively resistant to benzylpenicillin, gentamicin, cephalothin, polymyxin B, ofloxacin, metronidazole, furadonin.

Kluyvera cryocrescens are highly sensitive to piperacillin, amikacin, cefepime, chloramphenicol, ofloxacin, and trimethoprim. Kluyvera cryocrescens are relatively resistant to benzylpenicillin, streptomycin, cephalothin, polymyxin B, and metronidazole.

Yersinia enterocolitica is sensitive to piperacillin, gentamicin, cefepime, doxycycline, chloramphenicol, ciprofloxacin, nalidixic acid, trimethoprim, furadonin. Yersinia enterocolitica is relatively resistant to ampicillin, benzylpenicillin, streptomycin, cephalothin, tetracycline, polymyxin B, ofloxacin, metronidazole.

4 Conclusion

During the study, we isolated 575 strains of enterobacteria from various animal species belonging to the genus Escherichia, Shigella, Salmonella, Klebsiella, Proteus, Providencia, Hafnia, Morganella, Enterobacter, Citrobacter, Serratia, Erwinia, Kluyvera, Yersinia. In agricultural animals, the number of transient pathogenic enterobacteria Salmonella enteritidis and Yersinia enterocolitica did not exceed 0.22–0.12%, respectively, and Shigella dysenteriae and Shigella flexneri were not isolated. Among wild animals, the proportion of pathogenic Salmonella enteritidis and Yersinia enterocolitica did not exceed 0.24–0.11%, respectively, while Shigella dysenteriae and Shigella flexneri were isolated in a small concentration of 0.01%. In the feces of zoo animals, the amount of Salmonella enteritidis and Yersinia enterocolitica did not exceed 0.21-0.10%, respectively, while Shigella dysenteriae and Shigella flexneri were detected at the level of 0.01% of the total concentration of enterobacteria. The proportion of Salmonella enteritidis and Yersinia enterocolitica in domestic animals did not exceed 0.012-0.04%, respectively, and was the smallest compared to other animal species. At the same time, Shigella dysenteriae and Shigella flexneri were found in domestic animals at a low concentration of 0.003 and 0.005% of the total number of enterobacteria compared to wild and zoo animals.

Persistence factors in Salmonella enteritidis, Shigella dysenteriae and Shigella flexneri, Klebsiella oxytoca, and Yersinia enterocolitica were observed to be the highest among all isolated enterobacteria. Enterobacteria, having persistence factors, can survive (survive) in the macroorganism of humans and animals for an unlimited time, protecting themselves from cellular and humoral factors of nonspecific defense of the macroorganism. At the same time, the presence of persistence factors in enterobacteria allows exhibiting pathogenic properties.

Enterobacteria showed high resistance to benzylpenicillin from the group of natural penicillins, streptomycin, cephalothin from the group of cephalosporins of the first generation, to polymyxin B, to ofloxacin (tarivid) and metronidazole. Carbenicillin from the group of carboxypenicillins and piperacillin from the group of ureidopenicillins, kanamycin, amikacin, and gentamicin, cefepime from the group of IV generation cephalosporins, tetracycline, doxycycline and chloramphenicol, nalidixic acid, trimethoprim showed the highest antimicrobial activity against all cultures of enterobacteria isolated by us.

References

1. A. A. Galiakbarova, Emergency Doctor **1**, 56–65 (2021)
2. V. V. Ermakov, N. S. Titov, BIO Web Conf. **37** 00063 (2021)
3. V. V. Ermakov, N. S. Titov in *Actual Problems Of Veterinary Medicine, Biotechnology And Morphology: Collection Of Scientific Papers Of The National Scientific And Practical Conference* Kinel, Russia, 11–13 June 2021 136–140 (2021)
4. A. B. Kononenko, D. A. Bannikova, I. B. Pavlova, S. V. Britova in *Proceedings of the All-Russian Research Institute of Experimental Veterinary Medicine. Ya. R. Kovalenko* **80 (1),** 238–245 (2018)
5. A. I. Lusevich, E. S. Shcherbakova, Alley of Science **1, 7 (46),** 242–245 (2020)

6. E. M. Lenchenko, D. A. Blumenkrantz, Veterinary **1,** 25–29 (2020)
7. U. M. Nemchenko, E. A. Kungurtseva, M. V. Savelkaeva, E. V. Grigorova, E. I. Ivanova, T. V. Tunick, L. V. Rychkova, Experimental and Clinical Gastroenterology **4 (176),** 59–64 (2020)
8. E. N. Ruchko, V. I. Pleshakova, in *Modern Trends In The Development Of Veterinary Science And Practice: Proceedings Of The National (All-Russian) Scientific And Practical Conference*, Omsk, Russia, 15–19 April (2021)
9. M. V. Smirnova, S. D. Artemuk, E. I. Belkova, A. A. Meltzer, O. Yu. Timirbaeva, D. A. Kugotova, *Medicine: Theory and Practice* **4 (3),** 91–98 (2019)
10. I. F. Shishlo, S. A. Krasny, Vu. N. Dolgina, *Oncology and Radiology of Kazakhstan* **2 (60),** 4–8 (2021)
11. I. S. Shulga, M. E. Ostyakova, L. A. Lavrushina, I. E. Gorbacheva, *Veterinary Doctor* **1,** 65–73 (2022)
12. I. V. Shipitsyna, E. V. Osipova, *Clinical Laboratory Diagnostics* **67 (3),** 158–162 (2022)

Improved technology elements of table carrot cultivation in the Volgograd region

A I Belyaev[1], *A M* Pugacheva[1], *E P* Borovoy[2], *N Yu* Petrov[2], and *M P* Aksenov[2,*]

[1] Federal Scientific Centre of Agroecology Complex Meliorations and Protective Afforestation Russian Academy of Sciences, 97, University Avenue, Volgograd, 400062, Russian Federation
[2] Volgograd State Agrarian University, 26, University Ave., Volgograd, 400002, Russian Federation

Abstract. During the period of field research to find ways to improve the elements of table carrot cultivation technology, we proposed a comprehensive resource-saving technology for the cultivation of carrot root crops on drip irrigation in the Lower Volga region. A hypothesis was put forward to increase the productivity and quality of table carrots through the use of new generation biological products and fertigation of crops using drip irrigation. The research program included a comparative assessment of the zoned table carrot variety Shantane (control) and promising Red Core and Santa Cruz hybrids, which belong to the same early maturity group. For this, mineral fertilizers were calculated in doses for the planned yields of 90, 110, and 130 t/ha, using drip irrigation. Biological preparations Agriful (concentration 3 and 4 l/ha) and Fertigrain Foliar (concentration 1.3 l/ha) were used, in addition, to their complex use together with mineral fertilizers.

1 Introduction

The soil and climatic resources of the Volgograd region make it possible to grow vegetable products in gross volumes not only for domestic needs but also for export to other regions of the country [1–3]. Nevertheless, the biological potential of the studied region is currently exploited very inefficiently, only by 25–38%. The state of the old arable land fund of the region can be characterized as satisfactory but the state of reclamation systems is declining catastrophically [4].

Based on the formulated hypothesis, the fastest application of the latest agrotechnological elements with constant drip irrigation [5–7], the introduction of seeds of hybrids and carrot varieties of domestic and foreign producers [8], will significantly accelerate the growth of average yields of more than 100 t/ha.

The most important condition for increasing the efficiency of irrigated agriculture at the present stage is the rational use and implementation of modern intensive agricultural technologies [9–10]. The main component of these elements will be actively implemented drip irrigation regimes. Such an application will allow to form stable and high yields and to switch to modern industrial technologies for growing table carrots.

The development of agricultural production at the present stage is becoming one of the most important strategic tasks of the government of the country in the field of agrarian policy, aimed at the formation and further development of agricultural production. These activities are primarily necessary to ensure the country's food security, as well as the development and promotion of products in the world food markets.

The conditions for worthy and successful competition of domestic products in the world food market are high quality, satisfying the diverse taste preferences of consumers, and an acceptable purchase price. Therefore, in the program of proposed activities that we are considering, a special place is given to ecological and biological farming systems. These activities will ensure the formation of stable and high yields with excellent and good quality values [11–13].

Consequently, the observance of all elements of agricultural technology for growing table carrots, the water, and nutrient regimes of the soil, will basically allow for growing consistently high yields of carrot root crops in the conditions of the Volgograd region. However, the following issues need to be studied: new types of biological products, varieties and hybrids that appear on the market, and modes, and methods for implementing drip irrigation for a given crop.

It should be borne in mind that the vegetable growing complex is a rather complex system of interconnected technological links of agro-industrial production, processing, and storage. Therefore, it is realistic to reduce losses only under the condition of systematic integrated development of all its component techniques [14, 15].

2 Materials and methods

Experimental studies were carried out on the collective farm "V. A. Zaitsev". Gorodishchensky district, Volgograd region since 2019–2021 For research, a zoned variety of table carrot Shantane (control) and new promising hybrids Santa Cruz and Red Core belonging

* Corresponding author: aksenovmp@mail.ru

to the same early maturity group were selected. The dosage of mineral fertilizers was calculated for the planned yield of 90, 110, and 130 t/ha. Drip irrigation kept the humidity threshold at 75–80% HB.

Biological preparations Agriful (with a concentration of 3 and 4 l/ha) and Fertigrain Foliar (with a concentration of 1.3 l/ha) were chosen for the study, in addition, their complex effect with mineral fertilizers on the formation of the yield of table carrots of the Champagne variety, a hybrid of Santa Cruz and Hybrid Ed. Cor. Dosages of biological products from the manufacturer.

Carrots are very sensitive to the fertility of the soil, as well as to the nutrients contained in it in a balanced ratio. With a yield of 1 ton of carrots per hectare, the removal of microelements is as follows:
- K2 is 5.0 – 6.5 kg, depending on the grown hybrid or variety;
- P2O5 is 1.0 – 1.7 kg depending on the cultivated hybrid or variety;
- N is 2.2 – 4.0 kg depending on the cultivated hybrid or variety.

Table 1. Scheme of applying mineral fertilizers for carrot crops.

Indicator	Planned yield, t/ha								
	90			110			130		
	N	P_2O_5	K_2O	N	P_2O_5	K_2O	N	P_2O_5	K_2O
Required	240	120	440	300	150	550	360	180	660
For basic processing	120	90	220	150	90	280	180	90	330
1st top dressing (phase 3 sheets)	80	15	120	105	30	170	90	45	200
2nd dressing	40	15	100	45	30	10	90	45	130

Foliar top dressing was applied in three terms:
- phase 3 – 5 leaves with Agriful dilution (4 l/ha + Fertigrain Foliar (1.3 l/ha);
- after 10 days with Agriful dilution (3 l/ha + Fertigrain Foliar (1.3 l/ha);

We adopted the technology for growing carrot roots recommended for this zone. The repetition of the experiment is 4-fold. A 4-line sowing scheme was used: 0.05 + 0.30 + 0.30 + 0.05. The seeding rate was 1.2 million viable seeds per hectare. Sowing was carried out by a seeder – Maple. The area of the accounting plot was 36 m^2. The location of the plots was systematic.

3 Results and discussion

The experiments were carried out on typical medium loamy chestnut soils, which are most common in the Volga-Don interfluve. The presence of a bulk density of 24.7 t/m^3 and a total porosity of 22.08% indicates the presence of humus in the arable horizon was 1.8%. that these soils are very favorable for the growth of table carrots. The experimental soils were rich in exchangeable potassium (up to 310 mg/kg of soil). The absorption capacity of the plow horizon was within 22.8 mg-eq.

Achieving a stable and high yield of carrots is possible only with the introduction of drip irrigation. When conducting research, 10-12 days before the start of the harvest of root crops, the last watering was carried out. Harvesting began with the onset of the first autumn frosts, which contributes to the better preservation of root crops during storage. Not significant autumn frosts lead to deep chemical transformations in carrot roots. Under the influence of frost, complex molecular compounds are transformed into simple ones. These processes influenced the increase in the total sugar content in root crops (19 – 20%) and had a positive effect on the safety of root crops. The difference in the losses of carrot root crops put into storage, subjected to frost, and harvested before the onset of frost was 5–10% in favor of frozen root crops. These indicators also affect the overall quality characteristics of marketable root crops in storage.

Another feature of the late harvest is the fact that there is a fairly intensive increase in the mass of root crops, especially in autumn. Based on these provisions, early harvesting actually reduces the potential yield and is considered impractical. When harvesting, both the biological characteristics of the cultivated variety or hybrid and the weather conditions that existed during the growing season are considered. Table 2, Table 3, and Table 4 contain information on the obtained harvesting yield.

Table 2. Dependence of the yield of carrot roots on the application of mineral fertilizers, t / ha, (average value for 2019–2021).

N	Hybrid name, varieties	Experience option	Planned yield	Actual yield
1	Shantane	Control	–	88.37
		$N_{240}P_{120}K_{440}$	90	101.09

		$N_{300}P_{150}K_{550}$	110	100.34
		$N_{360}P_{180}K_{660}$	130	97.83
2	Santa Cruz	Control	–	98.71
		$N_{240}P_{120}K_{440}$	90	115.32
		$N_{300}P_{150}K_{550}$	110	139.25
		$N_{360}P_{180}K_{660}$	130	142.51
3	Red Cor	Control	–	103.73
		$N_{240}P_{120}K_{440}$	90	129.42
		$N_{300}P_{150}K_{550}$	110	148.89
		$N_{360}P_{180}K_{660}$	130	157.31

The analysis of the presented material allows us to state that in the aspect of applying the calculated amount of mineral fertilizers for the planned table carrot yields of 110 and 130 t/ha, it is possible when applying $N_{300}P_{150}K_{550}$ and $N_{360}P_{180}K_{660}$ on modern hybrids of Santa Cruz and Red Cor. While on the released variety Shantane, when applying similar doses of mineral fertilizers for the planned corresponding yields, it formed 100.34 and 97.83 t/ha. Against the background of natural fertility, the yield, depending on the variety or hybrid, varied from 88.37 t/ha (for the Shantane variety) to 103.73 t/ha (Red Core hybrid). That is, the introduction of the planned doses of mineral fertilizers allows you to form a given yield.

Table 3. The yield of carrot roots depending on the use of biological products, t / ha (average value for 2019–2021).

N	Hybrid name, varieties	Experience option	Planned yield	Actual yield
1	Shantane	Control	–	87.45
		1st top dressing	90	93.57
		2nd top dressing	110	102.81
2	Santa Cruz	Control	–	98.71
		1st top dressing	90	113.63
		2nd top dressing	110	130.92
3	Red Cor	Control	–	101.08
		1st top dressing	90	122.68
		2nd top dressing	110	141.39

A statement of the results of the experiments presented in Table 2 allows drawing the following conclusion: on the variants of natural soil fertility, neither the zoned variety nor modern hybrids (Santa Cruz and Red Core) are able to form the programmed yield according to the given program. The picture is completely different after the introduction of the new generation of biological products Agriful and Fertigrain Foliar. With the introduction of one top dressing, a research program was carried out to obtain 90 t/ha (for the Shatane variety – 93.57 t/ha, for the Santa Cruz hybrid – 113.63 t/ha and for the Red Core hybrid – 122.64 t/ha). Applying an additional second top dressing with biological products led to a further increase in the yield of table carrots. As a result, the Shantane variety had a yield of 102.81 t/ha, the Santa Cruz hybrid – 130.92 t/ha, and the Red Core hybrid – 141.39 t/ha. All this indicates that modern hybrids were more responsive to the use of new generation biological products.

Table 4. Dependence of the yield of carrot roots on the complex impact of biological products and mineral nutrition, t / ha (average for 2019–2021).

N	Hybrid name, varieties	Experience option	Planned yield	Actual yield
1	Shantane	Control	–	89.74
		$N_{240}P_{120}K_{440}$ + 1st top dressing	90	1121.67
		$N_{300}P_{150}K_{550}$ + 2nd top dressing	110	123.45

		$N_{360}P_{180}K_{660}$	130	98.11
2	Santa Cruz	Control	–	97.69
		$N_{240}P_{120}K_{440}$ + 1st top dressing	90	131.73
		$N_{300}P_{150}K_{550}$ + 2nd top dressing	110	158.61
		$N_{360}P_{180}K_{660}$	130	141.36
3	Red Cor	Control	–	104.37
		$N_{240}P_{120}K_{440}$ + 1st top dressing	90	141.56
		$N_{300}P_{150}K_{550}$ + 2nd top dressing	110	167.60
		$N_{360}P_{180}K_{660}$	130	172.08

The analysis of the experimental data showed that the maximum yield of carrot roots was obtained with the complex application of biological products, mineral fertilizers $N_{300}P_{150}K_{550}$ at the calculated dosage using the 2nd top dressing. The yield on the Red Core hybrid was 167.7 t/ha. Increasing the dosage of mineral fertilizers to $N_{360}P_{180}K_{660}$ increased the yield on this hybrid up to 172.08 t/ha. The combined use of new generation biological products in conjunction with the applied mineral fertilizers made it possible to raise the yield of table carrots to a higher level. The excess of the actual carrot yield of the Red Core hybrid over the planned yield level (130 t/ha) was 42.08 t/ha.

However, the cost of applied fertilizers was incomparable with the yield increase obtained. Against the background of natural fertility, the lowest root crop yield of 87.45 t/ha was obtained on the zoned variety Shantane. The Santa Cruz hybrid showed intermediate yield values between the released Shantane variety and the promising Red Core hybrid. The maximum yield on this hybrid, on average over three years, was obtained with the variant $N_{300}P_{150}K_{550}$ + 2nd top dressing and it was 158.61 t/ha, while in the control variant it was 97.69 t/ha. That is, the Red Core hybrid proved to be more responsive to the use of mineral fertilizers and the use of biological products.

4 Conclusion

The results of the research over a three-year period allow concluding that the recommendations for commodity producers of the Volgograd region on drip irrigation of chestnut soils for the cultivation of the Red Core hybrid against the background of the application of calculated doses of mineral fertilizers N300P150K550 + 2nd top dressing.

References

1. A. D. Akhmedov, E. E. Dzhamaletdinova, Scientific journal of the Russian Research Institute of Land Reclamation Problems **4 (36),** 1–16 (2019)
2. V. V. Borodychev, A. A. Martynova, M. N. Lytov *Cultivation of carrots under irrigation: from experiment to technology* (Volgograd: Volgograd State Agrarian University) (2019)
3. B. M. Kizyaev, V. V. Borodychev, A. A. Martynova, Vegetables of Russia **3,** 51–56 (2020)
4. A. S. Ovchinnikov, S. A. Lisichenko, V. V. Borodychev, A. A. Martynova, Fertility **3,** 30–32 (2015)
5. M. A. Likhomanova, O. A. Solovieva, *Proceedings of the Nizhnevolzhsky agrouniversity complex* **1 (61),** 161–173 (2021)
6. N. N. Dubenok, R. I. Shumakova, A. A. Martynova, *Proceedings of the Nizhnevolzhsky agrouniversity complex* **4 (60),** 27–39 (2020)
7. A. S. Ovchinnikov, S. A. Dusar, V. V. Borodychev, A. A. Martynova, *Proceedings of the Nizhnevolzhsky agro-university complex: science and higher professional education* **1 (45),** 9–20 (2017)
8. A. N. Khovrin, M. A. Kosenko, A. V. Kornev, A. M. Sokolova, Potatoes and vegetables **7** 32 (2019)
9. N. I. Matveeva et al., IOP Conf. Ser.: Earth Environ. Sci. **659,** 012067 (2021)
10. N. Yu. Petrov et al., IOP Conf. Ser.: Earth Environ. Sci. **786 (1),** 012002 (2021)
11. D. F. de Carvalho et al., Revista Brasileira de engenharia agricola e ambiental **22 (7),** 445–450 (2018)
12. B. B. Longwe, V. B. Wali, Indian journal of economics and development **12 (4),** 819–822 (2016)
13. Yu. N. Pleskachev, O. G. Chamurliev, L. V. Gubina, Bulletin of RUDN University. Series: Agronomy and animal husbandry **4** (2018)
14. V. J. da Silva et al., Bioscience Journal **6 (27),** 954–963 (2011)
15. D. F. de Carvalho et al., Ciencia Rural **46 (7),** 1145–1150 (2016)

Qualitative indicators of protein concentrates from pea and chickpea flour

Valentina Kolpakova[1,*], *Ruzaliya* Ulanova[1 2], *Denis* Kulikov[1], and *Vladimir* Bessonov[3]

[1] All-Russian Research Institute of Starch and Starch-containing Raw Materials Processing – Branch of Russian Potato Research Centre, Nekrasova str., 11, Kraskovo, Moscow region, 140051, Russia
[2] S.N. Vinogradsky Institute of Microbiology, Federal Research Center "Fundamental Foundations of Biotechnology", Russian Academy of Sciences, Moscow, 60th anniversary of October ave., 107143, Russia
[3] Federal Research Center for Nutrition, Biotechnology and Food Safety, Ustyinsky proezd, 2/14, Moscow, 109240, Russia

Abstract. A comparative analysis of the qualitative indicators of food and feed protein concentrates (PC) from pea and chickpea flour was carried out. The chickpea PC contains more protein than the pea PC: 83.22±0.35 and 71.78±0.35% on dry matter (DM), respectively, the biological value adjusted for protein digestibility (PDCAAS) in the pea PC (96%) is higher than that in the chickpea PC (76%). The PCs differed in the content of essential amino acids, copper, cobalt, manganese, nickel, the amount of flavonoids and foaming ability. Higher foaming capacity and lower foam stability in the chickpea PC correlated with higher flavonoid content and percent parallel β-structure and anti-parallel 3_{10}-helix proteins. A fodder biomass with a protein content of 61.68-64.10% and a biomass with a cultural liquid with 50.60-53.56% protein on DM were obtained. Biologically valuable concentrates differed in the mass fraction of fat, soluble, insoluble carbohydrates, potassium, magnesium, cobalt, manganese, sodium and the ratio of saturated:unsaturated fatty acids. A correlation was found between the amount of flavonoids, the optical density at D_{590} nm, and the color of preparations (correlation coefficient R=0.895). It is recommended to use the PCs for food purposes, serum concentrates, in feed for various animals.

1 Introduction

In recent years, in the country and abroad, interest in legumes for their use in food and feed products has increased due to the high nutritional value of protein and relatively low cost [1]. Protein concentrates and isolates are used in the production of bakery, pasta, imitation dairy products, curds, yoghurts, meat analogues, baby products, sports nutrition, etc. Therefore, the development of new technologies for concentrated forms of plant protein with high nutritional value and good functional properties is an urgent task [2]. Usually, proteins from plant materials are extracted with alkali solutions, which form new cross-linked covalent bonds, D-isomers of amino acids, etc. These chemical bonds have a negative impact on human and animal health. Safe technologies with enzyme preparations (EP) for obtaining a high yield of biologically valuable pea and chickpea proteins with appropriate functional properties are little known. At the same time, in the production of protein products, liquid by-products are always formed. These are serum waters. Along with the growth of world demand for protein-containing food products and feed from plant raw materials, the volumes of liquid effluents and waste that pollute the environment are also increasing [3]. One of the directions of the scientific and technological development of the world community for the coming years is the rational use of natural resources. Therefore, an important problem is the development of environmentally friendly technologies for the disposal of all types of waste and increasing the level of environmental culture during the deep processing of organic raw materials. Waste is a substrate for the synthesis of biomass of microorganisms [2, 4, 5]. Biomass is a part of the diet of farm animals to increase their productivity and an ingredient in food as a protein, lipid, carbohydrate source or a technical substance [6-8]. For example, a product obtained by fermentation of corn stalks with *Saccharomycetes* or by the consortium of *Saccharomycetes*, *Lactobacillus plantarum* and *Lactobacillus casei* has a positive effect on the animal organism and on the environment [9–11]. *Saccharomyces cerevisiae* yeast in feed recipes for broilers in an amount of 0.8% by weight of the diet also increased the efficiency of its use [12]. The study of the microbiota of fecal samples on days 21 and 42 using polymerase reaction (PMR) revealed a positive effect of supplementation with the yeast *S. cerevisiae* on the microflora of broilers and ruminants, increased fiber digestibility and the population of rumen cellulolytic bacteria *Ruminococcus flavefaciens* [13]. The addition of *S. cerevisiae* and/or *Aspergillus oryzae* biomass to the diet of cattle increased milk yield and fat content [14–16]. Also there are feed additives from coffee sludge [17], from alcohol stillage with wheat bran, which are obtained by growing the yeast *Saccharomyces*

* Corresponding author: val-kolpakova@rambler.ru

diastaticus and carotene-forming yeast *Rhodosporidium species* to increase the content of essential amino acids in them [18]. A feed additive with carotenoids and lipids was synthesized with the yeast *Rhodotorula glutinis*, *Rhodotorula mucilaginosa* and *Rhodotorula gracilis* on a medium from wastewater during potato processing together with waste from the production of glycerin [19]. At a sugar cane processing plant, *Aspergillus niger* mushroom biomass with lipids for biodiesel fuel was obtained from wine waste [20]. A microbial-plant concentrate with the yeast *S. cerevisiae* with a mass fraction of protein of 25.2±2.1%, fat - 22.1±3.2%, carbohydrates - 40.8±1.6% was synthesized on the basis of triticale grain extract for feeding pond fish. This concentrate contains potassium, calcium, cobalt, iron, zinc, molybdenum, nickel and other mineral elements.

According to FAO, today the annual protein deficit in the world is estimated at tens of millions of tons and by 2050 will reach 30 million tons [3]. Since ancient times, legumes and their processed products have been used as a source of protein in the diet of people, including those who do not eat meat [21, 22]. Also, little attention is paid to the processes of processing by-products of leguminous crops, which are formed during the production of protein concentrates using bioconversion. For example, it is known that with by-products from the extraction of pea protein, a food vegan protein concentrate is obtained to replace meat. The studies were carried out with 5 strains of filamentous fungi: *Neurospora intermedia*, *A. oryzae*, *Rhizopus oryzae*, *Fusarium venenatum*, *Monascus purpureus*, which were grown at 35±2 °C for 48 h until the protein content in the biomass was 43.13-59.74% on dry matter (DM). For every ton of by-product, about 680 kg of mushroom biomass can be obtained with 38% additional protein [5]. The synthesis of feed concentrates with cultures of *S. cerevisiae* and *Geotrichum candidum* was obtained with a protein mass fraction of 61.68-70.48% from serum, which remains after precipitation of pea protein [6].

Recently, the industry has been particularly interested in chickpea - a legume crop, which makes it possible to create technologies for protein concentrates (PC), isolates and a number of other useful products [2, 3]. This crop is resistant to hot climates and is a promising addition to soybeans and peas in conditions of recurring drought and global warming [1, 23–25]. The gross harvest of chickpeas in Russia in 2019 amounted to 506.3 thousand tons, but it decreased by 18.4% compared to the record harvest of chickpeas in 2018 (620.4 thousand tons). Since 2014, the chickpea harvest has been on an upward trend, increasing by 15.0%. In recent years, chickpea sown areas have increased by 355.2 thousand hectares (71.6%), over 5 years – by 178.1 thousand hectares, which amounted to 26.5%. These data indicate the stability of the raw material base of chickpeas and peas in Russia, which is promising to be used for the production of protein preparations of various compositions and purposes [2]. The urgency of the problem of providing the population with protein from this type of crop dictates the development of various technologies for protein preparations, therefore, the creation of methods for the utilization of liquid secondary products for future introduction into production is in demand. The purpose of this work is a comparative description of the qualitative indicators and properties of protein concentrates obtained by a biotechnological method for food purposes, and feed concentrates synthesized with the help of microorganisms from serum remaining after the extraction of the bulk of the dietary protein from pea and chickpea flour.

2 Materials and methods

As objects, we used pea flour from grain of the Yamal variety grown in the Altai Territory and chickpea flour from grain of the Volzhanin variety grown in the Volgograd region in 2019-2020. The chemical composition of pea flour in % in dry matter in mass fraction of protein (N×6.25) – 24.30±1.40, ash – 2.87±0.20, fat – 1.58±0.12, carbohydrates – 73.02±3.66. The chemical composition of chickpea flour in % in dry matter in mass fraction of protein (N×6.25) – 24.54±0.23, ash – 2.91±0.02, fat – 4.89±0.31, carbohydrates – 67.66±0.56. The amount of total nitrogenous substances in flour and in concentrates was determined by the Kjeldahl method, the mass fraction of moisture, ash and fat according to Russian Standards (GOST 10846-91, GOST 13586.5-93, GOST 10847-2019 and GOST 29033-91). Carbohydrates were determined by the difference between 100% and the sum of other components.

For isolation from PC and the secondary product (serum), we used enzyme preparations (EP) from Novozymes (Denmark): Shearzym 500 L from *Aspergillus aculeatus* with a xylanase activity of 500 U/g and optimal conditions of action 65–75 °C, pH 4.5–5.5. As a source of cellulase, carboxymethyl cellulase and β-glucanase activity, Viscoferm L was used, produced by *Trichoderma* and *Aspergillus* strains with a cytolytic activity of 600 U/g of raw material and an optimum of action at a temperature of 50–60 °C and pH 4.8–5.8. As a source of α-amylase, we used Fungamyl 800 L from the mold *A. oryzae*, which cleaves α-1,4 glucosidic bonds to form maltodextrins and maltose (50-60 °C, pH 5.0-6.5). The amyloglucosidase-containing EP was AMG 300 L 2500, isolated from the fungus *A. niger*, which cleaves both α-1,4 and α-1,6 glucosidic bonds in starch, dextrins and oligosaccharides to form glucose. The optimum action lay in the region of 55-60 °C, pH - 4.5-5.5. Alcalase 2.4 L EP from *Bacillus licheniformis* with a protease activity of 2.4 units/g was used as a source of proteases. We determined the enzymatic activity of the preparations according to Russian Standards (GOST P54330-2011, GOST R 53974-2010 and GOST R 55302-2012). All reagents were chemically pure. Ultrasound treatment of the protein suspension was performed on a Soniprep 150 ME device (MseLtd., UK).

To obtain feed microbial-plant concentrates (FMPC) from the collection of the Institute of Microbiology named after S.N. Vinogradsky (Moscow) we used the yeast *S. cerevisiae* 121 and the fungus *G. candidum* 977, the phylogenetic position of which was determined at the Research Institute of Genetics (Russia). Museum cultures from agar wort were subcultured into a test tube with pea or chickpea serum, which was pre-sterilized at a pressure of 0.1 MPa and cooled. Microorganisms were grown for 24 h, after which they were subcultured into 300 cm^3 flasks with 50 cm^3 of serum (pH 6.0-6.5). Cultivation was carried out on a shaker at a flask rotation speed of 150 min^{-1} and a temperature of 27±1 °C for 24–48 h. The suspension was inactivated at a temperature of 95±5 °C for 10-15 min and cooled to a temperature of 22±2 °C. The biomass was separated by centrifugation at 4000 min^{-1} for 10 min, dried separately or together with the culture liquid in a Hochvacuum HVDTG-50 lyophilizer (Germany) in a vacuum at –80 °C, and feed microbial-plant concentrates (FMPC-1 and FMPC-2) were obtained [26].

The amount of protein in solutions was determined by the Lowry method, soluble and insoluble carbohydrates in concentrates according to the method described in book [4]. The method is based on the enzymatic hydrolysis of starch and protein compounds with amylase, protease and amyloglucosidase to mono-, di-, oligosaccharides and peptides. For the hydrolysis of proteins and carbohydrates, α-amylase Fungamyl 800 L, protease Alcalase 2.4 L, and amyloglucosidase AMG from Novozymes (Denmark) were used. Soluble dietary fiber was precipitated with 4 volumes of 95% (v/v) ethanol for 2 hours at 4 °C, washed 2 times with 95% ethanol. The amount of dried mass was determined by the gravimetric method, the mass fraction of carbohydrates was expressed as a percentage (g/100 g) [27]. The digestibility of feed concentrates was determined according to Russian GOST 24230-80.

The carbohydrate composition of serum and culture liquid was studied on a gas chromatograph model GCMS-QP 2010 (Japan, Shimadzu Corporation) with a ReproGel Na column (9 μm, 8x300 mm). The amino acid composition of proteins was determined on a Hitachi model L-8800 chromatograph (Japan) with a cation exchanger (sulfonated copolymer of styrene with divinylbenzene) and a stepwise gradient of sodium citrate buffer solutions with increasing pH and molarity. The data were processed in the on-line system "MultiChrom 1.52" (Russia). A ninhydrin reagent was used to detect eluted amino acids. The instrument was calibrated with standard amino acid mixtures after dilution. The reproducibility in terms of release time was 0.3%, in terms of peak area – 1%, the lower limit of sensitivity – 3 pmol (signal-to-noise ratio = 2 for Asp). For acid hydrolysis, 3–5 mg of a sample was placed in a molybdenum glass ampoule and 0.3 cm^3 of a hydrolyzing mixture (concentrated hydrochloric and trifluoroacetic acids in a ratio of 2:1 with 0.1% β-mercaptoethanol) was added. The sample was frozen in liquid nitrogen, evacuated, and melted in a glass ampoule. Hydrolysis was carried out at 155°C for 1 hour, after which the ampoule was opened, the contents were transferred into a plastic tube (Eppendorf, Germany) and the hydrolyzing mixture was removed on a CentriVap Concentrator LABCONCO (US). 0.1 N HCl was added to the dry residue of the hydrolyzate, vigorously mixed in a closed plastic tube, and centrifuged for 5 min at 8000 min^{-1} in a Microfuge 22R centrifuge (Beckman-Coulter, USA). When calculating the amino acid score of the concentrates, the FAO/WHO reference protein scale was used [28]. Lipids were isolated from FMPC by the Folch method. After the lipids were evaporated in a rotary evaporator, chloroform, methanol hydrochloride (SupelcoMethanolic-HCl 0.5 N) were added to them, and the mixture was heated for 1 h at 90°C. The fatty acid composition of lipids was studied on a chromatograph with a Simadzu GCMS-QP 2010 Ultra mass detector at a temperature of 120°C, an injector at 200°C; interface – 205°C, detector – 200°C on an SLB-IL82 column (30 m, 0.20 μm, d = 0.25 mm) with a gel carrier at a flow rate of 35.6 cm/sec and its division 1:10. The gradient mode was changed from 120°C to 260°C at a rate of 5 °C/min for 2 minutes.

The content of lead and cadmium in the concentrates was determined according to GOST 30178-96, the remaining macro- and microelements were determined after dry ashing by the atomic absorption method on a Hitachi instrument (model Z 5300) in an air-acetylene flame with Zeeman correction. The functional properties of PC were studied according to the methods described in the works [29], flavonoids - by the spectrophotometric method with light absorption at 276 nm [30].

The molecular weights of flour proteins, extracts isolated under the action of enzyme preparations (EP), and serum were determined using gel electrophoresis under denaturing conditions (SDS-PAGE). Protein samples in an amount of about 50 μg were mixed with buffer in a ratio of 1:1. To prepare the buffer, 60 cm^3 of glycerol and 1 mg of bromophenol blue were poured into a beaker, and the pH was adjusted to 6.8 with concentrated HCL. 5 cm^3 of ß-mercaptoethanol was added to the solution, and the volume of the solution was adjusted to 100 cm^3. Samples and solutions of proteins markers in the amount of 102 μg/30 μl were heated for 2 min at 95–100°C. To prepare a 15% separating gel, 4.5 cm^3 of an AB solution were mixed (a sample of 29.6 g of acrylamide was dissolved in a small amount of water, 0.4 g of bisacrylamide was added and the volume was adjusted to 100 cm^3), 2.5 cm^3 of Tris-HCl buffer with pH 8.8, 3 cm^3 of water, 20 μl of TEMED and 160 μl ammonium persulfate (PSA). The solution was shaken and poured between glasses. To prepare a concentrating gel, 1 cm^3 of AB solution, 1 cm^3 of Tris-HCl buffer, pH 6.8, 3 cm^3 of water, 20 μl of TEMED, and 160 μl of PSA were mixed. The contents were shaken and poured into a device with electrophoresis plates. A comb was placed between the glasses and wells for protein samples were formed. Electrophoresis was carried out at a voltage of 50-60 V for the entry of samples into the gel, then at 120 V to separate the components. Ready electrophoregrams were stained with Coomassie bright blue and scanned. The experimental data were processed using TableCurve 2D 5.1, TableCurve 3D 4.0, Mathematica 10.3, and Statistica 10 programs. The confidence interval of the arithmetic mean was calculated using the significance level $p = 0.05$.

3 Results and discussion

3.1 Protein concentrates for food purposes

The extraction of proteins from pea and chickpea flour was carried out with hydrolytic EPs: cellulase, xylanase, amylase and protease at 3 stages according to the scheme shown in Figure 1. To determine the optimal parameters of protein extraction, an experiment planning matrix was compiled from 25 experiments on the dependence of protein solubility (R) on the EP concentration (sv), hydromodule (gl), and extraction duration (t). After solving the tasks, the proteins were transferred into the solution at optimal values of the parameters: for pea flour: sv – 1.50%, t – 4.2 h, gl – 15; for chickpea flour: sv – 1.81%, t – 5 h, gl – 25 at a mixing speed of both suspensions of 200 min^{-1}. Protein solubility was increased by 23±1% to 84±1% with a 3-minute ultrasonic treatment (UST) of the protein extract and a wave amplitude of 10 μm.

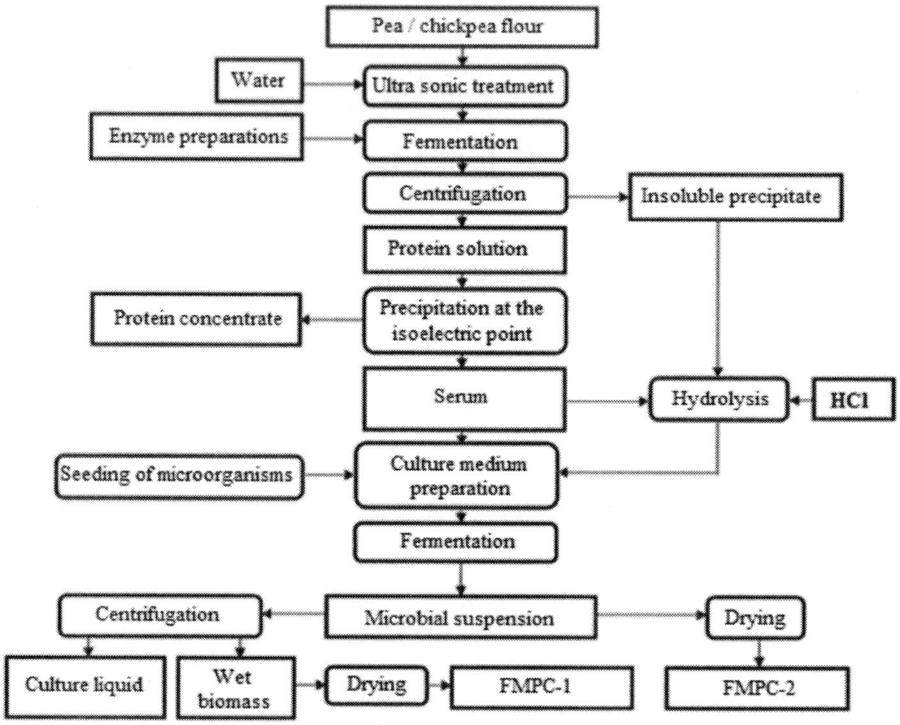

Fig. 1. Technological scheme for processing pea and chickpea flour into food and feed concentrates.

Proteins from the solution were precipitated with 10% HCL at the isoelectric point (pH 4.2), followed by centrifugation of the suspension at 5500 min^{-1} and separation of the precipitate from the serum. The precipitate was washed twice with water, freeze-dried, and PCs were obtained with the chemical composition presented in Table 1. The yield of proteins was 65–70% of their content in the raw material. Both concentrates have protein amino acid scores of 100% or more, with the exception of the valine score of the chickpea PC.

Table 1. Chemical composition of pea (PPC) and chickpea (CPC) protein concentrates.

<table>
<tr><th colspan="2" rowspan="3">Moisture, %</th><th colspan="7">Mass fraction, % on DM</th></tr>
<tr><th rowspan="2">Protein (Nx6.25)</th><th rowspan="2">Lipids</th><th rowspan="2">Ash</th><th colspan="4">Alimentary fiber</th></tr>
<tr><th>Soluble</th><th colspan="3">Insoluble</th></tr>
<tr><td colspan="2" rowspan="2">3.86±0.20</td><td colspan="7">PPC</td></tr>
<tr><td>71.78±0.35</td><td>4.47±0.27</td><td>1.80±0.27</td><td>9.67±0.76</td><td colspan="3">7.57±0.26</td></tr>
<tr><td colspan="2" rowspan="2">9.76±0.11</td><td colspan="7">CPC</td></tr>
<tr><td>83.22±0.35</td><td>2.23±0.31</td><td>1.11±0.38</td><td>8.01±0.70</td><td colspan="3">5.81±0.48</td></tr>
<tr><td colspan="9">Amino acid score, %</td></tr>
<tr><td>Val</td><td>His</td><td>Ile</td><td>Leu</td><td>Lys</td><td>Met+Cys</td><td>Thr</td><td>Trp</td><td>Phe+Tyr</td></tr>
</table>

 https://doi.org/10.1051/bioconf/20225200007

PPC								
110	144	166	144	145	109	160	212	183
CPC								
86	120	120	111	100	100	136	184	195

The amount of essential amino acids in the PPC did not differ much from the amount of such acids in the CPC and was equal to 256.21 and 247.9 mg/100 g, respectively (Figure 2). Adjusted for protein digestibility (PDCAAS) (88%) the biological value of the PPC is 96%, and that of the CPC is 76%.

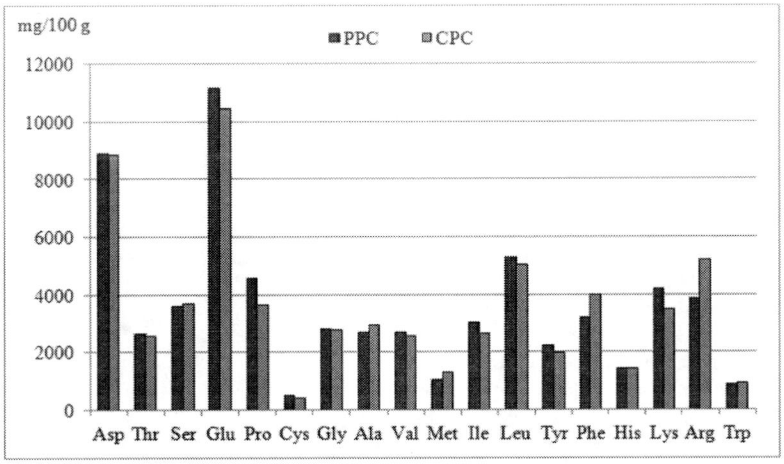

Fig. 2. Amino acid score of PPC and CPC.

The composition of macro- and microelements of PCs was characterized in Table 2. In the CPC, the mass fraction of iron was 2.1 times lower, and zinc was 1.4 times less as compared to the PPC, while copper, cobalt, manganese and nickel, on the contrary, were 1.2-4.8 times more than that in the PPC.

Table 2. The content of macro- and microelements in grains and protein concentrates.

No	Macro- and microelements	Pea		Chickpea	
		Grain	Protein concentrate	Grain	Protein concentrate
1	Sodium, mg/100g	33.0±2.5	103±7.0	4.93±0.49	109±7.0
2	Potassium, mg/100g	872±10.0	259±14.0	795.4±95.4	263±15.0
3	Calcium, mg/100g	115.2±1.5	219±14.0	96.5±9.7	207±18.0
4	Magnesium, mg/100g	106.5±2.0	10.3±0.7	122.9±1.23	10.7±0.6
5	Iron, mg/100g	6.80±1.3	114±8.0	5.16±1.33	53.9±4.0
6	Zinc, mg/100g	3.18±0.31	3.10±0.25	2.54±0.25	2.20±0.20
7	Copper, mg/100g	0.74±0.06	0.36±0.02	0.66±0.06	1.73±0.05
8	Manganese, mg/100g	1.70±0.08	0.51±0.04	2.91±0.29	0.92±0.07
9	Cobalt, mcg/100g	13.0±1.0	92.0±2.0	9.0±1.30	109.0±3.0
10	Nickel, mcg/100g	241.6±11.0	190±16.0	356±36.0	340±25.0
11	Cadmium, mg/kg	0.026±0.006	0.086±0.005	0.030±0.007	0.071±0.003
12	Chromium, mg/kg	0.007±0.002	≤0.005±0.001	0.008±0.001	≤0.005±0.002
13	Molybdenum, mg/kg	0.074±0.001	≤0.040±0.001	0.050±0.001	≤0.040±0.001

Table 3. Functional properties and amount of flavonoids in PCs.

Sample of PCs	WAC, g/g	FC, %	FS, %	OAC, g/g	EA, %	ES, %	PS, %	Flavonoids, mg/g of protein
PPC cream	2.79±0.04	42±1	32±1	2.24±0.01	52±2	47±9	11.60±0.25	2.78±0.24
PPC brown	2.44±0.03	91±1	10±1	2.25±0.03	56±3	51±3	13.10±0.15	15.05±0.71
CPC light yellow	2.82±0.05	85±0	12±0	2.26±0.03	55±1	52±1	12.20±0.50	14.06±0.41

Note: WAC – water absorption capacity; FC - foaming capacity; FS - foaming stability; OAC - oil absorption capacity; EA - emulsifying activity; ES - emulsifying stability; PS – protein solubility.

The values of functional properties (Table 3) are at the level of the values of the functional properties of cereal PCs obtained from wheat, rice, barley, rye, etc.

Table 4 presents data on the elements of the secondary structure of proteins of light powder of the PPC and the CPC, obtained on the basis of circular dichroism (CD) spectra. The measurements were carried out in a solution of 0.05 M acetic acid at 20°C and protein concentrations from 0.10 to 0.16 mg/cm^3.

Table 4. Elements of the secondary structure of proteins in the PPC and CPC, % of the sum of structures.

PPC cream		**CPC light yellow**	
α – helix -**7.2±0.3**	Regular - 3.0±0.1	α – helix – **0.1±0.005**	Regular – 0.1±0.0
	Distorted - 4.2±0.2		Distorted – 0.00
Antiparallel 3_{10}-helix - **26.9±0.5**	Left-twisted - 0.8±0.1	Antiparallel 3_{10}-helix – **39.2±0.3**	Left-twisted - 4.7±0.1
	Relaxed - 12.6±0.4		Relaxed - 17.6±0.4
	Right-twisted - 13.4±0.5		Right-twisted – 16.9±0.5
Parallel β - structure - **7.1±0.3**		Parallel β - structure - **3.0±0.2**	
Turn β – curves – 14.5±0.6		Turn β – curves – 14.8±0.7	
Others - 44.3±0.7		Others - 42.9±0.6	

The CPC proteins differed from the PPC proteins by the 7 time-less regular and irregular (distorted) α-helix and 2 time-less parallel β-structure. However, the CPC contains 1.26-6 times more proteins with an antiparallel 3_{10}-helix: left-twisted, right-twisted and relaxed. Turn β-curves and other types of the secondary structure are present in equal amounts. Consequently, the high FC, but the low value of FS in the PCs are due to the parallel β-structure, all types of proteins of the antiparallel 3_{10}-helix and flavonoids.

3.2 Feed microbial-plant concentrates from pea and chickpea serum

On serum after precipitation of pea and chickpea proteins from extracts that were isolated using hydrolytic EP, feed protein concentrates were synthesized with a consortium of yeast *S. cerevisiae* 121 and micromycete *G. candidum* 977 at a weight ratio of 1:1. Plans for the experiment were drawn up in the form of Latin squares, in which the function was the mass fraction of biomass (g/dm^3), factors - pH of the substrate (*pH*), temperature (*t*) and the amount of seed (*cm*) during the growth of microorganisms for 4 days. Through the listing of the solution, the coefficients and optimal values of the factors influencing the maximum yield of the mass fraction of biomass *md*, g/dm^3 were determined. The equations for the dependence of the biomass yield (*md*) on the influencing factors for chickpea serum had the form (1) for pea serum (2):

$$md = 3.798\,(0.44 + 0.00585cm^2)\,e^{-332624e\text{-}t}\,(0.3402 + 5.5403\,/\,pH^2) \qquad (1)$$

$$md = -2.94 + 0.544\,pH - 0.0356\,pH^2 + 0.181\,t - 0.003\,t^2 - 0.147\,cm + 0.0276\,cm^2 - 0.00447\,pH\,t. \qquad (2)$$

Note: md is the mass fraction of biomass, cm is the amount of seed, t is temperature, pH is the pH of the medium.

The correlation coefficients for equation 1 and equation 2 were significant ($p \leq 0.05$), they were equal to R=0.9644 and R=0.9869, respectively, which indicated that they adequately described the experimental data. From the equations, the optimal values of the factors providing the maximum biomass yield *md* = 0.82 g/dm^3 for pea serum were pH (*pH*) – 6.03, temperature (*t*) – 25.7°C, the amount of seed (*cm*) – 2–3%; for chickpea serum: *md* = 0.88 g/dm^3 with pH (*pH*) – 5.0, temperature (*t*) – 16.96°C, the amount of seed (*cm*) is 4%.

In the process of biomass synthesis from both types of serum, microorganisms completely assimilated glucose, xylose, arabinose, galactose, fructose, in total their number decreased more than 10 times. Raffinose, stachyose and maltose were practically not assimilated, the appearance of sucrose in the culture liquid was possibly due to the hydrolysis of the alpha 1 → 6-glycosidic bond between the residues of galactose and sucrose in raffinose. The amount of high molecular compounds (HMC) in the liquid increased due to the quantitative redistribution of fractions (Table 5).

Table 5. Carbohydrate composition of serum and culture liquid during biomass growth.

Days of growth of the biomass	**Carbohydrate composition of serum and culture liquid, % of total amount**						
	HMC	**Raffinose, stachyose**	**Sucrose**	**Maltose**	**Glucose**	**Xylose, galactose**	**Arabinose, fructose**
	Serum						
	29.75±0.43	17.65±1.2	0.0	4.75±2.3	2.29±1.2	39.07±1.6	6.49±0.2
	Cultural liquid during biomass growth						
1	57.56±0.10	26.00±0.81	4.73	8.21±0.07	0.0	0.0	3.51±0.41

2	53.78±0.09	33.30±0.70	0.20	8.07±0.06	0.0	4.86±0.13
3	55.88±0.08	28.28±1.20	3.05	7.79±0.08	0.0	5.02±0.05
4	58.47±1.10	27.04±0.92	0.25	10.05±0.10	0.0	4.45±0.33

Efficient accumulation of biomass was also due to the presence of relatively low molecular weight nitrogenous components compared to the native flour. The molecular weight (MW) of the components was determined by electrophoresis in SDS-PAGE with and without mercaptoethanol to break disulfide bonds in proteins. If the flour contains multichain components with MM from 15 to >250 kDa, which decomposed into single-chain polypeptides with MM (from 10 to 150 kDa) under the action of mercaptoethanol, then the serum contains components with a lower MM: from 10 to 25 kDa (Figure 3).

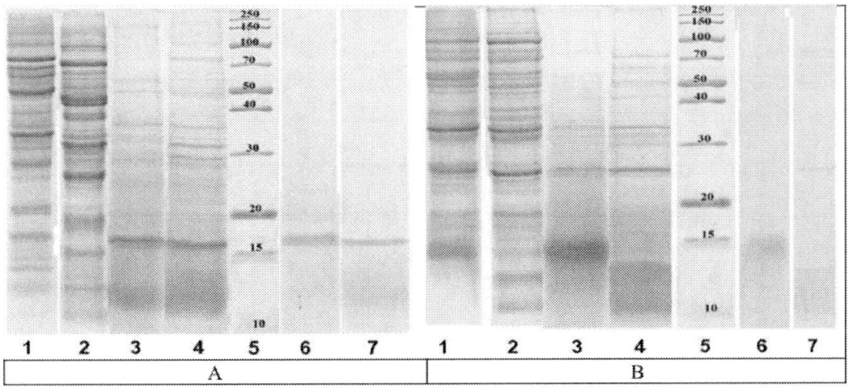

Fig. 3. Protein components in PAAG from: A – pea; B - chickpea; without mercaptoethanol: 1 – flour, 3 – extract of the 1st stage, 5 – markers, 6 – serum; with mercaptoethanol: 2 – flour, 4 – extract of the 1st stage, 7 – serum.

Carbohydrate and protein compositions of the nutrient medium already on the second day of growth (24 hours) of microorganisms on serum ensured the formation of a consortium of yeast and micromycetes, which positively affects the cell morphology (Figure 4 - D) and the chemical composition of the finished concentrates (Table 6).

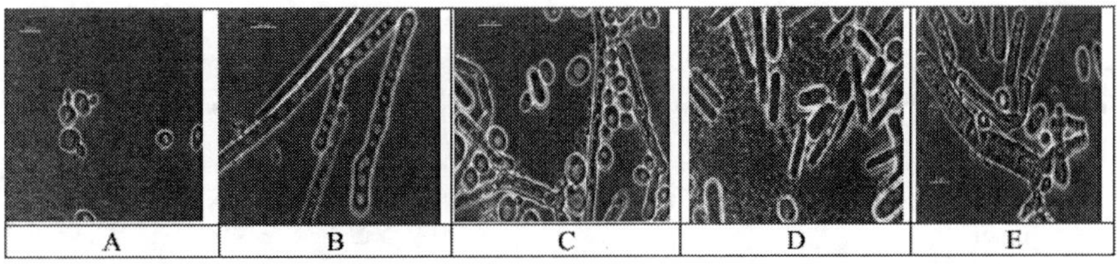

Fig. 4. Cells of monocultures and their consortium: pea serum: A - *S. cerevisiae* 121, B - *G. candidum* 977, C - consortium (48 hours of growth); chickpea serum: D - consortium (24 hours of growth), E - consortium (48 hours of growth)

Table 6. Chemical composition of microbial-plant concentrates from biomass with culture liquid (FMPC-1) and from biomass (FMPC-2).

Product	**Moisture, %**	**Mass fraction, % on DM**			**Carbohydrates**	
		Protein (Nx6.25)	**Ash**	**Lipids**	**soluble**	**insoluble**
FMPC from chickpea serum						
FMPC-1	7.20±0.26	50.60±0.58	12.27±0.05	2.65±0.01	15.22±0.64	20.31±0.33
FMPC-2	9.10±0.50	64.10±0.88	8.93±0.04	9.57±0.42	8.80±0.72	8.60±0.27
FMPC from pea serum						
FMPC-1	6.81±0.40	53.56±0.37	8.60±0.03	2.04±0.19	10.51±0.55	20.48±0.35
FMPC-2	6.81±0.41	61.68±0.47	8.60±0.04	8.31±0.36	7.13±0.55	14.27±0.44

The mass fraction of protein and lipids in the FMPC-2 from biomass is greater than the mass fraction of these compounds in FMPC-1 from biomass with culture liquid in both cultures by 15-27% and is 3.6-4.1 times, respectively. The sum of soluble and insoluble carbohydrates, on the contrary, is 1.4–2.0 times less. Differences in the chemical composition were also revealed for FMPC obtained on different serum. Differences in the mass fraction of protein amounted to 5-8%, in both types of chickpea FMPCs there are 15-29% more lipids than that in pea concentrates and 20-45% more soluble carbohydrates. The mass fraction of insoluble carbohydrates is almost the same in pea and chickpea FMPC-1, while in the FMPC-2 obtained from pea biomass, their amount is 66.5% that is more than that in the FMPC-2 from chickpea serum. A higher content of insoluble carbohydrates in the pea FMPC-2 (by 66%) compared to the chickpea FMPC-2 corresponded to a slightly lower value of their *in vitro* digestibility with EP: 84.41±0.32% and 88.46±1.30%, respectively. The differences are probably due to the characteristics of the chemical composition of the grain of the cultures, because the technology of serum bioconversion is the same.

Proteins of FMPCs contain 18 amino acids: glutamic, aspartic acids, glycine, proline, lysine, etc. (Figure 5).

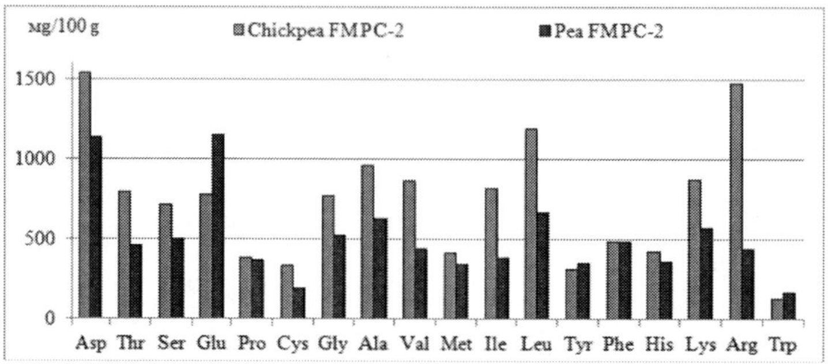

Fig. 5. Amino acid composition of feed microbial-plant concentrates from biomass (FMPC-2).

The score for all essential amino acids of the FMPC-2 from biomass on both types of serum is above 100% (Table 7), which indicates a high biological value of these concentrates.

Table 7. Amino acid score of feed microbial-plant concentrates from biomass (FMPC-2).

FMPC-2 from:	Score of essential amino acids, %								
	Val	His	Ile	Leu	Lys	Met+ Cys	Thr	Trp	Phe+Tyr
Pea serum	107	219	124	107	116	226	179	247	197
Chickpea serum	151	188	197	136	127	225	221	137	135

Table 8. Fatty acid composition of FMPC from pea and chickpea serum.

No	Fatty acid composition	Mass fraction, %		No	Fatty acid composition	Mass fraction, %	
		Pea serum	Chickpea serum			Pea serum	Chickpea serum
1	Decanoic acid Capric acid $C_{10:0}$	0.10	-	17	7-hexadecenoic acid Hypogeic acid $C_{16:1(7)}$	0.56	-
2	Undecanoic acid $C_{11:0}$	0.05	-	18	Hexadecanic acid Palmitic acid $C_{16:0}$	15.03	20.09
3	(R)-3,4-methyledimoxymethamphetamine $C_{11}H_{15}NO_2$	0.17	-	19	Trans-9-hexadecenoic acid Palmitoleic acid $C_{16:1(9)}$	3.65	8.26
4	Dodecanoic acid Lauric acid $C_{12:0}$	0.28	-	20	10-heptadecenoic acid $C_{17:1(10)}$	0.63	-
5	Nonandioic acid Azelaic acid $C_9H_{16}O_4$	0.09	-	21	Heptadecanoic acid Margaric acid $C_{17:0}$	0.52	-
6	Lauric aldehyde $C_{12}H_{24}O$	0.05	-	22	Octadecadiene- 9Z,12Z acid Linoleic acid $C_{18:2(9,12)}$	19.73	-
7	1-nonadecene $C_{19:1(9)}$	0.81	-	23	9-octadecenoic acid Oleic acid $C_{18:1(9)}$	40.43	16.56
8	10-methyldodecanoic acid $C_{13}H_{26}O_2$	0.05	-	24	6-octadecenoic acid Petroselic acid $C_{18:1(6)}$	4.31	1.01
9	Diphenolketone $(C_6H_5)_2CO$	0.08	-	25	Octadecanoic acid Stearic acid $C_{18:0}$	7.10	1.82

10	3-phenyl-2-butyl ester propenoic acid $C_{13}H_{16}O_2$	-	-	26	7-hexadecenoic acid Hypogeic acid $C_{16:1(7)}$	0.56	-
11	9-tetradecenoic acid $C_{14:1(9)}$ Myristolic acid	0.25	-	27	Trans-9-hexadecenoic acid $C_{16:1(9)}$ Palmitoleic acid	3.65	-
12	Tetradecanoic acid Myristic acid $C_{14:0}$	1.36	1.35	28	Hexadecanic acid Palmitic acid $C_{16:0}$	-	-
13	Pentadecanoic acid $C_{15:0}$	0.45	2.14	29	10-heptadecenoic acid $C_{17:1(10)}$	0.63	-
14	n-Hexyl ester of benzoic acid $C_{13}H_{18}O_2$	0.41	-	30	Heptadecanoic acid Margaric acid $C_{17:0}$	0.52	-
15	Heptyl ester of benzoic acid $C_{14}H_{20}O_2$	0.31	-	31	Octadecadiene-9Z,12Z acid Linoleic acid $C_{18:2(9,12)}$	19.73	41.26
16	1,4-benzenedicarboxylic acid diethyl ester	-	4.08	32	Nonadecanol-1 $C_{19}H_{39}OH$	-	3.42

However, for the FMPC-2 from chickpea serum, the score for threonine, lysine, leucine, isoleucine, and valine is higher than that for FMPC-2 from pea serum and less for aromatic amino acids and tryptophan. The score for sulfur-containing amino acids is the same.

The fatty acid composition of FMPC-1 from chickpea serum is represented by 10 components, from pea serum - 30 compounds (Table 8), among which the first concentrate has: 92.5% fatty acids of vegetable oils and animal fats, 7.50% esters and alcohols with the properties of flavors, essential oils and metabolites, in the second concentrate - 97.0% and 3.0%, respectively.

In the chickpea concentrate, the ratio of the sum of saturated (25.40%) and unsaturated fatty acids (67.09%) is 1:2.6, the content of omega-6 fatty acids (linoleic acid) is 41.26%. In the pea concentrate, respectively, the ratio is 1:3 (23.5% / 71.67%) and omega-6 fatty acids are 19.73%. The content of cis-isomers in the concentrate is 91.1%, trans-isomers is 5.1%. Thus, in terms of the composition and type of fatty acids, both FMPC-1s are close to edible oils and fats, while chickpea FMPC-1 contained 4.5% less as compared to these acids and 21.5% more omega-6 fatty acids (linoleic). The total amount of unsaturated fatty acids in pea FMPC is higher than that in chickpea FMPC (the ratio of saturated:unsaturated fatty acids is 1:3 and 1:2.6).

The mineral composition of FMPC-1 is represented by 14 macro- and microelements (Table 9). Chickpea FMPC-1 contains 1.8 times more potassium, magnesium, and cobalt, 10 times more manganese, and 2 times more sodium than pea FMPC-1 does. The amount of calcium, iron, and zinc is almost the same in both preparations.

Table 9. The content of macro- and microelements in FMPC-1 from chickpea and pea serum.

No	Macro- and microelements	FMPC-1 from:	
		chickpea serum	**pea serum**
1	Sodium, mg/100g	2460±172	1163±81
2	Potassium, mg/100g	3377±200	1844±100
3	Calcium, mg/100g	2010±156	2000±120
4	Magnesium, mg/100g	222±10	121±8
5	Iron, mg/100g	5.1±0.35	6.3±0.46
6	Zinc, mg/100g	11.4±0.90	14.0±1.2
7	Copper, mg/100g	1.50±0.04	1.12±0,04
8	Manganese, mg/100g	12.0±0.56	1,56±0,08
9	Cobalt, mcg/100g	107±3.00	57±2.00
10	Nickel, mcg/100g	210±15	440±36
11	Plumbum, mg/kg	≤0,001	≤0,001
12	Cadmium, mg/kg	0.151±0.007	0.171±0.009
13	Chromium, mg/kg	≤0.005	≤0.005
14	Molybdenum, mg/kg	≤0.04	≤0.04

The results must be taken into account when formulating feed and supplement recipes for different groups of animals.

One of the BA samples obtained from pea flour (Sample A) has a dark brown color (Figure 6).

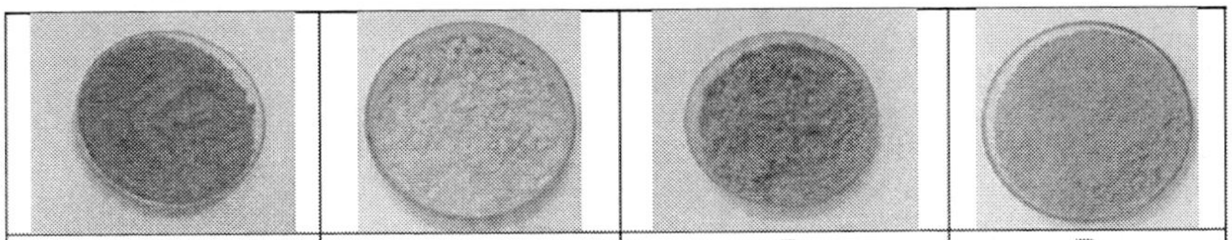

A B C D

Fig. 6. Appearance of PCs from: A - pea flour-1; B - pea flour-2; C - chickpea flour; D - FMPC from chickpea flour.

The color may be related to the reaction of melanoidin formation between the carbonyl groups of reducing sugars and the amino groups of proteins and amino acids, the formation of melanins with the participation of the amino acid tyrosine and enzyme tyrosinase, and the oxidation of −OH groups of flavonoids. The first of two reasons has not received experimental confirmation, while the relationship between the amount of flavonoids in samples and color shades has been established. The mass fraction of flavonoids in flours and PCs, expressed in mg/g of protein, correlated with the color shades of concentrates, while the amount of % of them on DM and in mg/g of the product did not reflect these features (Table 8).

Table 10. Mass fraction of flavonoids in flours and protein concentrates.

Product	The color of product	Optical density, D_{590} nm	Mass fraction of protein, % on DM	Mass fraction of flavonoids		
				% on DM	mg/g of product	mg/g of protein
Pea flour and concentrates						
Flour-1	Yellow	0.390	20.38	1.14	12.70	56.00
PC-1	Dark brown	0.080	71.78	1.08	11.22	15.05
FMPC-1	Light brown	0.100	42.50	1.11	12.23	39.14
Flour-2	Light yellow	0.080	20.20	0.02	1.79	9.89
PC-2	Light yellow	0.040	72.07	0.02	1.88	2.78
FMPC-2	Light yellow	0.040	61.68	1.12	12.37	2.85
Chickpea flour and concentrates						
Flour-3	Yellow	0.380	19.27	1.11	12.24	54.49
PC-3	Dark yellow	0.080	83.22	1.17	12.84	14.06
FMPC-3	Dark yellow	0.085	47.15	1.11	12.28	26.68

The less flavonoids in the flour, the less flavonoids in the composition of PC. Light yellow PC-2, obtained from flour-2 with a 5.6 time-lower amount of flavonoids, compared to flour-1, also contains less flavonoids than dark brown PC-1 does (5.4 times). The same dependence is typical of FMPC. Light yellow FMPC-2 contains 1.7 times less flavonoids than light brown FMPC-1 does. The high content of flavonoids in chickpea flour is also accompanied by a large amount of them in the finished PC-3 with a dark yellow color. The value of the optical density of aqueous solutions of all the studied products, measured at D_{590} nm, highly correlates with the mass fraction of total flavonoids, expressed in mg/g of protein ($R = 0.895$).

4 Conclusions

A comparative analysis of the qualitative indicators of food and feed PC from leguminous crops obtained with hydrolytic EP and ultrasonic processing of raw materials with the achievement of the solubility of pea and chickpea proteins up to 84±1% and their yield of 65-70% was carried out. Chickpea PC contains more protein than pea PC does: 83.22±0.35 and 71.78±0.35% on DM, but the biological value index, adjusted for protein digestibility (PDCAAS) (88%), was higher in pea PC than that in chickpea PC: 96% and 76%, respectively. The amount of essential amino acids is higher in pea PC (256.21 mg/100 g) as compared to chickpea PC (247.9 mg/100 g). PCs differ in the content of copper, cobalt, manganese, nickel, the amount of flavonoids, foaming ability and secondary structure elements. The higher content of flavonoids and more parallel β-structure and anti-parallel 3_{10}-helices in chickpea PC corresponds to a higher foam ability and lower foam stability compared to pea PC. Twisted β-curves and other types of the secondary structure do not affect the values of the functional properties of PCs.

No differences were found in the assimilation of carbohydrates by the yeast *S. cerevisiae* 121 and the micromycete *G. candidum* 977 from different types of serum. From carbohydrates and proteins with MM from 10 to 25 kDa, on the second day of growth, a consortium of microorganisms is formed with a mass fraction of protein of 50.60–53.56% on DM in biomass together with the cultural liquid and 61.68–64.10 in one biomass. Differences were found in feed PC from pea and chickpea serum biomass in the mass fraction of fat, soluble and insoluble carbohydrates. The amino acid score of all essential amino acids in concentrates from both types of serum is above 100%. The amount of unsaturated fatty acids in pea FMPC-1 is higher than that in chickpea FMPC-1 (the ratio of saturated: unsaturated fatty acids is 1:3 and 1:2.6), but chickpea FMPC-1 contained 21.5% more omega-6 fatty acids (linoleic) and more potassium, magnesium, cobalt, manganese, sodium. For all types of concentrates, a high correlation was found between the mass fraction of flavonoids in raw materials, concentrates and the color of dry preparations. The correlation coefficient R between the optical density of aqueous solutions measured at D_{590} nm and the mass fraction of flavonoids (in mg/g of protein) is 0.895. PCs of both types are recommended to be used for food purposes, FMPCs should be in animal feed, taking into account the characteristics of their chemical composition and quality indicators.

References

1. V. Zotikov, T. Naumkina, N. Gryadunova, V. Sidorenko, V. Naumkin, Legumes and Cereals, **17(1),** 6–13 (2016)
2. V. Kolpakova, D. Kulikov, R. Ulanova, L. Chumikina, *Food Processing: Techniques and Technology* **51(2),** 333–348 (2021) doi: 10.21603/2074-9414-2021-2-333-348
3. World Health Organization, Food and Agriculture Organization of the United Nations. Driving commitment for nutrition within the UN Decade of Action on Nutrition. WHO/NMH/NHD/17.11, FAO/CA1340EN/1/09.18. (2018)
4. D. Kulikov, V. Kolpakova, R. Ulanova, L. Chumikina, V. Bessonov, *Biotechnology* **36 (4),** 49–58 (2020) doi: 10.21519/0234-2758-2020-36-4-49-58
5. F. Pedro, R. Nair, D. Andersson, P. Lennartsson, M. Taherzadeh, *Fungal Biology and Biotechnology* **5,** 5–8 (2018) doi: 10.1186/s40694-018-0050-9
6. J. Xu, M. Zhang, T. He, H. Luo, K. Peng, X. Huang, J. Liu, *Bioresource Technology* **293,** 122032 (2019) doi: 10.1016/j.biortech.2019.122032
7. D. Sarris, Z. Sampani, A. Rapti, S. Papanikolaou, *Current pharmaceutical biotechnology* **20 (10),** 881–894 (2019) doi: 10.2174/1389201020666190211145215
8. W. R. C. Machado, L. G. Silva, W. R. M. Vanzela, D. Bianchi, *International food research journal* **26 (4),** 1259–1267 (2019)
9. X. Zhou, Z. Ouyang, X. Zhang, Y. Wei, S. Tang, Z. Ma, Z. Tan, N. Zhu, T. Teklebrhan, X. Han, *Animals* **9 (9),** 598 (2019) doi: 10.3390/9090598
10. G. Rychen et al., *Efsa Journal,* **15 (2),** 5700 (2017) doi: 10.2903/j.efsa.2017.4704
11. Z. Sun, T. Wang, N. Aschalew, W. Zhao, X. Chen, X. F. Zhang, Y. G. Zhen, G. X. Qin, *Journal of animal physiology and animal nutrition* **15** (2019) doi: 10.1111/jpn.13241
12. Y. G. Zhen, W. Zhao, X. Chen, L. J. Li, H. G. Lee, X. F. Zhang, T. Wang, *South African Journal of Animal Science* **49 (1),** 99–108 (2019) doi: 10.4314/SAJAS.V49I1.12
13. D. Sousa, C. A. Oliveira, A. V. Velasquez, J. M. Souza, E. Chevaux, L. Jose Mari, L. F. P. Silva, *Animal feed science and technology* **236,** 149–158 (2018) doi: 10.1016/J. ANIFEEDSCI.2017.12.015
14. M. I. Anjum, S. Javaid, M. S. Ansar, A. Ghaffar, *Iranian Journal of veterinary research* **19 (2),** 96–100 (2018) doi: 10.22099/IJVR.2018.4852
15. S. Ghazanfar, M. Qubtia, I. Ahmed, F. Hasan, M. I. Anjum, M. Imran, Journal of animal and plant sciences **28 (2),** 407–420 (2018)
16. S. Sallam, M. Abdelmalek, Z. Kholif, S.M. Zahran, M.H. Ahmed, H.S. Zeweil, M.F.A. Attia, O.H. Matloup, O.A. Olafadehan, Animal *Biotechnology* **17,** 1–7 (2019) doi: 10.1080/10495398.2019.1625783
17. E. Bashashkina, E. Pashinina, A. Pashinin, N. Suyasov, *Advances in chemistry and chemical technology* **22 (13),** 38–40 (2008)
18. E. Serba, E. Sokolova, N. Fursova, G. Volkova, Y. Borshcheva, E. Kurbatova, E. Kuksova, *Storage and processing of agricultural raw materials* **2,** 74–79 (2018)
19. A. Kot, S. Błażejak, M. Kieliszek, I. Gientka, B. Joanna, L. Reczek, K. Pobiega, *World Journal of Microbiology and Biotechnology* **35,** 150 (2019) doi: 10.1007/s11274-019-2732-8
20. G. Chuppa-Tostain, et al., *Fungal Biology and Biotechnology* **5,** 1–6 (2018) doi: 10.1186/s40694-018-0045-6
21. J. Soderberg, *Functional properties of legume proteins compared to egg proteins and their potential as egg replacers in vegan food* (2013)
22. A. Singhal, A. C. Karaca, R. Tyler, M. Nickerson, *Grain Legumes* (2016) doi: 10.5772/64020
23. A. Bondarenko, *Agrarian Russia* **1,** 24–26, (2019) doi: 10.30906/1999-5636-2019-1-24-26
24. A. Mudryj, N. Yu, H. Aukema, *Applied Physiology* Nutrition and Metabolism, **39 (11),** 1197–1204 (2014) doi: 10.1139/apnm-2013-0557
25. I. Revilla, *Composition, Processing and Impact on Active Components in Food,* pp 337–343 (2015) doi: 10.1016/B978-0-12-404699-3.00040-8
26. G. Semenov, *Vacuum freeze drying* (2013)
27. J. BeMiller, *Food Analysis,* pp 147–177 (2010) doi: 10.1007/978-1-4419-1478-1_10
28. S. Leser, *Nutrition Bulletin* **38,** 421–428 (2013) doi: 10.1111/nbu.12063
29. V. Kolpakova, L. Chumikina, L. Arabova, D. Lukin, A. Topunov, E. Titov, *Foods and Raw Materials* **4 (2),** 48–57 (2016) doi: 10.21179/2308-4057-2016-2-48-57
30. M. Gavrilin, O. Popova, E. Gubanova, *Chemistry of plant raw materials* **4,** 99–104 (2010).

Development and research of an information-measuring system for quality control of agricultural products

Stanislav Mishukov*, *Igor* Vorotnikov, *Maxim* Mastepanenko, and *Shaliko* Gabrielyan

Federal State Budgetary Educational Institution of Higher Education "Stavropol State Agrarian University", 12, Zootechnicheskiy Ln, Stavropol, 355017, Russian Federation

Abstract. The article presents the result of an information-measuring system development for quality control of agricultural products based on the assessment of their moisture content, chemical composition, presence of organic and mineral impurities. It also substantiates the relevance of the use of such systems by the agro-industrial complex enterprises. The proposed information-measuring system is based on the author's dielcometric method for determining the electrophysical parameters of capacitive sensors, the principle of which is to decompose the measuring process into two stages. The study of this method was carried out using simulation modeling in the SimInTech environment, while custom operational amplifier blocks and sample-and-hold circuit were built. As a result of the operational amplifier model research, graphs of the dependency of the output voltage on the change in the integration time constant and amplification gain were plotted, on the basis of which the optimal values of these quantities were selected. Studies of the constructed models of the first and second measurement stages using the proposed method showed a high accuracy in determining the electrophysical parameters of capacitive sensors filled with agricultural products, with a relative error not exceeding ±0.2%.

1 Introduction

The most important indicators of the agricultural products quality are moisture and the presence of impurities which have a significant impact on their safety, nutritional and consumer value, suitability for storage and processing, cost, etc. The presence of moisture in products leads to the acceleration of negative physicochemical and biological processes, such as germination, self-heating, respiration, the development of microorganisms, insects, mites, etc. [1].

The respiration of agricultural products is associated with the dissimilation process of complex organic substances to simpler ones - water, carbon dioxide and ammonia, the intensity of which directly depends on the initial moisture content and material contamination. For example, with a wheat moisture content of 15.5% or more, 0.05–0.2% of dry matter is lost per day, and the release of carbon dioxide CO_2 can reach 18 mg per 100 g of the product, which is associated with aerobic respiration, leading to an increase in heat release up to 16.7 MW/kg and self-heating material. This, in turn, creates conditions for the appearance of free moisture, which will sharply accelerate the microorganisms' development (bacteria, fungi, mold, etc.), insects and mites, which reduce the quality and change the chemical composition of products [2].

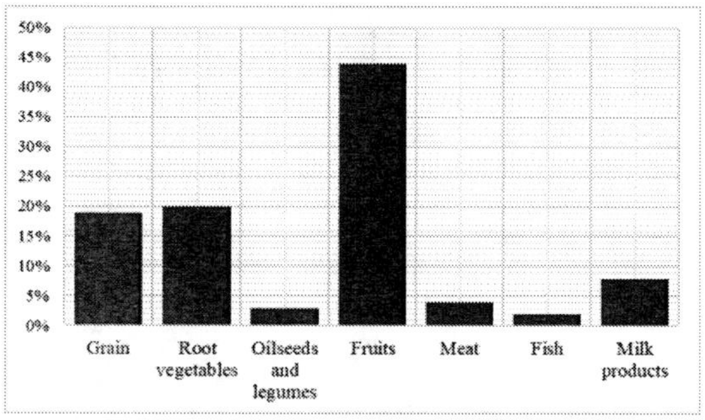

Fig. 1. The share of global losses of agricultural products (% of total losses).

As a result of these processes development, the loss of agricultural products can reach 1.3 billion tons per year (Fig. 1), which is a serious global problem that threatens food security and nutritional satisfaction of the population [3].

One of the ways to reduce losses is the rapid and accurate determination of moisture and impurities in products using modern technical solutions, for example, the information-measuring systems (IMS), which provide the ability to quickly obtain information about the quality of agricultural products and taking measures to bring the indicators to the established standards using technological operations of drying or cleaning [4].

* Corresponding author: stas.mishukov.92@mail.ru

 https://doi.org/10.1051/bioconf/20225200008

2 Materials and methods

At the moment, the advance research and development direction of such systems are dielcometric methods for determining the parameters of capacitive sensors (CS), based on indirect measurements of the material electrophysical properties, which make it possible to determine its moisture content, the presence of organic and mineral impurities, chemical composition, etc.

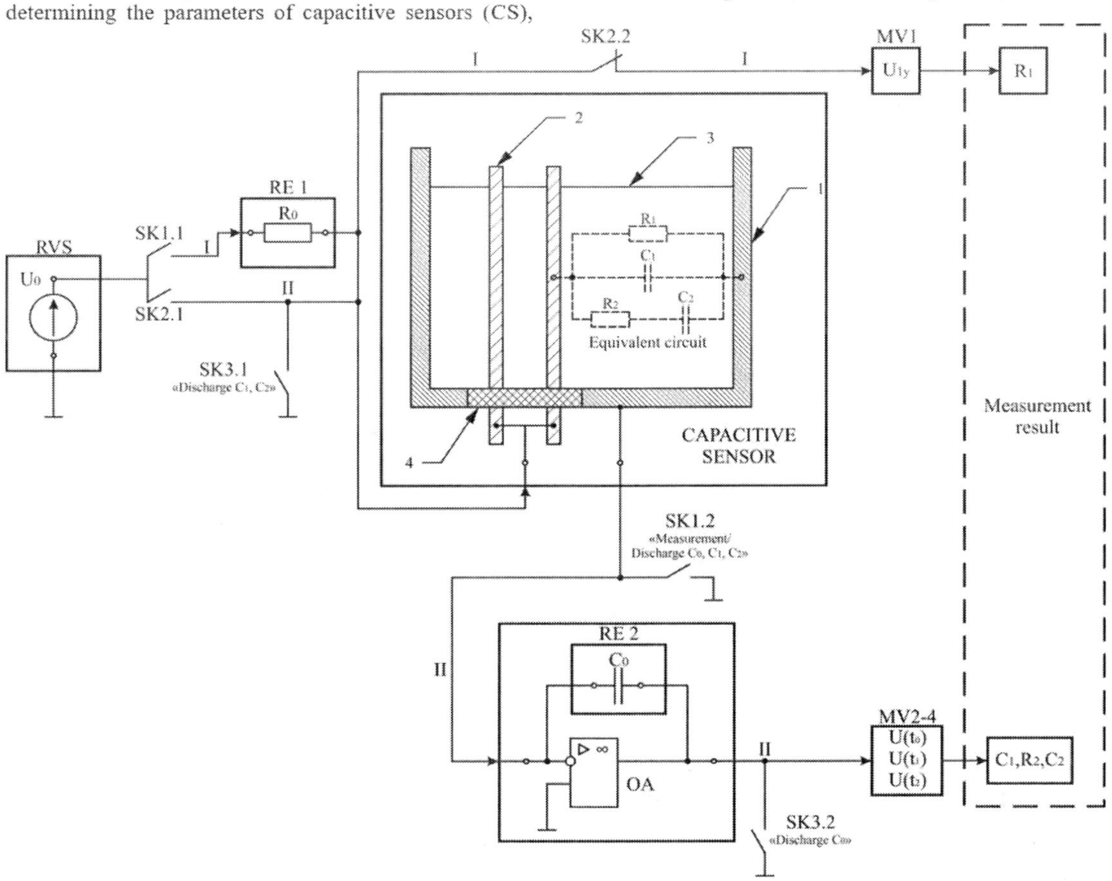

Fig. 2. Block scheme of the method for determining the electrophysical properties of agricultural products using capacitive sensors.

The use of capacitive sensors is due to their design simplicity and versatility in determining moisture and impurities in various agricultural products (wheat, rice, barley, corn, etc.), which provides unlimited opportunities for implementation in various technological processes of agricultural enterprises. For the system being developed, a coaxial capacitive sensor was used (Fig. 2), which is an aluminum container 1 filled with the test product 3 and is one of the electrodes in which internal aluminum rod electrodes 2 are built through a dielectric spacer 4.

Studies of similar capacitive sensors with agricultural materials placed in them in the works of Stuart O. Nelson [5] showed that their properties can be described by a four-element RC equivalent circuit (EC), including through resistance R_1 and capacitance C_1, relaxation resistance R_2 and capacitance C_2, while each of the elements characterizes individual electrophysical properties of the product. R_1 is organic impurities (leaves, weeds, stems, etc.); R_2 is mineral impurities (sand, pebbles, earth, etc.); C_1 is chemical composition (iron, phosphorus, sulfur, etc.); C_2 is the number of water particles and their sizes, i.e., humidity of the controlled environment.

The determination of the indicated properties of the material under study in the developed IMS can be implemented using the author's method [6], based on the decomposition of the measurement process into two stages. The first stage is performed in the steady operating conditions of the measuring circuit (MC) and allows determining the through resistance R_1. The second stage is carried out during a rapidly developing transient process in the MC and makes it possible to determine the rest of the electrophysical properties of the product - C_1, C_2 and R_2.

The block scheme of the proposed author's method is shown in Figure 2 and includes the following designations: RVS is a reference voltage source; RE1-2 is reference elements; SK1.1-SK3.2 are switching keys; OA is an operational amplifier; MV1-4 is a measured value; I-II are measurement stages.

To confirm the operability and relevance of using the proposed method for determining the electrophysical properties of agricultural products based on a capacitive sensor, it was simulated in the modern environment of model-based design SimInTech (Simulation In Technic) [7], which allows the construction and calculations of complex dynamic RC circuits in the real-time mode.

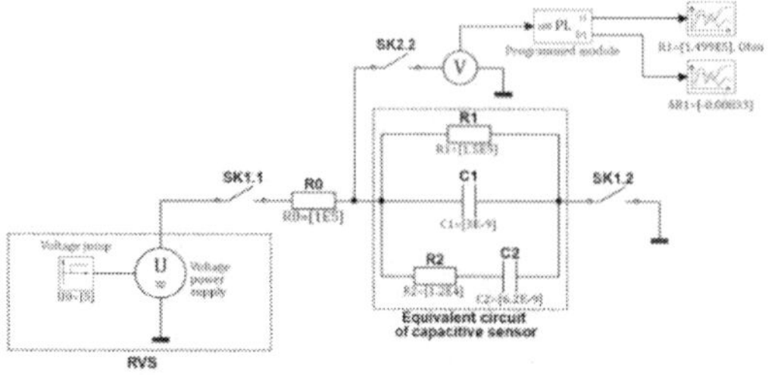

Fig. 3. MC model of the first stage of measurements.

The model of the first stage of measurements I is shown in Figure 3, while the measurement process is performed as follows. RVS, consisting of the standard blocks "Step" and "Controlled voltage source", forms a jump in DC voltage $U_0 = 5$ V, which through the switch SK1.1 and RE1 in the form of a resistor R_0, is supplied to the CS represented by a four-element EC grounded by the SK1.2. As a result of it the resistors R_1 and R_0 form a classic resistive divider, which makes it possible to determine the through resistance R_1 when measuring the voltage U_{1y} at the midpoint through the SK2.2 and the block "Voltmeter" according to the following expression:

$$R_1 = \frac{R_0 U_{1y}}{U_0 - U_{1y}}. \tag{1}$$

The calculation according to the specified expression is implemented using a programmed module which, based on a programming language similar to C/C ++, allows you to perform calculations of any complexity (Fig. 4).

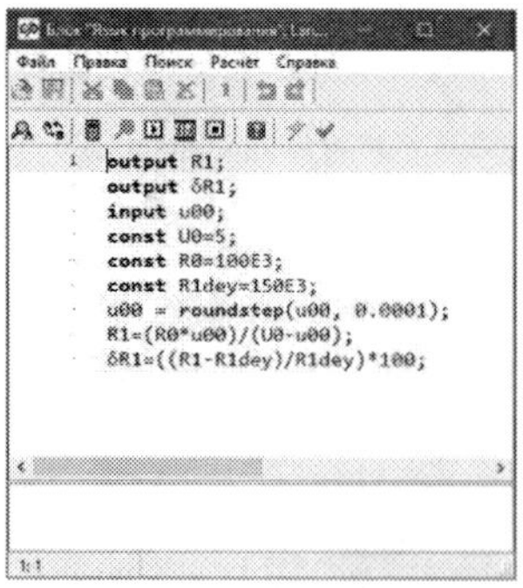

Fig. 4. Programmed module window.

To calculate the model, the values of the MC elements were taken from the known method for determining the parameters of bipolar circuits [8]: $U_0 = 5$ V; $R_0 = 100$ kOhm; $C_0 = 17$ nF; $R_1 = 150$ kOhm; $C_1 = 3$ nF; $R_2 = 12$ kOhm; $C_2 = 6.2$ nF; $\tau = 74.4$ μs. As a result of the calculations we obtained $R_1 = 149$ kOhm with a relative error $\delta R_1 = -0.0083\%$, which indicates the high accuracy of the proposed method at the first stage of measurements.

When building the MC model of the second stage of measurements II, it was necessary to develop an OA model, since such a block is not available in the standard libraries of the SimInTech environment. For this, blocks "Aperiodic link of the 1st order" 1, a comparing device 2, voltmeters with additional ports of mathematical communication for outputting the instantaneous value of the measured voltage 3-4, a voltage supply with the ability to set voltage function through port 5 or, in other words, voltage-controlled voltage source (VCVS) (Fig. 5).

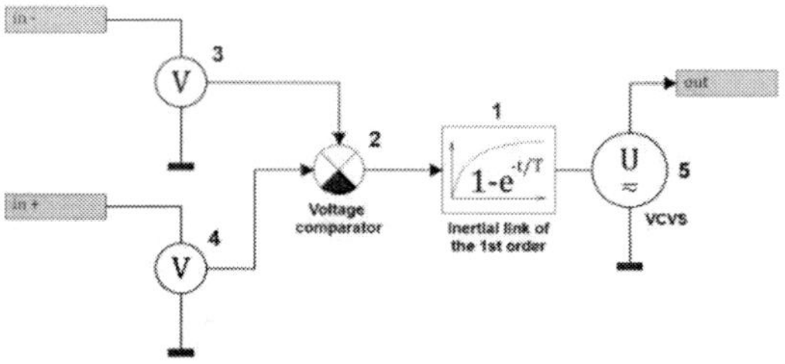

Fig. 5. Operational amplifier model.

The most important task in the construction of the OA model is the choice of the 1st order aperiodic link properties: amplification gain (transfer factor) at a constant input voltage k and time constant τ. The 1st order aperiodic link is the transfer function of the RC low pass filter and the transfer function of the integrator.

The complex amplification gain (transfer factor) can be represented as follows:

$$\dot{K} = \frac{K_0}{1 + j\omega\tau} \tag{2}$$

where K_0 is the DC amplification gain of the operational amplifier;

ω is the angular frequency of the input signal, rad/s;

τ is the time constant of the low-pass RC filter, corresponding to the integration time constant of the OA.

Based on this, the modulus of the gain transfer function will be equal to:

$$\left|\dot{K}\right| = \frac{K_0}{\sqrt{1+\omega^2\tau^2}} = \frac{K_0}{\sqrt{1+4\pi^2 f^2 \tau^2}}. \tag{3}$$

Then the dependence of the integration constant τ_{OY} on the frequency of unity gain ω_1 of the OA model at $\left|\dot{K}\right|$ =1:

$$\tau_{OA} = \sqrt{\frac{K_0^2 - 1}{\omega_1}}. \tag{4}$$

Therefore, the complex amplification gain (transfer factor) according to expression (2) can be represented as:

$$\dot{K} = \frac{K_0}{1 + j\omega\frac{K_0}{\omega_1}} = \frac{K_0}{1 + j\omega\frac{K_0}{2\pi f_1}}. \quad (5)$$

To select the correct properties of the 1st order aperiodic link, which describes the frequency characteristics of the operational amplifier, it is extremely important to establish the relationship between the value of the integration time constant τ_{OY} and the time constant of the measuring RC-circuit (MC) τ_{MC} connected to the inverting input of the operational amplifier (Fig. 6). When choosing these properties, one should take into account the transient processes occurring both in the MC and in the operational amplifier model itself, operating in the mode of a summing integrator connected according to an inverse circuit.

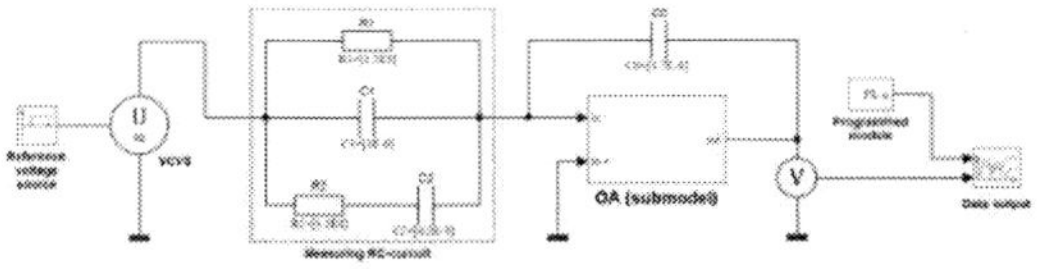

Fig. 6. Scheme for research of the OA model.

It is known that the output voltage at the operational amplifier can be described by the following expression [9]:

$$U_{out}(t) = \frac{U_0 C_1}{C_0} + \frac{U_0}{R_1 C_0} t + \frac{U_0 C_2}{C_0}\left(1 - e^{-\frac{t}{R_2 C_2}}\right) \quad (6)$$

where U_0 is the reference voltage, V;

C_0 is the standard capacitor, pF;

R_1, R_2 are MC elements in the form of resistances, Ohm;

C_1, C_2 are MC elements in the form of capacitances, pF.

The above expression is obtained on the basis of the conversion function of the inverting amplifier:

$$U_{out}(t) = -U_{input}(t) \cdot \frac{Z_{feedback}}{Z_{input}} \quad (7)$$

where $Z_{feedback}$ is the feedback resistance OA, Ohm;

Z_{input} is the resistance at the inverting input OA, Ohm.

The input signal is formed by step action on the input circuit and is an aperiodic function. In such a situation, when building an OA model, it is advisable to consider not the frequency characteristics, but the rate of aperiodic function change at the output, which has the following form:

$$\frac{dU_{out}}{dt} = \frac{U_0}{R_1 C_0} + \frac{U_0 C_2}{C_0} \cdot \frac{1}{\tau} e^{-\frac{t}{\tau}}. \quad (8)$$

From which it follows that the output voltage is inversely proportional to the time constant $\tau = R_2 C_2$. The above reasoning allows us to formulate the following recommendation for choosing the properties of the 1st order aperiodic link of the OA model - for a given practically significant gain $K_0 = 10^5$, the integration time constant τ_{OA} of the model can be determined using the expression:

$$\tau_{OA} = \frac{1}{K_0 \cdot \tau_{MC.calc}} \quad (9)$$

where K_0 is the given amplification gain of the OA;

$\tau_{MC\ calc.}$ is the calculated value of the MC time constant, established on the basis of preliminary data on the object of research and refined during the modeling process.

As an example, let's calculate τ_{OA} for the simulated circuit at $K_0 = 10^5$ and time constant $\tau_{MC} = 100$ µs:

$$\tau_{OA} = \frac{1}{10^5 \cdot 10^{-4}} = 0.1\ \text{s}. \quad (10)$$

With the indicated values, the amplitude-frequency characteristic (AFC) of the OA model will be as follows (Fig. 7). For clarity, the AFC is built on a logarithmic scale, on which the cutoff frequency is marked $\omega_{slice} = \frac{1}{\tau}$

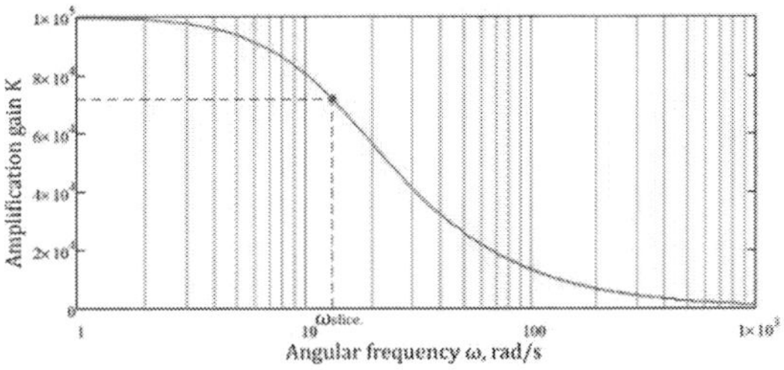

Fig. 7. AFC of the constructed OA model.

In order to confirm the operability of the constructed OA model, we will research the influence of the integration time constant and the amplification gain on the output voltage value $U_{out}(t)$ with the output of data in the form of appropriate graphs with a variation range for $\tau_{OA} = 0.1 \div 9.0$ s and $K = 10^5 \div 10^1$.

To do this, it will be necessary to add four additional circuits in parallel to the circuit for research and set the indicated parameters in the properties of the 1st order aperiodic link (Fig. 8).

As a result of the simulation, it was found that a change in the value of the integration time constant τ_{OA} leads to a change in the output signal $U_{out}(t)$ of the OA model in a wide range (Fig. 9), while the most optimal value is 0.1 s, which is confirmed by the maximum proximity of the constructed curve at the specified value to the ideal design curve obtained as a result of calculations in the programmed module. The constructed curves at values $\tau_{OA} = 0.5 \div 1.0$ s are quite close to the calculated ones and have a deviation not exceeding 0.4 - 0.6 V at each integration step, which is not applicable within the framework of the developed OA model due to the accumulation of additional error, leading to degradation of further calculation results in the MC

 https://doi.org/10.1051/bioconf/20225200008

models. The values τ_{OA} = 5.0÷9.0 s lead to an unacceptable deviation of the curves from the calculated one and will not be applied to the developed OA model.

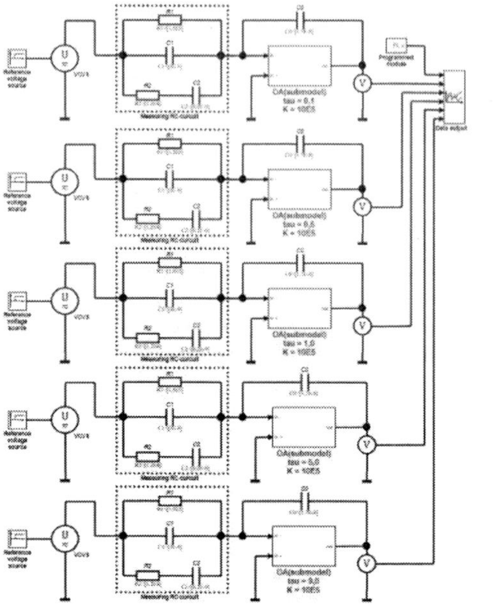

Fig. 8. Scheme of studying the influence of the integration time constant τ_{OA} on the OA model.

By analogy with the circuit in Figure 8, a circuit was built to assess the effect of the amplification gain K on the OA model, as a result of which it was found that changing the value of the gain K has a significant effect on the output voltage U out(t) of the OA model, while in comparison with the previous research, the range of change is more pronounced, and changes in the exponent shape of the transient process have cardinal differences from the calculated curve. The optimal value of the amplification gain K for the developed OA model is 10^5, which is confirmed by the superposition of the curve with this gain and the calculated curve. Invalid gain values are 10^4÷10^1, as evidenced by the deviations of the output voltage values from the ideal curve in the range from 0.45 to 2.5 V, while the last two curves have a linear voltage change, which does not completely correspond to the waveform of the output signal on the OA.

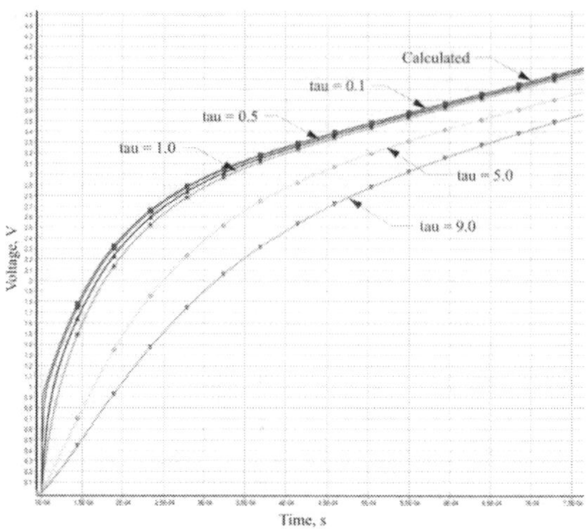

Fig. 9. Set of curves in the study of τ_{OA} influence on the OA model.

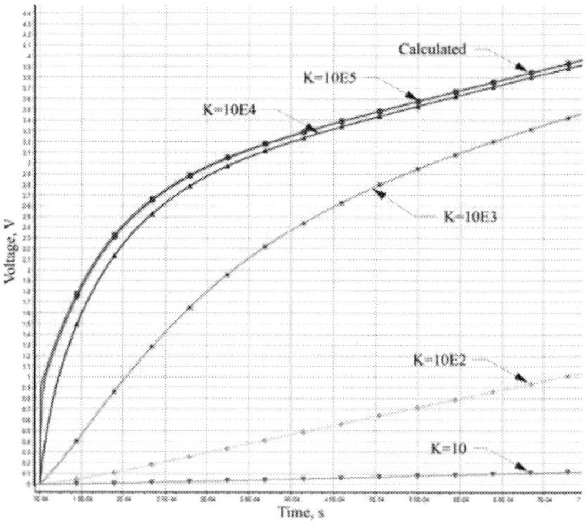

Fig. 10. Set of curves in the study of K influence on the OA model.

After confirming the performance and selecting the parameters of the OA model (Fig. 5), it was transferred to the constructed MC model of the second stage of measurements II according to the proposed method for determining the electrophysical parameters of the CS in the form of a submodel (Fig. 11).

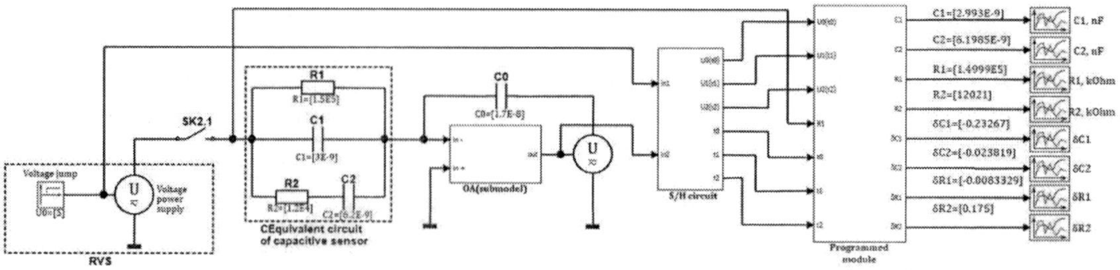

Fig. 11. MC model of the second stage of measurements II.

To obtain accurate time readings, signal sampling and accurate voltage values $U(t_0)$, $U(t_1)$, $U(t_2)$ during the developing transient process, the MC model implements a sample-and-hold circuit (S/H circuit), which is built using blocks «Impulse», «Turn-on delay» and «Storing the signal value», while similarly to the previous cases, the S/H circuit is combined into a separate submodel.

The measurement process at the second stage is as follows: a DC voltage jump is supplied from the RVS, which, through the SK2.1 goes to the CS connected to the inverting input of the OA, the negative feedback of which is implemented by the RE2 in the form of the capacitor C_0, after which the signal is sent to the S/H circuit, performing voltage readings $U(t_0)$, $U(t_1)$ and $U(t_2)$ at fixed times and transferring them to a programmed module that performs the calculation based on the following system of equations:

$$\begin{cases} U(t_0) = A_0 + A_1 t_0 + A_3\left(1 - e^{-\frac{t_0}{\tau}}\right) \\ U(t_1) = A_0 + A_1 t_1 + A_3\left(1 - e^{-\frac{t_1}{\tau}}\right) \\ U(t_2) = A_0 + A_1 t_2 + A_3\left(1 - e^{-\frac{t_2}{\tau}}\right) \end{cases} \tag{11}$$

The solution of this equations system, taking into account the through resistance R_1 obtained at the first stage, allows us to establish the ratios of the coefficients A_0, A_1, A_3 and the time constant τ:

$$A_0 = U(t_0) \tag{12}$$

$$A_1 = \frac{U_0}{R_1 C_0} \tag{13}$$

$$A_3 = \frac{\left(U(t_1) - A_0 - A_1 t_1\right)^2}{2U(t_1) - A_0 - U(t_2)} \tag{14}$$

$$\tau = \frac{t_1}{\ln\left(\frac{U(t_1) - A_0 - A_1 t_1}{U(t_2) - U(t_1) - A_1 t_1}\right)}. \tag{15}$$

Based on these parameters, the search values of C_1, C_2, R_2 of the studied CS, represented by a four-element EC, can be found using the following expressions:

$$C_1 = \frac{A_0 C_0}{U_0} \tag{16}$$

$$C_2 = \frac{A_3 C_0}{U_0} \tag{17}$$

$$R_2 = \frac{\tau}{C_2}. \tag{18}$$

3 Results and discussion

The calculation results of the MC model of the second measurements stage in the programmed module were (Fig. 11): C_1 = 2.993 nF; C_2 = 6.198 nF; R_2 = 12.021 kOhm. In this case, the relative error was: $\delta C_1 = \pm 0.232\%$; $\delta C_2 = \pm 0.023\%$; $\delta R_2 = \pm 0.175\%$. The positive results of modeling the proposed method in the SimInTech environment allow us to conclude that it is appropriate to use it in the developed quality control system for agricultural products based on the determination of their moisture content and impurities.

In this regard, the next stage in the development of IMS was the construction of its functional diagram (Fig. 12) in accordance with the known requirements for such systems [10], the operating principle of which is as follows. In one container of the CS we placed agricultural product with unknown moisture content W_x and the amount of impurities AdS_x, acting as a sample or object of study. The same product is placed in another container, pre-dried to the set value W_0 and cleaned of impurities AdS_0, which is a standard or measure of comparison. They, according to a given method determination of the electrophysical parameters of the CS, are connected to the measuring circuits (MC) 1-2, which include MC of the first and second stages of measurements with reference elements in the form of a resistor R_0 and a capacitor C_0. As a result of it, the parameters R_{1x}, C_{1x}, R_{2x} and C_{2x} of the research object are determined and R_{10}, C_{10}, R_{20} and C_{20} of the standard, which enter the comparison links in the form of comparators, where the input signals are compared and their ratio is established. They are transmitted through data links R_1, C_1, R_2 and C_2 to the data analysis unit, which performs the final calculations of product humidity W_{res} and impurities AdS_{res} based on the information received from the comparators and the calibration characteristics recorded in the long-term memory ROM.

Such IMS scheme makes it possible to further improve the accuracy of determining the product quality indicators by comparing with the measure, as well as reducing constant systematic and multiplicative errors. However, in real conditions of agricultural objects, the calibration characteristics stored in the system memory will act as a standard.

As a «Data analysis» block in the proposed system, it is recommended to use modern AVR microcontrollers. They have versatility, the required speed, built-in analog-to-digital converters (ADC), digital-to-analog converters (DAC), timers, a sufficient number of input / output ports for control switching in the MC, low power consumption, large amount of RAM and Flash memory, easy programming and customization flexibility.

4 Conclusion

Thus, the proposed quality control system for agricultural products has improved accuracy characteristics, due to the use of an improved method for determining the electrophysical parameters of CS with a low relative error of measurement results, not exceeding

±0.2% and confirmed by modeling in the SimInTech environment. And it also meets modern requirements versatility and speed, due to the use of a simple hardware-software implementation based on high-speed AVR microcontrollers, operational amplifiers, comparators and analog-to-digital converters.

The use of this IMS will allow agricultural enterprises to minimize losses by reducing the response time to product quality deviations such as excessive moisture, an increased amount of organic and mineral impurities, chemical composition. At the same time, due to the simplicity of the components design used in the system, in particular the CS, it can be concluded that there is a great potential for its implementation in the technological processes of harvesting, transporting, processing and storing agricultural products.

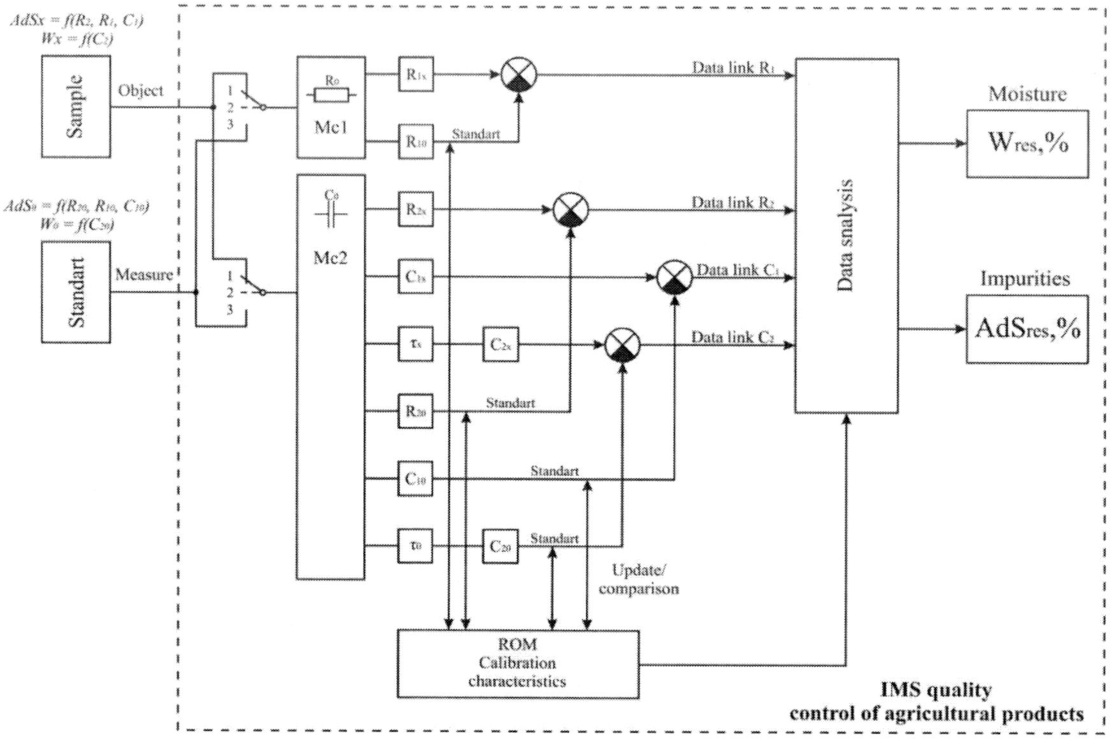

Fig. 12. Functional diagram of IMS quality control of agricultural products.

References

1. S.V. Egorova, V.G. Kulakov, E. M. Yutyusheva, E. S. Shumova, V.V. Andryushchenko, *Food Industry* **1,** 31–33 (2020).
2. M. B. Khokonova, *Biology in agriculture* **2 (31),** 39–42 (2021).
3. V.V. Kim, E. A. Galaktionova, K. V. Antonevich, *International Agricultural Journal* **4 (63),** 1 (2020).
4. I. N. Vorotnikov, M. A. Mastepanenko, S. Z. Gabrielyan, S. V. Mishukov, *IOP Conf. Ser.: Mater. Sci. Eng.* **643,** 012041 (2019).
5. Stuart O. Nelson, D. Sc. Samir Trabelsi, *Journal of microwave power and electromagnetic energy* **4 (50),** 237–268 (2016).
6. I. N. Vorotnikov, M. A. Mastepanenko, S. Z. Gabrielyan, S. V. Mishukov, *IOP Conference Series Earth and Environmental Science* **488 (1),** 012028 (2019).
7. F. I. Baum, O. S. Kozlov, I. A. Parshikov, V. N. Petukhov, K.A. Timofeev, A.M. Shchekaturov, *Atomic energy* **6 (113),** 354–357 (2012).
8. A.V. Knyazkov, A. S. Koldov, N. V. Rodionova, A.V. Svetlov, *Measurement. Monitoring. Management. Control*. **3 (25),** 69–78 (2018).
9. A. S. Koldov, *Proceedings of the International symposium "Reliability and quality"* **2,** 71–74 (2018).
10. M. G. Rodionov, A. A. Gorshenkov, *Information and measurement systems: theory of systems and system analysis* (2011).

Theoretical foundations and practical methods of disposal of manure drains of livestock complexes in the Republic of Tatarstan

Faik Safiollin[1,*], *Salavat* Suleymanov[1], *Nikolai* Trofimov[1], *Anas* Lukmanov[2], and *Ilfat* Gabbasov[1]

[1]Kazan State Agrarian University, 65, K. Marx Str., 420015, Kazan, Republic of Tatarstan, Russian Federation
[2]Tatarsky Centre of Agrochemical Service,210, Orenburgsky Tract, 420059, Kazan, Republic of Tatarstan, Russian Federation

Abstract. Irrigation creates a unique way to dispose manure effluents from livestock complexes in specially designated agricultural sewage farms (ASF). Theoretical foundations and practical methods of using manure runoff and selecting crops taking into account their incomplete dehelminthization were developed by the team of authors of this study. It was established that the use of manure effluents after their stepwise preparation ensures the production of 56.9 t/ha of green mass of perennial grass with production profitability from 58 to 86%, 57.0 t/ha of corn and 46.7 t/ha of feed mixtures against 33.8 t/ha on the control version of the experiment. These cultures go through the conservation stage. In this regard, they are environmentally friendly sources of feed in the diets of various groups of animals.

1 Introduction

Modern industrial production of milk and meat provides for the bedding-free livestock keeping, but the use of straw, peat or dry sawdust as a bedding material becomes the basis for an increase in the cost of livestock production [1–3].

On the other hand, the removal of animal excrement by hydraulic washing, the use of water for the care of abdomen, the washing of milk lines, and milking plants, sharply increase the volume of liquid manure, effluents and wastewater, which is 4 times higher than the volume of water of domestic and industrial origin [4–6].

Besides, manure runoff poses a real threat to environmental pollution, spread of infectious diseases of farm animals and humans [7–9].

Therefore, the development of highly efficient methods for the disposal of manure effluents of livestock complexes of the Russian Federation with a content of 2.2 million tons of nitrogen, one million tons of phosphorus and potassium [10] is extremely important from both ecological and practical perspective.

2 Materials and methods

The studies were carried out in 2018-2021 on the fields of Saba LLC Sabinsky and Integrated Agricultural Production Center named after Vakhitov of Kukmorsky municipal districts of the Republic of Tatarstan on typical gray forest soils. The research method corresponded to the method of the All-Russian Williams Fodder Research Institute) (2008). Wastewater was used in a ratio of 1:15. The norms and terms of irrigation were determined depending on the biological characteristics of the cultivated crops in the link of the by-farm crop rotation: perennial grass– corn – feed mixtures and the actual moisture content of the soils determined using the Dniester 1 soil moisture meter.

3 Results and discussion

In both farms, the density of livestock per 100 hectares of agricultural land exceeds the national average by 1.5–2.0 times. So, the number of cattle in the Integrated Agricultural Production Center named after Vakhitov is 3.180 heads, including 1.180 dairy cows. In addition to cattle, the farm is engaged in the production of pork – 2.757 heads and horse meat – 37 foals and horses. Large volumes of liquid manure, slurry and wastewater are formed while removing manure with hydraulic washing, washing of milking plants, a milk tank and sanitary care for animals. They are accumulated in huge specially built 3 liquid manure tanks (Figure 1).

The results of the studies showed that the most acceptable way to dispose of this waste is to ensure their stepwise preparation. At the same time, the polydisperse mass of slurry, liquid manure, wastewater enters the slurry collector through concrete pipers and, depending on density, is naturally divided into 3 parts. Particles with a density of more than one settle at the bottom of the slurry collector, a layer of concentrated liquid is formed above them, then part of the organic substance accumulates on its surface with a density less than one (feed remains, etc.).

* Corresponding author: dusai@mail.ru

Fig. 1. Satellite image of the Vakhitov Agricultural Complex with three livestock wastewater receivers.

After that, the liquid layer is pumped out into a special tank for clarification, and the dense precipitate is burned out using loaders and transported for application to a separate section of the agricultural irrigation field.

The analysis of samples of dense manure sediment carried out at the Tatarsky Center of Agrochemical Service showed that the nitrogen content, depending on its origin, ranges from 0.26 (cattle waste) to 0.39% (horse origin). The same pattern is typical for potassium (0.38 and 0.58%, respectively). However, in terms of phosphorus, this ratio changes fundamentally, since in dense horse manure its content is 5 times lower than potassium. In this regard, in our opinion, additional application of phosphorus fertilizers should be considered a mandatory agro-technical technique in agricultural sewage farms.

The final stage of treatment – clarification of wastewater – is carried out in horizontal tanks. After that, 1 m^3 of wastewater is diluted with clean water (15 m^3) and supplied to the agricultural sewage farms during the irrigation of crops.

Incomplete deworming of other disease-causing pathogens under the influence of solar radiation during the step-by-step treatment of manure effluents was also revealed in the course of study. In this regard, we can say with great confidence that it is necessary to cultivate only those crops in agricultural sewage farms that pass through the conservation stage (silage, haylage making or grass flour production) and through the body of animals. Only in this it is possible to exclude human infections. These are, first of all, maize for silage, perennial grass and feed mixture for silage (Table 1).

Table 1. Productivity of agricultural sewage farms (2018–2021).

Crop	Green yield, т/га t/ha	Increase		Feed unit content	Feed unit gross yield, t/ha
		t/ha	%		
Introduction of dense effluent of cattle manure, 40 t/ha					
Control (no fertilizer)	33.8	–	47.3	0.19	6.42
Perennial grass per sowing year	56.9	23.1	73.4	0.20	11.38
Corn	57.6	23.8	68.1	0.18	10.39
Feed mixes	46.7	12.9	49.9	0.20	9.34
HCP_{05}	0.32				
Effluent application					
Control (no fertilizer)	32.4	–	–	0.19	6.16
Perennial grass per sowing year	48.5	16.1	49.7	0.20	9.70
Corn	52.0	19.6	60.5	0.18	9.36
Feed mixes	45.1	12.7	39.2	0.20	9.02
HCP_{05}	0.29				

The use of liquid manure and clarified wastewater is a very effective way to increase crop yields.

The following patterns were revealed based on the results of yield accounting:

- without the introduction of manure effluents the productivity of irrigated gray-forest soils does not ensure the regulatory indicators – 6.16 and 6.42 t/ha of fodder units against the regulatory indicator of 6.50 t/ha;
- against the background of the use of both wastewater and dense sediment of cattle manure the highest yielding crop was maize for silage with a green mass yield of 52.0 and 57.6 t/ha, and in terms of gross yield of feed units there were no equal to perennial grass– 9.70 and 11.38 t/ha, respectively;
- feed mixtures (peas + oats + sunflower) occupied an intermediate position between control and perennial grass in terms of green mass yield and gross yield of feed units.

Fig 2. Corn cobs in milky ripeness against the background of the introduction of 40 t/ha of dense sediment of cattle.

Fig 3. Perennial grass with the introduction of wastewater.

Consequently, corn and perennial grassin agricultural sewage farms differ in the rate of biomass accumulation, which are profitable to cultivate from an economic point of view.

The cost of gross production (CGP) was determined by the formula:

CGP = Y · Feed unit · 8 thousand rubles, where

Y– green mass yield;

Feed units – content of feed units in green mass;

8 thousand rubles – sale price of 1 ton of oat grain in 2019 (1 ton of feed units nutritionally equal to 1 ton of oat grain – the world standard).

Calculations show a close correlation between the cost of gross production and the gross yield of feed units: the higher the first indicator, the larger the second value. Thus, perennial grass with the introduction of 40 t/ha of dense sediment of cattle manure runoff ensured the yield of 11.38 t/ha of feed units in the amount of 91.4 thousand rubles, which is 78 % higher than the control (Table 2).

Table 2. Economic indicators of the use of manure effluents in agricultural sewage farms(at 2021 prices).

Crop	CGP, RUB, in thousands/ha	TC, RUB, in thousands/ha	CNI, RUB, in thousands/ha	Profitability, %	Feed unit cost, RUB, in thousands/t
Introduction of dense effluent of cattle manure, 40 t/ha					
Control (no fertilizer)	51.4	43.8	7.6	17	6.8
Perennial grass per sowing year	91.4	58.1	33.3	57	5.1
Maize for silage	83.1	60.4	22.7	38	5.8
Feed mixes for haylage	74,7	56.3	18.4	33	6.0
Effluent application					
Control (no fertilizer)	49.3	38.4	10.9	28	6.2
Perennial grass per sowing year	77.6	41.8	35.8	86	4.3
Maize for silage	74.9	45.6	29.3	64	4.8
Feed mixes for haylage	72.2	50.9	21.3	42	5.6

The same pattern was characteristic in the block of studies using clarified wastewater: 77.6 and 57%, respectively.

The total costs (TC) are calculated according to the process chart. For the cultivation of perennial grass they turned out to be much less compared to corn and feed mixtures: in the first block of experiment the analyzed value was 58.1 thousand rubles/ha against 60.4 thousand rubles/ha when cultivating corn. This is due to the fact that perennial grass is sown once, but used 3-4 years. The cost of perennial grass tending is also less, since unlike corn, it does not require inter-row cultivation. Most importantly, the nodule bacteria of perennial grass with alfalfa content additionally absorb air nitrogen and provide the largest gross yield of feed units.

The reduction in the total cost of cultivating all studied crops using clarified wastewater should also be emphasized, since in this process there are no additional costs for their use: wastewater is delivered together with irrigation water. As a result, conditionally net income (CNI) determined by the formula CNI = CGP-TC on crops of perennial grass is 35.8 thousand rubles/ha, corn – 29.3 and feed mixtures – 21.3 thousand rubles/ha against 33.3; 22.7; 18.4 thousand rubles/ha on options with the introduction of dense sediment of manure runoff of livestock complexes using special spreaders terra variant 585.

In economic calculations, the most important indicator is the profitability (P) of production of a particular product. It is determined by the formula: $P = \frac{CNI}{TC} \cdot 100\%$. According to this indicator, our studies distinguish 2 crops: perennial grass – 57 and 86 % depending on the methods of using cattle manure and maize for silage with cobs in milk ripening – 38 and 64 %. For comparison it should be noted that in rich feed production of the Republic of Tatarstan the profitability of these crops does not exceed 25-30%, which is a significant proof of the high economic efficiency of irrigation of feed crops with the introduction of manure effluents from livestock complexes.

The calculations of the economic efficiency require the cost of production using the following formula: $C = \frac{TC}{Feedunitgrossyield}$,since it shows how much money is spent from the production of 1 ton of feed units. Depending on the culture and methods of manure disposal, the cost of feed production from perennial grass ranged from 4.3 to 5.1 thousand rub/t, corn – from 4.8 to 5.8 and feed mixtures – from 5.6 to 6.0 thousand rub/t. With a conditional sale price of 1 ton of feed units of perennial grass of 8 thousand rubles, the farm would receive from 2.9 to 3.8 thousand rubles of cash from perennial gras (8-4.3 and 8-5.1), corn – from 2.2 to 3.2 and feed mixtures – from 2.0 to 2.4 thousand rubles.

In addition to compliance with the recommended technology for the use of manure and the choice of crops in obtaining high economic returns, the territories of agricultural sewage farms are particularly important, provided they comply with the following ules developed by us:

- the length of the field must correspond to the length of the acquired sprinkler;
- the field configuration for front wide-grip sprinklers should be rectangular with a length-to-width ratio of 3:1 or 2:1;
- it is desirable to place the long side of the field across the north-west winds;
- it is strictly forbidden to place the ASF above the main and distribution pipelines;
- pipelines shall be located along the boundaries of fields. The number of pipelines shall be minimum, economically reasonable;
- sprinkler machines shall operate sequentially from one culture to another according to the rational process system;
- circular sprinkler machines shall provide no more than two crops with irrigation water, the transition to another irrigation area is carried out mechanically;
- the edges of irrigated area are sown with other mesophytic or xerophytic cultures to form rectangular fields forcircular sprinkler machines.

Most importantly, special attention should be paid to the forestry arrangement of the ASF territory. The forest strip created along the perimeter of the ASF should help reduce wind strength, increase relative air humidity, reduce moisture evaporation and spread ammonia odor and act as a biological drainage to prevent waterlogging of irrigated lands. In order to solve the latter problem, it is necessary to plant moisture-loving tree species (poplar, birch, aspen, and tall shrubs (willow, alder, willow) with a planting scheme of 3x1 m (3m distance between rows and 1m in rows), with a total width of 9–12 m and an openwork structure.

4 Conclusion

The disposal of manure effluents of livestock complexes in agricultural sewage farms and the forestry development of the ASF territory are a promising direction of increasing the volume of feed production with low cost and high profitability (up to 86%).On the other hand, it is of great environmental importance, especially for the Republic of Tatarstan with a cattle density of more than 1 million heads.

References

1. V. V. Ustyukova, *Agriculture* **4,** 12–20 (2020).
2. S. Bereziuk, I. Zubar, *Scientific Heritage* **46-7 (46),** 3–10 (2020).
3. E. V. Andreeva, *Environmental Safety in the Agro-Industrial Complex* **2,** 577 (2007).
4. S. Suleymanov, R. Safiollin, N. Loginov, L. Vafina, *BIO Web of Conf.* **37,** P00177 (2021).
5. V. F. Moiseichenko, et al. *Basics of Scientific Research in Agronomy* (Moscow: Kolos, 1996).
6. B. A. Dospekhov, *Field Experience Methodology* (Moscow: Agropromizdat, 1985).
7. E. S. Bryukhanova, G. V. Ushakov, A. G. Ushakov, *International scientific journal "Alternative Energy and Ecology"* **5 (85),** 71–82 (2010).

8. F. N. Safiollin, M. M. Khismatullin, *System of reclamation agriculture in the Republic of Tatarstan (general issues of land reclamation and agricultural cultivation of agricultural crops on watering)* (Kazan: Center for Innovation Technologies, 2015).
9. M. M. Khismatullin, R. Ullakh, *Regional Economics: Theory and Practice* **20-1 (496),** 168–185 (2022).
10. A. V. Yudanova, *Engineering and Technical Support of the Agro-Industrial Complex* **3,** 852 (2002).

Autonomous power supply of objects in rural settlements of Pavlodar region using alternative energy sources

Altai Alimgazin[1,*], *Ildar* Sultanguzin[2], *Irina* Akhmetova[3], *Askhat* Berguzinov[4], *Baurzhan* Zhakisihev[1], and *Oleg* Ushchekin[3]

[1] Eurasian National University named after L.N. Gumilyov, 2, Satbaev St., Nur-Sultan, 010000, Kazakhstan
[2] Moscow Power Engineering Institute Institute, 14, Krasnokazarmennaya St., Moscow, 111250, Russian Federation
[3] Kazan State Power Engineering University, 51, Krasnoselskaya St., Kazan, Republic of Tatarstan, 420066, Russian Federation
[4] Toraighyrov University, 64, Lomov Street, Pavlodar, 140000, Kazakhstan

Abstract. The issues of practical application at pilot facilities in the village of Konstantinovka, Uspensky district, Pavlodar region of a specific technology are considered. It is based on an autonomous power supply complex (APSC) using geothermal wells of the region with temperatures from +25°C to +42°C as a low-grade heat source. The results of integrated calculations of the economic effect resulting from the introduction of the AEC supply heat to a number of social facilities (school, hospital, kindergarten, 24-apartment building, sports complex) with a total heat load of 0.683 Gcal/hour (data from the akimat of the village of Konstantinovka) are presented.

1 Introduction

The transition to low-carbon development and decarbonization of the national economy of Kazakhstan involves the adoption of cardinal measures to improve energy efficiency and the introduction of new low-carbon technologies [1-3].

The Republic of Kazakhstan has also joined this global process by signing and ratifying the Paris Climate Agreement and declaring at the UN Climate Ambitions Summit in December 2020 its intention to achieve carbon neutrality by 2060.

The need to take drastic measures is dictated by technological and environmental risks, the solution of which does not require delay.

According to the latest data, Kazakhstan ranks among thirty polluting countries in the global ranking [1-3].

According to The Global Carbon Atlas, Kazakhstan's contribution to global carbon dioxide emissions by the end of 2019 amounted to 314 megatons of CO2.

In terms of carbon dioxide emissions, Kazakhstan ranked 21st among more than 221 countries.

Today, coal generation accounts for over 70% of our country's generation structure. And in the next 15 years this figure is planned to be reduced to 50% through the use of the best available technologies (BAT), the active introduction of renewable energy and hydrogen energy technologies, the commissioning of new gas turbine power plants, the development of decentralized energy supply, etc.

As world experience shows, one of the most promising technologies contributing to the effective decarbonization of the national economy includes technologies related to the use of steam compression and absorption heat transformers (heat pumps, cooling machines) [2-10].

The analysis of world-famous examples of the use of heat transformers in agriculture has shown broad prospects for improving the efficiency of processes in various areas of this industry [11–14].

The technological processes of agriculture are associated with a large consumption of heat, which is largely satisfied by electricity. Agriculture also has its own large secondary thermal resources, but due to their low-temperature potential, they are poorly and not fully used.

The use of reversible heat pumps (TN) in agricultural technological processes makes it possible to use low-potential waste heat to generate the necessary heat.

On livestock farms of the agro-industrial complex, the use of refrigerating machines (HMM) is required to cool milk. Also, there is a need for hot water, which is used for technological purposes and watering livestock. It is also necessary to provide heating for individual premises of the farm, including the calf house in the maternity ward. All these operations can be performed using equipment and energy-saving technologies.

The use of new innovative energy-saving technologies using heat transformers (TN, HM) is a real way to increase the efficiency of milk cooling and reduce energy consumption by optimizing and improving the design and technological scheme of milk cooling and water heating [12, 13].

In the food industry, the use of XM is very often required for the implementation of technological processes [12–14].

* Corresponding author: alimgazin_altai@mail.ru

For example, many breweries, meat and dairy plants and sausage factories have very large centralized HMS.

On the other hand, there is a great need for hot water used for various types of cleaning throughout the year. It is also necessary to provide heating of the premises.

Thus, there are all conditions for the beneficial use of TN and XM. However, we know only a few cases of their use in this area [12–14].

An example is the possibility of using TN with the simultaneous use of heat and cold during the pasteurization of liquids with their subsequent cooling.

The use of TN for the combined production of cold and heat in the food industry is also provided during the implementation of maturation processes, for example, cheeses or sausages [4, 5, 13].

For example, on dairy farms, the main share of energy consumption is the cost of electricity for the electric drive of HMS cooling freshly milked milk and for heating water for sanitary and technological needs.

The use of TN in this area allows you to simultaneously obtain the necessary cold and sanitary hot water with minimal energy consumption, which significantly affects the final cost of production.

The use of TN in dining rooms also significantly reduces the consumption of electricity for ventilation.

TN effectively solve the problems of heating greenhouses and farms. The range of application of TN is extremely wide, since their use is possible in greenhouses (for growing flowers, vegetables, ornamental plants, seedlings and seedlings), at livestock facilities and poultry farms [4, 5, 13, 14].

The use of TN in greenhouses [4, 5, 14] will dramatically reduce the cost of agricultural products.

The principle of using modern energy-saving heating and cooling technologies based on heat transformers for houses, greenhouses, animal and water complexes allows you to reduce the costs associated with fuel.

2 Materials and methods

Currently, in many rural settlements of the Republic of Kazakhstan, the problem of providing heat and electricity to both social facilities and agricultural sector facilities, small and medium-sized businesses is acute.

The combination of heating and cooling systems using steam compression and absorption heat transformers [2, 3] is very promising in agricultural enterprises of the Republic of Kazakhstan (farms, vegetable storages, meat and dairy complexes, greenhouses, etc.).

In accordance with the instructions of the President of Kazakhstan, the Ministry of Agriculture of the country has been successfully implementing a new special project "Auyl – El besigi" ("Aul – Cradle" within the framework of the "Spiritual Radiance" program since 2019 [15].

The villages of the Republic of Kazakhstan should have good roads, healthcare, decent education, sports centers, housing. It is planned to select 1–1.5 thousand villages from the currently existing 6,600 rural settlements and turn them into mini-towns with high standards of quality of life, which is quite possible.

The aim of the project is to modernize the socio-economic infrastructure of rural areas, which should ensure that by 2030 at least 80% live in settlements that meet modern standards of quality of life.

In many countries of the world, new heat supply systems based on thermal waters from TN have been developed and are actively used. Thermal waters can be used for heating buildings, for heating greenhouses, water in swimming pools, growing vegetables, fruits and for other purposes [4–7].

They are especially appropriate for the needs of heat supply, since thermal waters with temperatures below 45–50°C are most common, while additional high-temperature heat sources are provided, which are turned on at low outdoor temperatures. They can be peak boilers for additional heating of thermal waters, TN, electric heating installations.

In the Republic of Kazakhstan, according to hydrogeological studies, in a number of regions of Pavlodar, Almaty, Turkestan and other regions of the Republic of Kazakhstan there are a large number of artesian and geothermal wells with sufficient debit of self-draining water and temperatures from +200°C to 450°C.

Over the past few years, scientists of the L.N. Gumilev ENU, together with Russian colleagues (NRU "MEI", KGEU), have been conducting research on the prospects for the use of TN using both the heat of groundwater and reservoirs. There is geothermal heat of water from wells, and the heat of sewage treatment plants, ventilation emissions of buildings, waste heat of a number of industrial enterprises [9, 10].

As a result of carrying out in 2010–2021 at the Research Institute of "Energy Saving and Energy Efficient Technologies" (hereinafter – the Research Institute "Ei ET") L.N. Gumilyov ENU offers a specific result, a sample of a new product – TN in a block – modular layout. It is one of the main components of an autonomous power supply complex (AEC) using alternative energy sources, for the implementation of a number of state-funded and contractual research works (R&D).

The AEC is based on the basic heat pump technology for converting low-potential heat from low-temperature energy resources (waste heat from self-discharging geothermal wells in a number of regions of the Pavlodar region and other regions of Kazakhstan) into high-potential thermal energy suitable for practical use. It is not another modernization of traditional energy sources, but the introduction of a new, progressive, highly efficient and environmentally friendly method of obtaining heat.

The most important prerequisite for the use of AEC are sufficiently large volumes of low-potential heat lost with water from self-draining geothermal wells in various regions of the Republic of Kazakhstan.

For the first time, a specific technology and schemes for the use of AEC with the use of geothermal wells in the region at temperatures from +250°C to +450°C as a source of heat energy are proposed for practical implementation at social and agricultural sector facilities in pilot rural settlements of the Pavlodar region.

AEC using alternative energy sources in a stationary or modular configuration is a metal insulated container.

The practical implementation of the proposed energy-saving and environmentally friendly technology complies with the Laws of the Republic of Kazakhstan "On Energy Conservation and Energy Efficiency Improvement" (2012), "On Ratification of the Kyoto Protocol to the United Nations Framework Convention on Climate Change" (2009). There is the Law of the Republic of Kazakhstan "On Alternative Energy Sources" (2022), the Doctrine of (Strategies) of the Republic of Kazakhstan on achieving carbon neutrality (2022), the tasks of the transition of the Pavlodar region to the introduction of "green" technologies in various sectors of the economy.

Studies conducted in the period from summer 2018 to February 2022 have shown that the use of UEC will ensure reliable and uninterrupted power supply of facilities in rural areas.

For example, in the village of Konstantinovka of the Uspensky district of the Pavlodar region, the millionaire collective farm "30 years of the Kazakh SSR", previously known in the USSR, was located with a centralized heat supply system for the entire infrastructure of the village from the boiler house. Currently there are a number of peasant and farm farms for processing agricultural products, a number of public sector facilities (school, hospital, kindergarten, sports complex, 24-apartment residential building), which have serious problems with heating and hot water supply.

At the same time, according to the akimat of the Ravnopolsky rural district, which includes the village of Konstantinovka, there are 13 self-emptying artesian wells with a large water debit and a sufficiently high temperature in the district.

In 2018-2021 (according to preliminary agreements with regional akimats and orders of small and medium-sized businesses) it is proposed to use AEC for the use of waste heat from geothermal wells on a number of real objects in rural settlements of the Pavlodar region. They significantly reduce thermal pollution of the environment, obtain heat of high parameters for reliable and uninterrupted heat supply of objects in rural settlements.

When introducing AEC in the agro-industrial complex, housing and communal services, public sector facilities of rural settlements of the Republic of Kazakhstan, it is planned to obtain a significant economic and environmental effect, primarily by replacing traditional heat sources (solid, liquid fuel boilers, electric boilers), reducing greenhouse gas emissions into the atmosphere.

The use of autonomous technologies based on AEC will make it possible to return significant volumes of irretrievably lost heat to the heat supply system of objects (agro-industrial sector, social facilities, housing and communal sector, small and medium-sized businesses) without additional burning of traditional fuels. It will allow one to obtain a significant economic effect, to abandon, in some cases, purchased energy from outside, reducing thermal pollution of the environment.

At the same time, all recycled waste heat can be directed to the thermal and scheme of enterprises for heating and hot water at the above facilities, which, in the end, will enable significant savings in heat and electric energy purchased from local boiler houses.

It should be noted that the cost of own heat pump heat will be 2–4 times less than the price of purchased heat, i.e., with the launch of the AEC, there is a real opportunity to meet the existing demand for heat and electric energy for rural commodity producers, social facilities in rural areas at significantly lower costs.

3 Results and discussion

Below are the results of the enlarged calculations of the economic effect of the introduction of AEC for social facilities (school, hospital, kindergarten, 24-apartment building), for which the total heat load is 0.683 Gcal/hour (data from Akimat the village of Konstantinovka).

Initial data:

1. The existing source of heat supply is a new modular boiler house on solid fuel (Ekibastuz coal)

The fuel consumption in the boiler house for the heating season, tons – 870.

The cost of fuel for the heating season, Stopl., thousand tenge - 8,127.00.

The price of fuel, Coal, tenge/ton – 8,127,000 / 870 = 9,341.

The duration of the heating period is 210 days.

The cost of 1 kW of electricity, tenge/kWh – 21.56.

The name of the objects that are heated from a modular boiler room:

– Konstantinovskaya secondary school, heated area of Sot. m^2 – 3,493.4;

– sports complex, Sot., m^2 – 1 617.3;

– nursery-garden "Aigolek", Sot., m^2 – 1 010.1;

– 24-apartment residential building, Sot. m^2 – 1 124.3; there is no heating, a project with a modular coal boiler is currently being developed.

The total heated area from the modular boiler house, Sot. m^2 – 6 120.8.

To compare the efficiency and performance of the existing modular coal boiler house for the heat supply of all the above facilities we proposed block-modular heat pump unit (BMTU) an alternative heat supply system using the heat of geothermal sources the village of Konstantinovka. An enlarged calculation of both options was carried out based on the initial data provided by the Customer for these objects

3.1 Modular coal boiler house

The average annual consolidated operating costs for the boiler house will be, tenge/year:

Saved per year = Fuel saved + Saved wages + Saved fee. = 14 728.62 thousand tenge/year.

Autonomous Power Supply Complex (AEC)

Based on the data provided on the heated areas of the facilities, 1 water-to-water heat pump GSHP-400 (PRC) with a thermal capacity of 472 kW was selected for the installation, which is sufficient to cover the specified thermal load of the facility.

The placement of the necessary basic technological and auxiliary equipment is offered in the BNTU on the basis of a 40-foot container.

In the proposed version of the project, as a low-potential source of heat, it is proposed to use the water heat of a geothermal well in the village of Konstantinovka with a temperature of about +280°C.

In the enlarged calculations, we assume that the total annual operating costs when using AEC based on BNTU are equal to:

Saved AEC = electricity saved + service is saved. = 5,837,430 tenge/year.

The annual savings from the use of AEC compared to the existing heat supply system, tenge / year, is defined as:

Saved = Saved per year – Saved AEC = 8,891 190 tenge/year.

In fact, the total annual effect will be greater if the actual operating costs of the coal boiler house are fully taken into account (taking into account the specified costs for the purchase of coal, equipment repair costs, personnel payments, environmental payments).

The data of the enlarged calculations compared to 2 sources of heat supply of social facilities in the village of Konstantinovka are shown below in Table 1.

Table 1. Comparison of annual operating costs of facilities in the village of Konstantinovka, Uspensky district, Pavlodar region for heat supply from various heat sources (thermal load - 0.683 Gcal/hour).

No.	Name of the heat source	The cost of annual operating costs, Saved per year, thousand tenge*	Cost 1 Gcal/hour, tenge/Gcal (including VAT)
1	Modular boiler room The village of Konstantinovka	14 728,62	12 742
2	Autonomous power supply complex	5 837,43	856

* The data are given for January 2022; the exchange rate is $1 = 432 tenge.

During recent meetings (February 15–16, 2022) of representatives of the Research Institute "Ei ET" with the leadership of the akimat of the Uspensky district, Ravnopolsky rural district, representatives of small and medium-sized businesses, the use of AEC using alternative energy sources was proposed. It is a solution to the problem of reliable and uninterrupted provision of heat supply to the village of Konstantinovka.

A Memorandum of cooperation, a Letter of Intent, and an agreement on co-financing of this project were signed with the Akimat of the Uspensky district.

Similar documents were signed with representatives of small businesses in the region (FH "Oynak") concerning energy supply with the use of AEC of a greenhouse complex under construction, a shift settlement of 8 houses in the above-mentioned rural district.

The technical solutions are proposed by the scientists of the Research Institute "Ei ET" of the University for a specific pilot region (Uspensky district – the border of the Kachir district) of the Pavlodar region. It will allow one to breathe life into these settlements in the shortest possible time, since there are dozens of geothermal wells in the region with a self-draining water temperature of about 250°C – 450°C and with a water debit of 15 m^3/hour up to 75 m^3/hour.

Based on the results of the previous research conducted by the staff of the Research Institute "EiET" of the L.N. Gumilev ENU, we can safely say the following. The large-scale use of AEC using alternative energy sources in stationary or modular configuration in various regions of the country will help to solve problems in rural settlements with heat cooling and electricity supply of the agricultural sector and social (budgetary) facilities spheres.

Thus, the source of profitability of projects when implementing the UEC at various facilities of the agricultural sector (greenhouses, meat and dairy enterprises, vegetable and fruit storage facilities) and public sector facilities (schools, hospitals, kindergartens, sports complexes, administrative buildings and structures, etc.) are:

1. Significant economic effect, i.e., the difference in the cost of thermal energy purchased by consumers from boiler houses and produced with the use of AEC (3–5 times).
2. A significant reduction in the operating costs of the above facilities using alternative energy sources for heat and cold supply.
3. Reduction of greenhouse gas emissions in rural areas and, as a result, reduction of environmental payments of economic entities.

4 Conclusion

This article discusses the issues of the application of AEC using the heat of geothermal sources to improve the efficiency of heat supply systems for social facilities, agro-industrial complex, small and medium-sized businesses in a particular region of the Republic of Kazakhstan.

A real technology and schemes for the application of AEC using geothermal wells with t from +250°C to +450°C as a source of heat energy are proposed for implementation at social and agricultural sector facilities in pilot rural settlements of the Pavlodar region and other regions of the Republic of Kazakhstan.

When applying the proposed technology in various regions of the Republic of Kazakhstan, a significant economic and environmental effect is expected

https://doi.org/10.1051/bioconf/20225200010

primarily due to the replacement of traditional heat sources (solid and liquid fuel boilers, electric boilers), reduction of greenhouse gas emissions into the atmosphere and improvement of the environmental situation.

The results of calculations of the economic effect of the introduction of TN for the heat supply of a number of social facilities (school, hospital, kindergarten, 24-apartment building, sports complex) of a pilot settlement in the Republic of Kazakhstan with a total heat load of 0.683 Gcal/hour are presented.

References

1. G. V. Safonov, B. Yesekina, Sh., Tuebekova, Decarbonization of the national economy: strategies and scenarios, *Proceedings of the National Academy of Sciences of the Republic of Kazakhstan. A series of social sciences and humanities,* **313 (3),** 13–22 (2017)
2. S. Zh. Daukeeva (Ed.) *Decarbonization of extractive industries of the Republic of Kazakhstan* (Nur-Sultan: Bi-PRINT, 2021).
3. S. Daukei (Ed.) *Decarbonization of Economy's Extractive Sectors of the Republic of Kazakhstan* (Nur-Sultan: Bi-PRINT, 2021)
4. D. Ray, D. McMichael, *Heat pumps* (Moscow: Energoizdat. 1982)
5. G. Heinrich et al. *Heat pump installations for heating and hot water supply* (Moscow: Stroyizdat, 1985)
6. G. P. Vasiliev, Geothermal heat pump systems of heat supply, *Thermal power engineering*, **6,** 33–41 (2004)
7. T. G. Kuzmina, Energy saving during trigeneration in technologies with circulating water supply, *Turbines and Diesels*, *July – August*, pp. 12-14, (2012)
8. A. Sh. Alimgazin, S. G. Alimgazina, Heat pump technologies for heat supply of various objects. Experience of the Republic of Kazakhstan, *AVOK, Energy saving*, **8,** 68–73, (2013)
9. A. Sh. Alimgazin, I. A. Sultanguzin, Yu. V. Yavorovsky, I.G. Akhmetova, A. N. Berguzinov, Prospects of using absorption heat transformers to improve energy efficiency of industrial enterprises of the Republic of Kazakhstan, *Bulletin of Toraighyrov University. The Energy series* **4,** 21–33 (2020).
10. K. V. Alexandryan, R. G. Hanoyan, The use of low-potential heat for heating greenhouses, **3,** 6–8 (2002)
11. E. V. Shushunova, Efficiency of using heat pumps for cooling milk and heating water at livestock complexes of the agro-industrial complex: PhD dissertation (Moscow, 2012)
12. Aerocond Moscow, Retrieved from: http://www.aerocond.ru
13. A. G. Pozdnyakov, V. A. Shchelkanov, A. A. Gridnev, Absorption bromistolithium refrigerating machine ABHM-1000VN in the trigeneration system of a greenhouse plant. *Energy saving* **7,** 56–64, (2009)
14. Annual Message to the people of Kazakhstan of the First President of the country on the launch of a new special project "Auyl-El besigi" within the framework of the program "Rukhani Zhangyru" (Nur-Sultan, 2019)

Role and value of small forms of farming in the agricultural production of the Republic of Tatarstan

Ilgizar Gainutdinov*, *Farit* Mukhametgaliev, *Marsel* Khismatullin, *Fayaz* Avkhadiev, and *Asia* Subaeva

Kazan State Agrarian University, Kazan, 65, K. Marx Str., 420015, Russian Federation

Abstract. Providing the country's population with high-quality food while ensuring effective agricultural production in rural areas should become one of the priority tasks of the state. Besides, creating conditions for the development of small forms of economic activity in cooperation with large commodity producers, thereby achieving the development sustainability of rural areas, providing additional jobs, through stimulating the development of small agricultural businesses, is an urgent task of modern times. The paper is devoted to the study of the role and value of small farming in rural areas on the example of the Republic of Tatarstan. The results of the impact of state support programs on their number and further development are shown. The emphasis is put on the need to expand this assistance in other areas. The volumes of agricultural production in peasant farms are shown. Priority areas to increase crop production were identified, and recommendations to ensure sustainable income growth of small farms were given. The practical relevance of research results includes the possibility of their use to design programs for the development of regional food markets thus achieving high targets.

1 Introduction

The share of small forms of farming, which commonly include farms and households of the population, in agricultural production is quite high. Thus, their share in the total volume of agricultural products of the Republic of Tatarstan in actual effective prices ranged from 51.8% (2016) to 49.8% (2020). The production indices in peasant farms in comparable prices ranged from 115.4% (2016) to 120.9% (2020) compared to the previous year, and in the households of the population, respectively: 103.8% and 100.9%. Thus, 40.6% of the total number of cattle is concentrated in small forms of farming of the republic, including 41.6% of cows, 94.7% of sheep and goats, 10.1% of pigs, 69.1% of horses and 23.5% of poultry.

Small forms of farming also serve the sector of employment for the rural population and job placement. Many works are devoted to the study of the essence of small forms of farming [1, 4, 5], as well as their role and value in increasing the efficiency of land use, creating sustainability in the development of rural areas [2, 3]. At the same time, research aimed at studying the further development of small forms of farming, identifying the promising areas of their production activities, measures of state support, the creation of peasant farms and their cooperation with large representatives of agricultural business, is a promising area of considerable practical importance.

2 Materials and Methods

The theoretical and methodological basis of the study includes the use of scientific statements and concepts of the economic theory. The works of scientists and practitioners in agricultural economics on this topic were considered as a theoretical justification of the problem based on the principles of historicism. Scientific articles based on the materials of international and national scientific and practical conferences devoted to the study of the role and importance of the development of small forms of farming, as well as the materials of the authors' research were analyzed.

The general methodological basis of the study includes the systemic analysis. Constructive, deterministic, retrospective, dynamic and statistical methods of studying socio-economic systems, as well as monographic, abstract-logical, calculation-constructive, economic-statistical methods were used in the study.

The official data of the Federal State Statistics Service of the Russian Federation, the Ministry of Agriculture of the Russian Federation, the Ministry of Agriculture and Food of the Republic of Tatarstan, materials contained in official Internet sources, expert opinions, data obtained during the author's analysis and calculations were used.

3 Results and Discussion

The households of the population occupy a significant niche in economic activity among small forms of farming. First of all, the households of the population refer to citizens having personal subsidiary farms. In this regard, unlike other representatives of large and medium-sized businesses in agriculture, a significant

* Corresponding author: ilgizar-gg@mail.ru

https://doi.org/10.1051/bioconf/20225200011

part of small forms of farming are not entrepreneurs and commercial organizations. The contradictions in the legal status of small agricultural business entities, the clarification of the essence of small agricultural business is given as: "... a type of activity carried out in agricultural production and in the market of agricultural products, raw materials and food by agricultural producers – small agricultural business entities that meet the criteria of the legislation" [4]. According to the authors, small forms of farming, regardless of the organizational and legal form and forms of ownership, equally need measures of state support and the creation of equal conditions for their economic activities.

The number of peasant farms in the republic has a stable growth trend (Fig. 1). The stability of land use and the average size of the land area of farms suggests that these farms are created as the family forms of business in the countryside.

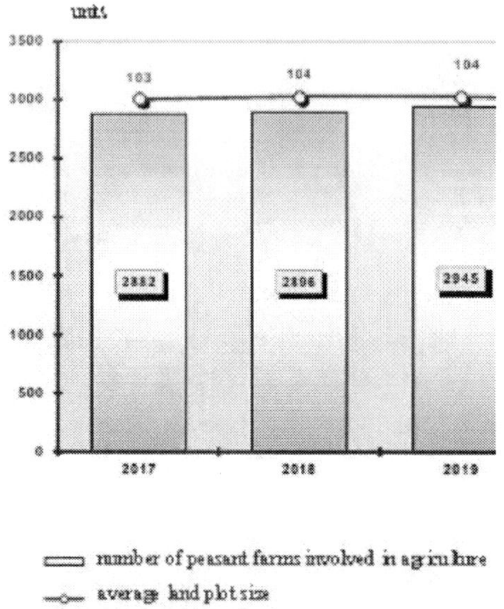

Fig. 1. Number of peasant farms and an average land plot size (according to the Office of the Federal Service for State Registration and Cartography in the Republic of Tatarstan).

Small forms of farming, including peasant farms, play an important role in providing food to the population of the republic, primarily potatoes, field vegetables, as well as milk and meat of cattle, lamb (Fig. 2).

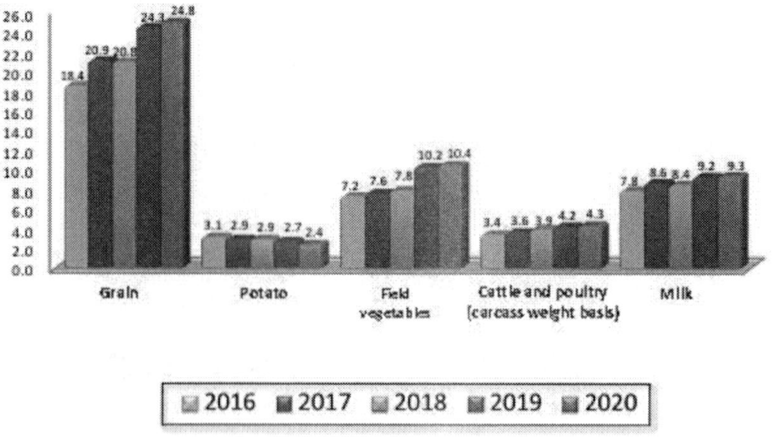

Fig. 2. Share of peasant farms in the total production of the main types of agricultural products (as a percentage of farms of all categories) (according to Tatarstanstat).

Most of animal species such as sheep and goats are mainly concentrated in the households of the population – 82.0%, in farms – 12.7%. If over 2016-2020 there is a decrease in the share of livestock and poultry meat production from 26.4% to 23.5%, milk – from 32.1% to 27.6%, eggs – from 26.6% to 20.9%, then in farms the share of livestock production is growing: 3.7 – 4.6%, 7.8 – 9.3%, 1.0% – 2.3%, respectively (Fig. 3).

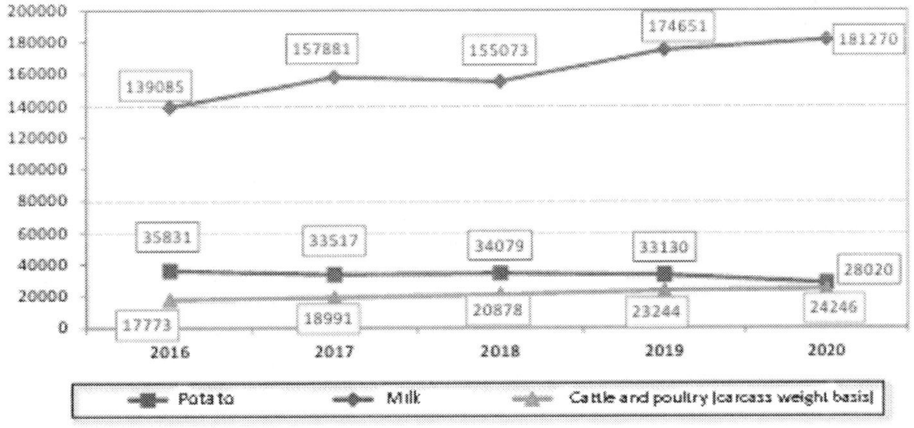

Fig. 3. Production of the main types of agricultural products in peasant farms (tons) (according to Tatarstanstat).

The share of farms in the sown areas of grain crops has also increased: from 19.0% in 2016 to 25.5% by 2020. Approximately proportionally to sowing areas, grain production is also growing – from 18.4% to 24.8% (Table 1).

Table 1. Share of crop production farms of the Republic of Tatarstan.

Indicator	2016	2017	2018	2019	2020	2020 to 2016, %
	Grain sowing areas, ha					
All categories of farms	1591763	1533971	1483999	1461709	1552490	97.5
Farms	301763	324467	326147	361715	395187	130.9
Share of farms,%	19.0	21.2	22.0	24.7	25.5	6.5 pp
	Production volume, tons					
All categories of farms	4115102	4879758	3657591	4167873	5200784	126.4
Farms	757012	1016656	759023	1010872	1290427	170.4
Share of farms,%	18.4	20.8	20.8	24.3	24.8	6.4 pp
	Potato sowing areas, ha					
All categories of farms	57991	55565	54160	52238	50862	87.7
Farms	2192	1533	1471	1243	1155	52.7
Share of farms,%	3.8	2.8	2.7	2.4	2.3	-1.5 pp
	Potato production volume, tons					
All categories of farms	1143743	1164297	1189580	1214133	1174207	102.6
Farms	35831	33517	34079	33130	28020	78.2
Share of farms,%	3.1	2.9	2.9	2.7	2.4	-0.7 pp
	Field vegetable sowing areas, ha					
All categories of farms	9390	10028	10011	10144	9947	105.9
Farms	702	881	1143	1246	1271	181.1
Share of farms,%	7.5	8.8	11.4	12.3	12.8	4.3 pp
	Volume of field vegetable production, tons					
All categories of farms	285062	286854	274648	286913	274292	96.2
Farms	20528	21639	21492	29296	28541	139.0
Share of farms,%	7.2	7.5	7.8	10.2	10.4	3.2 pp

The same trend is observed in the production of field vegetables. Although by 2020, the share of sown areas of field vegetables in farms increased from 7.5% to 12.8%, and production volumes amounted to only 10.4% of the total production of this product in all categories of farms. In this regard, farms should increase their production volumes in proportion to the occupied area.

During the analyzed period (2016-2020), the share of potato sown areas in farms decreased from 3.8% to 2.3%, i.e., by 1.5 percentage points. During the same period, potato production in this category of farms decreased from 3.1% to 2.4%, or by 0.7 pp. The decrease in potato sown areas for households and farms, combined with the crop failure of 2021, led to a significant decrease in gross yield, which led to a sharp increase in prices for potatoes and vegetables in 2021. Although in general, over 5 years (2016–2020), potato production in the republic increased from 1143.7 to 1174.2 thousand tons. However, there was a decrease in field vegetables from 330.9 to 325.8 thousand tons, respectively.

In general, despite some fluctuations in production volumes the production indices in farms in comparable prices as a percentage of the previous year, have always been higher compared to other forms of farms (Table 2).

Table 2. Production indices by farm categories (in comparable prices; as a percentage over the previous year).

	2016	2017	2018	2019	2020
Agricultural organizations	104.3	109.6	92.6	106.4	105.8
Households of the population	103.8	97.0	103.1	97.1	100.9
Peasant farms	115.4	119.5	91.1	120.9	109.8

The growth of the production index indicates the sustainable development of farms. This was also facilitated by the state support measures that were provided to the farmers of the republic. In 2021, small forms of farming received 675.9 million rubles in the form of grants from the budget, including in the following fields: *Development of Family Farms* – 315.9 million rubles; under the program *Development of Agricultural Consumer Cooperation* – 191.3 million rubles; *Agrostartup* – 161 million rubles and *Agroprogress* – 7.7 million rubles.

The results of competitions were summed up and the funds were allocated among the recipients with the conclusion of agreements in the Electronic Budget system. Eighty farmers received the grants for the development of family farms and the Agrostartup grants for 476.9 million rubles, including 30 farmers in dairy, 24 – in meat cattle breeding, 11 – in poultry farming, 4 – in horse breeding.

Herd expansion is expected over the next 2 years as a result of the full implementation of these projects: cows – by 1840 heads, feeder young cattle – by 1626 heads, horses – by 840 heads and poultry – by 540 thousand heads.

Starting from 2022 it is planned to ensure the state support for the development of agritourism in the rural areas of the Republic of Tatarstan. It is also planned to allocate grants in the amount of up to 10 million rubles for the construction of facilities to accommodate tourists and create the necessary infrastructure. Thus far about sixty entrepreneurs in the republic wish to fulfill their intentions in this area of business. Appropriate assistance is required from local executive authorities and professional consultants in choosing routes for tourists and in attracting clients. Two projects – *Rural Tourism with National Identity* of the Vysokogorsky district and *Bilyarsk Recreation Center with an Eco-Farm* of the Alekseevsky district have already been pre-selected by the Ministry of Agriculture of Russia for the Agrotourism grant in 2022.

However, the state support measures for small forms of farming cannot stop the decrease in the average number of employees in agriculture. Thus, the average number of employees of an agricultural organization in the Republic of Tatarstan by 2020 amounted to 41,192 people, or by 7,738 people (15.2%) less than in 2017. This is despite the fact that the total rural population decreased over the same period from 909.0 to 899.7 thousand people (a decrease of 1.1%), including working age population from 497.9 to 491.6 thousand people (a decrease of 1.3%), respectively. The reasons for the decrease in officially employed in agricultural production is well-known: low wages (19.4 thousand rubles in 2017, 27.3 and 32.0 thousand rubles in 2020 and 2021, respectively), insufficient financial resources, a significant part of which falls for the service and processing industries due to the lack of parity in prices for agricultural and industrial products. In this regard, the share of agriculture, forestry, hunting, fishing and fish farming in the structure of gross added value decreased from 6.8% in 2016 to 5.5% by 2019. Over 2016-2020 the prices sharply increased: for combine harvesters – by 22.9%, trucks – by 52.3%, diesel fuel – by 45.8%, electricity – 25.3%, and prices for mixed cattle feed – 2.2 times. At the same time, over the same period, the total consumer price index for food products increased by 17.5%, and the price index for agricultural producers only by 9.3%, including for crop products – 11.8%, and for livestock products – 9.4%.

The disparity in prices led to an increase in the cost of the entire range of agricultural products. For example, the production cost of grain in 2021 increased by 48% compared to the previous year, and as a result, the grain industry received less than 3 billion rubles in profit. Similar situation happened in the entire chain of crop production – sugar beets, potatoes, vegetables, and feed. The cost increase for feed will affect the increase in the cost of livestock production in the future.

Efficient agricultural production requires both digital and institutional transformation in agriculture in a timely manner [6, 7, 8]. Such measures include improving the legal framework for the functioning of small forms of farming. Along with measures of state financial support, significant assistance for the development of small and medium-sized businesses is provided in the legal sphere. So, on December 6, 2021, the Federal Law N407-FZ "On Amendments to Article 19 of the Federal Law "On Peasant (Farming) Economy" and certain legislative acts of the Russian Federation were adopted, thus allowing farms and agricultural consumer cooperatives to sell their agricultural products on the land used to carry out their activities. For these purposes, it will be possible to use capital construction facilities, non-capital structures, and non-stationary trading facilities. It is assumed that this will make it possible for farmers and consumer cooperatives to sell a significant part of agricultural products grown and processed on their own without unnecessary costs directly to consumers thus bypassing intermediaries. Besides, the Federal Law No. 299-FZ of July 2, 2021 was adopted, which allows farms building houses on agricultural land. The restrictions include the fact that the house should have not more than three floors, the area of the original land plot should not decrease. To do this the constituent entities of the Russian Federation should adopt the legislation that confirms this possibility. If approved and clarified by regional legislative bodies, the construction of houses on agricultural lands will make it possible to create peasant farms. All this will contribute to the development of significant territories that are currently deserted. These changes come into force on March 1, 2022. Thus, all these changes in legislation related to a certain extent to the activities of farms and small forms of farming will improve the efficiency of economic activity, develop social infrastructure, ensure employment and better development of empty rural areas.

4 Conclusion

The performed studies allow drawing the following conclusions:

1) The share of small forms of farming (households of the population and farms) in the Republic of Tatarstan that provide the population with agricultural products is significant in actual prices and ranges from 51.8% (2016) to 49.8% (2020). During the same period, the production index in peasant farms in a comparable estimate amounted to 115.4-120.9% compared to the previous year, and in the households of the population – 103.8 and 100.9%, respectively.

2) The number of cattle in the Republic of Tatarstan as a whole tends to decline slightly. A significant part of the livestock is concentrated in small forms of farming – about 40.6% of the total livestock, including 41.6% of cows, 94.7% of sheep and goats, 10.1% of pigs, 69.1% of horses and 23.5% of poultry.

3) The share of farms in the sown area of grain crops had an upward trend: from 19.0% in 2016 to 25.5% by 2020. There is also an increase in the production of grain products – from 18.4% to 24.8%. Farmers need to increase the production of potatoes and vegetables in proportion to the growth rate of sown areas.

4) All forms of state support for the development of small forms of business in the Republic of Tatarstan are required and they are successfully mastered and implemented in practice. The development of agritourism is planned for the future as an additional direction for the provision of services in rural areas.

5) In our opinion, the adoption of a number of laws at the federal level concerning the activities of peasant farms will contribute to the development of their infrastructure, the assimilation of new territories and the strengthening of the rural way of life.

References

1. L. Mikhailova, F. Avkhadiev, N. Asadullin, I. Gainutdinov, BIO Web Conf. **17,** 00095 (2020)
2. S. Sharipov. Agro-Industrial Complex: Economics, Management **3,** 53–56 (2008)
3. S. Sharipov, Economy of Agricultural and Processing Enterprises **8,** 51–54 (2013)
4. I. Gainutdinov. Achievements in Science and Technology of the Agro-Industrial Complex **2,** 6–9 (2012).
5. N. Yakushkin, S. Sharipov, I. Gainutdinov, *Small agricultural business in Tatarstan: issues of ethodology and practice* (Kazan) (2013)
6. A. Subaeva, Bulletin of Kazan State Agrarian University **4 (64),** 130–135 (2021)
7. A. Subaeva, Bulletin of Moscow State Agroengineering University named after V. P. Goryachkin **3 (85),** 47–53 (2018)
8. D. Khafizov, M. Khismatullin, E. Isaicheva, Bulletin of Kazan State Agrarian University **1 (27),** 51–54 (2013)

Efficiency of Giberelonand Cytodef-100 phytohormones in the crop structure and productivity of sunflower in edaphoclimatic conditions of the Republic of Tatarstan

Salavat Suleymanov*, *Faik* Safiollin, *Leonid* Egorov, *Svetlana* Sochneva, and *Nikolai* Loginov

Kazan State Agrarian University, 65, K. Marx St., Kazan, Republic of Tatarstan, 420015, Russian Federation

Abstract. The studies were carried out to assess the effect of foliar dressing with Cytodef-100 and Giberelonphytohormones on fruit elements and gross yield of sunflower oil in edaphoclimatic conditions of the Republic of Tatarstan. The study was conducted in the sunflower crops of the Avenger hybrid in the Republic of Tatarstan in 2019–2021. The agrochemical indicators of typical gray-forest soil were as follows: humus content according to Tyurin – 3.0%, mobile phosphorus – 160 and exchange potassium – 145 mg/kg of soil according to Kirsanov. The soil medium response was close to neutral (pH 6.6). The results of the studies show that the total diameter of sunflower heads increases from 10.9 cm in the control sample to 13.3 cm in the joint use of 3 agrochemicals in the phase 4-8 of sunflower leaves: Euro-Lightningherbicide (1.2 kg/ha) + Cytodef-100 (150 g/ha) + Giberelon (50 g/ha). At the same time, the diameter of the empty part of the heads has a clear downward trend: from 2.2 cm to 1.7 cm in the latter version. The increase in marketable products due to the combined use of Cytodef-100 antistress drug and Giberelonphytohormone solution in combination with chemical weeding was 15.2% compared to the control sample (spraying of sunflower crops with Euro-Lightning (1.2 kg/ha) without Agrosintez LLC products.

1 Introduction

Phytohormones belonging to the class of gibberellins have a wide range of biological activity: they stimulate the division and expansion of plant cells, which leads to rapid growth of stems and an increase in the number of productive branches; they regulate flowering and fruiting.

Giberelonhasadvantagesthatcansignificantlyexpandth escopeofapplicationofgibberellinsincropproduction, including in field crops for the following reasons:

- maximum effect of application is achieved within 2-3 weeks and is quite long;
- specially developed formulation of Giberelon allows maintaining the active substance in an active and stable state;
- minimal costs of use compared to similar expensive drugs;
- compatible with pesticides and various types of liquid fertilizers used in the same period.

Plants are often subjected to various stresses during their growth and development, which leads to significant losses in yield and product quality (for example, nutrient absorption through the root system is disrupted even with sufficient amounts of them in the soil). Additional energy is required to overcome stress against the main metabolic process. The introduction of Cytodef-100 phytohormone in vegetative plants can reduce stress, improve growth and formation processes, increase yield and quality of products.

2 Materials and Methods

Stationary field experiments were conducted in 2019–2021 on the basis of Agrobiotekhnopark (Narmonka village, Laishevsky municipal district of the Republic of Tatarstan) with coordinates: latitude – 55.5244865824 and longitude – 48.274901646, and laboratory tests – at the Center for Agroecological Research of Kazan SAU.

The agrochemical indicators of the typical gray-forest soil were as follows: humus content according to Tyurin – 3.0%, mobile phosphorus – 160 and exchanged potassium – 145 mg/kg of soil according to Kirsanov. The soil medium response was close to neutral (pH 6.6).

The air temperature in May and June 2019was higher by 3 and 2 °C, respectively, compared to long-term average annual data. During the critical period of moisture consumption (June-July)the precipitation rate was only 58 and 83%, respectively, which negatively affected the yield of the studied object.

The agrometeorological conditions of 2020 were characteristic of the Republic of Tatarstan. The

* Corresponding author: dusai@mail.ru

beginning of May was characterized by high temperature, and the second half – by a large amount of precipitation and average heat supply. A slight amount of precipitation fell in June, especially in the third decade(only 4% of the norm).

Unlike the long-term average annual indicators, July turned out to be cool, especially with low heat supply in the 1st and 2nd decades. At the beginning of August (1st and 2nd decades), low heat supply was combined with permanent rainfall (165 and 248% of the norm).

The agrometeorological conditions of 2021 were significantly different from the long-term average annual indicators: high average daily temperature in May (18.7 °C), June (23.4 °C), July (22.6 °C) and August (22.4 °C) versus 13.3; 18.1; 20.2 and 17.6 °C, respectively, of the norm. The growing season of 2021 was aggravated by moisture deficiency (53.3%). Despite this, sunflower, as a drought-resistant crop, was able to resist extremely unfavorable environmental factors and ensured 1.82 to 2.13 t/ha of commercial oilseeds.

The technology of cultivating sunflower was generally accepted in field experiments: the predecessor was winter rye for grain, which harvesting followed by a primary tillage; in spring – moisture closure, introduction of NPK to the planned yield of 2 t/ha; seeding on May 16 with a pneumatic seeder Spring 8 (Favorite) with the seeding depth of 6 cm, width – 70 cm between rows with a seed distance in rows – 26 cm (55 thousand pcs/ha) of germinating seeds. The experiment was carried out according to the methodology of Pustovoit All-Russian Research Institute of Oil Crops (2010).

3 Results and Discussion

The sunflower height and accumulation of its biomass does not fully reveal the advantages or disadvantages of the preparations, since the density of the plant stand before harvesting and the weed infestation of crops have a great influence on the formation of highly productive agrocenoses and production with the lowest costs for post-harvest sunflower oilseeds conditioning [1, 2, 3] (Table 1).

The high ground germination capacity and the ability of the sunflower to self-regulate the density of the plant stand by the end of the growing season should be emphasized before the analysis of data in Table 1. Of the 55 thousand pcs/ha of planted seeds, about 50 thousand/ha of seedlings were obtained (field germination – 90% against 80-85% for spring wheat). From 44.8 to 45.8 thousand pcs/ha remained for harvesting of this amount. Such a significant dyewood of plants (4.2–5.2 thousand pcs/ha) is explained by the fact that not all plants withstand the competition for sunlight, moisture, and nutrients).

Table 1. Preservation of plants for harvesting and weed infestation of crops.

Variant	Plant stand density, thousand pcs/ha	Weed infestation of crops	
		pcs m^2	g/m^2
Control without PPP treatment (background – zero tillage)	44.8	6.9	7.4
Spraying of plants with Cytodef-100 (200 g/ha) + herbicide in phase 4-8 of real leaves	45.2	5.4	6.1
Spraying of plants in 4-8 leaf rosette phase of real leaves with Giberelon (75 g/ha) + herbicide	45.0	5.7	6.6
Spraying of plants in phase 4-8 of real leaves with Cytodef-100 (150 g/ha) + Giberelon (50 g/ha) with herbicide	45.8	4.8	5.2
HCP_{05}	0.27		

Nevertheless, there is a direct relationship between the height and density of the sunflower stand with weed infestation of crops: the higher the height and density of the stand, the smaller both the total number of weeds and their dry weight [4, 5, 6]. Depending on the above factors, the total number of weeds decreases from 6.9 pcs/m^2in the control sample to 4.8 pcs/m^2in the version with spraying of crops with imidazoline herbicides in combination with additional feeding of plants with the anti-stress Cytodef-100 (150 g/ha) and Giberelonphytohormone (50 g/ha).

In general, the combination of chemical weeding with one inter-row cultivation, although they did not ensure the complete purity of sunflower crops from weeds according to the Isaev scale, they corresponded to the class of poorly weeded (less than 11 pcs/m^2).

The study of sunflower fruiting elements is of great practical importance, since the productivity of this crop depends on the parameters of the head (its total diameter, its productive area), the quantity and weight of productive seeds in the head and a thousand-seed weight [7, 8, 9] (Table 2).

Table 2. Change in sunflower fruiting elements under the action Cytodef-100 and Giberelon.

Variant	Head diameter, cm		Seed weight, g/head	Thousan d-seed weight, g	Biologicalyi eld, t/ha
	total	empty			

		part			
Control without PPP treatment (background – zero tillage)	10.9	2.2	41.1	37.9	1.84
Spraying of plants with Cytodef-100 (200 g/ha) + herbicide in phase 4-8 of real leaves	12.6	1.8	43.6	42.8	1.97
Spraying of plants in 4-8 leaf rosette phase of real leaves with Giberelon (75 g/ha) + herbicide	11.2	2.0	42.8	41.6	1.93
Spraying of plants in phase 4-8 of real leaves with Cytodef-100 (150 g/ha) + Giberelon (50 g/ha) with herbicide	13.3	1.7	44.7	43.7	2.05

Fig. 1. General view of field experiments with sunflower.

The results of the studies show that the total diameter of the heads increases from 10.9 cm in the control sample to 13.3 cm in case of the joint use of 3 agrochemicals in the phase 4-8 of sunflower leaves: Euro-Lightning (1.2 kg/ha) + Citadef -100 (150 g/ha) + Giberelon (50 g/ha).

At the same time, the diameter of the empty part of the heads has a clear downward trend: from 2.2 cm to 1.7 cm in the last version. There is a correlation relationship between the parameters of heads and the weight of seeds in one head, as well as between the thousand-seed weight. Large seeds with a thousand-seed weight of 43.7 g are formed in large heads, which exceeds the control sample by 15.3%. Most importantly, the productivity of each head increases from 41.1 to 44.7 g (a very significant increase in the productivity of sunflower heads under the action of preparations from Agrosintez LLC).

On the other hand, the structural analysis of the fruit elements allows calculating the biological yield of sunflower oilseed raw materials. To do this, it is enough to multiply the average density of the stem by the weight of seeds in one head. The calculations confirm the high efficiency of the variants in which the spraying was carried out with the anti-stress Cytodef-100.

For example, in the second version of the experiment (Euro-Lightning 1.2 kg/ha + Cytodef 200 g/ha), the biological yield of sunflower oilseeds was 1.97 t/ha versus 1.84 in the control and 1.93 t/ha in the combination of herbicidal treatment with the addition of Giberelon– 75 g/ha.

Fig. 2. Cytodef-100, VRP 150 g/ha + Giberelon, 50 g/h.

However, the highest results are obtained when they are used together. In this case, the biological yield of oilseeds increases to 2.05 t/ha, exceeding the control sample by 11.4% (yield increase – 0.2 t/ha).

It is known that the biological yield, as well as the bunker yield, do not provide for a clear conclusion on the studied problem. Thus, it is necessary to determine the volume of commercial products from 1 hectare of arable land, which is calculated taking into account the humidity and weed content of the final products according to GOST 22391-2015 (Table 3).

Table 3. Impact of Cytodef-100 and Giberelon phytohormone on gross oilseeds with baseline values.

Variant	Actual	Gross	Increase

	humidity, %	commercial yield, t/ha	t/ha	%
Control without PPP treatment (background – zero tillage)	10.3	1.78	–	–
Spraying of plants with Cytodef-100 (200 g/ha) + herbicide in phase 4-8 of real leaves	8.1	1.95	0.17	9.6
Spraying of plants in 4-8 leaf rosette phase of real leaves with Giberelon (75 g/ha) + herbicide	8.8	1.90	0.12	6.7
Spraying of plants in phase 4-8 of real leaves with Cytodef-100 (150 g/ha) + Giberelon (50 g/ha) with herbicide	7.0	2.05	0.27	15.2
HCP_{05}		0.20		

According to GOST 22391-2015,the purchasing centers accept sunflower oilseed raw materials and pay its cost based on humidity of 7%, weed impurity of 1% and oilseed impurity of 3%. In our studies, the weed and oilseed impurity in the obtained products corresponded to the basic indicators. Therefore, the only difference between the test variants was humidity, which ranged from 10.3% in the control sample to 7.0% in the last test variant. Therefore, the biological yield of the control sample determined at the test sites was reduced by 3.3%, in the second variant– by 1.1%, in the third – by 1.8, and in the last variant the actual humidity corresponded to GOST. Consequently, the studied preparations not only increase the productivity of arable land, but also accelerate the harvesting ripeness of sunflower. The above pattern was noted even in 2021 with a hot and arid growing season.

Table 4. Content of raw fat and gross yield of sunflower oil depending on the use of antistress and phytohormone preparations of Agrosintez LLC.

Variant	Rawfat, %	Gross yield of sunflower oil, kg/ha	Increase	
			kg/ha	%
Control without PPP treatment (background – zero tillage)	45.3	806.3	-	-
Spraying of plants with Cytodef-100 (200 g/ha) + herbicide in phase 4-8 of real leaves	46.0	897.0	90.7	11.2
Spraying of plants in 4-8 leaf rosette phase of real leaves with Giberelon (75 g/ha) + herbicide	45.8	870.2	63.9	7.9
Spraying of plants in phase 4-8 of real leaves with Cytodef-100 (150 g/ha) + Giberelon (50 g/ha) with herbicide	46.1	945.1	138.8	17.2

As a result, the increase in marketable products under the action of the combined use of Cytodef-100 anti-stress drug and Giberelonphytoharmon solution in combination with chemical weeding amounted to 15.2% compared to the control sample (spraying of sunflower crops with Euro-Lightning (1.2 kg/ha) without addition of Agrosintez LLC drugs.

It should also be noted a reliable increase in the actual yield with basic indicators and in the variants of the separate use of the studied drugs, especially the Cytodef-100 anti-stress preparation at the rate of 200 g/ha – an increase in commercial sunflower oilseed raw materials of 0.17 t/ha.

In terms of raw fat content, sunflower oilseeds are divided into 3 classes (GOST 22391-2015).

Class 1 – at least 50%;

Class 2 – at least 45%;

Class 3 – at least 40%;

According to this classification, our final products in all variants of the experiment corresponded to the second class (raw fat content – at least 45%). At the same time, we should note the dynamics of growth in the accumulation of raw fat as the yield of sunflower increases under the influence of Agrosintez LLC products.

4 Conclusion

The use of phytohormone preparations in order to reduce biotic and abiotic adverse environmental factors is a promising direction in modern crop production.

Thus, the results of 2-year studies showed that the use of Giberelon and Cytodef-100 phytohormone preparations on sunflower crops in the edaphoclimatic conditions of the Republic of Tatarstan increased plant preservation rate and reduced weed infestation of crops. Besides, these drugs had a positive effect on the structure of the sunflower crop. Thus, in the variant with the spraying of plants in phase 4-8 of the real leaves with Cytodef-100 (150 g/ha) + Giberelon (50 g/ha) together with the herbicide, the weight of seeds in 1 head was 44.7 g, and in the control – only 41.1 g. Ultimately, the positive effect of the studied preparations was reflected in the gross yield of vegetable oil. In case of the combined use of Cytodef-100 (150 g/ha) and Giberelon (50 g/ha), the gross vegetable oil yield was 945.1 kg/ha, which is 138.8 kg/ha or 17.2% higher than the control.

References

1. F. N. Safiollin, R. M. Nizamov, V. M. Panasyuk,S. R. Suleymanov, A. A. Akhmetzyanov *BIO Web of Conf.* **17**, 00241 (2020)
2. V. M. Lukomets, V. M. Penchukov, N. I. Zaitsev, *Bulletin of the agro-industrial complex of Stavropol* **S2,** 85–87 (2015)
3. B. N. Nasiev, A. S. Bushnev, *Oilseeds* **4 (180),** 45–49 (2019)
4. V. M. Lukomets, V. T. Piven, N. M. Tishkov, N. I. Bochkarev, S. A. Semerenko, *Plant Protection and Quarantine* **6,** 36–39 (2019)
5. D. T. Khaliullin, A. V. Dmitriev, *Modern science-intensive technologies* **12–2,** 272–276 (2015)
6. R. I. Safin, L. Z. Karimova, F. N. Safiollin, Sh. Z. Validov, B. G. Ziganshin, K. Z. Karimov, G. S. Minnullin, *E3S Web of Conferences*, **91,** 185–193 (2019)
7. V. F. Moiseichenko, et al., *Basics of Scientific Research in Agronomy* (Moscow: Kolos, 1996)
8. V. M. Lukomets, N. M. Tishkov, V. F. Baranov, *Procedure for Conducting Field Agrotechnical Experiments with Oilseeds* (Krasnodar, 2010)
9. B. A. Dospekhov, *Field Experience Methodology* (Moscow: Agropromizdat, 1985)

Evaluation of the gonadotropic preparation effect on the hormonal status of the infertile cows

Nataliya Bogdanova[1,*]

[1]PROVET LLC, 6B, Artyukhinoy Str., Moscow, 109390, Russian Federation

Abstract. To study of the effect of the gonadotropic preparation on the hormonal status of the infer-tile cows, 47 red-and-white cows belonging to the breeding farm (Voronezh Region, Russian Federation), having an ovarian hypofunction, at the age from 3 to 8 years old and with the live weight of 480-600 kg, were included in the experiment. The animals in the experimental group (n=27) received a single dose of placental gonadotropin, i.e. Follimag at the 1000 I.U. dose. The animals not injected with the preparation (n=20) formed the negative control group. Cows' blood samples were taken before injection, 4, 8 and 14 days after it. The hormone concentration (progesterone, es-tradiol-17β) was determined in blood serum by the immunoenzymometric analysis. Introducing the preparation led to the increase of the progesterone level in the blood of the experimental group cows on average of 2.85 times and estradiol-17β increased by 23.2% as compared with the background, progesterone increased 1.95 times and estradiol-17β by 9.0 % compared with the control on the 14th day of the experiment. It was established that the degree of the ovarian hormonepoietic reaction manifestation and its direction in prescribing the exogenous gonadotropins depends on the time of the preparation injection relative to natural waves of the follicles growth. Using the gonadotropic preparation provided a 21.6-time increase in the blood content of progesterone on the 14th day of the experiment and that of estradiol-17β of 1.29 times, when injected at the beginning of the follicles growth wave. When administered at the follicular atresia stage, it led to an increase in the progesterone level of 2.1 times and that of estradiol-17β of 1.29 times.

1 Introduction

Pathological alterations in the genital organs are the main cause of infertility in the cattle breeding stock, which leads to a lack of animal yield and reduction in the productive longevity of animals. High incidence of infertility provoking pathologies concerning reproductive organs is registered among the highly productive cows. Most often, sterility in cows appears during parturition and in the post-partum period [1, 2].

The most common causes of reproductive problems include the ovaries functional disorders in the form of their hypofunction, which account for about 50-60% or more of all the ovarian disorders [3, 4, 5]. External factors of the disease pathogenesis include inadequate rations and feeds of poor quality in terms of metabolic energy, plastic substances, vitamins and mineral elements, as well as organism intoxication, short-term exercises, year-round stall keeping and weak insolation. Parturition pathologies and problems with acute postpartum and chronic endometritis also increase likelihood of the hypofunction [4, 6, 7, 8, 9]. It was established that the ovarian hypofunction could appear against the background of elevated pro-inflammatory cytokines and leukocytes levels [9, 10]. Disorders in protein and lipid metabolism, hepatosis, weakening of the thyroid gland endocrine and of the hypothalamus-pituitary system, as well as the ovarian stroma are leading to a decrease in fertility [6, 11, 12, 13].

Correction of infertility in cows, which appears due to the gonads hypofunction, could be introduced in zoo-technical and veterinary practices using various methods, means and their combinations. A variety of approaches is a consequence of a large number of the ovarian dysfunction pathoetiological factors. Currently, biotechnical methods for normalizing the copulation function based on the introduction of gonadotropic preparations are widely used [8, 14, 15]. These methods induce the growth of follicles and ovulation in the ovaries of animals.

Hormone therapy of the cow infertility was being practiced for a long time. An accumulated extensive base of scientific research on the use of gonadotropic preparations indicates an ambiguous response of the ovaries to administering the exogenous hormones in different animals, and the expected clinical effect is not always achievable. In connection with the foregoing, the task of improving biotechnology in controlling the cows'

* Corresponding author: 2149137@list.ru

reproductive function through introduction of hormonal preparations is not reducing relevance of this problem for scientific community and for practical business.

We assume that the solution to the problems of ensuring the cows' fertility is impossible without studying the effect of exogenous gonadotropins on the ovaries hormone-producing function taking into account growth waves and follicular atresia in animals.

2 Materials and methods

The scientific research was carried out on the basis of the breeding farm located in the Pavlovsky district of the Voronezh Region, Russian Federation.

47 sterile cows with ovarian hypofunction aged 3-8 years old and weighing 480-600 kg of the red-and-white breed were selected for the experiment. Animals of the experimental group (n=27) were once injected with placental gonadotropin preparation, i.e. Follimag, of the 1000 I.U. dose. Animals of the control group (n=20) were not injected with the preparation and provided the negative control.

The ovaries functional state was determined by their surface structure, size and consistency. The absence of mature follicles and corpus luteum, decrease in size, smooth surface and hard consistency were considered to be characteristic signs of the hypo functional disorder. An additional sign was the uterine atony in cows. During the experiment, all animals were subjected to regular clinical observations and registration of the copulation cyclicity stages onset.

To analyze hormone levels using the enzyme immunoassay (EIA), i.e. progesterone and estradiol-17β, the venous blood was obtained from the animals before administering the preparation, then the procedure was repeated on the fourth, eighth and fourteenth days after administering the preparation. The whole blood was centrifuged, and the resulting serum was frozen for storage at –15°C. EIA was performed using test systems manufactured by Chema-Medica and the Uniplan enzyme-linked immunosorbent reactions analyzer. Results were processed by mathematical methods accepted in scientific research.

3 Results

Data on hormone concentrations in blood of cows of the experimental group were analyzed taking into account the ratio between the time of animals' treatment with the preparation and the copulation cycle stages, i.e. follicles growth and atresia in the ovaries.

Action of Follimag, which contained follicle-stimulating hormones (FSH) and luteinizing hormones (LH), manifested itself already in 4 days after administration (Table 1). Against the background of the lack of these hormones, production of copulation steroids by the ovaries was being suppressed. The level of progesterone in the blood reliably ($P<0.05$) increased from 1.68±0.269 to 4.83±1.274 nmol/l or 2.88 times. The response reaction in the form of increased progesterone synthesis up to the level of 4.77-13.75 nmol/l at that time was observed in 54.5% of the experimental animals. After 8 days, the average level of progesterone in cows decreased by 26.7% to 3.54±0.496 nmol/l (range from 1.78 to 7.06 nmol/l), although the response reaction to the preparation was already registered in 82.0% of the animals. After 14 days, the hormone concentration increased by 2.78 compared with the previous value and 5.85 times ($P<0.001$) compared with the initial value (up to 6.26-15.40 nmol/l). The response reaction to the preparation was registered in all treated cows; corpus luteum formation was registered in 75% of the animals.

Table 1. Follimag effect on the progesterone level (nmol/l) in the blood of cows with the ovarian hypofunction.

Indicators	Ovarian function state			
	Without accounting for copulation cyclicity		Follicles growth wave	Follicular atresia
Initial background	1.68±0.269		0.51±0.096	2.97±0.458
After preparation administration	4th day	4.83±1.274	6.97±1.751	3.05±0.425
	8th day	3.54±0.496	4.66±0.825	3.29±0.452
	14th day	9.83±2.148	11.02±2.129	6.26

Analysis of the same results in comparison with the ovaries functional activity stage showed that progesterone production by the gonads increased sharply in the case of introducing placental gonadotropin against the background of follicles growth. In this case, the progesterone level increased 13.7 times in 4 days after the Follimag injection compared to the background ($P<0.001$). After eight days the increase compared to the background level was 9.1 ($P<0.001$), and after 14 days – 21.6 times ($P<0.001$).

Results of each animal individual observations in the experiment demonstrated that best results of the Follimag therapy were registered, when the preparation was administered on the first day of the starting follicle growth wave. The progesterone content in the blood at this time reached 7.06-15.40 nmol/l. With Follimag introduction on the second day of follicular activity, the hormone content rouse up to 6.18-9.90 nmol/l, when administered on the third day - up to 4.77-8.51 nmol/l. Thus, there appeared a tendency of the decreasing effectiveness in the use of exogenous gonadotropins over time from the start of the follicle growth wave.

The lowest efficiency was shown by Follimag administration against the background of follicular atresia, the progesterone level altered by day 4 by 2.69%, and by day 8 – by 10.77%; the difference in results was not significant. On the 14th day, the hormone content increased 2.1 times in comparison with the background level, which indicated ovulation with the formation of the corpus luteum. A similar result could be a consequence of gonadotropins endogenous synthesis activation in the animals.

Table 2. Follimag effect on the estradiol-17β level (nmol/l) in the blood of cows with the ovarian hypofunction.

Indicators		Ovarian function state		
		Without accounting for copulation cyclicity	Follicles growth wave	Follicular atresia
Initial background		0.69±0.032	0.63±0.056	0.77±0.028
After preparation administration	4th day	0.66±0.027	0.66±0.057	0.67±0.022
	8th day	0.85±0.091	0.87±0.079	0.84±0.176
	14th day	0.85±0.103	0.81±0.136	0.99

In contrast to the progesterone content effect in the blood, preparation action on the estradiol-17β level started to be traced only on the 8th day after the animals' treatment (Table 2). By that time, the average increase in the concentration of the hormone was of 23.2%. Increase in the estradiol-17β content was more visible with the Follimag administration at the beginning of the follicles growth wave, i.e. 38.1%, compared with the background. When administered at the follicular atresia stage, the increase was 9.1%. After 14 days, the progesterone level under Follimag introduction at follicles growth range was by 6.9% lower, and with introduction at the atresia stage, it was by 17.9% higher than that on the 8th day. Compared to the background indicator, increase in the estradiol-17β concentration by the 14th day occurred in 28.6% of both cases.

Thus, the gonadotropic preparation effect on the state of animal ovaries morphological and functional structures state with their hypofunction was more indicated in alteration of the progesterone synthesis, and, to a lesser extent, in the estradiol-17β. Alterations in the hormones ratio index are presented in Table 3.

Table 3. Alteration in the progesterone and estradiol-17β ratio indices during administration of Follimag to infertile cows.

Indicators		Ovarian function state		
		Without accounting for copulation cyclicity	Follicles growth wave	Follicular atresia
Initial background		2.44	0.81	3.86
After preparation administration	4th day	7.32	10.56	4.55
	8th day	4.16	5.36	3.92
	14th day	11.56	13.60	6.32

All presented options of the Follimag introduction demonstrated evident alteration in the progesterone and estradiol-17β ratio index towards its increase. Most significant alterations could be observed with introducing the preparation at the follicles growth stage. On the 4th day, in comparison with the background, the index increased 13.0 times, on the 8th day, it was 6.6 times higher, and on the 14th day, it was 16.8 times higher. Using the preparation against the background of follicular atresia led to the index increase on the 4th day of 1.2 times, on the 8th day, the index approached the original, and on the 14th day, the index increased 1.6 times. Without taking into account copulation cyclicity, increase in the progesterone to estradiol-17β ratio was 3.0, 1.7 and 4.7 in the corresponding periods. That once again confirmed high effectiveness of using the gonadotropic preparation to correct the ovarian hypofunction during the follicles growth wave.

Data provided in Table 4 show that the introduction of the Follimag preparation contributes to an increase in the progesterone level in 4 days after administration to animals of 2.45 times in comparison with the control indicator, and after 14 days – 1.95 times. The estradiol content in the blood, on the contrary, decreased in the experimental animals after 4 days by 9.6% compared to the control group, and after 14 days it was higher by 9.0%.

Table 4. Sex steroids content (nmol/l) in the blood of the cows during Follimag administration.

Indicator		Hormones		
		Progesterone	Estradiol-17β	Progesterone / estradiol
Initial background		1.68±0.269	0.69±0.032	2.44
In 4 days	Control	1.97±1.511	0.73±0.043	2.70
	Experiment	4.83±1.274	0.66±0.027	7.32
In 14 days	Control	5.04±1.610	0.78±0.053	6.46
	Experiment	9.83±2.148	0.85±0.103	11.56

Increase in the progesterone and estradiol ratio index was also noted. However, the indicator exceeded the control group value 2.7 times on the 4th day, 1.8 times on the 14th in the cows of the experimental group after Follimag introduction.

4 Discussion

Increase in the progesterone and estradiol ratio index was also noted. However, the indicator exceeded 2.7 times the control group value on the 4th day, 1.8 times on the 14th in the cows of the experimental group after Follimag introduction. It is known that FSH and LH secreted adenohypophysis control follicles growth in the ovaries, their maturation and ovulation. Their release cyclicity coincides with the follicles growth waves [16]. In most cows, 2 or 3 follicles growth waves are observed during the copulation cycle, less often - 1 or 4 [17, 18]. If three peaks of the gonadotropic hormones secretion and follicles growth waves are present in the copulation cycle, then they cover the period before ovulation, the 4th –6th and 11th –13th days after ovulation and are interconnected with action of the steroid hormones positive and negative feedback [19].

FSH stimulates the follicles growth and differentiation and promotes the steroid hormones synthesis. The LH role lies in ensuring maturation of follicles, ovulation and ovule maturation; androgens and prostaglandins are synthesized under the hormone action [19]. Weakening of the ovarian response reaction to gonadotropins under the corticosteroids effect in addition to the LH and FSH reduced production by the adenohypophysis leads to violation of the follicles maturation. High cortisol levels associated with stressful exposure are able to suppress the LH wave-like production [20, 21]. Based on this, it could be concluded that administering the gonadotropin preparation simulates the endogenous FSH and LH secretion waves activating copulation hormones synthesis in the ovaries, follicle morphogenesis and indirectly affecting hormonopoiesis in the anterior pituitary gland.

Due to the natural wave character of the gonadotropins secretion, synchronization of the exogenous gonadotropins administration synchronization in order to correct ovarian hypofunction with endogenous release of FSH and LH becomes of utmost importance. The emerging synergism of exogenous and endogenous hormones leads to effective restoration of the copulation cyclicity.

5 Conclusion

Administration of Follimag to infertile cows with the ovarian hypofunction leads to an increase in the progesterone concentration in blood after 4 days of 2.45 times in comparison with the control indicator, and after 14 days - 1.95 times. The estradiol concentration during the preparation administration decreased in the experimental group after 4 days by 9.6% compared with the control indicators, and after 14 days, it was higher by 9.0%.

The conducted study shows that for the most effective treatment of the ovaries hypofunctional state it is necessary to synchronize the gonadotropic preparation effect with the follicles growth wave. The term of drug administration should be optimally timed with the first or second day of the new follicles growth wave.

References

1. G. V. Nebogatikov, *Innovative technologies in eliminating infertility in cows* (Volgograd: Volgograd State Agricultural Academy, 2008).
2. I. S. Koba, M. B. Lattice, M. S. Dubovikova, Bulletin of Altai State Agrarian University **2 (136),** 103–106 (2016).
3. V. D. Kocharyan, G. S. Chizhova, M. A. Nikitina, *Bulletin of the Nizhnevolzhsky AgroUniversity Complex* **3 (27),** 132–135 (2012).
4. N. N. Omelyanenko, V. N. Prus, V. L. Shnaider, *UO VGAVM Scholarly Notes* **50 (2 (1),** 201–204 (2014).
5. V. Ya. Nikitin, N. V. Belugin, N. A. Pisarenko, V. S. Skripkin, N. V. Fedota, *Bulletin of the Stavropol AIC* **(1),** 19–22 (2015).
6. I. Ventsova, V. Safonov, *American Journal of Animal and Veterinary Sciences* **16 (1),** 7–14 (2021).
7. V. A. Safonov, *Agricultural Biology* **43 (4),** 64–67 (2008).
8. V. M. Breslavets, A. V. Khokhlov, *Bulletin of Orenburg State University* **3 (41),** 252–254 (2013).
9. S. H. Cheong, O. G. Sá Filho, V. A. Absalon-Medina, A. Schneider, W. R. Butler, R. O. Gilbert, *PLoS ONE* **12 (5),** 0177356 (2017).
10. V. Mikhalev, V. Safonov, S. Shabunin, P. Parshin, A. Sineva, V. Lukina, *Advances in Animal and Veterinary Sciences* **10 (1),** 20–26 (2021).
11. E. Skovorodin, R. Mustafin, S. Bogoliuk, G. Bazekin, V. Gimranov, *Veterinary World* **13 (4),** 774–781 (2020).

12. J. F. Silva, N. M. Ocarino, R. Serakides, *Biology of Reproduction* **99 (5),** 907–921 (2018).
13. V. Safonov, *Biol. Trace Elem. Res.* **Jul 24** (2021).
14. A. Lishchuk, N. Malakhova, O. Piskunova, *IOP Conference Series: Earth and Environmental Science* **341,** 012046 (2019).
15. D. M. Bekenov, A. A. Spanov, N. S. Kenchinbayev, A. D. Baimukanov, *News of the National Academy of Sciences of the Republic of Kazakhstan. Series of agrarian sciences* **5 (53),** 83–87 (2019).
16. V. A. Anzorov, S. V. Moryakina, *Bulletin of Gorsky State Agrarian University* **52 (1),** 262–267 (2015).
17. N. I. Gavrichenko, L. N. Turchanova, *Actual Problems of Veterinary Obstetrics and Animal Reproduction* (Gorki, 2013), pp 522–527 (2013).
18. B. T. Khetagurova, A. V. Pankratova, F. N. Nasibov, L. E. Verdieva, *Young Scientists in Solving Actual Problems of Science* (Vladikavkaz, 2014), pp 325–327 (2014).
19. I. B. Manukhin, L. G. Tumilovich, M. A. Gevorkyan, E. I. Manukhina, *Gynecological Endocrinology. Clinical Lectures* (GEOTAR-Media, Moscow, 2020).
20. K. M. Breen, F. J. Karsch, *Frontiers in Neuroendocrinology* **27,** 233–245 (2006).
21. C. R. Ralph, M. N. Lehman, R. L. Goodman, A. J. Tilbrook, *Society for Reproduction and Fertility* **152,** R1–R14 (2016).

Simulation modeling of the machine and tractor fleet to improve the technical base of the agro-industrial complex

Boris Dokin [1], *Anna* Aletdinova [2 3] *, and *Alexander* Zheshko [4]

[1] Siberian Research Institute of Mechanization and Electrification of Agriculture, 630501, Krasnoobsk, Russia
[2] Novosibirsk State Technical University, Automated Control System Department, 630073, Novosibirsk, Russia
[2] National University of Oil and Gas «Gubkin University», Automated Control System Department, 119991, Moscow, Russia
[2] RUE "SPC NAS OF Belarus For Agricultural Mechanization", 220049 Minsk Republic of Belarus

Abstract. Modernization of the technical base of the agro-industrial base of the agro-industrial complex should be based on a scientific approach and the use of modern tools. In the article, the authors consider the possibilities of simulation modeling as an addition to the implementation of the task of optimizing the composition of the machine and tractor fleet, for example, to smooth out the peaks of loading of agricultural aggregates. The advantage of simulation models is the possibility of taking into account a large number of factors, the use of nonlinear dependencies between variables, its numerous runs when varying parameters, determining the degree of influence of predicted deviations of production conditions on the value of resource potential at the basic values of the system parameters. This provides estimates of the stability of these parameters and their adequacy to the functioning environment. The authors highlighted the advantages of using simulation modeling tools in the modernization and formation of an optimal agricultural machinery and tractor fleet of an agricultural organization. Modern software products can be divided according to the type of simulation model implementation: system dynamics, discrete-event and agent modeling. When considering the general requirements for the simulation model of the formation of a machine and tractor fleet, special attention is paid to verification. This implies the process of verifying the use of the formed composition of equipment by means of running a simulation model. Adequacy, integrity, informativeness, simplicity, scalability are the main requirements for the simulation model. The authors have identified the input and output parameters of the run of the simulation model of the formation of a machine and tractor fleet. Verification guarantees that the model is checked for compliance with the general requirements for a set of machines. Running the simulation model will allow you to adjust the production plan depending on the technological and technical availability, as well as the selected level of production intensification.

1 Introduction

With the development of information and communication technologies, there are more and more tools that allow you to create automated systems for building crop production technologies, databases of technologies and technical means, solve optimization problems, build simulation models.

The task of determining the optimal composition and structure of the machine and tractor fleet for agricultural organizations remains relevant at the present time, both for new enterprises and for existing ones in order to adjust the existing fleet. The optimization task is implemented using the linear programming method [1-5] and the method of end-to-end viewing of annual work packages (when smoothing the peaks of machine loading) [6, 7]. Simulation modeling is also popular and allows you to run a model with a large number of factors, which gives more accurate results. For agriculture, we wanted to give examples of implementation by the following scientists: K. Zhichkin, V. Nosov, L. Zhichkina, N. Fomenko, V. N. Shepel, N. V. Speshilova, M. V. Kitaeva, I. V. Zhupley [8-11].

The relevance of this study is to highlight the advantages of simulation modeling and the formation of general requirements for the simulation model within the framework of the optimization problem and the formation of the machine and tractor fleet of agricultural organizations.

2 Materials and methods

The main emphasis in the study is on the possibility and necessity of using simulation models in the formation of the composition of the machine and tractor fleet of an agricultural organization. The authors highlighted the

* Corresponding author: aletdinova@corp.nstu.ru

advantages of this approach and the necessary main stages of their construction, which is discussed in more detail below.

2.1. Materials

As the literature review presented by us has shown, there are two main directions for solving the problem of forming a machine and tractor fleet of an agricultural organization: graphical, based on the construction and smoothing of peak loading of machines; based on economic and mathematical modeling. The basis of the first approach is the calculation of technological maps, their analysis, determination of the number of machines of a certain brand and type, peak loads for specific agricultural time periods. It is simple, intuitive and accessible to a wide range of users. This approach makes it possible to solve most of the problems associated with the formation and optimization of the composition of the machine and tractor fleet. We recommend it for use in the modernization of the technical base and justification of the promising composition of equipment. It helps in evaluating the effectiveness of the existing composition of the machine and tractor fleet at an agricultural enterprise. One of the minor disadvantages of this approach is that for a large number of types of work and equipment, the construction of these maps is complicated.

The second approach is based on economic and mathematical modeling. During its implementation, models of the formation of the composition of the machine and tractor fleet are built. So a stochastic model can be reduced to a linear programming problem by varying the most appropriate optimality criterion, which is expressed quantitatively, usually tends to a minimum or maximum. The advantage of stochastic optimization models of the machine and tractor fleet is the property of the adequacy of the display of real production conditions. The decision-maker makes the choice of the decision. But it is necessary to note the disadvantages of this approach. Some researchers point out that due to the large number of external and internal factors, it complicates optimization, which leads to the need to reduce the number of variables, and therefore low efficiency in practice of the final solution [11, 12].

Therefore, considering the process of forming a machine and tractor fleet, it is advisable to combine these two approaches. Which leads to the need to use optimization and simulation models in one study. Simulation models are complementary, they allow us to take into account nonlinear dependencies between variables, limit the level of influence of variation in predicted production conditions on the size of the resource potential, and, due to their characteristics, evaluate the verification of the model. As we showed in our study [13], at an acceptable level of risk, if the level of stability of the system is close to critical, it is possible to determine new optimal basic parameters, adjust the input data for the optimization model in accordance with the conditions obtained as a result of running the simulation model.

Our analysis has shown that simulation modeling has the following advantages for research:

- Easy ability to add and analyze discrete variables and their nonlinear properties.
- Construction and research of complex systems containing stochastic parameters.
- Rapid modification of simulation models.
- Getting a set of alternative solutions.
- The ability to choose the most optimal for the given conditions or an effective solution.
- The use of model time for setting up experiments, studying the functioning of the system [14].
- Obtaining probabilistic estimates as a result of running the model.
- Evaluation of system indicators part with varying operating conditions.

2.2. Modeling methods

For any type of the simulation model, there is a set of general rules and stages of construction. These stages are shown in Figure 1 and form a cyclic model.

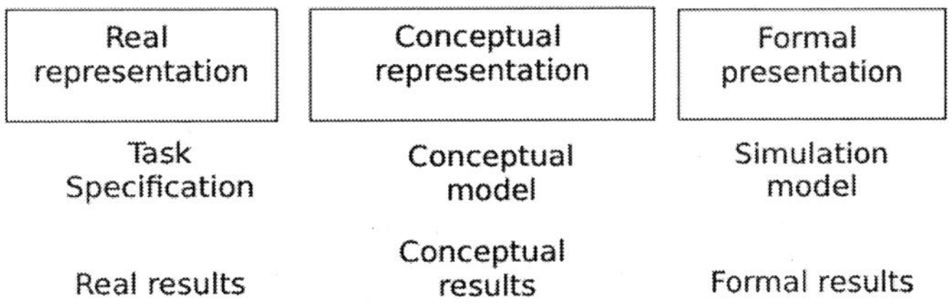

Fig. 1. Simulation cycle.

It reflects the real, conceptual and formal representation of the elements. At the first stage, the definition and formulation of the problem for which the simulation experiment is conducted takes place. Then it is necessary to collect, structure and classify the available and used data and information flows. After the data analysis is completed, the stage of forming a conceptual model begins, and then a simulation model. Next, the model is run, a series of experiments is set up, and a set of output data is collected. At the end, the analysis of the data obtained is carried out, conclusions are formulated and recommendations are made about the modeling object.

Thus, the simulation process is a cyclical process. Its results should be compared with the conceptual and real model. The presence of iterations during the simulation model run allows it to become approximating to the real system. At the same time, the main difficulty, in our opinion, is in collecting a large amount of input data.

3 Results and discussion

3.1. Software for the implementation of simulation models

Currently, there is a wide range of software products available; they allow you to implement simulation models using not only programming languages, but also visual programming. Let us consider modern tools that allow you to create any simulation models (Table 1).

Table 1. Software products for the implementation of simulation methods.

System dynamics	Discrete-event models	Agent-based modeling
AnyLogic	AnyLogic	AnyLogic
Pilgrim	Arena	NetLogo
Simulink	SIMSCRIPT	MASON
Berkeley Madonna	SLAM	Repast
VisSim	SIMAN	EcoLab
	AweSim	Cormas
	GPSS	

The existing tools are focused on a specific approach, i.e. for discrete dynamics, discrete-event models and agent-based modeling. For some of them, the choice is more diverse, this is explained by the peculiarities of each of the approaches, the presence of limitations in the construction of the model and the scope of the result. For agent-based modeling, the software appeared later than others and was presented in the early stages through libraries in Java or C++.

The analysis of the table demonstrates that the AnyLogic tool allows them to create simulation models for any type needed by the researcher. As you know, this software product is based on languages and methods used to create large information systems. The user receives a variety of tools (compared to other software presented in Table 1), which is not limited to the types of simulation research methods. The following advantages of AnyLogic should be noted: it is a convenient interface and libraries for building models, an intuitive way of setting and introducing probabilistic parameters. Thanks to the use of Java, the simulation model can be created as an independent application that runs independently of the software of a particular user.

3.2. General requirements for the simulation model for the task of forming a machine and tractor fleet

In the task of forming a machine-transport fleet, it is necessary to carry out actions that allow checking for errors in the generated plan for the use of equipment. As practice shows, verification is used for this, the procedure of which checks whether the output data and results meet the previously set implementation requirements. The verification process is based on the criteria applied to the software, the actions of the standard, the elimination of errors and shortcomings. There are approaches to solving the verification problem. The first approach suggests using methods of simulation modeling and other modeling, hardware emulation. But you can use the second approach. It is based on the methodology of the verification process, i.e. on the transfer of these procedures to more initial stages of design. As practice shows, researchers with a combination of these two approaches [15] obtain the optimal solution.

For the considered task of forming a machine and tractor fleet of an agricultural organization, we consider verification as a process of verifying the use of the formed equipment composition based on iterations of the simulation model.

Therefore, for our task, the verification procedure consists of the following stages: checking the formed duration of the periods of the agricultural year, ranking the distribution of technical means between technological operations for specific acreage; loading of the machine and tractor fleet, including modeling the load shares for the periods of the agricultural year for the formed complex of technical means and the block formation of an alternative plan and adjustments, taking into account verification carried out. Conceptually, the simulation model should correspond to the first three stages and allow the user to analyze the graphical representation of the data.

For our task, the verification procedure consists of the following stages: checking the duration of the agricultural year, the ranked distribution of the machine and tractor fleet between technological operations for agricultural operations; checking the workload of the machine and tractor fleet, including modeling the load shares for the periods of the agricultural year, the formation of adjustments and alternatives. There is a plan for the use of equipment at the entrance. And at the output we will get an alternative production plan or an adjustment of the plan taking into account the verification carried out. Conceptually, the simulation model should correspond to the first three stages and allow the user to analyze the graphical representation of the data.

Let us highlight the general requirements for simulation and analytical models. This is adequacy, integrity, informativeness, simplicity, scalability. Let us consider them more specifically for the task of forming a machine and tractor fleet.

To identify and formalize the requirements for the simulation model, which should correspond to the process under consideration, we will use the UML diagram of use cases (Fig. 2). The figure explicitly reflects the purpose of the simulation model being created, the necessary functional links between the projected model and external operating sources, both users making decisions and external data stores or other information systems.

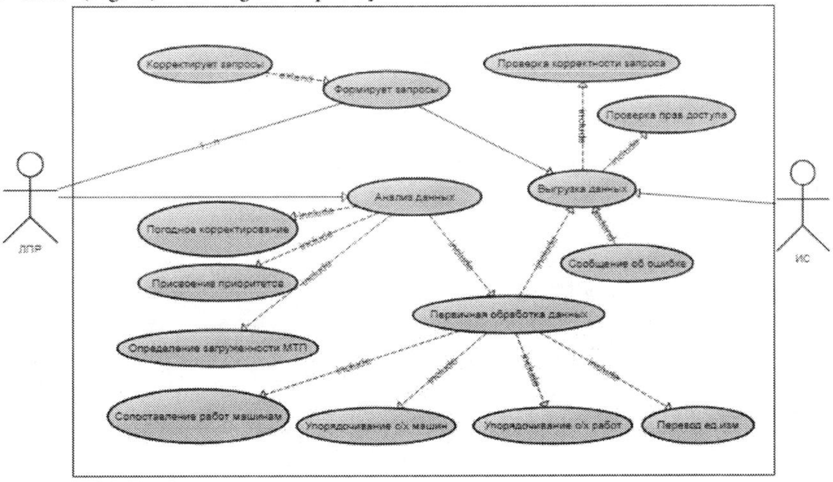

Fig. 2. Diagram of use cases.

After correcting the data, you can perform the rest of the calculations. As we have already noted, the periods of agricultural operations are set, depending on the natural and climatic zones, the weather and climatic conditions of the region, the availability of labor reserves. Researchers can vary the values of indicators of hourly loading of machines, the duration of technological operations, coefficients determining heat and water availability, the cost of agricultural machinery, the timing of ripening of crops. The function of the total costs of agricultural production in the model should tend to a minimum. At the next step, for a given set of equipment, productivity, reliability, and work readiness coefficients for each technical means, the workload for agricultural operations is calculated under conditions of limitation by the values of the size of annual work volumes and time periods. After running the simulation model, it is possible to estimate the level of workload of technical means, the probability of performing the specified volumes of technological operations, taking into account differences in cultivation technology, capacity and duration of work.

The information obtained after running the model allows the user to understand and analyze the efficiency of using equipment, the level of workload when changing the duration of technological operations, as well as to detect a shortage of technical, labor resources and adjust the initial data [6, 16].

Thus, the implementation of verification for the task of forming a machine and tractor fleet is the same as checking for compliance with the general requirements for a set of machines. When implementing verification, the adequacy, integrity, informativeness, simplicity and scalability of the model are checked. The user who applies simulation modeling can adjust the agricultural production plan. Existing software products allow implementing methods of system dynamics, discrete-event models, agent modeling. We consider AnyLogic to be the most universal.

Simulation modeling, on the one hand, is an excellent addition to solving optimization problems. And on the other hand, it can be used separately, because in particular cases it will be enough to make decisions.

References

1. K. Zhichkin, V. Nosov, L. Zhichkina, Z. Dibrova, T. Cherepova, *IOP Conference Series: Earth and Environmental Science* **315 (2),** 022023 (2019).
2. M. Li, Q. Fu, V. P. Singh, Y. Ji, D. Liu, C. Zhang, T. Li, *Science of the Total Environment* **651,** 1416–1434 (2019).
3. I. Galiev, K. Khafizov, R. Khusainov, M. Faskhutdinov, Ensuring possibility of functioning of tractors in agricultural production taking into account residual resources of their units and systems, *Proceedings of the 19th International scientific conference engineering for rural development proceedings* (20–22 May, Jelgava, Latvia) (2020).

4. J. Lowenberg-DeBoer, I. Y. Huang, V. Grigoriadis, S. Blackmore, Precision Agriculture. **21 (2),** 278–299 (2020).
5. J. E. Lee, *Modeling farm decisions for policy analysis* (CRC Press, 2019).
6. B. D. Dokin, A. A. Aletdinova, *E3S Web of Conferences* **285,** 02010 (2021).
7. B. D. Dokin, O. V. Elkin, E. A. Balushkina, A. A. Aletdinova, *IOP Conference Series: Materials Science and Engineering* **941,** 012038 (2020).
8. K. Zhichkin, V. Nosov, L. Zhichkina, N. Fomenko, *E3S Web of Conferences* **285,** 01010 (2021).
9. V. N. Shepel, N. V. Speshilova, M. V. Kitaeva, *SHS Web of Conferences* **62,** 08004 (2019).
10. I. V. Zhupley, *IOP Conference Series: Earth and Environmental Science* **981 (3),** 032012 (2022).
11. R. M. Peart, W. D. Shoup, *Agricultural systems modeling and simulation* (CRC Press, 2018).
12. N. Gupta, S. Grover, *Engineering and Applied Sciences Research* **2 (4),** 45–50 (2013).
13. B. D. Dokin, N. M. Ivanov, O. V. Elkin, M. S. Chekusov Alternative variants of technologies and technical means for grain production in Siberia, *Achievements of Science and Technology of Agriculture*, pp 49–51 (2015).
14. A. V. Ulezko, A. P. Kurnosov, A. A. Tyutyunnikov, Simulation modeling as a tool for the study of agro-economic systems, *Economics of agricultural and processing enterprises*, **8,** 28–30 (2012).
15. Yu. Karpov, *Simulation modeling of systems. Introduction to modeling with AnyLogic* 5 (Saint-Petersburg: BHV-Petersburg, 2005).
16. A. Lokhov, Functional verification of VLSI in the light of Mentor Graphics solutions, *Electronics: Science, Technology, Business* **1,** 58–63, (2004).

Anti-erosion reclaim development and its economic efficiency

Marcel Khismatullin[1]*, *Farit* Mukhametgaliev[1], *Nail* Asadullin[1], *Fayaz* Avkhadiev[1], *Rakhim* Ullah[1], *Asiya* Subaeva[1], and *Mars* Khismatullin[2]

[1] Kazan State Agrarian University, 65, Karl Marx St., Kazan, Republic of Tatarstan, 420015, Russian Federation
[2] Management "Tatmeliovodhoz", 15, Gvardeyskaya St., Kazan, Republic of Tatarstan, 420073, Russian Federation

Abstract. This paper considers the issues of the efficiency of agricultural production in the context of the globalization of the markets for agricultural raw materials and food and the increased competition in this regard. Modern realities and the adopted new Doctrine of Food Security of the Russian Federation, adopted in 2020, require an immediate solution to the problems of ensuring the country's food security by increasing the production of high-quality, environmentally friendly, and, most importantly, economically competitive agricultural products, which is not possible without intensifying production processes. However, the pursuit of high yields and high efficiency of agricultural commodity production often leads to the degradation of the prickly factor of production, namely, land. The paper presents the results of the analysis of the level of land degradation in the Republic of Tatarstan and identifies priority areas for combating erosion in agriculture. Measures to combat soil erosion are experimentally substantiated. The practical significance of the research results lies in the fact that they can not only contribute to the development of the agro-industrial complex and other sectors of the economy, improve soil fertility, the dynamic growth of agricultural products, reduce the disposal of agricultural land and create new jobs but also ensure the preservation and restoration of the Volga River. The results show that the ecological efficiency of one anti-erosion structure in combination with soil protection measures protects an average of 120 hectares of land from further development of erosion processes and is economically paid off by the results of two years of economic activity on protected lands.

1 Introduction

The efficiency of the functioning of the agro-industrial complex depends not only on the entrepreneurial activity and the ability of entities managing the land but also on the state of agricultural production factors. Earth, according to A.A. Zotov, F.N. Safiollin, E.S. Isaicheva, D.F. Khafizov, M.M. Khismatullin was, is, will be the main wealth of any country in the world. Therefore, the creation of a system of efficient, economical land use in terms of importance and relevance should be one of the leading places in the political and socio-economic life of the Russian Federation, including the Republic of Tatarstan [1–6].

Unfortunately, the productivity and ability of soils to self-heal have declined significantly in recent years. The alienation of nutrients from the harvest, dehumidification and erosion of soils, pollution and littering of land, an increase in anthropogenic pressure, and high plowing of agricultural land lead to the destruction of natural landscapes and ecosystems. This fact is reflected in the works of F. N. Safiollin [7], A. A. Zotov [4], I. M. Minnekhametova [8], F. N. Mukhametgaliev [9].

Irrigation and drainage reclamation, cultural engineering, and anti-erosion work in combination with agrochemical measures are the important factors for ensuring high efficiency of agricultural production, as well as the reproduction of soil fertility (A. R. Zakirova) [10, 11]. Leading scientists and practitioners in the field of agriculture I. A. Kayumov [12], G. S. Klychova [13, 14], A. M. Sabirov [15], Khismatullin M. M. [16] take into account the current market conditions and make a number of recommendations. They recommend that on reclaimed lands, first of all, cultivate vegetables, feed for livestock, develop seed production, and other most valuable and high-margin crops.

2 Materials and methods

The study is based on a theoretical and methodological basis that allows using the possibilities of interaction and complementarity between the fundamental provisions and the concept of land reclamation science. The basis of the theoretical and methodological base is the works of foreign and domestic scientists, practitioners A. A. Zotov [1], F. N. Safiollin [2], V. G. Grebennikova, I. A. Shipilov [17], D. I. Faizrakhmanov [18], V. A. Aksanov [19], D. F. Khafizova [20], I. A. Kayumov [7]. This study is based on international and domestic legal norms governing relations arising in the process of carrying out anti-erosion work, and the implementation of agricultural policy.

The methodological basis of the study is system analysis, within which the following approaches are applied: a constructive, deterministic, retrospective, and statistical.

* Corresponding author: marselmansurovic@mail.ru

https://doi.org/10.1051/bioconf/20225200015

During the study, the following were used: monographic, abstract-logical, calculation-constructive, and economic-statistical methods. The information base of the study was the official data of the Federal State Statistics Service, the Ministry of Agriculture of the Russian Federation, the Ministry of Agriculture and Food of the Republic of Tatarstan, the Federal State Institution "Management of Tatmeliovodkhoz", planning and reporting documents of agricultural organizations, materials of scientific conferences, materials contained in monographs, reports Research Institute, data obtained in the course of the author's analysis and calculations.

3 Results and discussion

Under the influence of natural, climatic, and anthropogenic factors, the scale of water erosion is increasing every year, causing enormous damage not only to agriculture but also to large and small rivers (ecology) of the region. The territory of the Republic of Tatarstan is characterized by great dissection, which is the basis of soil erosion. The development of soil erosion is facilitated by such a factor as a high degree of plowing of farmland (76.6%) with a low forest cover of arable land (3.3%, with optimal values of 4.7–7%).

The studies of the dynamics of changes in the quality indicators of lands revealed a clear trend towards a decrease in soil fertility and deterioration of the general environmental situation in the agro-industrial complex of the Republic of Tatarstan.

The scale of water erosion, the progressive degradation of the soil cover, and the damage caused by erosion processes to the agriculture of the republic have reached impressive proportions. So, in the Republic of Tatarstan, there are more than 21 thousand active ravines, their total length exceeds 29 thousand kilometers. As a result of the erosion of the fertile soil layer, more than one thousand hectares of agricultural land are annually taken out of circulation (S. R. Suleimanov) [21].

Compared with 1960, only due to soil erosion, the area of arable land has decreased by 247.5 thousand hectares. This is the territory of three regions of the Republic.

The areas of agricultural land subject to water erosion increased by 925 thousand hectares, the number of active peaks of ravines by 7450 pieces, and they occupy an area of 114 thousand hectares (V. A. Aksanov) [19].

The area of arable land requiring anti-erosion measures has more than doubled over the past 40 years and reached 2 million hectares. Snowmelt and stormwater from each hectare are washed off an average of 22 tons of fertile soil per year. The loss of humus for this reason in the whole country is about 700 thousand tons annually, which is equivalent to the introduction of 10 million tons of organic fertilizers. In other words, almost half of the applied mineral fertilizers are used for covering losses from flushing.

The area of erosion-hazardous agricultural land is 2263.2 thousand hectares, of which 1390 thousand hectares of arable land are subject to water erosion processes – 42%, including 6.7 of a strong degree, 254 of an average degree, and 1,129 thousand of a weak degree. ha. There are 21 thousand ravines with a total length of more than 29 thousand km, of which more than 21 thousand km are in the water area of the Volga River [22].

In the Republic of Tatarstan, more than 70% of the agricultural land area is located on slopes of various steepness: including arable land on slopes steepness:

- up to 1 degree – 42.4%,
- 1–3 degrees – 52.0%,
- 3–5 degrees – 5.6%.

According to F. N. Salakhutdinov [23], L. F. Sitdikova [24], N. M. Asadullina [25], a positive role in reducing erosion processes is played by the construction of anti-erosion structures for reclamation purposes, some of which, in turn, are a source of water for irrigated lands.

In the Republic of Tatarstan in different years, more than 880 anti-erosion and reclamation hydraulic structures (dams) with a total volume of more than 360 million m^3 (useful – 283.6 million m^3) were built and put into operation. The main part of these hydraulic structures was built in accordance with the approved design and estimate documentation, considering the volume of water accumulation: up to 500 thousand m^3 – 395 pieces; up to 1 million m^3 – 166 pieces; more than 1 million m^3 – 78 pieces (A. K. Subaeva) [26]. Tatarstan ranks first among the regions of the Volga Federal District in terms of the total area of lakes and reservoirs. In addition to reclamation purposes and watering of territories, they serve as a reliable tool designed to improve the ecological situation in the river basin of the Volga basin by reducing by at least 80% the volume of diffusion flows caused by erosion processes on sloping agricultural lands.

According to the state statistics authorities since 1968, 1681 reservoirs have disappeared on the territory of the republic for natural and anthropogenic reasons, including siltation [11]. Ponds and lakes built by damming ravines located on agricultural lands subject to erosional washout are exposed to the most active siltation. On them, the siltation rate reaches 10 to 50 mm per year, which is significantly higher than the average sedimentation rate typical for the territory of the Republic of Tatarstan [11].

Since 2013, as part of the implementation of the long-term republican target program for the overhaul of hydraulic structures, more than 450 facilities of 887 existing structures have been brought into the standard technical condition (Fig. 1). They accumulate more than 150 million m^3 of water, it also allows saving agricultural

 https://doi.org/10.1051/bioconf/20225200015

Fig. 1. Hydraulic structures in the Republic of Tatarstan.

land from the effects of water erosion on an area of 44.7 thousand hectares. During the spring passage of flood waters, the protection of the population and objects that fall into the zone of possible flooding is ensured. Through repair and restoration work, an increase in the technical reliability and stability of hydraulic structures was achieved. Thus, the threat of emergency situations at dams during the spring passage of flood waters through hydraulic structures is reduced [12].

The efficiency of anti-erosion measures is an aggregated assessment of environmental and economic damage prevented because of the implementation of environmental anti-erosion measures. Environmental efficiency (prevented environmental damage) is determined depending on the degree of reduction in the intensity of the above negative processes because of the implementation of anti-erosion measures. The environmental efficiency of the measures is the follows.

1 anti-erosion structure in combination with soil protection measures protects an average of 120 hectares of land from further development of erosion processes.

The cost of a complex of anti-erosion measures, including an anti-erosion pond is 2.5 million rubles.

The cost of production from 120 hectares with a yield of 30 centners of grain from 1 hectare:

120 x 30 x 1500 rubles/c = 5.4 million rubles. in year.

The payback period is two years.

One of the effective methods for reducing erosion processes is agroforestry. As part of the implementation of targeted programs for the development of land reclamation, agroforestry work is carried out in the republic, more than 2.5 thousand hectares of protective strips are planted annually. From 2016 to 2021, work was carried out on an area of about 11 thousand hectares. in the amount of more than 800 million rubles (about 150–160 million rubles per year).

Table 1. Financial support for the implementation of reclamation programs implemented in the Republic of Tatarstan in 2018–2020 (the table was compiled by the authors based on the analysis of internal documents of the Federal State Budgetary Institution "Department "Tatmeliovodkhoz").

Directions of funding	Years			
	2018	2019	2020	Total
For the construction, and reconstruction of reclamation systems and hydraulic structures from the budget of the Russian Federation for federal facilities million rubles	142.1	113.0	117.7	372.8
Capital repairs of ponds and hydraulic structures from the budget of the Republic of Tatarstan, million rubles	92.7	58.1	89.8	240.6
Construction, reconstruction, and overhaul of irrigated and drained lands, million rubles	62.3	46.9	70.2	179.4
Establishment of anti-erosion and field forest plantations, million rubles	150	186.5	150	486.5
In total for 3 years involved in land reclamation	447.1	404.5	427.7	1279.3

At the same time, there is an objective need in the republic to continue restoration work on hydraulic structures. The reason is that many facilities remain at risk of emergencies caused by the destruction of dams and flooding of nearby areas. In addition, a serious environmental pollutant, including rivers, is the use of pesticides and mineral fertilizers. About 2,000 tons of pesticides and more than 190,000 tons of mineral fertilizers are used annually on farmland in the republic. In addition to preventing river silting, erosion control structures delay the runoff of chemicals and pesticides used in agricultural formations. Thus, according to the results of a joint inventory of the Ministry of Emergency Situations of the Republic of Tatarstan and specialists from the Ministry of Agriculture and Food of the Republic of Tatarstan and the Tatmelioratsia company, as of January 1, 2021, 160 additional hydraulic structures require repair work. According to expert estimates, more than 840 million rubles of funds at current prices are required to bring them into a standard state.

The fact of siltation of many reservoirs and the development of degradation processes of the water body in connection with this, which necessitates their cleaning from bottom sediments, is also alarming. The conducted studies indicate that the low-fertile lands of the Predkama zone, which are most prone to erosion, can be protected by the correct selection and cultivation of legume-cereal perennial grasses on a large scale. In this case, it is possible to stop the development of erosion processes, and radically improve the fertility and structure of the gray forest, light gray, soddy podzolic, and dark gray soils of the analyzed region (V. G. Grebenkov, I. A. Shipilov) [17]. This method of combating the manifestations of soil erosion is less costly and affordable for a wide range of agricultural formations.

Also, this method helps to reduce the negative impact of drought on the performance of economic activities of agricultural organizations. For example, sowing perennial ryegrass from the legume family is the best way to reduce the negative effects of moisture deficiency. The reason is that the tap root system of alfalfa, goat's rue, and red clover draws water from such deep layers of soil, which is completely inaccessible to multi-cut ryegrass. In the same way, legumes start growing earlier in the spring and shade perennial grasses in the first cut.

As a result, the application of this method makes it possible to increase the economic efficiency of agricultural production both in crop production and animal husbandry. This result is confirmed by several studies by L. T. Vafin [27], V. M. Kosolapov [28], M. M. Khismatullin [29], and A. R. Valiev [30]. This method practically does not require a large amount of public investment. It is enough for agricultural formations to form their crop rotations widely scientifically reasonably, including legume-cereal perennial grasses.

4 Conclusion

In connection with the foregoing, we propose the following measures to improve the technical reliability and stability of hydraulic structures that are in pre-emergency and emergency conditions. We consider it expedient to include in the program for the protection of the Volga measures for the repair and cleaning of bottom sediments of hydraulic structures located in its water area and a set of rehabilitation measures aimed at restoring the ecological status of this reservoir.

We also consider it important to include in the program the construction of new anti-erosion structures on ravine-beam systems located in the Volga basin. This set of measures is necessary in order to significantly reduce polluted flood runoff and fix the growth points of ravines.

The construction of anti-erosion hydraulic structures will provide water for irrigation of about 57 thousand hectares of arable land. On the whole, according to expert estimates, in order to guarantee the provision of the republic's population with essential foodstuffs, it is necessary to have at least 100.000 hectares of irrigated land.

Taking this into account, we consider it necessary to include in the developed program of measures to create annually at least 5 thousand hectares of shelterbelts, afforestation of ravines, and steep slopes located in the Volga water area with annual funding of 300 million rubles.

The implementation of the proposed measures will ensure the preservation and restoration of the Volga River and will contribute to the development of the agro-industrial complex and other sectors of the economy, will ensure an increase in soil fertility, a dynamic increase in agricultural production, a reduction in the disposal of agricultural land and the creation of new jobs.

References

1. A. A. Zotov, M. M. Khismatullin, *Improvement and use of natural hayfields and pastures of the Middle Volga region* (Kazan: Zur Kazan) (2014)
2. F. N. Safiollin, M. M. Khismatullin, *The system of reclamation agriculture in the Republic of Tatarstan (general issues of land reclamation and features of irrigated crop cultivation)* (Kazan: Kazan State Agrarian University) (2015)
3. E. S. Isaichev, D. F. Khafizov, M. M. Khismatullin, M. M. Khismatullin, Bulletin of the Kazan State Agrarian University **6 (1 (19)** 82–84 (2011)
4. D. F. Khafizov, E. S. Isaichev, M. M. Khismatullin, Bulletin of the Kazan State Agrarian University **8 (1 (27),** 55–58 (2013)
5. D. F. Khafizov, E. S. Isaichev, M. M. Khismatullin, Bulletin of the Kazan State Agrarian University **5 (3 (17),** 77–78 (2010)
6. M. M. Khismatullin, D. F. Khafizov, Bulletin of the Kazan State Agrarian University **9 (3 (33),** 66–71 (2014)
7. F. N. Safiollin, M. M. Khismatullin, F. N. Mukhametgaliev et al., Financial business **9 (219),** 156–160 (2021)
8. I. M. Minnehametova, L. F. Gafiullina, M. M. Khismatullin, *Conditions and factors of development of agricultural consumer cooperatives. Cooperation and sustainable development. Lecture notes in networks and systems* **245** (2022)
9. F. N. Mukhametgaliev, M. M. Khismatullin, R. G. Khisamov, Bulletin of the Kazan State Agrarian University **10 (2 (36),** 31–35 (2015)
10. A. R. Zakirova, G. S. Klychova, A. R. Yusupova, R. Ullah, Collection of scientific papers based on the materials of the IX International scientific and practical conference dedicated to the memory of Professor V. P. Petrov (Kazan: Kazan State Agrarian University)70–74 (2021)
11. Decree of the Cabinet of Ministers of the Republic of Tatarstan of December 28, 2013, No 1083 On approval of the State program "Environmental protection, reproduction and use of natural resources of the Republic of Tatarstan", retrieved from: https://docs.cntd.ru/document/469122554
12. I. A. Kayumov, M. M. Khismatullin, M. M. Khismatullin, Eurasian Scientific Association **1 (7) 29),** 70–72 (2017)
13. A. S. Klychova, G. S. Klychova, I. V. Makunin, Accounting in Agriculture **6** 21–28 (2014)
14. G. Klychova, A. Zakirova, Sh. Khusainov et al., E3S Web of Conferences **157,** (2020)
15. A. M. Sabirov, D. I. Fayzrakhmanov, M. M. Khismatullin et al. Bulletin of the Kazan State Agrarian University **10 (4 (38)** 10–15 (2015)
16. M. M. Khismatullin, R. Ullah, M. M. Khismatullin, Regional Economy: Theory and Practice **20 (1 (496),** 168–185 (2022)
17. I. A. Shipilov, V. G. Grebennikov, *Collection of scientific papers of the Stavropol Research Institute of Animal Husbandry and Forage Production* **3 (1),** 68–71 (2010)
18. D. I. Fayzrakhmanov, F. N. Mukhametgaliev, A. R. Valiev et al., *Organizational and economic aspects of improving the efficiency of agrarian business* (Kazan: Kazan (Volga Region) Federal University) (2021)
19. V. A. Aksanov, Sh. A. Aliev, *Ways of mobilization of biological resources to increase the productivity of arable land, energy-saving, and the production of competitive agricultural products* (Kazan: Foliant) 266–269 (2005)
20. D. F. Khafizov, M. M. Khismatullin, M. M. Khismatullin, *Entrepreneurship in the agricultural sector* (Kazan: publishing house of Kazan University) (2007)
21. S. R. Suleimanov, F. N. Safiollin, M. M. Khismatullin *Proceedings of the international scientific and practical conference of the Kazan State Agrarian University dedicated to the 95th anniversary of the Faculty of Agronomy* (Kazan: Publishing House of the Kazan State Agrarian University) (2014)
22. S. A. Sharipov, Economics of Agricultural and Processing Enterprises **4,** 52–56 (2017)
23. F. N. Salakhutdinov, M. M. Khismatullin, I. R. Iskhakov, Bulletin of the Kazan State Agrarian University **6 (2),** 52–54 (2011)
24. L. Sitdikova, F. Mukhametgaliev, M. Khismatullin, et al., BIO Web of Conf. **17** (2020)

25. M. M. Khismatullin, F. N. Mukhametgaliev, N. M. Asadullin, F. N. Avkhadiev, M. M. Khismatullin, U. Raheem, BIO Web of Conf. **37** (2021)
26. A. K. Subaeva, L. M. Mavlieva, N. R. Aleksandrova, M. M. Nizamutdinov, Bulletin of the Kazan State Agrarian University **12 (3),** 124–127 (2017)
27. M. M. Khismatullin, M. M. Khismatullin, L. T. Vafina, F. N. Safiollin, IOP Conference Series: Earth and Environmental Science **341** (2019)
28. A. A. Zotov, V. M. Kosolapov, N. V. Panferov et al, *Improvement and use of floodplain meadows* (Moscow: Russian Academy of Agricultural Sciences) (2013)
29. M. M. Khismatullin, Bulletin of the Kazan State Agrarian University **14 (2 (53)** 64–67 (2019)
30. A. R. Valiev, M. M. Khismatullin, M. M. Khismatullin, et al., Bulletin of the Kazan State Agrarian University **16 (3 (63)** 160–166 (2021)

Import substitution and self-sufficiency in basic foodstuffs

Farit Mukhametgaliev*, *Landysh* Sitdikova, *Ilgizar* Gainutdinov, *Fayaz* Avkhadiev, and *Nail* Asadullin
Kazan State Agrarian University, 65, Karl Marx St., Kazan, Republic of Tatarstan, 420015, Russian Federation

Abstract. The relevance of the topic of the article is caused by the need to study the current situation in the food market and determine the level of development of the production potential of the domestic agri-food complex in the context of import substitution and fierce competition in the global food market. The purpose of the article is to assess the processes of development of the agro-food complex in the context of import substitution, identify trends, and substantiate priorities in the implementation of the country's food security strategy. The article carried out a comprehensive study of the current state of development of agricultural production, analyzed the current state of the food supply of the country's population, identified trends, identified priority strategic directions for the implementation of the import substitution policy, and ensured the country's food security. The article concludes on the long-term nature of the implementation of the import substitution strategy and ensuring food security. There is a need for an interconnected and balanced organization of import and export activities of enterprises of the agro-food complex. The authors substantiate the need for reorientation of enterprises after reaching the target indicators of food self-sufficiency from domestic sales of products to the development of international food markets. The theoretical provisions and practical recommendations proposed in the article make it possible to solve the problem of stable self-sufficiency of the country with competitive food products and the implementation of a food security strategy.

1 Introduction

The implementation of the strategy of self-sufficiency in basic food products of own production remains a priority for the development of the Russian agri-food complex. Along with this, the acceleration of innovative development, changes in the food market conditions, the need for a balanced and proportional development of agricultural production sectors, and the aggravation of regional and international competition have a huge impact on the dynamic changes in the country's agro-food complex enterprises. This implies filling the food security strategy with new content. Over the past eight years, global economic and political processes have taken place in the world food market. These changes contributed to Russia's transition to the implementation of an import substitution strategy. This strategy is focused on meeting domestic food needs with food at affordable prices for consumers through the development of the domestic agri-food business and the maximum reduction in import receipts of food products to the level determined by the Food Security Doctrine of the Russian Federation [1, 2, 3]. As a result of the transition to a new strategy of self-sufficiency in food supply for the population of the country, the following results were achieved. The volume of agricultural production has increased. The agrarian sector of the economy was diversified, and the subjects of the agrarian business were activated based on strengthening state support for the implementation of targeted programs for the development of individual sectors of agriculture. Increased consumption of basic foodstuffs. The continuing growth of agricultural production leads to a glut of certain segments of the domestic food market, and a decrease in the profitability of domestic producers, which is confirmed by the studies of Russian scientists [4, 5]. At the same time, new problems have emerged, significant among which are the following:

- the diversion of significant budget funds to address the issues of the agricultural business to create complete food chicks of the completed cycle without the participation of international partners,
- restriction in the development of world-class technical and technological innovations,
- removal from new trends in the development of the world economy for robotization and digitalization of production processes,
- decreased competitiveness of products in the global food market.

All these processes, of course, show the relevance of the problem of choosing priority strategic directions for the further development of the agricultural sector of the economy. There is a need to further study the problems of food self-sufficiency, determine the directions for the development of the import substitution policy with the subsequent achievement of food security and increase the export potential of the country's agro-industrial complex.

* Corresponding author: fem59@mail.ru

2 Materials and methods

The methodological basis for the study was the earlier work of domestic researchers, including the authors, devoted to the analysis of the level of development of the domestic agricultural sector, as well as the problems of implementing the strategy for ensuring the country's food security. The general methodological basis of the study is system analysis. The analysis of the indicators used was carried out by comparing the changes in values by years for 2017–2020 to identify fluctuations in these indicators in each subsequent year and improve the accuracy of calculations in the dynamic range of indicators. The information base for the study was the statistical data of the Federal State Statistics Service of the Russian Federation, the Ministry of Agriculture of the Russian Federation, data obtained during the analysis and calculations of the authors.

3 Results and discussion

Food security is currently a priority area of state support for the agro-food complex and the main direction of the state agrarian policy of Russia. The basis of food security is food independence, which is determined by the level of domestic production, and self-sufficiency in food products [6]. Domestic agricultural production is greatly influenced by the implementation of an export-oriented strategy for the development of the national economy. The introduction of an embargo on the import of foreign products after the events of 2014 accelerated the processes of import substitution in the food market. The implementation of the measures taken affected the structure and dynamics of imports of agricultural products and foodstuffs. Under the import, bans are meat and dairy products, vegetables, and fruits. As a result of the measures taken, by 2020, the targets for the level of self-sufficiency in basic food products were achieved or exceeded [7].

An analysis of the level of development of the agrarian sector of the economy over the past four years showed that the growth rate of cost indicators of production in comparable prices in general for agriculture in 2020 compared to 2017 reached 105.7%. This indicator means the fulfillment of the indicators established by the State Program [8] and exceeds the targets by 1.9 percentage points. The sustainability of the economic development of the agro-food complex in recent years has been maintained due to the positive dynamics of economic growth, the complexity of the structure of the complex, the diversification of its structure, the emergence of new industries because of the implementation of the import substitution policy.

Outpacing growth in the sectors of the agro-food complex has become a characteristic feature of the Russian economy in recent times. Although in 2018, a slight decrease of 0.2% in the growth rate of agricultural products was allowed. An analysis of the indicators for 2017-2020 shows that the gross value added in basic prices in the agricultural sectors increased by 9.0% (including in 2020 by 0.5%), and in the food industry by 109.2% (in 2020 by 0.3%), which is higher than the growth but for the economy as a whole (4.3%, and in 2020 a drop of 2.7% was recorded) [9].

In 2020, the index of the physical volume of gross value added by the sectors of the agro-industrial complex also grew and amounted to 100.5%. In the production of food products and beverages, the index of the physical volume of gross value added by the branches of the agro-industrial complex is 100.3%. During this time, GDP and gross value added for the economy as a whole were declining; respectively, 97.0% and 97.3% to the level of 2019. In 2020, for the first time in recent Russian history, food imports did not exceed its exports.

The increase in crop production in general for all categories of producers in 2020 compared to 2017 amounted to 5.1%, which exceeds the indicative value included in the program by 3.7 percentage points. Along with this, the growth rate of livestock production, although it increased in 2020 compared to 2017, was 105.1%.

Table 1. Production of the main types of import-substituting food products in the Russian Federation, thousand tons.

Types of products	Years				Indicators of 2020 in relation to 2019, %
	2017	2018	2019	2020	
Cattle meat	261.7	298.4	309	342.3	110.8
Pork	2404	2669	2819	3186	113.0
Poultry meat	4839	4877	4847	4804	99.1
Sausages	2259	2282	2282	2355	103.2
Vegetables and mushrooms	96.1	93	116.3	141	121.2
Fruits, berries, and nuts	15.6	16.8	22.2	28.2	126.7
Milk	5390	5457	5378	5626	104.6
Cream	133	150	163	195	119.5
Curd cheese	486	501	468	487	103.9
Butter	270	267	269	277	103.1
Cheese	464	467	540	572	105.8
Condensed milk products, million cans	837	806	717	720	100.4

Dairy products	2896	2819	2792	2745	98.3

But the value of the planned indicator remained unfulfilled by 0.4 percentage points. This indicator can be assessed as one of the limiting factors for sustainable agricultural growth. The growing potential of agricultural production, subsequently, led to an increase in the index of food production. The index of food production in 2020 (in comparable prices) compared to 2017, % amounted to 111.7%, which exceeds the target by 2.1%. There was an increase in the production of the main types of import-substituting food products (Table 1) [7, 9].

Table 1 shows that over the past four years, production has increased for almost all types of food products. The production of fruits, berries, vegetables, pork, and cattle meat is characterized by particularly high rates. The current import substitution trend has allowed Russia to increase its position in the World Food Security Ranking from 40th place in 2013 to 24th place in 2020 out of 113 countries assessed. In 2020, the global food security index of the Russian Federation amounted to 73.7 points and was 8.1 points higher than in 2013 [9].

The solution to the problems of ensuring the country's food security is associated with an increase in the level of consumption for a number of products and the achievement of target self-sufficiency in accordance with the requirements formulated in the Food Security Doctrine of the Russian Federation. In 2000, per capita consumption was below the standard level for vegetables and melons (108 kg against the norm of 140 kg), fruits and berries (62 kg against the norm of 100 kg), milk and dairy products (234 kg against the norm of 325 kg), as well as red meat and potatoes [9].

According to the goal of ensuring the sustainable development of agricultural business in the Russian Federation, the agro-industrial complex of the country is set an ambitious goal of increasing the export of products and food to $45 billion by 2025. This item is a fundamental position in the development and implementation of the state program and its subprograms. This position is because the results of the development of the national economy directly depend on the foreign economic activity of agricultural business entities to increase the export of food and agricultural raw materials [10, 11]. In this regard, improving the activities of agricultural business entities to increase exports is recognized as one of the priority areas. The reason for this is that the export potential and active foreign economic activity of agricultural business entities act as a catalyst in regulating equilibrium and balance in the organization of international trade in the country.

Table 2. Export of the main types of products of the agro-industrial complex of Russia, thousand tons [12].

Indicators	Years				Indicators of 2020 in relation to 2019, %
	2017	2018	2019	2020	
Food products and agricultural raw materials, $ billion	21.6	25.8	25.8	30.7	119.0
Meat	307.4	354.4	415.3	609.0	146.6
Milk	607.6	576.3	611.0	707.2	115.7
Butter	3.6	3.4	2.8	2.6	92.9
Sunflower oil	2326.9	2109.5	3098.1	3268.3	105.5
Flakes	43230.4	54895.9	39482.5	48739.0	123.4
Flour and cereals	250.3	305.0	368.6	375.5	101.9
Oilseeds	1737.4	2087.2	2721.9	3574.0	131.3
Sugar	559.8	387.5	687.0	712.0	103.6

An analysis of the indicators in Table 2 gives grounds to assert that the main commodity items in the structure of exports of agricultural products are cereals, oilseeds, and sunflower oil. The main export destinations are such countries as China, Turkey, Kazakhstan, the Republic of Korea, Egypt, Belarus, the Netherlands, and Iran. For specific types of agricultural products and products of their processing, one can note an increase in 2020 compared to 2019 in meat exports by 46.6%, grain by 23.4%, oilseeds by 31.3%, and milk by 15.7%. sunflower oil by 5.5%, sugar by 3.6%, flour and cereals by 1.9%. Along with this, a reduction in the export of butter by 7.1% was allowed. In general, exports in 2020 exceeded food imports for the first time and have a positive balance of $1.0 billion [7, 9]. Table 3 shows the main indicators of imports of basic food products. Table 3 shows the following trends. Russia continues to import large volumes of animal products. And imports of goods related to crop production are declining. The analysis of foreign economic activity showed that the subjects of the agricultural business in the development of exports have significant problems associated with the lack of development of the organizational structure and forms of institutions for export activities, low competitiveness of products in terms of price, and compliance with quality standards. This situation hinders the development of export activities of agricultural business formations and reduces the volume of exports of agricultural products in general [13, 14].

Table 3. Imports of basic food products to Russia, thousand tons [12].

Indicators	Years				Indicators of 2020 in relation to 2019, %
	2017	2018	2019	2020	
Food products and agricultural raw materials, $ billion	28.9	29.8	30.0	29.7	99.0
Meat	1084.6	879.7	771.8	648.0	84.0
Milk	6996.9	6493.0	6727.8	7044.4	104.7
Butter	99.7	90.7	119.7	11.2	93.1
Sunflower oil	25.7	25.7	2.9	2.7	93.1
Flakes	746.3	629.5	515.5	489.0	94.9
Flour and cereals	128.6	89.4	88.3	88.0	99.7
Oilseeds	2556.5	2521.4	2338.0	2296.2	98.2
Sugar	226.3	324.8	241.3	208.6	86.4

Food security is defined as the level of self-sufficiency as a percentage and is calculated as the ratio of the volume of domestic production of agricultural products, raw materials, and food to the volume of their domestic consumption. After the Western European countries and the United States announced sanctions against Russia, agriculture and the processing industry were given the task of providing the population of the country with products of their own production at affordable prices for the bulk of consumers. In a short period of time, the set level of self-sufficiency in basic foodstuffs was achieved (Table 4).

Table 4. Self-sufficiency in basic foodstuffs, % [7.9].

Indicators	Years					Threshold of the Food Security Doctrine of the Russian Federation
	2017	2018	2019	2020		
				plan	actual	
Corn	170.6	147.2	155.6	95	155.5	95
Vegetable oil	153.5	157.3	178.8	90	175.9	90
Sugar made from sugar beets	115.1	108	126.8	90	125.4	90
Potato	91.1	95.3	95.1	95	94.9	95
Milk and dairy products (in terms of milk)	82.3	83.9	83.9	85	84.4	90
Meat and meat products (in terms of meat)	93.5	95.7	97.4	85	96.7	85
Vegetables and gourds	86.8	87.4	87.7	87	88.4	90
Fruits and berries	32.5	32.5	38.8	40	39.5	60

Table 4 shows that in 2020 the threshold values of indicators of food independence (self-sufficiency) of the Russian Federation were reached or exceeded: grain – 155.5%, which is 1.6 times higher than the threshold set by the Food Security Doctrine, sugar – 125.4%, which is 1.4 times higher than the threshold value; vegetable oil – 175.9%, which is almost 2 times higher than the threshold value; meat and meat products – 96.7%, which is 11.7 percentage points above the threshold. Self-sufficiency remains below the thresholds of the Food Security Doctrine: milk and dairy products – 84.4%, which is 5.6 percentage points below the threshold; vegetables and melons – 88.4%, which is 1.6 percentage points below the threshold; fruits and berries – 39.5%, which is 1.5 times lower than the threshold value. Self-sufficiency in potatoes remained at the level of the previous year – 94.9%, which is 0.1 percentage points below the threshold value [9,12]. An assessment of the level of self-sufficiency in basic food products shows that the indicators correspond to their criteria for ensuring food security.

4 Conclusion

The agrarian policy on import substitution, carried out after 2014, contributed to the growth of agricultural production and the development of processing industries. Ultimately, by 2020, a positive trend has been achieved in the production of the main types of import-substituting food products and self-sufficiency in food for almost all major food groups. At the same time, the problem of self-sufficiency persists in milk and dairy products, fruits, and berries. After the

Western European countries and the United States announced sanctions against Russia, agriculture and the processing industry were given the task of providing the population of the country with products of their own production at affordable prices for the bulk of consumers. As in all cases of history, Russian agriculture and the agro-industrial complex emerged with dignity from the current situation. In conditions of limited material, technical and labor resources, the set level of self-sufficiency in basic food products was achieved in a short period of time [15].

Russia is now confidently becoming one of the key exporters in the global agri-food markets. However, the increased attractiveness of supplies to the foreign market can lead to an increase in the aggravation of contradictions between the growth opportunities of export-oriented industries and the need to ensure national food security, primarily due to the growth of physical and economic accessibility of food for the population with low and middle incomes. State program regulation plays a major role in the implementation of sustainable development priorities and the growth of the competitiveness of the agri-food complex. A feature of the current stage of development of the state support system is the combination of the use of program-target methods aimed at the implementation of targeted programs for the development of sectors of the agro-industrial complex and system-wide measures aimed at the development of rural areas. Ultimately, the positive trends that have developed in the agri-food sector as a result of the implementation of the import-substituting development strategy made it possible to move to an export-oriented strategy and implement a strategy to ensure the country's food security.

References

1. R. Valiev, B. G. Ziganshin, A. S. Lukin, et al., *Financial Business* **7,** 62–66 (2021)
2. S. Lukin, F. N. Mukhametgaliev, F. F. Mukhametgalieva, et al. Features of modern Russian agrarian policy, *Financial Business* **2,** 65–67 (2021)
3. A. Battalova, *BIO Web of Conf.* **17,** 00114 (2020)
4. R. Battalova, O. A. Ignatjeva, F. N. Mukhametgaliev, L. F. Sitdikova, *International Journal on Emerging Technologies* **10 (2),** 112–116 (2021)
5. D. I. Faizrakhmanov, A. R. Valiev, A. S. Lukin, et al., *Financial Business* **6,** 171–175 (2021)
6. *Doctrine of food security of the Russian Federation: approved. Decree of the President of the Russian Federation of January 21, 2020 No 20,* retrieved from: http://www.consultant.ru/document/cons_doc_LAW_343386/
7. National report *"On the progress and results of the implementation in 2020 of the state program for the development of agriculture and the regulation of markets for agricultural products, raw materials, and food"*, retrieved from: https//mcx.gov.ru/upload/iblock/98a/953ee7405fb0ebba38a6031a13ec0021.pdf
8. *State Program for the Development of Agriculture and Regulation of Agricultural Products, Raw Materials and Food Markets for 2013–2020* (Moscow) (2012)
9. *Federal State Statistics Service of the Russian Federation*, retrieved from: http://www.gks.ru/
10. D. I. Fayzrakhmanov, A. R. Valiev, B. G. Ziganshin, et al., *Management of mechanisms for improving the efficiency of labor resources in agriculture* (Kazan: Kazan State Agrarian University, 2021)
11. R. S. Tukhvatullin, *International Journal on Emerging Technologies* **10 (2),** 133–136 (2019)
12. *Balances of food resources*, retrieved from: http://old.gks.ru/wps/wcm/connect/rosstat_main-/rosstat/ru/statistics/enterprise/
13. L. F. Sitdikova, *Financial Business* **9,** 166–169 (2021)
14. I. G. Gainutdinov, F. N. Avkhadiev, *Bulletin of Kazan State Agrarian University* **2,** 86–95 (2021)
15. D. I. Fayzrakhmanov, *Grain Economy of Russia* **2,** 64–68 (2014)

https://doi.org/10.1051/bioconf/20225200017

Implementation of rural economy sustainable development policy

Landysh Sitdikova, *Farit* Mukhametgaliev, *Marcel* Khismatullin, *Farida* Zakirova, and *Lilia* Mikhailova
Kazan State Agrarian University, 65, Karl Marx St., Kazan, Republic of Tatarstan, 420015, Russian Federation

Abstract. The rural economy in modern conditions of the country's food security doctrine and import substitution policy is one of the main directions of socio-economic development of any society. The problems of sustainable development of the rural economy and the increase in growth indicators of the subjects of agrarian business are particularly interesting as a fundamental direction that ensures the socio-economic stability in rural areas. The paper is identifies modern systemic issues that restrain the implementation of the sustainable development policy of the rural economy, and develops priority directions to ensure the efficiency of the agricultural sector of the economy in modern conditions of external and internal socio-political challenges. The paper contains the results of the analysis of volumetric indicators and growth rates of agricultural production, the level of performance of the state target program indicators for recent years, identifies the development trends, priority directions that ensure the efficiency of the rural economy, proposes organizational and economic measures to guarantee the production growth and increase the efficiency of agrarian production, intensify investment activities and strengthen state support for the rural economy.

1 Introduction

The Russian economy in modern socio-political conditions desperately needs the aggressive development rates in all production and servicing sectors of the national economy. One of the most important directions in achieving the socio-economic stability of the country is the sustainable progressive development of the rural economy, which provides the population with basic necessities and forms the country's food security. With the development of human society accompanied by the population growth, there is an urgent need for a continuous increase in the volume of food products and goods directly related to the growth of the agricultural potential as a raw material base of the economy for the processing industry. Currently, agricultural business is one of the priority areas that strengthens the rural economy and develops rural areas. It performs socially important functions to ensure socio-political stability in rural areas based on increased business activity and employment of the rural population. The problems of agricultural business development are fully demonstrated at all levels of management of the socio-economic processes of the modern society. This situation forces to give the development of the agrarian economy the status of one of the priority national projects aimed at implementing promising tasks not only for the development of agricultural production, but also turning the agricultural sector into an attractive sphere of activity thus creating favorable living conditions in rural areas. The priority areas for the development of the agricultural sphere of life, first of all, imply the steady provision of increased growth rates of volumetric indicators in the production sectors of agricultural raw materials, the achievement of financially sustainable work of agricultural business entities. Indeed, the achievement of high indicators of agricultural production efficiency causes the need to identify and solve a whole range of external and internal challenges of our time, among which a special place is occupied by the problems of formation and creation of socio-political, organizational and economic prerequisites for further development and successful functioning of agricultural business entities adapted to new economic conditions, corresponding to specific conditions of production and financial activities. All this clearly shows the relevance of the problem of further progressive development of the agricultural sector of the economy, which causes the need for additional study of this issue, the development of practical recommendations for the implementation of measures for the effective functioning of the agricultural business.

2 Materials and methods

The methodological basis for the study included earlier work of domestic researchers, including the authors, devoted to the analysis of the level of development of the domestic agricultural sector, as well as the problems of implementing the country's food security strategy. The general methodological basis of the study is the systemic analysis. The analysis of the used indicators was carried out by comparing the changes in values by years over 2017-2020 in order to identify fluctuations in these indicators in each subsequent year and increase the correctness of calculations in a dynamic range of indicators. The information base for the study was statistics from the Federal State Statistics Service of the Russian Federation, the Ministry of Agriculture of the Russian Federation, data obtained during the analysis and calculations of the authors.

3 Results and discussion

The solution to the problem of increasing the efficiency of agricultural business entities is included in the general context of the development of the country's economy as a whole and in the concept of specific features of the rural economy related to natural and climatic factors and the natural and biological laws of crop production and animal husbandry. The most important component in this complex combination is the achievement of an increase in crop yields and animal productivity, as well as the expansion of integration and deepening of cooperation in building relationships between all related areas of the agro-industrial complex and agricultural sectors. The level of development of the agricultural business in modern conditions is determined by the achieved final indicators of one of the fundamental sectors of agriculture – crop production. For many years the value expression of the products of this industry has occupied almost half of the cost of gross production of the entire agricultural sector of the economy and has been largely involved in the formation of general indicators of the socio-economic development of rural areas. The products of crop industries are of great importance for the development of other sectors of the country's economies. First, crop production provides the country's population with food products such as bread and bread products, vegetable oil, sugar, potatoes, vegetables, melons, fruits and grapes. Second, crop production, like grain and industrial crops, forms the basis of the export potential of the agro-industrial complex. Third, it acts as a driver for the development of animal husbandry as a supplier of grain and feed necessary to obtain such animal husbandry products as meat, milk, egg, wool. Fourth, plants such as textile crops, oilseeds and partially cereals, fruits, vegetables and grapes, sugar beets are necessary raw materials for processing industry enterprises [1, 2].

The development of crop production in general and grain production in particular largely depends on the supply of livestock products to the population, which is an important direction in ensuring the country's food security. This goal can be achieved only with the stable development of the entire agricultural industry, the main direction of which is to increase gross grain production and improve the economic efficiency of its production through the rational organization of the production system. Grain farming forms the basis for agricultural production and has a deterministic influence on the development of all industries [3, 4].

The food security doctrine of the country and the improvement of the well-being of its population can be implemented only with an increase in crop production in all constituent entities of the Russian Federation. Besides, the sustainable development of this industry requires a whole range of conditions and factors. First of all, this concerns the widespread increase in farming standards, optimization of the size of land ownership of producers, increase in the economic fertility of soils and efficiency of the use of land resources. In modern conditions, the development of modern farming systems based on the use of innovative achievements covering technical, technological and organizational-economic measures to increase the volumetric production indicators and improve its quality serve the levers of management and organization of crop production [5].

The analysis of the level of development of the agricultural sector of the economy over the past four years showed that the growth rates of production cost indicators in comparable prices for agriculture as a whole in 2020 compared to 2017 reached 105.7%, which means that the state program indicators [6] exceed the targets by 1.9 percentage points (hereinafter – p.p.). Although in 2018 there was a slight decrease in the growth rate of agricultural products – by 0.2%. The performance versus targets of the program as a whole is the main indicator of the implementation of the agricultural development program of the country, which is provided for by the import substitution policy of the country's government regarding staple food and food supplies. The performance of planned and target indicators of the State Program are given in Table 1.

The table shows that the growth of agricultural production is not accompanied by the growing volumes of investments, their growth rates are lower than planned, i.e. they are falling behind, and this also contradict the proportionality of the balanced development of all mutually related indicators of industry development. This imbalance cannot be eliminated by the increase in added value created in the field of production activities of agricultural producers, which in 2020 amounted to 3472.7 billion rubles, or overfulfilment amounting to 8.2%. This phenomenon causes concern in the long term and threatens to reduce the capacity of the production potential of agricultural business entities. Against this background, at first glance, one can look with optimism at the export potential growth rate indicator, which is 41.2%, as a source of foreign exchange infusion ensuring the development of the agricultural sector of the economy. But here it is necessary to take into account a high share in the export revenues of intermediary enterprises, which buy products from agricultural producers at average domestic market prices. Due to the lack of experience in foreign markets and export transactions, the lack of development of market infrastructure and logistics, agricultural producers cannot independently enter into transactions with foreign importers of food products and are forced to sell their products through intermediaries. In this case, a significant part of

 https://doi.org/10.1051/bioconf/20225200017

export revenues remains with intermediaries and does not reach the subjects of the agricultural market and fails to act as investment sources for the development of agricultural production [8].

Table 1. Implementation of planned indicators stipulated by the State Program [7].

Index	2017	2018	2019	2020		
				plan	actual	performance, %
Index of agricultural production in farms of all categories (in comparable prices) by 2017, %	100	99.8	104.1	103.8	105.7	101.8
Index of actual volume of capital investments in fixed assets of agriculture by 2017, %	100	103.4	104.6	98.2	97.3	99.0
Production value added in agriculture, bln roubles	2896.8	3101.3	3352	3209.2	3472.7	108.2
Growth rate of agricultural exports by 2017, %	100	119.3	118.4	115.7	141.2	122.1
Index of crop production in farms of all categories (in comparable prices) by 2017, %	100	98.5	105	102.4	106.1	103.6
Index of livestock production in farms of all categories (in comparable prices) by 2017, %	100	101.1	103	105.5	105.1	99.6
Food production index (in comparable prices) by 2017, %	100	103.6	107.8	109.6	111.7	101.9
Profitability of agricultural enterprises (including subsidies), %	12.0	12.5	13.3	12.0	21.0	175

This situation affected the achievements of agricultural producers in terms of investment in fixed assets, which decreased in 2020 compared to the achieved indicators in 2017 by 2.7 p.p. and in terms of the growth rates did not reach 1.1% of the target indicator of the State Program [6]. The reason for this situation was, first of all, the excessive debt burden of the majority of agricultural business entities, an increase in the difference in the growth rate of prices for goods, services in the serving sectors of agriculture and agricultural products. Besides, recently much attention has been paid to the processing of agricultural products and bringing food products to consumers in a ready-to-eat form, which requires reorienting significant amounts of investment to agricultural processing industries and to the development of food industry enterprises [9].

Table 1 also shows that the growth of crop production in all categories of producers in 2020 compared to 2017 amounted to 6.1%, which exceeds the indicative value of the program by 3.7 p.p. The natural and climatic conditions of 2020 have recently been the most successful for farmers, which contribute to the growth and harvesting of a large volume of products for the main types of crops, often exceeding the record numbers of recent times in some crop industries.

At the same time, although increased in 2020 compared to 2017 in the range of 105.1% the growth rates of livestock production in terms of the planned indicator remained unachieved by 0.4 p.p., which can be estimated as one of the restraining factors of sustainable agricultural growth [10]. The reason for this was the reduction in the production of certain types of livestock products. From the point of view of strengthening the sustainable balanced growth of all agricultural sectors, this ratio of growth rates of crop production and animal husbandry is extremely undesirable since the sale of crop production in its original form is always not profitable for business in the domestic agricultural market. Therefore, it is desirable to sell crop production through livestock production through herd expansion and increased productivity, by using the growing potential of crop production in the form of grain and feed crops.

The growing potential of agricultural production subsequently led to an increase in the food production index in 2020 compared to 2017 by 11.7 p.p. and exceeded the targets of the State Program by 2.1%. The level of profitability of the production and economic activities of agricultural organizations, taking into account the amount of state subsidies, increased by 9 p.p. over the past four years. It exceeded the planned value by 9.0 p.p. The total level of profitability (21.0%) for modern agricultural organizations is not sufficient for the implementation of expanded reproduction, especially since intensive modernization of agricultural production is required everywhere [10, 11].

Modernization and technical re-equipment of the agricultural sector of the economy, given its starting position and the socio-economic importance of the results and development prospects, requires large investments. Over the past four years, the volume of investments in fixed assets in the agro-industrial complex has increased by 135.1 billion rubles in actual prices or by 23.1% through a progressive-increasing order. But this process is increasingly being passed on to agricultural producers. The share of the federal budget in the development of the fixed capital of the agro-industrial complex is only 0.1%, and the share of budget funds of the constituent entities of the Russian Federation in this process is 0.7% [12]. Here it is also necessary to take into account the following distribution of budgetary funds for the renewal of fixed capital: the first part is the renewal of the fixed capital of the budgetary service aimed at maintaining the governing bodies responsible for phytosanitary, high-quality, technical and other control functions in the food chain: maintaining the structure providing consulting, scientific, information and other types of services to enterprises of the agro-industrial complex. Only the second part of the budget funds is used to renew the fixed capital of enterprises and divisions of the agro-industrial complex engaged in direct crop production and animal husbandry. 63.4% of the fixed capital of agricultural business entities is renewed at their sole funds, another 35.8% – at the expense of borrowed funds of enterprises, organizations and other sources [13]. It should be noted that the share of borrowed funds is increasing from year to year, which indicates a reduction in the ability of enterprises of the agro-industrial complex to carry out expanded reproduction on their own. It is known that borrowed funds must be returned with interest. The more borrowed funds are involved in the turnover, the more difficult it will be to return them, they will increase apace and in the near future producers will be forced to withdraw active capital from circulation to return debts and limit funds for further development, which may increase in multiples due to the upcoming digitalization of the rural economy and automation of production processes. The commissioning of production facilities is carried out in all sectors of agriculture [14]. The largest renovation of premises for cattle was carried out in 2017 when 275.1 thousand cattle stalls were put into operation. In the following years the rate of commissioning accelerates in approximately the same size. The same situation is observed in the pig industry. This can be explained by the fact that after the announcement of sanctions by Western countries, an import substitution program was adopted, and the buildings started after that were ready for commissioning in one and a half to two years. Given the weak position of sheep farming, the growth rates of construction and commissioning of premises for this industry is increasing. Over four years, the growth rates have increased 3.7 times. The premises for poultry farming are being built at an accelerated pace. In the egg industry, the growth rates increased 1.5 times, in meat industry – 4.5 times. Poultry production is developing at a high pace not only to meet domestic needs, but also to supply poultry products for export.

Ultimately, the share of food products and agricultural raw materials in the commodity structure of exports of the Russian Federation in 2019 amounted to 8.8%. The total amount of food turnover of the Russian Federation with foreign countries in 2020 increased by 8.6% compared to 2019, with a decrease in imports by 0.5% and an increase in exports by 19.7%. The trade balance reached a positive value of US $886.3 mln (Table 2).

The increase in export supplies was facilitated by an increase in food volumes: meat (cattle – by 16.2 thousand tons (170.7%), pork – by 200.4 thousand tons (86.6%), poultry – 294.8 thousand tons (40.3%); sugar – by 1.2 million tons (76.2%); oilseeds – by 3.5 million tons (70.8%)); grain (barley) – by up to 6 million tons (51.7%).

In order to ensure the sustainable development of agricultural business in the Russian Federation, the country's agro-industrial complex sets an ambitious task to increase exports of products and food to $45 billion by 2025, which is a fundamental position in the development and implementation of the state program and its subroutines.

Table 2. Food turnover of the Russian Federation with foreign countries, million US$ [7].

Index	2018	2019	2020	2020/2019 growth rates	
				%	+/-
Turnover	55559	55532.2	60432.1	108.6	4774.2
Import	29757	29924.2	29772.9	99.5	-151.3
Export	25802	25608.1	30659.2	119.7	505.1
Trade balance	-3955	-4316.1	886.3	x	5202.4

This position is related to the fact that the results of the development of the national economy directly depend on the foreign economic activity of agricultural business entities to increase the export of food and agricultural raw materials. In this regard, improving the activities of agricultural business entities to increase exports is recognized as one of the priority areas. The reason for this is that the export potential and active foreign economic activity of agricultural business entities act as a catalyst in regulating the equilibrium and balance in the organization of international trade in the country as a whole.

4 Conclusion

After the Western European countries and the United States announced sanctions against Russia, the agricultural sector and the processing industry were tasked with providing the country's population with their own products at prices affordable to the majority of consumers. As in all cases of history, Russian agriculture and the agro-industrial complex with dignity got out of the current situation in the conditions of limited material, technical and labor resources. In a short period of time they reached the level of self-sufficiency in the staple foods. According to the consolidated reporting on the results of the activities of agricultural organizations receiving state support for 2020, the profitability (including subsidies) amounted to 20.0%, which is higher than the target (12%) by 9.0 p.p., and is also higher than the level of the previous 2019 by 7.7 p.p. (13.3%). In 2020, before-tax profit (including subsidies) was obtained in the agricultural sector of business in the amount of 624.2 billion rubles, which is 76.4% higher in 2019. At the same time, the share of profitable farms in the total number of agricultural business entities amounted to 86.1% (2 p.p. higher than the previous year) with a decrease in their total number. Excluding

 https://doi.org/10.1051/bioconf/20225200017

subsidies, the profitability is 16.3% versus 8% in 2019. The profitability level is not high, and taking into account the level of annual inflation, it turns out to be low. Such a level of payback of invested funds for agricultural organizations as dynamically developing entities of agricultural business does not allow conducting innovative and investment activities for the development of entrepreneurship in rural areas. Besides, the agricultural business plays a key role as a socially responsible enterprise, which took on the task of infrastructural development of rural areas in order to create favorable living conditions for the rural population, increase the attractiveness and depth of the rural lifestyle and agricultural labor.

References

1. A. S. Lukin, F. N. Mukhametgaliev, F. F. Mukhametgalieva et al., Features of modern Russian agrarian policy, *Financial Business,* **2,** 65–67 (2021)
2. A. R. Valiev, B. G. Ziganshin, A. S. Lukin et al., Organizational and economic problems of the development of the agricultural sector of the economy, *Financial Business,* **7,** 62–66 (2021)
3. F. N. Avkhadiev, Increasing the stability of grain production (based on materials of the Republic of Tatarstan), *Kazan State Agrarian University,* **4,** 104–108 (2016)
4. F. N. Avkhadiev, I. G. Gainutdinov, V. Ya Petrova et. al., Problems of regional grain market development, *BIO Web of Conf.* **17,** 00082 (2020)
5. A. R. Battalova, F. N. Mukhametgaliev, F. F. Mukhametgalieva, L. F. Sitdikova, Issues on increasing efficiency of agricultural business in the Republic of Tatarstan, *Journal of Environmental Treatment Techniques* **7,** 930–934 (2019).
6. *State Program for the Development of Agriculture and Regulation of Agricultural Products, Raw Materials and Food Markets for 2013–2020* (Moscow, 2012)
7. National report "On the progress and results of the 2020 state program for the development of agriculture and the regulation of markets for agricultural products, raw materials and food products", retrieved from: https//mcx.gov.ru/upload/iblock/98a/953ee7405fb0ebba38a6031a13ec0021.pdf
8. F. N. Avkhadiev, Main areas of technical modernization of agriculture of the Republic of Tatarstan, *Machinery and equipment for the village* **4,** 46–48 (2017)
9. R. S. Tukhvatullin, Priority areas of development of agricultural entrepreneurship in the regions of the Russian Federation, *International Journal on Emerging Technologies* **10 (2),** 133–136 (2019)
10. Federal State Statistics Service of the Russian Federation, retrieved from: http://www.gks.ru/.
11. A. Battalova, Development of integration processes in the agricultural sector, *BIO Web of Conf.* **17,** 00116 (2020)
12. D.I. Fayzrakhmanov, A.R. Valiev, B.G. Ziganshin et al., Management of mechanisms for increasing the efficiency of labor resources in agriculture, (Kazan: Publishing House of Kazan SAU, 2021)
13. I. G. Gainutdinov, F. N. Avkhadiev, State and peculiarities of the development of livestock industries in Russia and abroad, *Bulletin of Kazan State Agrarian University* **2,** 86–95 (2021)
14. A. R. Battalova, R. S. Tukhvatullin, Tendency of investment economy formation. *International Journal of Criminology and Sociology* **9,** 252–257 (2020)

Development of methodological tools for forecasting the activities of agricultural organizations to ensure food security in the region

Albert Iskhakov[1*], *Guzaliya* Klychova[1], *Gamlet* Ostaev[2], *Elena* Konina[2], *Ekaterina* Gainutdinova[2], and *Yelena* Milova[3]

[1] Kazan State Agrarian University, 65, Karl Marx St., Kazan, Republic of Tatarstan, 420015, Russian Federation
[2] Izhevsk State Agricultural Academy, 11, Studencheskaya St., Izhevsk, Udmurt republic, 426069, Russian Federation
[3] Korkyt Ata Kyzylorda University, 29A, Aiteke bi St., Kyzylorda, 120014, Kazakhstan

Abstract. The study of the issues of forecasting the activities of agricultural organizations in modern conditions is necessary and relevant. Development of forecasting tools in the management system of modern agricultural organization has a significant impact on improving the efficiency and effectiveness of its production and economic activity. In our opinion, the methodology of assessing the sustainability and effectiveness of agricultural organizations should be aimed at combining static and dynamic approaches, which is demonstrated in the present study. In economic practice, the possibility of managing the factors of socio-economic phenomena determines the need to measure their relationships. Multifactor correlation and regression analysis allows to estimate the measure of influence on the studied result indicator of each of the factors included in the model at a fixed position of other factors, as well as at any possible combinations of factors with a certain degree of accuracy. Correlation and regression analysis allows taking into account influence of various factors in forecast calculations. The research allows to create the forecasting variants of economical-mathematical models of activity of enterprise that allows to choose the trajectories of the most effective economic development. As an analysis and application of forecasting tools we chose a sub-sector of agriculture of the Republic of Tatarstan - dairy cattle breeding. The analysis of dairy cattle breeding of Russia and the Republic of Tatarstan for the last 30 years is given in the article, the food security issues are considered, the multifactorial correlation-regression model of economic activity of cattle breeding enterprises of the Republic of Tatarstan is built. As a result, based on the regression equation, the average annual milk yield per 1 cow was predicted. Thus, on the basis of the obtained results, it is possible to develop further action plan to improve the production-economic activity of the enterprise and search for reserves in increasing its efficiency.

1 Introduction

Agriculture and food industry are the basis in ensuring food security of the country. Consequently, food security is important for the economy of the country as a whole, because the life, health and well-being of citizens depend on it.

In conditions of extreme economic and political instability, the problem of food security becomes the most urgent for absolutely every state.

In the medium term, food security and provision of the population of the country is one of the main directions of national security of the country. As part of this task, special normative legal acts, state programs are adopted, which determine the main directions of development of agricultural production as a strategic sector of the economy.

Over the past 30 years, agriculture in Russia as a whole has demonstrated a trend of growth in indices and volume of products produced, including through government support and implementation:

- development concepts and strategies;
- federal target programs since 1995;
- the priority national project "Development of the Agro-Industrial Complex (AIC)" in three areas: accelerated development of animal husbandry, promotion of small-scale farming and provision of affordable housing for young rural specialists (2006-2007);
- state agricultural development programs since 2008;

* Corresponding author: iat20@yandex.ru

– priority projects from 2016 to 2018;
– national projects since 2019, including "International cooperation and export" (one of the federal level projects "Export of agricultural products").

Dairy products are included in the list of Doctrine of Food Security of RF and the share of domestic production of milk and dairy products (in terms of milk) should be at least 90%. The annual output of milk in Russia should reach 48 million tons or not less than 90% of the Doctrine threshold - 43 million tons. According to the forecast values, in 2022 in Russia it is planned to produce 32.5 million tons of milk, which is by 24% or 10.5 million tons less than the value established by the Doctrine.

In order to solve the tasks set to ensure food security, it is necessary to continue the implementation of state programs to support and develop agriculture, and agricultural producers to improve the efficiency of their activities, including through effective management methods [1].

2 Materials and methods

To achieve the goal and objectives of the study we chose deductive and inductive methods and techniques with elements of statistical analysis. These methods make it possible to assess the degree and nature of the relationship between the result and the factors that condition it. The results of statistical analysis can be used for forecasting the results of economic activity of agricultural organizations.

The main objectives of the state in dairy cattle breeding are to ensure food security of the population of the country, increasing the productivity of farm animals, including through effective breeding methods, innovation and technological development of the industry, development of the fodder base.

Dairy productivity of cows depends on many internal and external factors. As part of the analysis, it is necessary to identify the main of them and the extent of their influence with the possibility of applying the results of the analysis in the management and forecasting of the industry.

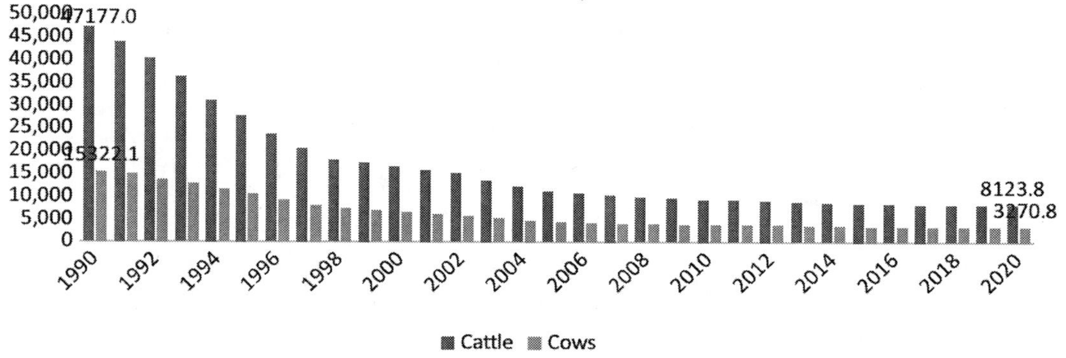

Fig. 1. The number of cattle and cows in agricultural organizations in Russia, thousand heads.

Over the period from 1990 to 2020 the number of cattle in agricultural organizations of the Russian Federation declined by 82.8%, including cows - 78.7%. Milk production decreased by 57.9% (from 42 452.1 tons to 17879.9 tons) [2].

In farms of all categories in the country the reduction of cattle in the analyzed period was 68.4%, of cows – 61.6%, the total volume of milk production decreased by 42.2% (from 55715.3 thousand tons to 32225.5 thousand tons) [2].

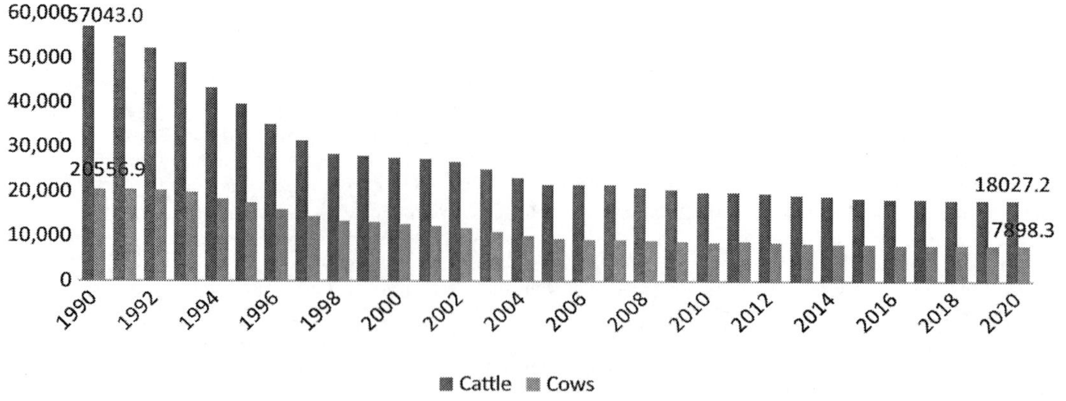

Fig. 2. The number of cattle and cows in all categories of farms of the Russian Federation, thousand heads (1990–2020).

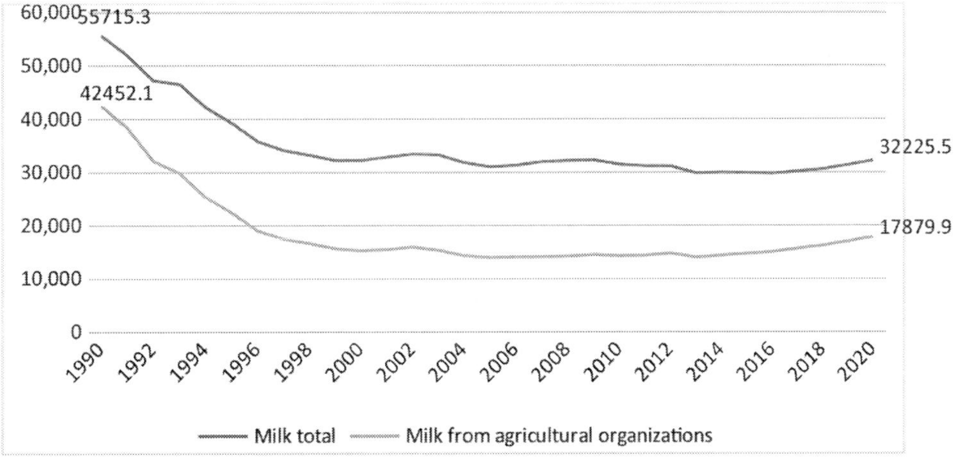

Fig. 3 - Milk production in Russia, thousand tons (1990–2020).

The processes of reducing the number of cows and providing the country's population with milk and dairy products are smoothed by intensifying milk production. The level of milk yields per cow in the country as a whole in farms of all categories during the analyzed period increased by 77.2%, in agricultural organizations – 2.4 times to 6728 kg [2].

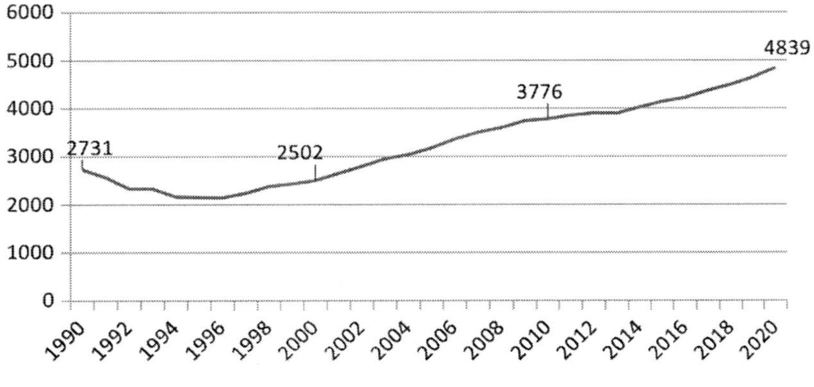

Fig. 4. Milk yields per cow on farms of all categories in Russia, kg (1990–2020).

In the Russian Federation among 85 subjects of the country only 17 regions (20%) the level of milk and dairy products production exceeds the required minimum standards of consumption (325 kg per year per person, according to the recommendations of the Russian Ministry of Health on rational norms of food consumption). The largest volume of milk production in the country in 2020 belongs to the Volga Federal District and accounts for 31% of the total Russian production. The leader in milk production in the Volga Federal District is the Republic of Tatarstan, so according to the results of 2020 the republic produced 1.9 million tons of milk, which is 2.5% more than in 2019. The share of agricultural organizations of the republic to the total production volume in 2020 was 63.1% or 1.2 million tons, which is 4.3% more than that in 2019.

Fig. 5. Milk production in the Republic of Tatarstan, thousand tons (1990–2020).

In addition to the growth in the production of raw milk, the volume of its processing has also increased in the republic. 322.7 thousand tons of drinking milk were produced in 2020, which is 29% more than in 2019; fermented milk products – 88.8 thousand tons, an increase of 6%; cheese, cheese products and cottage cheese produced 47.9 thousand tons with an increase of 8%.

According to the Dairy Market Intelligence Center (DIA), Tatarstan leads among the regions of the country by the number of plants: there are 8 enterprises operating in RT, and in 2020, 59.3% of milk and dairy products production comes from Zelenodolsky Dairy Plant, Mamadysh Cheese and Dairy Plant and "Alabuga Sote".

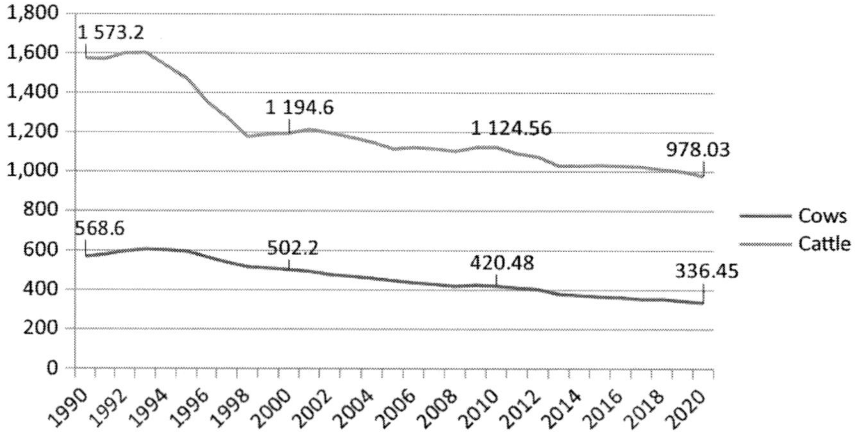

Fig. 6. The number of cattle and cows in the Republic of Tatarstan, thousand heads (1990–2020).

Despite the growth in milk production, there is a reduction in the number of cattle and cows, so for the analyzed period since 1990 the number of cows decreased by 40.8% and amounted to 336.45 thousand heads in 2020. Thus, the main factor in the growth of milk production is currently the milk productivity of cows. However, the reserve of milk production efficiency growth in the next years in Russia and in the republic will be an increase in the number of the main milking herd with a further increase in its productivity.

https://doi.org/10.1051/bioconf/20225200018

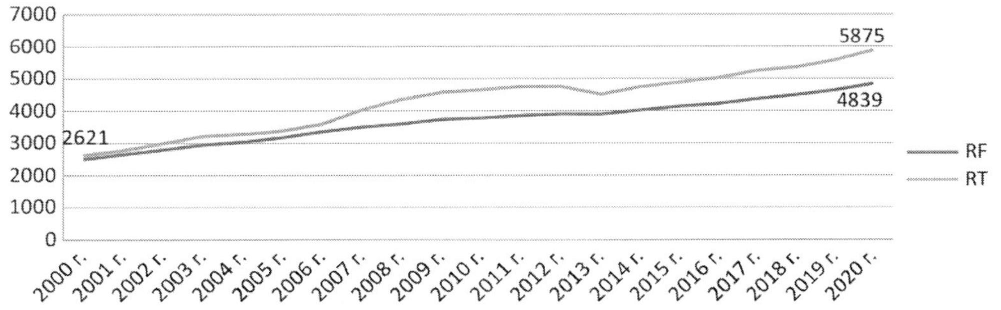

Fig. 7. Milk yield per cow in Russia and in the Republic of Tatarstan, kg.

Table 1. Large producers of milk in the Republic of Tatarstan in 2020.

Company	Volume, thousand tons	Number of cattle, thousand heads	Average yield per 1 cow, tons
JSC Agroholding «Krasny Vostok»	115.1	64.6	5.6
Ak Bars «HC JSC»	106.6	67.8	5.1
JSC «Agrosila»	67	31.9	5.8
LLC «APC Food Program» (Mamadysh)	55.4	20.8	10.9
AE Vakhitova (Kukmor)	34.3	7.9	11
LLC Management Company «AgroInvest»	31.6	9.7	6.2
LLC "Soyuz-Agro"	31	10.7	8.6
APC «Uralb» (Kukmor)	26.5	6.3	10.7
LLC «Orsis-Agro» (Nizhnekamsk)	17.8	4.2	9.7
LLC «Khuzangayevskoye»	17	10	3
LLC «Service-Agro»	12.9	5.8	7.7
LLC Agro-Osnova (Novosheshminsk)	12.5	8.2	4.2
LLC «MVP Sabinsky»	12	1.5	12.8
LLC «AF Karmaly» (Nizhnekamsk)	10.3	3.9	5.9
LLC «Nurlat Sete» (Nurlat)	9.8	5.4	5
JSC Tokarlikov (Almetyevsk)	8.9	2.9	8.9
LLC «August-Muslum»	8	12.5	5.2
LLC «Chistay-Agro» (Chistopol)	7.6	5.7	3.6
PSC JSC «TATAgrokhim»	6.3	3.3	5.6
CC Civil Code "Fortex"	6.2	2.3	6.3
AE LLC «Northern»	6	3.2	7.7
LLC AF «Chistopolskaya»	5.9	2.4	5.2
JSC «Racine»	5.6	1.1	4.1
LLC Agricultural Enterprise Lutfullin Z. R.	4.9	2.7	5.7
LLC «R-Agro» (Tatplodovoschprom)	4.4	1.9	4.4
JSC «Tatagroleasing»	3.5	1	3.7
JSC «Avangard»	3.3	2	4.3
JSC «Chelny-Khleb»	2.8	1	7
Total	633.2	300.7	X

Over the past 20 years the growth of milk productivity of cows in the republic has increased more than 2-fold and amounted to 5.9 thousand tons per year, which is associated with the improvement of the breed composition of the main herd, diet, conditions and technology of housing.

At present, the world leader in cow productivity is Israel, with an average indicator of 12 thousand tons of milk per year from one cow, which is 2 times higher than the republican value.

Let us consider the main production indicators of large producers of raw milk in the Republic of Tatarstan, which are given in Table 1.

Large producers of raw milk in the Republic of Tatarstan account for 32.6% or 633.2 thousand tons. The leaders in cow productivity are the following companies LLC LLC "MVP Sabinsky" (Sabinsky district) – 12.8

tons of milk from one cow per year, AE "Vahitova" (Kukmorsky district) – 11 tons, "APC Food Program" Ltd. (Mamadyshsky district) – 10.9 tons, "Ural" (Kukmorsky district) – 10.7 tons, "Orsis-Agro" Ltd. (Nizhnekamsky district) – 9.7 tons, Tokarlikova JSC (Almetyevsky district) – 8.9 tons [3].

Economic efficiency of dairy cattle breeding is characterized by some of the following indicators:

- average milk yield from 1 cow;
- costs per cow;
- Feed consumption per cow;
- cost of production;
- profit, level of profitability and others.

In economic practice the possibility of managing the factors of socio-economic phenomena determines the necessity to measure their interrelations.

To estimate the degree and nature of the relationship between milk productivity of cows and the factors, which determine it, we conducted correlation and regression analysis by solving the problem of multiple correlation.

For this purpose a multifactorial correlation and regression model of economic activity of livestock enterprises for 2019, covering a set of 269 units of agricultural organizations of all municipal districts of the Republic of Tatarstan was built.

The resultant (y) and factor (xi) signs are designated in this model. In economic processes, the resultant factors include indicators that directly or indirectly reflect the results of production activities. In this connection, the indicator characterizing the level of milk productivity of cows serves as an outcome factor.

Factor indicators, in turn, should reflect the level of their impact on the production process:

y – average annual milk yield per 1 head, cwt;

x1 – labor costs per 1 dairy head, thousand rubles;

x2 – cost of fodder per 1 goal, thousand rubles;

x3 – amount of subsidies per 1 goal, thousand rub;

x4 – density of cows per 100 hectares of agricultural land, heads.

Multifactor correlation and regression analysis allows one to estimate the measure of influence on the studied result indicator of each of the factors included in the model at a fixed position of the other factors, as well as at any possible combinations of factors with a certain degree of accuracy. In this case an important condition is the absence of a functional relationship between the factors.

Assessment of the parameters of the results of the solution is reduced to the comparison of obtained parameters with the norms of mathematical statistics. If the parameters meet the standards, then proceed to the interpretation of the solution results. Otherwise, the correlation and regression model is corrected by screening out observations with gross (anomalous) errors, and by excluding or replacing factor attributes according to mathematical restrictions imposed on their selection. The corrected model is used to obtain a new solution, which also evaluates the calculated parameters [4, 5, 6].

3 Results and discussion

According to the results of the solution of four-factor model, which includes 146 observation units after adjustment (Table 2), the tightness of the connection between the resultant and factor attributes selected in the correlation model is strong enough, as evidenced by the numerical value of the multiple correlation coefficient (*R* = 0.696).

Table 2. Four-factor correlation and regression model.

Number of observations	146				
Number of factors	4				
Result and factors	Y	X1	X2	X3	X4
Errors of correlation coefficients	0.043	0.069	0.058	0.071	0.064
Reliability of correlation coefficients	16.300	5.917	9.595	5.439	7.542
Arithmetic mean values	60.380	7.719	16.617	22.352	10.305
Mean square deviations	18.491	3.036	6.474	8.899	4.046
Coefficients of variation, %	30.625	39.333	38.960	39.814	39.263
Regression coefficients	11.157	0.818	0.808	0.635	1.483
Beta coefficients		0.134	0.283	0.306	0.325
Determination coefficients, %	48.418	16.657	30.479	14.733	23.064
Coefficients of individual determination	48.418	5.483	15.616	11.735	15.585
Multiple correlation coefficients	0.696				
Paired correlation coefficients	Y	0.408	0.552	0.384	0.480
	X1		0.404	0.082	0.414
	X2			0.282	0.396
	X3				-0.039

The absolute values of rYXj reflect the following strength of the relationship:

rYXj = 0 - no relationship;

0.00 < rYXj < 0.20 - insignificant, unreliable relationship;

0.20 < rYXj < 0.40 – weak relationship;

0.40 ≤ rYXj < 0.60 – average connection;

0.60 ≤ rYXj < 0.80 – strong connection;

0.80 <= rYXj < 1.00 – very strong relationship;

rYXj = r,00 - functional linear relationship

The error of the correlation coefficient (OR = 0.043) and the reliability of the correlation coefficient (TR = 16.300) confirm the presence of a strong relationship between the resultant and factor attributes and indicate sufficient reliability of the multiple correlation coefficient.

The interpretation of the results of the solution involves the study of coefficients of determination, regression coefficients and coefficients of separate determination.

The multiple determination coefficient (D = 48.418) characterizes the percentage measure of dependence between the factors included in the model and the result. It follows, that together the four factors of the correlation-regression model influence the change of milk productivity of cows by 48.418%, including the change of labour compensation by 5.483%, feed expenses by 15.616%, change of subsidies by 11.735%, population density by 15.585%.

The coefficients of the regression equation are the coefficients of the regression equation:

Y1 = 11.157 + 0.818x1 + 0.808x2 + 0.635x3 + 1.483x4

If we substitute the arithmetic mean values of the factor attributes – x1, x2, x3, x4 into the equation, then the calculated value of the result indicator – Y will also be the arithmetic mean.

Parameters of the linear regression equation at the factor attribute show, to what change in the average value of the resultant attribute a change in the factor attribute by one unit leads.

That is, if the average annual labor costs per 1 head are increased by 1 thousand rubles, then the average annual milk yield will increase by 0.818 ct. If the feed expenses per 1 head are increased by 1 000 rubles, then annual average milk productivity of cows will increase by 0.808 ct. The increase of the subsidies per 1 thous. rub. per 1 cow will increase the milk production level by 0.635 ct, and at the expense of the larger density of cows per 1 cow per 100 ha of agricultural land the annual average milk production will increase by 1.483 ct.

Thus, by changing the values of factor attributes it is possible to reach the maximum (or minimum) possible value of the resultant attribute. However, it is legitimate to change the factor attributes to decrease or increase only within the variation range or coefficient of variation of their arithmetic mean. Y1 = 11.157 + 0.818*1max + 0.808*2max + 0.635* max3 + 1.483*4max =

Y1 = 11,157 + 0.818*10,76 + 0.808*23.09 + 0.635*31.25 + 1.483*14.35 = 70.1 ct

It means that by changing the measure of influence of these factors, the average annual milk yield can be brought up to 70.2 centners per 1 cow, which is higher than the average level of the population under study.

It should be noted that the majority of agricultural enterprises do not use statistical methods of analysis, including correlation and regression analysis, due to the difficulties arising in the collection of digital data, their processing and interpretation. However, as this study shows, their use makes it possible to predict the value of certain basic indicators of financial and economic activity of the enterprise, taking into account the degree of influence of a number of factors on them. For a comprehensive assessment, forecasting and development of an action plan (roadmap) to improve the efficiency of financial and economic activity of an agricultural enterprise it is advisable to determine a number of key indicators with subsequent assessment of the degree of influence of certain factors on them. As an example, the study assessed and forecasted the productivity of farm animals.

In order to widely disseminate the practical application of this method in the management system of enterprises we recommend to form at the regional level the databases of the main production and financial indicators of agricultural producers (an option without naming them is possible, or to provide limited access to this information through the relevant regional ministries responsible for agriculture) [3].

4 Conclusion

The proposed tools can be used to forecast the results of the economic activity of agricultural organizations, to develop further action plans to improve the production and economic activity of the organization and to search for reserves to improve its efficiency.

References

1. Sh. M. Gazetdinov, M. Kh. Gazetdinov, O. S. Semicheva, F. F. Gatina, Reserves for improving the efficiency of integrated formations, *BIO Web of Conf.* **17,** 00026 (2020) doi: 10.1051/bioconf/20201700026
2. Official website of the Federal State Statistics Service. Retrieved from: https://rosstat.gov.ru/
3. N. Goloburdova, *"White oil" of Tatarstan: Khairullin Jr. is the main milkman, Mutigullin is the gray cardinal* (2021). Retrieved from: https://kam.business-gazeta.ru/article/507649
4. G. Klychova, A. Zakirova, R. Mannapova, K. Pinina, Y. Ryazanova, Assessment of the efficiency of investing activities of organizations, *E3S Web of Conferences* **110,** 02075 (2019) https://doi.org/10.1051/e3sconf/201911002075
5. G. Klychova, A. Zakirova, S. Alibekov, A. Klychova, V. Morunov, U. Raheem, Improvement of the tool of strategic management accounting, *Advances in Intelligent Systems and Computing 1258 AISC*, pp 669–686 (2021) https://doi.org/10.1007/978-3-030-57450-5_58
6. F. N. Mukhametgaliev, L. F. Sitdikova, M. M. Khismatullin, N.M. Asadullin, L.V. Mikhailova, Prospects of agricultural business in the Republic of Tatarstan, *BIO Web of Conferences* **17,** 00083 (2020) doi: 10.1051/bioconf/20201700083.
7. A. Zakirova, G. Klychova, K. Mukhamedzyanov, Z. Zakirov, A. Nigmetzyanov, A. Yusupova, Information and analytical system of strategic management of activities of enterprises, *Advances in Intelligent Systems and Computing 1258 AISC*

pp 687–707 (2021) https://doi.org/10.1007/978-3-030-57450-5_59

Effect of Mephosphonee on methane generation in organic waste processing

I Kh Gayfullin*, *B G* Ziganshin, *I N* Safiullin, *B L* Ivanov, and *R K* Khusainov

Kazan State Agrarian University, 65, Karl Marx St., Kazan, Republic of Tatarstan, 420015, Russian Federation

Abstract. The purpose of the study is to determine an effective technological method for the complete anaerobic processing of agricultural waste. It also aims to create and implement an effective technological method of sufficiently complete anaerobic processing of agricultural waste, including taking into account biologically active additives. A new technology for the disposal of organic agricultural waste and the influence of Mephosphone on the process of methane formation using a biogas unit at a processing facility developed at Kazan State Agrarian University is considered. The results of theoretical and experimental studies are presented, which made it possible to significantly reduce the utilization time, and reduce the cost of electricity and fuel. The experiments were carried out in a low-capacity biogas unit (LCBU) without air access in a heat-insulated reactor filled by 2/3. The results of numerical studies of the products of conversion of cattle manure are presented. Manure was kept in the reactor for 7 days. The use of Mephosphone in ultra-low concentrations (10-4) made it possible to obtain a useful product in the form of high-quality organic biofertilizer and biogas. The results of microbiological analysis of experiments in the test center of the Tatar Interregional Veterinary Laboratory are shown.

1 Introduction

Modern agriculture causes major damage to the environment similar to industrial enterprises. Animals consume up to 45% of the world grain harvest, emit 16% of methane generated in nature into the atmosphere, which affects the global warming more than carbon dioxide. The introduction of organic waste processing facilities will reduce methane emissions into the atmosphere. Non-controlled methane affects the greenhouse effect up to 20 times more strongly than carbon dioxide and is present in the atmosphere during 12 years. The use of new technologies for the collection of methane emitted into the atmosphere will make it possible to prevent the global warming [1].

The human impact on the ecosystem increases daily. People cultivate lands and create agricultural land. Annually, there are new species and varieties of plants in crop production and animal breeds in animal husbandry. This is a conscious human activity. Human activity makes rivers shallow, affects the development of soil erosion, extinction of plant and animal species, pollution of air, groundwater, destruction of useful insects (bees) etc. [2, 3].

The environment is polluted, especially with the development of technological progress. At the same time, poisonous substances cause great damage to nature when they enter water, soil, and air. Along with the above sources, debris, animal manure and the exhaust gases of improperly operating vehicles serve a significant source of natural pollution.

Every day there is a shortage of fossil organic fuels, the cost of their extraction and transportation increases significantly. The search for alternative, renewable energy sources and the creation of new energy-saving technologies is quite relevant [4]. One of the most important areas for renewable energy sources is the use of organic waste. Special attention is paid to the development of technologies for the production of biogas during the processing of organic waste. During the processing of all generated organic waste from fifty cattle heads it is possible to obtain more than 120 m3 of biogas, respectively 3.2 tons of biofertilizers per day [5].

Biogas is a valuable source of energy, which is especially promising in small peasant farms.

The limited reserves of fossil carbohydrate raw materials forced many European countries to intensify research in the field of alternative technologies. Currently, there are more than 500 energy complexes in Russia, including 5 large ones. One of them (Fig. 1) is located in the Republic of Tatarstan in the Aznakaevsky district, Aktyuba settlement [6].

The biogas unit is located on the basis of a livestock complex and processes cattle manure and other waste (grain waste, substandard corn silage, etc.). The generated electricity serves the own needs of the nearby farm and other facilities of the unit owner. With the current installed capacity of 0.5 MW, the unit has the prospect of expansion to 1.5 MW (provided the electric energy is supplied to the network).

* Corresponding author: ilnur-gai@yandex.ru

Fig. 1. Aktyuba biogas unit.

2 Materials and Methods

In terms of their design all biogas units are quite similar. The main element of the unit is a sealed container – a methane tank. Depending on the geometric parameters, the methane is made of glass, metal, plastic, concrete or a combination of both. The ratio of the height of the methane tank must be equal to its diameter. Methane tanks produced in Europe are usually more elongated than those made in the United States. Their filling capacity is 2/3.

The biogas unit (Fig. 2) developed at Kazan SAU consists of a reactor with a double housing, which acts as a cooling box. Hot steam or air can be supplied to the cooling box, and the necessary temperature regime of the substrate is heated and maintained in the reactor using a heating element. The reactor size ratio is rational ensuring the maximum production of biogas. The advantage of the plant is that it operates in various modes, including mesophilic and thermophilic, moreover, its operation is fully automated, thereby ensuring continuous control of biogas and fertilizer production. The biogas unit has a mechanism for substrate mixing. The mixing mechanism is simple. Mixing contributes to the activation of methane-forming bacteria, which facilitate the production of biogas. The novelty of the technical solution is protected by the patents of the Russian Federation for utility model No. 150764, No. 208815 [7, 8].

The loading of processed substrate and unloading of obtained biofertilizers from a methane tank is carried out by a pump through a branch pipe.

A certain temperature must be maintained in a methane tank to accelerate the formation and production of high-quality organic fertilizers. The basis of anaerobic fermentation of organic waste is the energy balance of technological losses and the preservation of heat in the methane tank during anaerobic fermentation [9]. There are three temperature regimes for biogas units, such as psychrophilic, mesophilic and thermophilic. In a psychrophilic temperature regime a temperature in the range of 15 to 20°C is considered rational. It is recommended to use the psychrophilic regime in summer. Since, for example, in the Republic of Tatarstan, the average annual temperature is about 3 °C, the impossibility of using this regime is obvious year-round. The mesophilic regime operates within the temperature range from 20–35 °C, and the thermophilic regime from 35–55 °C [10, 11].

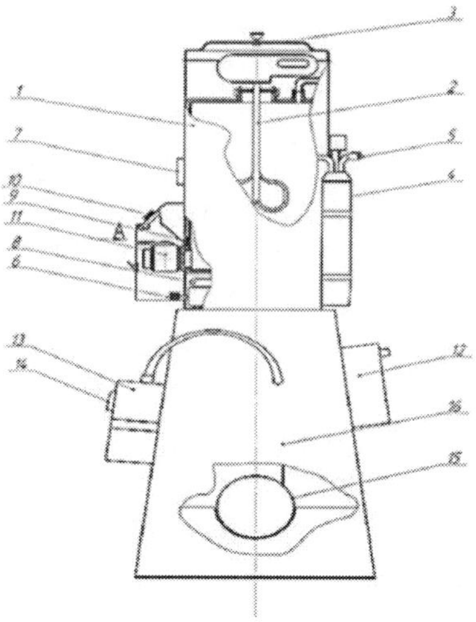

Fig. 2. Biogas unit scheme: 1 – methane tank; 2 – mixing device; 3 – methane tank cover; 4 – small gasholder; 5 – pressure gauge; 6 – filter bowl; 7 – thermometer; 8 – heating element; 9 – thermocouple; 10 – temperature controller; 11 – drainage tap of biofertilizers; 12 – settler; 13 – biogas counter, 14 – panel, 15 – low-speed compressor, 16 – housing support.

Besides, the production of biogas depends on many other factors. The amount of biogas produced from organic waste depends on the quality of the substrate and the bacterial composition. The smaller the particle sizes of the organic components of the initial substrate, the larger their specific surface area and, accordingly, the fermentation processes occur more intensively [12]. Various additives, including chemical preparations, organic additives change the composition of the starting substrate, and may function as a catalyst or moderator. There are more than a thousand strains of bacteria involved in the process of producing biogas alone. Chemical composition and physical conditions of the initial substrate also play an important role [13, 14, 15].

Table 1 shows the chemical composition of organic waste.

Table 1. Chemical composition of organic waste, %.

Composition	Cattle manure	Poultry manure	Horse manure
Carbon (C)	61	83	77
Oxygen (O)	12	9	31
Sulfur (S)	1	0.9	0.07
Hydrogen (H)	8	0.1	9
Nitrogen (N)	81	7	0.75

We studied the effect of Mephosphone on the degradation of cattle manure. The structural formula of the drug used is shown in Figure 3 [16].

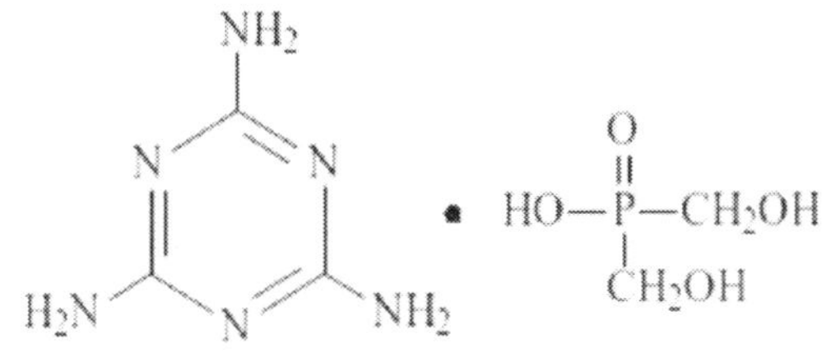

Fig. 3. Mephosphone structural formula.

When Mephosphone is added to the substrate, it decomposes into water and non-toxic nitrogen and phosphorus compounds.

To describe it in the form of a chemical reaction, it is written as follows:

$$C_{0,61}H_{0,08}O_{0,12}N_{0,18}S_{0,01} + 0.77H_2O + C_5H_{13}N_6O_4P \rightarrow 0,310\ NH_4^{+¿¿} + 0.276HCO_3^{-¿¿} + 0,495CO_2 + 0.678CH_4 + 0,011H_2S$$

Analyzing the resulting chemical reaction, we can say that 0.495 g of CO_2 and 0.678 g of CH_4 will be released from 1 g of ashless substance.

To determine the reliability of theoretical results, field experiments were carried out at the designed biogas unit. The biogas unit is shown in Figure 4.

Fig. 4. Biogas unit.

The loaded processing substrate was mixed periodically throughout the cycle. At the beginning, biogas accumulates in the methane tank and while passing through the hydraulic lock, it enters the small gasholder. From a small gas, at a certain pressure, gas is pumped through filters to a gas cylinder.

3 Results and Discussion

The influence of Mephosphone on biogas production was determined in accordance with the Technological Regulations developed by scientists of Kazan SAU [17]. Temperature and mixing frequency were monitored during the experiments [18]. The result of the experiments is shown in Figure 5.

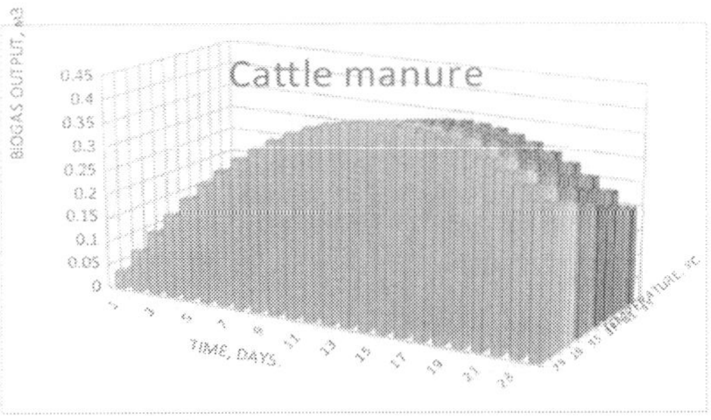

Fig. 5. Results of biogas production from cattle manure.

The diagram shows that while providing rational conditions for methane-forming bacteria, it is possible to obtain a higher output of biogas and organic fertilizers, which is very significant for a country without its own energy resources.

Various indicators were determined [19, 20] such as microbiological, agrochemical, quality, sanitary and parasitological indicators of fertilizers obtained in the test center of the Tatar Interregional Veterenary Laboratory. The results are presented in Table 2.

The composition of the manure substrate with the addition of Mephosphone.

Table 2. Results of the experiments with microbiological indicators.

Microbiological indicators			
Indicator	Measuring unit	Test result	Norm
Coliform bacteria		1	1–9
Enterococci		1	1–9
Clostridium	g	N/D	Not allowed
Pathogenic microorganisms	g	N/D	Not allowed
Agrochemical indicators			
Indicator	Measuring unit	Test result	Norm
Salt extract pH	pH	6	6–8.5
Mass fraction of organic matter	%	79	At least 70
Quality indicators			
Indicator	Measuring unit	Test result	Norm
Mass fraction of total nitrogen	%	0.5	At least 0.1
Mass fraction of potassium	%	0.51	At least 0.05
Mass fraction of total phosphorus	%	2.1	At least 0.05
Sanitary-parasitological indicators			
Indicator	Measuring unit	Test result	Norm
Cysts of enteric pathogenic protozoa microorganisms	Exs/100 g	N/D	Not allowed
Helmint eggs and larvae	pcs/kg	N/D	Not allowed
Sanitary-epidemiological indicators			
Indicator	Measuring unit	Test result	Norm

Larvae L and case K of flies		N/D	Not allowed

The microbiological analysis of the experiments showed that in the experiment with Mephosphone the content of enterococci and bacteria of the E. coli group was lower than in the experiment without Mephosphone.

In the methane tank, organic waste is processed using bacteria. Mephosphone makes it possible to increase the biomass concentration of these bacteria, as well as their vital products. The biofertilizer obtained in the biogas unit acquires new properties. The final nitrogen content of the Mephosphone treated substrate is higher than that of the untreated substrate. Acidity in untreated substrate is pH 6.7 and in a treated substrate – 7.5.

The studies showed that when Mephosphone is added to the substrate, the release of biogas begins on day 1-2 and lasts up to 7 days. In this case, the output of biogas increases to 25% compared to the use of a control substrate without the use of Mephosphone.

4 Conclusion

The above makes it possible to conclude that our country has a significant supply of renewable energy resources, with the use of which it is possible to fully satisfy the energy needs of the republic. The correct approach and rational use of existing energy resources together with the development and introduction of new technologies will allow our republic to avoid energy, environmental and economic crises. The widespread introduction of renewable energy plants will allow abandoning imported fuel and switching to fuel and energy self-sufficiency.

The disposal of organic agricultural waste using Mephosphone significantly reduced the utilization time and cost of maintaining the methane generation process.

References

1. I. Kh. Gayfullin, Z. M. Khaliullina, B. G. Ziganshin, Effect of Mephosphone on the efficiency of obtaining biogas and recycling carbon-containing waste, *Bulletin of Kazan State Agrarian University* **16-3 (63),** 19–26 doi: 10.12737/2073-0462-2021-19-26 (2021)
2. I. Kh. Gayfullin, B. G. Ziganshin, A. I. Rudakov, Problems of utilization and processing of organic agricultural waste, *Agricultural machinery 2018: VI international scientific congress* (Bulgaria, Burgas, June 25–28) pp 201–202 (Burgas: Scientific-Technical Union of Mechanical Engineering INDUSTRY 4.0, 2018)
3. I. Maksimov, N. Adigamov, A. Mustafin, D. Khaliullin, I. Gayaziev, A. Matyashin, R. Lukmanov, Theoretical fundamentals for determining soil erosion potential (energy concept): Part 1, *Quimica Periodical* **16 (31),** 540–557 (2019)
4. L. Xue, P. Zhang, H. Shu, C. Chang, R. Wang, S. Zhang, Agricultural waste *Water Environment Research* **8-8 (10),** 1334–1373 (2016)
5. B. G. Ziganshin, Yu. Kh. Shogenov, I. Kh. Gayfullin, I. I. Kashapov, *Modern Energy-Saving Solutions in Agriculture* (Kazan: Kazan State Agrarian University, 2018)
6. S. M. Naiman, Yu. A. Tunakova, Possibility of using biogas technologies for processing organic waste in Tatarstan. Production of biogas and energy, *Bulletin of Kazan Technological University* **16 (19),** 227–234 (2013)
7. F. S. Sibagatullin, Z. M. Khaliullina, R. R. Minnikhanov, Results of practical use of fertilizers from chicken manure in winter wheat cultivation, *BIO Web of Conferences* **37,** 00048 (2021)
8. F. F. Sitdikov, B. G. Ziganshin, R. R. Shaidullin, A. B. Moskvicheva, Use of modern technologies in dairy farming, *Bulletin of Kazan State Agrarian University* **15-1 (57),** 81–87 doi: 10.12737/2073-0462-2020-81-87 (2020)
9. V. Petric, Selimbasic, Composting of poultry manure and wheat strew in closed reactor: optimum mixture ratio and evplotion of parameters, *Biodegradation* **11,** 53–63 (2006)
10. Q. Zhang, J. Hu, D. J. Lee, Biogas from anaerobic digestion processes: research updates, *Renewable Energy* **98,** 108–119 (2016)
11. S. Sakar, K. Yetilmezsoy, E. Kocak, Anaerobic digestion technology in poultry and livestock waste treatment - a literature review, *Waste Management & Research* **27 (1),** 3–18 (2009)
12. D. Y. Suslov, P. S. Sedyh, Experimental studies of the process of obtaining biogas from wastes from agricultural enterprises, in *International Science and Technology Conference "EastConf"*, 8725328 (2019)
13. D. E. Belostotskiy, V. A. Miluykov, E. E. Ziganshina, A. M. Ziganshin, M. Siniagina, E. A. Boulygina, Impact of the substrate loading regime and phosphoric acid supplementation on performance of biogas reactors and microbial community dynamics during anaerobic digestion of chicken wastes, *Bioresource Technology* **193,** 42–52 (2015).
14. F. S. Sibagatullin, Z. M. Khaliullina, A. M. Petrov, K. O. Sinyashin, The prospects of using various commercial formulations to accelerate the process of "maturation" of chicken manure, *Bulletin of Kazan State Agrarian University* **14-1 (52),** 53–57 (2019)
15. R. Sabirov, A. Valiev, L. Karimova, A. Dmitriev, D. Khaliullin, Influence of physical factors on viability of microorganisms for plant protection, *Engineering for Rural Development* **18** doi: 10.22616/ERDev2019.18.N211 (2019)
16. F. S. Sibagatullin, Z. M. Khaliullina, A. R. Safiullina, Fermentation processes studying of chicken manure under the influence of a biologically active supplement "Mephosphone", *Bulletin of Kazan State Agrarian University* **2 (49),** 42–47 (2018)

17. State Duma Deputy of the Russian Federation, *Process regulations "Reducing the hazard class (neutralization) of animal waste with the use of a biologically-active preparation – Mephosphone"* (Kazan, 2017).
18. A. Sun, W. Cao, C. J. Banks, S. Heaven, R. Liu, Biogas production from undiluted chicken manure and maize silage: a study of ammonia inhibition in high solids anaerobic digestion, *Bioresour. Technol.* **218,** 1215–1223 (2016)
19. C. A. Pierce, R. J. Smernik, T. M. Mc Beath, Phosporous availability in chicken manure lower with increased stockpiling period, despite a lager orthophosphate content, *Plant and Soil* **373 (1–2),** 359–372 (2013)
20. E. E. Ziganshina, E. M. Ibragimov, P. Y. Vankov, A. M. Ziganshin, V. A. Miluykov, Comparison of anaerobic digestion strategies of nitrogen-rich substrates: performance of anaerobic reactors and microbial community diversity, *Waste Management* **59,** 160–171 (2017)

Evaluation of digital maturity for agricultural organizations based on DMPTSC analysis

Niyaz A. Safiullin*, *Chulpan M.* Kurakova, *Gulnara A.* Valeeva, and *Luiza T.* Yachina

Kazan State Agrarian University, 65, Karl Marx St., Kazan, Republic of Tatarstan, 420015, Russian Federation

Abstract. It is impossible to perceive modern agricultural production without information and communication technologies. The authors analyze the modern state of agriculture in Russia and provide data confirming that agricultural organizations are the main producers of goods. The paper describes the departmental project "Digital Agriculture", which is designed to increase the efficiency of agricultural production based on the introduction of digital technologies. The digital gap in rural areas is proved based on indicators of the use of information and communication technologies in urban and rural areas. The term "digital maturity" is defined in the aspect of digital transformation of agricultural organizations. The paper proposes the methodology to assess the digital maturity by groups of factors. A survey of the heads of large agricultural organizations was conducted according to the DMPTSC analysis, and measures were proposed to increase the efficiency of digital transformation of agricultural enterprises.

1 Introduction

Agriculture is one of the most dynamic sectors of the national economy. It provides the population with food, and processing enterprises with the necessary raw materials. One of the functions of agriculture is to ensure the country's food security. Agricultural production mainly includes crop and livestock industries, which are concentrated in three categories of farms: agricultural organizations, population farms and peasant farms.

Over 20 years agricultural production in Russia has grown significantly in all categories of farms (Figure 1).

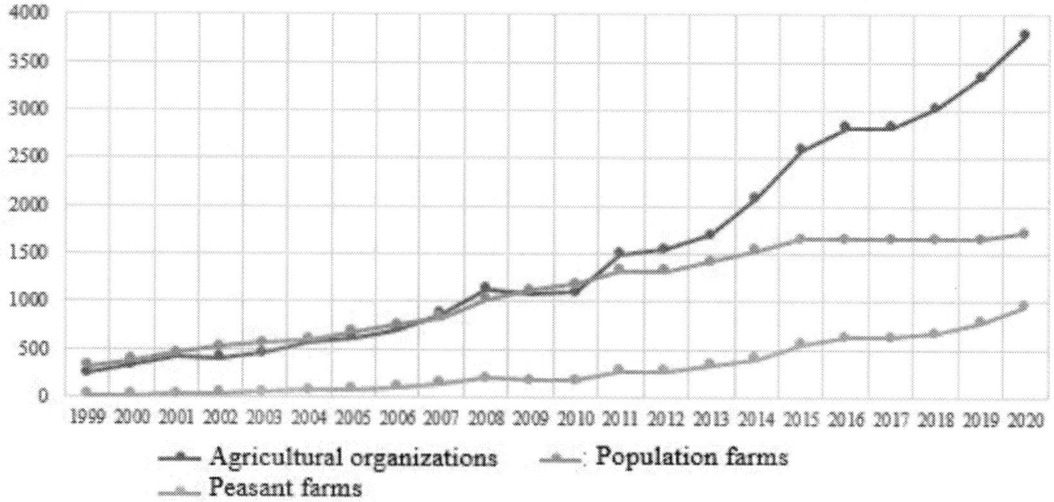

Fig. 1. Agricultural products by categories of farms in actual operating prices, bln rub.

The data shows that from 1999 to 2020 the volume of agricultural production in all categories of farms increased 11 times. Peasant farms demonstrate the highest growth rates (by 62 times), since this entrepreneurial activity appeared only in the early 90s of the 20th century and was strongly supported by the state in the late 2000s. At the same time, in 2020, 58.5% of agricultural products were produced in agricultural organizations, 26.6% – in population farms, and 14.9% – in peasant farms. Thus, agricultural organizations should become priority enterprises that should improve their

* Corresponding author: nsafiullin@outlook.com

efficiency through innovation and information technology.

The shift to a digital economy is seen as a key driver of the economic growth [1]. First of all, agricultural organizations must actively introduce modern digital technologies into their activities.

In recent years, all kinds of digital technologies have been increasingly used in agriculture abroad [2, 115]. For example, the *Food and Farm 2020* European Internet project developed in Europe ensures the compatibility of gages and sensors used by agricultural machinery. In the United States, about 70% of agricultural enterprises have digitalized their production [3, 281].

Currently, our country is implementing the *Digital Agriculture* project developed by the Ministry of Agriculture of the Russian Federation. According to the project, as a result of the industry digitalization, the investments in economic activities should increase to 5.9 billion rubles and revenue volumes – to 45 billion rubles by 2024. The departmental project includes three stages (Figure 1).

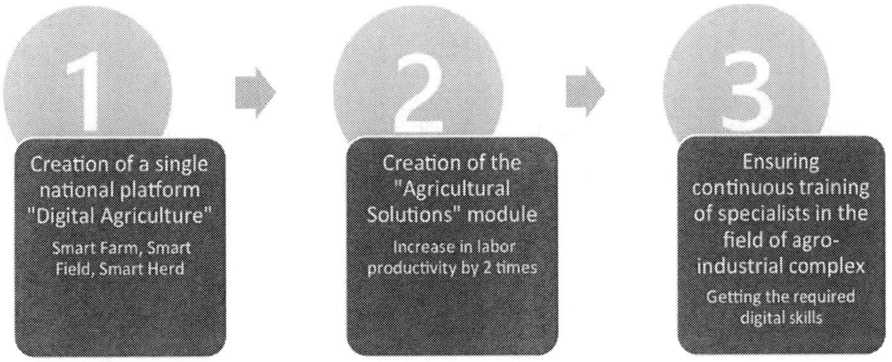

Fig. 2. "Digital Agriculture" stages.

Currently, Russia's digital agriculture is developing in five main areas indicated by the Ministry of Agriculture: effective hectare, smart contracts, agroexport "from field to port", agroindustrial solutions for agribusiness and "Land of Knowledge" [3, p. 00055]. Before starting to digitalize production activities, it is important to assess the readiness of the enterprise for digital transformation. This is a multilevel study that includes the analysis of the growth factors of an organization through the introduction of information and communication technologies. Thus, the digital transformation of agricultural organizations should begin with the assessment of digital maturity [4, 1].

Digital maturity is assessed in terms of the readiness of the organizational culture of an enterprise and its personnel for transformation, the current state of work with information and the effectiveness of production management. The assessment of digital maturity should contribute to the necessary action plans for the functioning of the enterprise in the digital economy and its transition to a fundamentally new level of development.

There is currently a so-called digital gap between urban and rural population. It is expressed in a substantial difference in access to digital technologies and in the level of digital competencies. In turn, this has an impact on economic activities in rural areas, i.e. employees of agricultural organizations have less knowledge, abilities and skills that are relevant in the digital economy (Fig. 3).

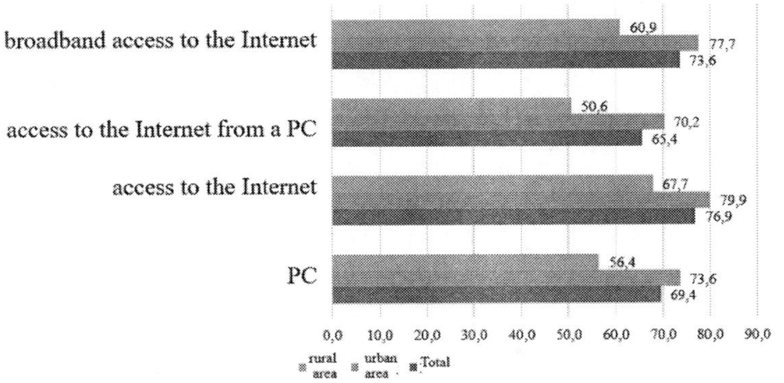

Fig. 3. Use of information and communication technologies in urban and rural areas in 2019, %.

2 Materials and methods

In January 2022, the authors conducted a survey among the heads of agricultural organizations and experts in the industry on the degree of readiness of enterprises for digital transformation. The survey was based on the DMPTSC analysis.

The DMPTSC analysis is the author's methodology that includes 6 groups of factors of digital maturity of agricultural organizations:

1. Data Utilization (D – Data)
2. Process Management (M – Management)
3. Product Management (P – Product)
4. Personnel Policy (S – Skills)
5. Use of end-to-end technologies (T – Technology)
6. Organizational Structure (C – Culture).

The analysis establishes the degree of influence of each factor from 1 to 3 points, where 1 – minimum degree of influence and 3 – maximum degree. Then, the heads of organizations and experts had to evaluate four indicators of digital maturity in each factor on a five-point scale (Fig. 3). After the survey, the author calculated a real estimate of the influence of factors using the following formula:

$$D_i = \frac{\overline{\text{Э}} \times I_i}{\sum I}$$

where D_i – real estimate of the impact of the digital maturity indicator,

$\overline{\text{Э}}$ – average value of expert assessments of the digital maturity indicator, points

I_i – degree of influence of the indicator on the efficiency of digital transformation, points.

Data Utilization (D – Data)

Assessment of data inventory quality	Assessment of data collection, analysis and visualization	Assessment of the efficiency of decision-making based on data	Assessment of data security

Use of end-to-end technologies (T – Technology)

Assessment of product management efficiency	Assessment of the degree of introduction of the customer-centric approach	Assessment of the level of design thinking	Assessment of the efficiency of the product management flexible approach

Process Management (M – Management)

Assessment of the efficiency of current process management	Assessment of the process inventory quality	Assessment of the degree of document development regulating the process	Assessment of process control quality

Product Management (P – Product)

Assessment of the degree of end-to-end technologies use	Assessment of employee competencies when utilizing end-to-end technologies	Assessment of regulatory support of the end-to-end technologies	Assessment of the level of customer satisfaction with services based on the end-to-end technologies

Personnel Policy (S – Skills)

Assessment of the degree of baseline competencies development	Assessment of the degree of personal competencies development	Assessment of the degree of professional competencies development	Assessment of the efficiency of digital transformation managers

Organizational Structure (C – Culture)

Assessment of the current state of the organizational culture	Assessment of the degree of staff opposition to organizational changes	Assessment of the quality of regular management	Assessment of the feedback degree

Fig. 4. Indicators of the DMPTSC analysis of digital maturity of agricultural organizations.

Specialists of the AIC Food Program and teachers of Kazan SAU participated in the digital maturity survey. The agricultural organization AIC Food Program is engaged in the breeding of dairy cattle and the production of raw milk. At the beginning of 2021, the organization employed 701 people. Over the past five years, investments in fixed assets in the group of companies amounted to more than 5 billion rubles. The group's enterprises are actively introducing digital technologies. So, for example, the "Herd Management" program reduced the service period from 145 to 109 days, reduced the disease rate from 12 to 2, and increased the pregnancy ratio from 18% to 28% by year. After the introduction of herd management and feeding systems, we were able to increase the milk yield from 26.5 to 34 kilograms of milk per dairy cow [5].

Table 1. Results of the DMPTSC analysis of the AIC Food Program.

	Degree of impact	Expert 1	Expert 2	Expert 3	Expert 4	Real estimate
Data Utilization (D – Data)						
Assessment of data inventory quality	3	5	4	5	3	**0.18**
Assessment of data collection, analysis and visualization	2	4	5	5	4	**0.16**
Assessment of the efficiency of decision-making based on data	3	4	5	5	3	**0.18**
Assessment of data security	2	4	5	5	4	**0.16**
Process Management (M – Management)						
Assessment of the efficiency of current process management	2	4	4	5	5	**0.20**
Assessment of the process inventory quality	2	5	5	5	3	**0.12**
Assessment of the degree of document development regulating the process	1	5	5	5	3	**0.06**
Assessment of process control quality	2	4	5	5	4	**0.16**
Use of end-to-end technologies (T – Technology)						
Assessment of product management efficiency	2	4	4	5	4	**0.16**
Assessment of the degree of introduction of the customer-centric approach	2	4	4	5	3	**0.12**
Assessment of the level of design thinking	1	4	4	5	4	**0.08**
Assessment of the efficiency of the product management flexible approach	2	4	4	5	5	**0.20**
Product Management (P – Product)						
Assessment of the degree of end-to-end technologies use	3	4	4	5	4	**0.24**
Assessment of employee competencies when utilizing end-to-end technologies	3	4	4	5	4	**0.24**
Assessment of regulatory support of the end-to-end technologies	2	4	4	5	5	**0.20**
Assessment of the level of customer satisfaction with services based on the end-to-end technologies	2	5	4	5	3	**0.12**
Digital Competencies (S – Skills)						
Assessment of the degree of baseline competencies development	2	4	5	5	3	**0.12**
Assessment of the degree of personal competencies development	1	5	4	5	3	**0.06**
Assessment of the degree of professional competencies development	3	4	4	5	4	**0.24**
Assessment of the efficiency of digital transformation managers	2	4	4	5	3	**0.12**
Organizational Structure (C – Culture)						
Assessment of the current state of the organizational culture	2	5	4	5	4	**0.16**
Assessment of the degree of staff opposition to organizational changes	2	5	4	5	5	**0.20**
Assessment of the quality of regular management	3	4	4	5	3	**0.18**
Assessment of the feedback degree	2	4	4	5	4	**0.16**

3 Results and Discussion

The data shows that people responsible for the digital transformation in AIC Food Program should pay special attention to the following aspects during its implementation:

1. Data collection, analysis and visualization. The most important at the initial stage of digitalization of any organization is the high-quality receipt and processing of primary data from various sources. In this case, it becomes important to establish contacts with stakeholders, the feedback of which serves valuable information for the agricultural organization. Stakeholders include consumers, suppliers, competitors and employees. Moreover, the more data, the more difficult it is to highlight the main aspects [6, p. 10]. Therefore, the leaders of transformation must have the skills to visualize them.
2. The development of documents regulating digital transformation is a complex and capacious stage of digitalization. It must take into account the interests of all concerned. It should contain indicators of the efficiency of digital transformations and algorithms for their implementation based on the peculiarities of agricultural production in the Agro-Industrial Complex Food Program.

3. The rate of production intensification and the growth of labor productivity depend on the level of development of the technical potential of agricultural production [7, p. 48]. The principles of design thinking when using end-to-end technologies are poorly developed in the studied organization. It is possible to significantly increase the competitiveness and attractiveness of products using this methodology to solve technical and business problems.
4. The assessment of customer satisfaction when receiving end-to-end services is the job of the marketing department. Such feedback can be obtained when delivering or transferring marketable products, as a result of monitoring social networks, or when directly communicating with the customers of an organization. The obtained data make it possible to adjust the quality characteristics of the produced goods.
5. The digital competencies of employees are the most important factor in the efficiency of digital transformations. The transition of the agricultural sector to an innovative development model is largely determined by the readiness of personnel to perceive and introduce modern technologies into agricultural production [7, p. 148]. The employees of the AIC Food Program must have not only basic and professional skills, but also develop personal ones. Soft skills are personality skills that can help employees work with other people and develop relationships with customers and partners.
6. Without changing the current state of organizational culture in the studied group of companies, the efficiency of digital transformation can decrease sharply. When determining the degree of development of the organizational culture, it is necessary to adhere to the principles of spiral dynamics. According to the authors, a culture of power with its mobilization nature prevails in the AIC Food Program. In a digital economy, when there is a continuous change in the external and internal environment of an organization, there must be a synthesis culture in which there is no clear hierarchy, and employees work in project teams created to address a specific task.

4 Conclusion

Information and telecommunication technologies are not only an integral part of the daily life of a modern person, but also a necessary technological platform for modern business processes [4, p. 00050]. The introduction of these technologies in the agricultural organization requires the consideration of all factors of the macro- and micro-environment. The DMPTSC digital maturity analysis proposed by the authors takes into account almost all aspects of digitalization and is the initial stage that evaluates the degree of readiness of the organization for deep digital transformations.

The conducted digital maturity study of the AIC Food Program based on the DMPTSC analysis reflects the current state of the organization and identifies the main areas that the digitalization specialists need to pay attention to during the digital transformation.

It is necessary to increase production volumes not by all means, but only by those that are economically feasible [10]. This methodology can be used to evaluate large agricultural organizations that are starting to introduce digital tools into their production and whose managers want to change the management principles using information and communication technologies.

References

1. G. D. Krupina et al., BIO Web of Conf. **17,** 00230 (2020)
2. Ch. M. Kurakova, N.A. Safiullin, Bulletin of Kazan State Agrarian University **15 (3 (59),** 114–120 (2020)
3. E. G. Erlygina, A. D. Vasilyeva, Bulletin of Science and Practice **6 (12)** (2020).
4. E. F. Amirova, M. G. Kuznetsov, E. G. Khakimova, A. V. Tolmacheva, BIO Web of Conf. **27,** 00055 (2020) doi: 10.1051/bioconf/20202700055.
5. N. A. Safiullin et al., BIO Web of Conf. **37,** 00160 (2021)
6. R. Minnikhanov, *Business Online,* retrieved from: https://kam.business-gazeta.ru/article/498804
7. K. N. Nafdiq, *Data: visualize, tell, use. Storytelling in analytics* (Moscow: Mann, Ivanov, and Ferber, 2020)
8. N. R. Alexandrova, A. K. Subaeva, L. M. Mavlieva, N. L. Titov, Bulletin of Kazan State Agrarian University **15 (1 (57),** 94–98 (2020) doi: 10.12737/2073-0462-2020-94-98
9. A. K. Subaeva, L. M. Mavlieva, Bulletin of Kazan State Agrarian University **13 (3 (50),** 147–150, (2018) doi: 10.12737/article_5bcf5799dbe707.03578820
10. A. K. Subaeva, *Bulletin of Moscow State Agro-*Engineering University named after V.P. Goryachkin **3 (85),** 47–53 (2018) doi: 10.26897/1728-7936-2018-3-47-53
11. E. F. Amirova, O. V. Kirillova, M. G. Kuznetsov, Sh. M. Gazetdinov BIO Web of Conf. **17,** 00050 (2020)
12. M. M. Khismatullin, M. M. Khismatullin, L. T. Vafina, F. N. Safiollin, *IOP Conference Series: Earth and Environmental Science* **341,** 012109 (2020) doi: 10.1088/1755-1315/341/1/012109

Effect of the Bacillus subtilis-based drug on the morphobiochemical and productive parameters of young goats

Galina Molyanova[1] and *Oksana* Semkina[2]

[1] Samara State Agrarian University, 2 Uchebnaya St., Kinel, Samara region, 446442, Russian Federation
[2] IP "Tsirulev E. P.", Head of the Peasant (Farm) Economy, Samara region, 445560, Russian Federation

Abstract. The effect of Extrasol based on Bacillus subtilis (Bisolby-Inter LLC) on the physiological, biochemical and productive parameters of young goats of the Zaanen breed was studied. The research and production experience were carried out on the 20 Zaanen goats on the Peasant Farm "Semkina O.V." in the Samara Region. The biological effect of the drug Extrasol is due to adsorption properties, the ability to enhance the activity of enzyme systems and increase intestinal and general immunity of the body. The use of the drug increases the intensity of anabolic processes in the body of animals: the amount of total protein in the blood serum of goats of the experimental group at 120 and daily age was higher than 9.7% (r≤0,05), albumins - by 33% (r≤0,001), compared with control animals. The blood glucose content of 120-day-old goats of the experimental group was 11.2% higher (r≤0,05) relative to control data.

1 Introduction

In recent years, dairy and meat goat breeding has become widespread in the Russian Federation. According to the research of domestic and foreign authors, dairy goat breeding is one of the most promising and profitable areas of animal husbandry. Balanced feeding of goats of dairy productivity makes it possible to fully realize the genetic potential inherent in their breed to obtain high-quality dairy products. Scientists have proved that the use of probiotics and other biologically active substances in addition to the basic diet leads to an increase in metabolism in the body of farm animals, an increase in nutrient absorption and a reduction in feed costs per unit of live weight gain [1-4].

Researchers Dementieva E.P. and Lobozhina Z.V. note in their scientific work that the use of probiotics in animal husbandry becomes an integral part of this process, since the growth and health of animals depend on the work of the intestines, the synthesis of vitamins and amino acids, the digestion of food and the assimilation of all necessary vitamins and minerals from feed, the utilization of metabolic products [5].

Plotnikova E.Y. in a scientific paper writes that among the most important biochemical properties inherent in Bacillus subtilis, it is necessary to highlight the ability of acidification of the medium, as well as the production of antibiotics. It is thanks to these properties that the hay wand from the genus of bacilli is able to reduce the impact of various conditionally pathogenic, as well as pathogenic microorganisms. Bacillus subtilis is an antagonist for yeast fungi, salmonella, proteus, streptococci, staphylococci. Other important properties of Bacillus subtilis include: synthesis of vitamins, amino acids and immunoactive factors; active production of enzymes capable of removing products of putrefactive decay of tissues [6].

Author Volodkina G. M. found that the use of the probiotic "Vetom 1.1", the active component of which is Bacillus subtilis, has a hematopoietic and immunostimulating effect, positively affects the growth rate of young cattle and reduces feed costs per unit of production [7].

Scientists Chelnokov V.A. and Sein O.B. studied the effect of the probiotic drug Vetom 1.1 on the physiological status and meat productivity of red-mottled bulls. Submitted to the experiment in experimental animals would have a higher hematocrit index (40.0±3.7%), the content of erythrocytes ($7.2 \pm 0{,}11 \cdot 10^{12/l}$) and hemoglobin at (120.0±6.4 g/l). At the same time, the differences in the content of erythrocytes were statistically significant (p<0.05-0.01). By the end of fattening, the bulls of the experimental group exceeded the animals of the control group by an average of 33.0 kg in slaughter weight. Carcasses of bulls taking the drug Vetom 1.1 differed in the content of the most valuable (69%) nutritionally components (muscles, fat), and the proportion of inedible parts (tendons, fascia, ligaments, bones) was less (31%) than in control animals (63% and 37%) [8].

Scientists Egorova T.A. and co-authors studied the effect of the drug Cellobacterin-T (contains a live culture of Bacillus subtilis bacteria and filler) on the productivity of broilers. Cellobacterin-T has a high probiotic activity, having a positive effect on the composition of the bacterial community in the blind

 https://doi.org/10.1051/bioconf/20225200021

processes of the intestine in broiler chickens. As a result of feeding the drug, there was a significant increase in the number of normoflora, as well as a decrease in the number of microorganisms traditionally associated with intestinal dysbiosis in humans and animals. In the group that consumed Cellobacterin-T, chickens were characterized by the greatest increase in live weight, accelerated growth; in addition, they had higher digestibility of feed nutrients and accumulation of vitamins A, E, B_2 and carotenoids in the liver [9].

Skvortsov E.V. and colleagues studied the effect of the probiotic drug Bacillus subtilis emulsified in palmitic acid on the physiological state and growth rate of laboratory rats. The results showed the presence of a positive effect of the use of emulsified bacteria on their survival during the passage of the gastrointestinal tract, that is, emulsified bacteria are resistant to low gastric pH and high concentrations of bile in the duodenum. An increase in the number of bacteria B. subtilis was accompanied by a decrease in the number of Bifidobacterium sp., Lactobacillus sp. and E. coli in the intestine. The ability of B. subtilis bacteria to have a positive effect on the body weight gain of rats has been shown. This means that emulsified Bacillus subtilis preparations show signs of effective probiotics and can be recommended for use in animal husbandry [10].

Afanasyev E.N. and colleagues have developed a probiotic drug for farm animals and poultry, which contains a mixture of suspensions of bacterial strains and a protective environment. After taking the drug daily for 18 days in the experimental group of calves, the total weight gain (by 25 heads) per day was 15 kg (0.6 kg per 1 head), and 270 kg per 18 days. The difference in the daily weight gain of calves for the experimental and control groups was 3.7 kg, and for the entire time of the experiment 67.5 kg. In the experimental group of piglets receiving this biopreparation, the total weight gain (for 20 heads) per day was 5.5 kg (0.275 kg per head), and for 18 days 99 kg. The difference in the daily weight gain of piglets for the experimental and control groups was 3.1 kg, and for the entire time of the experiment 55.8 kg. The use of the drug increased the total weight gain of poultry by 25% and reduced the incidence from 10% to 4% compared to the control [11].

In connection with the above, the use of modern domestic biologically active substances that improve the physiological and productive indicators of farm animals is an urgent research topic.

The purpose of the research is to study the effect of the drug Extrasol based on Bacillus subtilis H-13 1.5×10^8 (Bisolbi-Inter LLC) on the physiological, biochemical and productive indicators of young goats of the Zaanen breed.

2 Material and methods

The scientific work was conducted at the Department of "Epidemiology, pathology and pharmacology", test research laboratory, Department of biotechnology and veterinary medicine FSBEI HE "Samara State Agrarian University" and the dairy farm of KFX "Semkina O.V." of Privolzhsky district, Samara region.

The research and production experience was conducted on goats of 60 days of age, selected according to the principle of analogues, taking into account the breed, age, sex, body weight, physiological and clinical condition and physiological maturity. The animals in the control group received the basic diet (BD), balanced on the major indicators of nutritional quality in accordance with the rules of the RAAS. Animals of the experimental group BD with the addition of a solution of Extrasol based on Bacillus subtilis 5 ml per head once a day for 15-20 minutes before feeding. Solutions were administered to goats through a drencher in compliance with the rules of asepsis and antiseptics. The composition includes Bacillus subtilis, strain H-13 and metabolites obtained during the cultivation of the strain in a concentration of at least 100 million mg. CFU / ml on a mineral-silica powder filler. The drug has a positive expert opinion on the toxicological and hygienic assessment of the strain of the drug Extrasol based on Bacillus subtilis H-13 1.5×10^8 (LLC "Bisolbi-Inter") from 30.03.2010 from the Research Center for Toxicology and Hygienic regulation of Biological products (FSUN SIC TBP). The biological effect of the drug is provided by its high adsorption properties and surface activity. That allows it to adsorb a wide range of mycotoxins, pesticides, toxic metals, radionuclides contained in feed and simultaneously inhibit the development of pathogenic and conditionally pathogenic microorganisms, creating favorable conditions for the development of lacto- and bifidobacteria in the digestive tract. This generally reduces the toxic load on the body and simultaneously increases the activity of a number of enzyme systems of the animal body [12].

3 Results and discussion

Zoogenic indicators in the livestock premises of KFH "Semkina O.V." comply with the standards for the maintenance of small cattle. The microclimate in the room was characterized by the following indicators: the average air temperature was 18.60±0,30 ^{0}C, the relative humidity was 73.80±1.50%.

The chemical composition and nutritional content of the feed used in the farm for feeding goats corresponds to all-Russian data (Table 1). Felucene - salt for small cattle (MRS) was added freely to the feeders.

General physiological indicators: body temperature, pulse rate, respiration in goats from 60 to 120 days of age varied evenly and corresponded to their calendar day of development. The body temperature of the animals ranged from 39.2 to 41.250 C; pulse rate from 94 to 99 beats per minute; respiratory rate from 18 to 21 respiratory movements per minute.

Table 1. Structure of the main diet of goats, %.

stern	moisture	protein	fat	cellulose	BEV	ashes	dry matter	Nutritiousness, g	
								feed units	digestible protein
Grain-and-grass hay	17.0	8.4	0.9	23.4	41.1	10.0	83.0	0.46	41
grassy grass	64.8	4.2	1.3	10.8	18.9	1.53	35.2	0.21	15
Compound feed Megamix for calves	12.0	18.3	4.5	4.7	54.8	9.5	88.0	1.3	116

The effect of the drug Extrasol based on Bacillus subtilis H-13 1.5×10^8 on the biochemical parameters of young goats of the Zaanen breed is shown in Table 2.

Table 2. Clinical and physiological parameters of goats' blood serum.

Indicators	Group	
	control	experienced
60 days		
Red blood cells, $10^{12/l}$	7.32±0.12	7.56±0.25
White blood cells, $10^{9/l}$	10.72±0.31	10.35±0.28
Hemoglobin, %	125.3±5.01	122.9±4.36
Total protein, g/l	64.18±2.12	66.86±2.23
Albumins, g/l	24.34±0.97	25.26±0.76
Globulins, g/l	39.84±1.45	40.60±1.32
Glucose, mmol/l	5.41±0.26	5.68±0.17
120 days		
Red blood cells, $10^{12/l}$	7.31±0.12	7.82±0.19*
White blood cells, $10^{9/l}$	12.54±0.36	12.10±0.32
Hemoglobin, %	126.4±5.15	130.6±5.01
Total protein, g/l	64.28±2.01	70.54±2.16*
Albumins, g/l	25.71±0.91	31.74±1.25***
Globulins, g/l	38.57±1.21	38.80±1.15
Glucose, mmol/l	5.43±0.20	6.04±0.19*

Here and further note: * - $p\leq0.05$, ** – $p\leq0.01$, *** – $p\leq0,001$ – relative to the control data.

All the studied indicators were within the physiological norm. The content of erythrocytes in the peripheral blood of experimental animals at the age of 120 days after 60 days of daily administration of the drug was 6.8% higher ($p\leq0.05$) compared with similar data of control animals. The main function of hemoglobin is the transfer of oxygen to tissues. In the goats who took the drug Extrasol in addition to the main diet, there was an increase in hemoglobin compared to the control by 3.3%, which indicates a higher intensity of metabolic processes in the body of the goats.

The state of protein metabolism is determined by the total protein content. In the body, protein performs the following functions: participates in blood clotting, maintains blood pH constancy, performs a transport function, participates in immune reactions and many other functions. The amount of total protein in the blood serum of the experimental group of goats at 120 days of age was 9.7% higher ($p\leq0.05$), albumins by 23% ($p\leq0.001$) compared with the data of control animals. The use of the drug Extrasol increases the intensity of anabolic processes in the body of animals, as evidenced by a higher content of total protein and specifically its albumin fraction.

Glucose supports the normal functioning of individual cells, organs and the body as a whole. The results of the studies showed that the glucose content in the blood of 120-day-old goats of the experimental group was 11.2% higher ($p\leq0.05$) and was 5.68±0.31 mmol/l relative to the data of the control group and was 6.04±0.19 mmol/l. The use of the drug Extrasol to goats for 2 months contributed to an increase in the intensity of carbohydrate-lipid metabolism.

Table 3. Dynamics of growth and weight indicators of calves.

Indicators,	Group	
	control	experienced
1	2	3
60 days		
Body weight, kg	11,64±1,16	11,82±1,23
Body temperature, °C	40,23±1,02	40,13±0,92
Heart rate, beat / min	98,56±3,41	96,55±3,37
The respiratory rate and respiratory movements /min	19,06±1,21	19,34±1,01
120 days		
Body weight, kg	21,32±1,79	23,25±2,58*
Average daily gain, kg	0,161±0,02	0,190±0,04*
Body temperature, °C	40,02±1,25	39,9±1,22

Heart rate, beat / min	82,12±1,23	83,41±1,24
The respiratory rate and respiratory movements /min	16,32±1,23	17,36±1,14

All clinical and physiological parameters corresponded to the norm, but in the experimental group during the experiment, the indicators were higher, which indicates metabolic processes in the body of goats that are more intensive. The body weight of the goats in the control group at 120 days of age was 21.32±1.79 kg, in the experimental group - 23.25±2.58 kg, which was significantly higher by 1.93 kg relative to control. The average daily increase in the experimental group was significantly higher by 0.029 kg. The administration of the drug Extrasol based on Bacillus subtilis to goats increased the weight indicators of the animals.

Calculations of economic efficiency were carried out taking into account the costs of meat production and the proceeds from its sale at a price of 117 rubles per kg according to the website https://rosstat.gov.ru/storage/mediabank/cena_sx .xlsx, as well as the contractual cost of the drug based on Bacillus subtilis 350 rubles per 1 liter. The norm of feeding according to the experimental scheme is 0.5% of the preparation based on Bacillus subtilis from the dry substance OR 5 ml in liquid form, taking into account the recalculation of the active substance. The scientific experiment was carried out for 60 days. For one animal during the period of the scientific experiment, 300 ml of the drug was spent on preventive measures, which corresponds to the amount of 105 rubles, as well as 50 rubles for additional maintenance of the animal. The economic effect of the additional profit received at the end of the research and production experience was calculated on the basis of the data obtained. The use of the drug Extrasol contributed to an increase in the live weight of the animal, the possible profit from the sale of meat of young goats of the experimental group was higher by 70.81 rubles from each head.

4 Conclusion

The administration of the drug Extrasol based on Bacillus subtilis H-13 (Bisolbi-Inter LLC) to the young goats of the Zaanen breed in a dose of 5 ml per head daily from the age of 60 days allowed one to optimize the intensity of metabolic processes in the body of animals and increase the average daily gain. Based on the conducted studies, it is possible to recommend the drug Extrasol as a biologically active additive for young goats in order to increase the productive performance of animals and to obtain conditionally additional profit of 70.81 rubles per head.

References

1. A. S. Muhammadieva, R. S. Muhammadiev, R. S. Muhammadiev, L. R. Valiullin, Isolation and study of morphological and biochemical properties of new strains of lactic acid bacteria promising for creation of probiotic preparations, *Veterinarian* **3**, 39–46 (2020)
2. D. G. Poghosyan, R. S. Hajimusaev, Physiological features of young sheep with early intensive fattening, *Problems of the biology of productive animals* **3**, 77–86 (2017)
3. G. Molyanova, M. Nogotkov, N. Chigina, Effect of the Bacillus subtilis-based drug on the morphobiochemical and productive parameters of calves, *E3S Web of Conferences* **273**, 02011 (2021) https://doi.org/10.1051/e3sconf/202127302011
4. L. S. Samolhina, V. I. Ganina, I. I. Ionova, G. S. Komolova, M. A. Golovin, Bifidogenic properties of lactoferrin peptides of cow milk *Problems of Biological, Medical and Chemistry* **2**, 43–47 (2016)
5. E. P. Dementiev, J. V. Lobozhina, E. V. Tsepeleva, Influence of aeroionization and probiotic "lactobacterin" on calf microclimate, hematological indicators and calf growth intensity, *International Journal of Experimental Education* **11 (1)**, 113–155 (2015)
6. E. Y. Plotnikova, Effects of active metabolites of Bacillus subtilis in a new generation probiotic product. RMW, *Medical Review* **26 (3)**, 39–44 (2018)
7. G. M. Volodkina, I. S. Kurov, Effectiveness of probiotic preparations based on bacillus subtilis in raising young cattle, *Agri-food sector of the country's economy in the conditions of globalization and integration: materials of the international scientific and practical conference*, pp 72–76 (2016)
8. V. A. Chelnokov, O. B. Sein, M. V. Besedin, Physiological status and meat productivity in gobies when feeding probiotic drugs, *Bulletin of Kursk State Agricultural Academy* **7**, 69–70 (2020)
9. T. A. Egorova, T. N. Lenkova, L. A. Ilyina, Effects of probiotics based on Sacharomyces sp. and Bacillus subtilis on the bacterial community of blind intestinal processes and the productivity of broiler chickens, *Agricultural Biology* **51 (6)**, 891–902 (2016)
10. E. V. Skvortsov, R. S. Muhammadiev, R. S Muhammadiev, Preparation of the emulsified probiotic preparation Bacillus subtilis and its effect on laboratory rat intestinal microflora, *Scientific notes of Kazan State Academy of Veterinary Medicine named after N.E. Bauman* **241 (1)**, 187–192 (2020)
11. E. N. Afanasyev, I. S. Tyumentseva, N. E. Afanasyev et al., Patent No. 2326692, Russian Federation, *Probiotic drug for farm animals and poultry* (20.06.2008)

https://doi.org/10.1051/bioconf/20225200021

12. V. E. Ulitko, L. A. Pykhtin, Yu. V. Semenov et al., Reservation and use of nutrients by sows when enriching their diets with the pre-probiotic additive Bisolbi, *Agrarian science and education at the modern stage of development: experience, problems and ways to solve them: scient-prac. conf.* pp 99–104 (Ulyanovsk, 2017)

Anti-erosion reclamation and its economic efficiency

Guzaliya Klychova[1,*], *Alsou* Zakirova[1], *Marsel* Khismatullin[1], *Mars* Khismatullin[1], *Ullah* Raheem[1], and *Anatoly* Osipov[2]

[1] Kazan State Agrarian University, 65, Karl Marx St., Kazan, Republic of Tatarstan, 420015, Russian Federation
[2] Izhevsk State Agricultural Academy, 11, Studencheskaya St., Izhevsk, Udmurt republic, 426069, Russian Federation

Abstract. One of the most acute problems of agriculture in some regions of the Russian Federation, including the Republic of Tatarstan at present is a progressive degradation of the soil cover. Every year the scale of water erosion increases, causing huge damage not only to agriculture but also to large and small rivers of the region. In this regard, there is a need to analyze the impact of soil erosion on the efficiency of agricultural production and the development of science-based anti-erosion measures to combat soil erosion in agriculture of the Republic of Tatarstan. The results of the carried out research testify that ecological efficiency of one anti-erosion construction in complex with soil protection measures protects on the average 120 ha of lands from further development of erosion processes, economically pays back by the results of two years of economic activity on the protected lands. The practical significance of the research results lies in the fact that they can not only contribute to the development of the agro-industrial complex and other sectors of the economy, provide an increase in soil fertility, a dynamic increase in agricultural production, reduction of retirement of agricultural land and the creation of new jobs. They also ensure the conservation and restoration of water resources.

1 Introduction

The efficiency of the agro-industrial complex depends on not only the entrepreneurial activity and the ability of the subjects that manage the land, but also on the state of the factors of agricultural production. Land was, is, and will be the main wealth of any country in the world, so the creation of an effective, economical land use by its importance and relevance should be one of the leading places in the political and socio-economic life of the Russian Federation, including the Republic of Tatarstan [1, 2].

Unfortunately, the productivity and the ability of soils to self-repair in recent years have decreased significantly. Alienation of nutrients with crops, degumification and erosion of soils, pollution and littering of lands, increasing anthropogenic load, high plowing of agricultural lands, lead to the destruction of natural landscapes and ecosystems [3–6].

Timely erosion control measures increase erosion resistance and soil absorption capacity, evenly moisten the soil, reduce the volume and intensity of runoff and soil washout on sloping lands, prevent the concentration of runoff on arable land, contribute to safe discharge of excessive volumes of melt and rainwater, reduce the process of soil washout to environmentally acceptable values [7–12].

Counter-erosion measures are accepted to be divided into the following groups depending on the nature of the impact and purpose of agronomic techniques:

1. The first group includes agrotechnical methods, which improve water and physical properties of soil, increase water permeability of soil, deepen arable layer, create a fertile soil layer on natural undisturbed areas;

2. The second group includes agrotechnical methods, which promote surface water retention. These methods include transversal and contour plowing of furrow, trenching, intermittent furrowing, diking, micro-limans;

The third group includes agrotechnical methods, which provide a high level of anti-erosion stability of soil. These methods assume surface tillage, flat-tillage, mulching of the surface soil layer.

4. The fourth group includes agrotechnical methods, which regulate snow deposition and snow melting. These methods imply snow retention, strip exposing of

* Corresponding author: kgaukgs@mail.ru

soil or blackening of snow with ash, peat, cavity compaction of snow, plowing of snow.

The carried out erosion control measures must be systematic and take into account the specific natural and economic conditions of each economic entity [13–15]. In order to ensure that erosion control measures are carried out effectively, it is necessary to:

- properly organize the territory for efficient and rational implementation of erosion control measures [16, 17];
- acquire erosion control equipment to organize the daily protection of soils and increase their fertility [18, 19];
- conduct forest reclamation activities [20];
- organize the work of hydraulic engineering structures to prevent soil erosion. With the help of anti-erosion hydraulic structures, critical masses of surface runoff are regulated in case the intensity of rain exceeds the absorption capacity of the soil [21, 22].

Counter-erosion and cultural technical works in combination with agrochemical measures, irrigation and drainage melioration, are one of the important factors in ensuring high efficiency of agricultural production, as well as the reproduction of soil fertility [23, 24].

In order to increase the economic efficiency of land reclamation, one should:

- increase the level of meliorative land use in agricultural production;
- ensure targeted financing of land reclamation works;
- strengthen the ecologization of land reclamation activities [25–28].

The effective implementation of these measures will ensure the growth of agricultural production and increase the sustainability of agricultural industries.

2 Materials and methods

The conducted research is built on the theoretical and methodological basis, allowing using possibilities of interaction and complementarity of fundamental provisions and concepts of reclamation science. The methodological basis of the research is the system analysis, within which the following approaches are applied: constructive, deterministic, retrospective, statistical. During the study, the following methods were used: monographic, abstract-logical, computational-constructive, economic-statistical methods.

The information base of the study was the official data of the Federal State Statistics Service, the Ministry of Agriculture of the Russian Federation, the Ministry of Agriculture and Food of the Republic of Tatarstan, the Federal State Institution "Management of Tatmeliovodkhoz", planning and reporting documents of agricultural organizations, data obtained in the course of the author's analysis and calculations.

3 Results and discussion

Under the influence of natural-climatic and anthropogenic factors, the scale of water erosion increases every year, causing huge damage not only to agriculture, but also to large and small rivers (ecology) of the region. The territory of the Republic of Tatarstan is characterized by a large partitioning, which is the basis of soil erosion. Such factor as high degree of ploughed farmland (76,6%) with low afforestation of arable land (3,3%, with optimal values of 4,7-7%) contributes to soil erosion development.

Studies of the dynamics of changes in the qualitative indicators of land have revealed a clear trend of declining soil fertility and deterioration of the overall environmental situation in the agro-industrial complex of the Republic of Tatarstan.

A scale of water erosion, progressive degradation of soil cover and damage caused by erosion processes for agriculture of the republic has reached impressive proportions. So in the Republic of Tatarstan there are more than 21 thousand active ravines, their total length exceeds 29 thousand kilometers. As a result of erosion of fertile soil layer more than one thousand hectares of agricultural land is annually withdrawn from the turnover.

In comparison with 1960 only because of soil erosion the area of arable land has decreased by 247.5 thousand hectares, which is equal to the territory of three districts of the Republic of Tatarstan.

The area of agricultural lands subject to water erosion has increased by 925 thousand ha, the number of active gully tops – by 7450 pieces, they occupy an area of 114 thousand hectares.

The area of arable land, which requires anti-erosion measures, has more than doubled in the last 40 years and reached 2 million hectares. Melt water and storm water washes away on average 22 tons of fertile soil per hectare per year. Losses of humus for this reason in the country as a whole amount to 700 thousand tons annually, which is equivalent to the introduction of 10 million tons of organic fertilizers. In other words, almost half of the applied mineral fertilizers are spent to cover losses from washing away.

The area of erosion-prone agricultural lands is 2263.2 thousand hectares, of which 1390 thousand hectares of arable lands are exposed to water erosion processes – 42%, including 6.7 thousand hectares of high degree, 254 thousand hectares of medium degree and 1129 thousand hectares of low degree. There are 21 thousand gullies with a total length of more than 29 thousand km, of which more than 21 thousand km fall to the water area of the Volga River.

More than 70% of all agricultural areas in Tatarstan are located on slopes of different steepness: including arable lands on slopes with steepness to 1 degree – 42.4%, 1–3 degrees – 52–0%, 3–5 degrees – 5.6%.

Construction of anti-erosion reclamation structures plays a positive role in erosion processes reduction, part of which, in turn, are the source of water for irrigated lands.

In the Republic of Tatarstan, in different years, more than 880 anti-erosion and reclamation hydraulic structures (dams) were built and commissioned with total volume of more than 360 million m^3 (useful volume - 283,6 million m^3). The main part of the above mentioned hydraulic structures was built according to the approved design and estimate documentation with account of water storage volume: up to 500 thousand m^3 – 395 structures; up to 1 million m^3 – 166 structures; over 1 million m^3 - 78 structures. Tatarstan ranks first among the regions of the Volga Federal District by the total area of lakes and reservoirs. In addition to ameliorative purposes and watering of territories, they serve as a reliable means designed to improve the ecological situation in the Volga Basin by reducing by at least 80% the volume of diffusion runoff caused by erosion processes on sloping lands of agricultural purposes.

According to the data of state statistics bodies, since 1968, 1681 water bodies have disappeared on the territory of the republic due to natural and anthropogenic reasons, including silting. Ponds and lakes constructed by damming of ravines, located on agricultural lands, subjected to erosion washout, are exposed to the most active silting. Their siltation rate reaches 10–50 mm per year, which is significantly higher than the average sedimentation rate typical of the territory of the Republic of Tatarstan.

Since 2013, within the framework of implementation of the long-term Republican Target Program on Capital Repair of Hydraulic Engineering Structures, more than 450 ponds out of 887 existing structures (Fig. 1) have been brought into the normative technical condition. They accumulate more than 150 million m^3 of water, it also allows saving agricultural lands from water erosion impact on the area of 44,7 thousand ha. By carrying out repair and rehabilitation works, the technical reliability and stability of hydraulic structures were improved, the threat of emergency situations at dams during the spring flood water passage through the hydraulic structures was reduced, and the protection of the population and objects falling into the zone of possible flooding was ensured.

The efficiency of erosion control measures is an aggregated assessment of the environmental and economic damage prevented as a result of the implementation of environmental erosion control measures. Ecological efficiency (prevented ecological damage) is determined depending on the degree of reduction of intensity of the above negative processes as a result of implementation of anti-erosion measures. Ecological efficiency of measures is determined as follows:

1) one erosion control structure in conjunction with soil conservation measures protects an average of 120 hectares of land from further development of erosive processes;
2) the cost of the complex of anti-erosion measures, including anti-erosion pond – 2.5 mln. rubles;
3) the cost of production from 120 ha with yields of 30 centners of grain from 1 ha is 120 ha x 30 centners x 1500 rubles/c = 5.4 mln. rubles per year;
4) the payback period of two years.

One of the effective methods of reducing erosion processes is agroforestry. As part of the implementation of targeted programs for the development of reclamation in the republic, agroforestry and reclamation works are carried out, more than 2.5 thousand hectares of protective strips are planted annually. In 2016–2021, work was carried out on the area of about 11 thousand hectares at a cost of more than 800 million rubles (about 150–160 million rubles per year).

Land reclamation programs implemented in the Republic of Tatarstan in 2018–2020 received the financial support from the state (Table 1).

https://doi.org/10.1051/bioconf/20225200022

Fig. 1. Hydraulic structures in the Republic of Tatarstan (source: authors' own photos).

Table 1. Financial support of reclamation programs implemented in the Republic of Tatarstan in 2018–2020.

Areas of funding	**Years**			
	2018	**2019**	**2020**	**Total**
Construction, reconstruction of reclamation systems and hydraulic structures from the RF budget for federal objects mln. rub	142.1	113.0	117.7	372.8
Overhaul of ponds and hydraulic structures from the RT budget, mln. rub	92.7	58.1	89.8	240.6
Construction, reconstruction and overhaul of irrigated and drained lands, mln. rub	62.3	46.9	70.2	179.4
Creation of anti-erosion and field forest plantations, mln. rub	150	186.5	150	486.5
Total for 3 years of the attracted to land reclamation	447.1	404.5	427.7	1279.3

As we can see from Table 1, despite the fact that the volume of financing of reclamation works in the period from 2018 to 2020 decreased by 4.3%, we can observe the increase of state aid for construction, reconstruction and overhaul of irrigated and drained lands by 12.7%, as well as significant increase by 24% in financing of anti-erosion and field forest plantation works in 2019 compared to 2018.

At the same time, in the Republic of Tatarstan there is an objective need to continue rehabilitation work on hydraulic structures, as in many structures there are still risks of emergencies caused by dam failure and flooding of nearby areas. In addition, the use of pesticides and mineral fertilizers is a serious pollutant of the environment, including rivers. In the republic, about 2 thousand tons of pesticides and more than 190 thousand tons of mineral fertilizers in active substance are used annually on agricultural lands. In addition to preventing siltation of rivers, erosion control structures retard runoff of chemicals and pesticides used in agricultural formations.

Thus, according to the results of joint inventory, carried out by the Ministry of Emergency Situations of the Republic of Tatarstan and specialists of the Ministry of Agriculture of the Republic of Tatarstan and «Tatmelioration» company as of 01.01.2021, 160 additional hydraulic structures require repair work. According to expert estimates, more than 840 million rubles in current prices are required to bring them to the normal condition.

The fact of siltation of many reservoirs and development in connection with this the processes of degradation of the water body, which causes the necessity of their cleaning from bottom sediments also raises concern.

The conducted studies indicate that low fertile lands of the Predkamskaya zone, most prone to erosion, can be protected by proper selection and cultivation of legume-grass perennial grasses on a large scale.

In this case it is possible to stop the development of erosion processes, radical improvement of fertility and structure of gray-forest, light gray, sod-podzolic, dark gray soils of the analyzed region. This method of combating the manifestations of soil erosion is less costly and affordable for a wide range of agricultural formations, and contributes to reducing the negative impact of drought on the performance of economic activity of agricultural organizations. For example, the sowing of polygonum ryegrass with perennial legumes is the best way to reduce the negative effects of moisture deficit, because the tap root system of alfalfa, goat-grass and meadow clover takes water from such deep soil layers, which are completely inaccessible to polygonum ryegrass. Legumes also start growing earlier in spring and give shade to cereal perennial grasses in the first cut, which increases the economic efficiency of agricultural production in both crop and livestock production. This method practically does not require large amounts of public investment. It is enough for agricultural formations to scientifically groundedly form their crop rotations widely including legume-grass perennial grasses.

At the present stage of development agrarian production of the Russian Federation is being modernized and transformed on the basis of consistent implementation of measures related to its digitalization. At all stages of production and financial activities of agricultural enterprises, innovative technological solutions are developed and applied, aimed at increasing productivity and production of competitive products. In particular, sensors, robotics, chipping, and others are actively used in the process of economic operations. Smart field technologies used in agricultural production, integrated with data management tools, contribute to efficient land management. The result is an agro-ecosystem that is highly adaptable to permanently changing economic conditions. This fact undoubtedly contributes to the reduction of food security risks in the country. In addition, the active introduction of digital technologies into the activities of agricultural producers enhances stakeholders' access to information about the financial and economic activities of the enterprise, for example, about the products produced, works performed, services rendered, pricing, investment, supply and marketing policy and other similar information. At the same time, digitalization of land reclamation activities has certain features, which are due to the specifics of agricultural production (seasonality of works performed, climatic risks, biological processes) and the nature of agricultural activities.

For the consistent modernization and transformation of agricultural production, the widespread implementation of measures related to this digitalization is necessary:

- at all stages of agricultural production innovative technologies are developed and applied, which ensure the effectiveness of financial and economic activities;
- the system of state regulation of agricultural production based on the formation of a unified information space is improved to make effective management decisions.

Digital agricultural land use is based on scientifically grounded, rational planning, economic-technological and information forecasting, effective management and legal support. In this regard, there is a need for a comprehensive solution to problems of a fundamental and applied nature and consisting in the development of methodological support for the management of agricultural production, including land use.

The factors influencing the efficiency of agricultural land tenure and land use can include:

- macroeconomic factors, including the forecasting of socio-economic development of both the individual region and the country as a whole;
- introduction of innovative mechanisms which allow evaluating management decisions;
- application of modern digital technologies, automation and robotics for characterizing agro-ecological and soil peculiarities of the lands being introduced into agricultural turnover
- formation and timely updating of the database on the state of agricultural land by means of digital land management;
- development of accounting and monitoring of the effective use and protection of agricultural land, taking into account the existing agro-potential of land use and existing risks arising in the process of land management;

- development of legislative and regulatory provisions governing the use of agricultural land;
- digitalization of design and survey works, business planning, documentary support, controlling, online consulting in land management system;
- improving the system of training and professional development in the field of agricultural land management and land use, taking into account the development of modern information technologies.

Digitalization of land reclamation activities is more related to the development of information technology, which allows solving a wide range of management tasks to improve the efficiency of agricultural production (Fig. 2).

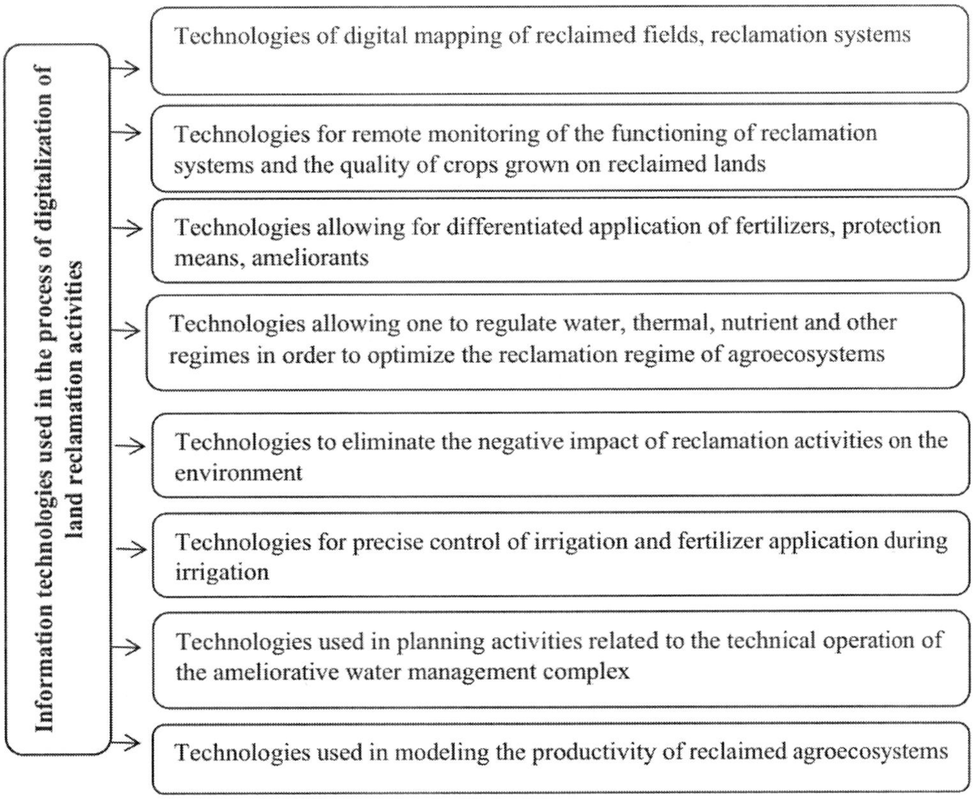

Fig. 2. Information technologies used in the process of digitalization of land reclamation activities.

Digitalization of land reclamation activity will allow implementing qualitatively new approaches aimed at increasing the efficiency of interaction between the subjects of the land reclamation sphere.

4 Conclusions

Thus, in the course of the study it was established that the efficiency of functioning of the agro-industrial complex depends on not only the entrepreneurial activity and the ability of subjects managing the land, but also on the state of agricultural production factors, which include land resources. To maintain the productivity and the ability of soils to self-repair, it is necessary to carry out timely anti-erosion measures. These measures are aimed at increasing anti-erosion stability and absorption capacity of soils, uniform moistening of soils, reducing the volume and intensity of runoff and soil washout on sloping lands, preventing the concentration of runoff on the arable land, reducing the process of soil washout to environmentally acceptable values. In the course of the conducted research, the tendencies of soil erosion development at the regional level were revealed and the necessary first-priority anti-erosion measures in agriculture of the Republic of Tatarstan were substantiated. The results of the analysis of the level of land degradation in the Republic of Tatarstan are given in the work, the priority directions of combating erosion phenomena in agriculture are determined. The measures to combat soil erosion are experimentally substantiated.

References

1. F. Mukhametgaliev, L. Sitdikova, M. Khismatullin, N. Asadullin, L. Mikhailova, Prospects of Agricultural Business in the Republic of Tatarstan, *BIO Web of Conferences* **17,** 00083 (2020) https://doi.org/10.1051/bioconf/20201700083
2. A. Klychova, G. Klychova, A. Zakirova, R. Sungatullina, K. Mukhamedzyanov, E. Philippova. Social development mechanism of an agricultural enterprise formation, *E3S Web of Conferences* **110,** 02072 (2019) doi.org/10.1051/e3sconf/201911002072
3. M. Safaei, H. Bashari, M. R. Mosaddeghi, R. Jafari, Assessing the impacts of land use and land cover changes on soil functions using landscape function analysis and soil quality indicators in semi-arid natural ecosystems, *CATENA* **177,** 260–271 (2019) https://doi.org/10.1016/j.catena.2019.02.021
4. M. Baude, B. C. Meyer, M. Schindewolf, Land use change in an agricultural landscape causing degradation of soil based ecosystem services, *Science of The Total Environment* **659,** 1526–1536 (2019) https://doi.org/10.1016/j.scitotenv.2018.12.45
5. H. E. M. El Azab, A. A. Afify, M. M. Soliman. The impact of informal irrigation practices on soil drainage condition, soil pollution and land suitability for agriculture in El Saf area of El Giza Governorate, *The Egyptian Journal of Remote Sensing and Space Science* **18,** 163–179 (2015) https://doi.org/10.1016/j.ejrs.2015.04.004
6. P. Tsymbarovich, G. Kust, M. Kumani, V. Golosov, O. Andreeva, Soil erosion: An important indicator for the assessment of land degradation neutrality in Russia, *International Soil and Water Conservation Research* **8,** 418–429 (2020) https://doi.org/10.1016/j.iswcr.2020.06.002
7. R. Ciampalini, P. Billi, G. Ferrari, L. Borselli, S. Follain, Soil erosion induced by land use changes as determined by plough marks and field evidence in the Aksum area (Ethiopia), *Agriculture, Ecosystems & Environment* **146,** 197–208 (2012) https://doi.org/10.1016/j.agee.2011.11.006
8. J. Han, W. Ge, Zh. Hei, Ch. Cong, Ch. Ma, M. Xie, B. Liu, W. Feng, F. Wang, J. Jiao, Agricultural land use and management weaken the soil erosion induced by extreme rainstorms, *Agriculture, Ecosystems & Environment* **301,** 107047 (2020) https://doi.org/10.1016/j.agee.2020.107047
9. B. Tang, J. Jiao, Y. Zhang, Y. Chen, N. Wang, L. Bai, The magnitude of soil erosion on hillslopes with different land use patterns under an extreme rainstorm on the Northern Loess Plateau, China, *Soil and Tillage Research* **204,** 104716 (2020) https://doi.org/10.1016/j.still.2020.104716
10. A. T. Barabanov, S. V. Dolgov, N. I. Koronkevich, V. I. Panov, A. I. Petel'ko, Surface runoff and snowmelt infiltration into the soil on plowlands in the forest-steppe and steppe zones of the east European plain, *Eurasian Soil Science* **51 (1),** 66–72 (2018) https://doi.org/10.1134/S1064229318010039
11. A. Chappell, N. P. Webb, J. F. Leys, C. M. Waters, S. Orgill, M. J. Eyres, Minimising soil organic carbon erosion by wind is critical for land degradation neutrality, *Environmental Science & Policy,* **93,** 43–52 (2019) https://doi.org/10.1016/j.envsci.2018.12.020
12. M. Crossland, L. Winowiecki, T. Pagella, K. Hadgu, F. Sinclair, Implications of variation in local perception of degradation and restoration processes for implementing land degradation neutrality, *Environ. Dev.* **28,** 1–13 (2018) https://doi.org/10.1016/j.envdev.2018.09.005
13. K. T. Lendering, S. N. Jonkman, P. H. A. J. M. van Gelder, D. J. Peters, Risk-based optimization of land reclamation, *Reliability Engineering & System Safety* **144,** 193–203 (2015) doi.org/10.1016/j.ress.2015.07.025
14. M. M. Khismatullin, F. N. Mukhametgaliev, N. M. Asadullin, F. N. Avkhadiev, M. M. Khismatullin, U. Raheem, Implementation of government support measures for reclamation as an incentive for the development of the agricultural industry: Experience of the Republic of Tatarstan, *BIO Web of Conferences* **37,** 00080 (2021) https://doi.org/10.1051/bioconf/20213700080

15. G. Klychova, A. Zakirova, S. Khusainov, A. Dyatlova, U. Raheem, E. Sagadeeva, Accounting and analytical support of internal management reporting on reclamation works, *E3S Web of Conferences* **157**, 04022 (2020) https://doi.org/10.1051/e3sconf/202015704022
16. W. Wang, H. Liu, Y. Li, J. Su, Development and management of land reclamation in China, *Ocean & Coastal Management* **102**, 415–425 (2014) https://doi.org/10.1016/j.ocecoaman.2014.03.009
17. H. Duan, H. Zhang, Q. Huang, Y. Zhang, J. Zhu, Characterization and environmental impact analysis of sea land reclamation activities in China, *Ocean & Coastal Management* **130**, 128–137 (2016) https://doi.org/10.1016/j.ocecoaman.2016.06.006
18. Q. Feng, W. Zhao, J. Wang, X. Zhang, M. Zhao, L. Zhong, Y. Liu, X. Fang, Effects of different land-use types on soil erosion under natural rainfall in the Loess Plateau, China. *Pedosphere* **26**, 243 https://doi.org/256 (2016) https://doi.org/10.1016/S1002-0160(15)60039-X
19. W. Wu, Zh. Yang, B. Tian, Y. Huang, Y. Zhou, T. Zhang, Impacts of coastal reclamation on wetlands: Loss, resilience, and sustainable management, *Estuarine, Coastal and Shelf Science* **210**, 153–161 (2018) https://doi.org/10.1016/j.ecss.2018.06.013
20. X. Li, Ch. Zhang, Effect of natural and artificial afforestation reclamation on soil properties and vegetation in coastal saline silt soils, *CATENA* **198**, 105066 (2021) https://doi.org/10.1016/j.catena.2020.105066
21. Q. Ke, K. Zhang, Interaction effects of rainfall and soil factors on runoff, erosion, and their predictions in different geographic regions, *Journal of Hydrology* **605**, 127291 (2022) https://doi.org/10.1016/j.jhydrol.2021.127291
22. A. V. Gusarov, The impact of contemporary changes in climate and land use/cover on tendencies in water flow, suspended sediment yield and erosion intensity in the northeastern part of the Don River basin, SW European Russia, *Environmental Research* **175**, 468-488 (2019) https://doi.org/10.1016/j.envres.2019.03.057
23. H. Jin, D. Shi, Y. B. Lou, J. Zhang, Q. Ye, N. Jiang, Evaluation of the quality of cultivated-layer soil based on different degrees of erosion in sloping farmland with purple soil in China, *CATENA* **198**, 105048 (2021) https://doi.org/10.1016/j.catena.2020.105048
24. H. Wang, T. Xie, X. Yu, Ch. Zhang, Simulation of soil loss under different climatic conditions and agricultural farming economic benefits: The example of Yulin City on Loess Plateau, *Agricultural Water Management* **244**, 106462 (2021) https://doi.org/10.1016/j.agwat.2020.106462
25. S. M. Cabrini, C. P. Calcaterra, Modeling economic-environmental decision making for agricultural land use in Argentinean Pampas, *Agricultural Systems* **143**, 183–194 (2016) https://doi.org/10.1016/j.agsy.2015.12.016
26. X. Chen, L. Yu, Zh. Du, Y. Xu, J. Zhao, H. Zhao, G. Zhang, D. Peng, P. Gong, Distribution of ecological restoration projects associated with land use and land cover change in China and their ecological impacts, *Science of The Total Environment* **17**, 153938 (2022) https://doi.org/10.1016/j.scitotenv.2022.153938
27. Dz. Faizrakhmanov, A. Zakirova, G. Klychova, A. Yusupova, A. Klychova. Formation and disclosure of information on social responsibility of agribusiness enterprises, *E3S Web of Conferences* **91**, 06004 (2019) https://doi.org/10.1051/e3sconf/20199106004
28. A. Dhar, M. A. Naeth, P. D. Jennings, M. G.l El-Din, Perspectives on environmental impacts and a land reclamation strategy for solar and wind energy systems, *Science of The Total Environment* **718**, 134602 (2020) https://doi.org/10.1016/j.scitotenv.2019.134602

To develop a methodology for finding the competencies of agricultural workers with a positive economic return

Anna Aletdinova[1 2 *], *Zoya* Kapelyuk[3], *Marina* Drozdova[3], and *Svetlana* Gorodkova[3]

[1] Novosibirsk State Technical University, Automated Control System Department, 630073, Novosibirsk, Russia
[2] National University of Oil and Gas «Gubkin University», Automated Control System Department, 119991, Moscow, Russia
[3] Siberian University of the Consumer Cooperation, Department of Theoretical and Applied Economics, 630087, Novosibirsk, Russia

Abstract. The paper describes a study by the example of processing the results of a survey of 200 employees of the agro-industrial complex on psychological tests, work experience and salary level. The authors identified the following groups of competencies: intellectual and educational, communicative, motivational and volitional, management. 15 psychological tests were used to assess their presence in respondents. After correlation and variance analyses, a statistically significant model of the dependence of the wage level on 12 indicators was obtained: ability and readiness for self-development and self-education, tolerance, high motivation for success, self-actualization of personality, strong interest in professional development, focus on interaction, high level of emotional intelligence. At the same time, it can be used in research only for the agricultural complex at the current time, i.e., at the beginning of the sixth technological order.

1 Introduction

An important factor in economic development, digitalization and the introduction of new technologies is the development of digital culture, the creation of a new educational environment, the development of technological infrastructure and the formation of human capital. This is also proved by the results of the Digital IQ study, which examines the impact of investments in digital technologies and systems. Any success of the transformation, the transition to a new technological order will depend on not only the intensification of technologies, the use of special sensors, algorithms, analytical tools and information systems, etc., but also on human capital. Organizations of the agro-industrial complex should attract, retain and train specialists of the "digital generation" who form the information society and possess the competencies that are most in demand in their field of activity.

The change in the functions of workers and the conditions of technological development of the agro-industrial complex will inevitably lead to the expansion of forms of employment, changes in working hours, the erasure of territorial boundaries, increased flexibility of labor, the transfer of economic and social activities to the Internet environment, the urbanization of agriculture.

We believe that in addition to classical factors such as education, seniority, place of residence, influencing the formation of human capital, a person's entrepreneurial abilities, high motivation for success and achievement of goals, free orientation on the Internet and possession of information and communication technologies, access to open knowledge can be significant. This is confirmed by the studies of N. Frederiksen, Tangney, V. Bouchard, S. Zaika, O. Gridin, D. Rusmingsih, E. L. Widarni, S. Bawono [1–5].

Drastic changes in the life of a modern person, increasing the pace of interaction in everyday life and production activities have influenced the paradigm of training specialists, i.e. there was a need to move from a qualification model to a competence model. Modern employees should continuously engage in self-education, expand and improve their competencies [6].

It is difficult for researchers to numerically assess most of the competencies of employees, because it is necessary to take into account the psychophysical state of a person, willpower, hereditary traits, so scientists suggest resorting to the use of psychodiagnostic methods. There are several areas of competence research at the interface with psychology:

- contradictions of professional competencies and intrapersonal conflict [7],
- assessment of cognitive characteristics of a person [8, 9],
- identification of success factors in the professional activity of an employee [10],
- economic and legal aspects of competencies, including assessment of economic impact [11–14].

As our analysis of scientific publications has shown, the least attention is paid to the creation of methods for psychodiagnostics of cognitive characteristics of an

* Corresponding author: aletdinova@corp.nstu.ru

employee that form his competencies. And such methods can be developed both for individual industries and be universal.
The purpose of the study is to select and substantiate a set of competencies/psychodiagnosis techniques to identify the competencies of agricultural workers in terms of informative features that affect the amount of wages.

2 Materials and methods

This section includes the main theoretical approaches and terms used in the study, as well as methodological justification and the main steps of scientific work.

2.1. Materials

Let us highlight the approaches and terms that will be used in the work.

First, let us introduce a restriction. We will consider competencies within the framework of a person's professional activity. Competencies are formally described requirements for personal, professional, etc. qualities of employees necessary for the successful functioning of the activities of entrepreneurs and/or organizations.

Professional competencies in Russia are determined by educational standards for educational institutions of different levels.

Critical competencies imply the competencies that are most in demand for this technological stage and this field of activity (and for our research for the agro-industrial complex) in the labor market. They are constantly changing in the process of economic development.

The set of competencies will be divided into the following groups: intellectual and educational, communicative, motivational and volitional, management.

A positive economic return implies an increase in the level of wages if the employee has the competence in question.

2.2. Research methods

To assess competencies, the authors chose the following methods of psychodiagnostics and labeled the results of them as variables:

- Assessment of the ability to self-development and self-education (SC1),
- Readiness for self-development (SC2),
- Creativity (SC3),
- Tolerance index (SC4),
- Communicative tolerance (SC5),
- Diagnosis of interference in establishing emotional contacts (SC6),
- The need to achieve a goal (SC7),
- Measuring tolerance (SC8),
- Resilience (SC9),
- Degree of risk tolerance (SC10),
- Motivation for success (SC11),
- The level of self-actualization of the individual (SC12),
- Morphological test of life values (SC13),
- Diagnosis of personality orientation (SC14),
- Emotional Intelligence (SC15).

The sample size of respondents is 200 employees of the agro-industrial complex. Correlation and variance analyses of the respondents' test results were used for the study.

The hypothesis in the study was that in the presence of the competencies in question, agricultural workers, with other identical characteristics have a higher salary level.

Let us introduce additional notation. The variable Y is salary, Ex – work experience, Scn – points obtained by psychodiagnostics methods for assessing competencies. Let us conduct a correlation analysis between the variables: SC1, SC2, SC3, SC4, SC5, SC6, SC7, SC8, SC9, SC10, SC11, SC12, SC13, SC14, SC15, Ex, Scn, Y.

3 Results and discussion

Let us construct a correlation matrix for the values: SC1, SC2, SC3, SC4, SC5, SC6, SC7, SC8, SC9, SC10, SC11, SC12, SC13, SC14, SC15, Ex, Scn, Y (Figure 1).

Variable	Correlations (Spreadsheet1.sta)															
	Ex	Sc1	Sc2	Sc3	Sc4	Sc5	Sc6	Sc7	Sc8	Sc9	Sc10	Sc11	Sc12	Sc13	Sc14	Sc15
Ex	1,000000	0,047183	0,170304	0,092099	0,074366	0,040908	0,006842	-0,016326	-0,048044	0,154161	-0,186372	0,096990	0,073491	0,188004	0,073661	0,126994
Sc1	0,047183	1,000000	0,139161	0,039356	0,096988	0,138504	0,092927	0,167893	0,119871	0,153907	0,028481	0,271680	0,343711	0,270739	0,163568	0,128322
Sc2	0,170304	0,139161	1,000000	-0,036054	0,033676	-0,096083	0,066945	0,135212	0,134274	0,171415	0,041675	0,224454	0,235648	0,216815	0,077466	0,120585
Sc3	0,092099	0,039356	-0,036054	1,000000	-0,014263	0,163357	-0,131310	0,075111	0,035610	0,059920	0,107189	-0,012923	-0,071226	0,012515	0,023235	0,047760
Sc4	0,074366	0,096988	0,033676	-0,014263	1,000000	0,152372	-0,011966	-0,018140	0,036009	0,190041	0,032636	0,103281	0,081165	0,216422	-0,044824	0,123077
Sc5	0,040908	0,138504	-0,096083	0,163357	0,152372	1,000000	-0,004160	0,162258	0,002959	0,118862	-0,019839	-0,001697	0,003284	0,019819	-0,014694	-0,088902
Sc6	0,006842	0,092927	0,066945	-0,131310	-0,011966	-0,004160	1,000000	0,052630	0,130591	0,042378	0,091283	0,027510	0,101005	-0,032549	0,059373	-0,009743
Sc7	-0,016326	0,167893	0,135212	0,075111	-0,018140	0,162258	0,052630	1,000000	0,090978	0,198081	-0,053921	0,147820	0,003849	0,027335	0,082424	0,072293
Sc8	-0,048044	0,119871	0,134274	0,035610	0,036009	0,002959	0,130591	0,090978	1,000000	0,144597	0,149935	0,088241	0,132236	0,121875	0,160746	0,031603
Sc9	0,154161	0,153907	0,171415	0,059920	0,190041	0,118862	0,042378	0,198081	0,144597	1,000000	0,022682	0,295426	0,292830	0,344737	0,275787	0,171419
Sc10	-0,186372	0,028481	0,041675	0,107189	0,032635	-0,019639	0,091283	-0,053921	0,149935	0,022682	1,000000	-0,021957	0,094371	-0,000425	0,040515	0,127369
Sc11	0,096990	0,271680	0,224454	-0,012923	0,103281	-0,001697	0,027510	0,147820	0,088241	0,295426	-0,021957	1,000000	0,350777	0,431834	0,208548	0,225724
Sc12	0,073491	0,343711	0,235648	-0,071226	0,081165	0,003284	0,101005	0,003849	0,132236	0,292830	0,094371	0,350777	1,000000	0,431536	0,241613	0,150247
Sc13	0,188004	0,270739	0,216815	0,012515	0,216422	0,019819	-0,032549	0,027335	0,121875	0,344737	-0,000425	0,431834	0,431536	1,000000	0,066395	0,329838
Sc14	0,073661	0,163568	0,077466	0,023235	-0,044824	-0,014694	0,059373	0,082424	0,160746	0,275787	0,040515	0,208548	0,241613	0,066395	1,000000	-0,002798
Sc15	0,126994	0,128322	0,120585	0,047760	0,123077	-0,088902	-0,009743	0,072293	0,031603	0,171419	0,127369	0,225724	0,150247	0,329838	-0,002798	1,000000
SL	0,194779	0,425411	0,352219	-0,011122	0,255196	0,069729	0,114075	0,168756	0,248407	0,567190	0,057575	0,625649	0,609094	0,623885	0,321021	0,369564

Fig. 1. Correlation matrix.

As we can see from the table, no factor forms a strong correlation. There is no multicorrelationality.

Let us output a table with the level of significance of the factors.

	Regression Summary for Dependent Variable: SL (Spreadsheet1.sta) R= ,88309730 R?= ,77986085 Adjusted R?= ,75798366 F(16,161)=35,647 p<0,0000 Std.Error of estimate: 8,4008					
N=178	b*	Std.Err. of b*	b	Std.Err. of b	t(161)	p-value
Intercept			-117,763	12,85009	-9,16440	0,000000
Ex	0,033216	0,039768	0,046	0,05475	0,83525	0,404814
Sc1	0,107864	0,041403	0,804	0,30845	2,60520	0,010042
Sc2	0,093989	0,040039	0,739	0,31471	2,34744	0,020119
Sc3	-0,021472	0,038826	-0,024	0,04386	-0,55304	0,581004
Sc4	0,087572	0,039038	0,078	0,03497	2,24323	0,026246
Sc5	0,031128	0,039591	0,014	0,01774	0,78623	0,432886
Sc6	0,045149	0,038271	0,081	0,06831	1,17972	0,239851
Sc7	0,018139	0,039765	0,159	0,34860	0,45616	0,648887
Sc8	0,083923	0,038881	1,402	0,64945	2,15847	0,032373
Sc9	0,229786	0,043214	0,324	0,06090	5,31743	0,000000
Sc10	-0,000818	0,039477	-0,001	0,02727	-0,02072	0,983494
Sc11	0,268531	0,043710	0,913	0,14864	6,14344	0,000000
Sc12	0,236365	0,045167	0,334	0,06379	5,23311	0,000001
Sc13	0,193070	0,047453	0,201	0,04934	4,06866	0,000074
Sc14	0,091964	0,040434	1,317	0,57902	2,27443	0,024261
Sc15	0,130825	0,040593	0,074	0,02308	3,22282	0,001537

Fig. 2. Significance levels of the model.

It is necessary to remove those factors whose significance level is $p \geq 0{,}05$. In our case, this refers to the variable "Work experience" and to competencies 3, 5, 6, 7, 10. It is necessary to consistently remove these factors from the regression equation, starting with the factor whose significance level is higher than the others are, each time rebuilding the model anew (Figure 3, 4, 5, 6. 7, 8).

	Regression Summary for Dependent Variable: SL (Spreadsheet1.sta) R= ,88309697 R?= ,77986026 Adjusted R?= ,75947695 F(15,162)=38,260 p<0,0000 Std.Error of estimate: 8,3748					
N=178	b*	Std.Err. of b*	b	Std.Err. of b	t(162)	p-value
Intercept			-117,735	12,73534	-9,24472	0,000000
Ex	0,033397	0,038672	0,046	0,05325	0,86360	0,389084
Sc1	0,107873	0,041273	0,804	0,30748	2,61364	0,009802
Sc2	0,093943	0,039852	0,738	0,31324	2,35729	0,019604
Sc3	-0,021590	0,038288	-0,024	0,04325	-0,56389	0,573610
Sc4	0,087541	0,038889	0,078	0,03484	2,25104	0,025728
Sc5	0,031131	0,039469	0,014	0,01769	0,78875	0,431414
Sc6	0,045077	0,037992	0,080	0,06781	1,18648	0,237170
Sc7	0,018215	0,039475	0,160	0,34606	0,46142	0,645116
Sc8	0,083831	0,038508	1,400	0,64322	2,17698	0,030929
Sc9	0,229781	0,043079	0,324	0,06071	5,33388	0,000000
Sc11	0,268584	0,043500	0,913	0,14793	6,17430	0,000000
Sc12	0,236283	0,044856	0,334	0,06335	5,26760	0,000000
Sc13	0,193118	0,047251	0,201	0,04913	4,08707	0,000069
Sc14	0,091940	0,040293	1,317	0,57700	2,28179	0,023803
Sc15	0,130694	0,039975	0,074	0,02273	3,26943	0,001316

Fig. 3. Significance levels of the model when excluding Sc10.

	Regression Summary for Dependent Variable: SL (Spreadsheet1.sta) R= ,88293315 R?= ,77957094 Adjusted R?= ,76063838 F(14,163)=41,176 p<0,0000 Std.Error of estimate: 8,3546					
N=178	b*	Std.Err. of b*	b	Std.Err. of b	t(163)	p-value
Intercept			-116,908	12,57827	-9,29446	0,000000
Ex	0,032183	0,038489	0,044	0,05299	0,83616	0,404289
Sc1	0,110289	0,040841	0,822	0,30426	2,70047	0,007655
Sc2	0,096231	0,039447	0,756	0,31005	2,43952	0,015779
Sc3	-0,020920	0,038168	-0,024	0,04311	-0,54811	0,584365
Sc4	0,086151	0,038679	0,077	0,03465	2,22735	0,027293
Sc5	0,033890	0,038919	0,015	0,01744	0,87080	0,385146
Sc6	0,045650	0,037880	0,081	0,06761	1,20512	0,229905
Sc8	0,084576	0,038381	1,413	0,64110	2,20359	0,028955
Sc9	0,232978	0,042416	0,328	0,05977	5,49270	0,000000
Sc11	0,270574	0,043181	0,920	0,14685	6,26597	0,000000
Sc12	0,233891	0,044448	0,330	0,06277	5,26217	0,000000
Sc13	0,191542	0,047013	0,199	0,04888	4,07420	0,000072
Sc14	0,092157	0,040193	1,320	0,57557	2,29286	0,023134
Sc15	0,131828	0,039802	0,075	0,02263	3,31206	0,001141

Fig. 4. Significance levels of the model when excluding Sc10, Sc7.

	Regression Summary for Dependent Variable: SL (Spreadsheet1.sta) R= ,88270304 R?= ,77916467 Adjusted R?= ,76165943 F(13,164)=44,510 p<0,0000 Std.Error of estimate: 8,3367					
N=178	b*	Std.Err. of b*	b	Std.Err. of b	t(164)	p-value
Intercept			-117,901	12,42068	-9,49228	0,000000
Ex	0,030338	0,038260	0,042	0,05268	0,79294	0,428960
Sc1	0,109263	0,040711	0,814	0,30329	2,68389	0,008024
Sc2	0,096843	0,039347	0,761	0,30927	2,46127	0,014880
Sc4	0,087375	0,038532	0,078	0,03452	2,26761	0,024659
Sc5	0,030576	0,038364	0,014	0,01719	0,79699	0,426608
Sc6	0,048465	0,037450	0,087	0,06685	1,29412	0,197444
Sc8	0,083341	0,038233	1,392	0,63863	2,17982	0,030696
Sc9	0,231891	0,042279	0,327	0,05958	5,48478	0,000000
Sc11	0,271051	0,043080	0,922	0,14650	6,29174	0,000000
Sc12	0,235983	0,044189	0,333	0,06241	5,34032	0,000000
Sc13	0,191551	0,046913	0,199	0,04878	4,08312	0,000069
Sc14	0,091655	0,040097	1,313	0,57419	2,28585	0,023543
Sc15	0,130502	0,039644	0,074	0,02254	3,29186	0,001219

Fig. 5. Significance levels of the model when excluding Sc10, Sc7, Sc3.

	Regression Summary for Dependent Variable: SL (Spreadsheet1.sta) R= ,88222334 R?= ,77831802 Adjusted R?= ,76219569 F(12,165)=48,276 p<0,0000 Std.Error of estimate: 8,3273					
N=178	b*	Std.Err. of b*	b	Std.Err. of b	t(165)	p-value
Intercept			-117,526	12,39771	-9,47963	0,000000
Sc1	0,108652	0,040658	0,809	0,30289	2,67236	0,008288
Sc2	0,101337	0,038892	0,797	0,30570	2,60559	0,010010
Sc4	0,088059	0,038479	0,079	0,03447	2,28851	0,023377
Sc5	0,032098	0,038273	0,014	0,01715	0,83868	0,402865
Sc6	0,048959	0,037403	0,087	0,06676	1,30897	0,192365
Sc8	0,080243	0,037990	1,340	0,63457	2,11221	0,036174
Sc9	0,234185	0,042133	0,330	0,05937	5,55829	0,000000
Sc11	0,270367	0,043023	0,919	0,14631	6,28420	0,000000
Sc12	0,234503	0,044100	0,331	0,06228	5,31755	0,000000
Sc13	0,195983	0,046526	0,204	0,04837	4,21229	0,000041
Sc14	0,093742	0,039965	1,342	0,57231	2,34560	0,020186
Sc15	0,132573	0,039513	0,075	0,02247	3,35514	0,000984

Fig. 6. Significance levels of the model when excluding Sc10, Sc7, Sc3 and work experience Ex.

	Regression Summary for Dependent Variable: SL (Spreadsheet1.sta) R= ,88168759 R?= ,77737301 Adjusted R?= ,76262062 F(11,166)=52,695 p<0,0000 Std.Error of estimate: 8,3199					
N=178	b*	Std.Err. of b*	b	Std.Err. of b	t(166)	p-value
Intercept			-118,995	12,26239	-9,70404	0,000000
Sc1	0,113975	0,040123	0,849	0,29891	2,84064	0,005066
Sc2	0,097724	0,038618	0,768	0,30354	2,53050	0,012320
Sc4	0,092325	0,038107	0,083	0,03414	2,42279	0,016477
Sc6	0,048637	0,037367	0,087	0,06670	1,30159	0,194861
Sc8	0,080047	0,037955	1,337	0,63399	2,10899	0,036446
Sc9	0,238822	0,041731	0,337	0,05881	5,72290	0,000000
Sc11	0,269678	0,042977	0,917	0,14615	6,27492	0,000000
Sc12	0,233366	0,044040	0,330	0,06220	5,29900	0,000000
Sc13	0,195670	0,046483	0,203	0,04833	4,20946	0,000042
Sc14	0,092071	0,039880	1,318	0,57109	2,30872	0,022192
Sc15	0,128580	0,039191	0,073	0,02228	3,28090	0,001261

Fig. 7. Significance levels of the model when excluding Sc10, Sc7, Sc3, Ex, Sc5.

	Regression Summary for Dependent Variable: SL (Spreadsheet1.sta) R= ,88039819 R?= ,77510098 Adjusted R?= ,76163397 F(10,167)=57,556 p<0,0000 Std.Error of estimate: 8,3372					
N=178	b*	Std.Err. of b*	b	Std.Err. of b	t(167)	p-value
Intercept			-117,242	12,21354	-9,59936	0,000000
Sc1	0,117361	0,040122	0,874	0,29890	2,92511	0,003922
Sc2	0,099740	0,038667	0,784	0,30393	2,57945	0,010757
Sc4	0,091862	0,038184	0,082	0,03421	2,40575	0,017231
Sc8	0,085644	0,037789	1,431	0,63122	2,26636	0,024713
Sc9	0,239903	0,041809	0,338	0,05892	5,73804	0,000000
Sc11	0,269782	0,043066	0,917	0,14645	6,26434	0,000000
Sc12	0,238134	0,043978	0,336	0,06211	5,41485	0,000000
Sc13	0,189743	0,046356	0,197	0,04820	4,09318	0,000066
Sc14	0,092249	0,039962	1,321	0,57227	2,30841	0,022203
Sc15	0,128339	0,039271	0,073	0,02233	3,26800	0,001315

Fig. 8. Significance levels of the model when excluding Sc10, Sc7, Sc3, Ex, Sc5, Sc6.

Let us construct a frequency histogram of the residuals (Figure 9).

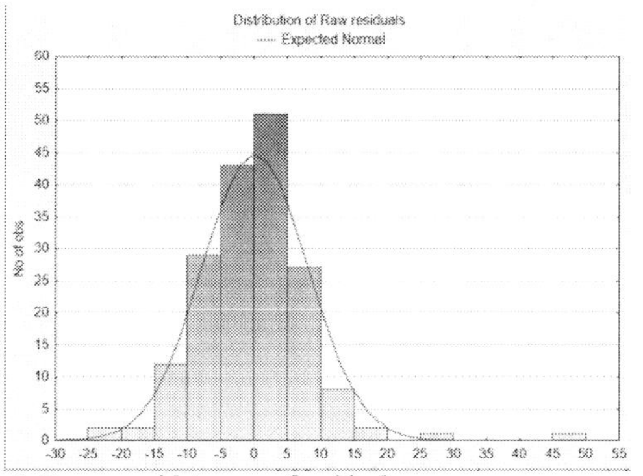

Fig. 9. Frequency histogram of residuals.

The histogram is distributed symmetrically, despite the outlier in the center, so the hypothesis of normality is not rejected.

Next, let us look at the normal probability graph (Figure 10).

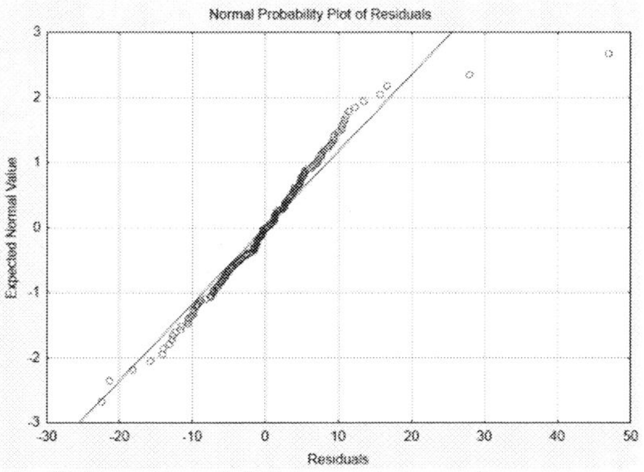

Fig. 10. Normally probabilistic remainder graph.

There are no systematic deviations from the theoretical normal curve, the residuals are distributed normally.

Let us look at the dependence of the predicted values and residuals (Figure 11).

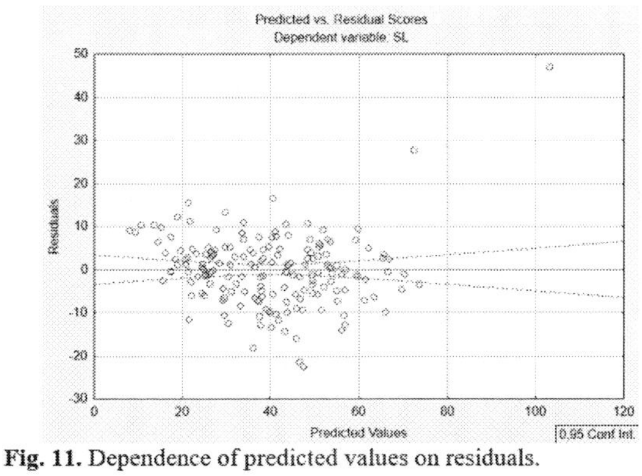

Fig. 11. Dependence of predicted values on residuals.

In the scattering diagram, we see the ratio of the residuals from the predicted values.

In the direction of the movement of points, there is a systematic pattern, expressed by the accumulation of points in the interval of residuals (-10;10) and predicted values in the interval (20;60). Based on this, we say that the residuals depend on the predicted values.

Let us evaluate the acceptability of the model as a whole by the analysis of variance. In the STATISTICA statistical package, it is designated as ANOVA. The analysis of variance is shown in Figure 12.

Effect	Analysis of Variance; DV: SL (Spreadsheet1.sta) Sums of Squares	df	Mean Squares	F	p-value
Regress.	40005,89	10	4000,589	57,55555	0,00
Residual	11607,89	167	69,508		
Total	51613,78				

Fig. 12. Variance analysis of the model.

By the level of significance, we can say that the model is acceptable and will work better than the forecast for average values.

The determination index is quite high, it indicates a good quality of the model (Figure 13).

Regression Summary for Dependent Variable: SL (Spreadsheet1.sta)
R= ,88039819 R?= ,77510098 Adjusted R?= ,76163397
F(10,167)=57,556 p<0,0000 Std.Error of estimate: 8,3372

Fig. 13. Assessment of the quality of the regression model.

Its value was 0.775, which means that approximately 77.5% of factors affect the response and are taken into account in our model.

Thus, when analyzing the impact of competencies on the salary level in the model, the following turned out to be informative: Sc1, Sc2, Sc4, Sc8, Sc9, Sc11, Sc12, Sc13, Sc14, Sc15. Consequently, abilities and readiness for self-development and self-education, tolerance, high motivation for success, self-actualization of personality, strong interest in professional development (based on the morphological test of life values), focus on interaction, a high level of emotional intelligence are not interrelated and affect the level of wages of employees of the agro-industrial complex. These competencies can be called critical, because employers are willing to pay employees more money for their availability. The set of critical competencies at different stages of economic development and for different fields of activity will be different. Therefore, research should be carried out separately for different industries, at different time periods.

References

1. J. P. Vesterager, P. Frederiksen, S. B. P. Kristensen, A. Vadineanu, V. Gaube, N. A. Geamana, A. G. Busck, *Land use policy,* **57** (2016)
2. J. P. Tangney, A. L. Boone, R. O. Y. F. Baumeister, *Self-Regulation and Self-Control* (Routledge, 2018)
3. V. Bouchard, A. Fayolle, *Corporate entrepreneurship* (Routledge, 2017)

4. S. Zaika, O. Gridin, Human Capital Development in the Agricultural Economy Sector. *Technology Audit and Production Reserves,* **1,** 4 (2020)
5. D. Rusmingsih, E. L. Widarni, S. Bawono, *Holistica–Journal of Business and Public Administration,* **12,** 1 (2021)
6. M. Kordaczuk-Wąs, S. Sosnowski, *Police Practice and Research,* **12,** 4 (2011)
7. V. M. Golyanich, A. F. Bondaruk, V. A. Shapoval, T. V. Tulupyeva, *Experimental Psychology (Russia)*, **11,** 3 (2018)
8. A. A. Aletdinova, A. I. Kiprisheva, *Journal of Phys.: Conference Series,* **1661,** 1 (2020)
9. J. Funke, A. Fischer, D. V. Holt, *Assessment and teaching of 21st century skills* (Springer, Cham, 2018)
10. R. A. Posthuma, M. Charles Campion, M. A. Campion, *European Journal of Work and Organizational Psychology,* **27,** 2 (2018)
11. G. Abuselidze, L. Mamaladze, *Journal of Physics: Conference Series,* **1840,** 1 (2021)
12. O. Kryvtsova, M. Karpa, K. Shvets, S. Lopatin, P. Yepryntsev, V. Kryvtsova, *AD ALTA: Journal of interdisciplinary research,* **11,** 2 (2021)
13. A. A. Aletdinova, A. V. Koritsky, *Regional Research of Russia,* **10,** 2 (2020)
14. J. Bratton, J. Gold, A. Bratton, L. Steele, *Human resource management* (Bloomsbury Publishing, 2021)
15. V. Sima, I. G. Gheorghe, J. Subić, D. Nancu, *Sustainability,* **12,** 10 (2020)

Problems of sustainable development of rural areas (on the materials of Samara region)

Anna Volkonskaya[1], *Oksana* Mamai[1,*], *Aleksandr* Kudryashov[2], and *Nikolai* Nikulnikov[2]

[1] Federal State Budgetary Educational Institution of Higher Education "Samara State Agrarian University", 2, Uchebnaya St., Kinel, urban-type settlement Ust-Kinelsky, Samara Region, 446442, Russian Federation
[2] Federal State Budgetary Educational Institution of Higher Education "Povolzhskiy State University of Telecommunications & Informatics", 23, L. Tolstoy St., Samara, Samara Region, 443010, Russian Federation

Abstract. The effectiveness of the development of the agricultural sector of the economy directly affects the socio-economic development of the country, rural areas. Among the factors of sustainable development of agricultural production, a special place has recently been given to state support. It allows not only to increase the efficiency of agricultural enterprises, but also contributes to the development of rural settlements and employment in them. Therefore, research in the field of integrated development of rural areas is currently relevant for many regions. The purpose of the study is to study the development trends of the agricultural sector of the economy as the basis for the development of rural areas, to identify problems that hinder this development. In the course of the study, various methods were used to study processes and phenomena in the field of organizational and economic relations: analysis and synthesis, systematic and logical approaches to studying trends in the development of the agricultural sector of the economy, the integrated development of rural areas, graphical interpretation of theoretical information and empirical data. As a result, shortcomings in the mechanism for implementing the state program for the development of rural areas were identified, and corrective measures for this mechanism were proposed.

1 Introduction

Rural areas are the most important productive, socio-demographic, cultural, ecological resources of the country that ensure its territorial integrity, as well as the unity of legal and economic space. Rural areas cover the population, land and other resources of the open landscape and small settlements outside the immediate economic areas of influence of major urban centers. Especially in comparison with urban regions rural areas are characterized by narrow relationships between industries and low population density. The predominant occupation of people in most rural regions is agriculture, there is lower socio-economic development, some labor activities, higher professional and social homogeneity of the population [1].

The state program "Comprehensive Development of Rural Areas" (CDRA state program) identifies rural areas as the territories of rural settlements and intersettlement territories. At the same time, small towns connected with rural areas through the joint use of infrastructure facilities and united by intensive economic, including labor, and social ties are considered the full participants of the CDRA state program.

The main purpose of this study is to identify the conditions and factors affecting the sustainable development of rural areas as the basis for ensuring the food security of the country and its regions.

2 Materials and methods

The research materials include the scientific works of domestic and foreign scientists in the field of sustainable development of rural areas. The study utilizes such methods as monographic, analytical, abstract-logical, economic-statistical, scientific deduction and induction, comparison, quantitative and qualitative analysis and others. The study is based on data from the Ministry of Economic Development of Russia, the Ministry of Agriculture of Russia and the Federal State Statistics Service.

3 Results

At the beginning of 2021, the rural population of Samara region amounted to 641.62 thousand people (20.2% of the region's population). The rural population lives in 27 municipal districts of Samara region, which are combined into 1284 rural settlements. On average, there are about 11 rural settlements, 48 rural communities per municipal district.

The vast majority of the rural population lives in rural settlements with a population of up to 10 thousand people. They differ from urban settlements by underdeveloped communication, transport and engineering infrastructure, do not have the necessary conditions for the development of entrepreneurship, especially small businesses, and also do not have a base that meets modern economic conditions [2]. Rural areas

* Corresponding author: mamai_ov@ssaa.ru

are characterized by high unemployment and low incomes.

According to sample surveys of labor, employment and unemployment in Samara region carried out by the territorial body of the Federal State Statistics Service for Samara region, the number of employed able-bodied population (men 16-59 years old, women 16-54 years old) living in rural areas in 2019 amounted to 260 thousand people, which is the minimum figure over the past 3 years. At the same time, against the background of a general decrease in the number of working-age population, its share in the total number of citizens of this category as a whole in Samara region (1443 thousand people) (18%) has not changed.

Compared to 2019, in 2020 in rural areas, in addition to a decrease in the number of employed able-bodied rural population by 5.8 thousand people (by 2.2%), there was a decrease in the number of unemployed by 1.6 thousand people (by 9.5%). Due to demographic factors, this decrease, which for the employed able-bodied population amounted to 1.5% in 2020, for the number of unemployed – 16%, is typical for the entire region.

At the same time, the number of rural working-age population who do not operate in the labor market for the production of goods, performance of works and the provision of services, i.e. which is not part of the labor force, increased by 6.9% over 2020-2021. The number of such a population amounted to 71.5 thousand people, which almost corresponds to the value of 2017 (71.2 thousand people). This increase is mainly caused by a 20% increase in the number of pensioners and citizens employed in personal subsidiary farms by 34%, as well as a 20% decrease in the number of full-time students and other reasons.

As of 01.01.2021, the registered unemployment rate in the municipal districts of Samara region was 1.14%, which is 0.27 p.p. higher than the regional value (0.87%) and 0.05 p.p. higher than the value for the same date in 2020.

Since the beginning of 2020, 8435 villagers, or 25.6% of all applicants, have applied to the employment service, 5691 people were recognized unemployed. In total, 5312 people were employed, or 63% of the total number of rural residents that applied to the employment service.

Traditionally, the main place of employment for the rural population is the agri-food cluster. The structure of the agri-food cluster consists of five main subsectors: crop production, animal husbandry, fish farming and fishing, traditional economic activities and food industry [3, 4].

The agricultural food cluster, although being one of the main clusters in Samara region, is far below in terms of employment of the region's population and income to other sectors of the economy (Fig. 1) [5].

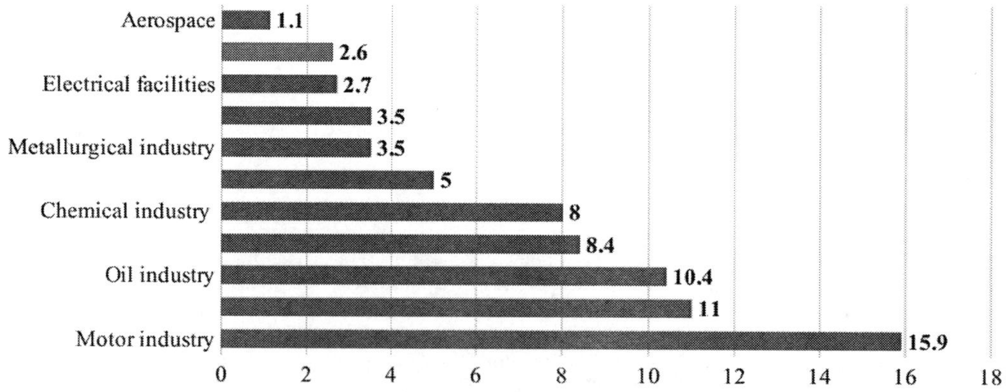

Fig. 1. Share of clusters in the total output of all clusters in Samara region, % (Compiled by the authors based on thematerials of the territorial body of the Federal State Statistics Service for Samara region, 2022).

Three global trends stand out in the agricultural sector around the world, each of which can open up new opportunities for the development of Russian enterprises [6, 7], including agricultural enterprises in Samara region:

– an increase in food consumption is associated with the population growth and increase in the share and income of the middle segments of the population;
– a large-scale use of genetic selection and engineering results plays an important role in the development of new technologies involving increased production in agriculture and improved properties of products;
– a growth in biofuel production associated with the development of its production technologies.
– The agricultural sector of the Russian Federation includes the following main trends, each of which creates new tasks for Russian companies:
– an increase in the export of grain and leguminous crops, any restrictions on grain export due to Russia's entry into the World Trade Organization are accompanied by an increase in prices in the domestic market;
– a ban on the import of meat with food additives prohibited in the Russian Federation and the development of animal husbandry.

Thus, the development of rural areas and the prosperity of the agricultural sector simply need the state support [8, 9]. The state program "Comprehensive

Development of Rural Areas" is directly aimed at the achievement of these objectives.

The state program for the comprehensive development of rural areas is aimed at achieving the following objectives:

- to commission (acquire) housing for citizens living in rural areas;
- to improve the living conditions of rural citizens, including by providing mortgages (loans) at a preferential rate;
- to implement socially significant projects for the improvement of rural areas;
- to increase the number of employees studying in federal state educational organizations of higher education within the Ministry of Agriculture of the Russian Federation;
- to increase the number of students studying in federal state educational organizations of higher education within the Ministry of Agriculture of the Russian Federation engaged in students' on-the-job training;
- to arrange engineering infrastructure facilities and improve sites located in rural areas meeting the requirements of compact housing development;
- to increase the level of gasification of residential buildings (apartments) with pipeline gas in rural areas;
- to increase the level of provision of drinking water to the rural population;
- to implement projects for the integrated development of rural areas within the framework of the departmental target program "Modern appearance of rural areas" of the state program of the Russian Federation "Comprehensive Development of Rural Areas";
- to increase the length of public roads of local importance in Samara region.

Program activities are carried out in 27 municipal districts, as well as in villages of 2 urban districts, 250 rural settlements. This will improve the quality of life of almost 600 thousand people, which is 93% of the total rural population of Samara region.

The program is being implemented in eight areas: water supply, gas supply, integrated housing, construction of housing provided under a housing lease agreement, improvement of housing conditions for citizens living in rural areas, improvement of rural areas, modern appearance of rural areas, promotion of rural employment.

As part of the direction for integrated housing development, residential complexes are being built in the villages of Bolshaya Chernigovka, Malaya Malyshevka, Kinelsky District, Surgut and Svetlodolsk villages, villages of Sergievsky District, while the work in Surgut village has been fully completed. 651 projects are being implemented in 27 municipal districts of Samara region and 2 urban districts. Currently, 83.4% of federal and regional funds have been spent.

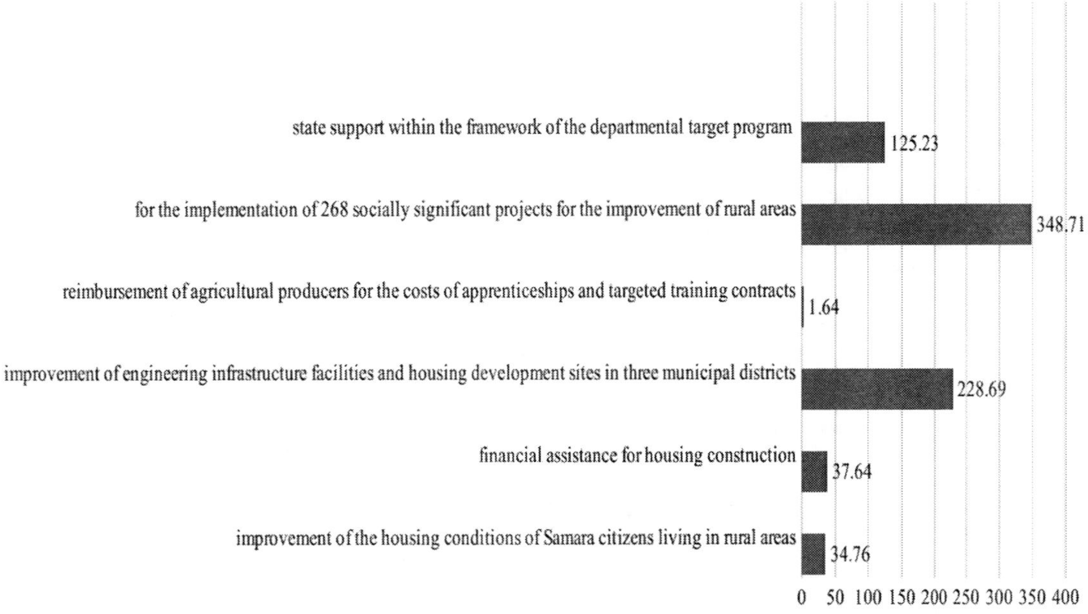

Fig. 2. Financial support for some areas of the "Comprehensive Development of Rural Areas" in Samara region, mln rub. (Compiled by the authors based on the materials of the territorial body of the Federal State Statistics Service for Samara region, 2022).

In accordance with the data of the Ministry of Agriculture of the Russian Federation in Samara region for 2021, more than 800 million rubles were allocated for the "Comprehensive Development of Rural Areas". Within the framework of the program and subroutines more than 98% of funds were spent (Fig. 2).

In this regard, Samara region became one of the regions that managed to attract funds for the next year.

In total, it is planned to build 723 facilities as part of measures aimed at the improvement of rural areas over the coming period before the end of the program.

Currently, the development of rural areas requires the government of Samara region to implement a set of the following measures:

- to foster efforts to the development of the agrarian economy;
- to support small businesses;
- to ensure the development of market infrastructure;
- to improve the standard of living of the rural population through the housing affordability.

4 Conclusion

The analysis of this program makes it possible to conclude that it is aimed at a whole list of development objects: development of the labor market (personnel potential), housing construction in rural areas, increased level of household improvement, formation of a modern appearance of rural areas, arrangement of engineering infrastructure facilities and improvement of sites located in rural areas. However, they are absent or presented in the minimum scope of the program for the development of agricultural production itself [10]. But it is production that determines the development of the entire system of rural areas.

Thus, the restoration and development of agricultural production can be added to all the above. Besides, it is worth considering different requirements for different territories, as well as changing the mechanism for implementing the rural development program (Fig. 3).

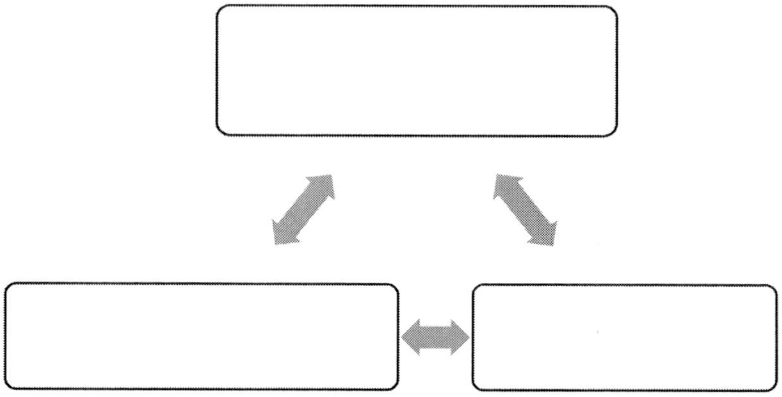

Fig. 3. Stages of the proposed rural development program implementation mechanism (Compiled by the authors).

At the same time, it is important to most accurately determine the needs of a particular territory and add some measures:

– to determine the powers of federal and regional state authorities, local authorities in the field of rural development;
– to establish the implementation stages of the state program for the medium-term development of rural areas, to determine its financial support and mechanisms for the implementation of specified measures;
– to ensure regular monitoring of rural development;
– to ensure the participation of economic actors, citizens, public organizations and associations in rural development.
– Besides, when adjusting the state programs of the Russian Federation in the fields of education, health care, culture, tourism, etc., it is necessary to include measures for the development of rural areas, as well as to form a pool of rural development projects.
– Thus, the integrated development of rural areas will solve three tasks:
– to increase the competitiveness of the agricultural sector – the main place of employment of the rural population;
– to ensure sustainable management of natural resources;
– to achieve a balanced territorial development of the rural economy, including the creation and sustainability of jobs.

References

1. M. Viccaro, S. Romano, C. Prete, M. Cozzi, *Land Use Policy* **111,** 105742 (2021)
2. A. V. Kostyukov, O. D. Pritulina, S. G. Davydova, *IOP Conf. Series: Earth and Environmental Science* **852 (1),** 012053 (2021)
3. O. N. Kusakina, E. G. Agalarova, E. A. Kosinova, Y. V. Rybasova, E. V. Skiperskaya, *Growth Poles of the Global Economy: Emergence, Changes and Future Perspectives: Lecture Notes in Networks and Systems* (Plekhanov Russian University of Economics; Luxembourg) pp 811–822 (2020)
4. B. B. Sharaldaev, V. G. Belomestnov, I. A. Sharaldaeva, V. V. Budazhapov, Y. Y. Grebnev, *IOP Conf. Series: Earth and Environmental Science* **885 (1),** 012010 (2021)
5. O. Kurlikov, A. Volkonskaya, O. Mamai, V. Parsova, *Proceedings of 20th International Scientific Conference: Engineering for rural development*, pp 464–470 (2021)

6. A. Chaplitskaya, W. Heijman, J. Van Ophem, O. Kusakina, Sustainability **13 (6),** 3509 (2021)
7. Y. Jiang, H. Long, C. Ives, R. Chen, Y. Zhang, *Habitat International* **121,** 102514 (2022)
8. L. Li, S. Ma, Y. Zheng, X. Xiao, *Land Use Policy* **114,** 105939 (2022)
9. A. V. Tolmachev, A. I. Trubilin, V. V. Smirnov, K. N. Shamrov, I. A. Papahchyan, *IOP Conf. Series: Earth and Environmental Science* **745,** 012003 (2021)
10. R. I. Safiullaeva, T. A. Neshchadimova, I. A. Demchenko, O. I. Solovieva, S. V. Zolotarev, *Lecture Notes in Networks and Systems* **205,** 229–238 (2021)

Ways to reduce carbon dioxide emissions from arable machinery and tractor units

Camill A. Khafizov, *Ramil* N. Khafizov, *Azat* A. Nurmiev*, and *Stanislav* A. Sinitsky

Kazan State Agrarian University, 65, K. Marx St., Kazan, Republic of Tatarstan, 420015, Russian Federation

Abstract. The article is devoted to the search for ways to reduce carbon dioxide emissions in the technological operation of plowing. Ways to reduce CO_2 emissions were determined on the basis of computational experiments conducted using a mathematical model of an arable unit compiled according to the efficiency criterion - the minimum integral emission of carbon dioxide. Integral CO_2 emission, in addition to direct and indirect CO_2 emissions over the life cycle of the arable unit, includes CO_2, which could be absorbed from the atmosphere by the crop lost due to the influence of the parameters of the unit. The influence of the parameters of the tractor and the unit on the emission of carbon dioxide during its manufacture, technical and production operation is analyzed. Computational experiments were conducted to identify the influence of some external factors (field area, seasonal load volume, etc.) on the efficiency of the arable unit.

1 Introduction

The world community was puzzled by a new problem – there is a global warming, the results of which can be catastrophic for many countries and peoples [1-4]. Many see the main reason for the coming changes in the active participation of people in the process of increasing carbon dioxide in the atmosphere, through the burning of fossil fuels, deforestation, increasing the number of livestock in order to feed the ever-increasing population on the globe. Mobile machines powered by fossil fuel energy are also widely used in agricultural production. One of the most energy-intensive technological operations is plowing, which cannot be completely abandoned for a number of reasons, including over compaction of the soil by tractor propulsion and the concomitant decrease in crop yields [5–8].

In this regard, the article considers ways to reduce the emission of integral carbon dioxide during this most energy-intensive, technological operation.

2 Materials and methods

To identify ways to reduce carbon dioxide emissions into the atmosphere, the method of mathematical modeling of the operation of the arable machine-tractor unit (MTU) based on the use of an integral criterion for optimizing the parameters of equipment is used - the total emission of carbon dioxide [9-11].

The amount of carbon dioxide emitted into the atmosphere during the manufacture, technical and production operation of arable units is a dependence:

$$CO_2 = CO_{2m.trac} + CO_{2m.am} + CO_{2m.tr} + CO_{2rmo} + CO_{2a.d.} + CO_{2dr} + CO_{2fuel} + CO_{2agr} + CO_{2s.c.} \rightarrow \min, \quad (1)$$

where CO_2 is the specific total carbon dioxide emission, kg/ha;

$CO_{2m.trac}$, $CO_{2m.am}$, $CO_{2m.tr.}$ are the amount of CO_2 emitted respectively in the manufacture of a tractor, agricultural machine, trailer, per 1 ha, kg/ha;

CO_{2rmo} is the amount of CO_2 emitted into the atmosphere during the repair and maintenance of a tractor, trailer and agricultural machine, kg/ha;

$CO_{2a.d.}$ is the amount of CO_2 emitted into the atmosphere during the assembly and disassembly of the unit, kg/ha;

CO_{2dr} is the amount of CO_2 emitted by the machine operator when driving a tractor, kg/ha;

CO_{2fuel} is the amount of CO_2 emitted during the combustion of fuel by the tractor engine, kg/ha;

CO_{2agr} is the amount of CO_2 from the atmosphere not bound by the harvest, due to its losses due to the violation of the agrotechnical terms of the technological operation, kg/ha;

$CO_{2s.c.}$ is the amount of CO_2 not bound by the harvest from the atmosphere due to its losses owing to soil compaction by tractor propulsion, kg/ha.

The proposed sum of the various components of the integral emission of carbon dioxide into the atmosphere will be the criterion for optimizing the parameters and operating modes of machine and tractor units in order to minimize harmful emissions.

When developing a mathematical model of an arable unit according to the specified indicator of the efficiency of its work, the relationship of each component of the optimization criterion with the parameters of the tractor and agricultural machine, as well as environmental

* Corresponding author: azat-nurmiev@mail.ru

factors, was revealed. In parallel, mathematical models of the arable unit were built on other indicators of the effectiveness of its work for comparative analysis.

3 Results and discussion

On the basis of the developed mathematical models of arable machine-tractor units according to the optimization criteria we identified minimum direct and indirect total energy costs, taking into account the energy of the lost crop; minimal emission of carbon dioxide at the stages of manufacture and operation of equipment. We take into account the amount of carbon dioxide that could be absorbed from the atmosphere, the crop yield lost during the operation of the unit; maximum capacity of the unit in one hour of shift time.The minimum fuel consumption per hectare of the area treated by the unit and the maximum traction efficiency of the tractor [12, 13], computational experiments were specified, some of the results of which are given in Table 1.

As can be seen from the table, the optimal values of the aggregate parameters, including the optimal basic parameters of the tractor, coincide according to three optimization criteria. This is due to the need to provide the arable unit with maximum productivity. In the case of the use of integral optimization criteria, maximum productivity ensures minimal crop losses due to violation of agrotechnical terms for the implementation of technological operations, which occupies a dominant position in the structure of total energy costs and carbon dioxide emissions, see Figure 1.

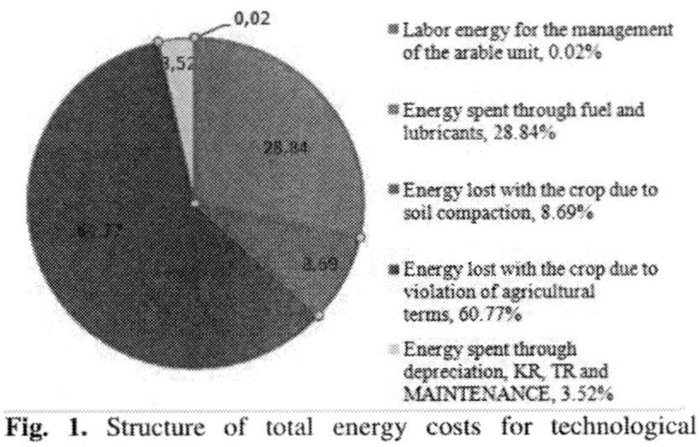

Fig. 1. Structure of total energy costs for technological ploughing operations.

To reduce the emission of carbon dioxide, it is necessary first of all to reduce crop losses due to violation of the agrotechnical terms of ploughing and fuel consumption for the implementation of the technological operation.

Figure 2 shows the dependence of carbon dioxide emissions on the weight of the tractor and the power of its engine.

The graph shows that CO_2 emissions are largely dependent on both the weight of the tractor and the power of its engine.

We will carry out the same calculations with an increased range of the main parameters of the tractor (the weight of the tractor is up to 260 instead of 220 kN and the power of its engine is up to 560 instead of 400 hp). The results of the calculation are presented in Figure 3.

The following results are obtained:

the optimal plough grip width (Bopt) = 5.25 m;

the optimal speed of the unit (Vopt) = 9.5 km / h;

the optimal tractor weight (Mtiopt) = 140 kN;

the optimum engine power (Neopt) = 520 hp;

the total emission of carbon dioxide (CO_{2min}) = 379.15 kg/ha.

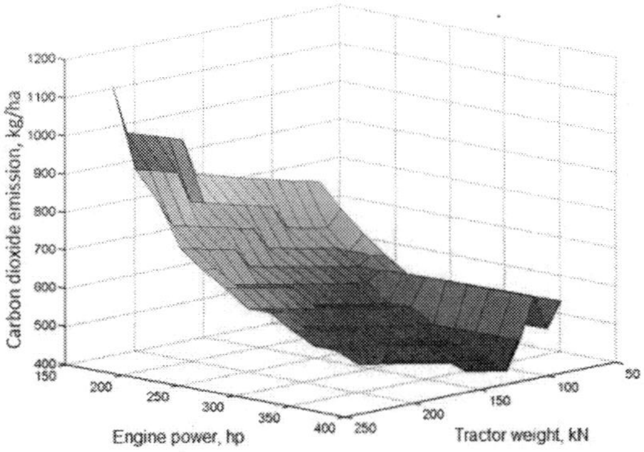

Fig. 2. Dependence of carbon dioxide emission on the main parameters of the tractor whenplowing.

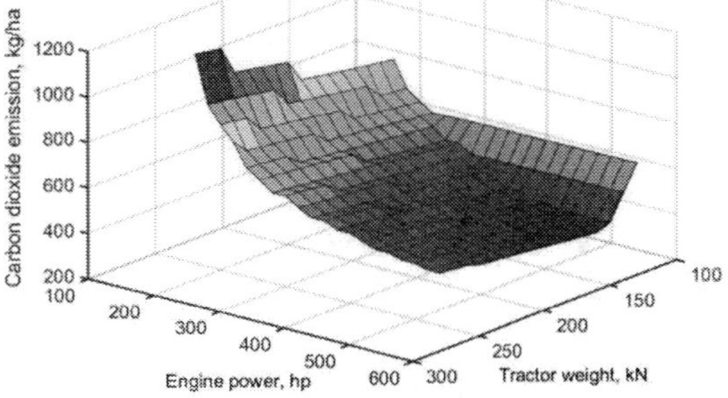

Fig. 3. Dependence of carbon dioxide emission on the main parameters of the tractor when plowing with an increase in their maximum limits.

As you can see, the optimal parameters of the tractor increased, CO_2 emissions decreased from 434.57 kg/ha to 379.15 kg/ha.

We will carry out the same calculations with an increase in the upper limit of the plough grip width from 5.25 to 5.95 m.

Calculation results are:

the optimal plough grip width (Bopt) = 5.95 m;

the optimal speed of the unit (Vopt) = 8.5 km/h;

the optimal tractor weight (Mtiopt) = 150 kN;

the optimum engine power (Neopt) = 520 hp;

the total emission of carbon dioxide (CO_{2min}) = 378.08 kg/ha.

As can be seen from the results of the calculations and Figure 4, the optimal parameters of the tractor and the arable unit have changed, while there is a slight decrease in carbon dioxide emissions compared to the previous calculation.

Table 1. Optimal parameters of the arable unit according to various optimization criteria for specified working conditions.

Options	Minimum energy consumption, MJ/ha	Minimum carbon dioxide emissions, kg/ha	Minimum fuel consumption, kg/ha	Maximum tractor efficiency	Maximum productivity, ha/h
Tractor weight, kN	130	130	100	110	130
Engine power, hp	398,11	398,11	214,55	266,79	398,11
Working width, m	5,25	5,25	4,55	4,2	5,25
Unit speed, km/h	7,5	7,5	5	6,5	7,5
The value of the optimal value of the optimization criterion	3405,2	434,57	18.87	0.68	2.32

Initial data for calculations:
Field area, = 100 hectares;
Rutting length = 1 km;
Moving distance of the unit = 3 km;
Seed density = 800 kg/m3;
Coefficient of strength of the bearing surface of the field = 0.9;
The amount of work performed by the unit = 500 hectares;
Number of tractors performing the operation = 1;
The number of hours of operation of the unit per day = 16 hours;
The planned yield of the main and by-products = 40 c/ha;
Tractor tire pressure = 0.16 MPa;
Number of wheels on one side of the tractor propulsion = 1;
Coefficient of adhesion of the wheels of the tractor to the soil = 0.7;
Coefficient of resistance to rolling wheels of the tractor = 0.12;
Plowing depth = 0.25 m;
Mass of CO_2 absorbed by culture = 243 kg/c
Maximum tractor weight = 220 kN;
Maximum engine power = 400 hp;
Maximum working width of the unit = 5.25 m;
Maximum speed of the unit = 10 km/h.

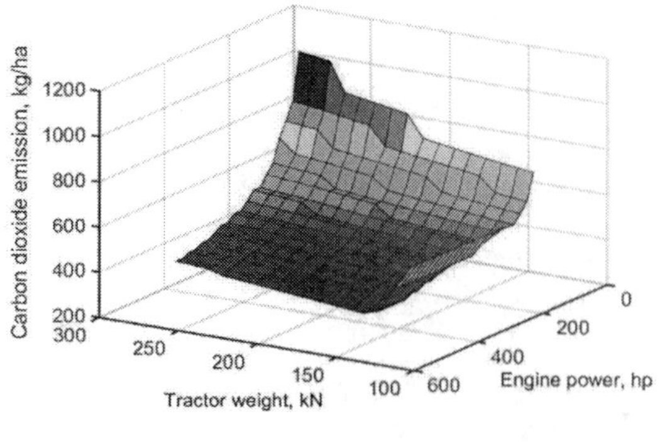

Fig. 4. Dependence of carbon dioxide emission on the main parameters of the tractor when plowing and increasing the maximum working width to 5.95 m.

Figure 5 shows the dependence of CO_2 emission on the grip width of the unit and its speed. The effect of increasing the plough's grip width on reducing CO_2 emissions was more intense than the effect of increasing the speed of the unit.

Let us calculate the optimal load on the arable unit for the autumn season of work for the given conditions of its work. The results of the calculations are shown in Figure 6.

As can be seen from Figure 6, the integral emission of carbon dioxide by the arable unit into the atmosphere largely depends on its seasonal load, reaching a minimum of 200 kg/ha at a load of 40-50 hectares.

We consider how the number of tractors engaged in plowing affects the emission of CO_2 when performing a seasonal load of 1500 hectares. The results of computational experiments are shown in Figure 7.

The optimal number of tractors for processing the field area of 1500 hectares is 3 pieces. At the same time, the minimum integral emission of carbon dioxide is 345 kg/ha. An increase and decrease in the number of occupied units leads to an increase in the integral emission of carbon dioxide.

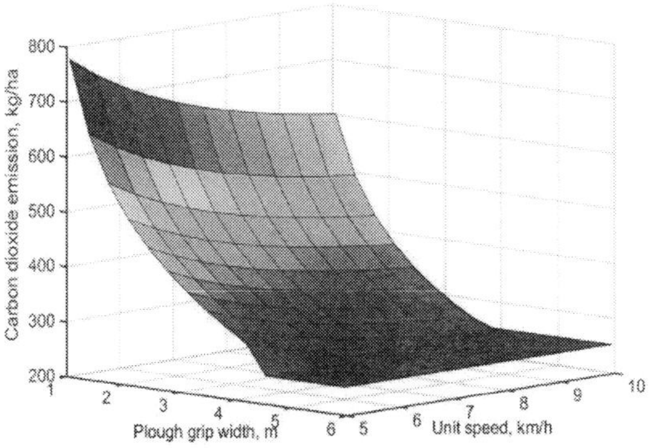

Fig. 5. Dependence of carbon dioxide emission on the operating speed and working width of the arable unit.

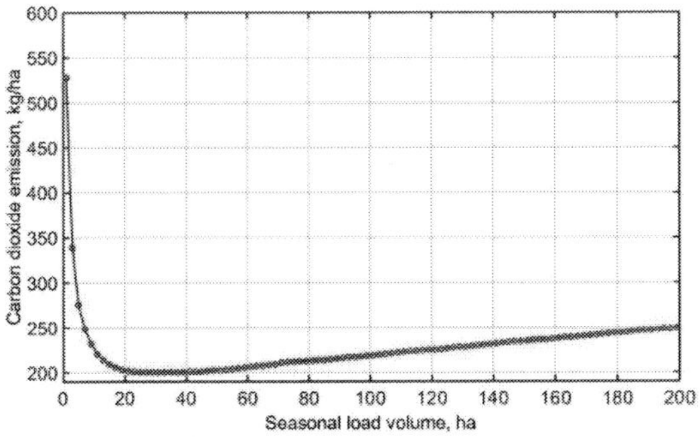

Fig. 6. Change in the emission of carbon dioxide into the atmosphere by the arable unit depending on the volume of its seasonal load.

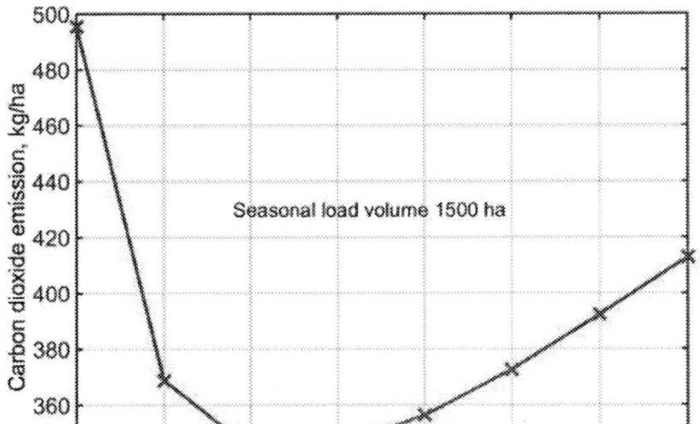

Fig. 7. Change in the emission of carbon dioxide into the atmosphere by the arable unit depending on the number of units engaged in the treatment of 1500 hectares of area.

As you can see from Figure 8, the lower the tire pressure of the tractor's wheels, the lower the emission of integral carbon dioxide. The minimum permissible tire pressure depends on the vertical load on the wheel and its values are given in the tire manufacturer's instruction manual [14–16].

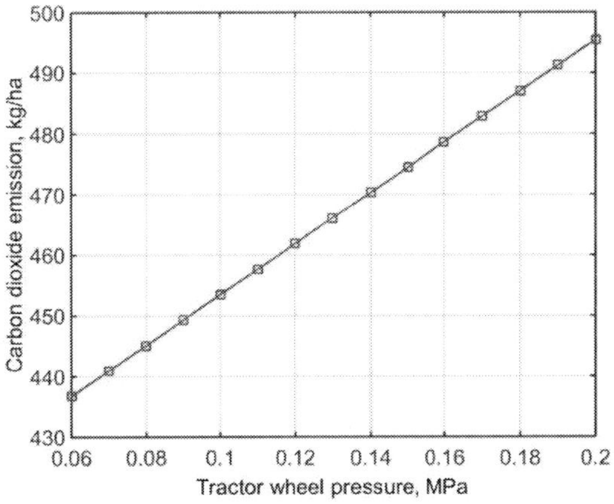

Fig. 8. Change in the emission of carbon dioxide into the atmosphere by the arable unit depending on the tire pressure of the wheels of the tractor.

The number of wheels on one side of the tractor also affects the emission of CO_2 into the atmosphere, as can be seen from Figure 9. If, with twin and built wheels of a tractor, a minimum pressure of 0.06 MPa is set in its tires, then working with built wheels becomes less profitable than with single and twin wheels on board, due to the increase in the weight of the tractor, as well as the area of the field compacted by the propulsion engines.

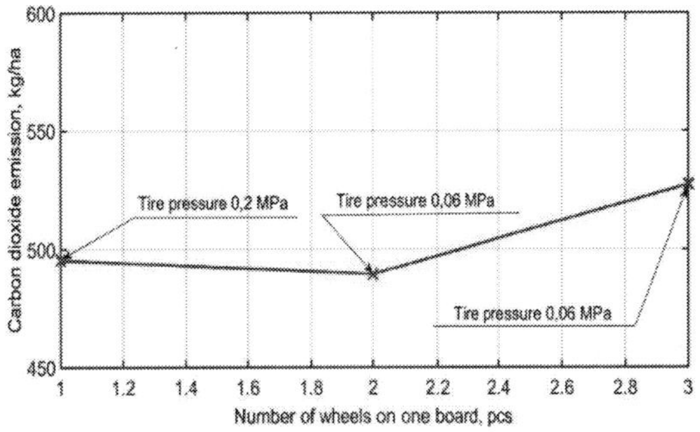

Fig. 9. Change in the emission of carbon dioxide into the atmosphere by the arable unit depending on the number of wheels of the tractor on one board.

To modernize tractors, sometimes it is necessary to change the brand and engine power [17–27]. The optimal engine power for a tractor weighing 156.8 kN (K-744R2) is in the range of 495–550 hp, as can be seen from the results of the calculations in Figure 10. With lower engine power, carbon dioxide emissions increase, and increasing engine power above the optimal value does not lead to a decrease in CO_2 emissions due to the speed limit of the arable unit within 10 km / h according to the conditions of the initial data.

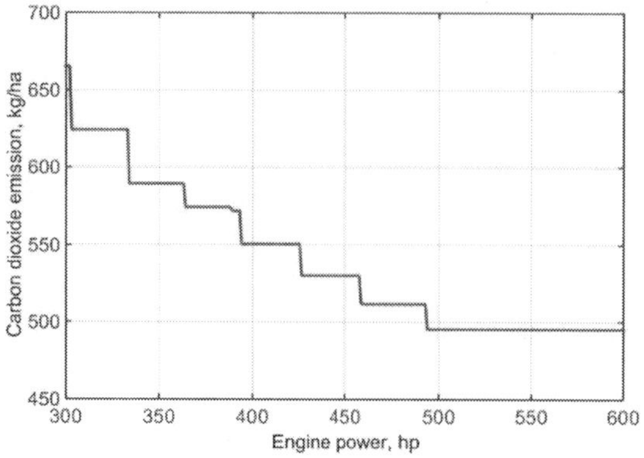

Fig. 10. Change in the emission of carbon dioxide into the atmosphere by an arable unit depending on the power of its engine at a weight of 156.8 kN.

In practice, usually the fields have different area and configuration. Figure 11 shows the dependence of carbon dioxide emissions into the atmosphere on the area of the field. From the figure it can be seen that arable tractors of high traction classes, in order to reduce CO_2 emissions, must be used in fields with an area of more than 10 hectares.

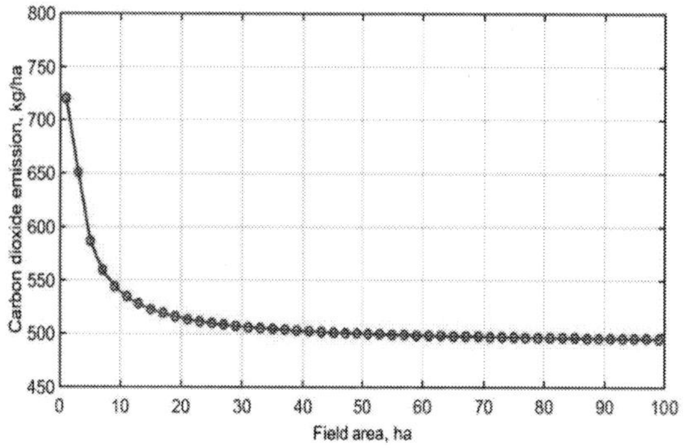

Fig. 11. Change in the emission of carbon dioxide into the atmosphere by the arable unit depending on the area of the treated field.

4 Conclusion

Computational experiments conducted on the basis of mathematical models of arable units compiled using various optimization criteria, including total energy costs, total carbon dioxide emissions, maximum productivity of the arable unit revealed a coincidence of optimal parameters for all three optimization criteria. The current picture is explained by the need to ensure the maximum productivity of the unit to minimize crop losses from violation of the agrotechnical terms of its work. The energy of the lost crop in the structure of total energy costs for ploughing occupies a dominant share and reaches 60% or more.

To reduce the integral emission of carbon dioxide into the atmosphere, the working arable unit must select the conditions of its operation, a unit with the optimal weight of the tractor, its engine power, grip width and operating speed. The ploughing unit should include a tractor with a weight in the range of 130-150 kN, with an engine power of 400-520 hp; a plough with a width of grip that maximizes the tractor's thrust load within the permissible towing limits to operate at a speed limited by engine power.

To reduce the emission of carbon dioxide, the pressure in the single tires of the tractor must be within the minimum value allowed by the manufacturer and selected on the basis of the vertical load on one wheel of the tractor. Twinning the wheels on the sides of the tractor is effective in reducing the pressure in the tires of the wheels to a minimum value, manifested, among other things, by increasing the force of adhesion of the wheels to the soil, increasing the tractive effort of the tractor with permissible slippage.

4. The volume of seasonal load on a ploughing tractor with these parameters should be at least 20 hectares. An increase in seasonal load leads to an increase in carbon dioxide emissions with an intensity of 322 grams per 1 hectare of load growth. On plowing, with a certain amount of seasonal load, there is always and it is possible to calculate the optimal number of tractors that provide minimal CO_2 emissions into the atmosphere. For a work volume of 1500 hectares, the optimal number of arable units is 3 pieces.

References

1. *Effects of global warming*. Retrieved from: https://www.nationalgeographic.com/environment/article/global-warming-effects.
2. L. Srivastava, *Catastrophic climate change*. Retrieved from: https://globalchallenges.org/catastrophic-climate-change
3. B. Parker. *Is Catastrophic Climate Change Inevitable?* Retrieved from: http://ccdatacenter.org/documents/IsCatastrophicClimateChangeInevitable.pdf).
4. E. G. Kolomyts, Ecological effects of modern global warming in forest geosystems of the Volga basin, *Geography and natural resources* **3 (162),** 43–54 (2020)
5. Y. A. Izotov, Combine plowing, *Innovative technologies and technical means for the agro-industrial complex: materials of the international scientific and practical conference of young scientists and specialists* (Voronezh State Agrarian University named after Emperor Peter I), pp 449–454 (2018)
6. M. R. Akhmetzyanov, Density of soil composition from the use of various methods of processing on crops of spring wheat in the conditions of the Republic of Tatarstan, *Agriculture and food safety: technologies, innovations, markets, personnel: Scientific works of the II International Scientific and Practical Conference dedicated to the 70th anniversary of the Institute of Mechanization and Technical Service and the 90th anniversary of the*

Kazan Zootechnical School (Kazan State Agrarian University), pp 268–274 (2020)

7. P. A. Postnikov, Density of the gray forest soil in dependence on the methods of basic processing, *Vladimirskiizemledelets* **4 (82),** 8–10 (2017)
8. D. T. Khaliullin, A. Belinsky, A. R Valiev, R. R Lukmanov, G. Bourges Optimization of plow adjustment, *BIO Web of Conferences* **27,** 00103 (2020)
9. K. A. Khafizov, Reduction of total energy costs on technological operations in the agro-industrial complex - the way to reduce greenhouse gas emissions in the atmosphere, *Bulletin of Kazan State Agrarian University* **3 (63),** 43–47 (2021)
10. R. N. Khafizov, Ways to reduce carbon dioxide emissions into the atmosphere in production processes in plant growing, *Bulletin of Kazan State Agrarian University* **3 (63),** 38–42 (2021)
11. K. A. Khafizov, System mathematical model of vehicles according to the optimization criterion - minimum emission of carbon dioxide into the atmosphere, *Dynamics of mechanical systems: materials of the II International Scientific and Practical Conference dedicated to the memory of Professor A. K. Yuldashev,* (Kazan State Agrarian University), pp 122–130 (2021).
12. R. N. Khafizov, Results of a multifactor experiment to determine the dependence of the maximum pressure of the tractor wheels on the soil on the parameters of the tractor and the physical and mechanical properties of the soil. Equation analysis, *Bulletin of Kazan State Agrarian University* **4(42),** 99–103 (2016)
13. K. A. Khafizov, Optimization of parameters and modes of operation of the MTA on the basis of energy analysis, *Tractors and Agricultural Machines* **7,** 32–34 (2006)
14. GOST 7463-2003 *Pneumatic tires for tractors and agricultural machines. Date of introduction* 2005-01-01.
15. *Tractor Belarus 3522.5 – Operating instructions* (3522.5-0000010 RE) (2011)
16. John Deere Operation manual tractors 8130, 8230, 8330, 8430 and 8530 (European edition). Omar259994 edition G6 (RUSSIAN). (Waterloo Works (European Edition) Printed in Germany, 2006)
17. A. A. Makushin, A. T. Kulakov, A. S. Denisov, B. P. Zagorodsk, Experience of modernization of the machine-tractor park during repair, *Machinery and equipment for the village* **1,** 41–44 (2009)
18. A. A. Makushin, A. T. Kulakov, N. I. Svetlichnichny, Restoration and modernization of machines at KAMAZ factories, *Tractors and Agricultural Machines* **12,** 46–48 (2007)
19. *Will MTZ tractors wait for powerful domestic engines?* Retrieved from: https://agronews.com/by/ru/news/analytics/2019-03-18/35363
20. I. S. Chernyavsky, A. V. Ustinenko, A. V. Bondarenko, Analysis of the loading ability of the transmission of the tractor t-150K when installing an engine with a capacity of 250 HP, *VisnikSevNTU* **148,** 57–63 (2014)
21. I. G. Galiev, K. A. Khafizov, F. Kh. Khaliullin, Modernization of the lubrication system of the bearing unit of the turbocharger of the auto-tractor engine, *Bulletin of Kazan State Agrarian University* **1 (52),** 71–76 (2019)
22. F. Kh. Khaliullin, G. V. Pikmullin, A. A. Nurmiev, M. A. Lushnov, Estimation of design parameters of the crank-connecting rod mechanism of engines for mobile agricultural machines, *BIO Web of Conferences* **37,** 00076 (2021)
23. F. Kh. Khaliullin, The influence of the design parameters of the crankshaft on its elastic-damping properties at torsional oscillations, *Innovations and Investments* **7,** 99–102 (2021)
24. F. Kh. Khaliullin, Determination of statistical data of conditional probabilities of the technical condition of internal combustion engines when compiling the Bayes diagnostic table, Bangkok, Thailand, *IOP Conf. Ser.: Mater. Sci. Eng.* **635,** 012017 (2019)
25. S. A. Sinitsky, et al., Investigation of the effect of air supply on the effective engine performance of a machine-tractor unit under unsteady load, *BIO Web of Conferences* **27,** 00025 (2020)
26. S. A. Sinitskiy, V. M. Medvedev, Influence of dynamic factors on the indicators of the MTA engine at an unstable load, *Technics and Equipment for the Village* **4 (274),** 16–19 (2020)
27. V. M. Medvedev, S. A. Sinitskiy, Mathematical model of evaluation of dynamic indicators of the MTA engine at an unestablished, *Bulletin of Kazan State Agrarian University* **2 (53),** 106–110 (2019)

Influence of the evenness of seeding using a disk-cell seeding mechanism on the productivity of scarlet amaranth

Evgeny I. Artamonov, *Sergey N.* Zhiltsov, *Maxim P.* Erzamaev*, *Dmitry S.* Sazonov, *Maxim S.* Prikazcnikov, and *Nikolai A.* Cherkashin

Samara State Agrarian University, 2, Uchebnaya St., Kinel, 446442, Russian Federation

Abstract. The paper presents the results of a field study of the effect of the evenness of seeding using a disk-cell seeding mechanism on the fresh yield and Scarlet amaranth grain of Kinelsky 254 variety. The problem of the amaranth evenness of seeding was solved by developing an experimental seeder of precision sowing for small-seeded crops. Field experiments were carried out on the fields of the Volga Research Institute of Selection and Seed Production named after P. N. Konstantinovin 2019-2021. To control the qualitative parameters of sowing, amaranth was sown with ballast (blind seeds) with a VS-4.2 vegetable seeder. The study revealed that the evenness of seed distribution in a row during sowing with an experimental seeder was 22%, for plants – 24.6%, control – 24.6 and 72.1%, respectively. The use of an experimental seeder made it possible to sow in accordance with agrotechnical requirements and create the best conditions for plant nutrition, which led to an increase in the fresh yield of Scarlet amaranthof Kinelsky 254 variety by 2.6 times, grain – by 1.6 times.

1 Introduction

The intensive development of animal husbandry associated with active import substitution in the whole country and in the Volga region in particular sharply raise the question of the need to increase the production and use of highly efficient feed [1, 2, 3]. One of the promising, unconventional forage crops with high yield and nutritional value is the Scarlet amaranth, the interest in which has grown in recent years [2, 3]. The Volga Research Institute of Selection and Seed Production named after P. N. Konstantinov(VRISSP named after P. N. Konstantinov) licensed the Kinelsky 254 amaranth and conducts the selection studies on its reproduction and creation of new varieties of fodder and grain [2, 3].

Scarlet amaranth belongs to the Amaranth family [4, 5, 6] – in most cases it is an annual large herbaceous plant with a dense spike-panicledinflorescence and small flowers. In different agricultural regions of Russia the plant can reach a height of up to 2.5 m, has a straight or branched stem, leafy to the base, a wide fleshy leaf. The weight of one plant can reach 8 kg, and the stem in the root part – 5 cm.

The nutritional value of Scarlet amaranth, in comparison with traditional legumes and cereal crops, is characterized by higher protein content and a well-balanced amino acid composition. Some basic valuable indicators of this culture are given in Tables 1 and 2 [4].

Table 1. Chemical composition of amaranth in the early flowering phase.

Indicator	In %of the dry weight	
	Varieties	
	Stere	Scarlet amaranth
Drymatter	13.5...20.2	17.1...19.0
Nitrogen-freeextractivesubstances	44.8...46.7	36.1...43.4
Protein	16.4...17.7	17.7...19.6
Cellulose	26.6...27.4	27.2...27.3
Ash	10.3...10.8	13.7...13.9
Sugar	4.7...6.5	5.3...7.9
Ascorbicacid, mg, %	30.8...36.2	28.3...28.7

* Corresponding author: erzamaev_mp@mail.ru

It is known that the green crop of amaranth can be used to prepare silage both in admixture with other crops, for example corn, and in pure form for the feed of cattle and pigs [4, 5]. The selection studies of V.F. Kazarin describe that one of the main conditions for obtaining the highest yield of amaranth, depending on the production purposes for which the sowing is carried out, is the optimal dotted distribution of seeds along the row. So, when cultivating the green crop of amaranth the distance between seeds should be 20–22 mm, and for grain cultivation – 40–45 mm respectively [2, 3]. The purpose of sowing amaranth determines the number of plants in a row per meter run, and it, in turn, is directly related to the sowing rate, germination of seeds and the chosen sowing method. Due to its agrobiological characteristics, the culture is very demanding on the food area, which sets high requirements for an even longitudinal distribution of seeds and plants in a row during sowing. The analysis of the domestic practice of cultivating Scarlet amaranth, as well as the structures of sowing devices and seeders, showed that due to a number of technical and technological limitations, it is not possible to fulfill the agronomic requirements for the longitudinal evenness of seeds in a row, which is caused by the low sowing rate, size, shape and low friction properties of seeds [7].

Table 2. Amaranth yield in comparison with leguminous crops and its nutritional value.

Indicator	Crop				
	Wheat	Corn	Rice	Beans	Amaranth
Biologicalyield, t/hha					
Freshyield, t/ha	–	32.0	–	–	56.4
Grainyield, t/ha	1.2–2.5	1.7	–	–	2.3
Substancecontent, %					
Protein	13.00	7.68	7.60	21.48	15.54-29
Fat	1.70	5.00	2.20	1.96	7.31
Cellulose	2.90	2.46	6.40	5.70	5.2
Ash	1.50	165	3.40	4.61	3.61
Copper	28.00	7.00	7.00	8.00	6.00
Sodium	0.01	0.01	0.01	0.02	0.12
Potassium	0.40	0.48	0.12	1.30	0.57
Phosphorus	0.41	0.27	0.18	0.41	0.54
Aminoacids, mg %					
Arginine	3.80	4.20	6.90	6.20	10.00
Lysine	1.20	0.70	3.80	0.00	8.00
Leucine	7.70	12.50	8.20	8.20	5.70

The studies revealed that the seeding rate may reach 0.1...0.9 kg/ha and depends on the established width of rows, while the depth of seeding may be 1...1.5 cm. For accurate or dotted seeding, seed germination should be at least 92% [2, 3, 6].

Amaranth is sown in the farms of Samara region by grain and vegetable seeders, and it is necessary to use some filler: calibrated river sand, blind amaranth seeds or seeds of other crops, granulated mineral fertilizers in a proportional ratio of 1:10, which significantly complicates and increases the cost of crop cultivation technology, and most importantly reduces the yield directly related to sowing quality indicators [3, 6, 7].

Such domestic grain seeders as SZT-3.6 equipped with a special device for sowing clover, melilot, rapeseed, SP-6 – for sowing rapeseed seeds, SLT-3.6 – for sowing in meadows and pastures, SZL-3.6 – grain seeder, to one degree or anotherperform the seeding of Scarlet amaranth in various ways: wide-row planting, row sowing and others. Seed dosing devices in most of these seeders are equipped with coils that do ensure small seeding rates on small-seeded crops, since they form a seed flow in portions of several pieces and throw them out simultaneously at certain intervals. This does not result in precise placement of seeds at the bottom of the furrow at the same inter-seed intervals, but in thickened continuous sowing. Besides, these seeders and their modern modernizations have quite large structural leakages along the entire route of seed movement from the bunker to the bottom of the furrow, where only on small seeds there is a phenomenon of "leakage" through gaps of seed material and, as a result, an increase in the seeding rate, thickening of crops, a decrease in grain yield and fresh yield [1].

In most grass and grain seeders their seeding and dispensing devices are equipped with dampers that are closed on the sowing of amaranth with a wide-row seeding method, however, the listed disadvantages are not eliminated by this operation [1].

The use of seeders by foreign manufacturers, such as: Monocenter– Germany, Unisem– France, Siklow-500– USA and many others, do not allow withstanding the agrotechnical requirements for sowing amaranth due to the fact that the pneumatic transportation of seed material leads to an increase in dispersion along the bottom of a furrow and along its walls and even in cases with pneumatic sowing of amaranth seeds – beyond the furrow, which is explained by the low weight of seeds (weight of 1000 seeds – 0.8 g) and high requirements for air flow pressure evenness. The engineering service specialists of Samara agricultural enterprises use various sealants and gasket material to prevent "leakage" of seed material through the gaps between the parts of the seeders in the dispensing devices and the transportation system. Additional operations to increase seeder tightness affect the seeding agro-termsand increase pre-seeding costs.

Besides, vegetable seeders VS-5.4 and VS-4.2 were also used to cultivate amaranth. Their main advantage is the creation of a compacted bed of the bottom of a furrow, a semi-shaped ploughshare, which allows improving the germination of seeds.

The main disadvantages of considered the seeders include the inability to ensure a small seeding rate in accordance with agricultural demands and to ensure longitudinal evenness of seeds in a row.

Based on the above-mentioned material it can be concluded that when Scarlet amaranth is cultivated for seeds and fresh yield, the best yield can be achieved at minimum sowing rates thus ensuring the even longitudinal distribution of seeds in a row and a wide-row sowing method. Today, the weaknessof the Scarlet amaranth sowing technology is the longitudinal evenness of the distribution of inter-seed intervals.

To study the effect of the evenness of sowing of Scarlet amaranth of Kinelsky 254 on grain and fresh yield.

2 Materials and methods

As a result of close scientific cooperation between the scientists of the Department of Technical Service of Samara State Autonomous University and VRISSP named after P. N. Konstantinov, the problem of even sowing of amaranth was solved by developing an experimental precision seeder for small-seeded crops (Figure 1).

Fig. 1. Experimental seeder (left) and seeding section (right).

The beet seeder SST-12B served the basis for the experimental seeder of precision seeding. For sowing seeds of Kinelsky 254 Scarlet amaranth, the units of the SST-12B seeder equipped with a disk-cell seeding mechanism were designed and manufactured [patent for the invention "seeding device" No. 2347349] with optimal design parameters of working organs justified during preliminary laboratory studies [8, 9].

The experiments to study the effect of the longitudinal evenness of the distribution of seeds and plants of Kinelsky 254 Scarlet amaranth in a row in terms offreshand grain yield were carried out on the fields of preliminary and main variety multiplication of the VRISSP named after P. N. Konstantinov from 2019 to 2021. Amaranth seeding with ballast was used to control the quality of seeding using the vegetable seeder VS-4.2. Blind seeds of the same culture were used as ballast. Seed material was used with a germination rate of 94% [10].

According to the established test method [11], the distribution of seeds during sowing was determined by the seedlings in three accounting sections of 2 m long and the width equal to the width of the seeder grip. A 2-meter ruler with single-centimeter divisions was applied at these sites, and the distance between individual seeds was measured on a cumulative basis (Figure 2) (the first seed was taken as zero).

Fig. 2. Evenness of distribution of amaranth seeds along the row using an experimental seeder.

Similarly, the distribution of plants along the row was determined after the complete emergence of seedlings at each of the three selected sites (Figure 3).

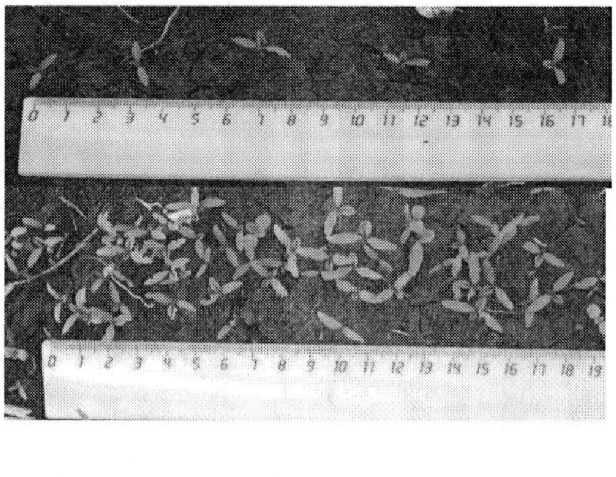

Fig. 3. Evenness of plant distribution along the row; using the experimental seeder VS-4.2.

The evenness of the distribution of seeds and plants was evaluated by the variation coefficient *v*.

3 Results and discussion

As a result of processing the obtained data to determine the longitudinal evenness, variation curves of the distribution of intervals between seeded seeds (Figure 4) and the obtained plant seedlings were built (Figure 5).

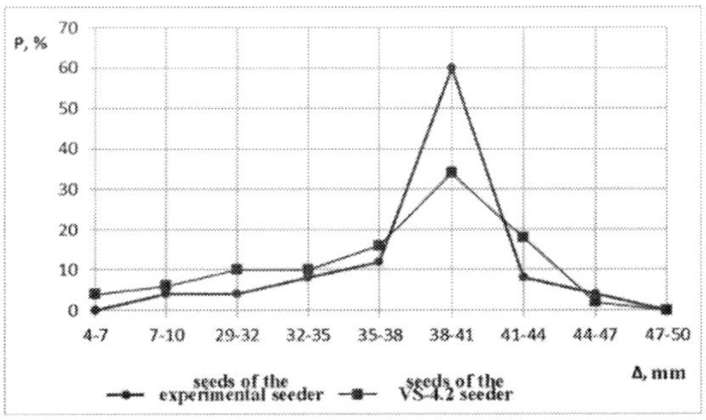

Fig. 4. Distribution of intervals between amaranth seeds in a row.

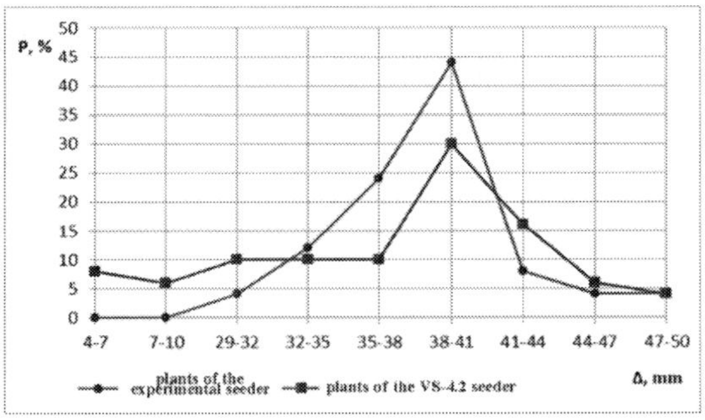

Fig. 5. Distribution of intervals between amaranth seedlings in a row.

The best indicators of the evenness of the distribution of seeds in a row were obtained during sowing with an experimental seeder. The variation coefficient of the intervals between the seeds was $v = 22\%$ for the crops planted with the experimental seeder and $v = 65\%$ for the crops planted by the seeder with a roller seeding mechanism.

The analysis of the variation curves (Figure 5) of the distribution of intervals between plants shows that when sowing with an experimental seeder, the variation coefficient of was $v = 24.6\%$, and with a control seeder – $v = 72.1\%$. A significant difference in evenness is explained by the fact that the seeding mechanismVS-4.2 supplies a mixture of seeds and ballast in portions. Besides, there is additional sowing of the furrow edges from the leakages of the seed transporting system, which leads to an additional increase in the seeding rate and significant thickening of crops. Intraspecific control in tall crops significantly reduces yields.

The uniform distribution of seeds when sowing amaranth using the experimental seeder made it possible to directly affect the crop yield. This is explained by high requirements for the food area of plants due to its size and high growth. The use of the experimental seeder at the VRISSP named after P. N. Konstantinov made it possible toincrease the grain yield on average by 1.6 times over the past three years, and the fresh yield increased 2.6 times compared to the crops planted with by the VS-4.2 (Table 3).

Table 3. Productivity of Kinelsky 254 amaranth using experimental seeder and control seeder VS-4.2(according to VRISSP named after P. N. Konstantinov).

Yield, t/ha	2019		2020		2021		Average	
	Experimental seeder	VS-4.2	Experimental seeder	VS-4.2	Experimental seeder	CO-4,2	Experimental seeder	VS-4.2
Fresh yield	30.0	11.5	55.0	21.2	84.4	32.4	56.3	21.7
Grain	0.9	0.6	1.2	0.75	2.7	1.6	1.6	0.98
Dry matter	7.2	2.7	13.2	5.1	19.4	7.6	13.3	5.1
Protein	0.8	0.56	1.5	0.7	2.2	1.49	1.5	0.92

Due to the best conditions for the evenness of sowing close to the agrotechnical requirements the sowing of Kinelsky 254 amaranth variety with an experimental seeder allowed achieving the maximum indicators for the fresh and grain yields, which had not previously been obtained in divisional selected cultivation or in preliminary reproduction. This made it possible to license the variety with the best indicators.

4 Conclusion

The field studies aimed at evaluating the effect of the evenness of longitudinal placement of seeds and plants of Scarlet amaranth in a row showed that the evenness of the distribution of seeds in a row during sowing with an experimental seeder was 22%, for plants – 24.6%, for control – 24.6% and 72.1%, respectively. This made it possible to perform sowing in accordance with agrotechnical requirements and create the best conditions for plant nutrition, which led to an increase in fresh yield by 2.6, and grain – by 1.6 times.

Thus, the scientific and industrial experiment of cultivating amaranth in Samara region allows recommending agricultural producers to use a wide-row sowing method with dotted placement of seeds to cultivate amaranth for various purposes.

References

1. E. I. Artamonov, *Sowing quality improvement of Scarlet amaranth seeds by improving technical means and technological process*: PhD thesis (Penza, 2013)
2. V. F. Kazarin, Agro-Inform**7,** 18–20 (2012)
3. A. V. Kazarina,Bulletin of Samara State Agricultural Academy**4,** 7–11 (2015)
4. P.F. Kononkov, *Amarant – promising culture of the 21stcentury*(Moscow: Publishing House of RUDN University, 1999)
5. *Amaranth in Bashkortostan*, retrieved from: http://amarant-rb.narod.ru/stat2.htm.
6. *Scarlet and red amaranth in Kabardino-Balkaria*, retrieved from: http://www.pandia.ru./text/77/393/78090.php.
7. N. V. Burlaka, *Improvement of the dosing technology and substantiation of the parameters of the air-jet seeding apparatus of the seeder for small-seeded crops*: PhD thesis (Saratov, 2004)
8. E. I. Artamonov, *Pat. No. 61981 of the Russian Federation, IPC A01S 7/04. Sowing device*Bulletin **9** (2007)
9. I. Yu. Galenko, Bulletin of Samara State Agricultural Academy**3,**13–18 (2013)
10. V. F. Kazarin, I. Yu. Galenko, E. I. Artamonov, Bulletin of Samara State Agricultural Academy**4** (2013)
11. *GOST 31345-2007. Tractor seeders. Test methods* (Moscow: Standardinform, 2007)

 https://doi.org/10.1051/bioconf/20225200027

Florfenicol 40% efficacy in piglets with respiratory pathologies

Sergey V. Engashev[1], *Aleksey V.* Savinkov[2,*], *Konstantin M.* Sadov[2], and *Pavel V.* Iliasov[3]

[1] Moscow State Academy of Veterinary Medicine and Biotechnology named after K.I. Skryabin, 109472, Moscow, Russian Federation
[2] Samara State Agrarian University, 2 Uchebnaya str., Ust-Kinelsky settlement, Kinel, Samara Region, 446442, Russian Federation
[3] Institute for Experimental Medicine and Biotechnologies, Samara State Medical University, Ministry of Health of the Russian Federation, 20 Gagarina str., Samara, 443079, Russian Federation

Abstract. Piglets with large-scale respiratory pathologies caused by a bacterial flora susceptible to phenicol antibacterial drugs received two doses of Florfenicol 40%. On day 4 after administration, this resulted in the complete reversal of clinical pattern, in the recovery of morphofunctional blood parameters, and in the reduction in prevalence of *Klebsiella pneumoniae* by 25%, *Streptococcus suis* by 50%, and *Staphylococcus haemolyticus* by 41.7%. The drug is tolerated with no adverse events. The results of this study allow recommending Florfenicol 40% as an antibacterial therapy in the acute infectious inflammatory respiratory pathologies in the store pigs.

1 Introduction

Respiratory diseases are the main causes of production losses in global pig breeding [6, 7, 9]. This study provides the therapeutic efficacy of a novel amphenicol chemotherapy, a florfenicol-based antibacterial veterinary drug for respiratory infectious pathologies in pigs. This drug is active against the chloramphenicol-resistant bacteria producing acetyl transferase. It inhibits the protein production in a bacterial cell at the ribosome level and has a bacteriostatic effect on the susceptible microorganisms [2, 10].

Florfenicol molecule contains a fluorine and a sulphonyl moiety, in contrast to chloramphenicol which is widely used in veterinary and has a similar structure with hydroxyl and nitro group, respectively. The advantages of this novel drug include: low toxicity for hematopoietic system (does not cause aplastic anemia), bacterial cell death due to peptidyl transferase inhibition and ribosomal protein production arrest, and also lack of resistant bacterial strains to date [5]. Florfenicol lacks a nitro group, an important molecular feature causing aplastic anemia. Therefore, it poses no risk for humans when consuming the animal products [1].

Florfenicol is more active than chloramphenicol or thiamphenicol, and may be more bactericidal towards some pathogens than previously thought. Florfenicol has a wide spectrum of antibacterial activity encompassing all chloramphenicol-susceptible organisms, e.g., Gram negative rods, Gram positive cocci and other atypical bacteria, such as micoplasma. Florfenicol is highly lactophylic which ensures sufficiently high concentrations to treat intracellular pathogens and cross some anatomical barriers [4, 5].

A high value of biological media is noted when defining microbiological activity and predicting the dosages for clinical use. Florfenicol has a concentration-dependent antimicrobial effect [1, 3, 8]. It maintains the therapeutically efficient concentrations, particularly in airway tissues which are the target organs [10]. When administered intramuscularly, florfenicol is rapidly and effectively absorbed (94.1%) and enters all organs and tissues. Its peak concentration persists for 48 h. Florfenicol and its metabolites are primarily eliminated with urine and, to a lesser extent, with feces [11].

Considering the above, Agrovetzashchita NVC LLC suggests using Florfenicol 40% drug in bacterial respiratory diseases in pigs.

Aims of the study:

- to investigate therapeutical efficacy of the Florfenicol 40% veterinary drug in bacterial respiratory diseases in pigs based on clinical, hematological and bacteriological findings;
- to investigate the safety and identify possible adverse effects of Florfenicol 40% when administered to pigs.

2 Materials and methods

The clinical study was conducted in post-weaning piglets aged 40-45 days, with the total of 24 animals owned by Myasoagroprom LLC, Krasnoyarsky district, Samara region. At the beginning of the study, test and control groups were formed, each containing 12 matched animals with clinical signs of respiratory pathology. All animals during the experiment were in the same housing and feeding conditions.

1 mL of the test drug, Florfenicol 40%, contains 400 mg of active substance florfenicol, and excipients. The

* Corresponding author: a_v_sav@mail.ru

drug is a clear colorless or pale-yellow liquid. Florfenicol contained in the drug is a thiamphenicol derivative whose molecule has a fluorine atom substituted for hydroxy moiety. It has a bacteriostatic effect against *Actinobacillus pleuropneumoniae*, *Pasteurella* spp., *Salmonella choleraesuis*, *Bordetella bronchiseptica*, *Haemophilus* spp., *Staphylococcus* spp., *Streptococcus* spp., *Klebciella pneumonia*, *Moraxella bovis*, and mycoplasma species, *M. hyopneumoniae* and *M. hyorhinis*, including chloramphenicol-resistant bacteria producing acetyl transferase. By binding 50S ribosome subunit in bacterial protoplasm, it inhibits peptidyl transferase slowing protein production at the ribosome level in the susceptible microorganisms. Florfenicol 40% is a moderately hazardous substance (GOST 12.2.007 category 3).

Florfenicol 40% was intramuscularly administered in test group animals in the dose of 1 mL per 30 kg of body weight, two times 48 hours apart. In the control group, we used Florox, a similar drug manufactured by NITA-FARM LLC, Russia. This drug was used according to the same regimen, intramuscularly in the dose of 1 mL per 25 kg of body weight, two times 48 hours apart.

The clinical study was conducted using conventional methods and design. The pathology status and Florfenicol 40% efficacy were assessed by the presence or absence of characteristic clinical signs and common ailment. The animal status was controlled throughout the study.

Blood counts considered RBC, WBC, eosinophils, basophils, lymphocytes, monocytes, platelets, hemoglobin level, and the erythrocyte sedimentation rate (ESR). Blood counts were performed on the Mindray BC-5300 hematological analyzer.

Florfenicol 40% antimicrobial activity was determined using the nasal washings from piglets having signs of bronchial pneumonia, by primary plating on 5% blood agar and universal chromogenic media (BioRad). Cultures were incubated for 2 days at 37 °C. All grown microorganisms were identified using MALDI-ToF mass spectrometry on Microflex LT (Bruker®) by direct application.

Efficacy was calculated using variation statistics and Student's t-test.

3 Results and discussion

At baseline, animals with respiratory pathologies had unsatisfactory appearance reflected by the following criteria: constrained posture (static forelimb astride posture or sitting dog posture, with craned neck), periodical cough, weak body type, poor nutritional status, sallow skin, pale dry conjunctiva and mouth mucosa, hollow eyes, catarrhal purulent nasal discharge. Feed refusal rate was 83.3% in Group 1 and 75% in Group 2. The other animal had poor appetite. All animals in both groups had slightly suprathreshold (above 40 °C) body temperature.

Next day after drug administration, both groups had a slight increase in temperature by 0.3-0.5 °C, however, as early as on day 2 mean body temperature was within the physiological range 39.7-39.9 °C. Hereafter, it dynamically decreased until the end of the experiment, and on day 5 after the end of the treatment, both groups had mean body temperature in the range of 39.2-39.4 °C. The body temperature decreased with the improvement of the overall health status. No differences in temperature by groups were noted.

On day 1 after start of treatment, there were no changes in appearance and activity of animals with respiratory pathologies. Cough was registered in 100% of cases. Imminent changes in appetites were noted. Complete feed refusal rate reduced to 50% in Group 1 and 58.3% in Group 2. On day 1 after start of treatment, all animals with the respiratory pathology had cough, however, changes in general health status were noted. The appearance and physical activity improved by 50% in Group 1 and by 58.3% in Group 2. No one piglet had complete feed refusal. 41.7% and 58.3% of animals in Group 1 and Group 2, respectively, had satisfactory appetites. All the animals still had cough next day after final administration. At the same time, there were no cases of poor appetites, 83.3% of piglets in Group 1 and Group 2 were characterized by satisfactory feed and water intake. The other animal had good appetites. Physical activity and general health improved in 75% and 83.3% animals in Group 1 and Group 2, respectively. Specific clinical parameters have changed on day 2 after final administration. There was no cough in 50% of animals in Group 1 and in 41.7% of animals in Group 2. The appetites significantly improved. 91.7% of animals in Group 1 and 83.3% of animals in Group 2 had good appetites, physical activity and appearance. On day 3 after the final administration, 100% of animals in both groups had good appetites, physical activity and appearance. There was no cough in 66.7% of animals in Group 1 and in 75% of animals in Group 2. 100% recovery of all estimated parameters and, therefore, complete recovery of piglets could be noted on day 4 after the final administration. No physiological abnormalities or clinical signs recurrence were revealed in piglets on day 5 after the final administration and hereafter.

Blood counts at the baseline have shown increase in WBC up to 23.30±1.270 $*10^9$/L in Group 1 and to 27.03±1.86$*10^9$/L in Group 2. By the end of study, WBC was normalized in these groups due to decrease by 24.1% ($P \leq 0.05$) and 27.3% ($P \leq 0.05$), respectively. Between-group differences in this parameter were insignificant.

WBC differential assessment allowed one to reveal the following trends. Band neutrophils level was insignificant at the baseline in all groups and slightly increased by the end of the experiment which had no effect on the general trend. Segmented neutrophils levels were slightly above upper limit of normal in both groups (56.75±1.403% in Group 1 and 53.91±0.958% in Group 2) which is a common phenomenon in the acute phase of inflammatory process. By the end of experiment, this parameter was normalized due to reduction by 16.5% ($P \leq 0.001$) and 13.6% ($P \leq 0.001$) in Group 1 and Group 2, respectively. No between-group differences were noted. On the other hand, lymphocytes showed an increase by 2.57% and 5.24%, respectively, which is indicative of positive dynamics towards the recovery. No differences between these parameters were noted within groups. At the baseline, eosinophil level was decreased being as small as 1% in both groups which was consistent with the general picture of the acute disease. By the end of experiment, this parameter was increased by 2.42% (P≤0.001) and 2.58% (P≤0.001) in Group 1 and Group 2, respectively. These parameters were comparable between groups. Note that the increase in eosinophils is a beneficial sign of recovery. Similar changes were noted for monocytes. Their level was low in all groups at the baseline, but by the end of experiment, this parameter was increased by 2.33% (P≤0.05) and 1.52% (P≤0.001) in Group 1 and Group 2, respectively. Between-group differences were insignificant.

Red blood parameters showed low hemoglobin level at the baseline in the both groups, however, it was above the lower limit of norm (87.45±0.908 g/L in Group 1 and 92.80±1.662 g/L in Group 2). Hemoglobin level increased at the end of the experiment by 20.2% (P≤0.001) in Group 1 and by 9.1% (P≤0.05) in the Group 2, respectively. There was insignificant difference in this parameter between groups, 3.9% in favor of Group 1. Hematocrit dynamics showed the same trends as the hemoglobin. At the end of the experiment, hematocrit increased by 19.5% (P≤0.001) in Group 1 and by 10.5% (P≤0.05) in the Group 2, respectively. These parameters were comparable in Group 1 and Group 2 (Table 1).

Table 1. CBC dynamics

Parameter	Treatment groups	
	1	2
ESR, mm/h		
Day 0	1.67±0.148	1.64±0.203
Day 9	1.42±0.202	1.83±0.216
WBC, 10^9/L		
Day 0	23.30±1.270	27.03±1.86*
Day 9	17.69±2.399*	19.66±1.350
RBC, 10^{12}/L		
Day 0	6.19±0.151	6.57±0.100
Day 9	6.53±0.095	6.38±0.114
Hemoglobin, g/L		
Day 0	87.45±0.908	92.80±1.662
Day 9	105.15±1.136***	101.22±2.572*
Hematocrit, %		
Day 0	33.69±0.450	36.73±0.625
Day 9	40.26±0.711***	40.58±1.18*
Platelets, 10^9/L		
Day 0	438.42±37.622	411.45±27.508
Day 9	338.50±20.939*	356.50±35.302
Bands, %		
Day 0	0.83±0.117	1.00±0.000
Day 9	2.92±0.505	3.25±0.605
Segmented, %		
Day 0	56.75±1.403	53.91±0.958
Day 9	40.26±0.711***	40.36±1.875***
Lymphocytes, %		
Day 0	39.25±1.343	42.09±1.004
Day 9	41.83±2.603	47.33±2.457
Eosinophils, %		
Day 0	1.00±0.000	1.00±0.000
Day 9	3.42±0.857*	3.58±0.395***
Monocytes, %		
Day 0	1.67±0.235	1.73±0.237
Day 9	4.00±0.704*	3.25±0.532*
Basophils, %		
Day 0	0.50±0.157	0.27±0.195
Day 9	0.42±0.202	1.50±0.375

*Note: *: P≤0.05; **: P≤0.01; ***: P≤0.001 vs. control; ˇ: P≤0.05; ˇˇ: P≤0.01; ˇˇˇ: P≤0.001 vs. baseline*

Platelet counts were significantly increased at the baseline (438.42±37.622 *10^9/L in Group 1 and 411.45±27.508 *10^9/L in Group 2, with the normal range of 180-300 *10^9/L). By the end of the experiment, this parameter reduced by 22.7% (P≤0.05) in Group 1 and by 13.4% in Group 2, respectively. These parameters were comparable to each other.

Analysis of susceptibility of bacteria isolated from nasal washings to Florfenicol 40% has shown that *Streptococcus suis, Moraxella canis, Moraxella pluranimalium, Enterococcus hirae* had the highest susceptibility of 0.5 µg/mL. The susceptibility of *Staphylococcus aureus* was of 1 µg/mL. The susceptibility of *E. coli* and *Staphylococcus haemolyticus* to the test drug was borderline and varied between 1-2 µg/mL. The susceptibility of *Klebsiella pneumoniae* varied within 2-8 µg/mL. *Pseudomonas fulva* had the lowest susceptibility among the cultures tested, which was 8 µg/mL.

The bacterial composition of nasal washings from sick piglets suggested that the disease was caused by a complex of opportunistic microorganisms due to decline in animal immune resistance. At the baseline, the most common microorganisms found in piglets in both groups include *E. coli, Klebsiella pneumonia, Rothia nasimurium, Streptococcus suis, Staphylococcus haemolyticus.* Thus, all bacteria commonly isolated from nasal washings were opportunistic microorganisms able to cause infectious inflammatory disorders with adverse outcomes, with *Rothia nasimurium* having the least rate of displaying its pathological properties.

The picture observed on day 1 of the study was as follows. In test Group 1, 41.7% of animals had *E. coli*, 91.7% had *Rothia nasimurium*; 58.3% had *Streptococcus suis*; and 50% had *Klebsiella pneumoniae and Staphylococcus haemolyticus. Staphylococcus chromogenes, Staphylococcus hominis* were identified in 16.7% of cases; and each of *Acinetobacter gerneri, Lactobacillus johnsonii, Staphylococcus warneri, Staphylococcus cohni, Moraxella canis, Bacillus megaterium* was only identified in 1 animal per group (8.3%).

In control Group 2, *Rothia nasimurium* was found in 66.7% of animals; *Klebsiella pneumoniae* in 58.3% of cases; *Staphylococcus haemolyticus and Klebsiella pneumoniae* in 33.3% of cases; *E. coli, Staphylococcus cohni* and *Streptococcus hyovaginalis* in 25% of cases; and *Enterococcus casseliflavus, Staphylococcus aureus* were found in 16.7% of cases. The following bacteria were found in one case per group (8.3%): *Aeromonas caviae, Stapylococcus saprophyticus, Acinetobacter lwoffii, Enterococcus hirae, Moraxella canis, Acinetobacter baumanii, Pseudomonas fulva, Aerococcus viridians, Staphylococcus muscae, Staphylococcus chromogenes, Staphylococcus xylosus, Staphylococcus hominis, Moraxella pluranimalium, Pasteurella aerogenes.*

After treatment by Florfenicol 40% antibacterial drug, prevalence of *E. coli* increased up to 91.7% in Group 1 and to 83.3% in Group 2. *Rothia nasimurium* also remained active. By the end of the experiment, its prevalence reduced to 50% in Group 1 and remained unchanged in Group 2. *Klebsiella pneumoniae* prevalence reduced by 25% in Group 1 and remained unchanged in control Group 2. In other cases, the antibacterial drug caused a frank depression of microbial flora. Thus, Group 1 had 50% reduction in prevalence of *Streptococcus suis* 50% (down to 8.3%; and 41.7% reduction in *Staphylococcus haemolyticus* (also down to 8.3%).

Other bacteria, namely *Staphylococcus chromogenes, Staphylococcus hominis; Acinetobacter gerneri, Lactobacillus johnsonii, Staphylococcus warneri, Staphylococcus cohni, Moraxella canis, Bacillus megaterium,* were not found at the end of the experiment.

By the end of the study, new microbial species were grown: *Enterococcus faecalis (50%); Acinetobacter lwoffii (41.7%); Acinetobacter johnsonii (33.3%); Enterococcus gallinarum, Arthrobacter histidinolovorans, Aerococcus viridians (16.7%); Bacillus mycoides, Citrobacter freundii, Staphylococcus sciuri, Acinetobacter calcoaceticus, Aeromonas caviae, Enterobacter cloacae, Lysinibacillus fusiformis, Acinetobacter towneri, Staphylococcus saprophyticus, Providencia alcalifaciens (8.3%).* These new microorganisms belong to obligate flora present in the process area of the facility.

In Group 2, the proportion of *Streptococcus suis* also reduced by 50%, down to 8.3%, and *Staphylococcus haemolyticus* was not found at the end of the experiment. By the end of the study, prevalence of *Enterococcus casseliflavus* reduced from 2 cases (16.7%) to 1 case (8.3%); *Aeromonas caviae* and *Stapylococcus saprophyticus* were also found in one case per group (8.3%) each; *Acinetobacter lwoffii* representation increased by 16.7%. *Staphylococcus haemolyticus, Staphylococcus cohni, Staphylococcus aureus, Streptococcus hyovaginalis* were not found at the end of the experiment.

At this time point, we registered new microbial species which were not found previously: *Enterococcus faecalis (41.7%); Acinetobacter gerneri (25%); Acinetobacter johnsonii, Acinetobacter radioresistens, Enterococcus gallinarum, Myroides odoratimimus (16.7%); Wautersiella falsenii, Proteus hauseri, Bacillus mycoides, Citrobacter koseri, Morganella morganii, Acinetobacter baumanii, Acinetobacter guillouiae, Enterobacter asburiae, Acinetobacter baylyi, Vagococcus fluvialis (8.3%),* see Table 2.

Table 2. Dynamics of bacterial flora in nasal washings

	Day 0		Day 9	
	n	%	n	%
Total number of animals	12	100	12	100.0
Treatment group 1				
E. coli	5	41.7	11	91.7
Rothia nasimurium	11	91.7	6	50.0
Streptococcus suis	7	58.3	1	8.3
Klebsiella pneumoniae	6	50.0	3	25.0
Staphylococcus haemolyticus	6	50.0	1	8.3
Streptococcus hyovaginalis	2	16.7	0	0.0
Acinetobacter gerneri	1	8.3	0	0.0
Staphylococcus chromogenes	2	16.7	0	0.0
Lactobacillus johnsonii	1	8.3	0	0.0
Staphylococcus warneri	1	8.3	0	0.0
Staphylococcus hominis	2	16.7	0	0.0
Staphylococcus cohni	1	8.3	0	0.0
Moraxella canis	1	8.3	0	0.0
Bacillus megaterium	1	8.3	0	0.0
Acinetobacter lwoffii	0	0.0	5	41.7
Bacillus mycoides	0	0.0	1	8.3
Citrobacter freundii	0	0.0	1	8.3
Staphylococcus sciuri	0	0.0	1	8.3
Enterococcus faecalis	0	0.0	6	50.0
Aerococcus viridans	0	0.0	2	16.7
Acinetobacter calcoaceticus	0	0.0	1	8.3
Aeromonas caviae	0	0.0	1	8.3
Acinetobacter johnsonii	0	0.0	4	33.3
Enterobacter cloacae	0	0.0	1	8.3
Arthrobacter histidinolovorans	0	0.0	2	16.7
Lysinibacillus fusiformis	0	0.0	1	8.3
Acinetobacter towneri	0	0.0	1	8.3
Staphylococcus saprophyticus	0	0.0	1	8.3
Providencia alcalifaciens	0	0.0	1	8.3
Enterococcus gallinarum	0	0.0	2	16.7
Treatment group 2				
E. coli	3	25.0	10	83.3
Rothia nasimurium	8	66.7	8	66.7
Klebsiella pneumoniae	4	33.3	4	33.3
Streptococcus suis	7	58.3	1	8.3
Enterococcus casseliflavus	2	16.7	1	8.3
Aeromonas caviae	1	8.3	1	8.3
Stapylococcus saprophyticus	1	8.3	1	8.3
Acinetobacter lwoffii	1	8.3	2	16.7
Staphylococcus haemolyticus	4	33.3	0	0.0
Staphylococcus cohni	3	25.0	0	0.0
Staphylococcus aureus	2	16.7	0	0.0
Streptococcus hyovaginalis	3	25.0	0	0.0
Enterococcus hirae	1	8.3	0	0.0
Moraxella canis	1	8.3	0	0.0
Acinetobacter baumanii	1	8.3	0	0.0
Pseudomonas fulva	1	8.3	0	0.0
Aerococcus viridans	1	8.3	0	0.0
Staphylococcus muscae	1	8.3	0	0.0
Staphylococcus chromogenes	1	8.3	0	0.0
Staphylococcus xylosus	1	8.3	0	0.0
Staphylococcus hominis	1	8.3	0	0.0
Moraxella pluranimalium	1	8.3	0	0.0
Pasteurella aerogenes	1	8.3	0	0.0
Acinetobacter johnsonii	0	0.0	2	16.7
Enterococcus faecalis	0	0.0	5	41.7
Acinetobacter radioresistens	0	0.0	2	16.7
Wautersiella falsenii	0	0.0	1	8.3
Proteus hauseri	0	0.0	1	8.3
Bacillus mycoides	0	0.0	1	8.3
Enterococcus gallinarum	0	0.0	2	16.7
Citrobacter koseri	0	0.0	1	8.3
Acinetobacter gerneri	0	0.0	3	25.0
Morganella morganii	0	0.0	1	8.3
Acinetobacter baumanii	0	0.0	1	8.3
Myroides odoratimimus	0	0.0	2	16.7
Acinetobacter guillouiae	0	0.0	1	8.3
Enterobacter asburiae	0	0.0	1	8.3
Acinetobacter baylyi	0	0.0	1	8.3
Vagococcus fluvialis	0	0.0	1	8.3

4 Conclusion

The objective findings obtained indicate that Florfenicol 40% has demonstrated high therapeutic efficacy in infectious inflammatory respiratory pathologies in post-weaned piglets caused by resident opportunistic flora which was supported by the animal clinical status, assessment of nasal washings, and blood counts.

When treated by the drug, complete recovery was achieved on day 4 after the final administration of the test and reference drugs. As a result, prevalence of *Klebsiella pneumoniae* reduced by 25%, *Streptococcus suis* by 50%, and *Staphylococcus haemolyticus* by 41.7%. The findings obtained in the reference group were comparable to the test group.

Therapy with Florfenicol 40% and reference antibacterial drug aids in normalization of WBC differential increase hemoglobin and hematocrit levels, reduce platelet count in animals with respiratory pathologies. WBC differential parameters in both groups suggest the trend to reduction in neutrophils, normalization of lymphocyte levels, increase in eosinophils and monocytes which is indicative of decline in acute phase processes mediated by while blood cells. Rapid significant increase in hemoglobin and hematocrit in the test group may indicate an activation of hemopoiesis due to elimination of infection. Animals with respiratory pathologies showed leukocytosis reversal. In general, this picture suggests recovery of piglets from the respiratory pathology.

Considering the above, we recommend using Florfenicol 40% in acute infectious inflammatory respiratory pathologies in store pigs caused by opportunistic resident flora, in two doses of 1 mL per 30 kg of the body weight 48 hours apart. The drug well tolerated with no adverse events.

References

1. E. Daeseleire, E. Van Pamel, C. Van Poucke et al. Veterinary Drug Residues in Foods, In *Woodhead Publishing Series in Food Science, Technology and Nutrition, Chemical Contaminants and Residues in Food (Second Edition)*, (Woodhead Publishing, 2017).
2. R. Del Pozo Sacristán, J. Thiry, K. Vranckx et al, *The Veterinary Journal* **194**, 420-422 (2012).
3. L. Denich, A. Farzan, R. Friendship et al, *Veterinary Microbiology* **257** (2021).
4. L. Dorey, S. Hobson, P. Lees, *Vet J.* **218**, 65-70 (2016).
5. Mark G. Papich. Florfenicol, In *Saunders Handbook of Veterinary Drugs (Fourth Edition)* 327-329 (2016).
6. P. Oba, M.M. Dione, B. Wieland, et al, *Porc Health Manag* **7**, 53 (2021).
7. P. Przyborowska-Zhalniarowicz, Y. Zhalniarovich, K. Wasowicz, *Vet Med-Czech* **66**, 242–247. (2021).
8. M. R. Qian, Q. Y. Wang, H. Yang et al, *Journal of veterinary pharmacology and therapeutics* **40**, 583-686 (2017).
9. G. Sarli, G. D'Annunzio, F. Gobbo et al, *Vet Sci.* **8(11)**, 256 (2021).
10. L. Jian-Zhong, F. Ki-Fai, C. Zhang-Liu et al, *European Journal of Drug Metabolism and Pharmacokinetics* **27**, 265-271 (2002).
11. L. Zhichang, R. Ting, Z. Dongping et al, *Journal of Veterinary Pharmacology and Therapeutics*, **41**, 719-725 (2018).

Modernization of the fuel supply system in the internal combustion engine by electronic control of the ring valve

Fanil Gabdrafikov[1], *Denis* Kharisov[1], *Ilgiz* Galiev[2]*, and *Rail* Khusainov[2]

[1] Bashkir State Agrarian University, 34, 50 years of October St., Ufa, Republic of Bashkortostan, 450001, Ufa, Russian Federation
[2] Kazan State Agrarian University, 65, K. Marx St., Kazan, 420015, Republic of Tatarstan, Russian Federation

Abstract. The research was carried out in order to develop a method for skipping fuel supplies (turning off individual piston strokes) at low frequencies of crankshaft rotations and at partial engine operating modes with direct-acting fuel supply systems to increase fuel efficiency. The developed method is easily implement using a ring-type discharge valve with electronic control. A valve made in the form of a split elastic ring and installed in a high-pressure line above the plunger pair controls the fuel supply. An electromagnet, located in the cavity of the annular valve, controls the valve, using an electronic regulator, acting at the right moment according to the signals coming from the sensors of the engine crankshaft speed, the volume of incoming air and the position of the piston. The proposed direct-acting fuel supply system with an electronically controlled ring valve reliably ensures the speed and load characteristics of the engine by affecting the number of cyclic feeds. With a decrease in the load and speed of the engine crankshaft, the number of cyclic fuel supplies decreases due to a decrease in signals sent to the electromagnet winding and, as a result, the valve does not attract the fuel and misses the fuel supply. Reducing the unevenness of the fuel supply by the dynamic component that occurs when the supply is switch off is ensured by reducing the inertia of the regulator due to the electronic control of the ring valve. The developed mathematical model of a direct-acting fuel supply system with an annular discharge valve allows us to reveal the relationship of fuel supply parameters with the design dimensions of the split ring. Such direct-acting fuel supply system with an electronically controlled discharge valve allows, due to the skipping of working strokes in low-load and idle modes, redistributing the provided amount of fuel to the working cylinders, significantly reducing fuel consumption.

* Corresponding author: drgali@mail.ru

1 Introduction

Agricultural production is one of the main sectors of the country's economy, which constantly needs to attract mobile energy resources to perform technological processes in crop production and animal husbandry. The most promising way to ensure the efficiency of the use of technical means in the agro-industrial complex is the modernization of the fuel supply system of the internal combustion engine [1].

Fuel supply systems of direct action of diesel engines have proven themselves well throughout the entire period of the long-term operation. Their wide distribution became possible thanks to the modified design of all elements of the system, unification, maintainability and unpretentiousness.

They work most effectively in modes close to nominal [2]. A study of tractor and combine diesels in operation has shown that most of the time in the balance is occupied by the work that requires only a partial load on the engine [3]. Similar results were shown by our earlier observations in conditions of a very high culture of equipment operation in the Berthold farm (Bavaria, Germany).

With a decrease in the load on the engine and the transition to low rotational speeds of its crankshaft and at idle, the efficiency of the fuel supply system deteriorates significantly due to a decrease in the value of the cyclic supply and an increase in its irregularities between sections and between cycles [4].

One of the options for preventing a decrease in the cyclic supply and the efficiency of diesel under partial loads is the method of turning off the cylinders [5]. This method proved itself when idling engines D-108, D-160 and D-180 of the Chelyabinsk Tractor Plant and A-01M of the Altai Engine Plant.

The positive effect of turning off the cylinders was obtained on the basis of the eight-cylinder engine of the Kamaz-740 car [6]. The disconnection of a part of the cylinders was carried out by connecting the high-pressure pump discharge line with the low-pressure line using a controlled electromagnetic valve in the fuel supply system.

Research is underway to analyse the turn-off of the cylinders of tractor and automobile diesels with simultaneous action on the valves of the gas distribution mechanism in partial load modes [7] to reduce losses on the piston pumping strokes in the suction and compression cycles. A reduction in the hourly fuel consumption of diesel up to 38% was found when some of the cylinders were switch off in low-load modes [8].

The company "MTU" (Germany) has modernized diesel engines 12V956TB, introducing the technology of sequential shutdown of cylinders [9].

Tula Technology [10] developed the method of disconnecting cylinders on an engine operating using a Miller cycle with an adjustable suction stroke through the control of the valve lift of the gas distribution mechanism. Studies have shown the possibility of reducing the engine fuel consumption by 23%.

The cylinder shut-off method is widely used on gasoline engines of passenger cars. Mercedes used the Active Cylinder Control (ACC) system to automatically turn off half of the cylinders on eight- and twelve-cylinder engines on its 500- and 600-series cars. The Multi-Displacement System (MDS) is used in Jeep, Dodge and Chrysler vehicles. General Motors has also implemented a Displacement on Demand (DoD) system in its cars. Honda has developed a Variable Cylinder Management (VCM) system. Since 2012, Volkswagen has been using a similar system called Active Cylinder Technology (ACT), but already in four-cylinder TSI engines with a volume of 1.4 litters. Audi also supplied its powerful V8 and V12 engines first, and then four-cylinder 1.4-liter engines with the COD (cylinder on demand) system.

The common disadvantages of the cylinder shutdown method on various engines are an increase in the required starting speed and the uneven rotation of their crankshaft.

The purpose of the scientific work is to study the method of skipping fuel supplies when operating a diesel engine under low loads with an electronically controlled ring valve and a mathematical description of its functioning.

2 Conditions, materials and methods of research

We have conducted research on the development of a method for skipping the fuel supply by alternating in all cylinders when operating a diesel engine with a partial load. This method will reduce the impact of the disadvantages of the cylinder shutdown method on the engine operation.

Preliminary calculations carried out for the D-240 four-cylinder tractor diesel engine showed that a consistent increase in the number of disconnected cylinders leads to an increase in the unevenness of rotation of its crankshaft and the conditional values of the starting revolutions. The calculation data are given in table 1.

From the data in the table, it can be seen that the values of δ are greater than the values of unevenness recommended for tractor diesels - 0.004...0.010, but slightly. To reduce δ to 0.01, it is necessary to start the engine when working using all cylinders or increase the moment of inertia of the flywheel.

Reducing the inertia of its mechanical regulator is essential for improving the quality of fuel supply. Electromechanical regulators with annular discharge valves, which can ensure the passage of fuel supplies, have shown themselves well in improving the quality of fuel supply in any modes [11].

The annular discharge valve is installed above the plunger pair of the high-pressure pump as shown in Figure 1.

* Corresponding author: drgali@mail.ru

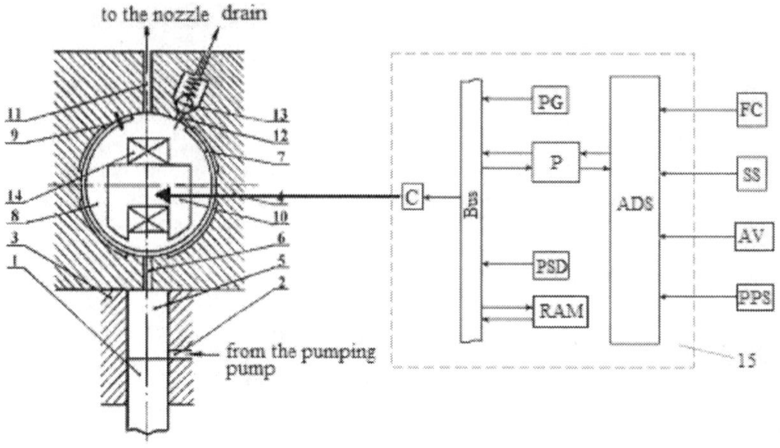

Fig. 1. Diagram of an electromechanical regulator with an annular discharge valve: 1 - plunger; 2, 6, 11, 12 - channels; 3 - sleeve; 4 - valve body; 5, 8 - cavities; 7 - split ring; 9 - pin; 10 - electromagnet; 13 - drain valve; 14 - electromagnet coil; 15 - electronic regulator; PSD - permanent storage device; PG - pulse generator; ADC - analog to digital converter; P – processor; RAM - random access memory; C – switchboard; PPS - piston position sensor; AV - incoming air volume sensor; SS - engine crankshaft speed sensor; FC - fuel supply control lever.

A split ring 7 is built into the valve body 1, fixed on one side with a pin 9. An electromagnet 10 with a coil 14 is installed in the valve cavity 8. The free side of the ring is capable of blocking the drain channel 12.

To supply fuel to the cylinder through the nozzle, the plunger 1 of the pump pumps fuel through the discharge valve by lifting the split ring 7, while the electromagnet 10 additionally attracts the ring, which covers the drain channel 12 with a free edge.

When the electromagnet is de-energized, the split ring is lower and the drain channel 12 opens. The high-pressure line connects to the drain and fuel injection into the cylinder stops. The residual fuel pressure in the high-pressure line is provided by the drain valve 13.

Thanks to the electronic regulator 15, the annular discharge valve can adjust the fuel injection advance angle and the feed duration (the value of the cyclic feed), determined by the moment and duration of the control pulse to the electromagnet.

The timely supply of a pulse to the electromagnet is provided in accordance with the data received from the sensors of the piston position, the volume of incoming air, the rotational speed of the engine crankshaft and the fuel supply control lever.

To skip the supply, when the plunger is injected with fuel, the control pulse is not applied to the electromagnet and the fuel is sent directly to the drain. The elasticity of the ring is calculated so that the height of its lifting and the movement of the free edge from the fuel pressure is not enough to close the drain channel 12.

Thus, an annular discharge valve with an electromagnet can regulate the operating modes of a diesel engine by passing fuel supplies depending on the load. When the load decreases, it is not the amount of cyclic feeds that decreases, but the number of missed feeds increases. In this case, the value of the cyclic feed will be equal to the nominal value. This will have a positive effect on the quality of fuel spraying, mixing and combustion. In this connection, it is expected to increase the efficiency of the engine.

3 Results of the study

A mathematical model of an electrically controlled fuel supply system with an annular discharge valve has been developed, taking into account the peculiarities of its operation, based on the use of the method of hydrodynamic calculation of existing direct-acting fuel supply systems.

The movement of fuel in the high-pressure nozzle was considered taking into account the pressure of waves coming towards at the speed of sound. The fuel flow rates in the injection and control channels were determined based on the continuity equation of the fluid flow.

To describe the ongoing process in a high-pressure pump, the equations of the volume of liquid (fuel) and the dynamic equilibrium of a ring-type discharge valve were used:

$$6 \cdot n_n \cdot \beta_a \cdot V_a \cdot \frac{dp_a}{d\phi} = \left(f_a \cdot s_p - \sigma_{iv} \cdot Q_a - \sigma_{vc} \cdot Q_1 - Q_3 - Q_4 \right). \quad (1)$$

$$6 \cdot n_n \cdot \beta_{vc} \cdot V_{vc} \cdot \frac{dp_s}{d\phi} = \left(\sigma_{vc} \cdot Q_1 - \sigma_{cc} \cdot Q_2 - f_{hp} \cdot s_{ci} \right) \cdot \sigma_2 \quad (2)$$

$$\frac{dV_{fv}}{d\phi} = -\left(\sigma_{vc} \cdot Q_1 - \sigma_{cc} \cdot Q_2 - f_{hp} \cdot s_{ci} \right) \cdot \frac{\sigma_3}{6 \cdot n_n}. \quad (3)$$

$$6 \cdot M_k \cdot n_n \frac{ds_{vc}}{d\phi} = (\left(f_i (p_a - p_s \right) + + f_i \cdot \rho_f \cdot s_p (s_p - s_{ci}) - p_{rp} \cdot b \cdot D - f_i \cdot p_s \cdot \sigma_c) \sigma_{vc}; \quad (5)$$

where p_a, p_s are the pressure of the supplied fuel in the cavities above the plunger and in the suction channel, Pa;

f_{cc}, f_i, f_{hp} are the cross–sectional areas of the control channels, injection and high pressure lines, mm^2;

s_p is the speed of the plunger, m/s;

s_{ci}, s_{cc}, s_{vc} are the fuel feed rates in the channels injection, control and in the valve cavity, m/s;

V_a, V_{vc}, V_{fv} are the volume in the cavities above the plunger, in the valve cavity and free volume, m^3;

ρ_f is the density of the supplied fuel, kg / m^3;

Q_0, Q_1, Q_2, Q_3, Q_4 are the fuel consumption in the system of equations in the intake channel, in the discharge channel, in the control channel, through the plunger gaps in the spool part and in the piston part, kg/s;

M_k is the weight of the ring with a cut, kg;

n_n and φ are the rotational speed (min^{-1}) and the angle of rotation (degree) of the cam shaft of the pump;

σ_{iw}, σ_{vc}, σ_{cc}, σ_2, σ_3, σ_c are the logical elements;

h_{rr} is the rise of the split ring, mm;

p_{rp} is the radial pressure of the ring, Pa;

b and D are the width and diameter of the ring, mm;

β_a, β_{vc} are the compressibility coefficient of fuel in the cavities above the plunger and in the valve cavity.

Formula (1) is the balance of the fuel volume in the cavity above the plunger. The following equations (2) and (3) are the balance of the fuel volume in the valve cavity. If there is no gap, the expression (2) is used for calculation, if there is a gap – (3).

The amount of fuel entering the cavity per unit of time with the ring valve raised in equation (2) is equal to the flow through the discharge channel, with the exception of fuel in the control channel and in the discharge pipe. When the control voltage is applied to the electromagnet, the step function is $\sigma_{cc} = 0$. Equation (3) (its left part) describes the volume changes (V_{fv}) per unit of time in the cavity (V_{vc}) of the ring. At the same time, the right part of this equation demonstrates an increase in the volume of fuel in the cavity of the V_{vc}, which leads to a decrease in the free volume of the V_{fv}.

The dynamic equilibrium of the annular discharge valve is described by formulas (4) and (5). In equation (4) the left part is the inertia force of the split valve ring, and the sum of the fuel pressure forces and the valve force in the right part.

The following logical elements were taken into account: with $p_a \geq p_{iw}$ $\sigma_{iw}=1$ was taken; with $p_a \leq p_{iw}$ $\sigma_{iw} = -1$ was taken; when the intake window was open, $\sigma_{iw} = 0$ was taken. When the ring was pressed against the body, $\sigma_c=1$ was taken; in other cases, $\sigma_c=0$; with $p_a \geq p_s$, $\sigma_{vc}=1$ was taken; with $p_a \leq p_s$, $\sigma_{vc} = -1$ was taken.

The order of use in the calculation of equations (2) and (3) is determined by the logical elements σ_2 and σ_3: so for $V_{vc} \geq 0$ and $p_{vc}=0$, $\sigma_2=0$ and $\sigma_3=1$ were taken; for $p_{vc}>0$ and $V_{vc}=0$ $\sigma_2=1$ and $\sigma_3=0$ were taken.

Injection pressure and fuel consumption are determined by the following expressions:

$$p^1_{\varphi} = \frac{(\mu f)^2_p}{(\mu_s \cdot f_h)^2}(p_f - p_g) + p_g, \quad (6)$$

$$Q_n = \mu_p \cdot f_h \cdot \sqrt{\frac{2}{\rho_f}} \cdot \sqrt{p_n - p_g}, \quad (7)$$

where $(\mu f)_p$ is the effective section of the atomizer, mm^2;

μ_p, μ_s is the fuel consumption coefficients in the plunger sections and spray holes;

p_g is the gas pressure in the cylinder, Pa;

p_f is the fuel pressure in front of the needle cone, Pa;

f_h is the cross-sectional area of the spray holes, mm^2.

For the calculation, the fuel injection process was divided into a number of stages, from the beginning of the plunger movement and the lifting of the annular discharge valve to the landing of the nozzle needle cone on the locking belt (completion of injection).

Calculations carried out using these equations made it possible to clarify the design dimensions of the developed fuel supply system with an electronically controlled annular discharge valve, in particular, the parameters determining the stiffness of the ring and the volume of the valve cavity: diameter (D), width (b) and thickness (t) of the split ring.

The adequacy of the developed mathematical model was checked by comparing the calculated data with the experimental results. Previously, the design dimensions of the annular discharge valve were assumed to be equal to D=20 mm, b=10 mm, t= 0.5 mm. The rotational speed of the camshaft of the fuel injection pump with a dimension of 1TN-9X10 is assumed to be n_n = 600 min^{-1}, the pressure in the nozzle is P_0 = 1.0 MPa with full cyclic fuel supply with the FD-22 nozzle.

Table 2 shows the calculated (C) and experimental (E) fuel pressure values in the nozzle and the values of the cyclic fuel supply.

Table 1. Calculated and experimental pressure values in the nozzle fitting and cyclic fuel supply

Indicators		Angle of rotation of the crankshaft, degree								
		4	6	8	10	12	14	16	18	20
Pressure in the nozzle fitting, MPa	**C**	0	25	45	50	52	53	20	0	0
	E	0	23	46	48	50	54	18	0	0
Cyclic fuel consumption, g/cycle	**C**	0	4.2	6.2	6.3	6.3	6.4	6.5	3	0
	E	0	5.1	6.0	6.2	6.3	6.4	6.5	3.2	0

A comparative analysis of the calculated and experimental data revealed that the deviation is no more than 3%, which confirms the adequacy of the mathematical model.

Fig. 2 shows the results of a study of the main performance indicators of the D-240 diesel engine among the number of cylinders operating simultaneously, i.e. when one, two and three cylinders are switched off, respectively. Experiments have shown that when the cylinders are switched off, the unevenness between the fuel supply sections also changes.

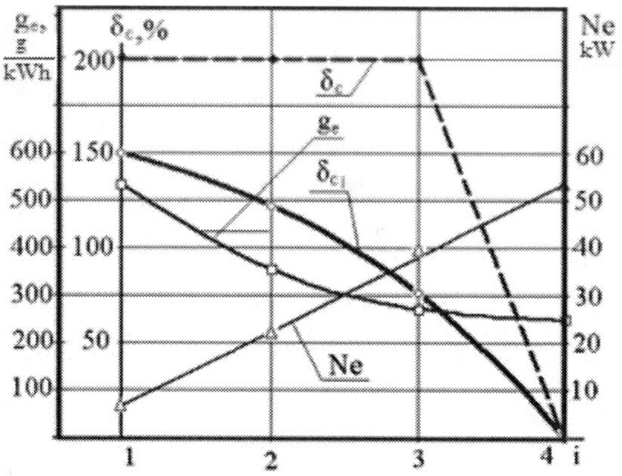

Fig. 2. Dependences of specific effective fuel consumption (g_e), power (N_e) and unevenness between sections (δ_c) of diesel D-240 on the number of working cylinders (i).

It is proven that the unevenness increases as the number of disconnected cylinders increases. The specific effective fuel consumption increases significantly and the effective power decreases (fig. 2).

Turning off the cylinders significantly affects the minimum-stable engine crankshaft speed. So when working using four cylinders, it was 750 min^{-1}, when using one – 700 min^{-1}.

Disconnecting one of the two cylinders of the D-21A engine with an effective power of N_e = 3 kW ensured a reduction in the specific fuel consumption from 700 to 420 g/kWh.

The calculated and experimental data obtained allow us to establish the possibility of regulating the engine crankshaft speed by passing fuel supplies depending on the load. For example, reducing the load by 1% makes it possible to skip every hundredth working stroke, by 10% – every tenth, by 20% – every fifth, etc. Such regulation makes it possible to inject the full fuel rate in each operating cycle, and therefore, as the load decreases, the specific fuel consumption will decrease, and the load characteristic itself will change smoothly.

Thus, turning off the cylinders, or even better, skipping the working strokes (fuel supply) as the load decreases are an effective way to increase the fuel efficiency of diesels.

4 Conclusion

The following conclusions can be made.

A direct-acting fuel supply system with an electronically controlled annular discharge valve ensures reliable regulation of the engine shaft rotation speed by passing fuel supplies while reducing the load.

The develop mathematical model of the fuel supply system with an electronically controlled ring valve allows you to establish the relationship and perform calculations of the design parameters of the fuel supply system (switching off the working strokes of the pistons) of the engine.

The effect of the proposed fuel supply system with an electronically controlled injection ring valve is manifested in an increase in the fuel efficiency of a diesel engine, especially in partial load and idle modes. So the diesel D-21A specific consumption decreased from 700 to 420 g/kW·h in the Ne = 3 kW mode.

References

1. I.V. Astakhov, L.N. Golubkov, V.I. Trusov et al. *Fuel systems and efficiency of diesels* (Mashinostroenie, 1990)
2. F.Z. Gabdrafikov, S.B. Shamukaev, Mechanization and electrification of agriculture, **9,** 22 (2010)
3. L.V. Grekhov, N.A. Ivashchenko, V.A. Markov, *Fuel equipment and diesel control systems: textbook for students of higher educational institutions* (Legion-Avtodata, 2005)
4. K.A. Khafizov, R.N. Khafizov, A.A. Nurmiev, I.G. Galiev, BIO Web Conf., **17**, 00022 (2020)
5. D.H. Valeev, V.A. Gergenreder, I.Y. Olesov, Engine building, **8**, 62 (1991)
6. A.N. Gots, V.F. Guskov Lecture Notes in Mechanical Engineering, 787 (2021)
7. A.V. Gritsenko, K.V. A.A. Glemba, Engineering Sciences, **33(1)**, 70 (2021)
8. Wenfu Sun, Xiaoqin Mo. IOP Conf. Series, **381**, 012065 (2018).
9. E. Ortiz-Soto, X. Yang, Ess Van, A. Joshi et al., Technology SAE Technical Papers (2021)
10. X. Zhu, S. Limbu, K. Cung, W. De Ojeda et al., SAE Technical Paper , 0855 (2016)
11. R.M. Bashirov, R.R. Galiullin, A.A. Ilyin, F.Z. Gabdrafikov, Patent for invention RU 2258823 (2005)

https://doi.org/10.1051/bioconf/20225200029

Accounting of heavy metals in agricultural land use

Mikhail Mikhailovich Geraskin[1*], *Zhanna Nikolaevna* Bakanova[1], *Vasily Ivanovich* Kargin[2], *Natalia Nikolaevna* Ivanova[2], and *Nikolay Nikolaevich* Neyaskin[3]

[1] State University for Land Administration, 15, Kazakova St., Moscow, 105064, Russian Federation
[2] Ogarev Mordovia State University,430005, 68 Bolshevist St., Saransk, Russian Federation
[3] State Center of Agrochemical Service "Mordovskiy", 430904, 35 Pionerskaya St., Yalga, Russian Federation

Abstract. People use many chemicals in their economic activities, which leads to the fact that they become involved in a cycle of anthropogenic transformations of the environment. It was proved that microelements and heavy metals are the most toxic among pollutants. Microelements in the arable layer of the soil depend on the type of soils, their location and content in soil-forming rocks. The studies revealed that the content of heavy metals in the soil in all cases is lower than the APC (MPC). The products obtained in the test areas are safe in relation to the content of heavy metals. Adaptive landscape farming systems can serve as an additional and significant step in optimizing the environmental situation when the soils are contaminated with heavy metals.

1 Introduction

People use many chemicals in their economic activities, which leads to the fact that they become involved in a cycle of anthropogenic transformations of the environment [1–2]. Often, due to imperfect cleaning systems, these substances enter the soil, water and air, thus contaminating and poisoning them.

It was proved that microelements and heavy metals are the most toxic among pollutants. They are the main indicator of anthropogenic impact on the environment, in particular on the soil [3]. It should be noted that these pollutants rank second in terms of hazard, second only to pesticides. In the future, they can become more dangerous than solid waste and waste from nuclear power plants.

Since metals are mobilized in the soil and various migratory forms are formed, it is in it that the main cycles of migration of heavy metals in the biosphere occur. Besides, the soil can serve as a source of secondary pollution of water and air. From the soil these substances are absorbed by plants, which in the worst way can get into human food.

All heavy metals are combined into a fairly large group of chemical elements, the main criterion is their atomic mass. Of these, three elements can be distinguished as especially toxic – mercury, lead and cadmium. The rest belong to microelements: zinc, cobalt, vanadium, copper, tin, molybdenum and nickel [8].

Most of the agricultural land is polluted by elements such as mercury, tin, bismuth, arsenic, lead, copper, which enter the soil with poisonous chemicals, biocides, plant growth stimulants, and structure-forming agents. The introduction of unconventional fertilizers derived from all kinds of waste, and alternative fertilizers made from various wastes into agricultural production often contain a significant set of pollutants in fairly high concentrations.

The purpose of the study is to monitor the content of microelements and heavy metals in the soils of the Republic of Mordovia and crops growing on them with the subsequent introduction of optimal crop rotations agricultural land planning in enterprises engaged in the production of food crops.

2 Materials and methods

Since 1993, the State Center of the Mordovian Agrochemical Service has been conducting a system of regular observations in 13 reference areas on the territory of the republic. By 2010, the number of reference plots reached 23 in 20 districts in 21 farms of the Republic of Mordovia. They are located in various natural and climatic zones and reflect the climatic and economic features of the main soil types (Table 1).

Conventional methods were used to study the accounting of the content of heavy metals in the soil: determination of the content of mobile and acid-soluble forms of heavy metals was carried out by the atomic absorption method; detection of heavy metals in plant tissues was carried out by atomic absorption method in flame of an air-acetylene burner.

3 Results and discussion

Microelements in the arable layer of the soil depend on the type of soils, their location and their content in soil-forming rocks. The content of boron and molybdenum in all reference areas ranges from medium to high (Table 2). Most

* Corresponding author: GeraskinMM@yandex.ru

of them are contained in reference plots No. 5, 7, 8. The content of zinc, manganese, cobalt is lower. Conversely, all reference areas contain high amount of copper in the arable layer. This pattern is observed over the entire study period [4].

Table 1. Topographic survey of reference plots and presence of pollutants.

Region	Nearest inhabited locality	RP No.	RP area, ha	Geographical coordinates							
				latitude	**longitude**	**farms**	**plants**	**airfields**	TPP	railroads	motorways
Leached chernozem											
Oktyabrsky	Nikolaevka	5	46	54° 10'	45° 15'	0.5	10	3	15	1	0.5
Oktyabrsky	Saransk	8	29	54° 10'	45° 15'	5	0.2	25	2	0.05	1.5
Lyambirsky	Lyambir	12	70	54° 15'	45° 07'	1.5	15	15	20	–	0.01
Chamzinsky	Sabur-Machkasy	21	112	54° 23'	45° 45'	0.3	3	–	–	1	2.0
Romodanovsky	Maloe Chufarovo	22	57	54°30'	45° 25'	1.5	20	–	22	2	0.2
Ardatovsky	Ardatov	28	27	54° 50'	46° 13'	1.5	–	–	–	–	0.001
Atyashevsky	Atyashevo	34	45	54° 43'	46° 18'	1.0	–	–	–	0.02	0.04
Podzolized chernozem											
Oktyabrsky	Nikolaevka	7	26	54° 08'	45° 08'	2.5	10	0.5	25	5	15.0
Ruzaevsky	Ruzayevka	13	71	54° 07'	44° 35'	2	1	–	–	0.08	0.5
Dubensky	Dubenki	36	65	54° 22'	45° 46'	5	–	–	–	–	0.001
Torbeevsky	Zhukovo	37	30	54° 18'	43° 39'	1.5	–	–	–	6	0.015
Dark-grey forest soil											
Zapadno-Polyansky	Akim-Sergeevka	17	27	54° 08'	42° 45'	–	–	–	–	–	0.03
Kovylkinsky	Panzha	27	18	54° 05'	43° 55'	0.2	–	–	–	–	0.1
Kochkurovsky	Kochkurovo	30	2	54° 03'	45° 26'	–	–	–	–	–	0.01
Krasnoslobodsky	Prigorodnoye	31	40	54° 25'	43° 45'	1	0.5	–	–	–	0.3
Kadoshkinsky	Kadoshkino	35	35	54° 08'	44° 21'	3.0	2.0	–	–	0.5	0.05
Grey forest soil											
Bolsheignatovsky	Varmazeika	29	40	55° 05'	45° 38'	0.8	–	–	–	–	0.03
Staroshaigovsky	Staroe Shaigovo	32	10	54° 14'	44° 28'	–	–	–	–	–	0.01
Atyuryevsky	Atyuryevo	33	47	54° 18'	43° 38'	0.8	–	–	–	–	0.01
Sod-podzolic soil											
Temnikovsky	Tat-Karaevo	14	70	54° 37'	43° 06'	1	–	10	–	–	0.1
Alluvial soil											
Bolshoy-Bereznikovsky	Bolshie Berezniki	3	22	54° 11'	46° 00'	0.5	–	–	–	–	0.001
Oktyabrsky	Nikolaevka	6	18	54° 10'	45° 09'	0.6	15	2	20	0.001	0.15
Ichalkovsky	Gulyaevo	10	10	54°40'	45°20'	5	–	–	–	0.01	–

The iron content in reference plots No. 6, 10, 29 is much higher than in other areas. The sulfur content in the soils of the reference areas is low and ranges from 1.6 to 6.0 mg/kg. Fluorine is the most active and reactive metalloid. Fluorine and its compounds in microdoses are necessary, in high doses – toxic to humans, animals, plants. The fluorine content in reference areas No. 5, 8, 21 is slightly higher compared to other areas – 0.83; 0,74; 0.76 mg/kg of soil, but the results obtained do not exceed the maximum permissible concentrations in soil.

An increased content of mobile forms of heavy metals and arsenic (Table 3) is observed at reference plots No. 7, 8, 10, 32, which are located near hazardous pollution sources. For example, plot No. 7 is located next to the runway of Saransk airport, area No. 8 is located near the rubber factory on one side and the railway bed on the other side.

The reference plot No. 10 is located in the flood meadows of the Alatyr River, toxic elements settle during the spring flood. Other reasons for the overestimated content of heavy metals are the location of reference plots near motorways (highways).

The mercury content is small in the arable horizon of all reference plots, i.e. compatible with the normal limits, but this element is more contained in RP No. 34 – 0.036 mg/kg of soil. In reference plot No. 32 the content of zinc, lead, nickel, arsenic is the highest – 1.81; 1.35; 1.25; 5.2 mg/kg soil. At the same time, the excess of the MPC in all areas is not observed.

Table 2. Characteristics of the arable layer by the content of mobile forms of microelements, sulfur, fluorine and iron.

Region	Year	RP No.	Mobile forms, mg/kg soil								
			B	**Mo**	**Cu**	**Zn**	**Co**	**Mn**	**Fe**	**S**	**F**
Leached and podzolized chernozem											
Oktyabrsky	2011	5	2.12	0.12	5.6	0.92	1.38	54	18.9	6.4	0.99
	2012		2.36	0.22	4.9	0.93	1.86	38.8	17.1	3.4	0.83
Oktyabrsky	2011	7	0.84	0.11	5.4	0.91	0.91	34	36.7	4.3	0.66
	2012		2.28	0.21	5.1	0.85	1.44	39.7	31.9	2.5	0.60
Oktyabrsky	2011	8	1.89	0.17	4.3	1.02	1.46	44	18.5	5.7	1.02
	2012		2.09	0.22	4.3	0.78	1.79	41.3	10.4	1.7	0.74

Lyambirsky	2011	12	1.25	0.37	8.1	1.39	1.05	41	20.3	4.0	0.39
	2012		1.25	0.21	6.8	0.68	1.07	48.2	6.8	3.3	0.57
Ruzaevsky	2011	13	2.01	0.40	7.8	0.74	1.31	67	31.6	5.6	0.47
	2012		1.87	0.29	4.8	0.73	1.00	63.5	18.9	1.7	0.56
Chamzinsky	2011	21	1.25	0.30	5.1	1.13	1.46	16	7.4	5.8	0.56
	2012		1.01	0.12	10.1	0.41	0.11	51.3	16.9	4.0	0.76
Romodanovsky	2011	22	1.32	0.21	6.3	0.98	1.07	35	22.5	5.2	0.36
	2012		0.92	0.22	5.7	1.61	1.22	55.5	12.2	3.7	0.66
Ardatovsky	2011	28	1.32	0.36	4.7	0.92	1.25	34	23.9	5.6	0.42
	2012		1.60	0.29	5.1	0.57	1.29	37.8	11.5	2.8	0.53
Atyashevsky	2011	34	1.41	0.27	9.6	1.13	1.05	66	17.0	4.8	0.44
	2012		2.08	0.15	7.8	0.65	1.58	54.1	32.8	6.0	0.59
Dubensky	2011	36	1.31	0.14	5.9	1.01	1.23	46	17.1	4.2	0.55
	2012		0.92	0.25	5.9	0.52	0.95	54.2	15.7	3.6	0.59
Torbeevsky	2011	37	1.02	0.16	6.9	0.93	1.46	55	17.1	4.4	0.48
	2012		1.51	0.15	7.1	0.32	1.15	59.3	14.2	1.7	0.55
Dark-grey forest soil											
Zapadno-Polyansky	2011	17	0.84	0.37	3.4	0.68	1.05	30	37.7	4.4	0.63
	2012		0.51	0.14	3.9	0.74	0.68	51.5	29.4	3.7	0.63
Kovylkinsky	2011	27	1.52	0.11	5.7	1.29	1.73	58	22.8	2.8	0.49
	2012		0.82	0.15	4.3	0.47	1.07	32.1	10.0	2.0	0.52
Kochkurovsky	2011	30	1.01	0.17	4.2	1.05	1.30	16	7.1	4.8	0.53
	2012		0.92	0.26	4.3	1.17	1.08	53.7	5.7	2.9	0.57
Krasnoslobodsky	2011	31	1.11	0.46	4.6	0.71	0.009	35	72.2	3.6	0.34
	2012		1.07	0.19	6.3	0.45	0.95	60.7	18.8	3.5	0.59
Kadoshkinsky	2011	35	1.30	0.15	6	0.83	0.09	33	34.2	5.3	0.54
	2012		0.75	0.21	5.1	0.65	0.95	50.6	27.7	1.6	0.59
Grey forest soil											
Bolsheignatovsky	2011	29	1.21	0.20	6	1.14	1.05	65	60.3	4.3	0.45
	2012		0.92	0.33	5.6	0.65	0.95	51.4	111.2	3.5	0.57
Staroshaigovsky	2011	32	1.40	0.20	5.8	1.12	1.38	44	100.5	5.1	0.57
	2012		0.78	0.17	4.2	1.81	1.28	39.5	81.3	2.7	0.44
Atyuryevsky	2011	33	0.95	0.16	6	0.59	0.09	45	68.8	4.1	0.38
	2012		0.59	0.15	6.2	0.53	0.75	38.3	40.4	1.6	0.61
Sod-podzolic soil											
Temnikovsky	2011	14	0.84	0.18	4.3	0.53	0.09	44	64.3	4.8	0.43
	2012		1.06	0.25	3.4	0.58	1.44	55.5	77.3	3.5	0.70
Alluvial soil											
Bolshoy-Bereznikovsky	2011	3	1.20	0.14	8	1.31	1.25	58	52.7	6.4	0.72
	2012		1.07	0.19	7.2	0.98	1.28	55.2	27.1	3.3	0.63
Oktyabrsky	2011	6	1.25	0.23	7.4	1.62	1.05	70	158.5	4.7	0.81
	2012		1.41	0.10	7.2	0.92	1.92	47.0	91.3	4.1	0.63
Ichalkovsky	2011	10	1.31	0.15	10.5	1.12	1.56	95	200.0	5.6	0.42
	2012		0.81	0.19	6.9	1.42	0.95	66.8	131.1	1.7	0.40
Average	2011		1.29	0.22	6.1	1.01	1.05	47	48.2	4.9	0.55
	2012		1.24	0.20	5.7	0.80	1.16	49.8	40.8	3.0	0.61

The collection of information on crops and ground adjustment of borders was carried out in those farms where RPs are laid. The measurements were made by height, density, weight from 1 m, fresh productivity in various phases of development, and crop accounting (Table 4). Weed types, their height, productivity, etc. were determined for seeding.

Table 3. Characteristics of the arable horizon of reference plots by the content of mobile forms of heavy metals and arsenic.

Region	RP No.	Observation (month)	Mobile forms (acetate-ammoniacal nuffer with pH 4.8), mg/kg							
			Cu	Zn	Cd	Pb	Ni	Cr	As	Hg
Leached and podzolized chernozem										
Oktyabrsky	5	22.04	0.20	0.93	0.060	0.77	0.45	–	3.8	0.020
Oktyabrsky	7	03.05	0.22	0.85	0.080	0.83	0.73	–	4.0	0.027
Oktyabrsky	8	26.04	0.26	0.78	0.050	1.12	0.58	–	5.4	0.019
Lambirsky	12	30.04	0.17	0.68	0.050	0.87	0.48	–	4.3	0.024
Ruzaevsky	13	04.05	0.20	0.73	0.080	0.62	1.03	–	4.6	0.025
Chamzinsky	21	29.04	0.38	1.01	0.060	0.96	0.68	–	3.8	0.020

Romodanovsky	22	24.04	0.28	1.61	0.060	0.93	0.63	–	4.5	0.027
Ardatovsky	28	03.05	0.22	0.57	0.080	0.89	0.83	–	3.1	0.018
Atyashevsky	34	03.05	0.15	0.65	0.040	1.27	0.85	–	5.4	0.036
Dubensky	36	28.04	0.27	0.52	0.040	1.14	0.68	–	3.9	0.022
Torbeevsky	37	27.04	0.16	0.32	0.070	1.35	0.52	–	3.5	0.026
Dark-grey forest soil										
Zapadno-Polyansky	17	27.04	0.23	0.74	0.050	0.93	0.72	–	2.7	0.014
Kovylkinsky	27	23.04	0.17	0.47	0.040	1.27	0.44	–	3.0	0.022
Kochkurovsky	30	25.04	0.25	1.17	0.060	0.94	0.92	–	3.9	0.018
Krasnoslobodsky	31	26.04	0.24	0.45	0.070	0.77	0.67	–	4.3	0.021
Kadoshkinsky	35	26.04	0.22	0.65	0.050	0.75	0.48	–	4.7	0.021
Grey forest soil										
Bolsheignatovsky	29	28.04	0.21	0.65	0.040	0.58	0.61	–	6.0	0.014
Staroshaigovsky	32	04.05	0.19	1.81	0.080	1.35	1.25	–	5.2	0.018
Atyuryevsky	33	26.04	0.25	0.53	0.080	0.70	0.41	–	4.0	0.022
Sod-podzolic soil										
Temnikovsky	14	25.04	0.31	0.58	0.080	1.24	0.65	–	1.9	0.017
Alluvial soil										
Bolshoy-Bereznikovsky	3	30.04	0.20	0.98	0.030	0.99	0.93	–	4.7	0.024
Oktyabrsky	6	22.04	0.19	0.97	0.040	0.75	0.67	–	6.5	0.022
Ichalkovsky	10	21.04	0.21	1.42	0.070	0.62	1.13	–	8.2	0.025
Average			0.22	0.80	0.060	0.94	0.71		4.4	0.022

Table 4. Crop chemical composition.

Region	RP No.	Crop	Product	Yield, t/ha	Total chemical elements, mg/kg of natural moisture products						Radiological indicators, Bq/kg	
					Cu	Zn	Pb	Hg	Cd	As	Sr-90	Cs-137
Leached and podzolized chernozem												
Oktyabrsky	5	winter wheat	grain	20.1	2.29	24.49	0.30	0.002	0.024	0.018	2.0	3.8
			straw	26.3	2.89	15.94	1.32	0.004	0.064	0.019	3.1	7.6
Oktyabrsky	7	oat	grain	18.1	1.52	20.18	0.42	0.002	0.014	0.019	1.9	4.0
			straw	22.3	1.63	9.71	1.71	0.004	0.039	0.019	3.3	4.2
Oktyabrsky	8	spring wheat	grain	20.4	3.06	21.38	0.27	0.002	0.015	0.018	1.9	4.9
			straw	26.1	0.97	15.08	0.51	0.003	0.064	0.020	3.9	5.9
Ruzaevsky	13	barley	grain	23.8	2.37	17.44	0.16	0.002	0.012	0.018	3.0	4.8
			straw	20.6	0.49	4.50	0.40	0.004	0.027	0.019	3.2	6.0
Chamzinsky	21	winter wheat	grain	27.4	1.89	17.3	0.38	0.004	0.038	0.02	3	5.9
			straw	35.2	0.54	5.06	1.31	0.006	0.034	0.021	3.8	6.5
Romodanovsky	22	corn	herbage	16.9	0.53	4.80	0.71	0.002	0.015	0.015	1.6	5.2
Ardatovsky	28	perennial grass	herbage	3.97	2.54	11.47	1.06	0.004	0.073	0.019	1.8	4.9
Atyashevsky	34	barley	grain	2.68	2.91	25.20	0.35	0.002	0.029	0.021	1.9	5.1
			straw	2.32	1.35	7.08	1.68	0.005	0.053	0.023	2.9	6.5
Dubensky	36	winter wheat	grain	21.34	30.53	0.33	0.002	0.047	0.016	1.5	3.6	42
			straw	1.31	7.50	1.17	0.004	0.042	0.017	3.0	5.4	30
Torbeevsky	37	winter wheat	grain	1.82	26.87	0.31	0.002	0.023	0.016	1.8	3.8	29
			straw	2.09	10.14	1.59	0.005	0.07	0.017	2.2	4.5	172
Dark-grey forest soil												
Zapadno-Polyansky	17	perennial grass	herbage	3.32	3.77	12.89	1.00	0.003	0.044	0.018	2.2	4.8
Kovylkinsky	27	barley	grain	11.51	17.90	0.67	0.002	0.010	0.019	2.4	4.3	37
			straw	21.67	7.35	1.09	0.006	0.031	0.023	3.2	5.6	40
Kochkurovsky	30	perennial grass	herbage	3.84	5.15	14.34	2.56	0.005	0.144	0.022	4.8	1.9
Krasnoslobodsky	31	annual grass	herbage	12.3	3.01	17.11	0.84	0.006	0.059	0.024	4.8	2.3

Chemicals are introduced mainly in those reference plots where grain crops are grown (winter and spring crops). They are used in most cases when sowing in rows (winter and other grains) and in the form of feeding in spring on winter crops. Organic fertilizers were not introduced at any reference plot during the entire observation period, and no chemical soil reclamation was carried out.

As mentioned above toxic elements emitted with spent products of motor engines, railway trains, flooding areas as a result of flood waters, pollution due to fuel ingress at gas stations, etc. settle in the soil and, as a rule, in crop products growing on them.

So at RP 7 there is a high content of lead, zinc in the by-products of oats, at RP 10 and 17 (perennial grass) – cadmium. Some elements are contained in greater quantities at RP 5, 8 than in other areas. These plots are located within the city, near the Rubber Products Plant, during the production of which the atmosphere, soil, and ultimately the agricultural culture growing on it, get polluted.

It should be noted that to determine the degree and completeness of dissolution of the studied heavy metals in soil, such indicators as the effective solubility product of sediments of these heavy metals, effective constants of instability of complexes existing in soil horizons and ion exchange constants in the soil-solution system are used [5].

As a result of the studies it can be stated that the content of heavy metals in the soil in all cases is lower than the APC (MPC). The products obtained in the tested areas are safe in relation to the content of heavy metals.

If metals contaminated the soil and vegetation, then the current situation can be corrected using a number of methods. First of all, it is necessary to promptly limit the flow of these substances to agricultural land. In the future, various methods should be applied that help to remove them from the root layer or bind them in the soil into low-dissociating compounds. The selection of crops and varieties that are less susceptible to them helps to reduce the accumulation of heavy metals in plants. Timely assessment of the soil for the presence of free forms of these elements and the compilation of crop rotations with the selection of different degrees of soil pollution for the cultivation of certain crops play an important role here.

In our case, we propose to consider the fifth more efficient method to develop environmental measures with the introduction of optimal crop rotations taking into account the established agrochemical indicators of soils and vegetation in agricultural enterprises of the Republic of Mordovia.

It has to be said that the first condition to restrict the entry of heavy metals into the soil must be observed as fundamental. Therefore, the restriction of doses for pollution of agricultural land due to environmental requirements is a prerequisite for the greening of agriculture.

Adaptive landscape systems of agriculture can serve as an additional and significant step in optimizing the environmental situation in case of soil pollution with heavy metals, where it is necessary to introduce optimal crop rotations developed in projects of internal land management on an agrarian landscape basis on these land arrays [6].

In these studies a reference plot means an entire field or a separately processed part of it, which are typical for a given region. The plot is not excluded from crop rotation and is in normal agricultural production conditions.

I. Leached chernozem

RP 5 – Nikolaevka village, Oktyabrsky region, 46 hectares (1994). Main pollutants (railway – 1 km, motorway – 0.5 km, airfield – 3 km). The studied areas were used in agricultural production at different times: grain crops were grown for 10 years, tilled crops – 4 years, perennial grass – 4 years, clean fallow – 1 year.

RP 8 – Greenhouse State Unitary Enterprise, Oktyabrsky region, 29 hectares (1994). Main pollutants (foundry – 3-3.5 km). Use of the plot: grain crops – 16 years, tilled crops – 1 year, vico-oat mixture – 2 years.

RP 12 – Cheremishevskoye LLC, Lambirsky region, 70 hectares (1994). Main pollutants (TPP-2 – 20 km, motorway – 10 m). Use of the plot: grain crops – 15 years, annual grass – 2 years, clean fallow – 2 years.

RP 21 – Saburmachkasskoye LLC, Chamzinsky region, 112 hectares (1994). Main pollutants (railway – 2 km, cement plant – 3 km, motorway – 2 km). Use of the plot: grain crops – 12 years, tilled crops – 3 years, vico-oat mixture – 1 year, perennial grass – 2 years, clean fallow – 1 year.

RP 22 – KZS and K, Romodanovsky region, 57 hectares (1994). Main pollutants (railway – 2 km, motorway – 0.2 km, industrial zone – 20 km, TPP-2 – 22 km). Use of the plot: grain crops – 12 years, tilled crops – 1 year, perennial grass – 3 years, clean fallow – 3 years.

II. Podzolized Chernozem

RP 7 – Lukhovskoye State Unitary Enterprise of the Republic of Mordovia, Oktyabrsky region, 26 hectares (1998). Main pollutants (railway – 5 km, motorway – 15 km, airport – 500 m). Use of the plot: grain crops – 4 years, perennial grass – 10 years, clean fallow – 1 year.

RP 13 – Demkin farm, Ruzaevsky region, 71 hectares (1994). Main pollutants (railway – 800 m, motorway – 50 m, industrial facilities – 1 km). Use of the plot: grain crops – 16 years, clean fallow – 3 years.

III. Dark-gray forest soil

RP 17 – Vector and K LLC, Zubovo-Polyansky region, 27 hectares (1999). Main pollutants (motorway – 30 m). Use of the plot: grain crops – 3 years, perennial grass – 11 years.

RP 27 – Samaevskoe LLC, Kovylkinsky region, 27 hectares (1994). Main pollutants (motorway – 100 m). Use of the plot: grain crops – 8 years, tilled crops – 6 years, annual grass – 1 year, clean fallow – 4 years.

IV. Sod-podzolic soil

RP 14 – Tarkhanovskoye LLC, Rodina branch, Temnikovsky region, 70 hectares (1996-2012). Main pollutants (motorway – 10 m). Use of the plot: grain crops – 3 years, tilled crops – 1 year, clean fallow – 1 year, perennial grass – 12 years.

V. Alluvial granular soil

RP 3 – Kirovskoye LLC, Bolshebereznikovsky region, 22 hectares (1996-2012). Main pollutants (motorway – 10 m, railway – 10 m, airfield – 2 km). Use of the plot: clean fallow – 3 years, perennial grass – 14 years.

RP 6 – Nikolaevka village, Oktyabrsky region, 18 hectares (1994-2012). Main pollutants (motorway – 150 m, railway – 10 m, airfield – 2 km). Use of the plot: vegetable and tilled crops – 11 years, grain vrops – 3 years, clean fallow – 3 years, perennial grass – 2 years.

RP 10 – Culture CJSC, Ichalkovsky region, 10 hectares (1994). Main pollutants (railway – 100 m – 2 km). Use of the plot: perennial grass (permanent).

4 Conclusion

In modern agriculture as a result of the increased man-made load the areas of soils contaminated with heavy metals increase, which makes it difficult to obtain high-quality agricultural products. They enter the agroecosystem in

https://doi.org/10.1051/bioconf/20225200029

various ways: from gas and dust emissions, with atmospheric precipitation, irrigation waters, etc. It should also be noted that it is during dispersion in ecosystem components that heavy metals enter the environment from technogenic sources, and the most important is the dispersion of aerotechnogenic emissions through the atmosphere. Currently, agrarian landscapes around large cities are subject to anthropogenic impact.

The environmental assessment should distinguish between pollution factors, and the content of heavy metals in suburban agricultural land represents the sum of the components of natural and man-made origin. Thus, it is possible to observe the expansion of the soil range with a reduced environmental value.

It should be noted that the decrease in the accumulation of humus, the decrease in the complexing and structure-forming ability, the biological activity of the soil solution is influenced by the accumulation of heavy metals and accordingly decreases the biological productivity of lands, and enzymatic and microbiological activity is inhibited. All listed processes lead to a decrease in soil resistance to pollution, and, accordingly, to a decrease in soil resistance to overcompaction and the development of gleyization. In fact, the accumulation of heavy metals in the soil leads to an increase in their mobility, to penetration into the deep layers of the soil profile, to pollution of vegetation, to an increase in the content of heavy metals in groundwater. It is also possible to note the unequal nature of transformation in various soils of heavy metals and, accordingly, characteristic buffering.

References

1. Ovchinnikova M. F., Agrochemical Bulletin **1**, 2–5 (2013)
2. Tolkach G. V., Bulletin of the Belarusian State Agricultural Academy **4**, 118–123 (2017)
3. Sokolov M. S., Agrochemistry **3**, 3–18 (2013)
4. Geraskin M. M., IOP Conference Series: Earth and Environmental Science **867**, 012060 (2021)
5. Vytnov A. I., Scientific notes of the Russian State Social University **4**, 222–227 (2011)
6. Geraskin M.M, Moscow Economic Journal **6**, doi: 10.24411/2413-046X-2020-10375

Modeling the efficiency of the use of production potential

Natalia Bondina[2], *Igor* Bondin[2], *Irina* Pavlova[2], and *Oksana* Mamai[1*]

[1] Samara State Agrarian University, 2, Uchebnaya St., Kinel, urban-type settlement Ust-Kinelsky, Samara Region, 446442, Russian Federation
[2] Penza State Agrarian University, 30, Botanical St., Penza, 440014, Russian Federation

Abstract. The efficiency of modern agricultural business depends on many economic factors, the key driver of which is the production potential. At the level of certain business entities, this is the maximum possible output of the finished product in the conditions of the achieved volume of production factors (labor, capital, land). The problems of stabilization and further production increase in the agro-industrial complex of Russia require an increase in the efficiency of the use of production potential by all participants in the agricultural business. At the same time, the major impact of digital transformation on all sectors of the country's economy changes the already established models of managing the production potential thus adapting them to modern information technologies. Therefore, in the current economic conditions, the success of the agribusiness entities largely depends on the ability to calculate, model the efficiency of the production potential use. The construction of an economic and mathematical model and the practical use of its results allows optimizing the production, material, personnel components of the production potential to ensure the maximum volume of finished products and achieve the best performance indicators.

1 Introduction

The production potential in the agricultural sector is the possibility of obtaining a certain amount of agricultural products in the conditions of the technologies in operation, and with the available amount of resources and their sufficient quality. It is very important that the resources are in an optimal combination in terms of quantitative and qualitative characteristics. The amount and size of resources must be sufficient to ensure production. They must be of the required quality as poor quality requires more resources. Structural proportionality of production potential elements shall also be maintained. At the same time, the general signs of digitalization of the country's economy lay the groundwork for improving not only technological processes for the production of goods, but also methods for ensuring its efficiency [1, 2]. Therefore, it is important to ensure the information support to manage the production potential. Thus, production, material, personnel, technical-technological and information components are critical in assessing the production potential of agricultural business participants.

The maximum efficiency of the production potential is achieved provided that all its components are balanced, which determines the need for the development of information technologies to manage the production potential on the basis of economic and mathematical modeling of its effective use [3, 4].

2 Materials and methods

The methodological basis of the study included the fundamental provisions of the scientific theory on the essence of production potential, modern research by domestic and foreign scientists. The study is based on methods used in economic science: general scientific (dialectical, analysis and synthesis, comparison and analogies), special (systemic, comparative analysis, economic and mathematical).

The information base of the study included official statistics; federal and regional regulatory legal acts; data from the Ministry of Agriculture of Penza region; author's research materials; data of Internet resources (scientific articles and works of practitioners, industry portals, economic reviews).

3 Results

The efficiency of the production potential of agricultural organizations depends on the efficiency use of its components. The main components of the production potential were analyzed based on the materials of agricultural organizations of Penza region [5].

An important condition for the efficiency of agricultural production is the provision of organizations with fixed production assets. Data on the structure of fixed assets are presented in Table 1.

* Corresponding author: mamai_ov@ssaa.ru

Table 1. Structure of the main production assets of agricultural organizations in Penza region, % *

Indicator	2001	2011	2015	2017	2018	2019	2020
Main agricultural production assets – total including:	100.0	100.0	100.0	100.0	100.0	100.0	100.0
buildings, structures and transfer devices	74.1	44.2	51.01	49.4	50.8	51.9	53.8
machinery and equipment	15.6	43.9	40.7	42.1	41.0	40.9	39.5
vehicles	3.8	6.1	4.1	4.2	4.4	4.1	4.0
working and productive livestock	3.9	5.2	3.7	3.7	3.3	2.8	2.3
other types of fixed assets	2.6	0.6	0.4	0.6	0.5	0.3	0.4

*As of the end of the year.

The table shows that in general there is a positive trend in the main indicators characterized as an active part of production assets. In comparison with the base period, their values for "machines and equipment" and "vehicles" increased for all years of the dynamic series. The ambiguous behavior of "draft and productive cattle" indicator ultimately confirms the tendency of a gradual decrease in the specific weight of livestock in the overall structure of production assets and the focus of agricultural production on crop production.

Despite the positive trends in the behavior of certain indicators in the structure of fixed assets, their dynamics in absolute terms is the opposite. The basic equipment fleet is decreasing every year (Table 2)

Table 2. Basic equipment fleet in agricultural organizations of Penza region, units*

Indicator	2001	2011	2015	2017	2018	2019	2020	2020 in % by 2001
Tractors	10412	2954	2885	2847	2643	2625	2720	26.1
Seeders	4460	1251	1103	1068	1224	1213	606	13.6
Mowing machines	938	399	379	331	336	311	301	32.1
Pickup presses	413	240	234	211	209	192	181	43.8
Windrowers	1561	249	211	230	226	260	227	14.5
Combines:								
grain-harvesting	3420	739	670	764	731	771	741	21.7
potato-harvesting	14	12	22	17	18	19	15	107.1
fodder harvesting	780	177	156	136	118	101	96	12.3
Beet harvesters	420	78	55	74	74	84	80	19.0
Sprinkler and irrigation machines and units	140	33	44	47	51	46	39	27.9

*As of the end of the year.

The exception is potato harvesters, the basic growth rate of which was 107.1%. For all other positions, there is a significant decrease in the number of pieces of equipment, which makes from 2 to 7 times. As a result, the production load per 1 unit of agricultural machinery increased (Fig. 1).

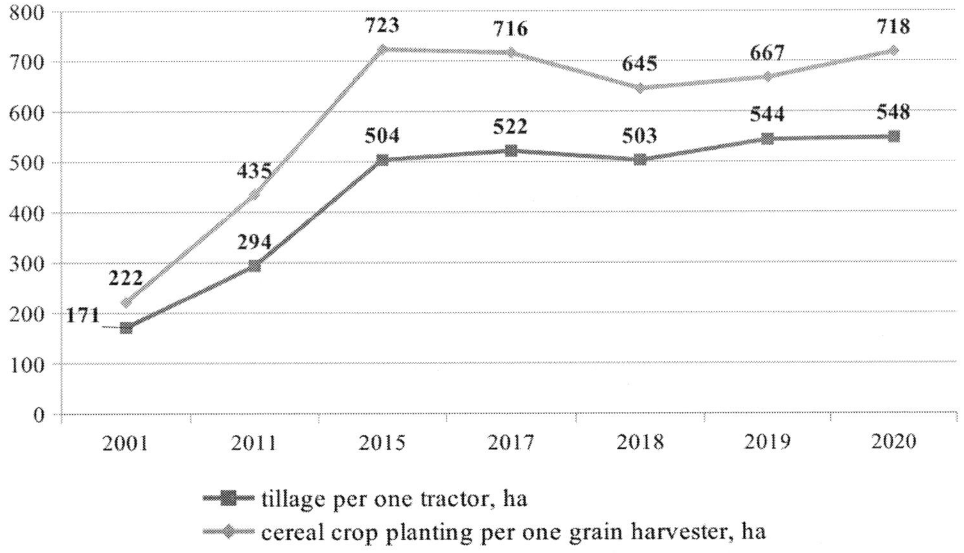

Fig. 1. Production load per 1 unit of agricultural machinery, ha

However, it should be noted that the growth of the processing area per 1 unit of agricultural machinery is caused not only by a decrease in the number of tractors and machines, but also by an increase in the area of agricultural land of Penza region over the analyzed period by 327.8 thousand hectares, as well as an increase in the productivity of the equipment itself, due to the acquisition of new modern units.

Its material basis has a great impact on the efficiency of the production potential [6]. The current assets of agricultural organizations are shown in Table 3.

Table 3. Current assets of agricultural organizations in Penza region *

Indicator	2001	2011	2015	2017	2018	2019	2020
Current assets coverage ratio	-0.66	-0.51	-0.44	-0.37	-0.71	-1.05	-1.21
Current to personal assets ratio	0.48	1.27	1.54	1.69	1.99	2.28	2.63
Material assets coverage ratio with personal current assets	-0.95	-0.97	-1.01	-0.77	-1.56	-2.28	-2.65
Cost factor of material current assets	0.20	0.20	0.21	0.25	0.21	0.19	0.18
Current assets to asset value ratio	0.27	0.31	0.48	0.51	0.45	0.4	0.39

*As of the end of the year.

In 2001-2020 there was an increase in the share of current assets in the total assets of business entities from 27% to 39 % in agricultural organizations of Penza region. At the same time, material reserves occupied a significant share, which in different years accounted for from 18 to 25% of the value of all property. However, throughout the analysis, the increase in the share of current assets was accompanied by a significant shortage of personal current assets. The situation as a whole is such that current assets cannot be financed from personal sources of funds. This is indicated by the actual level of the equity ratio of current assets, which during the entire analyzed period did not comply with its normative value (0.1). The negative value of this indicator indicates that organizations do not have their own funds to cover the need for current assets and widely use borrowed sources for this purpose. Thus, the improvement of current assets depends on the outstripping growth of the amount of current assets compared to the overall growth of current assets, inventories and personal sources of funds.

The labor resources of organizations determine the volume and timeliness of all work, affect the efficiency of the production potential [7]. Agricultural organizations of Penza region are characterized by labor shortage. During the study period, the number of permanent workers decreased by 46,630 people, including: tractor drivers – by 10,348 people, machine milking operators – by 4919 people, cattle workers – by 4728 people and so on. The labor resources for agricultural organizations in 2020 amounted to 19.5% of the 2001 level. On the one hand, this reduction was caused by the urbanization and mass migration of the agricultural population to the city, on the other hand – by the introduction of more advanced production technologies and a decrease in the share of manual labor [8, 9].

The reality is that high efficiency of the production potential can only be achieved by the rational interaction of all factors-resources of each level of production [10].

Therefore, in the current conditions, the problem of studying and modeling the relationship between the efficiency of the production potential and factors affecting its level becomes urgent, the role of the information component of the production potential increases.

To determine the degree of influence of each factor

and their combination on the efficiency of using the production potential, it is necessary to establish a close relationship between them based on the correlation-regression analysis. The main source in the formation of the information base was the official data of the financial statements. The analysis was carried out on the basis of 120 agricultural organizations of Penza region. The original information array was subjected to statistical analysis. As a result, a number of factors were identified, the impact of which on the effective indicator is most significant. The effective value (Y) was the cost of gross production per 100 hectares of agricultural land. Analytical procedures were carried out based on the indicators presented in Table 4.

Table 4. Factors influencing the efficiency of the production potential of agricultural organizations*

Factor	Factor name	Factor	Factor name
X1	capital-labor ratio, thousand rubles	X5	fertilizer costs, thousand rubles
X2	energy security, hp	X6	share of subsidies, thousand rubles
X3	labor security, people	X7	feed costs, thousand rubles
X4	current assets, thousand rubles	X8	soil quality of agricultural land, ball-bonitet

*Factors are presented per 100 hectares of agricultural land.

The resulting multiple regression equation has the following form:

$$Y=-1067+0.051X_1+0.11X_2+97X_3+0.21X_4+1.94X_5+0 \quad (1)$$

The multiple correlation coefficient R = 0.87 confirms a strong association between the performance indicators and the factorial features. The multiple determination coefficient R^2 = 0.75 indicates that 75% of variations in the performance indicators (Y) depends on the factors included in the model. The analysis of the obtained multiple regression equation showed that for the studied sample of agricultural organizations, the volume of gross production is most dependent on such a direct impact factor as labor security. Thus, with an increase in this factor by 100 rubles, the value of production potential as a whole will increase by 97 points. The remaining factors have a significant but somewhat smaller impact on the efficiency of the production potential.

A direct indicator of the degree of the production potential is the ratio of actually achieved results to its possible (normative) value.

$$Э=\frac{Y}{a^{¿}\tilde{Y}}=\frac{a}{a^{¿}} \quad (2)$$

$Y, \tilde{Y}$ – actual and theoretical levels of gross production, respectively;

$a, a^{¿}$ – actual and normative (maximum) integral indicators of resource efficiency, respectively.

The calculations showed that none of the organizations represented in the sample uses its production potential in full. Therefore, each of them can improve its performance and increase the efficiency of the production potential due to available internal reserves.

4 Conclusion

The modeling of the relationships of factors in determining the degree of the use of the production potential is one of the vector directions for the development of information technologies in the practice of agricultural organizations. Currently, the entire agricultural business is being transformed under the influence of information technologies. The introduction of modern methods of information processing expands the boundaries of using existing databases, contributes to improving the efficiency of organizations, and gives new competitive advantages to the agricultural business.

References

1. S. Alekseeva, G. Volkova, O. Sukhanova, E. Fudina Scientific Papers Series "*Management, Economic Engineering in Agriculture and Rural Development*" **21** (2), 19-25 (2021)
2. Y. Kolchugina, A. Chebotareva and E. Kobets Russian Economic Bulletin **2** (4), 14-17 (2019)
3. M. Dudin, K. Pavlova, E. Frolova, T. Samusenko, I. Popova Scientific Papers Series "*Management, Economic Engineering in Agriculture and Rural Development*" **18** (1), 143-152 (2018)
4. N. Kuznetsov, N. Ukolova, S. Monakhov, J. Shikhanova Scientific Papers Series "*Management, Economic Engineering in Agriculture and Rural Development*" **18** (1), 219-225 (2018)
5. Official website of the Ministry of Agriculture of the Penza Region, retireved from: https://mcx.pnzreg.ru/
6. N. Bondina, I. Bondin, I. Scientific Papers Series "*Management, Economic Engineering in Agriculture and Rural Development*" **21** (2), 111-117 (2021)
7. S. Gurfova 2018 Modern Economy Success **4**, 46-50 (2018)
8. E. Pozubenkova, O. Ulanova, T. Chuvorkina, O. Kadykova Scientific Papers Series "*Management, Economic Engineering in Agriculture and Rural Development*" **21** (2), 487-492 (2021)
9. D. Samygin, N. Baryshnikov, L. Mizjurkina Ekonomika regiona [Economy of region] **15** (3), 865-879 (2019)
10. D. Samygin, N. Studies on Russian Economic Development **26** (1) 59-62 (2015)

Application of special lubricating compositions to increase the efficiency of friction surface run-in

Sergey N. Zhiltsov[1]*, *Nikolay A.* Cherkashin[1], *Igor N.* Guzhin[1], *Maxim S.* Prikazchikov[1], and *Alexey L.* Khokhlov[2]

[1] Samara State Agrarian University, 2, Uchebnaya Str., Ust-Kinelskiy urban settlement, Kinel, 446442, Russian Federation
[2] Ulyanovsk State Agrarian University named after P. A. Stolypin, 14, sq., 69, Studentskaya Str., Oktyabrsky, 433430, Russian Federation

Abstract. The paper presents the results of laboratory studies of a number of additives introduced during operation to motor oil of internal combustion engines. It also evaluates their impact on tribological characteristics of friction pairs. It was found that all studied additives have a positive effect on the running-in quality. Reduced wear of friction pairs, reduced surface roughness compared to base motor oil is determined. The best indicators were recorded with the FENOM additive. The decrease in wear was 46% and roughness – 23.5% compared to the base oil.

1 Introduction

The analysis of the state of the machine tractor fleet of enterprises of the agro-industrial complex shows that farms still operate a large amount of equipment with a significant age and operating time. This situation means that a certain amount of complex types of repairs, including capital repairs, is required for this technology. It is worth noting that the service life of individual units and assemblies, as well as the entire machine after major repairs is much lower.

One of the most stressed units of tractors and cars is the engine. This unit is a resource-determining unit and may account for up to 65% of equipment failures after repair [1]. The quality of repair is determined by the technological preparation of the repair production, the availability of appropriate repair equipment and regulatory and technical documentation corresponding to the qualification of personnel (employees), incoming inspection of spare parts, quality control of the work performed, etc. The final stage of the engine's capital repair is its run-in. The purpose of the run-in is a complex run-in of friction surfaces to form optimal microgeometry and their relative position.

As is known, the running-in contributes to the formation of not only new roughness with certain parameters, but also the physicochemical properties of the surface layers of the material capable of perceiving the operating loads. In this case, the saturated contact of friction surfaces should as close as possible to the parameters of the saturated contact.

Under the action of chemical, mechanical, vibration, thermal, electrical processes the run-in of friction pairs causes changes in friction surface layers, which make it possible to prepare the transition for the unit to be able to hold the operating loads without excessive wear and jamming.

The efficiency indicators that determine the quality of run-in include the formation of optimal microgeometry (surface roughness) and a decrease in the run-in wear.

The works [2, 3] indicate that there are several areas of efficiency improvement that are used at various stages of equipment operation. Taking into account the current state of repair production, the direction for the use of special lubricating compositions for running in surfaces during the run-in period after repair is quite relevant. Many works [4, 5, 6] are devoted to the development of lubricating compositions and the technology of their application. As a rule, lubricating compositions represent a combination of commercial engine oils and additional components (additives) [7, 8]. The analysis of these works showed that additives, which are referred to as "metal conditioners" showed quite good results in the formation of high-quality friction surfaces.

2 Materials and methods

To evaluate the efficiency of using such additives in lubricating compositions, a number of additives were studied during the initial period of friction surface run-in.

Such additives as FENOM, Energy Release (ER), SMT-2 with similar composition and mechanism of action were selected for the preparation of lubricating compositions.

Comparative tests of lubricating compositions including various additives were performed according to the general procedure and research agenda.

* Corresponding author: 3204@mail.ru

The following lubricants were evaluated during the studies:

1. Base (motor oil M10-G2K, according to SAE 30 classification, according to API classification – CC class).
2. Base + FENOM additive.
3. Base + ER additive.
4. Base + SMT-2 additive.

Motor oil for the study was selected according to the recommendations of the lubrication chart for use in diesel engines.

The concentration of additives used in the lubricating composition was within the limits recommended by the manufacturer (Table 1).

Table 1. Composition of lubricants.

Composition of lubricants	Additive concentration, %
Base + ER	6%
Base + FENOM	4%
Base + SMT-2	6%

The experiments to determine the additive for obtaining the best tribological characteristics were conducted on a universal tribometer. The tribometer (Figure 1) is a research device that ensures the sliding friction mode according to the ring-disk scheme. Special cast iron discs were used for the study, a carbon steel ring was used as a counterbody. The samples were processed on a flat grinding machine. The surface roughness Ra of the samples was within 0.360 μm, which corresponds to the surface roughness of the cylinder liner after repair.

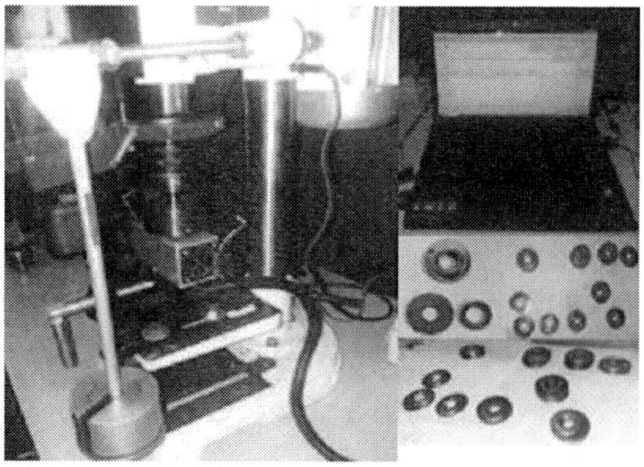

Fig. 1. Universal tribometer with software and samples for study.

The test procedure on the tribometer is not standardized, which allows the selection of the necessary modes and conditions within the technical characteristics.

The design of the tribometer allows simulating the operation of friction pairs in a sliding friction mode. It is equipped with devices for creating and implementing the selected test mode, as well as a system for monitoring and determining the friction parameters of the tested lubricants and samples. The software ensures primary processing of the received data.

The efficiency of lubricating compositions on friction pairs was evaluated according to the following parameters:

1) friction torque;

2) friction temperature;

3) wear of friction surface during testing was controlled by weight method using AND HR-200 scales;

4) surface roughness was determined using the Abris PM7M profile meter-profile recorder.

It is necessary to consider the peculiarities of operation when choosing a load and speed mode to simulate the studied transition. Calculations were made to assess the load and speed parameters for the studied unit.

Taking into account the results of calculations and the capabilities of the tribometer, the following test modes were adopted for the experiment: 1. Load – 450 N; 2. Speed – 1220 min^{-1}; 3. Time – 40 min.

3 Results

The study of tribological characteristics such as the friction torque (Fig. 2) and friction temperature (Fig. 3) shows that all tested lubricating compositions reduce the friction torque and temperature compared to base oil.
The effect of reducing the friction torque appears almost from the very beginning of the experiment in case of using lubricating compositions with additives. However, the dynamics of the friction torque change indicates certain processes occurring during the run-in period.

The action of the components included in the additives contributes to the creation of better tribological conditions for the interaction of friction surfaces compared to the base oil.

The temperature in the contact zone of the mating surfaces during testing increases, and here we can also see different temperature values. In this case, the use of additives in the lubricating composition contributes to the operation of the friction pair at a lower temperature. The comparison of these characteristics in lubricating compositions demonstrates better results of the lubricating composition containing the Energy Release additive.

The comparison of the tribological characteristics of the base oil and the lubricating composition containing the Energy Release additive with the best results shows that in absolute terms the friction torque decreased from 12.6 Nm to 8.08 Nm with the base oil and to 5.47 Nm with the lubricating composition containing the ER additive, which made a 21% reduction. The temperature reduction in the contact zone using the lubricant composition with the ER additive was 17% compared to the base oil.

The comparative wear tests showed that all lubricating compositions including additives have a certain anti-wear effect (Figure 4).

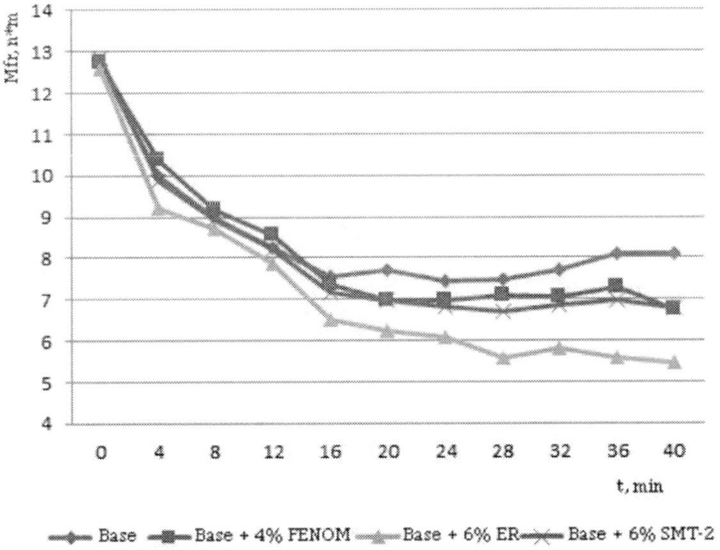

Fig. 2. Change in friction torque over the study period.

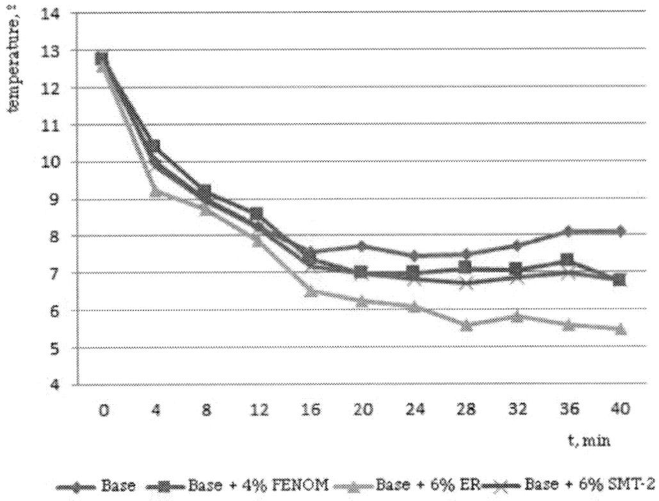

Fig. 3. Temperature change during the study period.

The wear evaluation of laboratory samples used in different lubricating compositions by weight method showed that the wear reduction compared to the base oil amounted to:

1. Base + FENOM – 26%
2. Base + ER – 46%
3. Base + SMT-2 – 37%

These results show that the lubricant composition with the ER additive has the greatest anti-wear effect.

The study of the roughness of sample surfaces before and after the tests made it possible to establish that the friction surface when lubricated with motor oil and oil with the addition of additives is also different (Table 2).

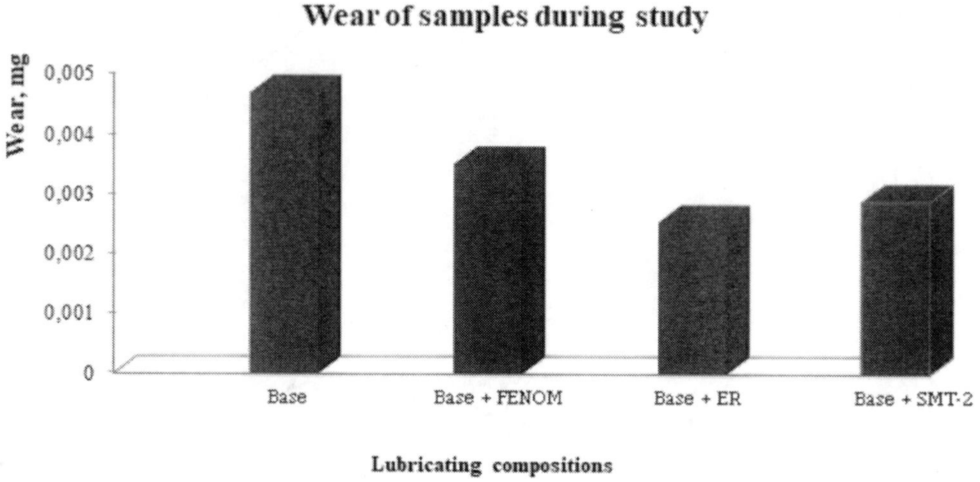

Fig. 4. Wear of samples when operating with various lubricating compositions.

Table 2. Surface roughness before and after tests.

Lubricant composition	Surface roughness, R_a, μm		Change, %
	Before tests (average)	After tests	
Base	0.358	0.326	8.9
Base + Energy Release (ER)		0.242	32.4
Base + FENOM"		0.270	24.5
Base + SMT-2		0.254	26

The roughness of the samples after base oil was Ra = 0.326 μm, and in a lubricating composition with ER additive – Ra = 0.242 μm. The roughness reduction made 32%.

4 Discussion

The comparative tests showed that all lubricating compositions provide higher tribological and anti-wear properties compared to the base oil.

The lubricant composition with the ER additive has the best characteristics among the test compositions. Its use allows reducing the friction torque and temperature in the contact zone by 21% and 17%, respectively.

The quality indicators of the running-in process are also better for this lubricating composition. Wear decreased by 46% and roughness by 23.5% compared to base oil.

The use of such lubricating compositions allows improving the quality of the run-in surfaces, which, with a decrease in wear during the run-in period, increases the post-repair service life.

References

1. V. V. Varnakov, A. L. Khokhlov, Use of additives for engine parts after repair, *Repair, Restoration, Modernization* **3,** 25–31 (2005)
2. V. V. Safonov, S. A. Shishurin, V. V. Venskaitis, K. V. Safonov, V.V. Ostrikov, Study of the effect of additives on the tribological characteristics of lubricants and the morphology of friction surfaces, *Agrarian Scientific Journal* **1,** 88–92 (2022)
3. V. A. Kochenov, K. E. Grunin, Increasing the durability and wear resistance of ICE tribological conjunction parts by changing their design features, *Tractors and Agricultural Machines* **8,** 47–52 (2017)
4. I. A. Tomilov, Study of the running-in of the sleeve-ring transition when running in oil with antifriction compositions, Bulletin of the International Academy of Agrarian Education **35,** 164–166 (2017)
5. O. S. Volodko, A. P. Bychenin Influence of remetallizants on the strength of oil in gearboxes, *Mechanization and automation of construction* pp 187–190 (Samara: Samara State Technical University, 2020)
6. Yu. B. Ramazanova, Study of antioxidant properties of multifunctional alkylphenolate additives and development of new lubricating compositions, *Oil refining and petrochemicals. Scientific and technical achievements and best practices* **8,** 49–51 (2018)
7. A. S. Nurutdinov, A. L. Khokhlov, I. R. Salakhutdinov, A. A. Glushchenko, M. M. Zamaldinov, Theoretical justification for the use of antifriction materials to reduce the wear of cylinder-piston group parts, *Bulletin of Saratov State Agricultural University named after N.I. Vavilov* **3,** 62–65 (2014)
8. V. I. Zhornik, M. A. Belotserkovsky, A. P. Yalovik, A. V. Dudan, Selection of the running-in

https://doi.org/10.1051/bioconf/20225200031

composition and determination of the modes of tribomechanical processing of steel gas thermal coatings, *Bulletin of Polotsk State University* Series B. Industry. Applied Sciences **3,** 85–91 (2017)

Designing structured food systems with a specified set of qualitative characteristics based on soy

Daria Kupchak [1,*], *Olga* Lyubimova[1], *Tatyana* Kalenik[2], *Sergey* Dotsenko[3], and *Olga* Skripko[4]

[1] Department of Food Technology, Khabarovsk State University of Economics and Law, 134, Pacific Street, Khabarovsk, 680042, Russian Federation
[2] Institute of Life Sciences and Biomedicine, Far Eastern Federal University, 10, Ajax, Russky Island, Vladivostok, 690922, Russian Federation
[3] Department of Service Technologies and General Technical Disciplines Amur State University, 21, Ignatievskoe Highway, Blagoveshchensk 675027, Russian Federation
[4] Department of Automation of Production Processes and Electrical Engineering Amur State University, 21, Ignatievskoe Highway, Blagoveshchensk 675027, Russian Federation

Abstract. The article proposes a method for inactivating the anti-nutritional factors of vegetable soy proteins by biological modification. It was carried out by germinating soy seeds in a mineralized medium, resulting in a decrease in trypsin-inhibiting activity and saturation of soy seeds with minerals. The change in the chemical composition of soy seeds is presented. Sprouted soybean seeds were later used for food production. Studies are presented on the basis of which a technology for obtaining structured food systems from soy has been developed, including operations for joint disintegration and extraction, pre-soaked soybean seeds and crushed fresh beet or pumpkin in a ratio of 1:1, separation of the resulting mixture, coagulation with 5% ascorbic acid solution, separation from whey, granulation and drying. The remaining whey was used to make drinks and sauces. Based on the analysis of mathematical models, the optimal parameters of coagulation of the protein-suspension system were determined. The mass fraction of ascorbic acid solution is 13.4-14.05%, the concentration of ascorbic acid solution – 4.7-5.4%, the duration of coagulation – 5.3-5.5 min, the optimal temperature of coagulation – 66.1-66.7 ° C and granulate drying: the initial moisture content of granules – 36.7-40.0%; drying temperature – 71.5-72.9 ° C; drying duration – 58-63 min., the strength of the obtained granules – 94.5-96.9%. Analysis of the chemical composition of the developed soy-vegetable food systems showed that the proposed technology allowed increasing the protein content to 26.4%, fat to 7%, dietary fiber to 8.6%, which exceeds similar indicators for dried vegetable concentrates existing on the market.

1 Introduction

An assessment of the current structure of food consumption of the population of the Russian Federation and a number of other countries indicates a violation of the nutritional adequacy of the diet and an increase in morbidity associated with alimentary factors [1,2].

Scientific developments of the nutriological profile and the use of innovative technological approaches in the creation of new generation products based on protein-rich crops contribute to the formation of a food assortment capable of ensuring the optimal alimentary status of the consumer [3,4].

A large number of factors, objects, technological parameters, technological modes when creating food components of the designed composition and properties necessitate a systematic approach and the use of mathematical modeling methods [5].

Today, soy products are considered as a means that can help a person avoid very common and formidable diseases that represent the main cause of morbidity and mortality in industrialized countries [5]. The number of diseases that soy products can play a certain role in the prevention of include, first of all, colon, breast and prostate cancer, atherosclerosis and cardiovascular diseases, hypertension, diabetes mellitus. However, the most interesting thing is that this highly beneficial effect of soy products for the body is associated today with the substances present in soy, in the past considered as anti-nutritional factors [6, 7, 8].

The combination of soy raw materials with vegetables, meat, fish in modern technological solutions makes it possible to create promising ingredients for the complex enrichment of biologically active substances of products included in the daily diet of the inhabitants of the country [6, 7, 8].

* Corresponding author: daria-kup@rambler.ru

Vegetable soy proteins are subject to requirements not only in terms of biological value, but also in terms of their digestibility, which is characterized by the ability of the protein to break down under the action of proteolytic enzymes of the digestive tract and be absorbed through the intestinal mucosa. This is due to the fact that among the biologically active substances of soy there are components that have a negative impact. These include trypsin inhibitors, substances of a protein nature that block proteolytic enzymes of the human gastrointestinal tract. The result of such blockade is a decrease in the absorption of protein substances in the diet.

The most obvious anti-nutritional effect of protease inhibitors represented in soy grains two varieties - trypsin inhibitors of the Kunitz type and the Bowman-Birk type. Quantitatively, the first type of inhibitor prevails. According to the mechanism of action, both belong to serine protease inhibitors. By suppressing the action of trypsin and chymotrypsin secreted by the pancreas, these substances disrupt the digestion of proteins in the intestine.

Heat treatment eliminates the effect of soy protease inhibitors by 80 – 90%. Soy oligosaccharides are represented by sucrose, raffinose and stachyose, whose content per dry weight of whole grain is 4.5, 1.1 and 3.7%, respectively. Raffinose and stachyosis are not digested by glycosidases of digestive juices in the small intestine and are fermented by anaerobic bacteria in the large intestine with the formation of gases - hydrogen, carbon dioxide and, to a lesser extent, methane, causing flatulence phenomena. Phytic acid (cyclic alcohol inositol hexaphosphate), binding divalent metal ions necessary for vital activity in the gastrointestinal tract, in particular iron, zinc, manganese, copper and calcium ions, can interfere with their absorption. Soy grain contains anti-vitamins - compounds that make it difficult for the body to use a number of vitamins: A, D, E, B_{12}. All of them, with the exception of the factor that prevents the utilization of vitamin A, are thermolabile. Lectins or phytohemagglutinins, which are very characteristic of various legumes, cause a hemagglutination reaction (gluing of erythrocytes) when injected into the blood or in vitro experiments and by chemical nature belong to glycoproteins. When ingested as part of food, lectins interact with specific receptors located on the surface of intestinal epithelial cells. The binding of lectins is accompanied either by a nonspecific suppression of the absorption of various nutrients, or by the penetration of pathogenic microorganisms from the intestinal cavity into the lymph and blood due to a violation of the protective mechanisms of epithelial cells. Soy saponins are complex compounds consisting of a nonpolar triterpenoid alcohol aglycone bound to a polar oligosaccharide residue. There are three types of aglycones in soybean saponins: sapogenols A, B, E. The complexity of the structure of soy saponins is due to variations in the oligosaccharide part and the possibility of adding additional groupings to the aglycone. Due to the amphiphilic nature, saponins have detergent properties. Saponins are classified as anti-nutritional factors of soy exclusively by analogy with saponins from other sources, which are, in fact, toxic. The inclusion of soy isoflavones in the group of anti-nutritional substances is due to their ability to cause an estrogenic reaction when administered to animals [9].

Protein modification makes it possible to increase the digestibility of soy products. Soy proteins that have undergone modification enter the human body already split into lower molecular weight peptides and individual amino acids. Despite the presence of substances in soy that are potentially capable of having an undesirable effect on the body, according to numerous studies, there is no reason to overestimate such a possibility and abandon soy foods. Foods made from raw soy and not modified to eliminate the activity of protease inhibitors, urease, lectins, and anti-vitamins by almost 90% are not suitable for food.

The aim of the work was to study the chemical composition of soy seeds in the process of biological modification in a mineralized medium, to develop and substantiate the technological parameters for obtaining structured food systems based on soy.

2 Materials and methods

The general scheme of the research included the structural and functional modification of proteins by limited proteolysis by its own enzyme complex. These are endoproteinases of germinating seeds with simultaneous saturation of soy raw materials with macro- and microelements (Na^{1+}, Ca^{2+}, Mg^{2+}), the development and optimization of the technology of soy-beet or soy-pumpkin compositions, coagulants and granulates which were the objects of studies and were evaluated by a combination of various indicators.

The studies were carried out using equipment that provides for the possibility of changing the parameters and modes of the processes under study: thermostats; shredder-extractor; heating boilers; press for pressing liquid fraction; pellet press; pump; spinning top; drying unit, etc.

In the experiments, 4 parallel definitions were reproduced. The numerical values indicated in the paper represent arithmetic averages, the reliability of which is P = 0.95, the confidence interval is ± 5%. The overall chemical composition, quality and safety indicators were determined by standard methods.

The processes under study were described by a second-order mathematical model. The construction of mathematical models and their analysis were carried out according to the Appol program, the Pareto-optimal solution method (CRS program).

3 Discussion of the results

For the process of biological modification of soybean seeds, a bicarbonate magnesium-calcium aqueous medium with a total mineralization of 1000-2500 g/dm "Amur" was used. The chemical composition of the mineralized aqueous medium is presented in Table 1.

Table 1. Chemical composition of the mineralized medium.

Element/ indicator	**Concentration mg-eq.%/value**
HCO_3	77-87
Mg	20-50
Ca	32-55
Na	12-28
pH	6.8-7.0

The rationale for the choice of soybean seeds of the Far Eastern selection was studied their chemical composition and individual biochemical parameters (Table 2).

Table 2. Chemical composition of individual soybean seeds of the Far Eastern selection.

Type	**Mass fraction, %**			**The content of the miner. substances mg/100g**
	proteins	**lipids**	**carbohydrates**	
Dauria	38.9±0.1	19.0±0.2	19.0±0.2	5.5
Garmonia	37.0±0.2	18.1±0.1	18.9±0.1	5.0
Lidia	39.3±0.2	17.5±0.3	18.1±0.3	5.0
Saltus	37.5±0.5	19.5±0.5	19.0±0.6	5.8
Venera	38.0±0.1	15.0±0.3	19.1±0.2	6.2
Primorskaya -13	40.0±0.2	18.0±0.4	17.0±0.4	0,1
Oktyabr' - 70	39.0±0.4	18.0±0.3	19.0±0.5	6.0

Long-term experimental data indicate that trypsin-inhibiting activity (TIA) negatively correlates with the protein content in soybean seeds (g = - 0.44 - 0.96). In varieties with a high protein content (44-47%), TIA is small and ranges from 11-18 mg/g with a reduced mass fraction of seed protein (32-38%), TIA is in the range of 24-32 mg/g [9, 10].

Biological modification of soybean seeds was carried out according to the study scheme (Figure 1) as follows: soybean seeds were inspected, damaged specimens were removed. They were treated with running water and soaked in a mineralized aqueous medium at a temperature of 18-20 ^{0}C for 24 hours. Soaking was performed to swell the seeds, soften the shell in order to remove it later and accelerate the germination process. The soybean seeds prepared in this way were placed in a container with a layer 15-20 cm high and installed in a thermostat. Here further germination of seeds was carried out at a temperature of 24-26 ^{0}C with mandatory periodic 5-fold replacement of the aqueous mineralized medium, observing the hydromodule 1:2, until the length of the shoots was 1.0-1.5 mm.

The content of mineral substances was determined during seed germination until the optimal size of the sprout was reached. They were compared with control samples, which were soybean seeds sprouted in a neutral aqueous medium in similar biotechnological parameters. The results are presented in table 3.

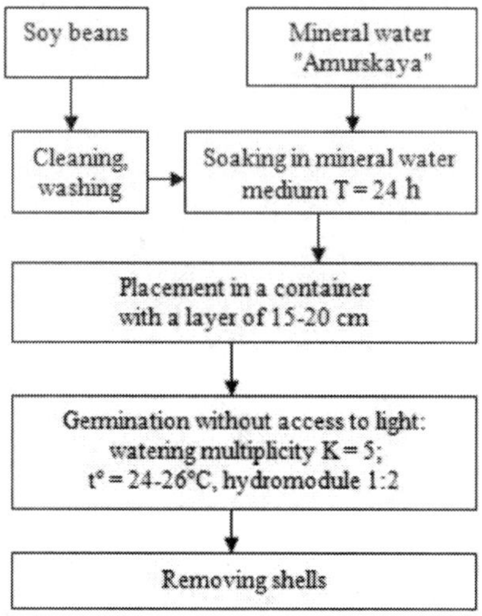

Fig. 1. Technological scheme of soybean seed biomodification.

Table 3. The content of the main mineral substances.

Name	**The content of macro-microelements, mg/100g**	
	Soy in a neutral medium	**Soy in a mineralized medium**
Ca^{2+}	277	550
Mg^{2+}	280	558
K^{1+}	1797	2500
PO_4^{3-}	704	1050
Na^{1+}	28	59
Mn^{2+}	400	780

Thus, in the process of biological modification, there is not only a decrease in the anti-nutritional components of soy proteins, but also the accumulation of minerals by soy seed sprouts during germination in a mineralized aqueous medium. As a result, the concentration of ions of the main mineral elements increases: Ca^{2+} 2 times, Mg^{2+} 2 times, K^{1+} 1.5 times, PO_4^{3-} 1.5 times, Na^{1+} 2 times, Mn^{2+} 2 times.

The sprouted soybean seeds were later used for the manufacture of food, after the necessary technological processing.

The production of structured forms of soy-based food systems was carried out according to the technological scheme presented in Figure 2, soybean seeds were inspected, washed in running water at t = 18-20 ° C, soaked in water for swelling and softening for 24 hours. Beets and pumpkins were sorted, washed, cleaned, rewashed, cut into small cubes. The prepared components were mixed, water was added, and joint disintegration of the components was carried out with simultaneous extraction of soluble substances of soy and vegetables in an aqueous medium. The resulting mixture was divided into liquid soluble (protein-vitamin dispersed system – PVDS) and insoluble fractions by filtration. The soluble fraction was injected with a 5% solution of ascorbic acid for the purpose of thermoacid protein coagulation. The resulting coagulate and the insoluble fraction were granulated, dried, and the serum remaining after coagulation was used to prepare sauces and drinks [11].

Evaluation and identification of a set of factors that allow purposefully shaping properties and composition of food systems showed that the most significant factors affecting the coagulation of protein substances in PVDS are the mass fraction of ascorbic acid solution, MK, %; ascorbic acid concentration, K, %; duration of coagulation, T, min. The optimization criterion is taken coagulation temperature, t, °C, and for the granulate production process: initial coagulate humidity, WH, %; drying temperature, t, °C; drying time, TS, min, optimization criterion – strength of finished granules, PR, %.

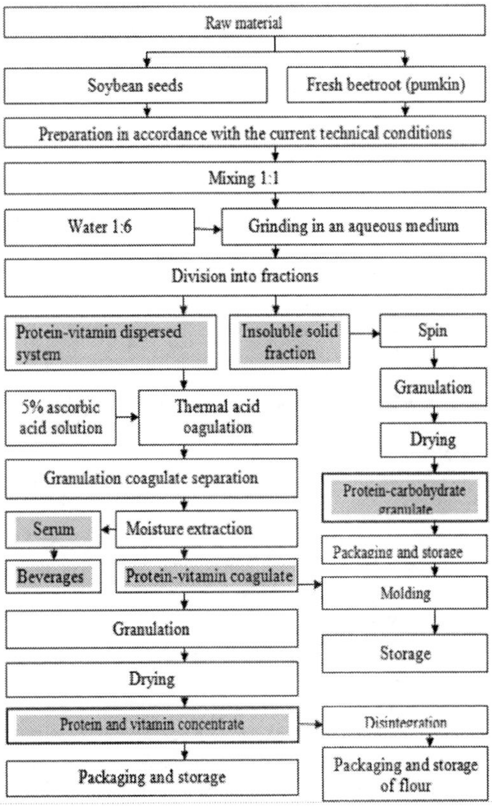

Fig.2. Technological scheme for obtaining soy-vegetable food systems.

Regression analysis of the dependencies allowed us to obtain mathematical models of the production of protein-vitamin coagulate and granulate based on fresh soy and table beet and fresh soy and pumpkin, using ascorbic acid (Table 4) [11].

For the studied processes, the regions of extreme values were determined. And a graphical analysis of the obtained three-dimensional dependencies was carried out, which shows that the optimal parameters of thermoacid coagulation of protein substances in PVDS are within the following limits: MD=13.4-14.05%; K=4.7-5.4%; T=5.3-5.5 minutes; t =66.1-66.7°C; soy-vegetable granulate production: WH=36.7-40.0%; t=71.5-72.9°C; TC =58-63min; PR=94.5-96.9%.

Table 4. Mathematical models of preparation of projected food systems.

Name of the food system	Mathematical models
	in encoded form
for PVDS based on soy-beet composition	$Y_1= 67.10 - 1.44X_1 - 0.71X_3 - 0.87X_1X_2 + 0.87X_2X_3 + 2.39X_1^2 + 1.04X_2^2 +1.71X_3^2$ opt
for PVDS based on soy-pumpkin composition	$Y_2 = 66.81 - 1.19X_1 - 1.33X_2 - 0.75X_3 - 2.62X_1X_3 + 0.87X_2X_3 + 2.09X_1^2 + 2.09X_2^2+ 2.09X_3^2$ opt
for soy-beet pellets	$Y_3 = 92.30 + 2.13X_1 - 1.29X_3 - 2.19X_1^2- 7.27X_2^2 - 4.90X_3^2$ 100 %
for soy-pumpkin pellets	$Y_4 = 95.52 - 3.76X_1 {}_+ 3.08X_2 + 2.45X_3 + 1.50X_2X_3 - 5.67X_1^2 - 6.01X_2^2 - 4.65X_3^2$ 100 %
	in decoded form
for PVDS based on soy-beet composition	$t_1^0= 176.39 - 8.41M_1 - 7.79K - 15.72T - 0.35M_kT + 0.43KT + 0.38M_kT + 0.43KT + 0.38M_k^2 + 0.26K^2 + 1.71T^2$ opt
for PVDS based on soy-pumpkin composition	$t_2^0 = 142.95 - 3.60M_1 - 8.08K - 10.74T - 1.05M_kT + 0.43KT + 0.33M_k^2 - 0.52K^2 + 2.09T^2$ opt
for soy-beet pellets	$Pr_1 = -481.30 + 1.75W_н + 10.40t^0 + 5.75T - 0.02W_н^2 – 0.07(t^0)^2 - 0.04T^2$ 100%
for soy-pumpkin pellets	$Pr_2 = -415.47 + 4.15W_н +7.82t^0 + 4.78T + 0.01t^0T - 0.05W_н^2 - 0.06(t^0)^2 - 0.04T^2$ 100%

The designed food systems have high organoleptic characteristics, a pronounced taste, color and aroma corresponding to the used raw materials (Table 5).

Table 5. Organoleptic characteristics of food systems based on soy and vegetables.

Indicator	Characteristic	
	Soy-vegetable paste	**Soy-vegetable granulate**
Appearance and consistency	Homogeneous paste-like mass, cheese-like in structure,	Dry granules with a rough surface, of the same size throughout the

	without foreign inclusions	volume with insignificant particles of crushed vegetable product (red-maroon or yellow-orange) Particles are porous, brittle, moderately brittle
Taste and smell	Moderately pronounced, characteristic of vegetable products, without extraneous tastes and odors	
Color	Red-burgundy or yellow-orange (depending on the vegetable component) and uniform throughout the mass	

The chemical composition of the developed granulates is shown in Figure 3.

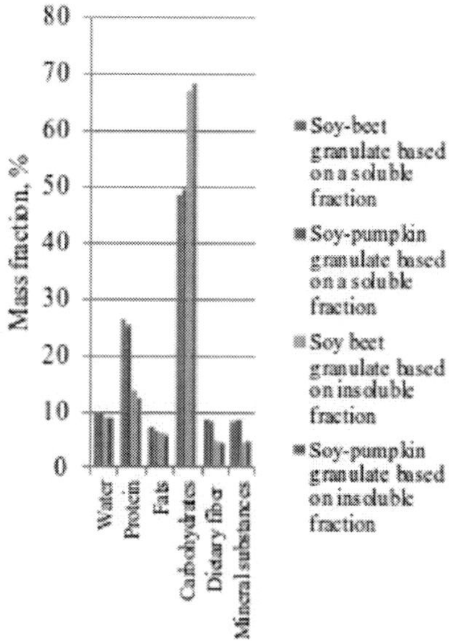

Fig.3. Chemical composition of the developed granules.

The analysis of the data presented in Figure 3 shows that the developed products based on soy and vegetables contain from 12.3 to 26.4% of protein, from 5.7 to 7.0% of fat, dietary fiber from 4.5 to 8.6%, which exceeds the content of these nutrients in the vegetable concentrates presented on the market in the form of powder or chips.

4 Conclusion

In the course of the conducted studies, the chemical composition of soybean seeds biomodified in a mineral aqueous medium was determined. As a result the concentration of ions of the main mineral elements increases: Ca^{2+} 2 times, Mg^{2+} 2 times, K^{1+} 1.5 times, PO_4^{3-} 1.5 times, Na^{1+} 2 times, Mn^{2+} 2 three times.

The technological parameters of biotechnology of new soy-vegetable systems containing functional food ingredients are mathematically substantiated and experimentally confirmed:

• duration of coagulation of soy protein at a temperature of 66.1-66.7 ° C in a 5% solution of ascorbic acid - 5.3–5.5 min; mass fraction of coagulant – 13.4-14.05%;

• duration of drying of the granulate to achieve the strength of the granules is 94.5-96.9% – 58-63min, at a temperature of 71.5–72.9 ° C and an initial humidity of 36.7–40.0%.

Taking into account the obtained parameters, the developed technology makes it possible to organize waste-free production of structured food systems based on soy, capable of enriching food products with full-fledged proteins, vitamins and ballast substances.

References

1. Methodological recommendations. MP 2.3.1.0253-21. Norms of physiological needs in energy and food substances for various population groups of the Russian Federation. (Moscow. 2021).
2. A. Baturin, A. Martinchik A. Kambarov. *Russian population nutrition structure at the turn of the 20th and 21st centuries.* Nutrition Issues. **4.** 60-70. (2020)
3. V. Tutelyan, D. Nikityuk, A. Baturin, A. Vasilyev, M. Gapparov, N. Zilina, V. Zhminchenko, A. Kambarov, V. Kodentsova, L. Kravchenko, S. Kulakova, N. Lashneva, V. Mazo, A. Sokolov, B. Sukhanov, S. Hotimchenko. *Nutriom as a direction of the "main blow": definition of physiological requirements in macro- and micronutrients, minor biologically active agents of food.* Nutrition Issues. **4.** 24-34. (2000).
4. A. Oyeyinka, J. Odukoya, Y. Adebayo *Nutritional composition and consumer acceptability of cheese analog from soy and cashew nut milk.* Journal of Food Processing and Preservation. **43(12)**. e14285. (2019).
5. A. Maitakov, A. Popov, N. Vetrova, L. Beryazeva, M. Zverikova. *Modeling of technologies for the production of multicomponent granular products.* Vestnik VGUIT **80(4)**. 63-68.(2018)
6. A. Sharikov, V. Stepanov, V. Ivanov, D. Polivanovskaya, M. Amelyakina. *Extrusion of mixtures of wheat and carrot pomace of high humidity in the technology of ready-to-use products.* Vestnik VGUIT **80(3).** 43-49. (2018).
7. N. Mayasti, M. Ushada, M. Ainuri. *Improvement of spaghetti composite quality based on local flours (mocaf, corn, rice and soybean).* AIP Conference Proceedings. **1(2175).** 020063. (2019).

8. D. Bánáti. *Veggie burgers, vegan meats? The ruling of the European parliament paved the way for meat substitutes with meat denominations.* Elelmiszervizsg Kozlemenyek. 66(4). 3159-3174. (2020).
9. O. Lyubimova. *Substantiation and development of technology of combined fish and vegetable products with by adding biologically modified soyan seeds.* PhD. Diss. Vladivostok, 2022. 206 p.
10. S. Dotsenko, O. Lyubimova a.o. *Soy grain processing method.* Patent RF, no. 2348179, (2007).
11. S. Dotsenko, D. Kupchak a.o. *Method of preparation of protein-carbohydrate products of functional orientation*. Patent RF, no 2482696, (2013).

Retrospective diagnosis of intrauterine diselementosis in newborn calves

Vladimir Safonov[1,*], *Emil* Salimzade[1], *Tatiana* Ermilova[1], and *Anton* Chernitskiy[2]

[1]Astrakhan State University, 20a, Tatischev St., Astrakhan, 414056, Russian Federation
[2]All-Russian Veterinary Research Institute of Pathology, 114B, Lomonosov St., Pharmacology and Therapy, Voronezh, 394087, Russian Federation

Abstract. One of the methods in diselementosis diagnosis, i.e. in determination of the trace elements content excess, deficiency or imbalance in the organism, is screening the hair elemental composition. It is known that the tail brush hair growth in cattle begins from the 7th month of gestation. The authors suggested that the tail brush hair elemental composition in newborn calves could be used for integral evaluation of the trace elements supply to the fetus in the last months of its intrauterine development. In order to elaborate criteria for retrospective diagnosis of the intrauterine diselementosis, 125 clinically healthy 1-day-old calves (67 male and 58 female animals) were examined; they were obtained from Simmental cows with a physiological course of gestation under biogeochemical conditions of the Central Black Earth region of the Russian Federation. The concentration of 11 essential (Co, Cr, Cu, Fe, I, Li, Mn, Se, Si, V, Zn) and 8 toxic (As, B, Cd, Hg, Ni, Pb, Sn, Sr) trace elements were studied in samples of unpigmented guard hairs from the animal tail brush by inductively coupled plasma mass spectrometry (Nexion 300D, Perkin Elmer, USA). The results were expressed as mean ± standard error of mean, median, minimum, and maximum. Reference values of essential and toxic trace elements concentration in the hair of newborns were determined using the percentile scales separately for the groups of male and female animals. Group comparison was conducted using the independent samples Mann-Whitney U test. Research results demonstrated that physiologically determined the range of the trace elements concentration in the hair of newborn calves corresponded to the interval between the 25th and 75th percentiles. Values beyond the specified interval boundaries were proposed to be considered as deviations corresponding to the diselementosis.

1 Introduction

The term "diselementosis" in biological literature is accepted to designate violations in the chemical elements homeostasis manifested by their content excess, deficiency or imbalance in the organism [1, 2]. Diselementosis in animals could course both latently and manifest itself in various clinically pronounced pathological conditions [2, 3].

Recent studies showed that intrauterine diselementosis in cattle is a risk factor provoking a number of diseases [3, 4] and metabolic disorders in the postnatal ontogenesis [5]. In this connection, evaluation of the fetus bioelemental status in cows, especially in the last trimester of pregnancy, acquires important diagnostic and prognostic value.

To evaluate the bioelemental status in cattle, hair, blood, urine, liver and other biopsy tissues are examined, in lactating animals – milk [6, 7, 8]. In this case, both the direct content of chemical element itself in organs and tissues [7, 8] and its metabolites [9, 10] could be evaluated in animals. For example, to assess the selenium status in cows, activity of the selenium-dependent glutathione peroxidase in blood is determined [11].

When selecting a biosubstrate for research, it is important to understand that the chemical elements concentration in the organism liquid media (blood, urine, milk) mainly reflects short-term and deep alterations in the bioelemental status [7, 8]. At the same time, the elemental composition of the solid tissues (hair and bones) characterizes disorders formed over a long period from several months to several years [12, 13]. For some chemical elements (arsenic, cadmium, lead, calcium, copper, zinc), its content determination is informative both in the organism liquid and in solid media [14, 15]. For others (aluminum, barium, boron, iron, magnesium, phosphorus, strontium), it is preferable to study their content in solid media, for example, in the hair [16, 17].

In the last 10–15 years, hair as the research object to evaluate the cattle elemental status gained significant popularity [18, 19]. Reference chemical elements

* Corresponding author: safonovbio@gmail.com

concentration ranges in the hair of heifers and cows of meat and dairy breeds were determined [20, 21]. Influence on the cattle hair elemental composition by sex and age [22, 23], productivity level [16, 24, 25], features of the region, season of the year and animal rearing technology [14, 19, 26] was studied.

Hair analysis in cattle diselementosis diagnosis has a number of advantages over traditional research objects, such as blood, urine and liver biopsy specimens [7, 8, 18]. The most important of them include the low level of stress load on animals during biomaterial sampling (compared to blood sampling or biopsy) and the elemental composition stability during long-term storage and transportation of the samples [12, 18, 27].

It was shown that chemical elements concentration in hair samples taken from the cattle body various parts were differing significantly [18]. The content of calcium, magnesium, copper, manganese, molybdenum, cobalt, nickel, strontium and cadmium in the tail brush guard hair was significantly higher, while zinc and potassium content, on the contrary, was lower in the samples taken from the tail brush compared to hair from other parts of the body (withers, dorsum) [18, 27]. High correlation was established between the chemical elements concentration in the tail brush integumentary hair of dairy cows and in the tissues of their parenchymal organs [27]. For lactating Black-and-White, Simmental and Holstein cows, indicative norms for the chemical elements content in the tail brush integumentary hair were elaborated [27]; sampling procedure, sample preparation, and analysis were described [21, 28].

Study of the chemical elements content in the tail brush integumentary hair in newborn calves is of great interest [5]. It is known that hair growth of the tail brush in cattle starts from the 7th month of gestation [29, 30]. In this connection, chemical elements concentration in the tail brush integumentary hair samples obtained from calves immediately after partition could be considered as integral indicators of the fetus supply with the studied elements in the last months of intrauterine development [5].

Objective of this study was to elaborate criteria for retrospective intrauterine diselementosis diagnosis in the Simmental newborn calves by chemical elements content in the tail brush integumentary hair.

2 Material and methods

2.1 Animals

Simmental cattle (*Bos taurus taurus*) was the research object. Research was conducted during the stall period (October-May) of keeping the animals in conditions of the Rechnoye Farm, Khlevensky District, Lipetsk Region of the Russian Federation. The sample included 125 clinically healthy 1-day-old calves (67 bulls and 58 heifers) weighing 43–55 kg and obtained from cows of 2–4 lactations with productivity of 5860–8338 kg. The trace elements content in the mother cows' ration generally met the animals' needs. It was not exceeding the maximum acceptable limits established by the USSR State Agriculture Committee [31] and the U.S. National Research Council [32, 33].

2.2 Sampling

Samples of integumentary unpigmented hair were obtained from the newborn calves tail brush using stainless steel scissors pre-treated with ethyl alcohol. From all the tail brush hairs cut as close as possible to the skin surface, an average sample weighing at least 0.5 g was taken. Hair samples were immersed in the clear for analysis acetone (Khimmed, Russia) for 15 minutes, then were washed three times with the deionized water (18 MΩ cm) and further were dried at the temperature of 60°C to the air-dry state followed by grinding into 2–3 mm long fragments.

2.3 Trace element analysis

Hair sample preparation and actual trace elements analysis were carried out, as described earlier [20], in the Micronutrients LLC accredited laboratory (Moscow, Russia) being the IUPAC associated company. 5.0 ml of high-purity concentrated nitric acid (Khimmed, Russia) was added to the 50.0 mg sample of hair. The mixture was decomposed in the Multiwave 3000 microwave device (PerkinElmer, Austria) in the following mode: 5 minutes – raising temperature to 200 °C, 5 minutes – stabilizing temperature at 200 °C, then cooling down to 45 °C. After that, the content was quantitatively transferred into the polypropylene tubes. The volume was adjusted to 15 ml by adding deionized water (18 MΩ cm) and thoroughly mixed by shaking in tubes closed with lids. The quantitative determination of chemical elements, including iron (Fe), zinc (Zn), copper (Cu), manganese (Mn), cobalt (Co), chromium (Cr), selenium (Se), iodine (I), boron (B), silicon (Si), nickel (Ni), vanadium (V), arsenic (As), lithium (Li), tin (Sn), strontium (Sr), lead (Pb), cadmium (Cd) and mercury (Hg), in the samples was carried out by inductively coupled plasma mass spectrometry on the Nexion 300D spectrometer (Perkin Elmer, USA). The instrument was calibrated using the monoelement Universal Data Acquisition Standards Kit (CT 06484, PerkinElmer, USA), and for internal standardization - the single-element pure yttrium standard of 10 µg/L (Y, Perkin Elmer, USA). The GBW09101 hair sample from the Shanghai Institute for Nuclear Research (China) was used as the reference. Extraction degree in the studied chemical elements varied from 90 to 110 %.

2.4 Statistical analysis

Experimental data was statistically processed using the IBM SPSS Statistics 20.0 software (IBM Corp., USA). The entire data were expressed as mean ± standard error of mean (SEM), median, minimum (Min) and maximum (Max). Reference chemical elements concentration values in the tail brush hair of newborn calves were determined by the 25th (lower limit) and 75th percentiles (upper limit) [20, 21]. Group comparisons were made using the independent samples Mann-Whitney U test; differences were considered significant at $P < 0.05$.

3 Results

Results of the essential and toxic trace elements quantitative determination in the tail brush integumentary hair of newborn calves are presented in Table 1 and Table 2, respectively.

Table 1. Hair essential trace element content (μg/g) in Simmental newborn calves.

Element	Mean±SEM	Median	Min–Max
Co	0.072±0.006 0.063±0.005	0.067 0.056	0.020–0.239 0.016–0.167
Cr	0.076±0.014 0.050±0.012	0.021 0.017	0.010–0.330 0.011–0.260
Cu	7.19±0.26 6.94±0.29	7.34 7.41	2.20–11.44 2.10–10.60
Fe	46.1±3.6 50.7±3.6	40.0 42.6	13.5–151.0 11.9–123.5
I	0.236±0.035 0.354±0.072	0.220 0.340	0.130–0.560 0.190–0.530
Li	0.025±0.005 0.019±0.003	0.020 0.020	0.010–0.080 0.010–0.030
Mn	8.17±0.29 9.64±0.40	8.43 8.69*	4.30–11.60 5.60–17.40
Se	0.298±0.018 0.263±0.016	0.270 0.260	0.060–0.734 0.080–0.526
Si	22.4±3.1 20.1±3.5	21.6 18.5	4.2–45.0 8.3–35.4
V	0.166±0.011 0.144±0.016	0.160 0.120	0.120–0.250 0.110–0.230
Zn	114.5±4.1 115.5±3.6	115.5 112.7	68.7–191.0 59.0–173.0

Above the line – group of male animals (n = 67), under the line – group of female animals (n = 58). * Differences between the groups are statistically significant at P = 0.026.

Table 2. Hair toxic trace element content (μg/g) in Simmental newborn calves.

Element	Mean±SEM	Median	Min–Max
As	0.074±0.009 0.068±0.010	0.057 0.057	0.006–0.233 0.002–0.211
B	0.881±0.346 0.446±0.080	0.510 0.450	0.270–4.93 0.260–0.860
Cd	0.013±0.002 0.014±0.003	0.010 0.010	0.002–0.082 0.002–0.077
Hg	0.034±0.004 0.029±0.004	0.030 0.030	0.020–0.060 0.020–0.050
Ni	0.585±0.087 0.873±0.138	0.350 0.500	0.060–2.79 0.090–3.11
Pb	0.522±0.105 0.634±0.141	0.220 0.480	0.030–2.00 0.040–1.73
Sn	0.029±0.004 0.028±0.008	0.020 0.020	0.006–0.060 0.009–0.070
Sr	0.740±0.071 0.820±0.089	0.600 0.600	0.320–2.93 0.320–2.80

Above the line – group of male animals (n = 67), under the line – group of female animals (n = 58).

Tables 3 and 4 present reference values of the studied trace elements concentration in the hair of newborn bulls and heifers determined on the basis of percentile scales [20, 21].

Table 3. The reference values* of hair essential trace element content (μg/g) in Simmental newborn calves.

Element	Males (n = 67)	Females (n = 58)
Co	0.048–0.098	0.040–0.084
Cr	0.015–0.153	0.015–0.275
Cu	5.55–9.08	5.45–8.63
Fe	30.1–53.1	31.7–58.2
I	0.140–0.295	0.190–0.530
Li	0.010–0.030	0.010–0.020
Mn	7.38–9.27	8.44–11.25
Se	0.200–0.405	0.160–0.367
Si	16.0–28.6	11.7–28.0
V	0.130–0.205	0.110–0.160
Zn	90.1–133.8	96.8–128.0

* Lower boundary corresponds to the 25th percentile, upper boundary – to the 75th percentile.

Table 4. The reference values* of the hair toxic trace element content (μg/g) in Simmental newborn calves.

Element	Males (n = 67)	Females (n = 58)
As	0.020–0.110	0.020–0.106
B	0.370–0.590	0.270–0.530
Cd	0.003–0.016	0.008–0.011
Hg	0.025–0.040	0.020–0.030
Ni	0.218–0.690	0.200–1.40
Pb	0.068–0.893	0.060–1.28
Sn	0.020–0.040	0.010–0.040
Sr	0.428–0.830	0.500–1.00

* Lower boundary corresponds to the 25th percentile, upper boundary – to the 75th percentile.

According to content in the hair of newborn calves (μg/g), trace elements under study could be ranked in the following descending series: Zn (114.7), Fe (41.3), Si (20.4), Mn (8.58), Cu (7.37), Sr (0.600), B (0.500), Ni (0.390), Se (0.260), Pb (0.235), I (0.230), V (0.150), Co (0.060), As (0.057), Hg (0.030), Li (0.020), Sn (0.020), Cr (0.019), Cd (0.010).

Female animals demonstrate higher concentrations of manganese in the hair than males do (9.64±0.40 versus 8.17±0.29 μg/g, $P = 0.026$). Hair concentrations of other essential and toxic trace elements did not differ statistically significant between the samples of male and female animals.

4 Discussion

For the first time, we carried out comprehensive evaluation of the 19 trace elements content in tail brush integumentary hair of the 1-day-old Simmental calves. Comparison of the obtained results with literature data [20, 21, 22, 27] showed that elemental profile of hair in newborn calves differed from that in the adult cattle in the lower content of iron, iodine, lithium, selenium, boron and strontium, and higher content of manganese, silicon, vanadium and mercury. Concentrations of other essential (cobalt, copper, zinc) and toxic (arsenic, cadmium, nickel, lead and tin) trace elements under study in the hair of newborns were comparable with the reference values described for the lactating cows [21, 27, 28].

Newborn Hereford and Holstein calves demonstrated manganese (2–4 μg/g) and iron (15–31 μg/g) concentrations in hair lower than our data (Table 1), but comparable in regard to copper (10–12 μg/g) and zinc (128–142 μg/g) [34]. The observed difference between the data published previously [34] and our data could be associated both with breed characteristics of the animals and with differences in biogeochemical conditions in their cultivation. Within the same biogeochemical zone in the Central Black Earth Region of the Russian Federation, hair concentrations of cobalt (0.057–0.097 μg/g), copper (7.40–10.60 μg/ g), iron (20.3–65.0 μg/g), manganese (8.11–12.80 μg/g), selenium (0.398–0.595 μg/g) and zinc (91.3–138.1 μg/g) in 1-day-old calves obtained from the Red-and-White cows with physiological course of pregnancy [5] were comparable with our results.

Table 1 shows that newborn heifers were characterized by a higher (by 18.0%, $P = 0.026$) concentration of manganese in the tail hair brush compared to the bulls. These results are of particular interest, since it was previously demonstrated that manganese concentration in the hair correlates with the level of physiological maturity in the newborns [5] and depends on the productivity direction [22]. It was established [22] that the manganese content in the dairy animals hair is significantly higher (by 48.6%, $P < 0.05$) compared to that of beef cattle.

S.A. Miroshnikov et al. (2015) found that the 5-8-months-old Simmental heifers, compared to bulls of the same age, were characterized by a higher content in their hair of not only magnesium, but also of cobalt, chromium, iodine, vanadium, arsenic, nickel, lead, strontium, and by a lower level of silicon [22]. Our study showed that statistically significant differences between the groups of 1-day-old male and female animals were found only in the content of magnesium in the hair, but not in other trace elements. Probably, characteristic features of the hair elemental profile in bulls and heifers are formed at the older age.

We used percentile scales in the present study, when determining reference values of the chemical elements concentration in the hair of newborn calves. M.G. Skalnaya et al. were the first to pioneer this approach (2003) [35]. The authors demonstrated that physiologically determined range of the trace elements concentration in human hair corresponded to the interval between the 25th and 75th percentiles in the representative sample, and values beyond the 25th and 75th percentiles indicated presence of the diselementosis [35]. Diselementosis diagnosed on the basis of these criteria could be latent in a patient (as a rule, if values are in the range from the 10th to the 25th percentile or from the 75th to the 90th percentile) or with characteristic clinical signs of deficiency, excess or imbalance in the trace elements content in the organism. As a rule, if the values go beyond the 10th or 90th percentile boundaries [1, 35]. Eligibility of using this approach in diselementosis diagnosis in cattle is confirmed by research results provided by the scientific group of Professor S.A. Miroshnikov [20, 21, 28, 36, 37].

In the present study, we evaluated the chemical elements concentration in the average sample from the tail brush guard hairs obtained from calves on the 1st day after birth. Since the tail brush hair growth in cattle begins from the 7th month of gestation [29, 30], results of this analysis make it possible to assess fetus supply with elements under study in the last months of intrauterine development, diagnose diselementosis, predict metabolic disorders [5, 9] and correct them in a timely manner.

We physiologically determined the concentration range of essential (Table 3) and toxic (Table 4) trace elements in the hair of newborns separately for the groups of male and female animals. This approach makes it possible to more accurately diagnose diselementosis in newborn calves taking into account their gender.

5 Conclusion

As a result of research using percentile scales, reference intervals were determined for essential (cobalt, chromium, copper, iron, iodine, lithium, manganese, selenium, silicon, vanadium, zinc) and toxic (arsenic, boron, cadmium, mercury, nickel, lead, tin, strontium) trace elements in the hair of newborn Simmental bulls and heifers in biogeochemical conditions of the Central Black Earth Region of the Russian Federation. Values beyond the 25th and 75th percentiles are proposed to be considered as deviations indicating intrauterine diselementosis.

The authors express their gratitude to V.A. Demidov, Laboratory Head, Micronutrients LLC (Moscow, Russia), and E.P. Serebryansky, Chief Analytical Chemist, for consultations and assistance in conducting research.

References

1. A.V. Skalny, *Bioelementology as an interdisciplinary integrative approach in life sciences: terminology, classification, perspectives*, J. Trace Elem. Med. Biol., **25**, S3–S10 (2011)
2. S.V. Sabunin, V.I. Belyayev, N.E. Papin, *Diselementosis - etiology, prophylaxis, treatment*, Veterinarnyy vrach, **3**, 39-43 (2014)
3. S. Shabunin, A. Nezhdanov, V. Mikhalev et al., *Diselementosis as a risk factor of embryo loss in lactating cows*, Turk. J. Vet. Anim. Sci., **41**, 453–459 (2017)
4. E. Kalaeva, V. Kalaev, A. Chernitskiy et al., *Incidence risk of bronchopneumonia in newborn calves associated with intrauterine diselementosis*, Vet. World, **13**, 987–995 (2020)
5. V.A. Safonov, V.I. Mikhalev, A.E. Chernitskiy, *Antioxidant status and functional condition of respiratory system of newborn calves with intrauterine growth retardation*, Agricultural Biology, **53**, 831–841 (2018)
6. M. Gabryszuk, K. Słoniewski, E. Meteraet al. *Content of mineral elements in milk and hair of cows from organic farms*, J. Elem., **15**, 259-267 (2010)
7. W.S. Swecker, *Trace mineral feeding and assessment*, Vet. Clin. North Am. Food Anim., **30**, 671–688 (2014)
8. A. Herold, A.-E. Müller, R. Staufenbiel et al., *Bovine trace mineral concentrations in different sample media with emphasis on fecal analysis*, Tierarztl Prax Ausg G Grosstiere Nutztiere, **48**, 5-14 (2020)
9. X.-J. Zhao, X.-Y. Wang, J.-H. Wang et al., *Oxidative stress and imbalance of mineral metabolism contribute to lameness in dairy cows*, Biol. Trace Elem. Res., **164**, 43–49 (2015)
10. F. Chen, J. Gao, D. Wu et al. *Clinical and pathologic features of a suspected selenium deficiency in captive plains zebras*, Biol. Trace Elem. Res., **176**, 114–119 (2017)
11. Y. Mehdi, I. Dufrasne, *Selenium in cattle: a review*, Molecules, **21**, 545 (2016)
12. D.K. Combs, *Hair analysis as an indicator of mineral status of livestock*, J. Anim. Sci., **65**, 1753–1758 (1987)
13. A. Przybylowicz, P. Chesy, M. Herman et al., *Examination of distribution of trace elements in hair, fingernails and toenails as alternative biological materials. Application of chemometric methods*, Cent. Eur. J. Chem., 10, 1590–1599 (2012).
14. R.C. Patra, D. Swarup, M.C. Sharma et al., *Trace mineral profile in blood and hair from cattle environmentally exposed to lead and cadmium around different industrial units*, Journal of Veterinary Medicine Series A, **53**, 511–517 (2006)
15. M. Spolders, M. Höltershinken, U. Meyer et al., *Assessment of reference values for copper and zinc in blood serum of first and second lactating dairy cows*, Vet. Med. Int., **2010**, 194656 (2010)
16. S. Miroshnikov, S. Notova, T. Kazakova et al. *The total accumulation of heavy metals in body in connection with the dairy productivity of cows*, Environ. Sci. Pollut. Res., **28**, 49852–49863 (2021)
17. L. Perillo, F. Arfuso, G. Piccione et al. *Quantification of some heavy metals in hair of dairy cows housed in different areas from Sicily as a bioindicator of environmental exposure—A preliminary study*, Animals, **11**, 2268 (2021)
18. S. Miroshnikov, A. Kharlamov, O. Zavyalov et al., *Method of sampling beef cattle hair for assessment of elemental profile*, Pak. J. Nutr., 14, 632–636 (2015)
19. M. Gabryszuk, J. Barszczewski, J. Dobrzynski, *The mineral elements content in hair of cows from conventional and organic farms*, Journal of Research and Applications in Agricultural Engineering, **63**, 54–56 (2018)
20. S.A. Miroshnikov, O.A. Zavyalov, A.N. Frolov et al. *The reference intervals of hair trace element content in Hereford cows and heifers (Bos taurus)*, Biol. Trace Elem. Res. **180**, 56–62 (2017)
21. S.A. Miroshnikov, A.V. Skalny, O.A. Zavyalov et al. *The reference values of hair content of trace elements in dairy cows of Holstein breed*, Biol. Trace Elem. Res. **194**, 145–151 (2020)
22. S.A. Miroshnikov, A.V. Kharlamov, O.A. Zavyalov et al., *Peculiarities of formation of cattle elemental composition due to efficiency and belonging to some gender and age group*, Herald of Beef Cattle Breeding, **92**, 94–99 (2015)
23. I. Sleptsov, V. Machakhtyrova, G. Machakhtyrov et al., *Age features of elemental status for the Kalmyk cattle breed under conditions of Yakutia*, Agrarian Bulletin of the Urals, **192**, 69–77 (2020)
24. A.N. Frolov, O.A. Zavyalov, A.V. Kharlamov, *Peculiarities of hair elemental composition and adaptability of imported heifers based on their productivity*, Herald of Beef Cattle Breeding, **94**, 39–44 (2015)
25. S. Miroshnikov, O. Zavyalov, A. Frolov et al. *The content of toxic elements in hair of dairy cows as an indicator of productivity and elemental status of animals*, Environ. Sci. Pollut. Res., **26**, 18554–18564 (2019)

26. D. Cygan-Szczegielniak, M. Stanek, E. Giernatowska et al., *Impact of breeding region and season on the content of some trace elements and heavy metals in the hair of cows*, Folia Biol., **62**, 163–169 (2014)
27. S.P. Zamana, *Determination of chemical element composition of the hair cover in cattle*, Sel'skokhozyaistvennaya Biologiya, **4**, 121–125 (2006)
28. S.A. Miroshnikov, O.A. Zavyalov, A.N. Frolov et al., *Reference ranges of concentrations of chemical elements in the wool of dairy cows*, Animal Husbandry and Fodder Production, **102**, 33–45 (2019)
29. W.H. Mao, E. Albrecht, F. Teuscher et al. *Growth- and breed-related changes of fetal development in cattle*, Asian-Aust. J. Anim. Sci., **21**, 640–647 (2008)
30. C.H. Krog, J.S. Agerholm, S.S. Nielsen, *Fetal age assessment for Holstein cattle*, PLoS ONE, **13**, e0207682 (2018).
31. Temporary maximum allowable levels of certain chemical elements and gossypol in feeds for farm animals and feed additives, *USSR State Agriculture Committee* (Gosagroprom USSR, Moscow, 1987)
32. Nutrient requirements of dairy cattle, *National Research council* (National Academy Press, Washington, DC, 2001)
33. Mineral tolerance of animals, *National Research council* (National Academy Press, Washington, DC, 2005)
34. C.C. O'Mary, M.C. Bell, N.N. Snead et al., *Influence of ration copper on minerals in the hair of Hereford and Holstein calves*, J. Anim. Sci. **31**, 626–632 (1970)
35. M.G. Skalnaya, V.A. Demidov, A.V. Skalny, *About the limits of physiological (normal) content of Ca, Mg, P, Fe, Zn and Cu in human hair*, Trace Elem. Med. (Moscow), **4**, 5–10 (2003)
36. S.A. Miroshnikov, G.K. Duskaev, O.A. Zav'yalov et al., *Wool composition centile parameters and detecting cattle elementosis*, Vestnik of the Russian agricultural science, **2**, 25–62 (2017)
37. S.A., Miroshnikov, A.V. Kharlamov, A.N. Frolov et al., *Experience of individual correction of elemental status of cows with reproductive disorder*, IOP Conf. Ser.: Earth Environ. Sci., **341**, 012080 (2019)

Impact of developed ruminant feed products on the surrounding ecosystem

Kristina Kondrashova*, *Ksenia* Inchagova, *Vitaly* Ryazanov, and *Galimzhan* Duskaev

Federal Research Centre of Biological Systems and Agrotechnologies of the RAS, 29, 9 Yanvarya St., Orenburg, 460000, Russia Federation

Abstract. The use of feed additives based on the so-called "phytobiotics" and a complex of trace elements is a promising direction in the nutrition of farm animals and birds. However, along with the positive characteristics associated with the increase in body weight, environmental friendliness of production, etc., these biologically active additives may have a negative effect on animals and humans associated with the toxicity of the used components. In this regard, the purpose of this study was to determine the toxic effect of such feed additives as phytochemicals and trace element metals using a recombinant luminescent strain *Escherichia coli* K12 MG1655 (pXen7) and a representative of ciliated protozoa *Stylonychia mytilus*. During the studies, the absolute toxicity of the Cu biocomplex and *Scutellaria baicalensis* was recorded in relation to the test organisms used. Against this background, *Artemisiae absinthil herba* and the Digestarom agent showed the smallest toxic effect. The toxic effect of other tested bio-additives was in high doses (0.625-10 mg/mL) and disappeared with a decrease in the concentration of the tested feed additive.

1 Introduction

In recent years there is a growing interest in the problem of the transition to a highly productive and environmentally friendly agricultural farming. Currently, research scientists working in the field of animal husbandry and feed production are actively developing and introducing new feeding schemes for farm animals, which make it possible to obtain large volumes of meat products [1]. Special attention is focused on the creation of feed additives, which are characterized by high indicators of the body weight gain, environmental friendliness, as well as minimum production costs [2]. However, along with the positive characteristics of the use of such bio-additives, there may be negative effects associated with the toxic effect of the used components on animals and humans [3]. Thus, there are data on high hepatotoxicity of plant extracts *Frangula alnus* [4], *Mentha pulegium* [5], *Chelidonium majus* associated with liver damage [6]; cardiovascular toxicity caused by *Aconitum carmichaeli* [7], *Curcuma longa* [8]; with respect to the digestive system – by *Mentha pulegium* [9]; neurotoxicity – by *W. somnifera* [10]. In turn, trace element metals used in animal husbandry and poultry farming also have a number of limitations for use, since these compounds have a dose-dependent effect [11]. All of the above paves the way for a more in-depth study of such bio-additives when used in the feeding schemes for farm animals and birds.

In this regard, the purpose of this study was to determine the toxic effect of feed additives – phytochemicals and trace element metals using recombinant luminescent strain *Escherichia coli* K12 MG1655 (pXen7) and a representative of ciliated protozoa *Stylonychia mytilus*.

2 Materials and Methods

2.1 Feed additives

The objects of study included the so-called "phytobiotics", namely *Artemisiae absinthil herba* (Phytopharm), *Salviae folia* (Krasnogorskleksredstva JSC), *Origanum vulgare* (Production Company KIMA Company LLC), *Scutellaria baicalensis* (Belovodye LLC), as well as *Inulae rhizomata et radices* (Krasnogorskleksredstva JSC): a microelement biocomplex represented by manganese (Mn), copper (Cu) and cobalt (Co); and Digestarom, which is a food additive improving the taste of feed and increasing the appetite of farm animals, including birds (Biomin LLC).

The samples were ground to a powdered state and dissolved in sterile distilled water for further biological analyses.

2.2 Bacterial strains and protista

The toxicity of the test feed additives was evaluated using two test objects. The first was represented by a recombinant luminescent strain *Escherichia coli* K12 MG1655 (pXen7) with a constitutive glow type. This biosensor was obtained by transforming the cells of the host strain with a hybrid plasmid *pUC18* with an embedded EcoRI DNA fragment of about 7 thousand bp

*Corresponding author: christinakondrashova94@yandex.ru

containing structural bioluminescence genes of the soil microorganism *Photorhabdus luminescens* ZM1 [12].

Another test organism for determining the toxic effect of the selected feed additives was the representative of the ciliated protozoa *Stylonychia mytilus*, which is a standard test object for determining the general toxicity of feed [13].

2.3 Toxicity study using *E. coli* K12 MG1655 (pXen7)

The *E. coli* K12 MG1655 (pXen7) biosensor was cultivated for a day on LB agar (Sigma, USA) at 37 °C, and the selective marker – ampicillin (100 µg/mL) – was added to the medium. The cells were then suspended in NaCl solution (0.9%) to an absorbance of 0.5 relative units at 450 nm, these measurements were made in plastic transparent wells using an Infinite 200 microplate photometer (Tecan, Austria). Thereafter, a 500 µL bacterial suspension was added to 1000 µL of LB broth. Then, two-fold dilutions of the test bio-additives were prepared at the final concentration of 0.001 mg/mL to 10 mg/mL. After that, 100 µL of diluted substances were added to the test wells, and 100 µL of water was added to the control wells, then 100 µL of the finished bacterial suspension was added to all wells. The plate was placed in the Infinite 200 Pro luminometer measuring unit (Tecan, Austria). The measurement was performed in a kinetic mode for 180 minutes at 37°C. The obtained results were primarily processed using the Magellan™ plate luminometer software, and further processing of the obtained data was carried out using the Excel 2010 computer program (Microsoft Inc.). The normalized luminescence was calculated using Formula 1:

$$I_{norm} = \frac{I_n^{test} \cdot I_0^{control}}{I_0^{test} \cdot I_n^{control}},$$

where I_{norm} – normalized luminescence at n-minute;

I_n^{test} – luminescence index of the test sample at n minute of the experiment;

I_0^{test} – luminescence index of the test sample at 0 minute of the experiment;

$I_n^{control}$ – luminescence index of the control sample at n minute of the experiment;

$I_0^{control}$ – luminescence index of the control sample at 0 minute of the experiment.

The graphs reflecting the dependence of normalized luminescence (relative units) on the concentration of test substances (mg/mL) were plotted on the basis of the obtained data.

2.4 Toxicity study using *S. mytilus*

The toxic effect of the test feed additives was also studied on the cell culture of the freshwater infusoria *S. mytilus*. It was prepared as follows. A day before the analysis, the mass of stylonychia was transplanted into a new Lozin-Lozinsky nutrient solution with the addition of yeast (*Saccharomyces cerevisiae*) as feed and placed in a thermostat at 25°C. Thus, the protist cells were in the exponential growth phase at the time of the experiment. The studied test functions included: survival, abundance (biomass).

The study was conducted in a concentration range of 0.039 to 10 mg/ml. Each sample of the test sample was studied five times. Transplantation and counting of stylonychia were performed using a light microscope (MT 5300L). The analysis was performed as follows. An automatic pipette was used to take 20 µl of the medium with stylonychia, which were placed in each of the five microaquariums. Then, 20 µL of the aqueous extract of the test feed additive prepared for biotesting was added with an automatic pipette. After 2 minutes, the number of stylonychia was counted in each microaquarium. After counting, 200 µl of the aqueous extract of the test sample was poured into each microaquarium and the start time of the test was marked. The toxicity of the test feed additives was determined by the survival of stylonychia after 1 and 3 hours of exposure, manifested as the movement cessation and change in the shape of cells from ellipsoid to rounded or violation of the integrity (lysis) of cells.

The survival rate of N stylonychia (%) was calculated using Formula 2:

$$N = \frac{N_2}{N_1} \cdot 100,$$

where N_2 – arithmetic mean (out of five tests) of the number of stylonychia at the end of the test after 1 hour of exposure, pcs;

N_1 – arithmetic mean (out of five tests) of the number of stylonychia at the beginning of the test, pcs;

100 – coefficient of conversion of the result into percentages.

The toxicity of the test feed additives was determined based on the following:

- from 70% to 100% survival of stylonychia – non-toxic sample;
- from 40% to 69% survival of stylonychia – slightly toxic sample;
- from 0% to 39% survival of stylonychia – toxic sample.

2.5 Statistical processing of study results

All experiments were performed in at least five replicates. The obtained results were processed by the methods of variation statistics in Excel for Windows 10.

3 Results

3.1 Evaluation of the toxic effect of test feed additives on recombinant luminescent strain *E. coli* K12 MG1655 (pXen7)

The toxicity assessment using a constitutively luminescent biosensor showed different activity of the test feed additives. Among all studied compounds, the *Inulae rhizomata et radices*, Digestarom, *Artemisiae absinthil herba*, *Salviae folia* and *Origanum vulgare* were not toxic, as evidenced by a slight level of luminescence relative to the control.

https://doi.org/10.1051/bioconf/20225200034

The biological complex of trace elements, as well as the *Scutellaria baicalensis* showed some toxic effect.

The Cu biocomplex completely inhibited the glow of a microorganism at maximum concentrations ranging from 1.25 to 10 mg/mL, with an increase in dilution from 0.625 to 0.313 mg/mL, the percentage of survived cells increased from 64% to 76% relative to the control, respectively. Further reduction in concentration did not cause a toxic effect on the test strain.

The Co biocomplex at the same concentrations as Cu had the toxic effects observed at concentrations of 5-10 mg/mL causing 100% quenching of bacterial luminescence. Lower concentrations of 1.25 and 2.5 mg/mL inhibited the biosensor glow by 64% and 86%, respectively, relative to the control, and with a decrease in concentration the effect was offset.

Table 1. EC_{20} and EC_{50} of the analyzed bio-additives for recombinant luminescent strain *E. coli* K12 MG1655 (pXen7), mg/ml.

Indicator	Co	Cu	Mn	Scutellaria baicalensis
EC_{50}	1.094	0.781	–	0.059
EC_{20}	0.729	0.273	7.5	0.004

*Note: The threshold concentration (EC_{20}) is defined as the concentration resulting in the smallest significant decrease in luminescence, i.e. suppression by 20%, and EC_{50} – suppression by 50%.

In the group of studied trace elements the Mn biocomplex showed the least toxic effect. Its maximum concentration had a minor toxic effect. The percentage of survived cells reached 80% (EC_{20}).

The evaluation of the concentration dependence of *Scutellaria baicalensis* showed its toxicity in a wide range of concentrations (0.005-10 mg/mL), and with a dose-dependent effect. Maximum concentrations were characterized by the greatest decline in bioluminescence, which was 88% (5 mg/mL) and 92% (10 mg/mL). With a subsequent decrease in concentrations (from 2.5 to 0.078 mg/mL), the toxic effect was reduced to 60%. At lower concentrations of 0.039-0.005 mg/mL, toxicity decreased to 25% and disappeared at subsequent dilutions (Table 1).

3.2 Biological effect of the test feed additives on the *S. mytilus* cell population

The analysis of the obtained data showed that the greatest toxic effect, as in the case of the *E. coli* K12 MG1655 (pXen7) strain, was the Cu biocomplex and *Scutellaria baicalensis*, the toxic effect of which was manifested after 1 hour of exposure in the entire range of studied concentrations of these substances, characterized by cell lysis.

In turn, the Mn biocomplex and *Salviae folia* showed a toxic effect in the concentration range of 1.25-10 mg/ml, and in this case, the cells changed shape from ellipsoid to rounded, and when exposed to *Salviae folia* – cell lysis occurred.

Significant toxicity was observed in the *Inulae rhizomata et radices*, which appeared in the concentration range of 2.5-10 mg/mL and was accompanied by lysis of all cells of the test object.

Against this background the *Artemisiae absinthil herba* and the Digestarom had a minor toxic effect, which manifested itself only in the concentration of 10 mg/ml for the *Artemisiae absinthil herba* and 5-10 mg/ml for the Digestarom agent, and in the concentration of 5 mg/ml the survival of test organisms corresponded to 50%.

The *Origanum vulgare* and the Co biocomplex were characterized by a high level of survival of stylonychia manifested in the concentration range of 0.039-2.5 mg/mL (Table 2).

4 Discussion

The study of phytochemical substances and metals of trace elements as feed additives allows referring to some features, as well as restrictions on their use in feeding farm animals and birds. Thus, the Cu biocomplex and *Scutellaria baicalensis* had absolute toxic activity against the biosensors used, which confirms the results of the studies carried out on other test organisms [3]. Moreover, if we analyze the causes of such action of these bio-additives in the present study, the toxicity of the Cu biocomplex can be explained by the participation of this trace element in biological processes in the cells of bacteria and protozoa (interaction with enzymes or allosteric effectors during metabolism, interaction with the luciferase complex responsible for luminescence, etc.) [15]. As for the *Scutellaria baicalensis*, this effect can be associated with its well-known antibacterial effect on many pathogenic bacteria, and in the case of stylonychiae, with the large-needle structure of the solid parts of the plant, which when swallowed cause the cell lysis. It is also necessary to take into account the effect of substances contained in extracts [16, 17].

Analyzing the effect of the remaining feed additives, it should be noted that their use should be strictly regulated in the feed of farm animals and birds, since exceeding the safe dose of these bio-additives may lead to negative consequences [3].

5 Conclusion

Thus, the results of the study indicate the need for a detailed study of the above feed additives in multicellular organisms, since the data obtained in single-celled biotests cannot guarantee the absolute safety of their use in animal husbandry and poultry farming.

This research was performed with the financial support from the Russian Science Foundation (Project No. 21-76-10014).

Table 2. Biological effect of the test feed additives on the *S. mytilus*.

Feed additive	Exposure time, h	Concentration, mg/ml								
		0.039	0.078	0.156	0.313	0.625	1.25	2.5	5	10
Origanum vulgare	1	NOEC	NOEC	NOEC	NOEC	NOEC	NOEC	NOEC	Tox	Tox
	3	NOEC	NOEC	NOEC	NOEC	NOEC	NOEC	NOEC	Tox	Tox
Salviae folia	1	NOEC	NOEC	NOEC	NOEC	NOEC	Tox	Tox	Tox	Tox
	3	NOEC	NOEC	NOEC	NOEC	NOEC	Tox	Tox	Tox	Tox
Inulae rhizomata et radices	1	NOEC	NOEC	NOEC	NOEC	NOEC	NOEC	Tox	Tox	Tox
	3	NOEC	NOEC	NOEC	NOEC	NOEC	NOEC	Tox	Tox	Tox
Artemisiae absinthil herba	1	NOEC	NOEC	NOEC	NOEC	NOEC	NOEC	NOEC	NOEC	NOEC
	3	NOEC	NOEC	NOEC	NOEC	NOEC	NOEC	NOEC	NOEC	Tox
Scutellaria baicalensis	1	Tox	Tox	Tox	Tox	Tox	Tox	Tox	Tox	Tox
	3	Tox	Tox	Tox	Tox	Tox	Tox	Tox	Tox	Tox
Digestarom	1	NOEC	NOEC	NOEC	NOEC	NOEC	NOEC	NOEC	LOEC	Tox
	3	NOEC	NOEC	NOEC	NOEC	NOEC	NOEC	NOEC	LOEC	Tox
Cu	1	Tox	Tox	Tox	Tox	Tox	Tox	Tox	Tox	Tox
	3	Tox	Tox	Tox	Tox	Tox	Tox	Tox	Tox	Tox
Co	1	NOEC	NOEC	NOEC	NOEC	NOEC	NOEC	NOEC	Tox	Tox
	3	NOEC	NOEC	NOEC	NOEC	NOEC	NOEC	NOEC	Tox	Tox
Mn	1	NOEC	NOEC	NOEC	NOEC	NOEC	Tox	Tox	Tox	Tox
	3	NOEC	NOEC	NOEC	NOEC	NOEC	Tox	Tox	Tox	Tox

Note: Tox – 0-39% survival rate of the test object; LOEC – 40-69% survival rate of the test object; NOEC – 70-100% survival rate of the test object [14].

References

1. D.K. Dhanasekaran, T.P. Dias-Silva, A.L. Filho, G.Z. Sakita, A.L. Abdalla, H. Louvandini, M.M., Agrofor. Syst., **94**, 1541-1553 (2020)
2. D. Yurin, N. Yurina, I. Tletseruk, M.I. Slozhenkina, A.A. Mosolov, A. Seidavi, A.V. Balyshev, IOP Conference Series: Earth and Environmental Science, **848**, 012083 (2021)
3. A. Hudson, E. Lopez, A.J. Almalki, A.L. Roe, A.I. Calderón, Planta Med., **84**, 613-626 (2018)
4. X. Dong, Phytother. Res., **30**, 1207-18 2016
5. K.M. Madyastha, C.P. Raj, Toxicology, **89**, 119-125 (1994)
6. R. Teschke, X. Glass, J. Schulze, Reg. Tox. Pharm., **61**, 282-291 (2011)
7. Z. Gardner, M. McGuffin *American Herbal Products Association's botanical Safety Handbook. Second edition Boca Raton* (FLa: CRC Press), pp 5-7 (2013)
8. J.L. Funk, J.B. Frye, J.N. Oyarzo, H. Zhang, B.N. Timmermann, J. Agric. Food Chem., **58**, 842-849 (2010)
9. A. Woolf, J. Toxicol. Clin. Toxicol., **37**, 721-727 (1999)
10. J. Wang, R. van der Heijden, S. Spruit, T. Hankermeier, K. Chan, J. van der Greef, G. Xu, M. Wang, J. Ethnopharm, **126**, 31-41 (2009)
11. V.I. Fisinin, S.A. Miroshnikov, E.A. Sizova, A.S. Ushakov, E.P. Miroshnikova, World's Poult. Sci. J., **74**, 523-540 (2018)
12. I.V. Manukhov, S.M. Rastorguev, G.E. Eroshnikov, A.P. Zarubina, G.B Zavilgelsky, Genetics, **36**, 322-330 (2000)
13. GOST 31674-2012 Feeds, compound feeds, material for compound feeds. Methods for the determination of common toxicity 2014 (Moscow: Publishing house of standards)
14. P. Jackson, N.R. Jacobsen, A. Baun, R. Deev, D. Kühnel, K.A. Jensen, U. Vogel, H. Wallin, Chemistry Central Journal, **7**, 154 (2013)
15. E. Fulladosa, J.C. Murat, M. Martínez, I. Villaescusa, Chemosphere, **60**, 43–48 (2005)
16. G.K. Duskaev, D.G. Deryabin, I.F. Karimov, D.B. Kosyan, S.V. Notova, J. Pharm. Sci. Res., **10**, 91-95 (2018)
17. I. Karimov, K. Kondrashova, G. Duskaev, O. Kvan, E3S Web of Conferences, **143**, 02034 (2020)

Theoretical justification of the parameters of the device for vertical measurement of soil hardness

Tatyana S. Gridneva*, *Sergey V.* Mashkov, *Sergey I.* Vasiliev, *Vladimir A.* Syrkin, and *Pavel* V. Kryuchin

Samara State Agrarian University, 2, Uchebnaya St., Kinel, Ust-Kinelsky, 446442, Russia Federation

Abstract. When evaluating tillage machines, one of the important places is occupied by agrotechnical assessment of the quality of the technological process of tillage. The properties of the soil as an object of cultivation, as well as the processes occurring during the interaction of the soil with any materials, are studied by agricultural mechanics, created by V.P. Goryachkin. One of the most important characteristics of the soil condition is the hardness of the soil, or the resistance provided by the soil to the penetration of any deformer into it. This article describes the forces and stresses that occur when the plunger penetrates into the soil. According to the theoretical provisions, friction force F, drag force Q, total force P, stresses σ1 and hardness T are calculated, theoretical dependences of soil hardness on the tip diameter on compacted and loose soil are given.

1 Introduction

When evaluating tillage machines, one of the important places is occupied by agrotechnical assessment of the quality of the technological process – tillage, and includes the determination of indicators of the state of the soil before processing, as well as indicators of the quality of the technological process, determined on already treated soil [1–3].

The properties of the soil as an object of cultivation, as well as the processes occurring during the interaction of the soil with any materials, are studied by agricultural mechanics, created by V.P. Goryachkin. One of the most important characteristics of the soil condition is the hardness of the soil, or the resistance provided by the soil to the penetration of any deformer into it [4, 5].

2 Materials and methods

To determine this indicator, devices for vertical measurement of soil hardness are used, the main working body of which is a plunger buried in the soil, and, as a rule, having a replaceable tip. Manual hardness testers used in tests have replaceable tips in the form of a flat disk with a base area of 1; 2 and 3 cm^2 with a plunger diameter of 1 cm. The size of the tip is selected depending on the density of the soil. However, the processes occurring in dense and loose soil are different, and the ratio of the diameter of the tip and the diameter of the plunger of the hardness tester is not justified.

If the tip has an equal diameter with the plunger or close to it, then the force required to bury the plunger will be greater, that is, the readings of the hardness tester will be overestimated, since the effect of soil exposure on the lateral surface of the rod is not taken into account.

Studies by a number of scientists [7-9] have proved that the soil has elastic properties, that is, the soil is characterized by partial recovery of volume when the applied load is removed.

We consider the penetration of a flat tip into the soil and its movement at a constant speed. If the depth is large enough in comparison with the width of the deformer, then the stress field in which the soil movement occurs and the stresses caused by it can be considered constant during the period of deepening. [9]

We consider two cases: the first is when the diameters of the plunger d and the tip *D* are equal, and the second is when the diameter of the tip *D* is greater than the diameter of the plunger *d*. Since we are interested in determining the effect of the ratio of the plunger and tip diameters on the accuracy of hardness determination, in the second case we will consider two options: immersion in loose soil and in untreated.

3 Results and discussion

The deepening of a flat cylindrical tip into the soil is accompanied by the appearance of a stress zone under it. Stresses cause the movement of structural soil particles, partial destruction and their mutual displacement. The theory of soil destruction assumes that when a certain load occurs on the surface, a certain array of stresses arises under it. A model formulated by Coulomb is known and used to describe the destruction of soil. According to this model, the load applied to the soil causes the destruction of the soil, the boundary of which is formed at an angle of $(45^0-\varphi/2)$ to the soil surface,

* Corresponding author: t-grid@mail.ru

where φ is the angle of internal friction of soil particles, deg. (fig. 1).

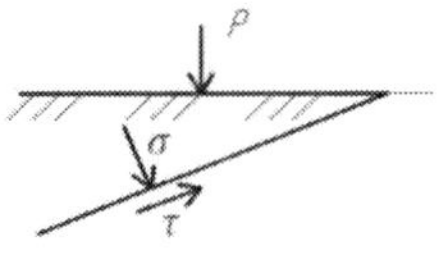

Fig. 1. Stresses arising in the soil sample during shear

According to Coulomb's law, the shear stress required for fracture must exceed the adhesion force of soil particles in the fracture plane and the friction force between surfaces at the fracture boundary. The calculated value of the required shear stress is:

$$\tau = c + \sigma_n tg\,\varphi, Pa, (1)$$

where c is the adhesion tension, Pa;

σ_n is the normal stress on the plane of destruction, Pa.

Adhesion tension c and angle of internal friction of soil particles φ depend on the mechanical composition of the soil, moisture content, porosity.

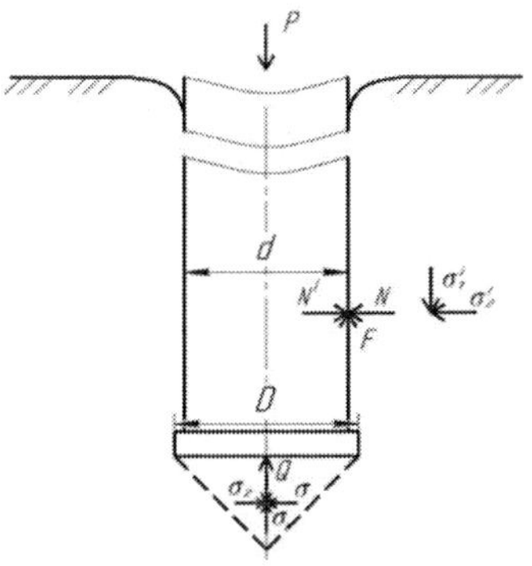

Fig. 2. Diagram of forces and stresses arising when the plunger penetrates into the soil

Let the tip of the hardness tester have a diameter D, and the diameter of the plunger d (fig. 2). Consider the case in which $D = d$. When the flat tip is immersed with the applied force P, according to the Coulomb-Mohr law, the normal stress in the stressed zone having the shape of a wedge (shown by the dotted line) is equal to:

$$\sigma_1 = \sigma_2 tg^2\left(45 - \frac{\varphi}{2}\right) + 2c\,tg\left(45 - \frac{\varphi}{2}\right), Pa, (2)$$

where σ_1 is the vertical stress acting on the tip surface, Pa;

σ_2 is the horizontal voltage, Pa.

The total force Q acting on the tip will be equal to:

$$Q = \sigma_1 S, N, (3)$$

$$S = \frac{\pi D^2}{4}, m^2, (4)$$

where S is the tip area, m^2.

$$Q = \sigma_1 \frac{\pi D^2}{4}. (5)$$

A cone-shaped zone with a base on the top surface appears on the surface of a solid material (tip) penetrating into an isotropic material. Such zone is called a compacted core with a taper angle $\gamma = 2\varphi$. When the tip moves, this compacted core acts on the soil and causes it to move sideways and compact under the action of the applied load. The stress σ_1 inside the core determines the force with which it acts on the surrounding masses of soil. If we take an element of the soil mass lying outside the zone of the compacted core at some depth h under the steady-state mode of sinking, then there will also be stresses arising due to the pressure of the overlying layers: σ_1' – vertical stress and σ_2' – horizontal stress.

The vertical stress acting on the soil element from the own weight of the soil is [11]:

$$\sigma_1' = ph, Pa. (6)$$

where p is the soil density, kg/m^3;

h is the location depth of the considered point, m.

The stress acting in the horizontal direction on the soil mass is equal to:

$$\sigma_2' = \xi \sigma_1', Pa. (7)$$

where ξ is the coefficient of lateral pressure in natural occurrence, depending on the type and condition of the soil.

Then:

$$\sigma_2' = \xi ph. (8)$$

During penetration, the side surfaces of the plunger and the tip from the side of the contacting soil are affected by the force of normal pressure N and the friction force F, directed against the direction of movement. The resultant of these forces R deviated from the normal by the angle of friction, this is the force with which the soil acts on the side surface of the plunger. Also, the plunger surface acts on the soil with the force R' directed in the opposite direction, and, accordingly, with the force N'. The force N' is equal to:

$$N' = \sigma_1' S', N. (9)$$

where S' is the area of the lateral surface of the plunger and tip, m^2.

Then the friction force will be equal to:

$$F = fN', N. (10)$$

where f is the friction coefficient of the soil on the material of the plunger.

$$F = f\,\xi ph\pi dl, N, (11)$$

where l is the length of the immersed part of the plunger, m.

Since l is the depth of the considered point of action of the force, the expression (11) can be represented as:

$$F=f\,\xi p h^2 \pi d, N.(12)$$

The friction force from the side of the soil will also act on the side surface of the tip, but since its height is rather small compared to the length of the plunger, it can be neglected.

Assuming that from the equilibrium condition, $\sigma_2=\sigma_2^{/}$, we get that:

$$\sigma_1=\sigma_2^{\text{¿}}\left[tg^2\left(45-\frac{\varphi}{2}\right)+2c\,tg\left(45-\frac{\varphi}{2}\right)\right]=\text{¿}$$

$$\text{¿}\,\xi ph\left[tg^2\left(45-\frac{\varphi}{2}\right)+2c\,tg\left(45-\frac{\varphi}{2}\right)\right].(13)$$

Then:

$$Q=\frac{\pi D^2}{4}\xi ph\left[tg^2\left(45-\frac{\varphi}{2}\right)+2c\,tg\left(45-\frac{\varphi}{2}\right)\right].(14)$$

Friction force F and the normal pressure force of compressive resistance (frontal resistance) Q balance the penetration force P, i.e.:

$$P=F+Q.(15)$$

$$P=f\,\xi p\,h^2\pi d+\frac{\pi D^2}{4}\xi ph\left[tg^2\left(45-\frac{\varphi}{2}\right)+2c\,tg\left(45-\frac{\varphi}{2}\right.\right.$$

$$P=\xi p\pi\left\{f\,h^2 d+\frac{\pi D^2}{4}\left[tg^2\left(45-\frac{\varphi}{2}\right)+2c\,tg\left(45-\frac{\varphi}{2}\right)\right]\right\}$$

Hardness (or soil resistivity) by its definition is the ratio of the resistance force to the area of the tip:

$$T=\frac{P}{S}, Pa,(16)$$

or

$$T=\frac{\sigma_1 S+\sigma_2 S^{\text{¿}}}{S}.(17)$$

$$T=\sigma_1+\frac{F}{S}.(18)$$

That is, σ_1 is the specific drag force, then it is a value that does not depend on the area S. Since when finding T, σ1 does not depend on S, the friction force F on the side surface of the plunger has a greater influence on the value of T.

Now we consider the plunger movement in the case when the plunger diameter is smaller than the tip diameter: $d<D$ (fig. 3).

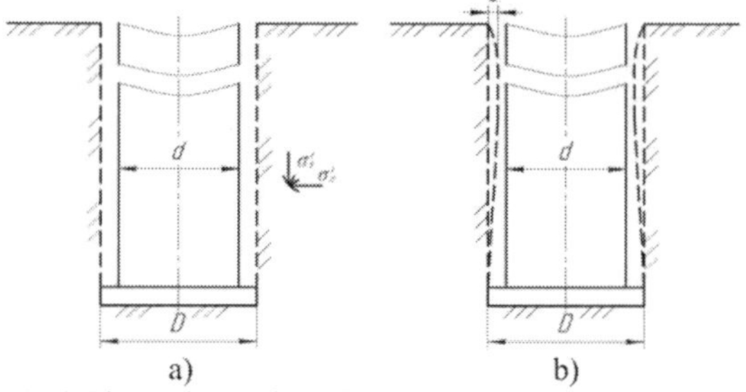

a) b)

Fig. 3. Plunger immersion scheme with a tip (D>d): a) in dense soil; b) loose soil

If the soil has a high density, then the angle of internal friction of the soil in this case will have a smaller value, and the adhesion force will have a larger one. The force transmitted by the tip as it passes will compress the walls of the hole. It has been proven [7–9] that the soil belongs to the media with elastic-plastic properties, that is, it is able to restore, partially or completely, its initial volume after the application and removal of a short-term load. The process of volume restoration after applying a certain load under any conditions lasts no more than 15 sec., and the time during which a significant part of the volume is restored is 3–6 sec. [12]. That is, this value is comparable to the time during which the process of immersion of the hardness tester plunger takes place. In this case, the nature of the volume recovery proceeds according to the law described by the logarithmic dependence:

$$p_i=p_{max}e^{-bt}, kg/m^3,(19)$$

where p_i is the value of soil density in the process of unloading, kg/m³;

p_{max} is the maximum value of density during loading, kg/m³;

b is the experimental coefficient;

t is the time of unloading, sec.

The change in the value of decompaction ε from 0 to ε_{max} of the soil sample subjected to the action of the load and compaction p_{max} occurs within 3...6 sec. Let us assume that the load applied to the walls of the hole with diameter D formed during the passage of the tip compacts the walls (ε=0). According to the law of elastic deformations, the value of ε will increase according to dependence (17), acquiring the value of ε_{max} within 3 sec.

It can be assumed that when the tip reaches the maximum depth in the upper part of the plunger at a depth h', the friction force F' will act. Taking into account the average immersion time and the distance traveled, it can be assumed that h'=¼h, in this section of the hole ε will be equal to ε_{max}, and in the lower layers it will be zero.

Thus, if the ratio D–d<2ε, then with an increase in ε in the upper layer, a friction force F' is equal to:

$$F^{\text{¿}}=f\,\xi p\left(h^{\text{¿}}\right)^2\pi d.(20)$$

In the case when $D–d>2\varepsilon$ (fig. 3), it is theoretically possible to take the friction force F equal to zero, since the hole wall does not act on the side surface of the plunger. The force of friction on the side surface of the tip is neglected.

In the second and third cases, we assume that with an increase in D, and, accordingly, S, σ_l remains constant, and P increases in proportion to S.

Now consider the movement of the tip in loose soil. The tip, plunging, compresses the soil and deforms it in lateral directions. In this case, a compacted core is created, which perceives the frontal resistance to penetration and tangential friction forces along the side surface of the tip. The soil from the impact of the tip is compacted, and a well with compacted walls is formed. The degree of their compaction depends on the initial compaction of the soil and its properties - the adhesion force and the angle of internal friction.

Theoretically, we assume that in the case of both $D=d$ and for $D>d$, the forces F, Q, P and the values σ_l and σ_2^{i} will be determined according to the above expressions. That is, if the soil is loose and the force N_l applied in the lateral direction is less than the cohesive forces of the soil, then soil particles, crumbling from the walls of the well, will act on the side surface of the plunger (fig. 4, b).

According to the theoretical provisions, the calculation of the friction force F, the frontal resistance Q, the total force P, the stress σ_l and the hardness T was carried out, with the following parameters corresponding to the moisture content of the medium loamy black soil of 24%: for movement in solid soil with $c=400\cdot10^3$ kg/m^2, $\varphi=22^0$; $\xi=0,7$, $p=1.2\cdot10^3$ kg/m^3. When calculating in the case of movement in loose soil, the density was taken equal to $0.9\cdot10^3$ kg/m^3, $c=300\cdot10^3$ kg/m^2, $\varphi=26^0$.

Calculations were carried out for the following points: $D=d=11$ mm; $D–d<2\varepsilon$ ($D=13$ mm, $d=11$ mm) (in this case, it was assumed that within 3 sec the precipitation ε reaches the value $\varepsilon>(D–d)/2$ in the upper 50 mm layer. So friction force F, and in the underlying ones – $F=0$, let's say that F will be recorded when the tip of the lower layer passes 30 cm); $D–d >2\varepsilon$ ($D>13$ mm), in this case $F=0$ was taken.

The results of calculating the soil hardness are shown in fig. 4, 5.

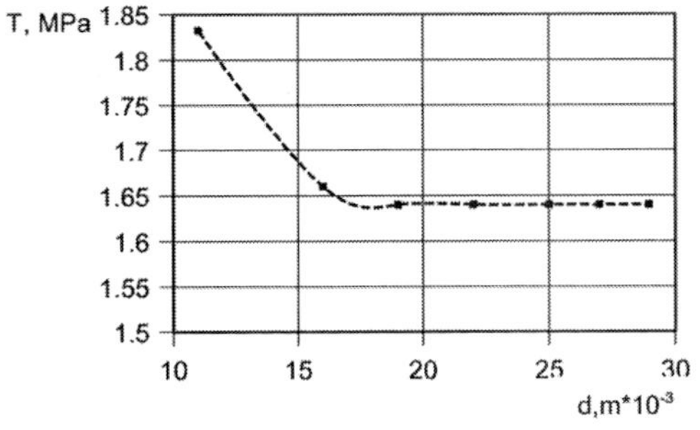

Fig. 4. Theoretical relationship between soil hardness and tip diameter (compacted soil)

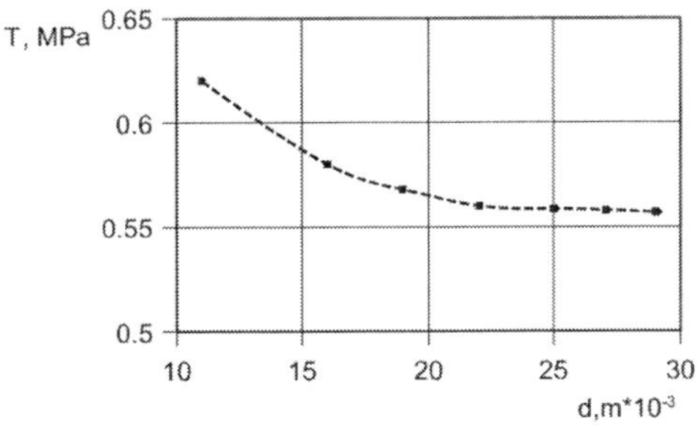

Fig. 5. Theoretical relationship between soil hardness and tip diameter (loose soil)

4 Conclusion

On the basis of the theoretical studies, the following conclusions can be drawn.

Calculations performed according to theoretical provisions showed that the decrease in soil hardness with an increase in the cross-sectional area of the measuring device tip is explained by the decrease in the F/S value in formula (1), due to the fact that the plunger diameter d remains constant, and the ratio D/d increases. The stress that occurs on the surface of the tip remains constant with a change in its diameter, which means that the total force P is more influenced by the friction force along the side surface of the plunger, which decreases with increasing D/d.

References

1. M.A. Kanaev, O.V. Karpov, S.A. Vasiliev, M.R. Fathutdinov. *Theoretical research of the scaffolder and soil interaction*, Bulletin of Samara State Agricultural Academy **3**, 19-23 (2017)

2. O.N. Chernikov, A.P. Bychenin, D.S. Sazonov, *Soil structure and hardness change resulting from the depth of a tractor track under different wheel drive towing strength*, Bulletin of Samara State Agricultural Academy **1**, 22-27 (2019)

3. M.P. Erzamaev, D.S. Sazonov, E.S. Nesterov, S. N. Zhil'tsov, *The influence of tillage longline process parameters for energy outlay*, Bulletin of Samara State Agricultural Academy **2**, 3-12 (2018)

4. G. Eudoxie, D. Phillips, R. Springer, *Surface Hardness as an Indicator of Soil Strength of Agricultural Soils*, Open Journal of Soil Science, 341…346 (2012)

5. T.S. Nugmanova, *Improvement of methods and technical means for determining soil condition indicators during testing of tillage machines*, Diss. cand. tech. sciences. Samara (2002)

6. V.F. Babkov, V.M. Bezruk, *Fundamentals of soil science and soil mechanics*. Moscow, Higher School (1986)

7. V.M. Matsepuro, *Investigation of the laws of soil and*

 https://doi.org/10.1051/bioconf/20225200035

soil resistance, Diss. cand. tech. sciences. Minsk (2002)

8. V.N. Savinykh, *Investigation of the regularities of stress relaxation and soil resistance to deformation as the basis of technological design of earthworking processes*, Diss. cand. tech. sciences. Minsk (1971)

9. S.S. Nugmanov, S.I. Vasiliev, T.S. Gridneva, *Methods and technical means for measuring soil hardness in coordinate agriculture*, Monograph (2009)

10. A. Kulen, H. Kuipers, *Modern agricultural mechanics*. Moscow, Agropromizdat (1986)

11. N.A. Tsytovich, *Mechanics of soils*. Moscow, Higher School (1979)

12. Yu.A. Saveliev, P.A. Ishkin, *Improving the efficiency of soil decompression using shallow autumn strip treatment*, Monograph (2017)

13. Yu.A. Savelyev, Yu.A. Kirov, P.A. Ishkin, M.A. Petrov. *Optimization of parameters of powered tillage tool with soil spikers*. Bulletin of Samara State Agricultural Academy **3**, 30-38 (2021)

Influence of a stopping angle of a spiral coil on the operation of a spiral-screw dispenser

Evgeny Lyalin*

Perm State Agro-Technological University named after Academician D.N. Pryanishnikov, 23, Petropavlovskaya St., Perm, 614990, Russia

Abstract. Dosing of dry concentrated feed to dairy cows is one of the most important tasks in the field of livestock mechanization. Devices that carry out volume dosing are more reliable and easier to be manufactured. Volumetric dispensers also include a spiral-screw dispenser – a special case of a screw dispenser. It is proposed to study the working process of the spiral-screw dispenser, namely the influence of a stopping angle of the last turn of the transporting spiral at the end of the discharge hole of the cylindrical casing. The paper indicates the method of determining and the value of the physical and mechanical properties of granulated feed. A laboratory installation on which it is possible to change the angle of installation of the spiral on the drive shaft is described. To study the effect of the stopping angle of the last turn of the transporting spiral on the accuracy of the formation of the feed portion, spirals with a diameter of 49, 73, 97 mm with different steps of 0.75, 1 and 1.25 of the diameter values were used. As a result of the experiments, it was revealed that the lowest coefficient of variation is 0.56 – 17.50%. And, accordingly, the greatest dosing accuracy is obtained when the last turn of the spiral stops at the position of 0°. That is, when the turn stops at the edge of the discharge neck in its lower part and this is typical for all sizes of spirals that were used.

1 Introduction

As noted earlier by many scientists and researchers, spiral-screw feeders – a special case of screw feeders – are promising in the field of volumetric dosing of bulk materials [1 – 3]. With regard to agriculture, specifically in animal husbandry, this is expressed in the dosing of dry concentrated feed to various farm animals [4 – 9].

In this industry, spiral-screw dispensers are increasingly used, since in dairy farming, according to zootechnical requirements, it is necessary to produce highly concentrated feed with high accuracy (at least 95%) [10 – 14].

Especially well proven spiral-screw dispensers are, in which the dosing of a portion of feed occurs according to the number of whole perfect revolutions of the spiral [5, 15]. When testing and studying the workflow of the dispenser, the question arose how and in what way the location of the last turn of the dosing spiral at the edge of the discharge neck would affect the operation of the dispenser.

In this regard, the goal was to investigate the process of dosing a portion of food at different angles of start and stopping the spiral.

2 Materials and methods

The task of the tests was to determine the accuracy of the formation of a portion of feed, depending on the angle of location of the last turn of the spiral at the end of the discharge neck.

The tests were carried out on granulated feed corresponding to GOST 9268-2015 with the following physical and mechanical properties: humidity – 6.1%; angle of natural slope – 42°; collapse angle – 48°; the angle of friction on the steel – 22°; bulk density – 658 kg/m^3. The sum of the Karr score is 81; flowability is good. These properties and criteria were determined according to GOST R 51850-2001, GOST R 57059-2016, GOST 13496.0-2016, GOST 28254-2014 [16].

To study the process of dosing and forming a portion of feed with a spiral-screw dispenser, a dispenser was designed, the scheme of which is presented in Figure 1. The basis of the spiral-screw dispenser includes a replaceable spiral and a casing. The spiral is driven by means of a coupling through a 16.3730 DC gear motor with a capacity of 100 W [5].

This motor gearbox allows stopping the spiral with a step of its rotation angle of 90°.

Figure 2 shows a general view of the laboratory spiral-screw dispenser

* Corresponding author: evgen159@list.ru

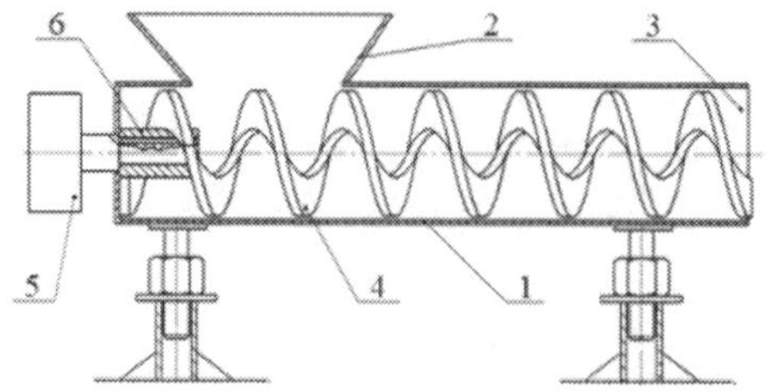

Fig. 1. Schematic diagram of spiral-screw dispenser: 1 – cylindrical casing; 2 – loading neck; 3 – discharge neck; 4 – spiral; 5 – drive (geared motor); 6 – coupling

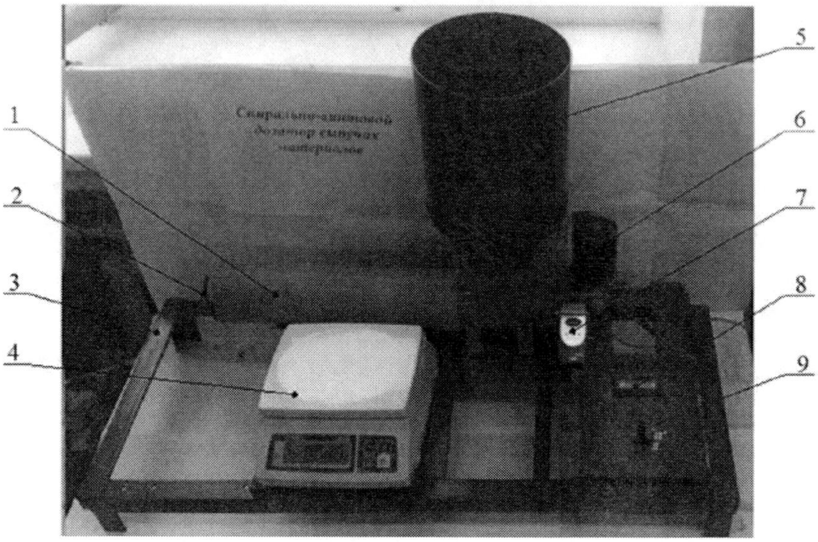

Fig. 2. General view of the experimental laboratory installation: 1 – cylindrical casing, 2 – spiral, 3 – frame, 4 – scales CAS SW-05, 5 – hopper with loading neck, 6 – motor reducer 16.3730, 7 – tachometer SM8238, 8 – measuring complex MYLB-G. T. Power RC 130A, 9 – switch

The estimation of the error in the formation of the feed portion was carried out on spirals of various diameters and pitches (Table 1), (Fig. 3), to ensure a gap of 5 mm between the spiral and the cylindrical casing, 3 round pipes with diameters of 59, 83 and 107 mm were selected.

Table 1. Parameters of the transport spirals used

Spiral diameter d, mm	Strip profile, mm	Spiral pitch, s	Spiral length l, mm
49; 73; 97	4x12	0.75d; 1d; 1.25d	600 ± 50

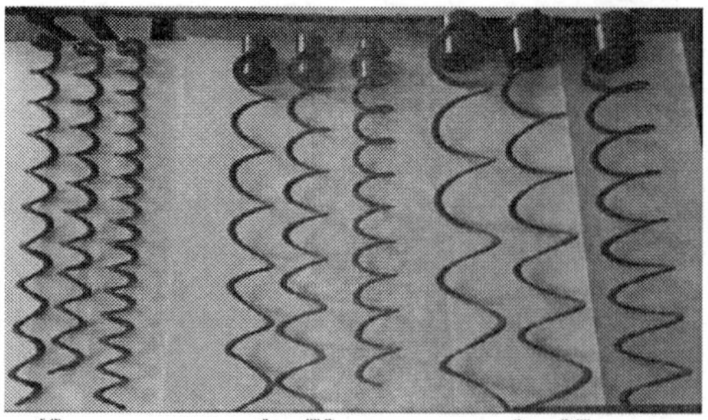

d = 49 mm d = 73 mm d = 97 mm

Fig. 3. Screw spirals for dispenser

The laboratory dispenser works as follows. When the actuator is switched on, the rotational motion of the drive shaft 8 through the shaft key connection to the rod is transmitted to the transport coil 4, thereby allowing the movement of the bulk material disposed at the bottom of the loading neck 2 in the direction of the discharge neck (window) 3.

The use of a geared motor with an eccentric as a spiral drive allows the spiral to be stopped in a position where the last turn is at the bottom of the discharge neck. The presence of a contact strip on the drive shaft allows you to change the angle of the last turn of the spiral with a step of 90° as shown in Figure. 4. That is, the spiral will constantly stop in one strictly defined position – 0, 90, 180 and 270 degrees, while making the x-number of revolutions.

The experiment was conducted as follows. On the control panel, the number of revolutions of the spiral was set, N = 2 turns. The spiral was installed in one of the 4 positions. After that, the power was supplied to the drive and the feed was issued to the receiving tank and weighed on the SCALES CAS SW-05 with an accuracy of two grams, each experiment was repeated in five time-repetition [16]. At the end of all the series of experiments, the data were structured and summarized in Table 2.

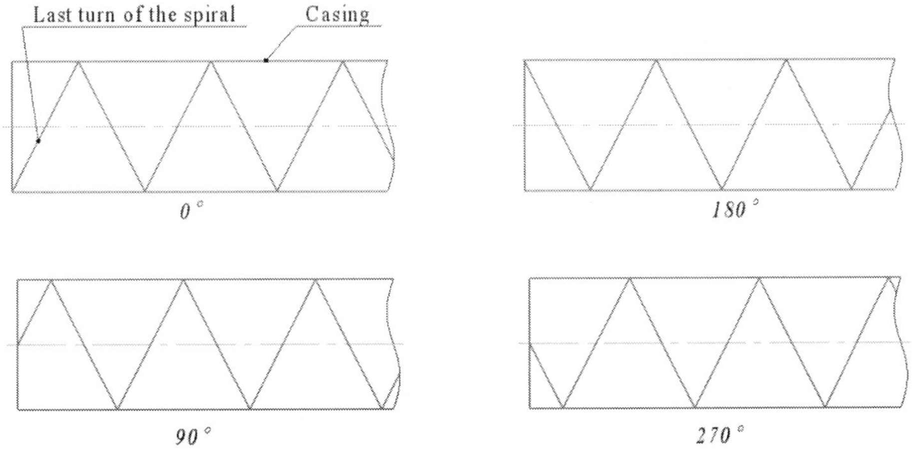

Fig. 4. Options for stopping angles of the last turn of the spiral at the end of the discharge neck

Table 1. Parameters of the transport spirals used

Angle of coil	D_k	d_{sp}	S	Weight of feed issued in experiment i					Σm_i, g	M_{av}, g	v, %
				m_1, g	m_2, g	m_3, g	m_4, g	m_5, g			
0°	109	97	1.25d	1352	1350	1326	1338	1372	6738	1347.6	**17.50**
90°				1356	1326	1334	1338	1382	6736	1347.2	29.64
180°				1322	1310	1374	1306	1364	6676	1335.2	59.87
270°				1324	1342	1328	1364	1358	6716	1343.2	18.65
0°	109	97	1d	1210	1202	1218	1220	1216	6066	1213.2	**3.51**
90°				1190	1222	1198	1230	1212	6052	1210.4	18.03
180°				1208	1198	1210	1216	1232	6064	1212.8	10.37
270°				1226	1190	1240	1222	1218	6096	1219.2	21.99
0°	109	97	0.75d	970	968	970	978	976	4862	972.4	**1.55**
90°				940	978	970	970	968	4826	965.2	17.67
180°				954	950	960	948	980	4792	958.4	13.92
270°				926	956	968	946	962	4758	951.6	22.77
0°	83	73	1.25d	600	598	594	612	607	3011	602.2	**6.93**
90°				614	580	614	618	590	3016	603.2	38.62
180°				600	612	617	602	622	3053	610.6	11.77
270°				590	610	578	594	607	2979	595.8	22.85
0°	83	73	1d	538	536	552	540	544	2710	542	**5.90**
90°				556	518	536	538	536	2684	536.8	27.00
180°				532	542	530	542	548	2694	538.8	8.49
270°				542	540	550	538	532	2702	540.4	6.34
0°	83	73	0.75d	410	404	402	420	402	2038	407.6	**11.54**
90°				392	412	412	412	386	2014	402.8	32.41
180°				402	412	396	422	396	2028	405.6	25.01
270°				398	392	404	402	414	2010	402	13.13
0°	59	49	1.25d	196	194	206	202	198	996	199.2	**9.32**
90°				196	198	194	198	208	994	198.8	11.75
180°				194	192	188	202	208	984	196.8	26.50
270°				198	188	196	204	196	982	196.4	13.36
0°	59	49	1d	170	172	170	170	172	854	170.8	**0.56**
90°				180	170	178	172	176	876	175.2	7.85
180°				170	178	176	178	172	874	174.8	6.04
270°				178	170	174	184	172	878	175.6	14.03
0°	59	49	0.75d	142	136	144	140	142	704	140.8	**5.23**
90°				148	148	142	144	152	734	146.8	8.28
180°				140	144	138	146	138	706	141.2	7.48
270°				136	132	142	138	136	684	136.8	7.72

3 Results and discussion

To assess the quality of the formation of a portion of feed, a coefficient of variation was adopted, which allows characterizing the dosing error with the lowest coefficient of variation, the greatest accuracy of feed coding will be observed.

In Table 2, the following symbols are adopted: D_k - diameter of the cylindrical housing, mm; d_s – spiral diameter, mm; S – pitch of the spiral in fractions of units of diameter; m_i – feed weight issued in experiment i, g; Σm_i – sum of feed masses for 5 experiments, g; m_{av} – average value of the mass of the feed issued, g; v – coefficient of variation, %.

Analysing the data obtained during the experiment, it can be noted that the lowest coefficient of variation is obtained when the last turn of the spiral stops at a position of 0°, that is, when the turn stops at the edge of the discharge neck in its lower part (Fig. 5). This phenomenon is observed on all studied standard sizes of spirals.

Fig. 5. Spiral stopped at 0°

Coefficients of variation at D_k = 83 mm, d_{sp} = 73 mm and S = 0.75...1.25d ranged from 1.55% to 59.87% and the lowest values of the coefficient of variation were 1.55%, 3.51%, 17.50% with the corresponding spiral steps obtained at 0°.

Coefficients of variation at D_k = 109 mm, d_{sp} = 97 mm and S = 0.75... 1.25d ranged from 5.90% to 38.62% and the lowest values of the coefficient of variation were 11.54%, 5.90%, 6.93% with the corresponding spiral steps also obtained at 0°

Coefficients of variation at D_k = 59 mm, d_{sp} = 49 mm and S = 0.75...1.25d ranged from 0.56% to 26.50% and the lowest values of the coefficient of variation were 5.23%, 0.56%, 9.32% at the corresponding spiral steps obtained similarly at 0°.

4 Conclusion

Thus, the previously put forward assumption was confirmed and the last turn of the spiral is able to perform the function of a cut-off and provide a more accurately formed portion of the feed by stopping at the bottom of the casing and counteracting the uncontrolled angle of collapse of the feed. And the results obtained are recommended to be taken into account when designing a spiral-screw dispenser and when using it in practice.

References

1. V.V. Konovalov, Substantiation of the parameters of the spiral-screw feeder of the concentrated feeds, Vestnik All-Russian Research Institute of Animal Husbandry Mechanization, **3(22)**, 68-73 (2011)
2. Y. Shimizu, P.A. Cundall, Three-dimensional DEM simulations of bulk handling by screw conveyors, Journal of Engineering Mechanics, **9(127)**, 864-872 (2001)
3. P.W. Cleary, DEM modelling of particulate flow in a screw feeder model description, Progressing Computational Fluid Dynamics, an International Journal, 2(7), 128-138 (2007)
4. P.S. Zolotarev Spiral-screw conveyor for bulk materials, Machinery and equipment for rural areas, **12**, 25-26 (2009)
5. E.A. Lyalin, M.A. Trutnev, Improving the accuracy of dosing concentrated feeds with a spiral-screw dispenser, Rural mechanizer, **1**, 26-27, 29 (2018)
6. P.A. Savinykh, Theoretical studies of feeding and unloading transporters, Achievements of Science and Technology of the Agro-Industrial Complex, **3**, 61-64 (2009)
7. L. P. Zaripova, F. S. Gibadullina, Sh. K. Shakirov et al., *Forage of the Republic of Tatarstan: composition, nutritional value and use* (Foliant, Kazan, 2010)
8. W. Brade, The stature of Holstein cows – a critical review, Tieraerztliche Umschau, **73(3)**, 74-84 (2017)
9. A.S. Zavgorodnyaya, I.G. Shashkova, V.S. Konkina et al., *Adaptive Management of the Agricultural Enterprise in the Conditions of Environmental Uncertainty*, J. of Advan. Res. in Dynamical and Control Syst., **7** (Special Issue), 2022–2031 (2018)
10. V.N. Vinogradov, M.P Kirilov, S.V. Kumarin, Modern approaches to the use of concentrated feed, Zootechny, **6**, 10-11 (2002)
11. V.D. Li Optimal mode of distribution of concentrates to highly productive cows, Zootechny, **7**, 16-17 (1996)

12. N.A. Morozkov, S.V. Tretiakov, V.A. Voloshin, *The system of full-fledged feeding of Black-and-White cattle at milk production complexes, which ensures an increase in milk productivity and an improvement in milk quality* (Perm, 2015)
13. V.G. Sofronov, A.S. Sajfullin, E.I. Yamaev, N.I. Danilova, SH.K. SHakirov, P.V. Sofronov, E.L. Kuznecova, Transactions of Kazan State Academy of veterinary medicine named after N.E. Bauman, **232**, 133–136 (2017)
14. B. Laurence, T. Pablo, R. Jacques, Y. Chilliarda, Animal Feed Sci. and Technol., **222**, 31–42 (2016)
15. E.A. Lyalin, M.A. Trutnev, Discrete method of dosing free-flowing concentrated feed with spiral-screwed feeder, COMETa 2018 "Conference on Mechanical Engineering Technologies and Applications" Proceedings, 401-406 (2018)
16. O.I. Polivaev, et al. *Test of agricultural machinery and energy-powered installations* (Lan, Saint Petersburg, 2016)

Research of the operating process of the mobile mixed fodder dispenser with a spiral-crewed dosing unit

Evgeny Lyalin*, *Mikhail* Trutnev, and *Nikolai* Trutnev

Perm State Agro-Technological University named after Academician D.N. Pryanishnikov, 23, Petropavlovskaya St., Perm, 614990, Russian Federation

Abstract. Highly productive dairy cows require additional distribution of concentrated fodder. Many farms of Russia and the Perm Krai use displacement feed propotioners, one of them is a spiral-crewed dosing unit. This dosing unit has good dosing accuracy and can be installed on the battery-driven dispenser. The device is equipped with the battery capacity of 55 A·h, a transporting spiral has the diameter of 97 mm with the pitch of 73 mm and the rotating frequency of 43 min-1. The dispenser was tested at several commercial dairy farms of the Perm Krai. Since the amount of concentrated fodder is individual for every cow, it must be calculated according to corresponding methodology taking into account the milk yield per day. In-process tests show operating capacity of the designed dispenser with a spiral-screwed dosing unit and its efficiency in mixed fodder distribution. The volume of dispensed fodder is equal to 552…604 g during the test. Taking into account the loss of time to move the dispenser between the animals, its productivity is equal to 3.11 kg/min, inequality in the dose of bulk bran does not exceed 4.18%, and 3.27% in the dose of granulated mixed fodder. Power consumption for fodder distribution is 45 … 47 W/h. In-process tests justify the value of earlier conducted scientific research.

1 Introduction

Important tasks of agricultural production are to increase the production of livestock products, increase labour efficiency and reduce the cost of the resulting products. The implementation of these tasks can be achieved only with the holistic mechanization of labour-intensive processes and the introduction of modern technologies. The efficiency of production and the volume of livestock products directly depend on the level and quality of animal feeding, as well as the balance of diets, taking into account the nutritional value of feed [1].

For highly productive cows with the annual milk yield equal to 6000 kg, consumption of concentrated fodders can reach 40% of daily nutritional value of animal's diet [1].

The general diet for groups of animals is based on their physiological state and daily milk yield. When feeding highly productive dairy cows, it is necessary to take into account the rate of feed intake during the milking process. It is known that dairy cows can eat up to 2.5 kg of loose and about 3 kg of granulated feed within 8–10 minutes [1, 2].

The redistribution of the volume of fed compound feeds over lactation periods (45...60% in the first period) increases the concentration of energy in the dry matter of the diet, and the adopted feeding system (2-3 times a day) causes poor eating of the main type of feed [3, 4].

The research of the Federal Research Center for Animal Husbandry (FRCAH) established that the optimal distribution of concentrated fodders is the following. Six-fold is in lactation phase I, three- or four-fold in lactation phase II, and two-fold in lactation phase III or 380…410, 290…360, and 140…240 g of concentrated fodder per 1 kg of milk yield. At the same time, the one-time delivery of concentrates shall not exceed 3 kg per one animal. When dairy cows are fed six times with compound feed, the optimal dispensing interval is 3 hours. At the same time, the maximum amount of compound feed given out during the day should not exceed 12...14 kg [2, 5].

Various types of dosing devices can be used for metered distribution of mixed fodder [6-9]. One of such devices is a spiral-screwed dosing unit that performs fodder dosing according to the number of full turns of the spiral. The dispenser is manufactured at the Department of Agricultural Machinery and Equipment of the Perm State Agro-Technological University [10, 11].

In the course of tests, the dosing unit yielded good results in the accuracy of dosing the dry mixed fodders [10]. Therefore, the aim of research is formulated as follows: to determine the efficiency of the mixed fodder dispenser with a spiral-screwed dosing unit in production conditions, to determine the inequality of fodder distribution, productivity and power consumption of the dispenser.

* Corresponding author: evgen159@list.ru

2 Materials and methods

The task of in-process tests of the spiral-screwed dosing unit is to study and estimate its operating process directly at a commercial dairy cattle farm. In order to solve this problem, the mobile dispenser of bulk concentrated fodder with a spiral-screwed dosing unit was designed and manufactured at the Perm SATU (Fig. 1) [10].

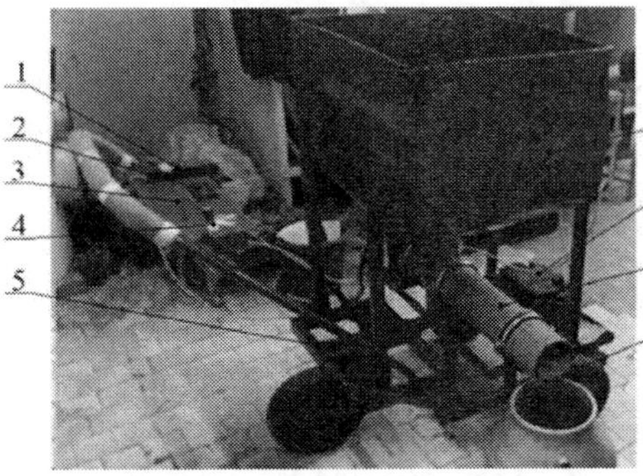

Fig. 1. Mobile fodder dispenser with a spiral-screwed dosing unit: 1 – SM8238 turn counter, 2 – MYLB-G.T. Power RC 130A measuring system, 3 – control panel, 4 – mode switch, 5 – handcart, 6 – tanker for concentrated fodder, 7 – battery, 8 – tubular enclosure, 9 – spiral.

The dispenser contains: a trolley with a pair of wheels and a handle for moving, a feed bin 6, in the lower part of which a spiral screw dispenser is installed. It consists of a cylindrical casing 8, a spiral 9 made of a rectangular section tape and an electric drive, which is controlled by a control panel 3. It includes sensor 1 of the counter of full revolutions of the spiral 9, a microprocessor and a "start" button 4, there is a battery 3 to power the electric drive-geared motor with appropriate control circuit. To prevent loss of feed, the spiral is set in such a way that the last turn when it stops is always in the lower position.

The distributor works as follows. Before starting work, the microprocessor of the distributor enters data on the norms for issuing compound feed to each animal, converted into the number of revolutions of the spiral 9, according to its serial number. Next, the feed is loaded into the hopper 6, after which the dispenser is moved manually by the handle to the places where the feed is dispensed to the animals. At the place of distribution of feed to the first animal, the operator presses the "start" button 4 on the control panel 3 and, due to the power supply from the battery 7, the electric drive starts the rotation of the dispenser spiral, dispensing feed to the animal. When the spiral 9 rotates, the revolution counter 1 transmits data to the microprocessor on the number of completed revolutions. Upon reaching the specified number, the microprocessor sends a signal to stop the dispenser electric drive, while the spiral 9 stops in such a way that its last coil is in the lower position at the end of the casing 8. After this, the operator moves the distributor to the next animal and presses the "start" button again. Thereby a signal is sent to the microprocessor that the distributor is located at the second animal and it is necessary to make another number of revolutions of the spiral 9 in order to dispense the appropriate portion of the compound feed. At the end of the feed distribution to the second animal, the operator moves the dispenser to the third and subsequent animals, and the process is repeated. In case of emptying hopper 6, it is filled again and the distribution process continues from the animal on which it was interrupted.

The tests of the dispenser were carried out as prescribed by the procedure guidelines according to the standard technique of the STO AIST 19.2-2008 (Industry Standard) [12, 13].

General and main technical specifications of the tested dispenser are represented in Table 1.

Table 1. Technical specification of the mixed fodder dispenser with a spiral-screwed dosing unit.

No.	Parameter	Measuring Unit	Value
1	Power of Electric Motor	W	50
2	Spiral Rotating Frequency	min^{-1}	41-45
3	Spiral Diameter	mm	97
4	Spiral Pitch	mm	73
5	Distribution on: - bulk bran - granulated mixed fodder	kg/min	 7.2 13.9
6	Tanker Capacity	m^3	0.14
7	Carrying capacity, max	kg	90
8	Mixed Fodder Distribution Height	m	0.41
9	Overall Dimensions, max - length - width - height	m	 1.3 0.75 1.1
10	Weight	kg	63
11	Battery Capacity	A-h	55

The amount of mixed fodder required for animals is determined as follows:

$$Q_{day} = m_{day} \cdot q_i, \quad (1)$$

where m_{day} is daily milk yield per cow; q_i is normal rate of mixed fodder distribution.

The normal rate of fodder distribution in turn depends on the period of lactation [1, 2].

The amount of one-time distribution of mixed fodder is defined by the following formula:

$$Q_{one} = Q_{day} / k_d, \quad (2)$$

where k_d is multiplicity of distribution.

To distribute the estimated amount of mixed fodder, the spiral shall make the following number of turns:

$$n_{turn} = Q_{one} / q_{turn}, \quad (3)$$

where q_{turn} is portion mass dispensed per one spiral turn.

Value q_{turn} can be determined by means of the graph in Fig. 2, where its values are given for various spirals used in the tested spiral-screwed dosing unit.

More correct mass of one portion can be determined experimentally by giving, for instance, 10 portions into a receiving container and the weighting it. Dividing this mass by 10, we get the mass of one dose. This check and calibration must be carried out for each new batch of mixed fodder as its properties change.

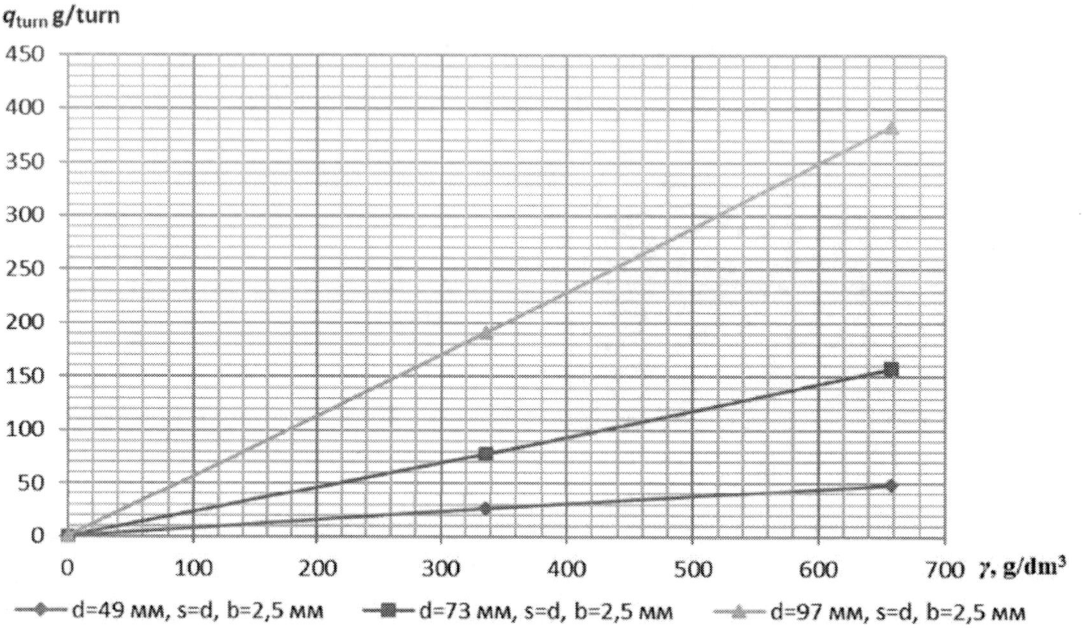

Fig. 2. Dependency graph of portion mass dispensed per one spiral turn (q_{turn}) depending on γ as the bulk density of mixed fodder

In the end of determining the amount of fodder required for each animal and a dose mass given out per one spiral turn, the distribution of mixed fodder is carried out to the entire range of animals with the preset multiplicity and interval. The volume of dispensed portions is weighted and mass values are entered in the test log. During the tests, the voltage value and current drain are recorded to determine the power consumption of spiral-screw dosing unit as well as the time of test start and its end to calculate the productivity of fodder dispenser.

Distribution of mixed fodder is carried out as follows. Fodder dispensers stop in front of the first feeding station and the operator switches on the dosing unit. When the intended amount of fodder is dispensed, as it is shown on the turn counter, the dosing unit is switched off and automatically set up to its initial position. After that, the dispenser moves to the next animal and the procedure is repeated.

3 Results and discussion

In-process tests of the fodder dispenser with a dosing unit were carried out at the "Rus" Lobanovsk dairy farm of the Perm District, where mixed fodders were distributed to heifers. Since there are no dairy cows, the distribution rate for all animals was the same. Taking into account the assigned distribution rate, the spiral with the diameter of d = 97 mm and the pitch of s = 0.75 d = 72.75 mm was installed in the fodder dispenser. The gap between the spiral and enclosure was b = 5 mm.

The following tests were carried out at the "Ocherskoe" commercial dairy farm (Fig. 3), where the one range of animals consisted of 42 cows was served with 42 portions of bulk or granulated mixed fodder per day.

 https://doi.org/10.1051/bioconf/20225200037

Fig. 3. Process of metered mixed fodder distribution with a dosing unit

The results of tests are shown in Table 2.

Table 2. Results of in-process tests of the mixed fodder dispenser with a dosing unit

No.	Bulk Mixed Fodder			Granulated Mixed Fodder		
	X_i, g	$X - X_{av}$, g	$(X - X_{av})^2$	X_i, g	$X - X_{av}$	$(X - X_{av})^2$
1	566	-5.8	33.64	544	-17.6	309.76
2	573	1.2	1.44	570	8.4	70.56
3	560	-11.8	139.24	572	10.4	108.16
4	584	12.2	148.84	544	-17.6	309.76
5	576	4.2	17.64	572	10.4	108.16
6	552	-19.8	392.04	556	-5.6	31.36
7	570	-1.8	3.24	572	10.4	108.16
8	602	30.2	912.04	560	-1.6	2.56
9	604	32.2	1036.84	570	8.4	70.56
10	584	12.2	148.84	562	0.4	0.16
11	594	22.2	492.84	580	18.4	338.56
12	571	-0.8	0.64	574	12.4	153.76
13	554	-17.8	316.84	570	8.4	70.56
14	566	-5.8	33.64	574	12.4	153.76
15	570	-1.8	3.24	548	-13.6	184.96
16	560	-11.8	139.24	562	0.4	0.16
17	552	-19.8	392.04	540	-21.6	466.56
18	554	-17.8	316.84	548	-13.6	184.96
19	568	-3.8	14.44	556	-5.6	31.36
20	574	2.2	4.84	572	10.4	108.16
21	576	4.2	17.64	552	-9.6	92.16
22	566	-5.8	33.64	550	-11.6	134.56
23	578	6.2	38.44	568	6.4	40.96
24	600	28.2	795.24	550	-11.6	134.56
25	570	-1.8	3.24	564	2.4	5.76
26	574	2.2	4.84	562	0.4	0.16
27	564	-7.8	60.84	570	8.4	70.56
28	560	-11.8	139.24	568	6.4	40.96
29	570	-1.8	3.24	554	-7.6	57.76
30	562	-9.8	96.04	564	2.4	5.76
Average	**571.8**	**0.00**	**14.070**	**561.6**	**0.00**	**10.820**
Standard Deviation	0.00			0.00		
Variation Coefficient	2.46			1.93		
Minimal Dose	552			540		
Maximal Dose	604			580		
Deviation	**±23.91**			**± 18.38**		
% of Inequality	**4.18**			**3.27**		

4 Conclusion

As a result of in-process tests using the mixed fodder dispenser with a spiral-screw dosing unit, we established the following.

1. Distribution inequality reconciles with the livestock requirements (±5%) and is equal to 4.18% with the bulk mixed fodder distribution, and 3.27% with the granulated mixed fodder distribution, whereas the amount of the portion accounted for 571.8 and 561.6 g, respectively.

2. The productivity of the dispenser is 3.11 kg/min with the dose mass of 552...604 g taking into account the time for its movement, and the power consumption is within the range of 45...47 W·h.

References

1. V.N. Vinogradov, M.P Kirilov, S.V. Kumarin, Modern approaches to the use of concentrated feed, Zootechny, **6**, 10-11 (2002)
2. N.A. Morozkov, S.V. Tretiakov, V.A. Voloshin, *The system of full-fledged feeding of Black-and-White cattle at milk production complexes, which ensures an increase in milk productivity and an improvement in milk quality* (Perm, 2015)
3. 7. L.P. Zaripova, F.S. Gibadullina, Sh. K. Shakirov et al., *Forage of the Republic of Tatarstan: composition, nutritional value and use* (Foliant, Kazan, 2010)
4. 8. W. Brade, The stature of Holstein cows – a critical review, Tieraerztliche Umschau, **73(3)**, 74-84 (2017)
5. 14. B. Laurence, T. Pablo, R. Jacques, Y. Chilliarda, Animal Feed Sci. and Technol., **222**, 31–42 (2016)
6. P.S. Zolotarev Spiral-screw conveyor for bulk materials, Machinery and equipment for rural areas, **12**, 25-26 (2009)
7. *Cart of auto fodder distribution for cows.* Retrieved from: https://agrovektor.ru/physical_product/510514-telezhka-avtomaticheskoy-razdachi-kormov-dlya-korov.html (Access date: 14.04.2022)
8. *Alicatt E 301* Retrieved from: http://www.alicatt.fr/e301.html (Access date: 14.04.2022)
9. Y. Shimizu, P.A. Cundall, Three-dimensional DEM simulations of bulk handling by screw conveyors, Journal of Engineering Mechanics, **9(127)**, 864-872 (2001)
10. E.A. Lyalin, M.A. Trutnev, Improving the accuracy of dosing concentrated feeds with a spiral-screw dispenser, Rural mechanizer, **1**, 26-27, 29 (2018)
11. E.A. Lyalin, M.A. Trutnev, Discrete method of dosing free-flowing concentrated feed with spiral-screwed feeder, COMETa 2018 "Conference on Mechanical Engineering Technologies and Applications" Proceedings, 401-406 (2018)
12. V.I. Pakhomov, M.A. Tischenko, S.V. Bragintsev, et al. *Engineering and calculation of processing line for preparation and distribution of graded feed mixtures by multi-function devices*. (Resselkhozacademia Zernograd, 2013)
13. STO AIST 19.2-2008 *Agricultural Machinery. Machines and equipment for fodder preparation. Procedure for determining the function performances*. (Minselkhozprod, Moscow, 2010)

Postfire dynamics of standing dead tree stock in northern boreal forests

Liudmila Mukhortova [1,*], *Leonid* Krivobokov [1], and *Dmitry* Schepaschenko [2,1]

[1] V. N. Sukachev Institute of Forest Siberian Branch of Russian Academy of Sciences, FIC KSC SB RAS, 50/28, Akademgorodok, Krasnoyarsk, 660036, Russian Federation
[2] International Institute for Applied Systems Analysis (IIASA), 1, Schlossplatz, Laxenburg, A-2361, Austria

Abstract. Wildfire is one of the main forest disturbing factors in the boreal zone of Siberia that can cause significant changes in tree stands dynamics. Tree mortality caused by fire can significantly increase a standing dead tree pool that is one of the poorly studied components of forest ecosystems. The aim of this study was assessing of post-fire changes in the standing dead tree pool in northern boreal larch forests of Central Siberia (Russia). We analyzed dynamics of the standing dead tree stock on experimental plots, which were affected by wildfire of moderate severity in 2013. The stock of standing dead trees was measured on these plots before and 1, 2, and 7 years after the fire. It was found that about half of the pre-fire standing dead trees fall down during the first year after the fire. At the same time, tree mortality caused by the fire significantly contributed to the total standing dead tree stock in these ecosystems. Our study showed that a significant part of the pre-fire standing dead trees and trees killed by fire can remain standing after the moderate severity fire. This standing dead wood conserves carbon for a long time.

1 Introduction

Forests cover more than 50% of the territory of Russia [1]. Assimilating a huge amount of atmospheric carbon dioxide they play a significant role in the stabilization of the Earth climate system [2, 3]. A significant part of the carbon assimilated by trees is sequestered in wood of the tree trunks. Tree stand growth and development usually are accompanied by partial tree mortality due to natural reasons or disturbances. Standing dead trees (snags) are an important component of the forest ecosystem providing wildlife habitat [4-7], serving as a long-term nutrient and carbon store, a contributor to long-term soil development [8, 9], and are essential for post-disturbance forest recovery [10].

In the natural undisturbed ecosystems, the stock of dead standing trees does not exceed 12-15% of the total woody biomass [11]. However, an impact of some exogenous factors such as wildfires, pests, drought and pollution can kill a significant part of the tree stand sufficiently increasing the stock of snags. As it was shown in many studies wildfires create a large pool of dead trees that increase future fuel loads, influence fire behaviour [12-15].

Stand replacing fires occur on 1.5 to 5.0 million ha of the forests on the territory of Russia [16].

Harvesting recently died trees (known as a salvage logging) may reduce fire risk and future fire disturbances [17-19]. This type of management practice is extensively implemented worldwide [6, 20], however in the northern regions of Siberia with extremely low population and limited accessibility, such salvage logging is problematic to implement.

An expected climate change can further increase spread and intensity of wildfire [21-23] so that fire and forest managers have prominent management concerns about the dead trees loading under changing climate [21, 24]. Management of post-fire vegetation is critical to avoid fuel accumulation and cascade fire events [24].

Fire-killed wood can be extracted and used in the southern part of Siberia, where the infrastructure is available. A snags stock in the low accessible northern regions is difficult to assess and use. It means that in the north the dead wood will be included in the natural biological turnover.

Lack of the data on "lifespan" of snags in forest ecosystems of Siberia and fire impact on snag stock dynamics leads to high uncertainties in estimates of carbon budget in forest ecosystems and their ecosystem services. In the global perspective estimates of snag carbon, the stock is critical for countries which report greenhouse gas emissions in the frame of the United Nations Framework Convention on Climate Change (UNFCCC) [25]. Woody detritus is one of five main carbon pools in forest ecosystems, monitoring which are required to report greenhouse gas emissions from the forest sector [26].

* Corresponding author: l.mukhortova@gmail.com

The temporal dynamics of woody debris during forest development and the response of this pool to the fire are also important for forest management and assessment [27, 28]. However, a few studies have examined the dynamics of coarse woody debris in northern boreal larch forests of Central Siberia.

The goal of this study was to improve our understanding of natural post-fire snag dynamics. Our objectives were to (1) describe patterns of snag dynamics following wildfires, (2) test the fire effect on pre-fire stock of snags, and (3) assess a possible contribution of snags to the total tree-stand C budget in larch (*Larix gmelinii*) forests of the northern boreal subzone of Central Siberia – region with a high fire frequency regime [29].

2 Materials and methods

2.1 Study area

The study was carried out in Central Evenkia (Evenk district of Krasnoyrsk territory (Central Siberia, Russia)) in the northern taiga subzone.

Sample plots were established near the Tura settlement (64° N, 100° E). This territory is a continuous permafrost zone. The climate is continental, moderately wet. Mean annual temperature is -8.9°C. The mean temperature of January is -36°C, in June +16°C, leading to the annual temperature range of 52°C. Accumulated degree days above +10°C are about 1000 °C. Mean annual precipitation is 370 mm. The distribution of precipitation by seasons is relatively even, height of the snow cover is 50–60 cm. A vegetation period is lasting 70–80 days [30]. The main forest forming tree species in northern taiga is larch (*Larix gmelinii* (Rupr.) Rupr.), which occupy 84% of the forested area. Low shrubs-greenmosses larch forests and sparse larch forests prevail on all locations [30].

The study area historically characterized a mixed-severity fire regime with mean fire-return intervals approximately 80 years [29].

2.2 Sample plots

In July of 2013 we established 3 sample plots on the southern slope. All live and dead trees taller than breast height (1.3 m) were measured and identified to species within these plots. Each of the dead standing trees was marked. An age of the tree stands on sample plots covered a range from 200 to 250 years. Initial tree stand characteristics on the sample plots are shown in Table 1.

In August of 2013 all three sample plots were damaged by the fire of moderate severity. In 2014 we accounted the amount of pre-fire snags that remained standing. The similar measurement of marked snags was made two and seven years after the fire. Seven years after the fire additionally all live and dead trees on the sample plots were measured. In 2013 and in 2020 for each dead tree we recorded their status. We used four stage classification of dead standing trees, which is the similar to stages from 3 to 6 in snag succession series according to Thomas et al. [31]:

1 stage – dead trees with bark, which kept all the twigs, including last (finest) order;

2 stage – trunks have partially kept first order branches;

3 stage – branches are completely lost, bark can be kept on the trunk or can be flaked;

4 stage – branches are lost, trunk is broken at the height more than 1.3 m, bark can be kept or absent.

Described stages can characterize the relative "age" of snags – time passed since the tree death, because thin twigs keep only on the recently died trees. After the tree death, twigs become fragile due to drying and easily break from the contact with twigs of surrounding trees, by wind or snow. Breaking the first order branches requires longer time.

2.3 Measurements of living trees and snags

All live trees and snags were measured on each sample plot. The diameter at breast height (DBH) and height of all trees above 2.5 cm DBH was measured. Based on the measured data, we estimated growth stock volume and stock volume of snags (m^3 ha^{-1}).

The Microsoft Excel was used to perform statistical analysis. We applied the "t-Test: Two-Sample Assuming Unequal Variances" from the Data Analysis ToolPak to estimate the significance of differences. The statistical significance of the results was verified at the significance level of 95%. Mean values in the text are given together with confidential interval (± standard deviation).

3 Results and discussion

The measured pre-fire live tree density on the studied sample plots ranged from 4267 to 5389 trees ha^{-1} with the total living growing stock from 123 to 200 m^3 ha^{-1}. The snag density was 1620-3945 trees per ha that is 25-42% of the total tree number on the sample plots. However, the stock of snags did not exceed 4-16% of the total tree stand stock because most of the snags had smaller size than living trees did. The average DBH of the standing dead trees was almost twice smaller and average height was 2-3 m lower compared to living trees. It means that the most part of snags was a result of natural tree mortality during tree stand development due to competition and genetics particularities.

Table 1. The main characteristics of tree stands on the sample plots.

Sample plot ID	Tree species	Living trees			Dead trees		
		Density, trees ha^{-1}	DBH, cm	Height, m	Density, trees ha^{-1}	DBH, cm	Height, m
SP1	*Larix gmelinii*	5389	8±3	8±3	3945	4±2	5±2
SP2	*Larix gmelinii*	4714	6±2	7±3	1620	3±1	5±2
SP3	*Larix gmelinii*	4267	7±3	9±4	2400	4±1	6±2

Inventory made 7 years after the fire showed that the range of post-fire mortality across the plots varied from 75 to 97%, so that the living growing stock decreased up to 3-25% of pre-fire stock. An amount of the snags on the sample plots increased by 35-125% and the snag's stock increased 4.5-9.5 fold due to the death of trees from the top forest layer, which have a higher volume than pre-fire snags do. As a result, 7 years after the fire, living trees contributed only 3-34% to the total tree stand stock in contrast to 84-96% in the pre-fire stands (Fig. 1).

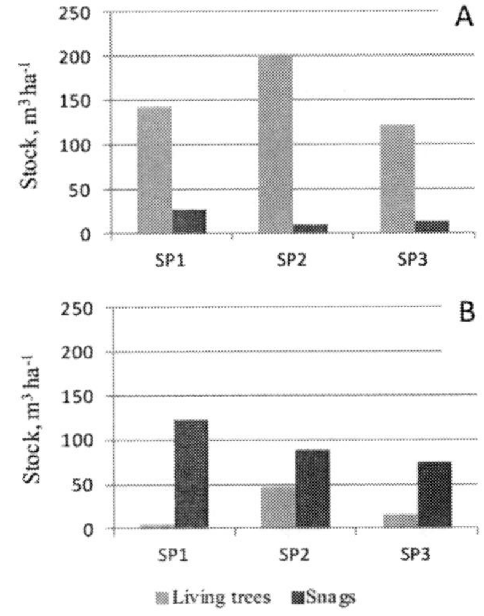

Fig. 1. Ratio between growing stock and snags volume on the sample plots before (A) and 7 years after the fire (B): SP1-SP3 – sample plots according to Table 1.

The comparison of the average size of remaining living trees with the pre-fire one indicated that only the largest trees could survive after the fire of moderate severity. The average DBH of the remaining living trees ranged from 9 to 12 cm, and their average height ranged from 9 to 16 m that is higher than the average parameters of the pre-fire tree stands (6-8 cm and 7-9 m, respectively). Preferred surviving of the larger trees was reported in other studies in different regions (e.g., [32]). The larger trees typically have thicker bark and taller crown-base heights that prevent tree from the damage by flame. In contrast, Dunn et al. [33] observed a positive correlation between fire severity and average snag density and height, so larger trees were killed. This supports the assumption that size-dependent mortality may vary among species of different fire tolerance [32].

Significant changes were observed in the structure of the snag stock. If before the fire most of the snags belonged to the 3rd stage (39.5-65.7%) (Fig.3A), the 7 years after the fire about 93-98% of snags was dead trees of the 1st stage (Fig.2B). This indicates that 7 years is too short for snags to lose twigs and branches (to shift the snags from the 1st to the 2nd stage). So organic matter (carbon) accumulated in these twigs is preserved from the fast decomposition at least for 7 years after fire and these fine woody debris do not supply to the ground fuel load.

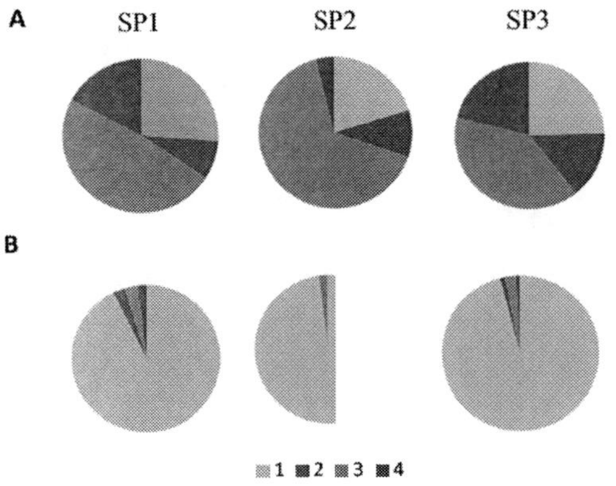

Fig. 2. Ratio between snag stages on the sample plots before (A) and 7 years after the fire (B): SP1-SP3 – sample plots according to Table 1.

The total stock of the tree stands decreased by 25-35% during 7 years after the fire. It means that the most part of the trees killed by fire of moderate intensity kept standing as snags and continue to hold carbon accumulated in the wood.

The stock of snags in the larch forests of northern taiga observed in our study was in the same range or exceeded one reported earlier for postfire larch forests in the same region [30, 34]. The stock of standing dead trees was reported in the range of 12.5-19.9 m^3 ha^{-1} in the tree stands of 178-290 years old [30]. At the same time, obtained in our study contribution of the snags to the total standing volume was similar to that reported for postfire ecosystems. The stock of snags reached 20-41% of total standing volume in the post-fire forests [30].

To assess the significance of snags as carbon depot the estimation of changes in pre-fire stock of snags is very important especially for the forests in regions with high fire frequency regime. It was found that one year after the fire about 50% of the pre-fire snags were kept as standing (Fig.3A). During the second year after the fire, the pre-fire snags lost from 6 to 14%. During the following 5 years, the rate of the snag fall decreased and averaged 3.9% per year (from 1.5 to 6.1%).

The stock of the pre-fire snags decreased by 30-63% during the first year after the fire, by 10-28% during the second year and during the following 5 years the average rate of the snag stock loss was 5.6% (from 4.2 to 7.0 %) per year (Fig. 3B). The similar trajectory of snag density and snag stock losses indicates unselective fall of the pre-fire snags.

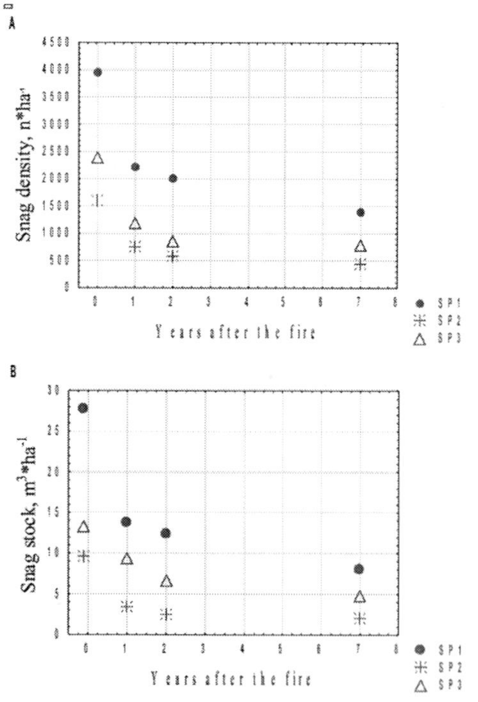

Fig. 3. Dynamics of the pre-fire snags density (A) and stock (B) on the sample plots: SP1-SP3 – sample plots according to Table 1.

Totally, during the fire and two years following the fire the snag stock lost on average 59±11% of the pre-fire stock. This was fall down of the most damaged by fire trees independently of the tree size. During the following 5 years, the rate of snags loss decreased to 5.6±0.56% per year. It means that remaining 40% of pre-fire snag stock can remain standing for a long time after the fire providing carbon deposition.

It is possible that our observation period (7 years) is too short to observe the maximal post-fire tree death and snag fall. Dunn et al. [33] reported that in ponderosa pine (*Pinus ponderosa* subsp. ponderosa) and dry-mixed conifer forests in USA the rapid drop in snag density began around 10 years post-fire and lasted next 10 to 15 years depending on fire severity. However, the snag fall dynamics observed in our study indicated that the rate of the pre-fire snag loss significantly decreased to the seventh year after the fire (Fig. 3). The main reason of the faster loss of damaged snags in northern taiga larch forests can be the smaller tree size and shallow root system.

Tree mortality and increasing stock of snags and downed dead wood can significantly alter carbon balance in such post-fire forests due to changes in the ratio between carbon assimilation and carbon release during wood decomposition, because the production part (living tree density) decreased more than 23-fold with simultaneous significant dead wood increase.

A low rate of snag wood decomposition in northern boreal forests provide long-term deposition of carbon accumulated in standing dead trees. The decomposition rate of downed dead wood in this region is 1.03-1.52% per year [35]. It means that estimated carbon flux from decomposition of dead trees can range from 49 to 200 kg C ha^{-1} per year that is about 13% of possible NPP in undisturbed larch forests in this region [36] or 63±22% NPP of the living trees remaining after fire.

Recently burned forest can be harvested and provide timber for wood-based products, while snags are suitable for bioenergy for relatively long period of time after the fire event. Harvesting damaged forest stands reduces risk of consequent wildfire, but this management option is infrastructure dependent.

4 Conclusions

There are four main conclusions can be made based on this study:

1. Wildfires can cause widespread tree mortality in northern boreal larch forests of Central Siberia. However, the most part of the killed trees remain standing contributing to the snags stock.
2. About half of the pre-fire stock of snags falls during the first year following the fire. The rate of these snag loss significantly decreases following years that allow remaining pre-fire snags serve as a carbon depot for the long time.
3. Carbon flux from dead wood decomposition probably does not exceed the rate of carbon assimilation even in the forest strongly disturbed by fire.
4. Low rate of snag's wood decomposition can provide relative carbon sequestration for the fire-damaged forest. However, on the other hand, the long-

term standing of the dead trees can increase coarse woody fuels loadings for these forests.

Results of this study can be used for assessment of carbon budget in post-fire forest ecosystems in northern boreal larch forests. Additionally, understanding the fuel trajectory in fire-killed forests is essential for forest managers.

The research (data processing and analyzing) was funded by RFBR, Krasnoyarsk Territory and Krasnoyarsk Regional Fund of Science, project number 20-44-240008, sample plots were established in the frame of State Assignment of V.N. Sukachev Institute of Forest SB RAS No 0287-2021-0008 (state registration number No. 121031500339-0).

References

1. FAO. *Global Forest Resources Assessment. Report. Russian Federation* (Rome, 2020)
2. Y. Pan, R.A. Birdsey, J. Fang et al., Science **333,** 988-993 (2011)
3. C. Le Quere, G.P. Peters, R.J. Andres et al., Earth Syst. Sci. Data Discuss **6,** 689-760 (2013)
4. M. E. Swanson, J. F. Franklin, R. L. Beschta, C. M. Crisafulli, D. A. DellaSala, R. L. Hutto, D. B. Lindenmayer, F. J. Swanson, Front. Ecol. Evniron. **9 (2),** 1-10 (2010)
5. C. J. Dunn, J. D. Bailey, For. Ecol. Manage. **276,** 71-81 (2012)
6. J. Castro, G. Moreno-Rueda, J. Hydar, Conserv. Biol. **24,** 810-819 (2010)
7. M. W. Ritchie, E. E. Knapp, C. N. Skinner, J. Forest. **287,** 113-122 (2013)
8. M. E. Harmon, et al., Adv. Ecol. Res. **15,** 133-302 (1986)
9. D. B. Tinker, D. H. Knight, Ecosystems **3,** 472-483 (2000)
10. D. B. Lindenmayer, P. J. Burton, J. F. Franklin, *Salvage logging and its ecological consequences.* (Island Press, Washington, D.C., USA, 2008)
11. L. V. Mukhortova, L .V. Krivobokov, D. G. Schepaschenko, A. A. Knorre, D. S. Sobachkin, *IOP Conference Series: Earth and Environmental Science* **875 (1),** 012059 (2021)
12. M. D. Passovoy, P. Z. Fulé, For. Ecol. Manage. **223 (1-3),** 237-246 (2006)
13. J. R. Thompson, T. A. Spies, L. M. Ganio, *Proc. Natl. Acad. Sci.* **104 (25),** 10743-10748 (2007)
14. D. Kulakowski, T. T. Veblen, Ecology **88 (3),** 759-769 (2007)
15. P. G. Monsanto, J. K. Agee, For. Ecol. Manage. **255 (12),** 3952-3961 (2008)
16. A. Z. Shvidenko, D. G. Shchepashchenko, E. A. Vaganov, A. I. Sukhinin, S. Maksyutov, I. McCallum, I. P. Lakyda, *Doklady Earth Sciences* **441 (2),** 1678-1682 (2011)
17. L. F. DeBano, D. G. Neary, P. F. Ffolliott, *Fire Effects on Ecosystems* (John Wiley & Sons, 1998)
18. D. W. Peterson, E. K. Dodson, R.J. Harrod, For. Ecol. Manage. **338,** 84-91(2015)
19. M. C. Johnson, M. C. Kennedy, S. C. Harrison, D. Churchill, J. Pass, P. W. Fischer, For. Ecol. Manage. **470,** 118190 (2020)
20. S. Thorn, C. Bässler, R. Brandl, P. J. Burton, R. Cahall, J. L. Campbell, J. Castro, C. Y. Choi, T. Cobb, D. C. Donato, E. Durska, J. Applied Ecol. **55 (1),** 279-289 (2018)
21. J. S. Littell, D. McKenzie, D. L. Peterson, A. L. Westerling, Ecol. Applications, **19 (4),** 1003-1021 (2009)
22. J. S. Littell, D. McKenzie, H. Y. Wan, S. A. Cushman, Earth's Future **6 (8),** 1097-1111(2018)
23. P. E. Dennison, S. C. Brewer, J. D. Arnold, M. A. Moritz, Geophys. Research Lett. **41 (8),** 2928-2933 (2014)
24. M. Coppoletta, K. E. Merriam, B. M. Collins, Ecol. applications **26 (3),** 686-699 (2016)
25. M. B. Russell, S. Fraver, T. Aakala, J.H. Gove, C. W. Woodall, A. W. D'Amato, M. J. Ducey, For. Ecol. Manage. **350,** 107-128 (2015)
26. U. Söderberg, S. Wulff, G. Ståhl, Scandinavian J. For. Res., **29 (3),** 252-258 (2014)
27. C. Woodall, B. Walters, S. Oswalt, G. Domke, C. Toney, A. Gray, For. Ecol. Manag. **305,** 48-59 (2013)
28. M. Garbarino, R. Marzano, J. D. Shaw, J. N. Long, Ecosphere **6 (3),** 1-24 (2015)
29. V. I. Kharuk, M. L. Dvinskaya, I. A. Petrov, S. T. Im, K. J. Ranson, Reg Environ Change **16,** 2389-2397 (2016)
30. *Permafrost Ecosystems: Siberian Larch Forests* pp 99-122 (Ecological Studies; Springer: Dordrecht, The Netherlands, 2010)
31. J. W. Thomas, R. G. Anderson, C. Maser, E. L. Bull, Agric. Handb **553,** 60-77 (1979)
32. R. T. Belote, A. J. Larson, M.S. Dietz, For. Ecol. Manage. **353,** 221-231 (2015)

33. C. J. Dunn, C. D. O'Connor, M. J. Reilly, D. E. Calkin, M. P. Thompson, For. Ecol. Manage. **441,** 202–214 (2019)
34. E. Köster, K. Köster, F. Berninger, A. Prokushkin, H. Aaltonen, X. Zhou, J. Pumpanen, J. Environ. Manage. **228,** 405–415 (2018)
35. L. Mukhortova, A. Kirdyanov, S. Evgrafova, O. Shapchenkova, L. Krivobokov, *Climate and permafrost ecosystems: Proceedings of IX International Symposium "C/H2O/energy balance and climate over the boreal and Arctic regions with special emphasis on Eastern Eurasia"* (Yakutsk, Russia. Nagoya: Nagoya University Publishing House, 2016)
36. T. Kajimoto, Y. Matsuura, M. A. Sofronov, A. V. Volokitina, S. Mori, A. Osawa, A. P. Abaimov, Tree Physiology, **19 (12),** 815-822 (1999)

 https://doi.org/10.1051/bioconf/20225200039

Effect of colistin sulfate in the health status and bacteriological gut flora in store pigs with nonspecific gastroenteritis

Sergei Engashev[1], *Aleksei* Savinkov[2], *Konstantin* Sadov[2], and *Pavel* Iliasov[3]

[1] K.I. Skryabin Moscow State Academy of Veterinary Medicine and Biotechnology (MVA) FSBEI HE, Moscow, Russia
[2] Samara State Agrarian University FSBEI HE, 2 Uchebnaya St., Kinelsky urban locality, Samara region, Russia
[3] Samara State Medical University FSBEI HE, Ministry of Health of the Russian Federation, Samara, Russia

Abstract. Piglets with the nonspecific gastroenteritis received Colistin 12000000 with drinking water for 5 days is the doses 0.6 g/100 kg in Group 1 and 0.8 g/100 kg in Group 2. This provided the complete therapeutic efficacy (the clinical recovery) on Day 4 in both treatment groups. The dynamics of the erythrocytes and leukocytes suggested the recovery of the morphofunctional homeostasis after the disease in both treatment groups. After the treatment with the Colistin 12000000, the prevalence of the *Streptococcus gallolyticus* reduced by 8.3% in the Group 1. *Klebsiella pneumoniae* prevalence reduced by 33.3%. In Group 2, the prevalence of *Streptococcus gallolyticus* reduced by 41.7% after the 5-day therapy and was 25.0% at the end of the study. *Klebsiella pneumoniae* prevalence reduced twice, and it was ultimately observed in 16.7% of cases. Other bacteria presented in the washings at the baseline in both groups were no more observed at the end of the treatment. The symbiotic flora was found in both groups at the end of the experiment in a sufficient number. Thus, the efficient dose of the test drug was 0.6 g/100 kg provided the desirable clinical effect comparable to the higher dosing regimen. The drug well tolerated, with no side effects or adverse events reported throughout the study.

1 Introduction

Non-infectious diseases related to the decrease in natural resistance of animals result in significant economic losses in animal production. Such diseases can be up to 90% of all animal disease cases in the farm [5, 6]. Therewith, gastrointestinal pathologies account for about 40% of the non-infectious diseases in terms of the incidence, general spread and significance of the losses. In most of cases, they are caused by the failures related to the animal feeding and housing [2, 3].

Currently, pig industry is on the top of the most efficient segments of the agrifood industry in the terms of the meat production. However, it is characterized by the highest incidence of mass gastrointestinal pathologies due to the use of the intensive hog growing technologies. Dyspepsia and gastroenteritis are the most common gastrointestinal diseases in the store pigs. It is of the importance that the gastrointestinal diseases of various etiology cause up to 70% of mortality in the pig farms [7]. They are extremely dangerous for piglets who frequently develop severe gastrointestinal diseases related to the decrease in the systemic immune reaction and caused by the external factors, including but not limited to bacteria [8, 9]. The advanced cases include complications reflected in the dystrophic and degenerative changes in the tissues and organs [10].

Since the most gastrointestinal pathologies are underlain by the changes in the gut flora, the antibacterial drugs remain the most efficient therapies for such diseases.

Currently, there are extensive investigations aimed at the search for the antibacterial drugs efficient in the infectious inflammatory gastrointestinal diseases, and the antibiotics based on the colistin sulfate are contemplated in this regard [1]. Colistin (polymyxin E) is a cyclic polypeptide antibiotic produced by certain *Paenibacillus polymyxa* strains. This polymyxin antibiotic has a bactericidal effect against most resting and dividing Gram negative bacteria due to the changes in the structure and the function of their surface and plasma membrane. It is used in the calves, goats, poultry, sheep and pigs for the gastrointestinal infections caused by the colistin-susceptible bacteria such as *E. coli*, *Hemophilus*, and *Salmonella* [3].

However, understanding the efficacy of such drugs requires a sufficient level of the process and clinical testing. Considering the above, Agrovetzashchita NVC LLC suggests using the Colistin 12000000 in the bacterial gastrointestinal diseases in the pigs.

This study is aimed at the investigation of the efficacy of the Colistin 12000000 in the gastrointestinal pathologies in the store pigs.

This requires addressing the following task: to determine the efficacy of the Colistin 12000000 in doses 0.6 g and 0.8 g per 100 kg (5-day therapy) in the bacterial gastrointestinal infections in the store pigs.

2 Materials and methods

The clinical trial was conducted in 35-45-day piglets at the pig holding. Two treatment groups were formed, wherein each group included twelve piglets with a gastrointestinal disorder and diarrheal syndrome.

Colistin 12000000 was administered orally with water for 5 days in the dose 0.6 g per 100 kg in Group 1 and in the dose 0.8 g per 100 kg in Group 2.

During the study, the animals were housed in separate cages. The diet was the same in all groups and corresponded to the normal diet for this animal species and age.

The efficacy of Colistin 12000000 was estimated based on the clinical signs of the gastrointestinal pathology at the end of the experiment as confirmed by the bacteriological analysis and blood counts.

The bacteriological samples were taken from the rectal mucosa with a sterile swab. The blood was drawn from the vena cava cranialis with a disposable sterile vacuum system into the anticoagulant tubes and clot activator tubes. Blood and nasal washings were taken on the Screening Day 1 before the initiation of the treatment and five days after the end of the treatment.

Bacteriological analysis of the rectal swabs was performed by the primary inoculation of 5% blood agar and universal chromogenic media (BioRad). The cultures were incubated for 2 days at 37 °C. All grown microorganisms were identified using the MALDI-ToF mass spectrometry on Microflex LT (Bruker®) by the direct application.

The blood counts considered the RBC, the WBC, the eosinophils, the basophils, the lymphocytes, the monocytes, the hemoglobin level, and the erythrocyte sedimentation rate (ESR). The blood counts were performed on Mindray BC-5300 hematological analyzer (P.R.China).

Digital experimental data were processed using the descriptive statistics to test the significance of the differences in the parameters compared with Student's t-test commonly used in the biology and animal science in Microsoft Excel.

3 Results and discussion

At the baseline, the general health parameters corresponded to an acute phase of the gastrointestinal disease and dehydration due to the diarrhea. The piglets showed a depressed state with poor appetites. 25.0% of animals in the Group 1 and 16.7% in the Groups 2 completely refused the feed. 75.0% and 91.7% of animals, respectively, had an unsatisfactory appearance. The motor activity was reduced in all animals. 100% of cases showed a diarrheal syndrome. The bowel evacuation was accompanied by the tenesmus and a characteristic sitting posture. The stool was liquid, foul-smelling, with a feed and mucus impure. The body temperature was within the age-related physiological normal range. Some individuals had the temperature above 40°C.

The bacterial composition of the rectal washings from the sick piglets suggested that the disease was caused by a complex of the opportunistic microorganisms due to the general sickness related to the negative effect of a number of the process parameters.

After recording the clinical parameters and taking the biomaterial for the laboratory tests, the animals were orally administered with the antibacterial drug and water according the study regimen by the groups.

As early as on Treatment Day 1, no one animal completely refused the feed, and most animals had shown the satisfactory appetites. The motor activity increased in 83.3% of the piglets in both groups, 66.7% of the animals in the Group 1 and 75% in the Groups 2 had a satisfactory appearance. 75.0% and 83.3% of the piglets, respectively, still had the diarrhea.

On Treatment Day 2, the piglets demonstrated 4.2% decrease in their body temperature ($P<0.001$). During the rest of the treatment period, the temperature reduced by 1-2.3% vs. the baseline ($P\leq0.05$ or $P\leq0.01$).

On Treatment Day 4, 100.0% of the piglets had no diarrhea, showed good appetites, a satisfactory appearance, and a good motor activity. During the follow-up, all the parameters corresponded to the physiological normal ranges. No diarrhea signs were noted.

The body weight analysis is provided in Table 1. The treatment resulted in the intensive growth of the animals yielding a higher average daily gain vs. the gain anticipated for the process in this holding (0.3 kg in the week 1 post weaning, 0.38 kg in the week 2, and 0.34 kg average gain).

Table 1. Piglet body weight at the baseline and at the end of the study, kg.

Parameter	Treatment Group 1	Treatment Group 2
Baseline		
Σ n=12	102.6	139.8
x±Sx	8.6±0.19	11.7±0.41
10 days after the end of the treatment		
Σ n=12	180	241.8
x±Sx	15.0±0.47	20.2±0.76
ΔΣ	77.4	102
Δ x	6.5	8.5
ΔΣ, %	175.4	173.0
Δ x, %	175.4	173.0
Average daily gain	0.403	0.531

Thus, the gain in Group 1 and in Group 2 was 18.5% and 32.4% higher than the in-process gain, respectively

This suggests a steady trend of high average daily gain vs. the in-process values, indicating a high efficacy of the Colistin 12000000 in the terms of the interruption of the bacterial infection and the intrinsic rapid recovery of the healthy vital signs.

The blood count data obtained before and five days after the treatment are provided in Table 2.

The study has shown that the erythrocyte sedimentation rate (ESR) was within the normal range (2-9 mm/h). At the same time, this parameter reduced by the end of the study. This reduction was 8.5% in the

Group 1 and 25.1% in the Group 2. A low confidence with high values of the differences is due to the high intrinsic variability of the parameter.

The pigs are characterized by a sufficiently broad WBC range with a high upper level (10-22 × 109/L). The piglets had normal WBC values at the baseline in both groups. However, by the end of the study this parameter increased by 29.6% in the Treatment Group 1 (P≤0.001) and by 80.5% in the Treatment Group 2 (P≤0.001). As a result, the WBC was slightly above the norm in the Group 2.

Table 2. CBC dynamics.

Parameter	Group	
	Group 1	Group 2
ESR, mm/h		
Baseline	2.00±0.405	2.67±0.397
End of study	1.83±0.252	2.00±0.270
WBC, 10^9/L		
Baseline	15.96±1.663	14.05±2.348
End of study	20.34±1.257***	25.36±2.448***
RBC, 10^{12}/L		
Baseline	6.11±0.494	6.72±0.396
End of study	6.42±0.130	6.43±0.080
Hemoglobin, g/L		
Baseline	71.91±5.183	71.83±4.835
End of study	79.25±2.235***	93.03±1.024***
Hematocrit, %		
Baseline	27.09±2.527	31.97±2.333
End of study	35.61±0.988***	38.71±0.715***
Bands, %		
Baseline	1.00±0.000	1.08±0.095
End of study	0.92±0.087	2.25±0.261
Segmented, %		
Baseline	52.50±1.388	45.75±2.002
End of study	49.83±2.545	43.58±1.545
Lymphocytes, %		
Baseline	42.67±1.313	50.25±0.324
End of study	45.00±2.562***	46.58±1.732
Eosinophils, %		
Baseline	1.09±0.091	1.50±0.000
End of study	2.42±0.300	3.27±0.449***
Monocytes, %		
Baseline	2.36±0.244	1.33±0.127
End of study	1.33±0.196	1.67±0.213
Basophils, %		
Baseline	0.17±0.117	0.00±0.00
End of study	0.42±0.155	2.58±0.489

*Note: *: P≤0.05; **: P≤0.01; ***: P≤0.001 vs, control*
Note: ˇ: P≤0.05; ˇˇ: P≤0.01; ˇˇˇ: P≤0.001 vs. baseline

Note that the piglets had no significant changes in the WBC differential characteristic for an inflammatory process. In the course of the disease, the changes in the WBC differential percent values were reported. The segmented neutrophils percentage reduced in both groups vs. the baseline. The reduction was 2.67% in the Group 1 and 2.17% in the Group 2. The bands content was small in both groups. The monocyte levels were small both at the baseline and at the end of experiment in all groups. By the end of the experiment, their levels wwere similar in both groups. The lymphocytes increased by 2.33 in the Group 1 (P≤0.001), and reduced by 3.67% in Group 2. At the same time, the lymphocytes level was similar in both groups. We noted an increase in the eosinophils by 1.03% in Group 1 and by 1.77% in the Group 2 (P≤0.01). We also noted an increase in the basophils by 0.25% in Group 1 and by 2.58% in the Group 2 (P≤0.01).

RBC estimation has shown that this parameter corresponded the normal range in both groups throughout the study and varied between 6.11-6.72 × 10^{12}/L. At the same time, the baseline hemoglobin levels were significantly below the normal range. In the course of the study, it improved by 10.2% in Group 1 (P≤0.001), and by 29.5% in the Group 2 (P≤0.001). Similar changes were reported for the hematocrit. At the baseline, it was at the lower reference limit, but increased by 35.1% in the Group 1 (P≤0.001) and by 21.1% in the Group 2 (P≤0.001) by the end of the experiment.

We also registered a decrease in the ESR indicative of an improvement in the inflammatory process. The WBC rise up to the upper limit of the norm and above it with no apparent adverse changes in the WBC differential suggesting an increase in the immune reactivity. The WBC differential parameters in both groups suggest the trend to the reduction in the neutrophils, the normalization of the lymphocyte levels, increase in the eosinophils which is indicative of the decline in the acute phase processes mediated by the white blood cells.

The most significant changes were related to the erythrocytes. Rapid significant increase in the hemoglobin and hematocrit in both groups may suggest an activation of the hemopoiesis due to the elimination of the adverse effects of the infection.

Therefore, the described changes related to the erythrocytes and leukocytes suggested the recovery of the morphofunctional homeostasis after the disease in both treatment groups.

At the baseline, we estimated the susceptibility of the microorganisms isolated from the rectal washings of the piglets with the gastrointestinal pathologies to Colistin 12000000. The results are provided in the Table 3.

The choice of the piglets was based on their group affiliation, the most intensive microbial growth in their samples and the apparent colony forming effect as well as on the presence of the most dangerous pathogens.

The analysis demonstrated that the susceptibility of *E. coli* and *Pseudomonas aeruginosa* to Colistin 12000000 was 0.5 µg/mL in both groups. *Klebsiella pneumoniae* had the same susceptibility. The susceptibility of *Pasteurella aerogenes* varied within 0.5-2 µg/mL.

Table 3. Analysis of the susceptibility of the microorganisms to Colistin 12000000, µg/L.

No.	Microorganism	MIC, µg/mL
1	*E. coli*	0.5 – S*
6	*Klebsiella pneumoniae*	0.5 – S*
7	*E. coli*	0.5 – S*
15	*Pasteurella aerogenes*	0.5 – S*
16	*E. coli*	0.5 – S*

16	*Klebsiella pneumoniae*	0.5 – S*
20	*Pasteurella aerogenes*	0.5 – S*
23	*Pasteurella aerogenes*	0.5 – S*
30	*Pasteurella aerogenes*	0.5 – S*
31	*E. coli*	0.5 – S*
35	*Pasteurella aerogenes*	0.5 – S*
38	*Klebsiella pneumoniae*	0.5 – S*
44	*Pasteurella aerogenes*	2 – S*
47	*Pasteurella aerogenes*	0.5 – S*
48	*Pseudomonas aeruginosa*	0.5 – S*
58	*E. coli*	0.5 – S*
57	*E. coli*	0.5 – S*
60	*Pasteurella aerogenes*	0.5 – S*

**S; susceptible; *R: resistant*

Thus, Colistin 12000000 was efficient against both Gram positive and Gram-negative flora. All the bacteria studied were susceptible to this drug which allows its use in the various bacterial infections. Of special note is the fact that the growth of *Escherichia coli*, one of the main causative agents of the infectious inflammatory diseases, was efficiently inhibited at the drug concentration of 0.5 µg/mL.

When estimating the species composition of the microbial flora in the nasal washings, the most common microorganisms found in both groups included: *E. coli, Streptococcus gallolyticus, Klebsiella pneumoniae.*

Table 4. Rectal microbial flora dynamics.

Microbial flora	Baseline		End of study	
	n	%	N	%
Total number of the piglets	12	100	12	100
Treatment Group 1				
E. coli	11	91.7	12	100.0
Streptococcus gallolyticus	8	66.7	7	58.3
Klebsiella pneumoniae	5	41.7	1	8.3
Pasteurella aerogenes	3	25.0	0	0.0
Enterococcus faecalis	2	16.7	0	0.0
Proteus mirabilis	1	8.3	0	0.0
Streptococcus suis	1	8.3	0	0.0
Acinetobacter towneri	1	8.3	0	0.0
Rothia nasimurium	1	8.3	0	0.0
Lactobacillus salivarius	0	0.0	5	41.7
Lysinibacillus fusiformis	0	0.0	3	25.0
Enterococcus hirae	0	0.0	3	25.0
Lactobacillus johnsonii	0	0.0	2	16.7
Acinetobacter radioresistens	0	0.0	1	8.3
Lysinibacillus boronitolerans	0	0.0	1	8.3
Treatment Group 2				
E. coli	12	100.0	12	100.0
Streptococcus gallolyticus	8	66.7	3	25.0
Klebsiella pneumoniae	4	33.3	2	16.7
Proteus mirabilis	1	8.3	1	8.3
Enterococcus villorum	1	8.3	0	0.0
Enterococcus faecalis	1	8.3	0	0.0
Aerococcus viridans	1	8.3	0	0.0
Citrobacter freundii	1	8.3	0	0.0
Streptococcus hyovaginalis	1	8.3	0	0.0
Pasteurella aerogenes	1	8.3	0	0.0
Streptococcus orisratti	1	8.3	0	0.0
Enterococcus hirae	0	0.0	8	66.7
Lysinibacillus fusiformis	0	0.0	3	25.0
Lactobacillus salivarius	0	0.0	1	8.3
Lactobacillus johnsonii	0	0.0	1	8.3
Staphylococcus saprophyticus	0	0.0	1	8.3
Bacillus megaterium	0	0.0	1	8.3
Comamonas kerstersii	0	0.0	1	8.3
Corynebacterium stationis	0	0.0	1	8.3

Escherichia coli is a rod-shaped opportunistic bacterium living and growing in the absence of the oxygen, mainly in the animal and human intestine. Some its strains belong to the symbiotic flora but can cause the enteric infections in adverse conditions.

Streptococcus gallolyticus is a Gram-positive opportunistic pathogen which can cause bacteriaemia and is predominantly found in the large intestine.

Klebsiella pneumoniae is a Gram-negative rod-shaped immobile capsule-forming bacteria belonging to *Klebsiella* genus which can also cause bacteriaemia and various organ system-related diseases.

Therefore, all three most common bacteria isolated from the intestine are opportunistic microorganisms, with *Klebsiella pneumoniae* having the most apparent pathogenicity.

In the Treatment Group 1, *E. coli* was found in 91.7% of the cases; *Streptococcus gallolyticus* in 66.7%; *Klebsiella pneumoniae* in 41.7%; *Pasteurella aerogenes* in 25%; *Enterococcus faec*alis in 16.7%; and *Proteus mirabilis, Streptococcus suis, Acinetobacter towneri* were found in 8.3% of the cases.

In the Treatment Group 2, *E. coli* was found in all cases; *Streptococcus gallolyticus* in 66.7%, *Klebsiella pneumoniae* in 33.3% of the cases. Other bacterial species (*Proteus mirabilis, Enterococcus villorum, Enterococcus faecalis, Aerococcus viridians, Citrobacter freundii, Streptococcus hyovaginalis, Pasteurella aerogenes, Streptococcus orisratti*) were found in single cases.

After the 5-day therapy, the prevalence of *Streptococcus gallolyticus* in the Group 1 reduced by 8.3% and ultimately was 58.3%. *Klebsiella pneumoniae* prevalence reduced by 33.3%, this bacterium was only found in one case, while the number of *E. coli* increased by 8.3%. Other previously isolated bacteria were no more found. At this time point, new bacteria were isolated, including *Lactobacillus salivarius* (41.7%); *Lysinibacillus fusiformis* and *Enterococcus hira*e (25.0%); *Lactobacillus johnsonii* (16.7%); *Acinetobacter radioresistens, Lysinibacillus boronitolerans* and *Lysinibacillus boronitolerans* (8.3%).

In the Group 2, prevalence of *Streptococcus gallolyticus* reduced by 41.7% and was 25.0% five days after the 5-day therapy. *Klebsiella pneumoniae* prevalence reduced twice, and it was ultimately registered in 16.7% of the cases. *E. coli* number remained unchanged. Other previously isolated bacteria were no more registered, except for the *Proteus mirabilis* (8.3%). New bacterial species were found, including *Enterococcus hirae* (66.7%); *Lysinibacillus fusiformis* (25.0%); *Lactobacillus salivarius, Lactobacillus johnsonii, Staphylococcus saprophyticus, Bacillus megaterium, Comamonas kerstersii, Corynebacterium stationis* (8.3%).

In both cases, Colistin 12000000 had no effect on *E. coli* that may be due to either high antibiotic resistance or symbiotic nature and a good sustainability of this microorganism in the gut flora. Given the good animal recovery dynamics when treated with the antibacterial drug, this bacterium likely had no key role as an etiological factor of the disease.

It is worth noting that among the new isolated microorganisms, *Lactobacillus salivarius, Lactobacillus johnsonii* and *Enterococcus hirae* are symbiotic species naturally presented in the intestinal lumen. Their presence is a positive parameter characterizing the digestive function. The amount of these microorganisms was highest at the end of the experiment. Other newly isolated bacteria can be considered commensal species, and some of them may be opportunistic ones. However, their amount was insignificant. Increase in the activity and emergence of new bacterial species after the treatment suggests their presence in the process premise environment and cannot be considered as an adverse factor as the animals completely recovered by this time point.

4 Conclusion

The study has shown that Colistin 12000000 has a high therapeutic efficacy in the infectious and inflammatory gastrointestinal pathologies caused by the resident opportunistic bacterial flora in the store pigs.

Oral administration of the drug in the piglets with the nonspecific gastroenteritis for 5 days in the doses 0.6 g/100 kg or 0.8 g/100 kg provided the complete therapeutic efficacy (the clinical recovery) on the Day 4 in both treatment groups. The dynamics of the erythrocytes and leukocytes suggest the recovery of the morphofunctional homeostasis after the disease in both treatment groups. After use of Colistin 12000000 in a dose 0.6 g/100 kg for 5 days, the prevalence of *Streptococcus gallolyticus* reduced by 8.3%. *Klebsiella pneumoniae* prevalence reduced by 33.3%. After the 5-day therapy with Colistin 12000000 in a dose 0.8 g/100 kg, the prevalence of *Streptococcus gallolyticus* reduced by 41.7% and was 25.0% at the end of the study. *Klebsiella pneumoniae* prevalence reduced twice, and it was ultimately registered in 16.7% of the cases. The other bacteria presented in the washings at the baseline in both groups were no more observed at the end of the treatment. The symbiotic flora was found in both groups at the end of the experiment in a sufficient number.

Thus, the efficient dose of the test drug was 0.6 g/100 kg as provided the desirable clinical effect comparable to the higher dosing regimen.

The drug well tolerated, with no side effects or adverse events reported throughout the study.

The results of the study of the efficacy of Colistin 12000000 give ground for recommending this drug dissolved in the drinking water for the treatment of the gastrointestinal diseases in the pigs with the diarrheal syndrome in the oral dose 0.6 g/100 kg for 5 days.

References

1. I. S. Koba, A. V. Pchelnikov, A. K. Shantyz, A. V. Luneva, L. M. Kashkovskaya, V. I. Dorozhkin, *Proceedings of Kuban State Agrarian University* **87,** 104–109 (2020)
2. O. N. Nikolaeva, M. S. Rodionova, *International Workshop and Conference Dedicated to 95th Anniversary of a Correspondent Member of RAAS, Honored Scientist of RSFSR and RF, Professor M.M. Dzhambulatov* (Makhachkala), pp 298–301 (2021)
3. M. S. Rodionova, *Proceedings of International Workshop and Conference Dedicated to 20th Anniversary of Veterinary Medicine Department of Izhevsk SAA* (Izhevsk), pp 156–158 (2020)
4. S. Y. Steblovskaya, E. P. Evglevskaya, *Proceedings of International Workshop and Conference*, pp 309–311 (2017)
5. V. M. Sorokoletova, N. N. Gorb, *Papers of Workshop and Conference of Academicians, Students, Master Students and Postgraduate Students Dedicated to 80th Anniversary of Novosibirsk SAU*, pp 408–412 (2016)
6. A. F. Makhmutov, G. N. Spiridonov, K. N. Makaev, M. T. Khuramshina, *Proceedings of International Workshop and Conference Dedicated to 90th Anniversary of Professor V.A. Kirshin*, pp 232–236 (2018)
7. M. E. Kehrli, Jr., J. Stasko, Kelly M. Lager, *Animal Frontiers January* **4,** 44–45 (2014)
8. A. T. Köhler et al., *Journal of Hospital Infection* **100,** 40–46 (2018)
9. Radmilo Colovic et al, *Toxins* **11** (2019)
10. L. Slivinska, B. Lukashchuk, *The Ukrainian Farmer*, pp 178–180 (2017)

Increasing the staff use efficiency in consumer cooperatives: current trends in conditions of digitization of agriculture

Marina Drozdova[1], *Lidia* Nagovitsyna[1], *Zoya* Kapelyuk[1,*], *Svetlana* Gorodkova[1], and *Anna* Aletdinova[2]

[1] Department of Theoretical and Applied Economics, Siberian University of the Consumer Cooperation, 26, Karl Marx Avenue, 630087, Novosibirsk, Russian Federation
[2] Automated Control System Department, National University of Oil and Gas "Gubkin University", 65, Leninsky Prospekt, 119991, Moscow, Russian Federation

Abstract. The purpose of the study is to identify current trends in increasing the efficiency of using the personnel of consumer cooperation organizations. This, along with other sectors of the economy, performs the functions of the third stage of the economic reproduction process (exchange), and the third sphere of the agro-industrial complex sectors that complete the production process in the sphere of commodity circulation. The personnel of economic entities carry out procurement, transportation, processing, storage and marketing of products to end consumers. In the context of existing demographic problems, the decline of the rural population, the challenge is to increase the efficiency of using the part of the economically active population employed in the diversified activities of consumer cooperation. A long-term study of the functioning of cooperatives made it possible to accomplish the purpose of this study. This concerns the development of ways to increase the efficiency of using the personnel of cooperative organizations – *preliminary measures*. Such measures include training and retraining of specialists, advanced training of workers in cooperative higher and secondary vocational educational institutions, and *subsequent* ones as a result of the creation of integrated complexes, using modern equipment, new technologies and digitalization advances, providing for the efficient use of personnel. As a new trend, the training of specialists who work with cooperative members is proposed. The economic participation of members will have a positive impact on the performance of their cooperative and increase the efficiency of the use of personnel of consumer cooperation organizations.

1 Introduction

Consumer cooperation is an integral part of the agro-industrial complex of the Russian Federation, and it is directly involved in the economic reproduction process performing the exchange (buying and selling) function. Despite the lingering systemic crisis, diversified multi-industry consumer cooperation continues to be a significant business player and has a positive impact not only on the regional economy within its operation area, but on the economy of the Russian Federation as a whole. In the face of increased competition, cooperatives seek to provide conditions for their staff professional development to stay competitive in the interests of their members.

The research is aimed at identifying the causes of the negative trends in the system of consumer cooperation, including the inefficient use of labor potential and all other elements of the resource potential. The results of the study show that one of the main negative factors is that members who are not employed by cooperatives are not involved in their activities. The importance of the economic participation of cooperative members is underestimated and frequently neglected.

Unlike enterprises and organizations of other legal forms, consumer cooperatives have a definite client base – their members, whose participation in the activities of their organization was determined in 1995 by one of the most important principles declared by the International Cooperative Alliance.

The participation of members in the activities of their cooperative improves their material well-being through mutually beneficial cooperation, enhances the performance of cooperatives in different sectors, and promotes the presence of consumer cooperation in the regional consumer market. Member economic participation ensures the efficiency of using the organization's resource potential and all its components, including staff potential.

In the context of this study, labour resources are considered to be the factor that determines the efficiency of the functioning of the rural economy, including

* Corresponding author: zkapelyuk@inbox.ru

agricultural enterprises, agricultural and consumer cooperatives. Labor resources are currently subject to requirements that meet the interests of enterprises and organizations in developing qualified personnel potential as well as related human capital characterized by professional competencies [1–8].

2 Materials and methods

In the course of the study, general scientific methods were used, including comparison, deduction, induction, generalization and interpretation of the obtained data, as well as the author's method: structural logical construction of chains of dependencies and relationships of socio-economic phenomena and processes.

3 Results and discussion

Increasing the staff use efficiency of enterprises is a challenge that has been dealt with since the emergence of machine production, mechanization and automation of production processes, which caused the need for the division of labour.

Until the 1990s, the five-year plans for socio-economic development set the task of increasing labour productivity, i.e. the efficiency of personnel use. This task did not lose its relevance at the end of the 20th century and became even more urgent in the 21st century, since the fall in the birth rate after 1990 led to a decrease in the working-age population in the 2000s. Its reduction can only be compensated by the growth of labour productivity and introducing the achievements of scientific and technological progress into the practice of enterprises in all spheres of economic activity, including consumer cooperation.

Advances in science and technology increasingly affect the integration and diversification of industries and activities, the creation of conglomerates, the improvement of federal state educational standards, forms and systems of remuneration, and the efficiency of the use of labour resources. In the system of Russian consumer cooperation, they also cause the development of modern trends in the use of human resources. The scientific novelty of the article lies in the adaptation of labour innovations to consumer cooperation.

In the context of the current shortage of qualified personnel, the trends for increasing the efficiency of rural labour potential use can be divided into two groups: that with the direct participation of cooperative organizations, which are the object of this study, or the one without such participation.

The main hypothesis is that cooperatives, as a form of association (conglomerate) of their organizations and enterprises, mainly related to small businesses, can and should become competitive and efficient due to the features and advantages of the cooperative business model.

A number of features, the most prominent of which is its dual nature, characterizes consumer cooperation as the largest economic system with almost two hundred years of functioning. It explains the opportunities for the interconnected development of a cooperative not only through entrepreneurial activity, but also through the socialization of cooperative-member relations – economic participation of cooperative members in the activities of their organizations as associations. The cooperatives that consider socialization as a condition for development, increasing labour productivity based on the benefits for cooperative members are successfully developing, and as a result, stay competitive and effectively use all of their resource potential, including the economic potential of their members.

The next essential factor is that the demand for consumer cooperation, its products and services for population entails the need for a timely adaptation of types and sectors of activity in order to meet new member and consumer needs.

Entrepreneurial activities of consumer cooperatives are carried out in the sectors which are most in demand in the rural areas. These are trade and public catering, procurement of agricultural products, non-timber forest products, medicinal and technical raw materials. They also include products of handicraft and fishing activities, primary processing and production of consumer essential goods, household products, building materials, and different services: educational, health, medical, transport, construction, household services. Cooperatives either set up enterprises operating within the same industry or establish concerns that combine interconnected industries acting as links in one chain and conglomerates involving non-production activities such as trade.

Currently, integrative transformation within cooperative organizations is the most cost-effective trend of organizational innovations implemented by consumer cooperation in the rural markets. Integrated entities in the cooperative economy are represented by procurement – production complexes, procurement – production – trade complexes and public catering complexes and are viewed by the authors as drivers of socio-economic development and communications of the rural population.

The competitiveness of complexes that combine procurement, trade, production and public catering is increasing significantly as a result of a shift in priorities in favour of these industries and limited opportunities of cooperatives to compete with retail chains.

The advantages of the above-mentioned complexes as large-scale associations of small and medium-sized cooperative businesses lie in the possibility of using the latest equipment and modern technologies. The efficient use of labour resources occurs through not only the mechanization and automation of production processes, increasing the number of equipment shifts and the level of activity, but also through attracting qualified personnel with specialized professional competencies. The use of workers with specialized training makes it possible to reduce the labour intensity of products made by cooperative manufacturing enterprises due to the rational use of time during the working day, the improvement of technological processes. And, consequently, it allows one to increase the efficiency of human capital of cooperatives.

The long-term advantage of complexes is economies of scale, which ensures lower costs per unit of output and, as a result, an increase in profits. Cooperative organizations that create a personnel training fund, in addition to it, can use part of the profits received for training, retraining, advanced training of workers in their own educational institutions, which will subsequently increase both the wages of workers and their labour productivity.

Famous Soviet economist S. G. Strumilin in his study "Problems of Labour Economics", back in the mid-1930s, theoretically substantiating the relationship between wages and labour productivity, noted, "Why does the school increase earnings so much? Obviously, only because a worker who has been to school thinks better, uses his strength more skilfully, and therefore improves the quality and productivity of his work" [9, p. 406].

Various ways of developing and improving the professional competencies of workers in traditional and new industries and types of consumer cooperation activities will make it possible to obtain a synergistic effect. It leads to a further increase in labour productivity and the emergence of properties in a cooperative system (complexes) that are not inherent in its elements separately, i.e., creation and institutionalization of a new industry "Membership Socialization".

We suppose that this industry will become an economic catalyst, accelerating the return of cooperative members to the sphere of consumer cooperation. It will require professional training of a new category of specialists - organizers of the economic participation of members. These specialists should be able to use digital technologies that reduce the complexity of accounting the contribution of members to the performance of a cooperative. These workers will be included into the executive personnel of the council chairman, heading the cooperative organization as an association of members. They should become holders of knowledge about the basics of cooperation, its advantages, its significance in the life of rural population and its role in providing an additional source of satisfaction of the material and other needs of members of the consumer society. In other words, such specialists should transform into cooperative officials who promote cooperative ideas and education among different age groups and categories of the rural population.

The result of their activities at the initial stage will be an increase in the number of members who are the social base of consumer cooperation. The need to meet the needs of the growing number of members of the cooperative will require an increase in the volume of production, trade and procurement activities, public catering and other sectors of the economy of the cooperative organization. And it will lead, albeit indirectly, to increasing the efficiency of using the labour potential and, first of all, labour productivity in a cooperative.

The problem of increasing the efficiency of using the labour potential of consumer cooperatives is extremely relevant due to a number of reasons. We consider the decline in rural population to be the main one.

In 1992, which is considered the year when market reforms began, the number of people living in rural areas was 38.9 million or 26.3% of the total population. In 2000, the proportion of the population living in the zone of consumer cooperation increased to 26.8%, but in subsequent years, the opposite negative vector of change transformed from a current trend into a common regular one. Thus, in 2000–2020, the number of village residents decreased by 5.9%, or by 2.3 million people, to 25.26% of the total population of the country [10]. This is due to the continuing low attractiveness of agricultural labour because of low wages, high labour intensity, bigger opportunities for the urban population, as well as the aggressive penetration of ideas of "easy money" (blogging, etc.) into the minds of young people, and other reasons. Under these conditions, the need to increase labour productivity is getting decisively important.

Consumer cooperation, which is recognized as a socially oriented system that conducts socially responsible business, voluntarily assumes the implementation of the 7th principle of the International Cooperative Alliance "Concern for Community". This principle implies the responsibility of cooperative organizations for the economic, social and cultural development of the area of activity, as well as the preservation of the environment. Concern for Community is manifested in the participation of cooperatives in the formation of local budgets and extra-budgetary funds, supplying the social sphere organizations, such as schools and preschool institutions, hospitals and other organizations, with high quality food products at fair prices, including those produced at cooperative enterprises.

In the context of this study, creation of jobs and the efficient use of the personnel potential of cooperative organizations are considered as an important aspect of implementing of the principle Concern for Community. The data in Table 1 indicates higher rates of growth in labour productivity in trade and public catering as the traditional sectors of consumer cooperation compared with similar indicators for the Russian Federation as a whole.

Table 1. Dynamics of labour productivity of trade and public catering workers in the Russian Federation and average indicators for consumer cooperation for 2019–2020 (as a percentage of the previous year).

Industries	2019	2020
Russian Federation, total	102.6	99.6
including :		
trade	101.5	101.9
public catering	100.2	83.0
Consumer cooperation of the Russian Federation, total	101.2	108.5
including		

trade	101.2	113.1
public catering	105.6	86.5

The employment figures can increase only in the sectors that are reasonably considered drivers of the socio-economic development of consumer cooperation – the procurement and processing of agricultural products and raw materials, wild mushrooms, berries, medicinal and technical raw materials and other non-timber forest products. Since chain retailing demonstrates higher competitiveness, the consequence of this is a decrease in the share of trade and public catering in the total volume of activities of consumer cooperatives.

The decline in the numbers of the rural population, its political and economic isolation, as well as the need to enhance food security and independence, necessitate the growth of labour productivity of personnel in all sectors of the rural economy through the introduction of all types of innovations. This will allow improving and expanding the number of competencies based on the use of digital technologies.

The consumer cooperation system has its own organizations of higher and secondary vocational education and successfully carries out training, retraining, and advanced training of workers. Cooperative education ensures the effective use of the economically active population, which can also be considered a significant trend in increasing the efficiency of staff use in the system of consumer cooperation.

The Russian cooperative education system is represented by three higher educational institutions: the Belgorod University of Cooperation, Economics and Law (Belgorod), the Russian University of Cooperation (Mytishchi, Moscow Region) and the Siberian University of Consumer Cooperation (Novosibirsk). All universities have a network of regional branches, i.e., institutes all over the country. The system also includes 38 educational organizations of secondary vocational education: technical schools and colleges.

In order to promote professional cooperative education, at the initiative of the Siberian University of Consumer Cooperation, in the mid-2000s, the Association of Higher and Secondary Specialized Educational Institutions of Siberia and the Far East was created, uniting cooperative technical schools and colleges of most Siberian and Far Eastern regions. The Association acted as an advocate of specific cooperative professional competencies, participated in developing the human capital and the labour potential of consumer cooperation organizations.

The dual nature is inherent not only in consumer cooperation as an economic system, but also in cooperative education and is expressed in the combination of professional and cooperative education. The development of professional competencies, combined with the knowledge in the sphere of cooperation, the features and benefits of its diversified activities in rural areas, should be considered as a renewed trend for increasing the efficiency of using labor resources that are currently limited in number and skill level.

The shortage of qualified personnel for the system of consumer cooperation is due to the gradual transformation of cooperative professional education into professional education in cooperative universities [11]. This negative process, which continues to the present, began in the 1990s. The main reason is limited financial resources, which are mostly used for investing in diversified activities to ensure moderate competitiveness. Striving to increase the overall economic and financial performance, universities change priorities and are not always efficient in staffing cooperatives.

4 Conclusion

The current state of the agro-industrial complex in Russia and the basic economic systems (branches) involved in its functioning – agriculture and consumer cooperation – requires taking organizational and economic measures aimed at increasing the staff use efficiency in enterprises and industries that ensure the country's food security.

Consumer cooperation, characterized by a dual nature, in accordance with the Law of the Russian Federation "On consumer cooperation (consumer societies, their unions) in the Russian Federation", also participates in ensuring food security, using the specific features of the system. These are the ability to create integrated complexes that combine procurement, production, trade, public catering and thus allow one to effectively use professional competencies of the system's personnel and introduce technological and digital advances.

References

1. A. A. Aletdinova, *Herald of the Belgorod University of Cooperation, Economics and Law* **6,** 164–175 (2018)
2. Yu. V. Meikshan, *Herald of the Siberian University of Consumer Cooperation* **4,** 34–39 (2019)
3. L. A. Poltarykhin, S. A Shelkovnikov, P. V. Zhuravlev, *Economy and Entrepreneurship* **9,** 483–487 (2021)
4. L. A. Poltarykhin, S. A Shelkovnikov, P. V. Zhuravlev, S. G. Golovina, *Economy and Entrepreneurship* **9,** 1107–1110 (2021)
5. I. G. Kuznetsova, M. S. Petukhova, S. A. Shelkovnikov, *Economics, Labor, Management in Agriculture* **1,** 23–28 (2020)
6. S. A. Shelkovnikov, E. V. Sharavina, I. G. Kuznetsova, M. S. Petukhova, *Transbaikal State University Journal* **2,** 114–122 (2020)
7. O. V. Ozhogova, S. A. Shelkovnikov, I. G. Kuznetsova, *Transbaikal State University Journal* **10,** 99–109 (2020)

 https://doi.org/10.1051/bioconf/20225200040

8. I. G. Chirkova, S. A. Shelkovnikov, M. A. Tikhonchuk, *Bulletin of Agrarian Science* **1,** 145–153 (2020)
9. S. G. Strumilin, *Problems of Labour Economics* (1982)
10. Russian Statistical Yearbook (2021). Retrieved from: https://rosstat.gov.ru
11. M. I. Drozdova, L. P. Nagovitsina, *Herald of the Belgorod University of Cooperation, Economics and Law* **3,** 84–95 (2017)

Prospects of personnel training for the agricultural sector in the conditions of digitalization

Zoya Kapelyuk[1,*], *Irina* Shchetinina[2], *Anna* Aletdinova[3], *Svetlana* Gorodkova[1], and *Yulia* Derevyanko[2]

[1] Department of Theoretical and Applied Economics, Siberian University of the Consumer Cooperation, 26, Karl Marx Avenue, 630087, Novosibirsk, Russian Federation
[2] Siberian Scientific Research Institute of Agricultural Economics of Siberian Federal Scientific Centre of Agro-BioTechnologies of the Russian Academy of Sciences, Centralnaya St., Presidium, Krasnoobsk, 630501, Russian Federation
[3] Automated Control System Department, National University of Oil and Gas "Gubkin University", 65, Leninsky Prospekt, 119991, Moscow, Russian Federation

Abstract. The article is devoted to identifying the main problems associated with the lack of young and qualified personnel in the agricultural sector of the economy. This includes financial, household and other problems of the rural areas and population, which reduces the prestige of agricultural jobs for young people, causes reluctance to get education related to agriculture, and after graduation to work in agricultural production. In order to solve these problems, it is necessary to change the system of personnel training, move to a crosscutting continuous agricultural education, form a positive image of a rural worker in children from preschool and school years. One must also improve university and secondary vocational education, and, based on the use of information technology, provide the opportunity for continuous mastering the latest scientific and practical innovative achievements throughout their working life.

1 Introduction

The modern agricultural sector of the agro-industrial complex, like the entire economy, is experiencing the rapid development of information technologies, the transition to digitalization, and the use of smart technologies in production. Leading farms actively cooperate with research, consulting and engineering companies. However, at many agricultural enterprises there is still a serious technical and technological lag, which is associated with financial and personnel problems. Therefore, government authorities and heads of agricultural enterprises need to pay serious attention to solving these problems.

2 Results and discussion

The studies conducted show that many agricultural educational institutions face significant problems with training personnel for the agro-industrial complex. The problems arise at the stage of student recruitment, since few students apply to do majors related to agriculture. Most young people do not consider jobs in agriculture to be prestigious and choose other areas of study for obtaining a profession.

In particular, in 2020 there were 34.7 thousand students pursuing degrees in *Oil and Gas Business*, 105.5 thousand people majoring in *Construction*. In contrast, the educational program of *Zootechnics* attracted 16.6 thousand students, and only about 1.1 thousand young people studied *Soil Science* (Table 1) [1].

Table 1. The number of students in Russia in some areas of education, 2020.

Areas of education (programs)	Program Code in Russia	Number of students
Construction	08.03.01	105457
Informatics and Computer Engineering	09.03.01	51017
Applied Informatics	09.03.03	48936
Information Systems and Technologies	09.03.02	45800
Oil and Gas Business	21.03.01	34661
Economics	38.04.01	33357
Soil Science (Bachelor's degree)	06.03.02	874
Soil Science (Master's degree)	06.04.02	272
Zootechnics	36.03.02	16568

Moreover, after having graduated from agricultural educational institutions, not all students are ready to work in the field of agriculture. According to the statistics of Novosibirsk State Agrarian University, only

* Corresponding author: zkapelyuk@inbox.ru

61–67% of graduates return to work in the village or choose a field of activity related to their profession.

As a result, a difficult situation is developing in Russia's agriculture. Over 40% of agricultural workers do not have special professional education, the number of young people in agricultural production is decreasing, and the proportion of people of pre-retirement and retirement age is growing. In the period from 2017 to 2020, the share of the rural population among workers aged 15–29 decreased by 8.1%, while those aged 55 and over increased by 7.9% (Fig. 1) [2; 3].

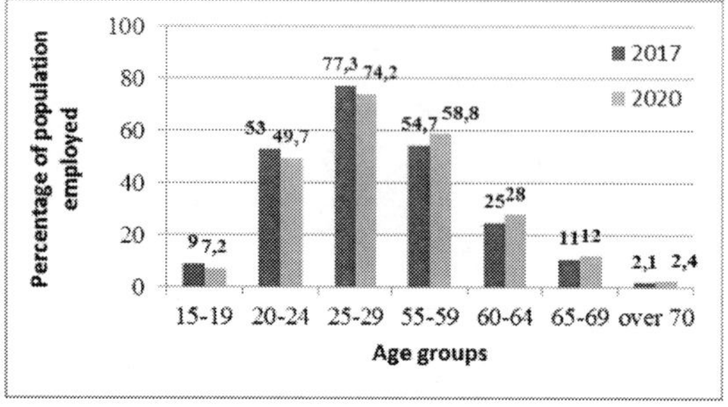

Fig. 1. Dynamics of employment of the rural population of Russia by age groups, % of the population of the corresponding age group.

The mass movement of population, including qualified specialists, from rural to urban areas leads to a slump in agricultural production and becomes a threat to the existence of rural settlements. Moreover, in Siberia, the decline in production and depopulation of villages is proceeding faster than in Russia as a whole. While in Russia, the number of rural settlements has decreased by an average of 10%, over the past 10 years, in Siberia this number fell by 16% [4]. The number of cattle over the past 30 years has dropped by an average of 3.2 times in Russia, while in Siberia this decrease is 15% higher. In some regions, the reduction of farm livestock population is even greater. For example, in the Novosibirsk region, the number of cattle fell by 3.7 times in 1990–2020. This situation requires the adoption of certain measures by the state and municipal authorities.

In 2012, the Federal Law "On Education in the Russian Federation" provided for the formation of a system of continuous education in the country. However, this kind of work in the agricultural sector began only in 2022; the Ministry of Science and Higher Education and the Ministry of Agriculture, together with the Innopraktika Foundation [5], carry it out.

In our opinion, in order to develop a crosscutting continuous educational system, it is necessary to involve rural children, starting from preschool age, and systematically continue working with them during school years, creating a positive image of an agricultural worker for them. This approach can be effective as it is in the kindergarten that a child begins to form the most important components of the perception of the world and its initial assessment [6; 7]. Material support for children's educational institutions should be provided at all levels to ensure the accessibility of education to all rural children, since in some regions up to 30% of children aged 3–7 years are not provided with places in preschool educational organizations [8].

In rural schools, efforts to develop children's interest in agricultural occupations should be continued on a constant basis. It is necessary to involve employees of agricultural enterprises, parents engaged in agricultural production, to provide career guidance so as to purposefully prepare schoolchildren for mastering the professions required in agriculture, taking into account the specifics of the regions, lifestyle and type of activity. This is due to the fact that many regions of Russia have their own significant features in the development of agriculture, the culture of the population, the organization of education, etc. For example, school education in the northern regions of Russia has distinct peculiarities [9]. In this connection, a dilemma arises: should schoolchildren study in small groups, preserving their traditional way of life, but to the detriment of the quality of education. Or should they study in boarding schools, where education is better, but there is a danger of children being separated from their usual way of life? In the latter case, they may no longer go back to rural labour in their region. In this case, it is important to find a balance and continue working with children in boarding schools, involving them, in accordance with their age, in rural labour traditional ones for the northern regions. A similar approach should be taken in other regions of the country, given the wide variety of natural and climatic conditions in Russia, the types and methods of agricultural production, the traditions of the rural population and their way of life in rural areas.

An important area of work with schoolchildren should be their focus on the future of innovative and high-tech agriculture, which needs to be developed in the digital economy, and which will require qualified employees. Therefore, in order to succeed in the labour market, it is necessary to receive vocational education. At the same time, teachers of rural schools need to understand that the future belongs to digitalization in all sectors of the economy, and everyone should be prepared for it [10].

The next stage of crosscutting (continuous) approach to the training personnel for the agrarian sector of the economy should include encouraging students to enter professional higher and secondary educational institutions of agricultural specialization. Ways of attracting young people may include increasing scholarships and providing benefits to pay for places in dormitories. In this connection, it is advisable to develop special programs to support students studying in agricultural areas. They should include cost compensation based on co-financing from the budgets of the federal and regional levels with the involvement of different structures of agro-industrial clusters: manufacturing enterprises, financial organizations (Rosselkhozbank, etc.), including insurance

https://doi.org/10.1051/bioconf/20225200041

organizations, retail chains and other stakeholders interested in the development of the agro-industrial complex and agriculture as its main component.

Important conditions for receiving increased scholarships throughout the entire period of study should be the progress of students (good and excellent marks) and the obligation to return to the village to work and use competencies in agriculture for a period of at least 3–5 years. This will provide agriculture with young, qualified employees for a certain period of time. For example, in the Novosibirsk region, up to 3,000 young specialists per year can be attracted to rural areas in this way.

The training of students in professional educational institutions should be focused not only on the needs of agricultural production at the currently prevailing technical and technological level, but also on the digitalization of agriculture. Some farms already have technologies based on various precision agriculture software, "smart" technologies, unmanned aerial vehicles, etc. [11, p. 82–98; 12, p. 96].

For example, automated technologies "Smart Land Use", "Smart Field" take into account the types of land, its location in a particular farm, municipality, etc., analyze the state of the soil, the environment, plants and other parameters. Based on this, digital planning, modeling with the use of geoinformation technologies (GPS, etc.) is carried out. "Smart Garden" system provides for keeping records of data on the areas occupied by fruit and berry crops, analyze the soil condition in these gardens, and monitor the quality of plants, which allows taking the necessary measures to improve the above-mentioned characteristics. The robotic complex "Smart Greenhouse", created on the basis of digital technologies, provides control of the indoor microclimate, lighting, plant nutrition, energy consumption and other indicators necessary to increase the efficiency of production, growth and development of plants, obtaining high yields. Automated vegetation systems use sensors that automatically measure the properties of the soil, the condition of plants by electrical, electromagnetic, radiometric, laser, acoustic, thermal and other parameters. They are designed to automatically control the mode of lighting, irrigation, ventilation, heating and other processes according to the program built into them.

The technological complex "Smart Farm" is designed for livestock husbandry, provides a reduction in the level of diseases of farm cattle, control of safety, animal productivity and quality of the final product. Automated dairy farm management systems with GPS/GLONASS and RFID (Radio Frequency Identification) technologies help to solve a set of production and management tasks, keep records of livestock and its vaccination, control movement and other indicators, optimize selection work, etc.

The demand for digital technologies will continue growing in the future. According to the Higher School of Economics, the achievements of robotics, the creation of autonomous robotic systems, smart greenhouses and smart farms should become a driver in agriculture [11, p. 83]. Foreign sources also note that by 2021 there were "75 million agricultural devices of the Internet of things in the world, and by 2050 the average farm will generate 4.1 million units of data per day" [13, p. 16]. Therefore, students of agriculture should be ready to work in these conditions and acquire the necessary knowledge and skills.

In addition, the Ministry of Science and Higher Education, the Ministry of Agriculture of Russia, as well as heads of regions and educational institutions should keep in mind that the educational program of IT specialists for the agricultural sector is to cover the specifics of production, product storage, marketing and logistics. This provides the management of enterprises with the necessary information for making effective decisions.

In the educational sphere of agrarian educational institutions, it is necessary to make wider use of the new opportunities brought by information technologies, which allow involving specialists and managers from the agro-industrial complex, manufacturing, research and other organizations of Russia and foreign countries, using online interaction options in the educational process. This will make it possible to establish a close connection of all educational institutions with advanced and innovative enterprises, research organizations, state and municipal authorities, and ensure the acquisition of new knowledge and best practices by both students and agricultural specialists (Fig. 2).

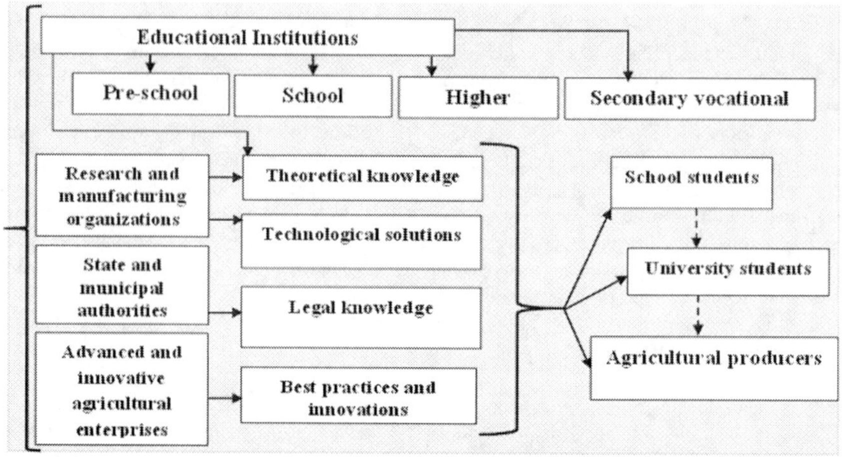

Fig. 2. The model of continuous acquisition of new agricultural knowledge and best practices in the context of digitalization.

At the same time, teachers of agricultural educational institutions should understand the importance of using not only traditional full-time teaching methods, but also the opening opportunities for online interaction with students and graduates who need consultation on their future work. Researchers highlight the above-mentioned ideas in relation to various areas of education [14].

The application of up-to-date information technologies is also essential for teaching distance education students in agricultural universities. For example, the process can be organized with the use of the Open Agricultural Education electronic platform developed by Digital Agro LLC. This platform provides students of all agricultural universities in Russia with access to distance learning. The professional knowledge platform participants are currently universities, academies and research organizations that offer hundreds of different educational programs for more than 300,000 students. Using this platform, any student or employee of the agricultural sector can connect to participating organizations, enter their libraries, use literature and learn about scientific developments and other activities of participating organizations. Those who apply through this platform can become both teaching and learning participants in any educational program. Students of all agricultural educational institutions should be encouraged to work with this electronic platform and taught how to use the information available.

In addition, as a number of scientists point out, the use of information technologies and the strategy “Education 4.0” will promote lifelong learning [15; 16; 17]. This means that agricultural specialists who have mastered advanced information technologies will be able to receive data on the latest scientific, technical and other achievements at any time as well as consultations of scientists and teachers of educational institutions they graduated from. It will help them introduce innovations.

Thus, a crosscutting agrarian education is a life-long educational program developed and implemented through all levels of education: from the preschool age of children till the retirement of an agricultural worker. Advanced information technologies will increasingly play a significant role in this process.

3 Conclusion

The rapid development of information technologies in the agrarian sector of the economy is a significant trend, and it should be taken into account at all levels: from federal government to on-farm management. At the same time, special attention should be paid to personnel training, since at present there are serious problems in agriculture related to the quantitative and qualitative composition of enterprise employees, the outflow of young people from rural areas.

The studies carried out made it possible to identify the main reasons for the lack of qualified personnel in the rural areas, who are necessary for the development of the latest technologies and the transition of the agrarian economy to an innovative development path.

To deal with the existing problems, it is necessary to switch to continuous education and to its most promising option – cross-cutting agricultural education aimed at building a favourable image of agricultural occupations for children and schoolchildren, starting from an early age. The should be a financial support of students of professional educational institutions, providing opportunities to master modern information technologies that make it possible to introduce innovations into production throughout their working life.

References

1. Ministry of Science and Higher Education of the Russian Federation Statistics (Number of Students). Retrieved from: https://www.minobrnauki.gov.ru/opendata/9710062939-svedeniya-o-chislennosti-studentov-po-kursam-napravleniyam-podgotovki-i-spetsialnostyam

2. Labor force, Employment and Unemployment in Russia. 2021: Statistical Digest of Rosstat (2021)
3. Labor and Employment in Russia. 2021: Statistical Digest of Rosstat (2021)
4. Regions of Russia. Socio-Economic Indicators. 2021: Statistical Digest of Rosstat (2021).
5. Ministry of Science and Higher Education of the Russian Federation on Agrarian Education. Retrieved from: https://minobrnauki.gov.ru/press-center/news/ ?ELEMENT_ID=46368
6. A. K. Belolutskaya, T. N. Levan, S. A. Zadadaev, O. A. Shiyan, I. B. Shiyan, *Educational Studies Moscow*. **3**, 237–259 (2021).
7. A. Diaz, N. Eisenberg, C. Valiente, S. VanSchyndel et al., *Journal of Research in Personality* **67**, 3–14 (2017)
8. L. Yu. Bedareva, E. A. Semionova, G. S. Tokareva, *Educational Studies Moscow* **2**, 60–82 (2020).
9. O. V. Vasilieva, V. E. Okhlopkov, *Educational Studies Moscow* **4**, 285–310 (2021)
10. A. A. Deryabin, I. E. Boitsov, A. A. Popov, P. D. Rabinovich, K. E. Zavedenskii, *Educational Studies Moscow* **3**, 212-236 (2021)
11. Digital Transformation of Industries: Starting Conditions and Priorities. HSE Report (2021).
12. K. O. Vishnevsky, L. M. Gohberg, et al., *Digital Technologies in the Russian Economy* (2021)
13. World Government Summit. Agriculture 4.0: The Future of Farming Technology (2018). Retrieved from: https://www.worldgovernmentsummit.org
14. D. M. Rogozin, O. B. Solodovnikova, A. A. Ipatova, *Educational Studies Moscow* **1**, 271–300 (2022)
15. V. A. Komarov, A. V. Sarafanov, *Business Informatics* **2**, 47–59 (2021)
16. A. A. Maltsev, V. A. Maltseva, *International Organisations Research Journal* **4**, 189–195 (2020)
17. A. A. Hussin, *International Journal of Education & Literacy Studies* **3**, 92–98 (2018)

Effect of the drug "Degelm-14" on spontaneously infected chickens with eimeriosis

Orazali Mullakaev, *Minsagit* Lutfullin, *Evgeny* Kirillov*, *Elena* Zaikina, and *Gulnar* Nizamova

Kazan State Academy of Veterinary Medicine named after N. E. Bauman, 35, Siberian tract St., Kazan, 420029, Republic of Tatarstan, Russian Federation

Abstract. The studies show that when using the preparation "Degelm-14" after infection of birds, the histological structure in most of the studied organs was gradually restored. In the glandular stomach and small intestine on the 22nd day after application of the studied preparation the microstructure of the organ recovered completely. Microstructural changes in the cloacal pouch and spleen after application of Degelm-14 indicated restoration of the proliferative activity of the cells of the germinal centers. In the liver, at the final stage of the study, there were no signs of metabolic disorders in hepatocytes, and the dark pink coloration of their cytoplasm indicated an increase in synthetic processes in the cells of the organ parenchyma. But in the kidneys there were still observed signs of impaired glomerular filtration, regulating the function of the organ.

1 Introduction

The introduction of digitalization in poultry farming has caused a change in the epizootic nature of the course of parasitic diseases. The use of modern poultry housing technology has led to the reduction or elimination of helminth infections, but the large concentration of livestock in limited areas has created favorable conditions for the development of protozoal diseases, especially Eimeriosis. Eimeriosis causes death of chickens, causing visible damage, but the main problem is hidden losses [1–3]. To control Eimeriosis, it is necessary to choose the right Eimeriostatic, while taking into account its effect on the host organism. Therefore, it is necessary to conduct histological and histochemical studies of host tissues after the use of drugs before introducing new antiparasitic drugs into practice [4–8].

2 Materials and methods

"Degelm-14" is a composition including H-tetradecyltributylphosphonium bromide, 5.7 – bis – (m – nitroanilino) – 4.6 – dinitrobenzofuroxane in a certain weight ratio and glucose. "Degelm-14" is a crystalline powder of light red color with Tpl – 198°C; it dissolves well in polar solvents such as dimethyl sulfoxide, dimethylformamide and in vegetable oil. Degelm-14 powder is poorly soluble in water and in non-polar solvents such as hexane, heptane, benzene, etc. The active substances of this composition are phosphonium salt that penetrates through cell plasmolemma and causes its destruction. This salt is a structural analog of phospholipid fragments of biological cell biomembranes; substituted nitrobenzofuroxane, a component generating six molecules of nitric oxide (NO-radical) per molecule in vivo due to phosphorylation processes. Glucose is an excipient at a weight ratio of the active components of 1:10.

Translated with www.DeepL.com/Translator (free version). The coccidiostatic effect of Degelm-14 was studied on three hundred spontaneously invaded broiler chickens of cross 15. During the period of the experiment, birds were under similar conditions of housing and feeding. "Degelm-14" was given to poultry in the morning by group method with half of concentrated feed for two consecutive days at a dose of 35 mg/kg. To study pathomorphological changes in experimental coccidiosis, experimental birds were slaughtered on 7 and 22 days after infection.

3 Results and discussion

The histological structure of the glandular stomach was preserved. In the mesohumeral glands, shortened during the preceding invasive process, the exocrinocytes of round-oval shape with a small volume of cytoplasm were preserved. The mesohumeral glands retained a small volume. In some glands there were detected areas with signs of desquamation and destruction of secretory cells, as well as cells with atrophy phenomena. The folds of epithelial tissue in the area of tube gland mouths were shortened and deformed. Subepithelial lymphoid tissue of the organ was represented by compact formations in which germinal centers were not marked. Muscular basis of glandular region of the stomach had signs of atrophy

* Corresponding author: kirilloff.eug@yandex.ru

and weakly expressed edema of intermuscular connective tissue (Fig. 1).

The connective tissue base of the reconstructed intestinal villi was represented by numerous reticulocytes, undifferentiated and lymphoid cells and single smooth muscle cells. The highest density of these cells was noted at the base of the villi. The submucosal base was poorly expressed and intermediately passed into the muscle plate, which was not of uniform thickness. The muscular layer retained signs of unfolding. Intermuscular nerve plexuses retained signs of neuroplasma swelling and pale colored nuclei. Eimeric schizonts when stained with SHIF reagent were not detected both on the surface and inside the cells of the epithelium of villi and crypts.

The histological pattern of the glandular gastric pattern on the 22nd day after infection was well preserved. Complex mesenteric glands were distinguished by the well-defined structure of labyrinth, formed of numerous glandular tubes, cells of which had a cubic shape and the presence of granulated secretory material in the supranuclear zone. Areas with signs of disruption of the epithelial monolayer of exocrine cells in the glandular tubes, characteristic of sick birds, were not detected. Under the mucosa of the organ appeared compact, dense clusters of lymphoid tissue, with a weakly marked structure of germinal centers. Lymphoid clusters were most pronounced near the mouths of complex glands. In the small intestine, on the seventh day after application of Degelm-14, the formed intestinal villi were characterized by a polymorphic structure

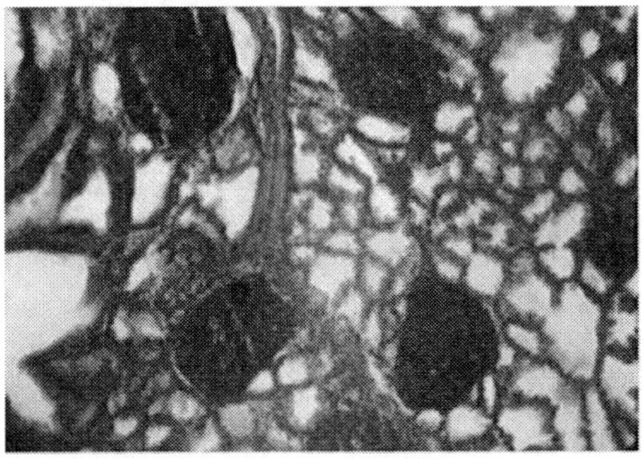

Fig. 1. Gastric gland. Formation of numerous subepithelial lymph nodes on the 7th day after application of Degelm-14. Hematoxylin and eosin staining. Ou.140.

Most of the intestinal villi had a wide base and narrowed apical part. The monolayer of columnar epitheliocytes was mainly located on the lateral surface of the villi. The apical region of individual villi was partially devoid of epithelial cover and partially due to the delayed renewal contained numerous enterocytes with signs of atrophy (Fig. 2, 3).

Fig. 2. The small intestine. Single mitotically active cells were detected in the regenerative area of crypts on the 7th day after application of Degelm-14. Staining with hematoxylin and eosin. 240.

On the 22nd day after application of the drug in the mucosa of the small intestine, there were villi with flattened, wide base. The prismatic epithelium of villi was characterized by the presence of numerous bocalovid cells with signs of mucus hypersecretion. Their greatest accumulation was concentrated on the lateral surface of the villi. No increased secretion was noted in the apical zone of villi. The connective tissue base of villi was characterized by hyperplasia of reticulocytes, fibroblasts, endothelial and adventitial cells of blood vessels. Most capillaries of the villi had marked wall profiles and were in a state of full-bloodedness. In the basal part of the crypts, the figures of mitosis were noted among the epitheliocytes (Fig. 4).

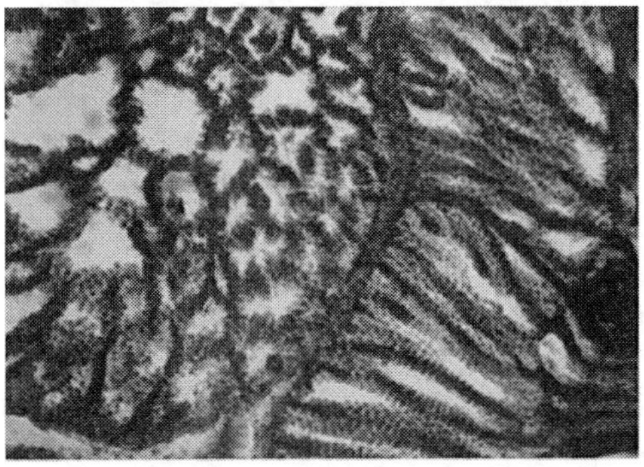

Fig. 3. Glandular stomach. Lymphoid clusters of the greatest expression were observed near the mouths of the complex glands on the 22nd day after application of Degelm-14. Hematoxylin and eosin staining. Stained with hematoxysin and eosin. 240

The nuclei of most crypt epithelial cells, as well as those of villi, were saturated with finely structured chromatin. The muscular layer was located continuously and was characterized by uneven thickness of the inner and outer layers of the muscular shell with weakly expressed signs of edema and disorganization of intermuscular connective tissue components with preservation of nerve plexus cell structures.

The connective-tissue base of the mucosa was sparse. Single mitotically active epitheliocytes were detected at the bottom of crypts. At the base of villi the connective-tissue base was represented by numerous reticulocytes, lymphoid cells and single myocytes. The submucosal base was weakly expressed and passed into the muscularis, which was not of uniform thickness. In the muscular layer, there were signs of fragmentation. Eimeric schizonts were not detected by SHIF reagent staining.

The pattern of the cloacal pouch structure in birds after seven days of using the preparation was well enough preserved. The lobules were of different size and angular shape. On the 7th day of the drug "Degelm-14" application the lymphoid tissue proliferative activity was observed, as a result the cortical and brain areas of the lobules were clearly manifested. The interlobular fibrous connective tissue was edematous and unfolded. In submucosa there was a formation of new lobules of the organ. It was manifested by appearance of small cell clusters formed by reticulocytes, macrophages and lymphoid cells. Epithelium of bursa mucosa at the top of folds had weakly expressed signs of vacuole dystrophy in some places.

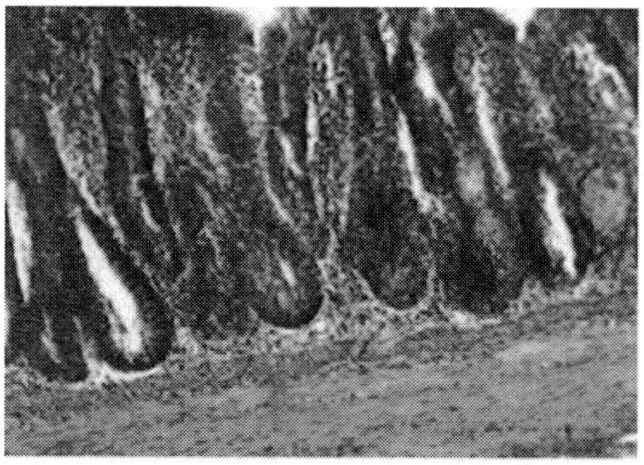

Fig. 4. The small intestine. In the regenerative area of crypts we detected single mitotically active cells on the 22nd day after application of Degelm-14. Staining with hematoxylin and eosin. Stained with hematoxysin and eosin. 240.

Fabricated bursa of birds on day 22 of the preparation had preserved mucosal epithelial layer, which was a cylindrical epithelium formed of equal height cells with a pronounced polar structure. In the folded areas, the height of epitheliocytes sharply increased. Oval, heterochromatin-saturated epithelial cell nuclei were shifted to the basal pole. Mucous membrane loose connective tissue contained numerous fibroblasts, reticular cells, between which there were bundles of collagenic and reticular fibers. Blood vessels had a well-defined endothelial structure. Lymph nodes were numerous oval, round-shaped structures with a sparse center and an increased concentration of lymphocytes along the periphery (Fig. 5). The generative center of the lymph nodes contained numerous blast cells with mitotic figures. In addition to the formalized lymph nodes, small clusters of lymphoid tissue were detected.

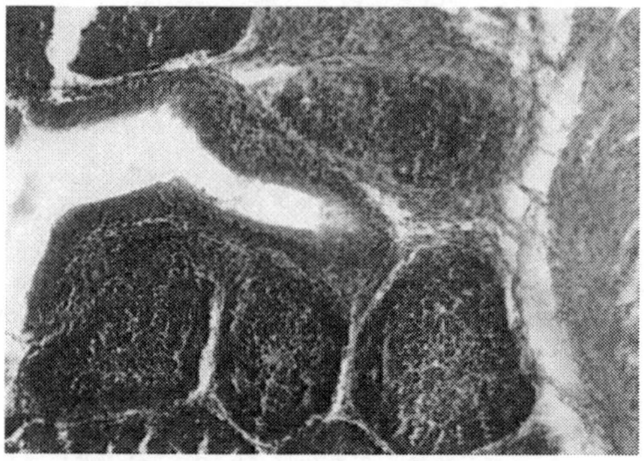

Fig. 5. Fabricated bursa. Significant accumulations of large lymphocytes, cortical and brain substance of nodules on the 22nd day after application of "Degelm-14" were noted. Hematoxylin and eosin staining. 240.

On the seventh day in the parenchyma of the spleen of the birds treated with Degelm-14, rounded lymph nodes of the white pulp were located irregularly. Individual nodules contained pronounced germinal centers containing reticulocytes with thickened processes surrounded by blastic, undifferentiated lymphocytes with mitosis figures. The mantle and marginal zones of the nodules were weakly expressed. The periarterial zone, located around the central artery of the lymph node, was a loose cluster of small lymphocytes.

In birds on the 22nd day of obtaining the studied preparation, the spleen structure pattern was well expressed. White pulp was represented by compact lymph nodes with small germinal centers containing blast cell forms, cells with single mitotic figures and large and medium lymphocytes. The mantle zone of the nodules consisted of small and medium-sized lymphocytes. The periarterial zone of the nodules was characterized by a large accumulation of small lymphocytes. The marginal zone contained dense clusters of lymphocytes, macrophages, and reticulocytes due to increased proliferative activity (Fig. 6). Surrounding lymph nodes, the moderately blood-filled red pulp was distinguished by the presence of compact clusters of small and medium-sized lymphocytes surrounded by reticular cells with swollen processes.

Microstructural changes in the spleen after application of "Degelm-14" indicated the restoration of proliferative activity of cells in the germinal centers of lymph nodes, promoted active formation of small lymphocytes, migration of these cells into the thymus dependent zone of the organ. Moderate and weakly blood-filled central arteries were distinguished by the absence of signs of connective tissue component disorganization and marked wall structures. Only in separate vessels we observed swelling of endothelial cells, large epithelioid-macrophage muffs had well-defined borders and blood vessel profiles. At the edges of the muffs there were compact non-uniformly thick clusters of macrophage lymphocytes

On the seventh day after exposure to Degelm-14, a significant number of hepatocytes with oxyphilic stained cytoplasm were found in the weakly expressed bar structure of liver parenchyma in birds; simultaneously, a population of small hepatocytes with pycnomorphic nuclei was present. Numerous glycogen granules were found in the cytoplasm of larger hepatocytes when stained with Schiff reagent, while in small liver cells these granules were much fewer. Stellate reticuloendotheliocytes were present in the wall of sinusoidal capillaries. Few clusters of lymphoid cells were found in underdeveloped connective tissue septa.

The liver of birds after seven days of using the preparation "Degelm-14" had a preserved histological structure of the organ. Most hepatocytes were distinguished by the polymorphic structure. Along with small hepatocytes in histological sections, there were clusters of large cells with nucleus with karyoplasm filled with chromatin. Sinusoidal capillaries were moderately blood-filled. Stellate reticuloendotheliocytes located in the walls had basophilically stained cytoplasm.

The histological structure of birds' kidneys after seven days of exposure to the studied preparation was characterized by signs of weak edema of the glomerular capsule cavity. Renal corpuscles were characterized by polymorphism and increased volume. Weakly marked capillary profiles, low number and atrophy of mesangial cells in the glomerular vasculature indicated impaired glomerular filtration. Epithelial cells of the wall of proximal sections of convoluted tubules showed signs of metabolic disorders in the form of granular dystrophy resulting in accumulation of oxyphilic protein mass in the lumen of some tubules of this section. Epithelium of tubules of the nephron loop and distal parts of tubules, as well as collecting tubes were characterized by the absence of destructive phenomena.

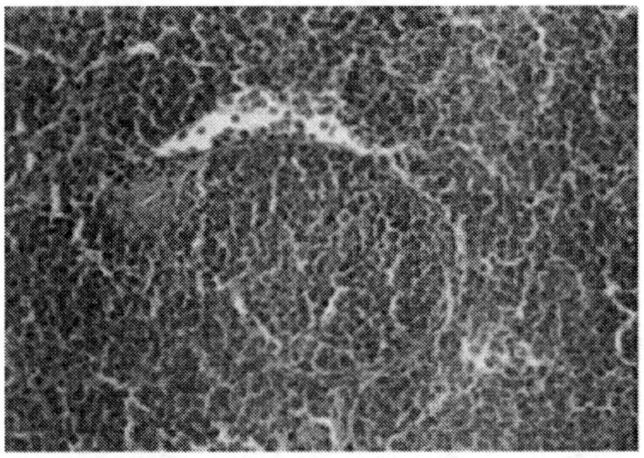

Fig. 6. Spleen. Lymph node with well-marked and structurally functional areas on the 22nd day after application of Degelm-14. Hematoxylin and eosin staining. 140.

The kidneys on the 22nd day of "Degelm-14" application had a well-preserved structure. The capsule cavity of the organ's glomeruli was characterized by slight edema. Nuclei of numerous podocytes were distinguished by low number of hyperchromic nuclei. Weak signs of protein dystrophy with the presence of oxyphilic masses in the tubule lumen were observed predominantly in the epitheliocytes of the proximal wall, creating an additional obstacle for primary urine outflow in some nephrons. Vascular glomeruli of juxtamedullary nephrons were characterized by more significant edema of capsule cavity with flattening of podocytes. Minimal manifestation of metabolic disorders was noted in the epitheliocytes of collecting tubes. Most of them had free lumen.

4 Conclusion

The conducted studies show that the application of the preparation "Degelm-14" after infection of birds, the histological structure in most of the studied organs was gradually restored. In the glandular stomach and small intestine on the 22nd day after application of the studied preparation the microstructure of the organ was completely restored. Microstructural changes in the cloacal pouch and spleen after application of Degelm-14 indicated restoration of the proliferative activity of the cells of the germinal centers. In the liver, at the final stage of the study, there were no signs of metabolic disorders in hepatocytes, and the dark pink coloration of their cytoplasm indicated an increase in synthetic processes in the cells of the organ parenchyma. But in the kidneys there were still observed the signs of glomerular filtration disturbance, regulating the organ function. Pathomorphological evaluation of the action of Degelm-14 showed the presence of active eimeriosis activity of this compound in relation to eimeriosis pathogens. During all periods of the study in the intestinal mucosa it was not possible to detect the presence of pathogenic schizonts. Effective purification of the body of sick birds was detected as early as on the 7th day after exposure to "Degelm-14". In the body, regenerative processes proceeded with increasing their expression both in the structures directly in contact with pathogens of invasion (gastrointestinal tract, liver, pharyngeal bursa), and in the organs which had no direct connection with pathogens of invasion (kidneys, spleen). It should be noted that the changes in the structure of organs and tissues of birds that received this compound were characterized by the lowest level of reactogenic action. Minimal manifestation of disorganization processes of connective and epithelial tissues at the initial period and their almost complete absence of expression on the 22nd day of experiment, against the background of regenerative and biosynthetic processes increase in the studied organs and tissues of birds treated with "Degelm-14", indicated the absence of reactogenic damaging effects of this tested agent. It should also be noted that the compound "Degelm-14" does not cause

https://doi.org/10.1051/bioconf/20225200042

side immunosuppressive effects on the structures of the birds' immune system organs, while other studied drugs and compounds acted with significant damage to the structure of spleen lymph nodes, subepithelial lymphoid tissue of gastrointestinal tract organs.

References

1. R. A. Asrutdinova, I. N. Zalyalov, E. G. Kirillov, *BIO Web of Conferences* **17,** 00184 (2020)
2. E. O. Kachanova, *Proceedings of VEEV* **80,** 177–182 (2018)
3. A. A. Mironova, *Actual problems and methodological approaches to the diagnosis, treatment and prevention of animal diseases: mater. all-Russian scientific and practical conference* (Persianovsky: Donskoy GAU) 128–133 (2017)
4. A. A. Mironova, L. P. Mironova, T. V. Molokova, Actual problems and methodological approaches to the diagnosis, treatment and prevention of animal diseases: mater. all-Russian scientific and practical conference (Persianovsky: Donskoy GAU) pp 128–133 (2017)
5. O. T. Mullakaev, I. N. Zalyalov, E. G. Kirillov, *BIO Web of Conferences* **17,** 00031 (2020)
6. R. T. Safiullin, *Scientific Notes of Kazan State Academy of Veterinary Medicine. N.E. Bauman.* **236,** 4-169, 174 (2018)
7. J. H. Oxford, *Selvaraj Journal of Applied Poultry Research* **28 (4),** 1279–1287 (2019)
8. K. Sidorenko, M. Mkrtchyan, Y. Kuznetsov, *Advances in Animal and Veterinary Sciences* **8,** 2.7, 11 (2020)
9. S. M. Silva, C. Ferreira, J. Paupério et al., *Acta-Parasitologica* **60 (2),** 350–355 (2015)

Analysing the staff supply of a dairy enterprise in the Samara region

O. F. Pyatova, *T. V.* Shumilina*, *T. G.* Lazareva, and *Yu. Yu.* Gazizyanova

Samara State Agrarian University, 2, Uchebnaya St., Kinel, urban-type settlement Ust-Kinelsky, Samara Region, 446442, Russian Federation

Abstract. The key objective of any enterprise's staff policy is to form a system of providing managers, specialists and workers of mass professions who can solve tasks of future development at a high professional level. The volume and timeliness of all work, the efficient use of fixed assets and, as a result, the volume of production, its cost, profit and several other economic indicators depend on the availability of labour resources and the efficiency of their use. The most important social and labour indicators in enterprises are ensuring full employment of labour resources and high labour productivity, creating good working conditions for staff and increasing wages, achieving proper economic growth and quality of life for workers. Labour has a major role to play in realizing these economic objectives. The article analyses the indicators characterizing the level of labour supply of the enterprise, their qualitative composition and the level of labour productivity.

1 Introduction

One of the main conditions for the successful production and social development of an enterprise is the selection and recruitment of staff [6]. Staff supply is an activity that aims at staffing a team of workers who are professionally trained and able to perform their duties effectively within the law and within their job description, at the level of up-to-date requirements. Staff policy is not only about providing effective staff and encouraging them to work constructively, but also about creating attractive working conditions, safety and opportunities for advancement [2]. During the recruitment process, the employer does not know the actual quality of the workers, but has access to several characteristics (education, age, gender, work experience), considered as signalling information about the capabilities and abilities of the worker [1]. Staff efficiency includes not only the attributes and characteristics of individual workers, but also the quality of interaction of these workers and their groups among themselves. It also involves the psychological climate, in which the release of everyone's energy has a synergetic effect. Thus, the staff of an enterprise and their changes have certain quantitative, qualitative and structural characteristics to be measured with lesser or greater certainty and reflected in the following absolute and relative indicators:

– the average number of workers of the enterprise and (or) its internal divisions for a certain period;

– the share of workers of individual divisions (groups, categories) in the total number of workers of the organization;

– the rate of growth (increase) in the number of workers of the organization for a certain period;

– the share of workers with higher or secondary vocational education in the total number of workers and (or) workers of the organization;

– the average work experience in the specialty of managers and specialists of the organization;

– the list and attendance number of workers of the organization and (or) its internal divisions, certain categories of groups on a certain date;

– the staff turnover.

2 Results and discussion

When analysing the structure of labour resources by category and profession, it is important to assess whether the enterprise has sufficient labour resources of a particular quality. The supply of appropriate labour resources depends on comparing the actual number of workers by category and profession with the planned need. Particular attention focuses on analysing the availability of human resources for the most important professions [3].

Table 1 gives an indication of the labour force availability of the enterprise.

Tabl e1. Labour supply of the enterprise in 2021, people.

* Corresponding author: tanyashum86@mail.ru

Category of workers	Planned	Actual	Supply, %
Average number of production staff	210	197	93.8
including workers	175	167	95.4
including butter shop workers	23	21	91.3
raw shop workers	40	42	105.0
whole-milk workers	50	48	96.0
support workers	62	56	90.3
Employees	35	30	85.7

The average number of workers decreased by 13 people, or 6.2%. This decrease was mainly due to auxiliary workers. There was also a 14.3% drop in the number of employees due to the dismissal of an accountant, an economist and posts are still vacant.

An important factor in the qualitative composition of the workforce is the optimum ratio of men to women in different positions and occupational groups. General trends in the participation of men and women in different positions are as follows: men are better able to cope with physically demanding work, line managers (directors of enterprises, heads of workshops, sites, shifts), while women have proven themselves in functional management positions (heads of departments, offices, sectors, groups) and specialists. Women also find it easier to cope with monotonous work and areas requiring tidiness. An analysis of the qualitative composition of staff with their education and qualifications involves determining the number of workers with different levels of education, the qualitative level of placement of workers in their positions, the extent to which specialists with higher education are used rationally, etc.

Analysis of the qualitative composition of the workforce involves studying workers by gender, age, education, qualifications, length of service and other socio-demographic characteristics (Table 2).

Table 2. The qualitative composition of the enterprise's workforce.

Group of workers	Number of workers, people		Specific gravity, %	
	in average for 2017–2020	for 2021	in average for 2017–2020	2021
Age composition, years: up to 20	8	5	3.9	2.5
from 20 to 30	98	87	48.3	44.2
from 30 to 40	44	60	21.7	30.5
from 40 to 60	47	41	23.2	20.8
above 60	6	4	2.9	2.0
Total	203	197	100	100
By education level: Secondary, secondary vocational	165	157	81.3	79.7
Higher	38	40	18.7	20.3
Total	203	197	100	100
By service length, years: Up to 5	8	10	3.9	5.1
from 5 to 10	90	76	44.3	38.6
from 10 to 15	40	45	19.7	22.8
from 15 to 20	47	51	23.2	25.9
over 20	18	15	8.9	7.6
Total:	203	197	100	100

Examining the groups of workers by age, we can see that the largest share of workers is in the 20-30 age group, which decreased by 3.9 p.p. in 2021. The share of working pensioners decreased from 2.9% to 2% in 2021 in the total share of workers. This is a positive development, as most of the workforce comprises people in the prime of their lives and energy. Young people entering the working age are the primary source of labour replenishment in the economy by replacing the jobs of older generations that are leaving.

Most workers have secondary and secondary vocational education – 81.3% of the total share of workers. A positive development is an increase in the share of workers with higher education by 1.6 points in 2021 compared to the average data for the last 4 years. This category of workers consists mainly of managers and specialists of the enterprise.

In terms of length of service, workers with between 5 and 10 years of service have the largest share. Although in 2021 the share of workers between 15 and 20 years of service decreased and the share of workers between 15 and 20 years of service increased. 7.6% of all employees have been with the company for over 20 years.

The number of workers and their professional and qualification structure can be determined through the production programme, the production rates, the planned increase in labour productivity and the structure of the work. The qualitative composition of the labour force in terms of qualifications needs to be analysed (Table 3).

Table 3. Qualification composition of the enterprise workers in 2017□2021.

Worker qualifications	Pay rate multiplier	Number of workers, people	
		in average for 2017–2020	for 2021
First rate	3.96	5	5
Second rate	4.11	12	13

Third rate	4.41	59	61
Fourth rate	4.69	87	80
Fifth rate	7.96	10	10
Sixth rate	10.06	30	28
Total	–	203	197
Average pay rate of workers	–	3.86	3.82
Average pay rate multiplier	–	5.51	5.48

The movement of workers has reduced their qualifications: the average pay grade and tariff coefficient have fallen by the end of 2021.

Since changes in the qualitative composition come as a result of the movement of the labour force, this issue receives a great deal of attention in the analysis (Table 4).

Table 4. Data on the movement of the workforce in the enterprise.

Indicators	In average for 2017-2020	for 2021
Average annual number at the beginning of the year, people	204	195
Employed, people	51	24
Outflow, people.	54	20
resigned by their own volition	26	10
dismissed for breach of work discipline	8	4
Number of staff at the end of the year, people	201	199
Average number of workers, people.	203	197
worker turnover rate	0.25	0.12
Worker attrition rate	0.12	0.10
staff turnover rate	0.17	0.07
staff consistency factor	0.77	0.81

The intake turnover rate in 2021 decreased by 0.13 points compared to previous years. The situation is similar for the number of departures. Mostly, people resign of their own volition. Only a small percentage, namely 2% of workers, was dismissed in 2021 for breaches of labour discipline. staff turnover is very low, over the period and has continued to decline. The staff permanence rate is increasing, which is also a positive development in this situation.

The rate of utilization of working time is the indicator of team efficiency because the amount of production is proportional to the amount of working time and inversely proportional to its labour intensity at a given level of labour productivity.

We analysed the use of working time based on timesheets, photographs of working time, sick leave, overtime, idle time and pay slips. Preliminarily, we analysed the use of labour resources by worker category in the enterprise (Table 5).

Table 5. Use of labour resources by worker category.

Category of workers	Average annual number, people.	Days worked per year by one worker	Hours worked per year by one worker	Average working hours, h	Total working time, hours
Workers in the butter shop					
on average for 2017-2020	23	214	1605	7.5	36915
for 2021	21	207	1656	8.0	34776
Cheese shop workers					
on average for 2017-2020	43	210	1638	7.8	70434
for 2021	42	207	1615	7.8	67830
Workers in the whole-milk shop					
on average for 2017-2020	51	204	1530	7.5	78030
for 2021	48	207	1656	8.0	79488
Support workers					
on average for 2017-2020	56	219	1577	7.2	88301
for 2021	56	207	1635	7.9	91577
Employees					
on average for 2017-2020	32	216	1620	7.5	51840
for 2021	30	207	1656	8.0	49680

The analysis shows that there are no fundamental changes in the use of working time at the enterprise. The decrease in numbers in 2021 has led to a reduction in the total working time of the workers in the butter and cheese shop and the workers. A reduction in the number of workers in the whole-milk shop was compensated for by an increase in working hours.

To identify the causes of daytime and shift losses, we compare the actual and planned balances of working time. Table 6 presents the analysis.

The enterprise lost the most unproductive time due to subjective factors: additional leave with the permission of the administration, absenteeism, and downtime, counting as unused resources to increase the working time fund. The company also has significant non-productive labour costs, which resulted from time spent on producing substandard products and correcting defects, as well as due to deviations from the technological process. They amount to 591 h.

Table 6. Analysis of the working time use in 2021, days.

Indicator	Per worker		For all staff	
	Planned	Actual	Planned	Actual
Calendar number of days	365	365	76650	71905
including holidays and weekends	120	120	25200	23640
Nominal working hours	245	245	51450	48265
Absences at work	36	46	7560	9062
including annual leave	24	24	5040	4728
study leave	2	3	420	591
additional leave of absence with the approval of the administration	4	5	840	985
diseases	6	11	1260	2167
absenteeism	–	2	–	394
down time	–	1	–	197
Working time available	209	199	43890	39203
Working shift duration, h	8.0	7.8	–	–
Working time budget, h	1672	1552.2	351120	305783.4
Pre-holiday shorter days, h	14	14	2940	2758
Intra-shift downtime, h	-	58.6	–	11544.2
Useful working time availble	1658	1980.6	348180	390178.2
Overtime worked, h	–	8	–	1576
Non-productive working time, h	–	3	–	591

The basis of society's development is the production of material goods, and the labour of the people who create these goods, along with nature, is the source of its social wealth. The more productive the labour, the richer the society, and vice versa.

Increasing productivity is a universal economic law.

To determine the average annual output per worker and to calculate the impact of the factors, see Table 7.

Table 7 shows a decrease in output per worker in 2021, compared to previous years, in the amount of 15 thousand rubles, while there is an increase in the output of 16 thousand rubles relative to the worker's intensity of work. There is also an increase in the average daily and average hourly output, both compared to the previous period and to the plan.

Determine the effects of factors such as the share of workers, the number of days worked by one worker per year, the length of the working day and the average hourly output of workers (using absolute differences) on the annual average output of one worker.

Table 7. Input data for factor analysis of labour productivity.

Indicator	In average for 2017-2020	for 2021		Deviation	
		planned	actual	from planned	from average
Average annual number of workers, people	203	210	197	–13	–6
including workers	171	175	167	–8	–6
Share of workers in the total number of workers	0.84	0.83	0.85	+0.02	+0.01
Days worked per year by one worker	211	209	199	-10	-12
Average working hours, h	7.9	8.0	7.8	-0.2	-0.1
Product output, thousand rubles	54738	56230	56200	-30	1462
Average annual output per worker, thousand rubles	270	268	285	+17	–15
Worker's output: annual average, thousand rubles.	320	321	336	+15	+16
average daily, rubles.	1516	1537	1691	+154	+175
average hourly, rubles	192	192	217	+25	+25

In 2021, the average annual output per worker increased by 17,000 rubles compared to the target:

a) the share of workers in the total number of company staff increased by 6.4 thousand rubles;
b) the number of days worked by one worker during the year decreased by 13.0 thousand rubles;
c) the working hours decreased by 6.5 thousand rubles;
d) the average hourly output of workers increased by 30.1 thousand rubles.

The application of computer technology in determining the grade of milk reduced labour output by 37,000 hours, or 8.4%. As a result, the average hourly output rose by 9.2%, or 142 rubles.

Unproductive labour costs due to defective work amounted to 591 hours. This reduced the average hourly output by 0.14%, or 0.22 rubles.

Modernization of existing equipment has reduced labour costs by 6,780 man-hours, or 1.55%, resulting in an increase in the average hourly output of 1.6%.

Changes in product structure have a significant impact on the average production level. As the proportion of more labour-intensive products increases, the labour cost of producing them increases. Table 8 shows the calculation.

Table 8. Calculation of the impact of the production structure on the change in working time in 2021.

Type of product	Labour costs per 1.000 rubles of marketable production, hours	Commodity production, thousand rubles		Product structure		
		planned	actual	planned	actual	change
Butter	9.8	16000	15453	0.365	0.330	–0.035
Fat cheese	10.0	15800	16007	0.361	0.340	–0.021
Whole-milk products	10.3	10600	14214	0.242	0.300	+0.058
Low-fat	12.5	70	64	0.002	0.001	–0.001
Non-fat cheese	10.1	1300	1547	0.030	0.029	–0.001
Total	10.0	43770	47285	1.0	1.0	–

We calculate the change in average labour intensity as follows:

$$[(-0.035) \times 9.8 + (-0.021) \times 10 + (0.058) \times 10.3 + (-0.001) \times 12.5 + (-0.001) \times 10.1] = 0.0218 \text{ h}$$

Thus, the increase in labour output per 1.000 rubles of production was 1.31 min.

For the entire production 1.031 man-hours.

Due to the increase in the share of more labour-intensive products, such as whole-milk products, the total labour costs increased by 1031 man-hours, but due to changes in the production structure, the production in value terms also increased by 11298 thousand rubles.

The return on staff indicator is of great importance for assessing the efficiency of the use of human resources in a company in a market economy.

Calculating these figures for 2021 and the average for the previous 4 years, we obtain 142.1% and 102.5% respectively. The structural-logical model of factor analysis of staff profitability allows us to establish how much the profit per worker has changed due to the level of profitability of sales, the share of revenue in total production and labour productivity. Table 9 presents the data for the analysis.

Table 9. Data for factor analysis of staff profitability.

Indicators	In average	for 2021	Deviation

	for 2017-2020		
Profit from product sales, thousand rubles	208	280	+72
Average number of workers, people.	203	197	-6
Profit per worker, rubles	1025	142 1	396

In Table 9, the profit per worker in 2021 is higher than the average for the past 4 years by 396 rubles, including the change:

a) labour productivity – increased by 81 rubles;
b) the share of sold products in total production – increased by 11 rubles;

c) profitability of sales – increased by 304 rubles.

Errors in the calculations are due to rounding.

A combination of favourable economic and social factors is necessary to optimise the use of the workforce. The study of their impact on the efficiency of labour force utilisation makes it possible to identify the reserves of savings of live labour due to individual factors, imbalances in the equipment of processing production with productive forces. It determines specific measures for the implementation of the final objectives in the near and more distant future.

The impact of individual factors on labour efficiency varies. We have found the closest relationships between labour productivity, labour efficiency, labour capital formation, labour supply and employment, labour force composition, remuneration level and labour regime, development of the nonproduction sphere [4, 5].

Depending on the nature of the interaction between the factor and the result, we have identified the following reserves for efficiency gains as a result of our calculations.

When calculating the labour supply of the mill, it was found that the enterprise has a labour supply of 93.8%. If, however, the actual number of workers is equal to the planned number of workers, the gross product will consequently increase.

The reserve for increasing production by employing additional labour can be determined by multiplying the increase by the actual average annual production per worker ($W \uparrow GP$ = 13 × 285 = 3705 thousand rubles.). As a result of the calculations, we obtain that the reserve for increasing production in 2021 is equal to 3,705 thousand rubles.

Reducing loss of working time is one of the reserves for increasing production. To calculate it, we need to multiply the loss of working time through the fault of the enterprise (whole-day and intra-shift downtime, reduction of holiday with the permission of the administration, absenteeism, unproductive work) by the planned average hourly production of production: ΔGP = 13711.2 × 192/1000 = 2633 thousand rubles.

In turn, the successful development of production in the enterprise will contribute to an increase in jobs [7].

3 Conclusion

The analysis has led to the following conclusions: the total number of workers is decreasing each year, indicating a decline in production. It is mainly the share of auxiliary workers and clerical workers that is decreasing. This indicates a fairly rational workforce allocation, as it is the proportion of key workers that directly affects the amount of production.

The educational structure of the workforce shows that there is a decreasing trend in the proportion of workers with higher and secondary vocational education. But unfortunately, workforce deployment is not entirely rational, i.e. the qualifications of a worker are not always the same as the work they do. The analysis revealed that the percentage of correspondence between the complexity of the work and the qualifications of the workers performing it is low, indicating the misuse of workers in the enterprise.

We can describe the sex and age composition as satisfactory. The average age of workers is 42-43 years, which is a favourable and capable age for workers.

The indicator for length of employment is also the most favourable, at 8 years, and characterises the workforce as stable.

An analysis of the movement of human resources in the company has revealed that there is a problem with staff turnover. This has led to a waste of human resources: increasing recruitment, hiring and adaptation costs.

An analysis of the use of working time revealed that the company's workforce is also under-utilised. In particular, the use of working time in 2021 was 95.2 % on average by one worker. The main reasons for these losses include increased absenteeism due to sickness, with the permission of the administration, and absenteeism.

To improve the efficiency of the company, we can suggest measures to improve the use of human resources:

- providing training and skills development for workers in the main trades. This activity is aimed at a more rational professional-qualification placement of workers.
- introducing additional payments and wage supplements as a method aimed at reducing staff turnover.

References

1. D. V. Bezvitelnov, N. V. Kotelnikova, Bulletin of Dimitrovgrad Engineering and Technological Institute 2d24() 75–79 (2021)
2. G. G. Vukovich, A.V. Nikitina, Society: Politics, Economics, Law 1) 34–38 (2018)
3. Yu. E. Galkina, I. V. Golovkin, Economics and Management of Innovative Technologies 1, (2018), retrieved from: https://ekonomika.snauka.ru/2018/10/16184

 https://doi.org/10.1051/bioconf/20225200043

4. B. M. Genkin, *Motivation and organization of effective work (theory and practice)* (Moscow: Norma: INFRA-M) (2020)
5. M. S. Gorina, Innovations and Investments **6)** 189–195 (2019)
6. E. A. Okunkova, Bulletin of the Altai Academy of Economics and Law **12dB()** 183–189 (2019)
7. O. F. Pyatova, T. V. Shumilina, O. V. Chumakova, S. M. Mirgazimova IOP Conference Series: Earth and Environmental Science **341d1()d**012014 (2019)

Development of technology for the production of thermally conductive carbon foams using microspheres and pitches

Tagir Vagizov[1,*], *Rishat* Akhmetzyanov[1], *Ilgiz* Galiev[1], *El'mira* Sharafutdinova[2], and *Nazirya* Galimova[2]

[1] Kazan State Agrarian University, 65, K. Marx St., Kazan, 420015, Republic of Tatarstan, Russian Federation
[2] Kazan National Research Technological University, 68, K. Marx St., Kazan, 420015, Republic of Tatarstan, Russian Federation

Abstract. The technological process of obtaining thermally conductive carbon foams based on microspheres and pitches, including the stages of mixing the initial components, pressing the mixture of components, obtaining semi-finished "green" foam, low-temperature carbonization, carbonization, high-temperature heat treatment, pyro-compaction, mechanical processing has been developed. It is shown that the structure of the resulting foam is characterized by a lower content of voids, the layered structure of the walls of microspheres. The obtained values of thermophysical and physical-mechanical characteristics provide the possibility to use the foams for work in extreme conditions

1 Introduction

1.1 Relevance

The development of many industries necessitates the development of new and unique materials capable of working under extreme conditions while retaining their technical properties during operation. In recent decades a lot of attention is paid to the research aimed at development of carbon materials of the foam structure [1-9]. The main peculiarity of carbon materials as a class is a possibility of directed development of the porous structure during realization of the technological process by selection of fillers, additives and binders, heat treatment regimes. At the same time it is possible to regulate the total porosity, specific surface area and pore-size distribution in rather a wide range. Parameters of the porous structure have a great influence on physical-mechanical, thermal-physical and other properties of carbon foams [10-15].

One of the promising directions in the field of modern materials science is the development of chemically and corrosion-resistant high-temperature materials, including carbon-based ones. It is known that carbon materials exist in a wide variety of forms: carbon, graphite, diamond, fullerenes, glass carbon, graphene, highly oriented pyrolytic graphite, thermally expanded graphite, graphite polyhedral crystals, nanofibers, nanorods, single-layer and multilayer nanotubes, graphite whiskers, carbon black, carbon fibers, carbon foams, etc. Among the wide variety of listed materials, foam structure materials, including syntactic carbon foams, occupy a special place in terms of efficiency and prospects of application. A distinctive feature of syntactic carbon foams is the presence of a well-developed regular macroporous structure, which determines the maximum level of technical properties compared to other foamy materials at the same density. In addition, they have high specific strength characteristics, a heat conductivity coefficient adjustable in a wide range, high heat and heat resistance, electrical conductivity, resistance to aggressive media and other special properties.

Analysis of the situation in the field of obtaining and using carbon foams has shown that the problem of developing basic technological solutions for creating a new generation of heat-insulating and heat-conducting syntactic carbon foams with a wide range of values of the thermal conductivity coefficient and physical and mechanical characteristics, as well as approaches to modeling technological processes and designing products based on them, is extremely relevant.

Syntactic foams are the types of composite materials formed by binding hollow microspheres with a metal, polymer or ceramic matrix [6]. In this context, the term "syntactic" means "taken together", and more precisely means the regularity of the structure, when an elementary cell can be isolated, the repetition of which can describe the structure of the foam macroform. The shells of hollow microspheres can be glass, polymer, carbon, ceramic, metal.

Basic requirements for microspheres are flowability, strength, defect-free, moisture and chemical resistance, hydrolytic strength, the possibility of changing the granulometric composition and volume filling coefficient within wide limits. Obtaining foams with optimal strength properties is possible only if the gas inclusions have a spherical shape. The diameter of microspheres, as a rule, is 1–500 microns, the wall thickness is 1–4 microns, the bulk density is 70–500 kg/m^3, the apparent density is 50–250 kg/m^3.

* Corresponding author: toha-174@mail.ru

Microspheres with a diameter of up to 40 mm are also used as hollow fillers of synthetic carbon foams.

The presence of hollow particles provides a low density of foams, a low coefficient of thermal expansion, and in some cases transparency for electromagnetic or acoustic waves. In the process of carbonization, oxidation, and solvent exposure, microspheres can transform to form foam with open cells. The structure of synthetic carbon foams provides for the presence of open pores, their connection with each other by a system of regular channels and a well-developed three-dimensional macroporous structure.

For operation of thermal-conductive foams in extreme conditions the following properties are required: bulk density – 0.1–1.0 g/cm^3; porosity – 60-97 %; thermal conductivity coefficient is from 10 to 180 W/m·K; thermal expansion coefficient – (2–8)· 10^{-6} K^{-1}; compressive strength – 0.3–20.0 MPa; fracture strength – 0.3–10.0 MPa; stability in inert environment – 2800°C, in air 600°C; carbon content is not less than 99%.

The aim of the work is to develop the technology, study the structure and operational properties of thermally conductive foams based on graphitizing, including mesophase-forming materials.

1.2 Input components

In order to obtain foams with high values of thermal conductivity coefficient we used raw materials that provide a high degree of graphitization during heat treatment. A necessary condition for this is the formation of a visco-plastic phase (mesophase as a liquid crystalline state of matter) during carbonization. This is ensured by the use of mesophase-forming coal and petroleum pitches with high values of the coefficient of thermal conductivity. In order to increase the physical and mechanical properties and the thermal conductivity coefficient of foams, it is effective to use additives of mesophase particles (microspheres), which helps to increase the degree of graphitization of the material. Carbon microspheres produced by carbonization of hollow phenol-formaldehyde resin (PFR) microspheres at 900°C were used as a filler. High-temperature coal tar pitch with a softening point of 120°C was used as a binder.

In carrying out this work, an integrated approach was used, according to which technologies for obtaining blanks of products with controlled properties based on syntactic carbon foams are considered as a set of interrelated processes for converting raw materials, which are consistently transformed at the stages of technological conversion into blanks of products, the properties of which are predetermined by the variety of relationships between the characteristics of the raw materials, as well as the operating parameters of the processes.

2 Materials and methods

Foams obtaining technology consists of the following stages: mixing of microspheres and pitch; formation of "green" foam by pressing a mixture of components; low-temperature carbonization of samples of "green" foam; high-temperature treatment; pyrocompaction of graphitized foam samples; mechanical processing.

2.1 Mixing of components

In a three-neck round-bottom flask of 500 ml we select and transfer portions of carbon microspheres (25.0±0.2 g), coal or oil pitch (25.0±0.2 g) in the form of plates, powder. A reflux condenser with chloralkalcium tube is attached to the flask and a stirring element of a mechanical stirrer with fluoroplastic working part is placed therein. While stirring at 60 rpm, 250 ml toluene is added to the contents and heated in an oil bath. The mixture was boiled under reflux while stirring at 100 rpm for 90 minutes, after which the contents of the flask was cooled to 30±5°C and portions of 150±50 ml were transferred into a 500 ml flask. The mixture is evaporated at a rotary evaporator (40°C, 250±50 mm Hg until complete removal of toluene). The mixture is then dried under vacuum while stirring on a rotary evaporator (120°C, 12±5 mm Hg, 60 min). The obtained granulate is removed from the flask and ground to a particle size of less than 250 microns to obtain a press powder.

The initial component is a coal or oil pitch with a mesophase yield of at least 20 wt. %. If necessary to work with pitches having a mesophase yield of less than 25 wt. %. Mesophase microspheres obtained by heat treatment of the initial pitch at a mesophase temperature (400–500 °C) followed by extraction of the soluble part in toluene can be added to increase the thermal conductivity of the foam. Mesophase microspheres are added to increase the mesophase yield from the pitch to 40–50 wt. %. Carbon micro- and nanoparticles in the form of finely ground carbon black, carbon nanotubes, natural graphite, graphene nanoparticles in the amount of 0.1–5.0 wt. % can be added to the pitch in order to increase the strength of the material.

The production of "green" foam. The powder is placed in a matrix with an inner diameter of 35 mm and shaped in a hydraulic press. Pressing regimes are loading speed of 0.25–0.35 tf/min, pressing pressure – 1.2 tf, holding at pressure for 3 minutes. The moulded billet is extracted, its overall dimensions, mass, apparent density are controlled.

Low temperature carbonization. Samples of "green" foam are wrapped in a double layer of paper, placed in a steel container, covered with a layer of graphitized swarf 5–10 cm high, the top is poured layer (0.5–1.5 cm) of high-temperature coal pitch, covered with a lid and placed in a muffle furnace. The samples in a container are heated at a rate of 2.5±0.5°C/min to the mesophase-forming temperature of the pitch (450 °C), held for 150 minutes, after which the furnace is turned off and cooled to 60±40°C for 12 hours. The samples are removed, the overall dimensions, shrinkage, apparent density, ultimate compressive strength and thermal conductivity coefficient are controlled.

Carbonization. Samples of "green" foam are wrapped in a double layer of paper, placed in a steel container, covered with a layer of graphitized swarf 5-10 cm high, on top of which we poured a layer of high-temperature coal pitch of height 2.5 mm, covered with a lid and placed in a muffle furnace. The samples in the container are heated at a rate of 2.5±0.5 °C/min to 900 °C, incubated for 120 minutes, then the furnace is turned off and cooled to 60±40 °C for 12 hours. The samples are removed and the overall dimensions, shrinkage, apparent density, ultimate compressive strength and thermal conductivity coefficient are checked.

High temperature heat treatment. Obtained samples are placed in graphite crucibles with lids; the filler is poured as waste material from artificial graphite production. Crucibles are placed in an electric vacuum furnace, the furnace interior is evacuated to a residual pressure of less than 1 mm Hg. The furnace is heated for 8 hours to 2100 °C, allowed to stand for 2 hours and then cooled naturally. The samples are removed and the loss of mass, apparent density, compressive strength and heat transfer coefficient are monitored. The obtained samples are placed in graphite crucibles and the waste material from artificial graphite production is poured onto the crucibles. The crucibles are covered with a lid, placed in a graphite furnace and heated to 2700 °C and incubated for 60 minutes. Then the samples are extracted, weight loss, apparent density and true density, ultimate strength and compressive modulus of elasticity, thermal conductivity and thermal expansion coefficients, thermal and oxidation resistance are controlled.

Pyro-compaction. Samples of graphitized foam are loaded into the chamber of pyro-foam oven, the oven is sealed, purged with argon, then the samples are heated and kept at 800-1100 °C, methane pressure of 8-12 mm Hg for 20-120 hours. After that the samples are extracted, weight gain, apparent density, ultimate compressive strength and heat transfer coefficient are controlled.

Mechanical processing. Foam samples can be machined by cutting, drilling, milling, grinding at low tool feed rates with a Mohs hardness greater than 7.

3 Results and discussion

The results of scanning electron microscopy (SEM) examination of foams samples are shown in the figure. It can be seen that the basis of the obtained foams is microspheres with holes or partially destroyed probably due to the volatiles release during carbonization (Fig. 1, a, b). The size of obtained microspheres is less than 90 microns (average – 60 microns), bound by dense interlayers of binder which is a product of carbonization of pitch. The foam structure is characterized by a reduced content of voids due to the release of volatiles from the binder, which may be associated with a higher carbon content in the stove compared to the PFR. The binder layers are markedly textured, indicating the formation of graphite with a jet structure (Fig. 1, c–e). It is necessary to note also the formation of small isometric particles with the size less than 10 microns and chains of particles (fig. g-k) that specifies the course of secondary processes in a gas phase on the mechanism “liquid – steam – crystal” with the formation of soot-like particles from light fractions of pitch or not removed toluene residues.

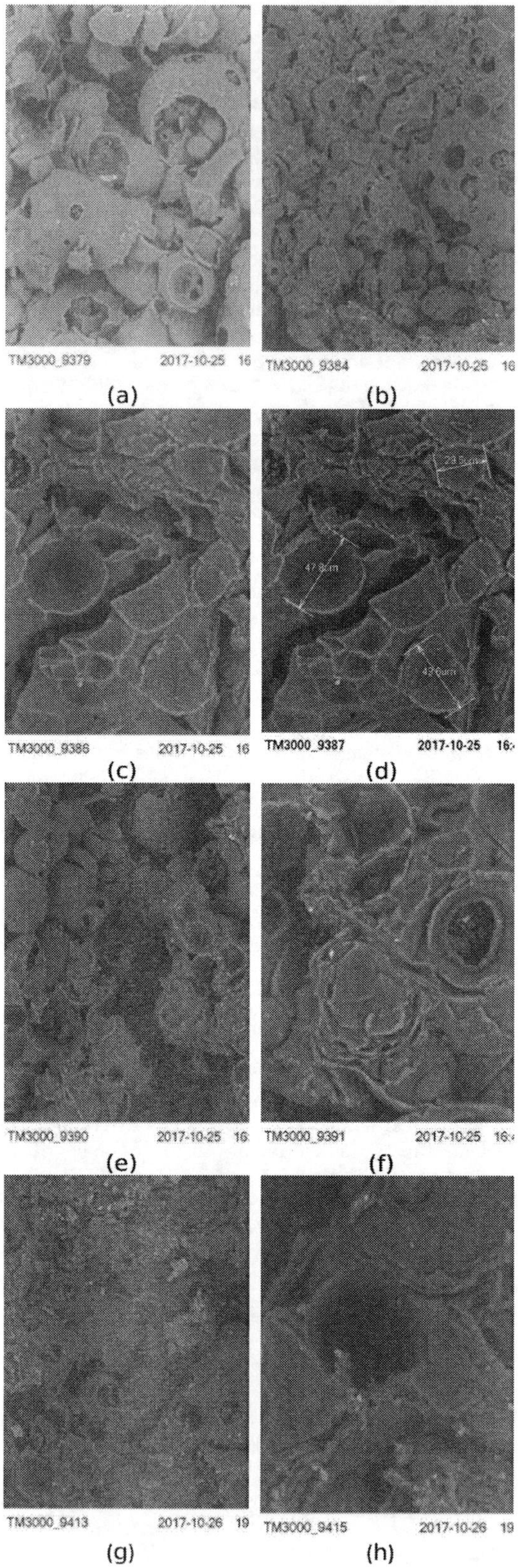

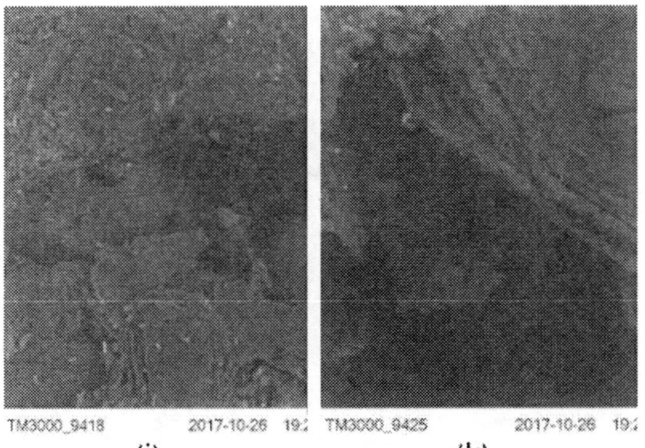

Fig. 1. SEM images of carbon foam samples.

A distinctive feature of the foam samples is the formation of a layered structure of the walls of the microspheres, indicating the course of coke formation, the subsequent graphitization of the binder on their surface, a change in the degree of graphitization of the walls material (Fig. 1, f, h, k).

The results of experimental studies of the basic technical properties of carbon foam samples obtained using the technology described above are shown in Figures 2–5.

Figure 2 shows the nature of the change in the apparent density of foams from the content of microspheres in the initial mixture. It can be seen that the dependence of the density is almost linear, which makes it possible to control the density of the material by changing the ratio of the initial components (filler – binder).

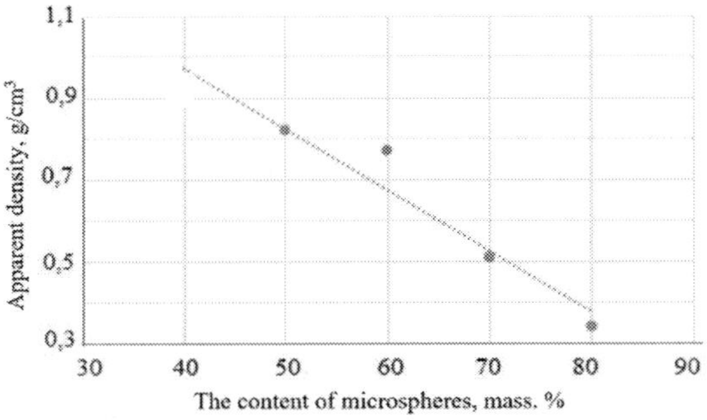

Fig. 2. Dependence of the apparent density samples on the content of microspheres.

Strength is the main requirement for structural materials, and the level of 20 MPa is sufficient for mechanical processing of the resulting foam samples. Figure 3 shows that the foam meets these requirements when the content of microspheres is 40 wt. % or less. In general, the strength naturally decreases with an increase in the content of microspheres, which corresponds to an increase in the porosity of the material.

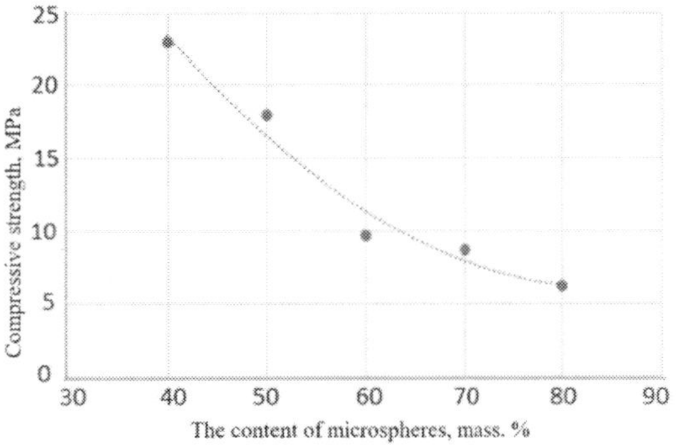

Fig. 3. Dependence of compressive strength samples on the content of microspheres.

Hardness is an important property that also determines the possibility of mechanical processing of foams. It can be seen that with an increase in the content of microspheres, the hardness of the foams decreases (Fig. 4) due to an increase in the porosity of the material.

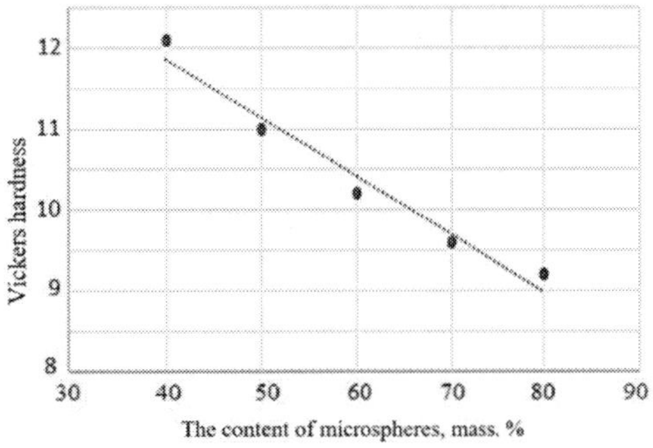

Fig. 4. Vickers hardness dependence foam on the content of microspheres.

Thermophysical properties determine the suitability of materials for use as high-temperature heat sink products. The tests carried out made it possible to establish that this class of materials satisfies the specified requirements in terms of temperature and oxidative resistance. The thermal conductivity coefficient clearly linearly depends on the content of microspheres (Fig. 5).

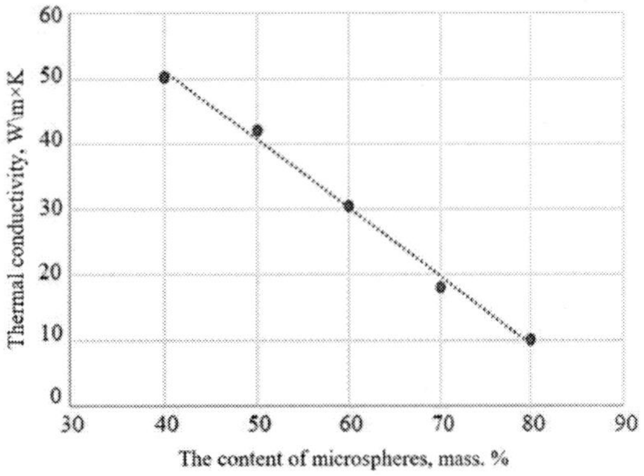

Fig. 5. The dependence of the coefficient of thermal conductivity foam on the content of microspheres.

4 Conclusion

Thus, a technological process of obtaining thermally conductive foams using carbon microspheres, coal or oil pitch was suggested and worked out. The process consists of the following stages: mixing the initial components in the form of carbon microspheres and coal or oil pitch in equal proportions; the formation of "green" foam by pressing the resulting mixture of components; low-temperature carbonization of samples of "green" foam; high-temperature heat treatment, pyrocracking of graphitized foam. The obtained values of the basic indices of the foams correspond to the predetermined characteristics, which makes it possible to use them for the manufacture of products operating under extreme conditions.

References

1. Y. Chen, B. Chen, X. Shi, H. Xu, W. Shang, Y. Yuan, L. Xiao, *Electrochimica Acta* **53,** (2008).
2. A. Fedden, *Thesis Air Force Institute of Technology AFIT/GA/ENY* (2006)
3. S. A. Nada, W. G. Alshaer, *Energy Conversion and Management* **105,** (2015).
4. F. A. Shutov, *Advances in Polymer Science* **73,** (1986)
5. M. Inagaki, J. Qiu, Q. Guo, *Carbon* **87,** (2015)
6. S.T. Benton, *Carbon* **10**, (1972)
7. E. Bruneton, *Carbon* **20,** (2002)
8. E. R. Galimov, T. N. Vagizov, A. V. Belyaev, *Solid State Phenomena* **284,** (2018)
9. E. E. R. Galimov, V. L. Fedyaev, V. M. Samoylov, E. A. Danilov, E. I. Timoshhuk, D. V. Ponomareva, A. V. Nakhodnova, M. A. Orlov, *Proc. 8th Int. Sc. and Tech. Conf. Innovative mechanical engineering technologies, equipment, materials* (2017)
10. V. L. Fedyaev, E. R. Galimov, V. M. Samoylov, *Proc. 8th Int. Sc. and Tech. Conf. "Innovative mechanical engineering technologies, equipment, materials 2017"* (2017)
11. E. R. Galimov, E. E. Tukbaev, V. L. Fedyaev, N. Ya. Galimova, V. M. Samoylov, M. A. Orlov, A. S. Borodullin, I. R. Mavliev, *Proc. 8th Int. Sc. and Tech. Conf. "Innovative mechanical engineering technologies, equipment, materials 2017"* (2017)
12. V. M. Samoylov, E. A. Danilov, E. R. Galimov, V. L. Fedyaev, N. Ya. Galimova, M. A. Orlov, *IOP Conf. Series: Materials Science and Engineering* **240** (2017)
13. G. Badun, M. Chernysheva, A. Grigorieva, E. Eremina, A. Egorov, *Radiochimica Acta* **104,** (2016)
14. N. Sorokina, D. Savchenko, S. Ionov, A. Tikhomirov, I. Nikol'skaya, V. Avdeev, *Journal of Physics and Chemistry of Solids* **71 (4)** (2010)
15. A. Grigorieva, E. Goodilin, K. Dubova, T. Anufrieva, L. Derlyukova, A. Vyacheslavov, Y. Tretyakov, *Solid State Sciences* **12 (6)** (2010)

Selection of the main parameters of tractors for direct sowing of grain crops according to various optimization criteria

Camill A. Khafizov, *Ramil N.* Khafizov, *Azat A.* Nurmiev* *and Ilnar N.* Gayaziev

Kazan State Agrarian University, 65, K. Marx St., Kazan, 420015, Republic of Tatarstan, Russian Federation

Abstract. The article discusses the questions of optimization of parameters and operating modes of the tractor as a part of the arable unit by various criteria of optimization. The following parameters have been chosen for consideration. These are minimum total energy costs, taking into account the energy of the lost crop; minimum emission of carbon dioxide into the atmosphere; minimum fuel consumption per 1 ha of crops; maximum performance; maximum traction efficiency of the tractor. The optimal weight of the tractor, the power of its engine, the width of the sowing unit and its working speed were determined by carrying out computational experiments with a change in the resistivity of the sowing complex. It was revealed that the optimal parameters of the tractor and the sowing unit do not coincide according to all optimization criteria with an increase in the specific resistance of the soil to the working bodies of the seeder, the optimal value of the tractor weight and the power of its engine grow. The optimal weight of the tractor should be 120 kN with the possibility of loading up to 150 kN, it is desirable to have a two-level engine of 500 and 600 hp. The working width of the seeder must be able to change from 14 to 18 m.

1 Introduction

In scientific articles and technical literature devoted to improving the efficiency of operation of machine and tractor units [1–11], different indicators are used in technological operations to characterize their work, various criteria for conducting optimization calculations, usually without justifying their choice. There is no comparative analysis of the results obtained using several optimization criteria (performance indicators). In this regard, the proposed article is devoted to the analysis of the optimal basic parameters of the tractor and sowing unit, obtained on the basis of calculations for a number of optimization criteria.

2 Materials and methods

To substantiate the main parameters of the tractor (tractor weight, engine power) and the parameters of the sowing unit (gripper width and operating speed), the method of mathematical modeling of the unit for direct sowing of grain crops according to optimization criteria was used. The criteria are minimum total energy costs, taking into account the energy of the lost crop; minimal emission of carbon dioxide into the atmosphere; minimum fuel consumption per 1 hectare of sowing. There is also maximum performance; the maximum traction efficiency of the tractor [1–4] and computational experiments carried out using the software application "Tractor Calculation", compiled on the basis of the use of mathematical models of sowing machine and tractor units.

3 Results and discussion

We will conduct computational experiments using the software application "Tractor Calculation" setting the operating conditions of the sowing unit and the values of external environmental factors of the tractor-operator-tool-field-soil-harvest (TOOPPU) system within the limits observed for the soil and climatic conditions of the Republic of Tatarstan [12–17]. Calculations were carried out in parallel according to all the criteria for optimizing the parameters of the sowing unit. For a comparative analysis of the obtained data, we will set two values of the specific resistance of the soil to the working bodies of the planter (4 kN/m and 6.25 kN/m). This factor most strongly affects the main parameters of the designed tractor, as well as the width of the gripper and the working speed of the planter for direct sowing. As a prototype of the projected seeder, a planter of the "Agromaster" type with cultivator type coulters was chosen [18–20].

Baseline data for calculations:

specific soil resistance per 1 m of the seeder = 4 kN/m;
field area = 100 ha;
cutting length = 1 km;
moving distance of the unit = 3 km;
seed density = 800 kg/m^3;
strength coefficient of the bearing surface of the field = 0.9;
amount of work performed by the unit = 500 hectares;
number of tractors performing the operation = 1;
number of operation hours of the unit per day = 16 hours;
planned yield of the main and by-products = 40 c/ha;
tractor tire pressure = 0.16 MPa;
number of wheels on one side of the tractor propulsion = 1;

* Corresponding author: azat-nurmiev@mail.ru

coefficient of adhesion of the wheels of the tractor with the soil = 0.65;
coefficient of resistance to rolling wheels of the tractor = 0.15;
mass of CO_2 absorbed by the crop (wheat) = 243 kg/c;
maximum tractor weight = 260 kN;
maximum engine power = 550 hp;
maximum working width of the unit = 18 m;
maximum speed of the unit = 12 km/h.

The results of the calculations are shown in Table 1 and in Figures 1-5. The table is divided into two parts according to the specific resistance of the seeder by 1 m of the width of the gripper of the unit. Each line corresponds to the results of calculations for one of the five optimization criteria and in them. In addition to the optimal parameters of the tractor and the unit for this criterion, the values of performance indicators corresponding to other optimization criteria are given.

As can be seen from Table 1 and Figures 1, 2, and 4, for the selected specific resistance of the seeder of 4 kN/m, the optimal values of the parameters: the weight of the designed tractor, the power of its engine, the width of the unit and its operating speed completely coincided according to three optimization criteria. These are the following optimization criteria: minimum total energy consumption (∑Ei), integral carbon dioxide emission ($\sum CO_{2i}$) and maximum unit capacity (W). The first two optimization criteria, unlike the rest of the criteria, take into account the influence of the parameters of the tractor and the unit on the size of the crop being formed, through the loss of a potential harvest due to an incorrectly selected tractor and seeder [21–25].

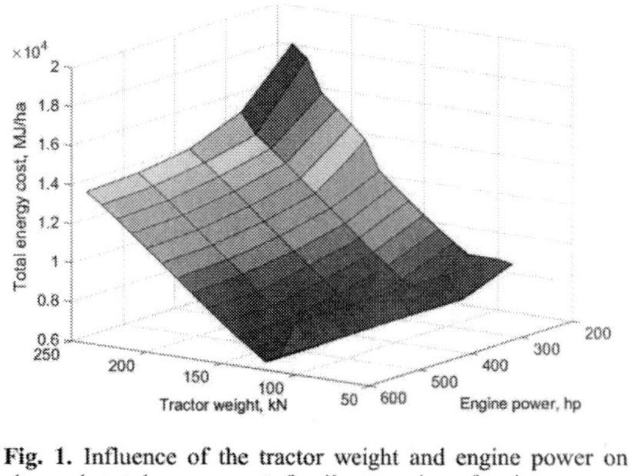

Fig. 1. Influence of the tractor weight and engine power on the change in total energy costs for direct sowing of grain crops.

For the accepted operating conditions for sowing units, the optimal ones are the weight of the tractor being 120 kN, the engine power being 490 hp, the width of the grip of the sowing unit being 18 m, the operating speed being 12 km/h. With such parameters of the tractor and the unit, the minimum total energy costs are ensured, taking into account the energy of the lost crop being 6399 MJ / ha, the minimum emission of carbon dioxide into the atmosphere being 30.4 kg/ha, the maximum capacity of the unit being 12.18 hectares/h.

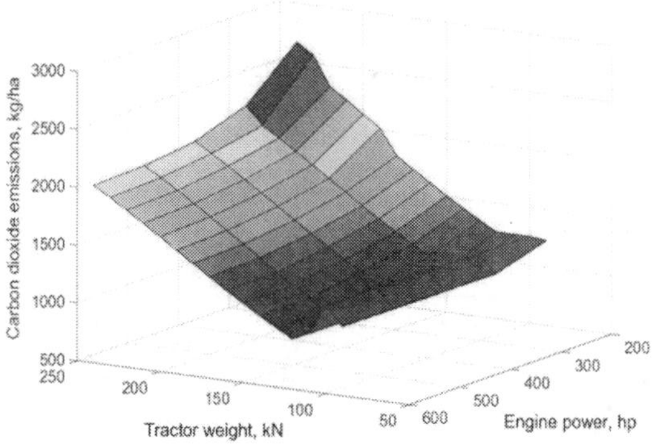

Fig. 2. Influence of the tractor weight and its engine power on the change in carbon dioxide emission into the atmosphere when directly sowing the crops.

To save only direct energy costs through fuel and lubricants, the tractor should have a lower weight of 110 kN, lower engine power of 192 hp. And it, accordingly, can perform work at a lower speed of 5 km/h. Natural fuel economy will be 0.47 kg/ha compared to the previous parameters of the unit, but the total (total) energy costs will increase by 444 MJ/ha, equivalent to 10 kg/ha of natural diesel fuel (see Table 1).

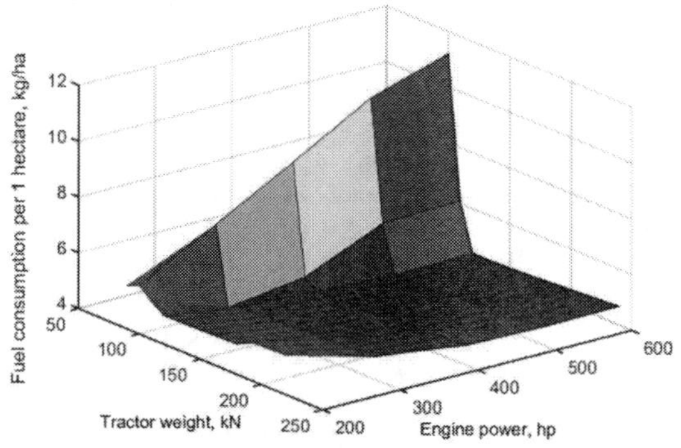

Fig. 3. Influence of the tractor weight and engine power on the change in per-hectare fuel consumption during direct sowing of grain crops.

The choice of the optimization criterion to focus on is the behalf of equipment holders. The optimization criterion is the maximum traction efficiency of the tractor not deserving special attention, because it is a purely specific criterion for optimizing the use of only the tractor itself and does not bring an effect on the scale of the TOOPPU.

 https://doi.org/10.1051/bioconf/20225200045

Table 1. Optimal parameters of the tractor and unit for direct sowing according to various optimization criteria.

Optimization criterion	Value of performance indicators					Optimal parameters of the tractor and the unit according to the optimization criterion			
	($\sum E_i$), MJ/ha	($\sum CO_{2i}$), kg/ha	(G), kg/ha	(W), ha/h	(η)	Tractor weight, kN	Engine power, hp	Working width of the unit, m	Operating speed of the unit, km/h
	Specific soil resistance of 4 kN/m								
Minimum total energy cost ($\sum E_i$)	6399	932,67	4.33	12.18	0.6970	120	491	18	12
Integral carbon dioxide emission ($\sum CO_{2i}$)	6399	932,67	4.33	12.18	0.6970	120	491	18	12
Minimum fuel consumption per 1 hectare (G)	6843	997,10	3.86	8.84	0.6959	110	192	18	5
Maximum aggregate performance (W)	6399	932,67	4.33	12.18	0.6970	120	491	18	12
Maximum tractor traction efficiency (η)	9214	1344	6.12	4.61	0.6985	100	159	15	5
	Specific soil resistance of 6,25 kN/m								
Minimum total energy cost ($\sum E_i$)	10713	1574,6	6.56	8.89	0.6981	150	596	14	12
Integral carbon dioxide emission ($\sum CO_{2i}$)	10713	1574,6	6.56	8.89	0.6981	150	596	14	12
Minimum fuel consumption per 1 hectare (G)	11234	1643,7	5.98	7.11	0.6951	170	300	18	5
Maximum aggregate performance (W)	11055	1621,7	6.36	9.67	0.6982	180	591	17	10
Maximum tractor traction efficiency (η)	14257	2093,6	8.04	3.42	0.6985	120	382	11	10

In addition, the maximum values of the tractor's traction efficiency are in the zone for the optimal parameters of the sowing unit according to the three main optimization criteria, the difference is 0.0015 (see Table 1), which is an insignificant difference.

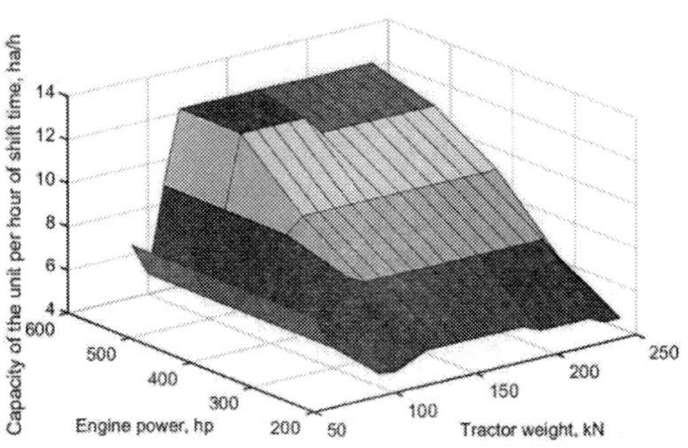

Fig. 4. Influence of the tractor weight and its engine power on the change in the replacement capacity of the unit during direct sowing of grain crops.

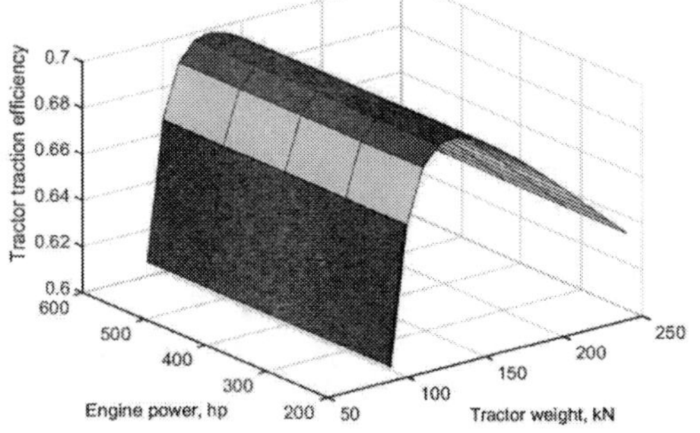

Fig. 5. Influence of the tractor weight and engine power on the change in tractor traction efficiency during direct sowing of grain crops.

Increasing the specific resistance of the seeder from 4 to 6.25 kN/m causes the need to increase the optimal weight of the tractor by 3 tons, engine power by 105 hp, reduce the working width by 4 m according to the optimization criteria: minimum total energy consumption ($\sum Ei$); integral carbon dioxide emission ($\sum CO_{2i}$). To obtain maximum productivity, the tractor must be 6 tons heavier compared to that with the previous calculations and this is explained by the fact that the performance optimization criterion does not take into account the impact of the tractor's weight on crop losses due to the compaction of the soil by its propulsion.

4 Conclusion

A systematic (TOOPPU system) mathematical model of units for direct sowing of grain crops according to various optimization criteria has been compiled. These are minimum total energy costs, taking into account the energy of the lost crop; minimal emission of carbon dioxide into the atmosphere, taking into account CO_2 not absorbed from the atmosphere by the lost crop; minimum fuel consumption per 1 hectare of sowing; maximum performance; maximum traction efficiency of the tractor. Computational experiments were conducted.

The optimal values of the tractor weight were 120 kN, the power of its engine was 481 hp, the width of the grip of the unit was 18 m and its operating speed was 12 km/h at low values of the specific traction resistance of the seeder (4 kN/m). They coincide according to three optimization criteria: minimum total energy costs, taking into account the energy of the lost crop; minimal emission of carbon dioxide into the atmosphere, taking into account CO_2 not absorbed from the atmosphere by the lost crop. The maximum productivity is due to the need to reduce crop losses from violation of the agricultural term of the technological operation, a unit with a maximum replaceable productivity of 12.18 hectares/h is selected.

Increasing the specific resistance of the seeder to 6.25 kN / m leads to an increase in energy costs for the execution of the technological operation, so the optimal values of the tractor weight and the power of its engine increase. According to the first two optimization criteria, the weight growth of the tractor is limited by the increase in crop losses due to soil compaction, so the optimal weight is 150 kN. The criterion for optimizing the maximum capacity of the unit for 1 hour of shift time does not take into account the influence of the tractor weight on soil compaction, so the optimal weight of the tractor for it is the largest being 180 kN. There is an increase in the weight of the tractor and the power of its engine and other optimization criteria.

A good desire to save direct energy costs through fuels and lubricants leads to an increase in total energy costs, an increase in carbon dioxide emissions into the atmosphere, and a decrease in the productivity of the sowing unit. For optimization calculations of the parameters of sowing units, it is necessary to use more complex, systemic optimization criteria, such as minimum total energy costs, taking into account the energy of the lost crop; minimal emission of carbon dioxide into the atmosphere, taking into account CO_2 not absorbed from the atmosphere by the lost crop.

Within the change in the specific resistance of the seeder from 4 to 6.25 kN/m, the optimal weight of the tractor should be 120 kN with the possibility of loading up to 150 kN. It is desirable to have a two-level engine of power of 500 and 600 hp.

References

1. D. V. Ushtyk, Criteria for optimizing the parameters of tillage aggregates and tractors. Resource-saving technologies in the agro-industrial complex of Russia, *Materials of the II International Scientific Conference* (Krasnoyarsk) pp 322–326 (2022)
2. A. M. Dementiev, Ensuring energy saving in technological processes of soil cultivation by optimal design of combined aggregates of block-modular structure: abstract of the PhD dissertation (2011)
3. Y. D. Pogulyaev, Increasing the productivity of machine-tractor aggregates on the basis of optimal and quasi-optimal control of energy modes: abstract of the PhD dissertation, (Chelyabinsk, 2006)
4. A. V. Filin, Selection of machine and tractor units for maximum productivity and minimum operating costs. *Energy-Efficient and Resource-Saving Technologies and Systems: Interuniversity Collection of Scientific Papers, Saransk: National Research Mordovian State University named after N.P. Ogarev* pp 235–237 (2016)
5. A. R. Valiev, et al., Agro-bio-techno park as an innovative factor of increasing competitiveness of agriculture under global challenges, *Rural development. Bioeconomy Challenges,* pp 1365–1368 (2017)
6. C. Khafizov, A. Nurmiev, R. Khafizov, N. Adigamov, Method of justification for parameters of tractor-implement unit with regard to their impact on crop productivity, *Engineering for Rural Development: Proceedings* pp 176–185 (2018)
7. I. Gainutdinov, Beet production efficiency and ways to increase it in case of negative market conditions in the commodity market, *BIO Web of Conferences* **17,** 00108 (2020)
8. I.G. Gainutdinov, R.R. Gadelshin, Prospects for the development of production and processing of potatoes in Russia, *Vector of economy*, **6(48)**, 30, (2020).
9. S. A. Sinitskiy, V. M. Medvedev, Influence of dynamic factors on the indicators of the MTA engine at an unstable load, *Technics and equipment for the village,* **4(274)**, 16-19, (2020)

10. I. G. Galiev, A. A. Mukhametshin, I. R. Iskhakov, A. R. Shamsutdinov, Increasing the efficiency of tractors in modern conditions, *Bulletin of Kazan State Agrarian University* **1 (12),** 169–172 (2009)
11. M. N. Kalimullin, R. K. Abdrakhmanov, I. G. Galiev. Perfection of potato cultivation technology, *Technics and equipment for the village* **4,** 6–9 (2017)
12. The farming system of the Republic of Tatarstan. Innovation based on tradition. Part 1. General aspects of agriculture. Center for Innovative Technologies, Kazan, 168 (2013)
13. The farming system of the Republic of Tatarstan. Part 2. Agricultural technology for the production of crops, Center for Innovative Technologies, Kazan, 292 (2014)
14. The farming system of the Republic of Tatarstan. Part 3. The system of organization and management of production in agriculture (agrarian management), Center for Innovative Technologies, Kazan, 280 (2014)
15. N. V. Trofimov, S. R. Suleimanov, S. V. Sochneva, N. A. Loginov, Adaptive landscape farming system - the basis for land conservation in the Republic of Tatarstan, *Bulletin of Kazan State Agrarian University* **13 (1 (48),** 69–73 (2018)
16. R. S. Shakirov, M. Sh. Tagirov, Increasing the efficiency of agriculture in the Republic of Tatarstan, *Agriculture* **7,** 9–12 (2014)
17. R. I. Safin, A. R. Valiev, R. V. Minikayev et al., Establishment of the precision farming system in the Republic of Tatarstan, *Bulletin of Kazan State Agrarian University* **5 (2 (16),** 153–156 (2010)
18. I. G. Galiev, Ensuring possibility of functioning of tractors in agricultural production taking Into account residual resources of their units and systems, *Engineering for Rural Development*, 48–53 (2020)
19. E. Alekseenkova, Integrated approach. Combined sowing machines market, *AgroForum* **5,** 56–62 (2021)
20. M. I. Sanzharovskaya, Agromaster – technology of the only passage through the stubble. Engineering and technical support of the agro-industrial complex, *Abstract Journal* **1,** 242 (2009)
21. R. N. Khafizov, Optimization of the main parameters of the tractor and aggregate for sowing grain crops taking into account their influence on the harvest: abstract of the PhD dissertation (2017)
22. K. A. Khafizov, Optimization of parameters and modes of operation of the MTA on the basis of energy analysis, *Tractors and Agricultural Machines* **7,** 32–34 (2006)
23. C. A. Khafizov, Optimization of main parameters of tractor and unit for plowing soil, taking into account their influence on yield of grain crops *Engineering for Rural Development*, pp 585–590 (2020)
24. A. Nurmiev, Optimization of main parameters of tractor working with soil-processing implement, *Engineering for Rural Development: Proceedings* (Jelgava), 161–167 (2018)
25. K. A. Khafizov, Reduction of total energy costs on technological operations in the agro-industrial complex - the way to reduce greenhouse gas emissions in the atmosphere, *Bulletin of Kazan State Agrarian University* **3 (63),** 43–47 (2021)

Synthesis of spatial five- and six-letter mechanisms with rotational pairs by the movement of the output link

Alexander P. Mudrov[1,2], *Fanil F* Khabibullin[2], *Gennadi V.* Pikmullin[1], and *Zurab D.* Gurgenidze[1]

[1] Kazan State Agrarian University, 65, K. Marx St., Kazan, 420015, Republic of Tatarstan, Russian Federation
[2] Kazan National Research Technological University, 68, K. Marx St., Kazan, 420015, Republic of Tatarstan, Russian Federation

Abstract. The article discusses spatial mechanisms with rotational pairs that allow the movement of the output link or working body with a given accuracy while ensuring the necessary rigidity and reliability of the structure. Its durability is also significantly increased, energy losses are reduced, since all hinge assemblies (rotational kinematic pairs) are made on rolling bearings and are reliably isolated from the external environment. The article is also devoted to the synthesis (design) of functional spatial six-branch mechanisms with rotational pairs reproducing with a given accuracy the linear function of the change in the angle of rotation of the output link, provided it is fully rotatable. The objective function and dependencies necessary for the design of such mechanisms are obtained and an algorithm for solving this problem in CAD is developed.

1 Introduction

Spatial mechanisms are quite widely used in agricultural engineering [1,2]. Basically, four-, five- and six-pin mechanisms are used in the structures, but there are very few of them whose hinges are made in the form of rotational kinematic pairs alone, despite their obvious advantages [3-5,12-15]. Spatial mechanisms with rotational pairs allow the movement of the output link or working body with a given accuracy while ensuring the necessary rigidity and reliability of the structure. Its durability is also significantly increased, energy losses are reduced, since all hinge assemblies (rotational kinematic pairs) are made on rolling bearings and are reliably isolated from the external environment. The tasks of synthesizing such mechanisms, despite their complexity, are undoubtedly important in the theory of mechanisms and machines.

2 Materials and methods

To synthesize spatial 6R mechanisms according to a given law of the output link, an analysis of structural schemes was carried out, the constructive implementation of which will accurately reproduce the required linear motion function. We showed that the reproduction of a linear function on some part of the turnover of the output link can be achieved using two-crank six-link spatial mechanisms (Figure 1a) or their five-link modification (Figure 1b). The structural parameters of these mechanisms are related by dependencies:

$$\alpha_2=180^0-2\alpha_6,\quad l_2=0; \tag{1a}$$

$$\alpha_3=\alpha_5-\alpha_1, l_3=l_5-l_1; \tag{1b}$$

$$\alpha_4=180^0-\alpha_6, l_4=l_6; \tag{1c}$$

$$\frac{l_1}{\sin\alpha_1}=\frac{l_5}{\sin\alpha_5}=\frac{l_6}{\sin\alpha_6}. \tag{1d}$$

Here and further, α_i is the angle of intersection of the geometric axes of the hinges of the link, counted counter clockwise, taking the axis of the hinge facing the observer as the starting point; l_i is the shortest distance between the axes of the hinges (the theoretical length of the link).

3 Results and discussion

The kinematics of the output link of spatial mechanisms with rotational pairs depends only on the angular parameters of the links [6]; therefore, they will be the determined parameters when designing mechanisms.

As for the linear parameters, when determining them, it is necessary to set the length of one of the links arbitrarily, based on the functional purpose of the mechanism and the design of the hinge assemblies, while other lengths are determined by the above formulas.

Let's first find the relationship between the rotation angles of the input (φ_1) and output (ψ_5) links of the mechanism. This dependency is defined by the following expression:

$$\psi_5=\arcsin\frac{\left(\cos\alpha_6-\cos\alpha_1\right)\sin\varphi_1}{1-\cos\alpha_1\cos\alpha_6-\sin\alpha_1\sin\alpha_6\cos\varphi_1}+¿$$

$$+\arcsin\frac{(\cos\alpha_6+\cos\alpha_1)\sin\varphi_1}{1+\cos\alpha_1\cos\alpha_6-\sin\alpha_1\sin\alpha_6\cos\varphi_1}+\dot{\iota}$$

$$+\arcsin\frac{-(\cos\alpha_6+\cos\alpha_5)\sin\varphi_1}{1+\cos\alpha_5\cos\alpha_6-\sin\alpha_5\sin\alpha_6\cos\varphi_1}. \quad (2)$$

Here α_6 is the angle of intersection of the axes of the hinges of the frame (rack), varying from 85^0 to 65^0 with an interval of 0.5^0. The angle α_5 of crossing the axes of the hinges of the driven link - crank is taken in the range from 213^0-α_6 to 183^0-α_6 with the same interval, and the angle α_1 of crossing the axes of the hinges of the driving crank should be less than ninety degrees.

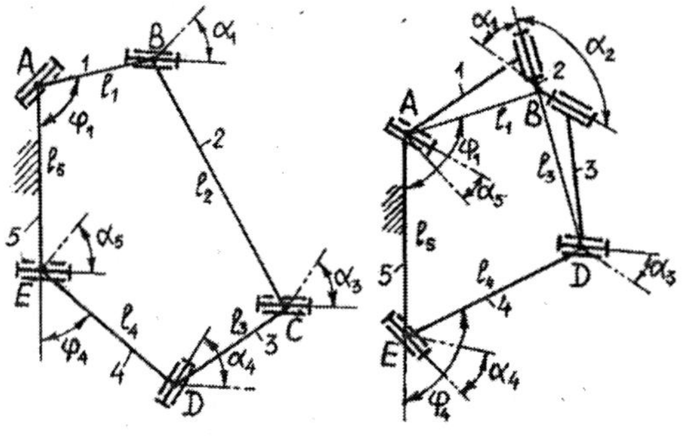

Fig. 1. Kinematic schemes of spatial hexagons (a) with the dependence of the angles of intersection of the geometric axes of the joints of the links α_4= 〖180〗 ^0-α_6; (b) a five-link, with one spherical kinematic pair, modification of the hexagon.

A graphical representation of this expression is given in Figure 2 in two coordinate systems - $\varphi_1 0 \psi_5$ and $\varphi C \psi$. Moreover, the second coordinate system is more convenient for carrying out further calculations and programming them on a computer.

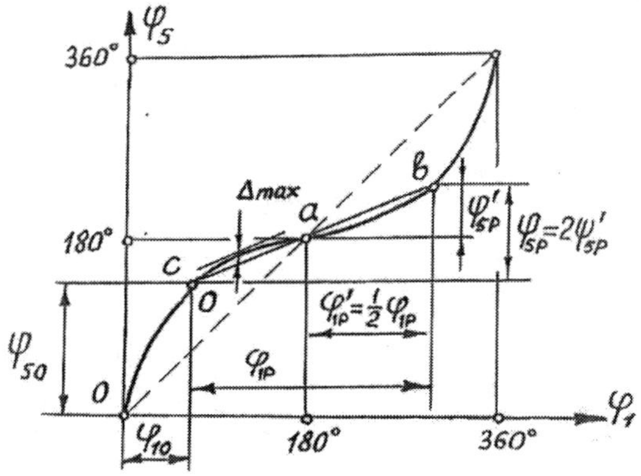

Fig. 2. Graph of the dependence of the angle of rotation of the output link on the angle of rotation of the input for the hexagon.

As can be seen from the above graph, the curve $\psi_5=f(\varphi_1)$ is symmetric with respect to point *a*. Therefore, the straight line segment *cb* = 2*ab* , as well as the angle $\varphi_p=2\varphi_p^{'}$ $(\psi_p=2\psi)$. Preliminary calculations determined that the value of the φ_p of the working angle of rotation of the input crank should be within 180^0 - 300^0. Then the corresponding working angle φ_p of the output crank rotation will be with it in the following relationship [7, 8]:

$$K=\frac{\psi_p}{\varphi_p}=\frac{\psi_p^{'}}{\varphi_p^{'}}=0{,}1\ldots0{,}5. \quad (3)$$

The synthesis of the mechanism is carried out so that the maximum deviation Δ_{max} of the curve $\psi_5=f(\varphi_1)$ in the section *cb* does not exceed the value limited by the relative error:

$$\Delta\psi^0=\frac{\Delta_{max}}{\psi_p}\cdot100\,\%\leq\Delta\psi. \quad (4)$$

It is desirable that the deviation be $\Delta\psi=0{,}2-0{,}5\%$.

The synthesis of the mechanism will be carried out by the method of the purposeful search from the minimization condition of the function Δ_{max} (or $\Delta\psi^0$) [9]. However, before proceeding to the definition of this function, we will make an equation linking the angle α_1 and other input and output parameters of the synthesis mechanism. It has the form:

$$\alpha_1=\arcsin\frac{A}{B}, \quad (5)$$

where:

$$A=\sin\alpha_6\left[\left(1-\cos\frac{\varphi_p}{2}\cos\frac{K\varphi_p}{2}\right)(1++\cos\alpha_5\cos\alpha_6)+\sin\frac{\varphi}{2}\right.$$

(6)

$$B=\left(\cos\frac{K\varphi_p}{2}-\cos\frac{\varphi_p}{2}\right)\left(1+\cos\alpha_1\cos\alpha_6++\sin\alpha_1\sin\alpha_6\text{co}\right.$$

(7)

Expression (5) is derived from the equation:

$$l_1\cos\varphi_1=l_6+l_5\cos\varphi_5-\dot{\iota}\,l_6(\cos\varphi_1\cos\varphi_5--\cos\alpha_5\sin\varphi_1\sin$$

$$\cdot\{[\cos\alpha_5\sin\varphi_1\cos\gamma_3-(\sin\alpha_5\sin\alpha_6--\cos\alpha_5\cos\alpha_6\cos\varphi_1$$

(8)

considering that:

https://doi.org/10.1051/bioconf/20225200046

$$\frac{l_1}{\sin\alpha_1}=\frac{l_5}{\sin\alpha_5}=\frac{l_6}{\sin\alpha_6},$$

$$\varphi_1=180^0+\varphi_p'=180^0+\frac{\varphi_p}{2},$$

$$\psi_5=180^0+\psi_p'=180^0+\frac{K\varphi_p}{2}$$

and

$$\sin\gamma_3=\frac{-(\cos\alpha_6-\cos\alpha_5)\sin\varphi_1}{1+\cos\alpha_5\cos\alpha_6-\sin\alpha_5\sin\alpha_6\cos\varphi_1},$$

$$\cos\gamma_3=\frac{(1+\cos\alpha_5\cos\alpha_6)\cos\varphi_1-\sin\alpha_5\sin\alpha_6}{1+\cos\alpha_5\cos\alpha_6-\sin\alpha_5\sin\alpha_6\cos\varphi_1}.$$

In turn, the dependence (8) is found by the solution of the vector equation compiled for the vector contour ABCD (see Figure 3):

$$\overline{AB}=\overline{AF}+\overline{FE}+\overline{ED}+\overline{DC} \qquad (9)$$

in the projection on the Y axis.

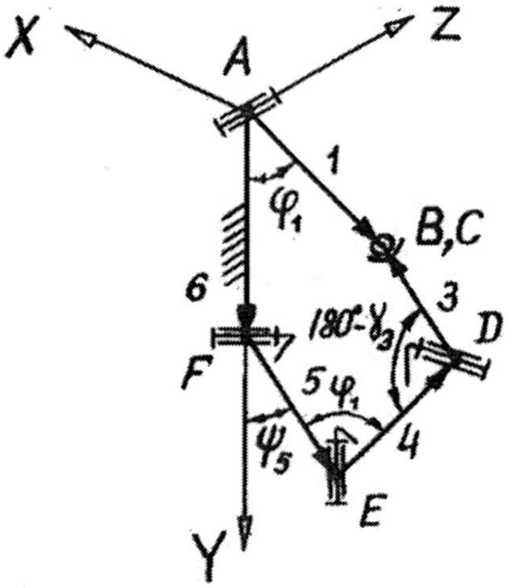

Fig. 3. Kinematic scheme of spatial hexagons with a vector contour AFEDBC.

Now let's move on to the Δ_{max} function. For the mechanism under consideration, it is as follows:

$$\Delta_{max}=\arcsin\frac{(\cos\alpha_6-\cos\alpha_1)\sin\varphi_K}{1-\cos\alpha_1\cos\alpha_6-\sin\alpha_1\sin\alpha_6\cos\varphi_K}+$$

$$+\arcsin\frac{(\cos\alpha_6+\cos\alpha_1)\sin\varphi_K}{1+\cos\alpha_1\cos\alpha_6-\sin\alpha_1\sin\alpha_6\cos\varphi_{1K}}+\text{¿}$$

$$-\arcsin\frac{(\cos\alpha_6+\cos\alpha_5)\sin\varphi_{1K}}{1+\cos\alpha_5\cos\alpha_6-\sin\alpha_5\sin\alpha_6\cos\varphi_{1K}}-K\varphi$$

(10)

The angle φ_K included here is found from the equality of the partial derivative (∂ψ_5)/(∂φ_1) to the angular coefficient K, that is:

$$\frac{\partial\psi_5}{\partial\varphi_1}=\frac{\cos\alpha_6-\cos\alpha_1}{1-\cos\alpha_1\cos\alpha_6-\sin\alpha_1\sin\alpha_6\cos\varphi_K'}+\text{¿}$$

$$\frac{+\cos\alpha_6+\cos\alpha_1}{1+\cos\alpha_1\cos\alpha_6-\sin\alpha_1\sin\alpha_6\cos\varphi_K'}-\text{¿}$$

$$\frac{-\cos\alpha_6-\cos\alpha_5}{1+\cos\alpha_5\cos\alpha_6-\sin\alpha_5\sin\alpha_6\cos\varphi_K'}=K, \qquad (11)$$

where $\varphi_K'=180^0+\varphi_K$

From this expression, after a series of transformations, we get:

$$x^3+ax^2+bx+c=0, \qquad (12)$$

where $x=\cos\varphi_K$.

$$a=\frac{1}{K\sin\alpha_1\sin\alpha_5\sin\alpha_6}\{\sin\alpha_1(\cos\alpha_6+\cos\alpha_5)--2\sin\alpha_5\text{co}$$

(13a)

$$b=\frac{1}{K\sin^2\alpha_1\sin^2\alpha_6\sin\alpha_5}\{2\sin\alpha_1(\cos\alpha_5\sin^2\alpha_6--\sin\alpha_1\text{sir}$$

$$+K[2\sin\alpha_1(1+\cos\alpha_1\cos\alpha_6)+\sin\alpha_5(1--\cos^2\alpha_1\cos^2\alpha_6)]\},$$

(13b)

$$c=\frac{1}{K\sin^2\alpha_1\sin^3\alpha_6\sin\alpha_5}\{(1+\cos\alpha_1\cos\alpha_6)[(1-\cos^2\alpha_1\text{cos}$$

$$\cdot(\cos\alpha_5+\cos^2\alpha_1\cos\alpha_6)\}. \qquad (13c)$$

To solve the cubic equation (12), we find the discriminant:

$$\Delta=\left(\frac{q}{2}\right)^2+\left(\frac{pq}{3}\right)^3, \qquad (14)$$

here

$$q=\frac{2a^3}{27}-\frac{ab}{3}+c, \qquad (15a)$$

$$p=\frac{-a^2}{3}+\text{¿}b. \qquad (15b)$$

If the discriminant Δ>0, then the real root of the equation is one:

$$x_1=\sqrt[3]{\frac{-q}{2}+\sqrt{\Delta}}+\sqrt[3]{\frac{-q}{2}-\sqrt{\Delta}}-\frac{a}{3} \qquad (16)$$

and, therefore,

$$\varphi_K=\arccos x_1.$$

If Δ = 0, and *p* and *q* are not equal to zero, then there are two real roots:

$$x_1=\frac{3a}{p}-\frac{a}{3}, x_2=-\left(\frac{3q}{2p}+\frac{a}{3}\right) \quad (17)$$

and

$$\varphi_{(1)K}=\arccos x_1, \varphi_{(2)K}=\arccos x_2.$$

If Δ<0, then there are three real roots:

$$x_1=2\left|\sqrt[3]{r}\right|\cos\frac{\beta}{3}-\frac{a}{3},$$

$$x_2=2\left|\sqrt[3]{r}\right|\cos\frac{\beta+360^0}{3}-\frac{a}{3}, \quad (18)$$

$$x_3=2\left|\sqrt[3]{r}\right|\cos\frac{\beta+720^0}{3}-\frac{a}{3}.$$

Here:

$$\left|\sqrt[3]{r}\right|=\left|\sqrt{\frac{-p}{3}}\right|; cos\beta=\frac{-q}{2r}; r=\left|\sqrt{\frac{-p^3}{27}}\right|.$$

Then:

$$\varphi_{(1)K}=\arccos x_1,$$
$$\varphi_{(2)K}=\arccos x_2,$$
$$¿\varphi_{(3)K}=\arccos x_3.$$

The value of the angle φ_K is chosen so that the condition is met $\varphi_K \le \varphi_p$.

Thus, the dependences necessary for the optimization synthesis by the objective function of a six-link mechanism (or its five-link modification), reproducing a linear function on some part of the output link turnover with its full turn ability, are obtained [10, 11].

Synthesis is carried out as follows. The working section φ_p of the angle of rotation of the input crank is set, as well as the extreme upper values of the coefficient K and the angles α_5 and α_6.After finding the angle α_1, the objective function $\Delta\psi^0$ is calculated, which is compared with the lower value $\Delta\psi$. In this case, the condition must be met:

$$\Delta\psi^0 \le \Delta\psi \quad (19a)$$

or

$$\Delta\psi^0_{i+1}-\Delta\psi^0_i \ge 0. \quad (19b)$$

If this is not done, then the values of the angles α_5 and α_6 are successively reduced by a certain amount and the process is repeated until one of the conditions (19) occurs. If all possible parameter values are K, α_5, α_6 are sorted out, and the condition $\Delta\psi^0 \le \Delta\psi$ is not obtained, then $\Delta\psi$ increases by a certain amount and the process repeats until the condition (19a) is reached.

Having determined the values K, α_1, α_5 and α_6 with the corresponding value $\Delta\psi$, they proceed to determine other parameters of the mechanism.

The value of the working (rectilinear) section of the law of change of the angle of rotation of the output crank:

$$\psi_p=K\varphi_p. \quad (20)$$

The origin of the coordinates of the new system φCψ is determined by the coordinates (see Figure 2):

$$\psi_{50}=180^0-\frac{\psi_p}{2}, \varphi_{10}=180^0-\frac{\varphi_p}{2}. \quad (21)$$

The angles α_2,α_3 and α_4, as well as the linear parameters of the links, for a given length of one of them, are determined by formulas (1).

4 Conclusion

Optimal structural schemes of spatial multi-link mechanisms with rotation-al pairs (as well as their five-link modification with a spherical kinematic pair) have been identified, which reproduce the required accuracy, the linear function during the rotational movement of the output link.

For these mechanisms, the objective function and the corresponding analytical dependencies are obtained, which can be used to automate the design of such spatial mechanisms in CAD.

In addition to the objective function, additional equations linking the desired parameters and design data have been compiled. The boundaries of their change have been established, which makes it possible to simplify the task of purposefully searching for parameters of the designed mechanisms.

References

1. G. D. Ananov, *Kinematics of spatial articulated mechanisms of agricultural machines* (Moscow; Leningrad: Mashgiz, 1963).
2. E. I. Vorobyev, F. M. Dimentberg, *Spatial hinge mechanisms. Closed and open kinematic circuits* (Moscow: Nauka, 1991)
3. A. V. Verkhovskiy, Six-dimensional spatial articulated mechanisms, *Bulletin of Tomsk Technological Institute* **61 (1),** 47-52 (1947)
4. V. A. Zinovyev, *Spatial mechanisms with lower pairs* (Moscow: Gostekhizdat, 1952)
5. A. N. Evgrafov, G. N. Petrov, Calculation of geometric and kinematic parameters of the spatial link mechanism with excessive links, *Problems of mechanical engineering and machine reliability* **3,** 3-8 (2013)
6. F. M. Dimenberg, *Theory of spatial articulated mechanisms* (Moscow: Nauka, 1982)
7. V. A. Glazunov, A. Sh. Koliscor, A. F. Krainev, *Spatial mechanisms of parallel structure* (Moscow: Nauka, 1991)

8. G. T. Bennett, A new mechanism, *Engineering* **76,** 777–778 (1903)
9. K. Brunnthaler, H. P. Schrocker, M. A. Husty, New method for the synthesis of Bennett mechanisms, *Proceedings of CK2005, International Workshop on Computational Kinematics* (Cassino) pp 53–61 (2005)
10. A. P. Mudrov, A. G. Mudrov, S. M. Yakhin, N. Z. Mingaleev, G. V. Pikmullin, Study of spatial hinge mechanisms and their use in agricultural machines, *BIO Web of Conferences* **27** 00012 (2020)
11. A. P. Mudrov, A. G. Mudrov, G. V. Pikmullin. Designing a spatial 5R mechanism according to a given law of motion of the output link, *Bulletin of Kazan State Agrarian University* **15-2 (58),** 107–113 (2020)
12. M. G. Yarullin, I. R. Isyanov. Kinematics of the connecting rod of a two mobility five-link space mechanism with a double crank, *Advances in Mechanical Engineering Selected Contributions from the Conference "Modern Engineering: Science and Education* (May, Saint Petersburg) (2018)
13. O. N. Levitskaya, N. I. Levitsky, *Course of theory of mechanisms and ma-chines.* (Moscow: Higher School, 1978)
14. I. M. Salakhov, A. V. Matyashin, N. F. Vafin, R K. Abdrakhmanov, Substantiation of the use of an oscillatory type of working organ for tillage, *Machinery and Equipment for the Village* **3,** 21–23 (2018)
15. B. L. Ivanov B. G. Ziganshin, A. V. Dmitriev et al., Droplet size of virocide disinfectant liquid from vortex injector sprayer under different operating conditions, *Engineering for Rural Development* (26–28 May, Jelgava) pp 564–571 (2021)
16. A. A. Oliveira Jr. J. C. M. Carvalho, Modeling of the Bennetts linkage as leg of a mobile robot, *12 th IFToMM World Congress* (Besancon) pp 1–6 (2007)
17. Y. Chen, E. J. Baker, Using a Bennett linkage as a connector between other Bennett loops, *Proc. IMechE* **219,** 177–185 (2004)
18. B. V. Shitikov, *Investigation of spatial hinge mechanisms. The manuscript. Report on research work* (Kazan: KhTI, 1957)
19. Sh. R. Galiullin, *Justification of the pa-rameters of a bar elevator with uneven movement of the blade to increase a separating ability of the potato har-vester: Thesis* (Moscow: VIM, 1989)

Modular microwave unit for heat treatment of raw materials at factory-farm enterprises

M. V. Prosviryakova[1], *B. G.* Ziganshin[1], *Y. K.* Shogenov[2], *O. V.* Mikhailova[3,*], and *A. A.* Tikhonov[4]

[1] Kazan State Agrarian University, 65, K. Marx St., Kazan, 420015, Republic of Tatarstan, Russian Federation
[2] Russian Academy of Sciences, 14, Leninsky prospect, Moscow, 119334, Russian Federation
[3] Nizhny Novgorod State Engineering-Economic University, 22, October St.,Knyaginino, 606340, Russian Federation
[4] Nizhny Novgorod State Agricultural Academy, 97, Gagarin Avenue, Nizhny Novgorod, 603107, Russian Federation

Abstract. The objective of the work is to develop a microwave unit with removable modules for continuous heat treatment and disinfection of animal-derived raw materials of different structures with resonators, providing electromagnetic safety without shielding and high electric field intensity with reduced operating costs while preserving the feeding value of the product. The study includes 3D modeling of resonators for each module using CST Microwave Studio software and multi-criteria evaluation of regression models of process parameters, allowing one to estimate the feeding value of the product. The first module is designed for heat treatment of secondary animal-derived raw materials, including frozen blocks of meat waste. There are modules can be installed simultaneously or separately in the microwave unit, depending on demand and available raw materials. All modules are removable and easy to clean after completion of heat treatment, and the source block is closed and is subject only to maintenance inspection. Studies of electromagnetic safety for the perforated cylindrical resonator were performed. The operating modes of the second module were found, the duration of defrosting and heating of the animal colostrum was 10–15 min with the generator power of 4.8 kW and energy costs of 0.18–0.20 kW·h/kg.

1 Introduction

Electrotechnology is developing more and more widely every year. We have found a fairly wide application in various industries, one of which is modern farming. The electromagnetic field energy is used for processing farm products, since microwave energy has the capability of resonant energy interaction with raw materials [29-33]. Many authors (Atabekova G.I., Baskakova S.I., Ginzburg A.S., Didenko A.N and others) have presented the fundamentals of electromagnetic waves influence on dielectric materials. There are many studies on the development of microwave units for heat treatment of raw materials of factory-farm enterprises by Borodina I.F., Strebkova D.S., Novikova G.V., Tsuglennok N.V. and others [13-27].

In livestock and poultry farming, it is necessary to use new technologies and developments to improve production efficiency, in particular, to use animal-derived supplements to balance the diet of animals. Therefore, the development of electrotechnology and devices that increase the feeding value of protein supplements is relevant.

From analysis of the earlier studies it was noted that developed units have high operating costs mainly due to batch operation, and continuous units do not ensure radio-tightness. The problem of increasing the feeding value of protein feeds by the action of the microwave electromagnetic field in the modular multi-generation continuous units with air-cooling magnetrons with the observance of electromagnetic safety in the conditions of farms remains unresolved.

The paper aims to develop a microwave unit with removable modules for continuous heat treatment and disinfection of animal-derived raw materials of different structures with resonators, providing electromagnetic safety without shielding and high electric field intensity with reduced operating costs while preserving the feeding value of the product.

The novelty of the study is the scientific and methodical basis of developing the modular continuous microwave unit, including methods of identifying effective resonator designs with justification of electrodynamic parameters of the «generator-resonator-load» system and operating modes, as well as assessments of electromagnetic safety through multi-criteria regression models.

2 Materials and methods

The development of the modular microwave unit with air-cooling magnetrons provides a step-by-step optimization of the design of the resonators in terms of their electrodynamic parameters and the analysis of mathematical models, describing the electromagnetic field distribution in them. The 3D modeling of

* Corresponding author: ds17823@yandex.ru

resonators was performed using CST Microwave Studio, SolidWorks. Multi-criteria evaluation of regression models, taking into account the organoleptic, biochemical, physicochemical, microbiological indicators of the product, characterizing the feeding value, and statistical analysis of experimental data were carried out using the MS Excel, Mathcad 14, and Statgraphics Plus for Windows.

For heat treatment and disinfection of animal-derived raw materials in farms, a three-module multi-generation microwave unit with resonators of different design has been developed, which provides electromagnetic safety and sufficient electric field intensity to save the feeding value of the product. Each module is designed for heat treatment of a specific type of secondary animal-derived raw material.

The first module serves for heat treatment of secondary animal-derived raw materials, including frozen blocks of meat waste. The module allows producing cooked and disinfected products from non-food meat waste, which can be used as a protein supplement for animals. Different designs of resonators have been investigated using air-cooling magnetrons under change of their location in order to avoid the mutual influence of generators on each other, to reduce heat loads and breakdowns. The power shift of magnetrons operating in impulse mode relative to each other at average radiation power is taken into account.

The second module is designed for thermal treatment of frozen animal-derived raw materials with phase transition, which coagulate at temperatures above 40 °C, for example, frozen blood, frozen cow colostrum, etc. Unfrozen and heated colostrum can be used to feed calves as a product of high feeding value, and defrosted blood from butchers can be used in the production of meat products. Dozens of modules for such raw materials have been developed and patented, the general part of these constructions is the sectioning of resonators based on the idea of separating the processes of defrosting and heating of raw materials due to a significant difference in its dielectric and thermophysical characteristics of these processes. The engineering principles of design and optimization of the unit are given.

The third module is necessary for heat treatment of viscous raw material, which provides the combination of dewatering of cut meat, fine grinding, and heat treatment of solid fraction for use as the protein supplement for animals. The key feature of the proposed module design is the use of a supercritical contraction, which ensures radio-tightness (no radiation from their resonator) on the one hand and the possibility of continuous loading of defrosted raw materials on the other hand. Since the contraction size must be at least two wavelengths in length, some conical resonators have metal meshes to eliminate radiation, while others have screw coils that limit radiation. Furthermore, mesh cells at a frequency of 2450 MHz are acceptable for unloading finished bulk material, and with a system of several meshes, the full reflection of waves inside the resonator can be provided.

All modules are removable and easy to clean after the heat treatment is completed. The source block is closed and subject to maintenance inspection only.

The analogues for each resonator are given below.

1. An analogue of the first module is a microwave unit with a quasi-resonator for the thermal treatment of ground fat-containing raw material in a continuous mode [1]. The apparatus comprises a torus and a capacitor part. The central part of the resonator has a non-ferrous receiving container in the form of a truncated cone. Magnetrons are mounted on the external lateral surface of the torus.
2. The analogue of the second module is a microwave unit for defrosting and heating cow colostrum with coaxial resonators [2]. The cylindrical perforated resonator, located on the horizontal axis of the spherical resonator, rotates from the electric drive.
3. The analogue of the third module is a microwave unit with a conical resonator for thermal treatment of non-food animal-derived waste [3].

The disadvantages of analogues are the difficulties of sanitizing the parts of the unit and the risk of the detergent entering into the magnetrons through the ventilation case.

The study is focused on the development of new technologies in animal farms, with their further implementation into modern production. Materials in the field of agrotechnical developments are given. The methodology for developing microwave heat treatment units for animal-derived raw materials includes the following order [4–12].

1. Analysis of electrophysical and thermophysical parameters of raw materials: dielectric permeability, dielectric loss angle tangent, dielectric loss factor, heat capacity, density depending on electromagnetic field frequency, temperature, humidity, fatness; calculating the depth of wave penetration into raw materials.
2. Identification of functional criteria: continuous mode setting, air cooling magnetrons, high electric field intensity, radio tightness without shielding, capacity of the unit according to the farm requirement, multi-generator resonators, multimodule unit, separate resonators for raw material with phase transition, etc.
3. Development of several microwave units with different designs of resonators for heat treatment of the same raw material: toroidal resonators with different capacitor zone locations, coaxial resonators, conical resonators, spherical and hemispherical resonators, cylindrical resonators, etc.
4. Justification of transportation mechanisms of raw materials at different aggregate state: screw contours, centrifugal disc, perforation, pump viscous fluid, cutters, electric vibrator, etc.
5. Comparative assessment of microwave units with different resonator designs according to the least deviation criteria: preservation of the feeding value of the product, reduction of heat treatment time of raw materials, observance of electromagnetic safety. Selection of efficient design of microwave continuous unit with resonators, which provide heat treatment and

disinfection of the raw material with preservation of the feeding value of the product.
6. Assessment of technical and economic indicators for the application of modular multiple-generation microwave unit in farms.

The modular microwave continuous unit for heat treatment of raw materials comprises modules located at an angle to the horizontal plane, and the block of sources with stationary non-ferromagnetic bases 2, 20, 32 of the modules (Fig. 1), with waveguides with air-cooling magnetrons, are located at 120 degrees from each other. A cooling fan 1 and electric drives 1 and 13 of modules are arranged in the source block inside the contact stationary bases, containing elements for fixing the demountable cases of resonators 3, 16, and 18.

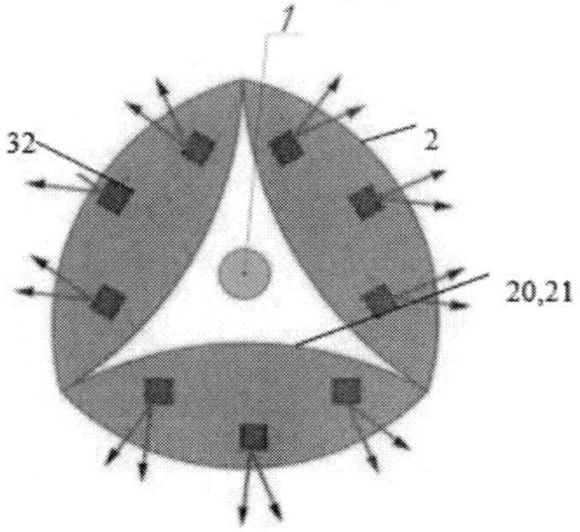

Fig. 1. Source block with contacting stationary non-ferromagnetic bases of modules: 1 – fan electric drive; 2 – stationary base of the first module; 20, 21 – stationary bases of the second module; 32 – stationary base of the third module.

The first module (Figs. 2a, 3a) has the form of a non-ferromagnetic toroidal resonator 3, a detachable body is attached to its stationary base, consisting of a toroidal part 7 and a capacitor part 6. In the gap of the capacitor part of the resonator, near the coil, there is a screen dielectric cylinder 5, inside which a spiral dielectric mixer 13 is mounted. The diameter of the screen cylinder is smaller than the diameter of the stationary base of the resonator 2. In the small-diameter capacitor plate, a mesh with a screw cutter 10 is installed, located in the center of the toroidal resonator 3. The shredder 9 is arranged in the receiving tank of the cutter. A non-ferrous magnetic pipe with a ball valve 12 is attached to the toroidal part 7 of the resonator. Above this pipe is a dielectric disc 11 with an electric drive.

The second module (Figs. 2b, 3b) is performed in the form of two cylindrical resonators of different diameters, placed one inside the other in such a way that the bases 20, 21 are in contact and form a space where the electric vibrator 17 and a receiving tank 14 with shutter are mounted. Inside the container there is a cutter 15 in the form of finned rolls. A stationary base 21 of internal resonators 18, where magnetrons with waveguides are mounted, locates close to the stationary base 20 of external resonator 16. The bases have elements for fixing a removable body. The lower part of the lateral surface of the internal cylindrical resonator 21 is perforated, and a pipe with a ball valve 19 is connected to the lateral surface of the external cylinder.

The third module (Figs. 2c, 3b) has a coaxial design, there are a screen dielectric truncated cone 26 without a base and a dielectric truncated rotor 24 inside a truncated conical non-ferromagnetic resonator 23. A dielectric pipe 22 is connected under the dielectric rotor 24 through the lateral surfaces of the resonator 23 and the screen dielectric truncated cone 26. Behind the resonator, this pipe is connected to a non-ferromagnetic pipe, which acts as a supercritical waveguide. A truncated dielectric cone 27 is mounted to the truncated part of the screen cone 26, inside which a dielectric screw 29 is coaxially placed, the turns of which are directed to the outlet pipe, located coaxially in the non-ferromagnetic pipe 31, linked with the truncated part of the resonator.

Each module is intended for a specific process.

The process in the first module (Fig. 2a), designed for the heat treatment of secondary animal-derived raw materials, is as follows. The electric drives of the dielectric mixer 13, cutter «Volchok» 10, shredder 9, and fan 1 located in the source block, are started. Non-food animal-derived waste is loaded into the receiving tank 8, then the shutter opens. The raw material is comminuted using the shredder 9, then it is pumped and finely ground by the cutter «Volchok» 10, and through the mesh it falls into the screen dielectric cylinder 5. Further, it is necessary to launch the generators 4, located on the stationary base 2 of the first module, at a given power. The magnetron emitters are directed into the capacitor part of the resonator. In toroidal resonator 3, a running wave (12.24 cm length, 2450 MHz frequency) occurs. The raw materials are heated due to the polarization currents, and cooked during the mixing process with the dielectric spiral mixer 13. The bulk cooked product is passed through a screen dielectric cylinder and, when the ball valve 12 is opened in the pipe (outlying waveguide), it is discharged from the toroidal part 7 of the resonator. At the end of the process, all sources should be turned off in a reverse sequence. Next, the removable module (toroidal resonator with mixer without electric motor) is detached from the base 2 and all components are sanitized.

The process in the second module (Fig. 2b), designed for thermal treatment of frozen animal-derived raw materials with phase transition, coagulating at temperatures above 40°C (frozen blood, frozen cow colostrum, etc.) occurs as follows. The ball valve 19 is closed, electric vibrator 17 and electric drive of the cutter with finned surface are started. The frozen product is loaded into the receiving tank 14 using the shutter. The raw material after grinding by the shredder 15 enters into the internal cylindrical resonator 18 and the generators on the stationary base 21 are started. The emitters from the magnetrons are directed into cylindrical resonators 16, 18.

The frozen raw material is defrosted by a microwave electromagnetic field. During the phase transition, the colostrum flows through the lower part of the lateral surface of the internal resonator 18 into the external cylindrical resonator 16, then activates the generators located on the stationary base of the second module 20 at a given power. In both cylindrical resonators, the microwave electromagnetic field of different doses is activated. Such technology of dielectric heating of raw material in different resonators at frozen and liquid states, i.e. during phase transition, is because of the change of dielectric loss factor of raw material is opposite, i.e. it increases in the negative temperature range during the defrosting process, but decreases in the positive temperature range during the heating process. The module that realizes these processes in different resonators reduces the defrosting and heating time of the colostrum, saving the required physical, chemical, and biological properties of the raw material while reducing the energy consumption. The expected annual economic effect from the use of microwave defroster in relation to defrosters of different design, operating on steam-water mixture, is 323350 RUB by reducing operating costs.

The process in the third module (Fig. 2c), designed for heat treatment of second viscous raw material, ensuring the combination of dewatering of cut meat, fine grinding, and heat treatment of solid fraction for use as the protein supplement for animals, occurs as follows. An electric drive 25 of the dielectric rotor 24, dielectric screw 29, and electrical drive for viscous liquid pump 22 are started.

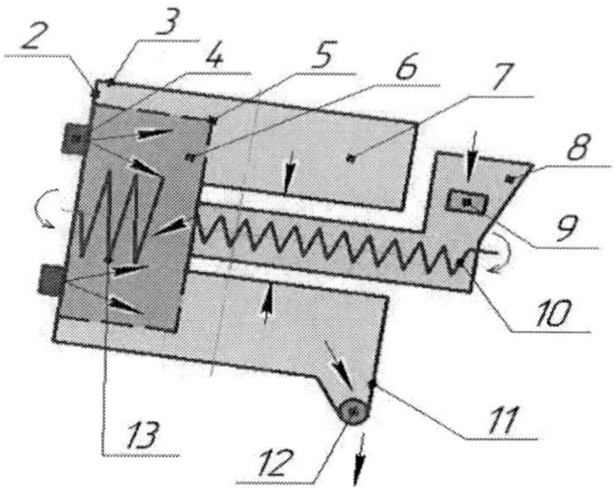

(a)

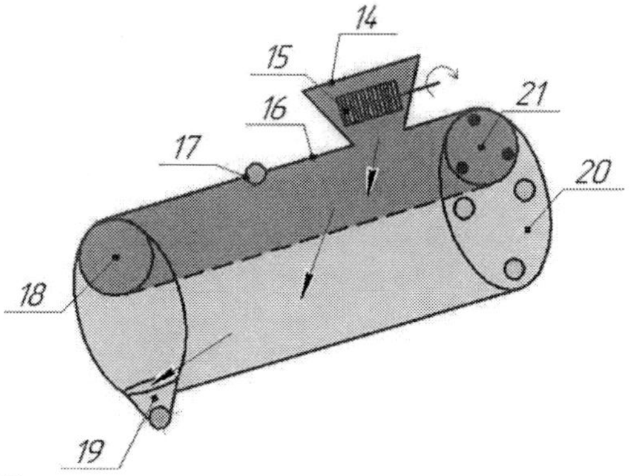

(b)

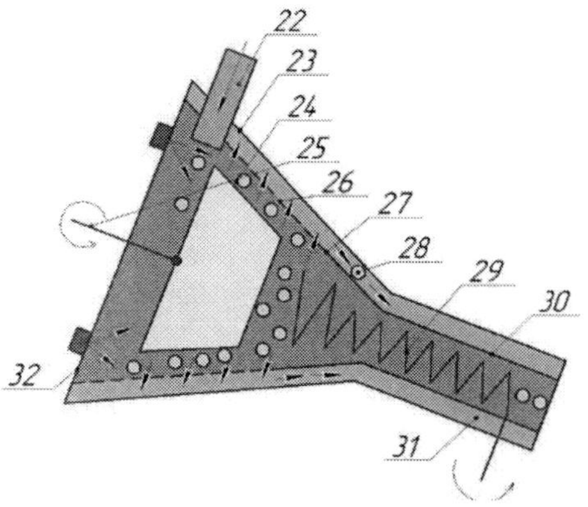

(c)

Figure 2. Scheme of microwave modules for the thermal treatment of raw materials: a) the first module with a toroidal resonator; b) the second module with coaxial cylindrical resonators; c) the third module with a conical resonator; 2 – stationary base of the toroidal resonator 2 (Fig. 1); 3 – toroidal resonator; 4 – magnetrons with waveguides; 5 – screen dielectric cylinder; 6 – capacitor part of the resonator; 7 – toroidal part of the resonator; 8, 14 – receiving tank; 9, 15 – shredder; 10 – cutter «Volchok»; 11 – disc dielectric with the electric drive; 12 – discharge pipe with the ball valve; 13 – spiral dielectric mixer; 16 – external cylindrical resonator; 17 – electric vibrator; 18 – internal cylindrical resonator; 19 – pipe with the ball valve; 20 – stationary base of the external cylindrical resonator with magnetrons; 21 – stationary base of the internal cylindrical resonator with magnetrons; 22 – dielectric pipe for the pumping viscous fluid; 23 – truncated conical non-ferromagnetic resonator; 24 – dielectric rotor in the shape of the truncated cone coated with the fine-grained abrasive material; 25 – electric drive; 26 – screen dielectric truncated cone without bases; 27 – truncated dielectric cone; 28 – flow channel for the extracted liquid; 29 – dielectric screw for the transporting cooked solid fraction of raw material; 30 – screw case with the electric drive; 31 – non-ferromagnetic liquid fraction pipe; 32 – stationary base of the truncated conical resonator (Fig. 1).

Further, the viscous secondary raw material enters the truncated conical resonator through the joined pipes (non-ferromagnetic resonator and dielectric resonator). The generators that are located in the source block on the stationary base 32, the emitters of which are directed into the resonator 23, are activated. The dielectric rotor 24, which coated with fine-grained abrasive material, rotates and finely grinds the raw material due to centrifugal force, the liquid fraction passes through the screen dielectric truncated cone 26, the raw material is dehydrated, the solid fraction is cooked and is removed from the resonator using the dielectric screw 29, 30. The screw is located in the non-ferromagnetic pipe 31, which acts as the supercritical waveguide. In addition, due to

https://doi.org/10.1051/bioconf/20225200047

the critical cross-section in the conical resonator [4], beyond which the waves do not propagate, but are reflected in the opposite direction, forming a standing wave, the radiation does not occur.

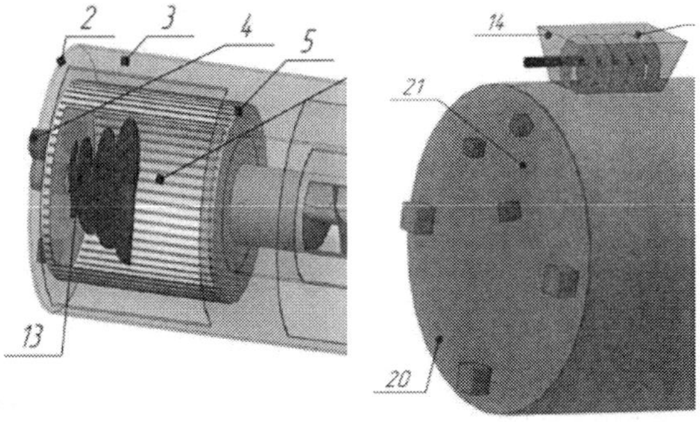

a) b)

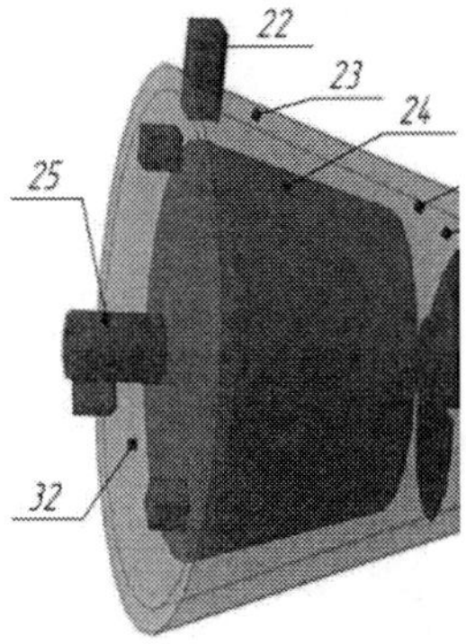

c)

Fig. 3. Scheme of microwave modules for heat treatment of raw materials: a) first module with the toroidal resonator; b) second module with coaxially arranged cylindrical resonators; c) third module with the conical resonator (the items are given in Figures 1–2)

3 Results and discussion

The electromagnetic safety of the cylindrical perforated resonator has been assessed. Table 1 shows the planning matrix.

Table 1. Active planning matrix of the 2^3 experiment for electromagnetic assessment of cylindrical perforated resonator safety.

No.	*Parameters*						*Optimization criteria*	
	Specific power of generator, W/cm³		*Free area ratio*		*Thickness of shielding case, mm*		*Radiation flow rate, μW/cm²*	*Safe operation time, h*
	x1	P_s	x2	*k*	x3	*d*	Y1(P_{tot})	Y2(*τ*)
1	+	0.195	+	0.4	+	4	420	2.5
2	+	0.195	-	0.04	-	2	400	2.6
3	-	0.05	+	0.4	-	2	390	2.65
4	-	0.05	-	0.04	+	4	140	4.1
5	0	0.1	0	0.22	0	3	160	3.9
6	-	0.05	0	0.22	0	3	120	4.5
7	+	0.15	0	0.22	0	3	200	3.4
8	0	0.1	-	0.04	0	3	170	3.6
9	0	0.1	+	0.4	0	3	220	3.3
10	0	0.1	0	0.22	-	2	200	3.4
11	0	0.1	0	0.22	+	4	190	3.5

3D Surface Plot of Y1 against X_1 and X_2

$$Y1 = 202.81 - 1823.11 \cdot x_1 - 131.49 \cdot x_2 + 15658.17 \cdot x_1^2 - 3942.65 \cdot x_1 \cdot x_2 + 2002.7 \cdot x_2^2$$

3D Surface Plot of Y1 against X_1 and X_3

$$Y1 = 1260.97 - 5178.97 \cdot x_1 - 524.72 \cdot x_3 + 15658.17 \cdot x_1^2 + 829.49 \cdot x_1 \cdot x_3 + 64.89 \cdot x_3^2$$

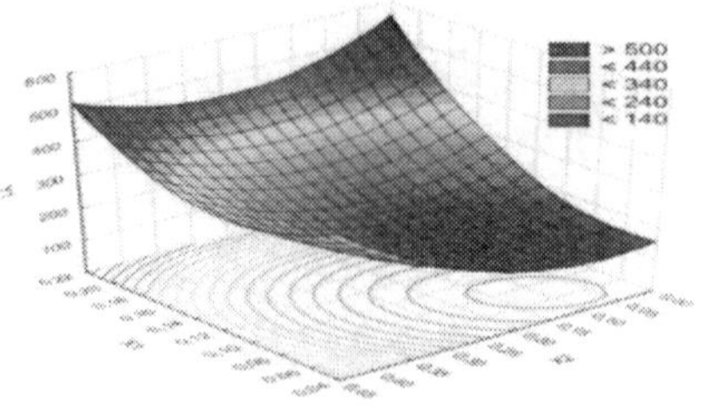

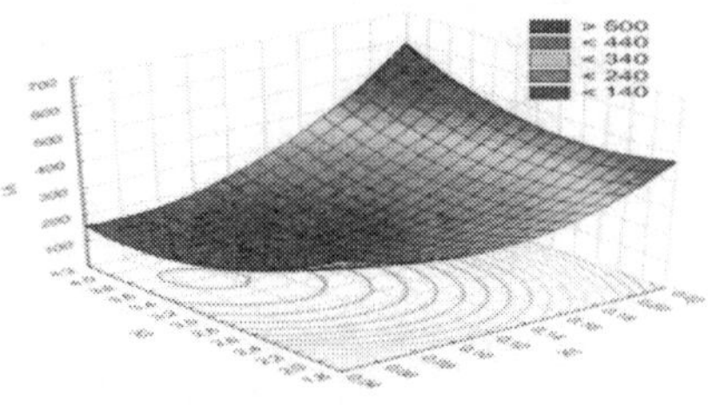

3D Surface Plot of Y1 against X_2 and X_3

$$Y1 = 1419.03 - 2155.70 \cdot x_2 - 678.08 \cdot x_3 + 2826.51 \cdot x_2^2 + 402.78 \cdot x_2 \cdot x_3 + 91.59 \cdot x_3^2$$

3D Surface Plot of Y2 against X_1 and X_2

$$Y2 = 4.62 - 11.45 \cdot x_1 + 2.43 \cdot x_2 - 3.1 \cdot x_1^2 + 23.23 \cdot x_1 \cdot x_2 - 15.49 \cdot x_2^2$$

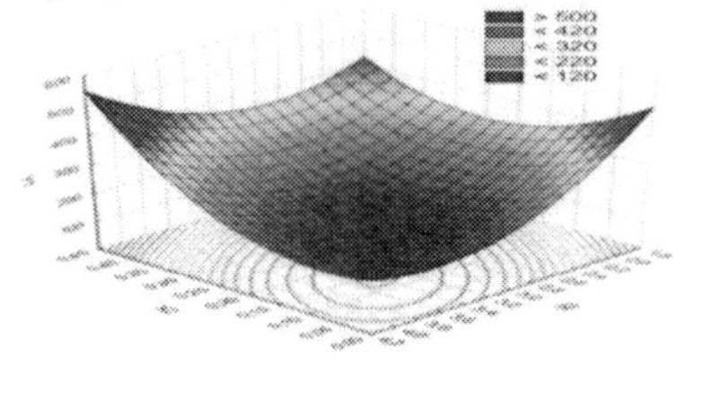

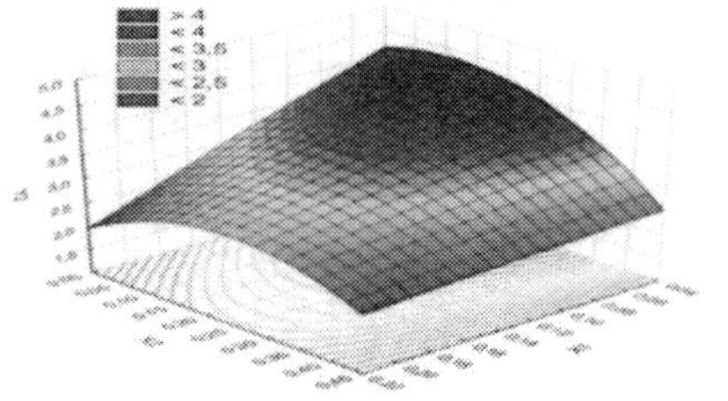

3D Surface Plot of Y2 against X_1 and X_3

$$Y2 = -2.48 + 8.0 \cdot x_1 + 3.80 \cdot x_3 - 3.1 \cdot x_1^2 - 4.78 \cdot x_1 \cdot x_3 - 0.5 \cdot x_3^2$$

3D Surface Plot of Y2 against X_2 and X_3

$$Y2 = -3.02 + 11.76 \cdot x_2 + 3.66 \cdot x_3 - 15 \cdot x_2^2 - 2.29 \cdot x_2 \cdot x_3 - 0.49 \cdot x_3^2$$

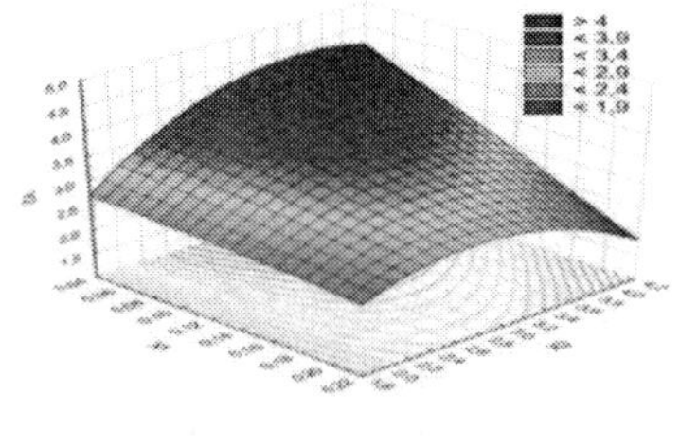

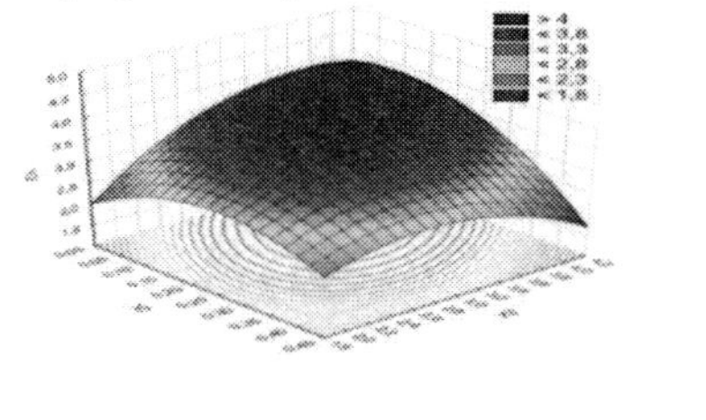

Fig. 4. Contour cross-sections and response surfaces of three-factor models for electromagnetic safety of the perforated cylindrical resonator.

The obtained regression models for evaluating the electromagnetic safety of the perforated cylindrical resonator:

$$Y1 = 202.81 - 1823.11 \cdot x_1 - 131.49 \cdot x_2 + 15658.17 \cdot x_1^2 - 3942.65 \cdot x_1 \cdot x_2 + 2002.7 \cdot x_2^2 (x_3 = 3\ mm)$$

$$Y1 = 1260.97 - 5178.97 \cdot x_1 - 524.72 \cdot x_3 + 15658.17 \cdot x_1^2 + 829.49 \cdot x_1 \cdot x_3 + 64.89 \cdot x_3^2 (x_2 = 0.22)$$

$$Y1 = 1419.03 - 2155.70 \cdot x_2 - 678.08 \cdot x_3 + 2826.51 \cdot x_2^2 + 402.78 \cdot x_2 \cdot x_3 + 91.59 \cdot x_3^2 (x_1 = 0.1 W/cm^3)$$

$$Y2 = 4.62 - 11.45 \cdot x_1 + 2.43 \cdot x_2 - 3.1 \cdot x_1^2 + 23.23 \cdot x_1 \cdot x_2 - 15.49 \cdot x_2^2 (x_3 = 3\ mm)$$

$$Y2 = -2.48 + 8.0 \cdot x_1 + 3.80 \cdot x_3 - 3.1 \cdot x_1^2 - 4.78 \cdot x_1 \cdot x_3 - 0.5 \cdot x_3^2 (x_2 = 0.22)$$

$$Y2 = -3.02 + 11.76 \cdot x_2 + 3.66 \cdot x_3 - 15 \cdot x_2^2 - 2.29 \cdot x_2 \cdot x_3 - 0.49 \cdot x_3^2 (x_1 = 0.1 W/cm^3)$$

where Y1 is the radiation flow rate, μW/cm^2; Y2 is the safe operation time, h; x_1 is the specific power of the generator, W/cm^3; x_2 is the resonator perforation area ratio; x_3 is the thickness of the shielding case, mm.

The results of the regression model study (Fig. 4) show that if the coaxial resonator is made of aluminum 3 mm thick at a specific power of 0.1 W/cm^3 and the resonator perforation area of the annular base of the coaxial resonator of 0.22, the radiation flux is 160 mW/cm^2. The unit can operate during 4 hours at this radiation level.

Modeling of 3D structures of electromagnetic fields in resonators with three magnetrons (CST Studio Suite software, Time Domain Solver) was performed. Below are the results of the study of electric (A/m) and magnetic (V/m) components, surface current (A/m) (Fig. 5). The results of the electrodynamic parameters study show that in order to ensure high electric field intensity (3–4 kV/cm), it is necessary to design toroidal resonators, conical resonators (third module) provide electromagnetic safety during continuous operation. In order to ensure the uniform distribution of the electromagnetic field, magnetrons are located at 120

https://doi.org/10.1051/bioconf/20225200047

degrees from each other along the perimeter of the resonator.

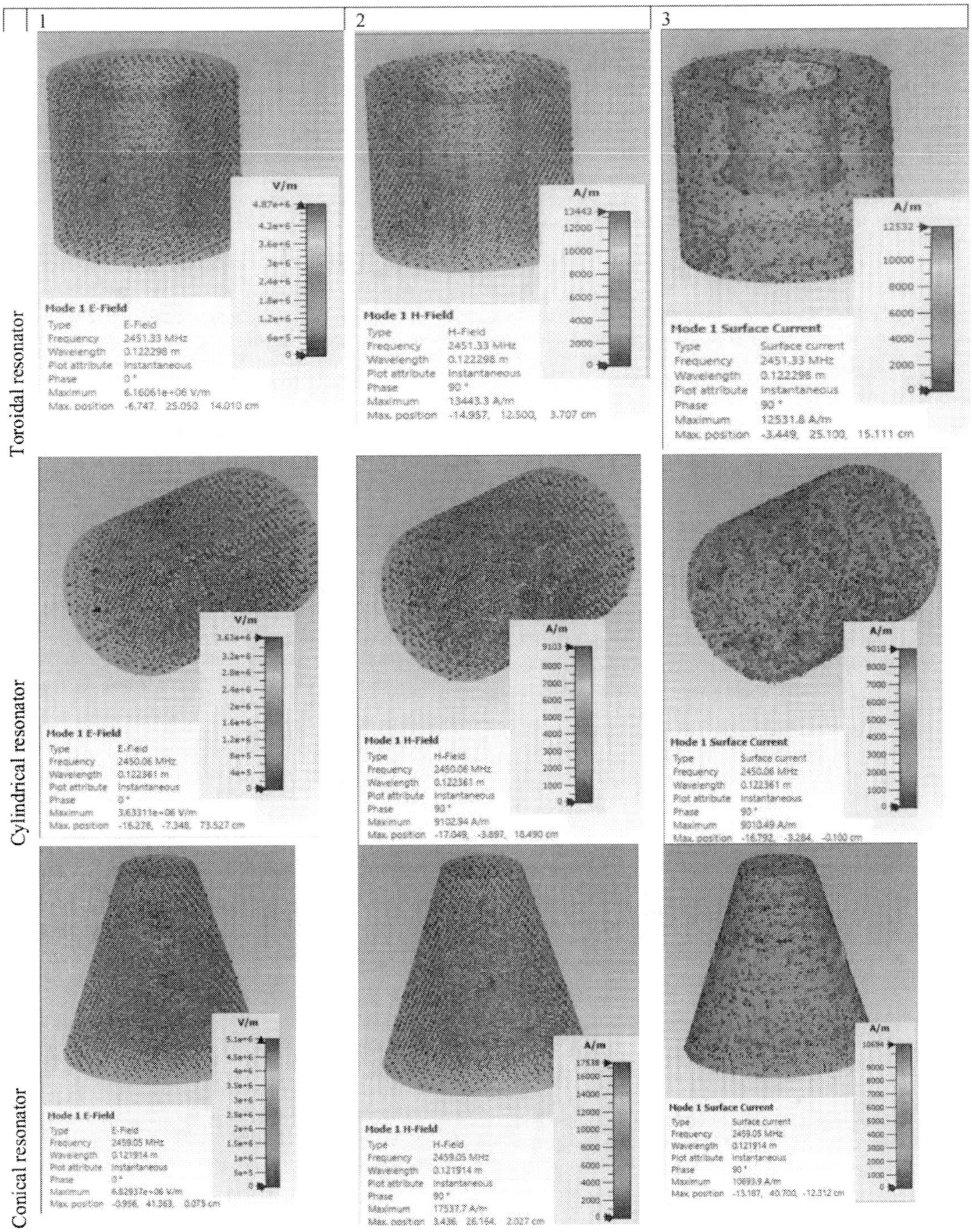

Fig. 5. Electrodynamic parameters of resonators (mode 1): a) toroidal (first module): b) cylindrical (second module); c) conical (third module); 1 – electric field intensity, V/m; 2 – magnetic field intensity, A/m; 3 - magnetic component intensity (surface current), A/m.

4 Conclusion

In the paper the microwave unit with removable modules for continuous heat treatment and disinfection of animal-derived raw materials of different structures with resonators was developed, providing electromagnetic safety without shielding and high electric field intensity with reduced operating costs while preserving the feeding value of the product. The electrotechnological process and the designs of the proposed units for treatment animal-derived materials were described.

The results of the study are justified modes of microwave module operation. The first module provides an increase in the feeding value of the protein product by 5 points (maximum is 15) compared to the traditional technology under influence of the microwave electromagnetic field in a continuous mode within electromagnetic safety of units, which estimated by energy flux up to 160 $\mu W/cm^2$ during no more than 4 hours per day. The effective mode of operation of the second module is with the duration of defrosting and heating of the animal colostrum is 10-15 min with the power of the generators of 4.8 kW and energy costs of 0.18-0.2 kW·h/kg. The third module provides dewatering and heat treatment of viscous raw materials, so the energy costs are associated with the heating of the solid fraction.

References

1. A. A. Tikhonov, A. V. Kazakov, M. V. Prosviryakova, G. V. Novikova, Patent of the Russian Federation No. 2726565. Microwave installation with a quasi-stationary toroidal resonator for rendering decontaminated fat from crushed fat-containing raw materials in a continuous mode **Bull. No. 20,** No. 2019122928 (2020)
2. G. V. Novikova, M. V. Belova, O. V. Mikhailova, D. V. Tarakanov, I. A. Sorokin, A. A. Tikhonov, A. V. Kazakov, Patent of the Russian Federation No. 2734593. Microwave unit for defrosting and heating cow colostrum with coaxially located resonators **Bull. No. 29,** No. 2020104252 (2020)
3. A. A. Tikhonov, A. V. Kazakov, G. V. Novikova, B. I. Gorbunov, Patent No. 2710154 RF. Microwave installation with a conical resonator for dehydration and heat treatment of meat waste **Bull. No. 36,** No. 2019118371 (2019)
4. O. O. Drobakhin, P. I. Zabolotny, E. N. Privalov, Resonant properties of axially symmetric microwave resonators with conical elements, *Radiophysics and Radio Astronomy* **1 (4),** 433–441 (2009)
5. V. N. Udalov, B. V. Vasiliev, G. V. Lysov, Yu. M. Egorov, Patent No. 1830197A3 RF, IPC N05V6/64. Microwave defroster **Bull. No. 27,** No. 4881731/09 (1993)
6. D. V. Poruchikov, A. N. Vasiliev, I. G. Ershova, G. V. Novikova, M. V. Belova, Patent No. 2694944 RF, IPC A47J39/00. Microwave unit for defrosting bovine colostrum **Bull. No. 20,** No. 2018143727 (2019)
7. G. V. Novikova, M. V. Prosviryakova, O. V. Mikhailova, I. G. Ershova, D. A. Tarakanov, A. A. Tikhonov, Patent No. 2752938 RF, IPC A47J.39/00. Two-module continuous-flow microwave unit for defrosting and heating cow colostrum **Bull. No. 23,** No. 2020141711 (2021)
8. G. V. Novikova, M. V. Prosviryakova, O. V. Mikhailova, I. G. Ershova, B. G. Zigan-Shin, D. A. Tarakanov, Patent No. 2752941 of the Russian Federation, MPK A47J.39/00. Radiotight multicavity unit for defrosting animal colostrum **Bull. No. 23,** No. 2020141715 (2021)
9. G. V. Novikova, V. F. Storchevoi M. V. Prosviryakova, O. V. Mikhailova, I. G. Ershova, D. A. Tarakanov, Patent No. 753424RF, IPC A47J.39/00. Continuous-flow microwave unit with quasi-stationary toroidal resonators for defrosting and heating animal colostrum **Bull. No. 9,** No. 2021104198 (2021)
10. M. V. Prosviryakova, O. V. Mikhailova, E. A. Shamin, G. V. Novikova, *Electromagnetic Fields and Microwave Technologies* (Knyaginino: GBOU VO NGIEU, 2020)
11. M. O. Binelo, V. Faoro, O. A. Kathatourian, B. Ziganshin, Airflow simulation and inlet pressure profile optimization of a grain storage bin aeration system, *Computers and Electronics in Agriculture* (2019)
12. G. V. Novikova, O. V. Mikhailova, S. P. Zaytsev, Development of microwave technology and installation for defrosting and heating of bovine colostrum, *Bulletin of the Chuvash Agricultural Academy* **3 (14),** 90–94 (2020)
13. O. O. Drobakhin, P. I. Zabolotny, N. B. Goreyev, D. Yu. Saltykov, Sensors based on biconic microwave resonators for monitoring the parameters of dielectrics *19th Crimean Conf. Microwave and Telecommunication Technology (CriMiCo)* (Sevastopol) **1418,** 775–776 (2009)
14. R. K. Lorenzoni, M. O. Binelo, M. De F B Binelo, B. G. Ziganshin, Quasi-2D simulation of soya beans flow in mixed flow dryer, *Journal of Stored Products Research* **89,** 101727 (2020)
15. A. N. Didenko, *Microwave Energy: Theory and Practice* (Moscow: Nauka, 2003)
16. A. V. Strekalov, Yu. A. Strekalov *Electromagnetic Fields and Waves* (Moscow: RIOR: INFRA-M, 2014)
17. *CST Studio Suite*, retrieved from: https://lorentz.ru/cst-studio-suite/?yclid=7063004186007886553
18. V. Gevorkyan, V. Kochemasov Volumetric dielectric resonators - main types, characteristics, manufacturers, *Electronics* **4 (00154),** 62–76 (2016)
19. O. O. Drobakhin, Resonant properties of axially symmetric microwave resonators with conical elements, *Radiophysics, Radio Astronomy* **14 (14),** 433–441 (2009)
20. I. A. Rogov, V. Ya. Adamenko, et al., *Electrophysical, Optical and Acoustic*

Characteristics of Food Products (Moscow: Light and food industry, 1981)
21. Yu. N. Pchelnikov, *Microwave Electronics*. (Moscow: Radio and communication, 1981)
22. O. V. Drogaytseva, *Increasing the level of heating uniformity of dielectric materials in microwave devices of waveguide and resonator types*: PhD Dissertation (Saratov: SGTU, 2011)
23. V. A. Kolomeitsev, V. V. Komarov, *Microwave Units with Uniform Volumetric Heating* (Saratov: SGTU, 2006)
24. N. Dubkova, V. Kharkov, B. Ziganshin, Effect of Mode Amplitude on Power Consumption in Vibrating Mixer, *6th International Conference on Industrial Engineering (ICIE 2020): Lecture Notes in Mechanical Engineering* (Sochi, Russia) pp 362–369 (Springer International Publishing, 2021)
25. V. N. Kopusov, On the issue of creating multi-magnetron equipment for modern technologies, *Microwave Engineering and Telecommunication Technologies,* 652–653 (Sevastopol: Weber, 2001)
26. D. L. Rakhmankulov, S. Yu. Shavshukova, I. N. Vikharev, Features of microwave installations for heating, *Bashkir Chemical Journal* **15 (1),** 57–61 (2008)
27. A. E. Semenov, *Microwave heating devices of the resonator type with an adjustable supply of electromagnetic power*: PhD Dissertation (Saratov: SGTU, 2008)
28. A.V. Tsygankov, *Electrotechnological microwave installations for uniform heating of dielectric materials on waveguides of complex sections*: PhD Dissertation (Saratov: SGTU, 2003)
29. O. V. Mikhailova, *Fundamentals of Patent Science and Scientific and Technical Developments for Implementation of Microwave Technology in the Processing of Agricultural Raw Materials* (Knyaginino: NGIEU, 2021)
30. G. Tushar, Z. Huacheng, K. Ashim, H. Kama Microwave drying of spheres: Coupled electromagnetics-multiphase transport modeling with experimentation. PartII: Model validation and simulation results, *Food Bioprod. Process* **326,** 96 (2015)
31. L. Donglei, T. Juming, D. Patrick, L. Frank, T. Zhongwei, Analysis of electric field distribution within a microwave assisted thermal sterilization (MATS) system by computer simulation, *J. Food Eng.* **188,** 87 (2016) doi: 10.1016/j.jfoodeng.2016.05.009
32. S. Tianyi, Z. Zhijun, H. Jingxue, Z. Shiwei, W. Xiaowei, Z. Wenqing, Sensitivity analysis of intermittent microwave convective drying based on multiphase porous media models, *Int. J. Therm. Sci.* **153,** 106344 (2020) doi: 10.1016/j.ijthermalsci.2020.106344
33. L. A. Campañone, N. E. Zaritzky, Mathematical analysis of microwave heating process, *J. Food Eng.* **69,** 359 (2005) doi: 10.1016/j.jfoodeng.2004.08.027
34. S. J. Guido, D. V. Martin, I. S. Andrzej, D. S. Georgios, On the effect of resonant microwave fields on temperature distribution in time and space. *Int. J. Heat Mass Transf.* **55 (13),** 3800 (2012) doi: 10.1016/j.ijheatmasstransfer.2012.02.065

Application of feed additive "BioPrimum dry" with probiotic effect in dairy farming

Vadim A. Ruin, *Anna A.* Kistina, *Yuri N.* Prytkov, and *Anna S.* Panfilova

National Research Mordovia State University (MRSU), 68, Bolshevistskaya St., Saransk, Republic of Mordovia, 430005, Russian Federation

Abstract. Currently, the search for alternative ways to replace the use of antibiotics in animal husbandry is being intensively conducted all over the world, including in Russia. One of the possibilities for the replacement of antibiotics in feed mixtures is the development and testing of probiotics in production conditions. They are biomass of bacteria in the vegetative or spore form with a clearly expressed antagonistic activity against the pathogenic and conditionally pathogenic microflora of the gastrointestinal tract chyme. The paper presents the results of a study on the use of the BioPrimum Sukhoy (dry) feed additive added in different dosages into cows' diets. It was found that the introduction of the feed additive in the composition of the cows' diets at a dosage of 75 mg/kg of dry matter of feed resulted in an increase of red blood cell count and hemoglobin content in the blood by 11.40 and 6.05%, respectively, compared with analogs of the control group. There was an increase in milk productivity by 13.8%, in the amount of milk fat by 14.8% and milk protein by 14.2%.

1 Introduction

To prevent and treat gastrointestinal diseases of bacterial etiology and reduce economic losses when changing feed, after courses of antibiotic therapy and vaccination, about 10 years ago, after numerous tests, several probiotic preparations of a wide spectrum of action were created. They were based on strains of bacteria B. Bacillus subtilis. They are producers of biologically active substances (BAS), such as enzymes, antibiotics, insecticides, and other substances that regulate and stimulate digestion and have anti-allergenic and antitoxic effects. Bacteria B. Bacillus subtilis are highly adaptable to various conditions of existence (presence or absence of oxygen, growth, and development in a significant temperature range, the use of various organic or inorganic compounds as food sources, etc.). They contribute to the spread of bacilli in soil, water, air, food and other objects of the external environment, as well as in the animal bodies [1–9].

B. Bacillus subtilis strains produce antagonistically active biomass against pathogens, as well as proteolytic, amylolytic, and lipolytic enzymes. The B. Bacillus subtilis strains used for the development of the preparation, multiplying in the intestines of animals, release BAS. Under the influence of BAS the digestion processes are activated, as a result of which the average daily gain in live weight increases, productivity increases together with the safety of the livestock, and feed cost per unit of production is reduced.

2 Materials and methods

The purpose of our study was to evaluate the effect of different levels of the BioPrimum Sukhoy feed additive on morphological and biochemical parameters of blood, quantitative and qualitative indicators of milk produced.

To achieve this goal, we conducted a scientific and economic experiment in the production conditions of Agrosoyuz LLC of the Ruzaevsky municipal district of the Republic of Mordovia. There were cows with the inclusion of different dosages of the BioPrimum Sukhoy feed additive in their diets. To experiment on the principle of pairs of analogs, taking into account the breed, age, live weight, and lactation period, 40 cows were selected on the twentieth day after calving, and 4 groups of 10 heads each were formed. All animals were clinically healthy and kept in the same conditions. The cows were fed three times a day during the experiment. The diets were compiled according to the recommended detailed norms of the Russian Academy of Agricultural Sciences (RASKhN) (2003), taking into account the live weight, physiological state, productivity of the cows, and the chemical composition of feed. The animals of the control group received the basic diet. The analogs of the 1st, 2nd, and 3rd experimental groups, in addition to the basic diet, were fed with the BioPrimum Sukhoy feed additive at the rate of 60; 75; 90 mg/kg of dry matter of the diet, respectively, or 1,200, 1,500 and 1,800 mg per head per day, respectively (Table 1). The BioPrimum Sukhoy feed additive contains a lyophilized microbial mass of spore-forming bacteria Bacillus subtilis WB3482 (VKPM B-1722) and Bacillus amyloliguefaciens 31 (VKPM B-2336) at least $2x10^9$ colony-forming units (CFU) in 1 g, 15% of chromium picolinate, and up to 100% of filler: calcium carbonate. The chromium content ranges from 0.018 to 0.02 g in 1 g of the preparation. It does not contain genetically modified organisms and products. The content of harmful impurities does not exceed the maximum permissible standards in force in the Russian Federation. In appearance, it is a loose powder from beige to cream color.

Table 1. Scheme of scientific and economic experiment.

Group	Number of heads per group	Level of feed additive in the diet, mg/kg of dry matter	Qty of feed additive per head per day, mg
Control	10	Basic diet	–
1st experimental	10	60	1.200
2nd experimental	10	75	1.500
3rd experimental	10	90	1.800

To control the physiological state and the course of biochemical processes in the body of dairy cows under the influence of different dosages of BioPrimum Sukhoy (Table 1), we conducted a study of the dynamics of morphological and biochemical parameters of the cows' blood at the beginning and end of lactation.

3 Results and discussion

According to the results of our study, we found that different dosages of the BioPrimum Sukhoy feed additive in the diets of cows at the beginning of lactation had a certain effect on hematological parameters. Thus, in the blood of cows of the 2nd experimental group, there was an increase in the content of red blood cells and hemoglobin by 11.40% and 6.05% compared with the analogs of the control group and by 6.05% and 2.78% compared to the peers of the 1st experimental group. An increase in the feed additive in the diets of cows of the 3rd experimental group to 90 mg/kg of the dry matter of the diet contributed to a slight decrease in the studied indicators compared to the 2nd experimental group. A decrease of 3.95% and 2.03%, respectively, in red blood cells and hemoglobin, was observed (Table 2).

A similar pattern was observed in the morphological parameters of blood at the end of lactation. Thus, in the blood of cows of the 2nd experimental group, there was an increase in the content of red blood cells and hemoglobin by 6.01% and 6.65% compared with analogs of the control group. The number of white blood cells almost did not change, with only a slight increase in the blood of animals of the control group. It can be assumed that all this indicated a more intensive course of redox processes in the body of animals.

Feeding cows with the feed additive "BioPrimum dry" contributed to the transformation of dry matter into milk. So, if the amount of dry matter in the milk of cows of the 2nd experimental group was 12.73%, then it was 0.26% ($P<0.01$) lower in the milk of analogues of the control group.

With a decrease and an increase in the dosage of the probiotic, there was a decrease in the amount of dry matter by 0.10–0.17%. However, the indicators were significantly higher than those that were in the milk of cows of the control group.

An increase of the dry matter in the milk of the cows in the experimental groups, who received the feed additive "BioPrimum dry" as part of the diet, was caused by the consumption of nutrients and mineral elements by cows along with feed mixtures of their digestibility, participation in milk formation.

Experimental studies confirm that the inclusion of a feed additive in the main diet gives an increase in milk sugar (lactose) in milk. The superiority in this value in the milk of cows of the 1st experimental group over the analogues of the control group was by 0.02%, the 2nd – 0.13%, the 3rd – 0.05% with a significant difference. It is important to emphasize that the use of the feed additive "BioPrimum dry" in feeding dairy cows activates the microflora of the rumen, stimulates scar digestion, increases the digestibility of nutrients in the diet improves the use of nutrients, carbohydrate metabolism.

According to the data obtained in our experiments, it was found that the level of calcium and phosphorus in the milk of the experimental groups exceeds the control. In the milk of cows of the 1st experimental group, the calcium content was higher by 0.03 g and phosphorus by 0.05 g, respectively, in the 2nd – by 0.04 g and 0.03 g, and in the 3rd – 0.03 g and 0.11 g.

Feeding a probiotic has a positive effect on the accumulation of carotene in the milk of cows of the 2nd experimental group prevails, who received the feed additive "BioPrimum dry" as part of the diet at a dose of 75 mg/kg of the dry matter of the diet.

The digestibility of nutrients, assimilation and use of mineral elements contributed to an increase in the energy value of milk. So, if the milk of cows of the 2nd experimental group had an energy value of 74.02 kcal per 100 g, which is 0.85 kcal or 1.16% more than that in the control group and 0.20% with the 1st experimental group.

When the dosage increased to 90 mg / kg of the dry matter, it contributed to a slight increase in the energy value of milk.

Thus, the use of a feed additive with a probiotic "BioPrimum dry" in the diet of red-mottled cows contributes to the improvement of the qualitative composition and nutritional value of milk.

The amount of milk fat in milk, the most important cost indicators in the sale of milk, characterizes the nutritional, biological value of milk. The percentage of fat in milk depends on the cost of milk on sale. The content of milk fat in milk during lactation changes under the influence of the physiological state, lactation period, balanced feeding.

During the lactation period, under the influence of different levels of feed additives, there are spikes and declines in the monthly dynamics of the fat content in milk. It was found that in the first month of lactation, the fat content in the milk of cows of all groups was almost at the same level and amounted to 3.81-3.86%.

Thus, due to different dosages of probiotic supplements, providing beneficial intestinal microflora, stimulating metabolic processes, milk formation, there is an increase in the milk and fat content.

Starting from the third month of lactation, there was a gradual decrease in the production of milk fat obtained from cows of all the experimental groups. This decrease depended primarily on the amount of milk received during a certain period of lactation. Since the dairy productivity of cows and the fat content of the milk obtained are inversely related, the lower the milk yield by the end of lactation, the higher the fat content of milk. The smallest amount of milk fat was obtained from the cows of all the experimental groups in the last month of lactation.

The use of different dosages of the probiotic supplement "BioPrimum dry" in the diets of dairy cows significantly contributes to an increase in the quantity and quality of milk fat. Along with an increase in the amount of milk fat, the yield of milk protein steadily increases, which directly depends on the protein intake from the feed.

It can be argued that the protein content tends to decline in the first months of lactation with a slight increase from the middle of lactation and an increase to a maximum by the end of lactation in the milk of cows of the studied groups. In the cows of the control group, the average protein content in milk for 305 days of lactation was 3.15%, which is 0.01–0.04% lower than that in the peers of the experimental groups.

An increase in the protein content in the milk of cows receiving a probiotic supplement is an indicator of stimulating protein metabolism in the body, which is confirmed by the biochemical status of the blood.

Among the cows of the experimental groups, the superiority in the amount of milk fat for the entire lactation period was on the side of the peers of the II and III-experimental groups, their superiority over the animals of the I-experimental group was 14.41 kg (6.81%; $P<0.001$) and 14.53 kg (6.87%; $P<0.001$), respectively.

It should be noted that the use of different doses of the feed probiotic supplement "Bioprimum dry" in the diets of dairy cows significantly contributes to an increase in the quantity and quality of milk fat.

So, in the period from the first to the third month inclusive, the decrease in protein concentration in the milk of cows of the control group was 0.02%, I experimental – 0.02%, II experimental – 0.01% and III experimental group – 0.02%. It should be noted that starting from the fifth month of lactation, there was an increase in the protein content in the milk of cows of all groups. So, in the period from the fifth month to the end of lactation in the milk of cows of the control group, the value of the studied indicator increased by 0.06%, the experimental groups – 0.05–0.08%.

Thus, it can be argued that reliable data were obtained on the protein content in milk during the lactation period of experimental cows of the red-mottled breed. They indicate the fruitful effect of the feed additive with the probiotic "Bioprimum dry" on the presented indicators.

Table 2. Morphological parameters of cows' blood.

Group	Red blood cells, 10^{12} g/l	White blood cells, 10^{9} g/l	Hemoglobin, g/l
Beginning of lactation			
Control	6.14±0.03	9.22±0.12	104.17 ±0.81
1st experimental	6.45±0.05	8.98±0.06	107.47 ±1.60
2nd experimental	6.84±0.04	8.32±0.10	110.46 ±1.31
3rd experimental	6.58±0.10	8.54±0.06	108.26 ±1.04
End of lactation			
Control	6.06±0.09	9.03±0.03	101.98 ±1.14
1st experimental	6.13±0.13	8.81±0.11	105.34 ±1.55
2nd experimental	6.42±0.10	8.32±0.10	108.77 ±0.90

3rd experimental	6.2 4±0.09	8.54±0. 07	106.37 ±1.12

An important indicator of the state of protein metabolism in the body is the content of total protein and its fractions in the blood. In our studies we used the BioPrimum Sukhoy feed additive in the diets of cows of the 2nd experimental group at the rate of 75 mg/kg of dry matter of the diet. This led to an increase in the amount of total protein in their blood compared to analogs from the control group by 5.87% and 4.25% compared to the 1st experimental group (P<0.001). A similar pattern was observed for the albumin and globulin count. In the blood of cows of the 2nd experimental group, the content of albumins is by 9.86% and globulins by 2.31% higher compared to the peers of the control group and, respectively, by 6.47 and 2.19% compared to the animals of the 1st experimental group. Being of the globulin fraction, gamma globulins occupy the largest share, and their lowest concentration in all age periods was observed in the animals of the 2nd experimental group. The number of alpha- and beta-globulins in the 2nd experimental group was also higher compared to the experimental animals of the control, 1st, and 3rd experimental groups.

As a result of the conducted study, it was found that the inclusion of different dosages of the BioPrimum Sukhoy feed additive in the diets of cows had a positive effect on the quantitative and qualitative indicators of milk. We found that during the first lactation, 8.806.6 kg of milk was obtained from the first-calf cows of the 2nd experimental group, which is 13.8% (P<0.001) and 5.0% (P<0.001) more compared to the analogs of the control and the 1st experimental group. The increase of the dosage of the BioPrimum Sukhoy feed additive to 90 mg/kg of the dry matter of the diet did not contribute to a further increase in milk productivity. However, the milk yield for the first lactation was higher by 443.2 kg or 5.7% (P<0.001) compared to the analogs in the control group [10–12].

The qualitative indicators of milk are directly dependent on the intake of nutrients and BAS into the body and their ratio in the diet and bioavailability.

In general, 326.7 kg of milk fat was obtained from cows of the 2nd experimental group during lactation, which is 42.1 or 14.8% (P<0.001) more than analogs of the control group and 13.1 or 4.2% (P<0.01) compared to the 1st experimental group.

A similar trend was observed in the yield of the amount of milk protein. Thus, from cows of the 2nd experimental group we obtained 295.9 kg of protein during the lactation, while the amount of protein obtained from peers of the control group was less than that by 14.2% (P <0.001) and 5.0% (P <0.01) less than from the 1st experimental group. Significant differences in the qualitative composition of milk were observed after the first days of feeding the BioPrimum Sukhoy feed additive to animals of the 2nd experimental group. Thus, the amount of dry matter increased by 0.58%, the non-fat milk solids (NFMS) by 0.37%, fat content by 0.03%, protein content by 0.01% compared to analogs of the control group [10–17].

4 Conclusion

Based on the above, it was found that the inclusion of the BioPrimum Sukhoy feed additive in the amount of 75 mg/kg of dry matter in the diets of cows contributed to the normalization of the blood composition and an increase in quantitative and qualitative indicators of milk productivity.

References

1. A. P. Kalashnikov, V. I. Fisinin, V. V. Shcheglov (Eds.), *Norms and rations of feeding of farm animals* (Moscow, 2003).
2. F. S. Khaziakhmetov, A. F. Khabirov, R. Kh. Avzalov, Results of the use of Vitafort probiotic in the diets of young farm animals, *Bulletin of Orenburg State University. Agrarian University* **3,** 140–143 (2019)
3. I. M. Mikolajchik, L. A. Morozova, I. V. Arzin, The effect of yeast probiotics on the digestibility of nutrients in the diet and the level of dairy productivity of cows, *Dairy and Beef Cattle Breeding* **7,** 28–31 (2017)
4. I. M. Mikolajchik, L. A. Morozova, N. M. Kostomakhin, I. V. Arzin, Features of digestion in highly productive cows when using yeast probiotic additives, *Chief Animal Technician* **12,** 27–33 (2017)
5. N. G. Makartsev, *Feeding of Farm Animals* (Kaluga: SUE Oblizdat, 1999)
6. Ya. Barta, G. Bergner, J. Buchko, et al., *Non-traditional feed in the diets of farm animals* (Moscow: Kolos, 1984)
7. L. K. Ernst, Z. M. Naumenko, *Fodder resources of the forest* (Moscow: RASKHN, 2006)
8. V. K. Pestis, M. A. Kavrus, A. N. Mikhalyuk, *Probiotics in animal husbandry and veterinary medicine* (Grodno: GGAU, 2006)
9. I. V. Petrukhin, *Feed and feed additives* (Moscow: Rosagropromizdat, 1989)
10. B. V. Tarakanov, On the types of fermentation in the rumen, *Animal Science* **6,** 8–9 (2001)
11. M. T. Taranov, *Biochemistry and productivity of animals* (Moscow: Kolos, 1976)
12. A. I. Ovsyannikov, *Fundamentals of experimental business in animal husbandry* (Moscow: Kolos, 1976)
13. G. I. Goncharova, A. B. Smolyanskaya, K. Ya. Sokolova, Microbial ecology in norm and pathology, *Antibiotics and Chemotherapy* **6,** 462–466 (1989)

14. N. V. Danilevskaya, V. V. Subbotin, Probiotics in calves' diets: animal health and food safety for humans, *Milk and feed. Management* **2**, 16–20 (2008)
15. A. A. Aliyev, *Metabolism in Ruminants* (Moscow: Engineer, 1997)
16. I. Artemov, R. Chernykh, V. Pepelina, The use of zeolite-containing rocks in the diets of cows, *Dairy and Beef Cattle Breeding* **2**, 22–24 (2011)

Current trends in the development and efficiency of irrigation systems

I. R. Nafikov, *R. R.* Lukmanov*, *M. A.* Lushnov, *R. K.* Khusainov, and *I. I.* Kashapov

Kazan State Agrarian University, 65, K. Marx Str., Kazan, 420015, Republic of Tatarstan, Russian Federation

Abstract. The article presents research methods for the developed irrigation systems, for increasing the efficiency of water use in the agricultural sector. Based on an analysis of literature and patent sources, the authors present ways of modifying irrigation systems, such as physical and hydraulic barriers and magnetic treatment of irrigation water. The article also presents the results of laboratory tests of the proposed drip irrigation system. It investigates five different types of drip trays to quantify hydraulic performance. Researchers obtained dependencies of uniformity of water flow through the drip trays when changing the operating pressure in the system.

1 Introduction

The main task of the agro-industrial complex is to provide the population with quality and inexpensive foodstuffs, and the processing industry with the necessary types of raw materials. The efficient use of both individual machine-tractor aggregates (MTAs) and the entire machine-tractor fleet (MTF) of agricultural enterprises plays a major role in solving this problem [1–3].

The efficient use of machinery and tractors means producing the required quantity and quality of agricultural products with low labour and monetary costs, which requires crop irrigation [4–6].

Agriculture is by far the largest consumer of fresh water, with about 70% supplied by rivers and lakes, and in some developing countries consumption is as high as 90%.

Due to continued population growth in developed countries, there is a growing irrigated area for food supply [7–9]. Underground drip irrigation helps the plant's root system absorb moisture better. The main objective of subsurface drip irrigation is to produce food with less water and energy. Hence, the application of technologies to save water in irrigated agriculture is an urgent task [10].

Subsurface drip irrigation (SDI) is the most modern method of irrigation. Underground drip irrigation of various crops is carried out through plastic drip tubes containing emitters (drip trays) placed at certain distances (Figure 1).

Using SDI systems improves the efficiency of irrigation water and fertiliser use. The SDI systems allow irrigation and application of nutrients directly to the root zone of the plant, not on the surface. This reduces the loss of moisture due to evaporation from the soil surface and achieves efficient use of fertilizer.

Fig. 1. Subsurface drip irrigation.

The advantages of subsurface irrigation also include longer system life due to the reduced effect of solar radiation; easier ploughing and other mechanised work; and fewer weeds and fungal diseases.

Despite the numerous benefits listed above and the high efficiency of water use, the use of SDI is very limited. This is mainly because existing underground drip irrigation systems, have serious drawbacks. These disadvantages include the higher cost of the system, clogging of the drip trays, system breakage due to penetration of plant roots, and the difficulty of detecting and repairing leaks.

Another disadvantage of subsurface drip irrigation is that the soil structure can reduce water flow through the drip trays due to the intra-soil pressure near the drip tray during irrigation. Changing the flow rate can seriously

* Corresponding author: look-rus@mail.ru

© T he Authors, published by EDP Sciences. This is an open access article distributed under the terms of the Creative Commons Attribution License 4.0 (http://creativecommons.org/licenses/by/4.0/).

https://doi.org/10.1051/bioconf/20225200049

affect irrigation uniformity. In addition, in loose soil, some moisture can seep below the root system due to the underground location of the drip lines.

At the seedling stage, the distance to the subsurface irrigation line is large and decreases as the root system of the plants increases. Therefore, there was a suggestion to put a waterproof polyethylene film below the drip lines. Polyethylene film 60 cm wide, 0.06 mm thick was laid to a depth of 30 to 40 cm. This method is called a physical barrier and significantly increases the amount of moisture retained in the root system area, either from drip line or rain, and limits deep percolation (seepage).

To improve soil moisture distribution in underground drip irrigation, a secondary drip line was laid below the primary drip line. This method depends on the factor that water moves faster into dry soil (due to a higher head gradient) than into wet soil. Thus, when the secondary drip line moistens the soil below the primary line and the water from the upper line is redistributed upwards, and this method is called a hydraulic barrier.

The results showed that when applied in the field, the hydraulic barrier increased potato yields by 12% and 48%, and the 'physical barrier' by 131% and 138% respectively. These results prove the advantage of using such methods to improve crop yields.

Figure 2 shows that below the soil surface there is film 1 (hydraulic barrier), then two drip lines 2 and 3 at 15 and 25 cm depth and a physical barrier 4 at 30 cm depth. There were vertical multi-sensor probes 5 used to measure soil moisture at distances of 4, 20 and 25 cm from the drip lines.

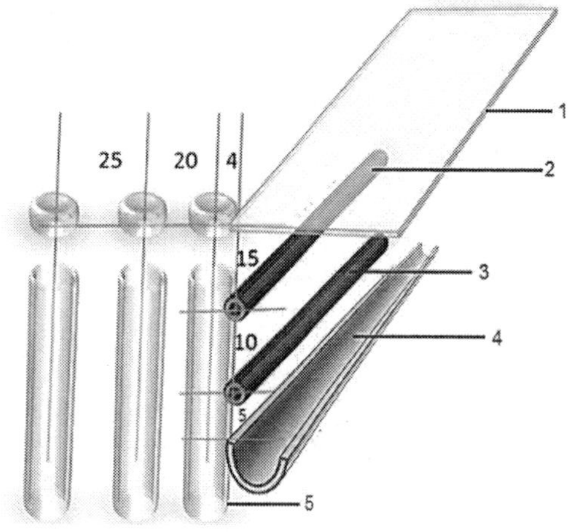

Fig. 2. Subsurface drip irrigation with physical and hydraulic barrier.

A capacitance-type moisture detector also has a use for measuring water movement in the soil (Figure 3). The soil moisture measurement system consists of vertical tube 1, three sensor electrodes 3, cable 2, which transmits the soil moisture information to the control unit. Soil moisture sensors 4, 5 and 6 of the EnviroSCAN brand (Australia) are placed at 10 cm intervals from each other [11].

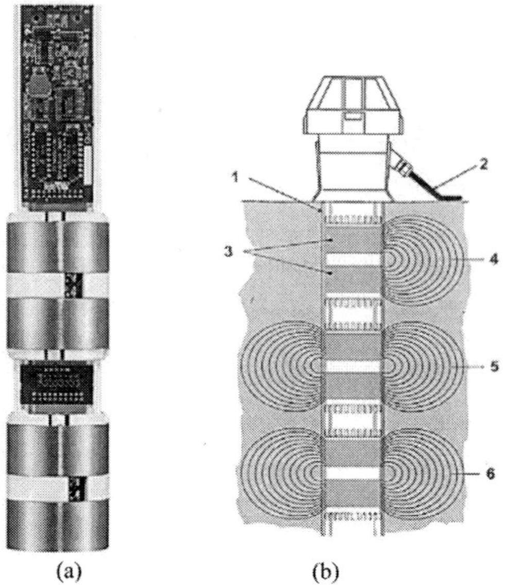

Fig. 3. EnviroSCAN sensors (a) – general view; (b) – diagram of the location and operation of the sensors at depth.

2 Materials and methods

2.1 An alternative method of subsurface drip irrigation

The developed alternative method of subsurface drip irrigation avoids most of the above disadvantages. This system is being introduced in the south of Spain to irrigate olive orchards. It consists of installing a water distribution network and drip trays on the soil surface. The system works as follows: drip trays positioned above the soil surface feed water into a perforated plastic pipe positioned vertically in the ground (Figure 4). In this way, it formed moisture distribution inside the soil, right at the bottom of the perforated tube. If the depth of the perforated tube is large and its diameter small enough, evaporation losses from the soil surface can be considered negligible.

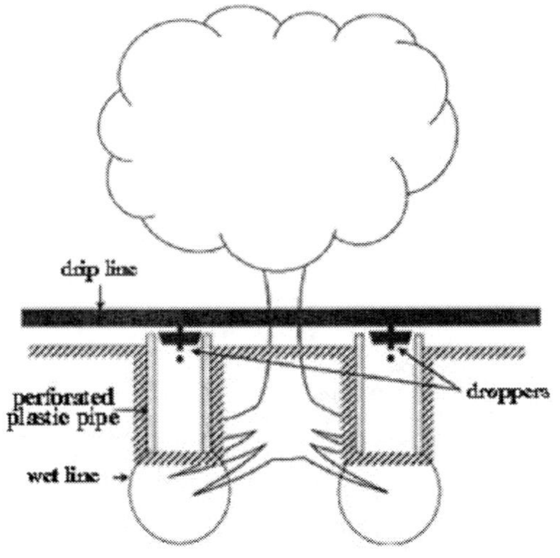

Fig. 4. Alternative method of subsurface drip irrigation.

The proposed alternative system has the following advantages This drip irrigation system is easy to install and relatively economical, with no clogging of the drip trays with soil.

The disadvantage of this system is the additional cost of installing perforated pipes under each drip tray. However, it avoids the cost of burying drip lines as with traditional subsurface irrigation.

Maintenance of the system is much easier than in underground irrigation systems, as clogged drip trays are easier to locate and replace at a lower cost.

To evaluate the effectiveness of the alternative subsurface drip irrigation method and to compare it with the surface drip irrigation method, we carried out a laboratory experiment. We have experimentally confirmed that the yield of fruit and berry crops is higher with the alternative irrigation method. The new system improved irrigation water efficiency compared to the traditional drip system and achieved water savings of up to 20% [12].

One of the most important directions of scientific and technological progress in mechanisation and electrification of agriculture, is the development of methods and technical means, considering the requirements of environmental safety, application of electrophysical ways of influence on biological objects to increase their yields. Using electromagnetic and magnetic fields in seed pre-treatment and plant exposure, as well as in preparing irrigation water in irrigation systems, can increase the yield of grain, vegetable and other crops, their resistance to disease and stimulate plant growth.

At present, there are results of experimental studies indicating a positive effect of magnetic influence of irrigation water on growth, development and yield of agricultural crops.

Piped water that passes through the magnetic field of a permanent magnet is magnetised (figure 5). The passage of water through a magnetic field changes its structure and some physical-mechanical, chemical and bacteriological properties such as density, brine capacity and the degree of sedimentation of solids.

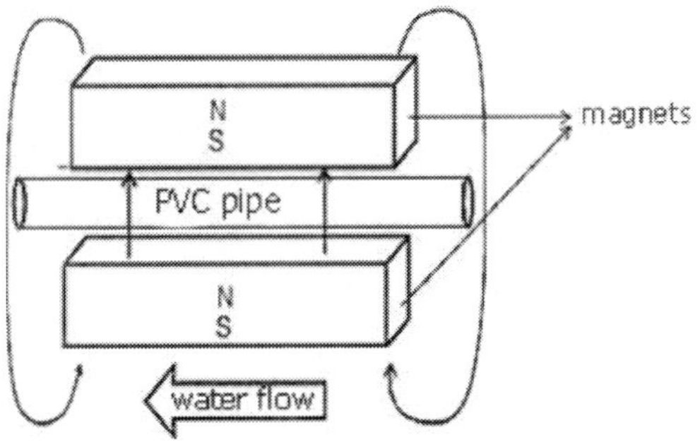

Fig. 5. Schematic diagram of magnetising water flow with magnets in a PVC pipe.

A device with a magnetic field in the range 3.5 to 136 mTl was used for magnetised irrigation water. Analysis of the data collected during the study shows that the effect of the magnetic treatment varied according to the type of plant and the type of irrigation water used. And there was also an increase in plant yield and plant water uptake.

3 Results and discussion

To improve the efficiency of drip irrigation systems, the drip irrigation unit was upgraded to include a soil moisture sensor to assess the uniformity of water distribution by drip trays at different pressures [13].

We selected five different types of Rivulis Eurodrip drip trays for the study (Figures 6 and 7), the specifications of which are shown in Table 1.

Table 1. Technical characteristics of the drip trays tested.

Type of drip tray	Series	Cover colour	Nominal flow rate, (l/h)
with pressure compensation	A	green	8
	B	black	3.85
	C	brown	2.2
without pressure compensation	D	black	4
	E	green	8

Drip trays with pressure compensation (Figure 6) compared to drip trays without pressure compensation (Figure 7) ensure uniform water supply to the active water use layer according to the manufacturer's specifications through each drip tray, and the water flow through such a drip tray does not change with pressure variations in the system.

Fig. 6. Pressure-compensated drip trays.

Fig. 7. Drip trays without pressure compensation.

Before starting the experiments, we vented the system and adjusted the inlet pressure of each main pipe to the required value. We placed 500 ml measuring vessels directly under each drip tray to determine the volume of water leaked in two minutes (Figure 8).

Experiments were carried out at three different pressures (P) – 0.05, 0.2 and 0.35 MPa, in triplicate.

Fig. 8. Experimental installation for drip irrigation.

The results of the experimental investigations carried out on all drip trays are shown in Figures 9 and 10 [17].

By increasing the operating pressure from 0.05 to 0.35 MPa, the volumetric flow rate of the A series droppers increased from 7.4 to 8.4 l/h, B from 3.6 to 4.9 l/h, C from 2.1 to 2.4 l/h, D from 3.1 to 8.2 l/h and E from 5.7 to 15.2 l/h (Table 2).

Table 2 shows that droppers with pressure compensation are less sensitive to changes in flow when pressure increases than droppers without compensation.

Table 2. Fluid flow rates at different pressures and deviations from the specified nominal flow rate.

Type of drip tray	Nominal flow rate (l/h)	Fluid flow rate at different pressures					
		0.05 MPa	Δ q %	0.2 MPa	Δ q %	0.35 MPa	Δ q %
A	8	7.4	-7.5*	7.9	-1.3	8.4	5.0
B	3.85	3.6	-6.5	4.3	11.7	4.9	27.3
C	2.2	2.1	-4.5	2.2	0.0	2.4	9.1
D	4	3.1	-22.5	5.9	47.5	8.2	105.0
E	8	5.7	-28.8	11.5	43.8	15.2	90.0

* A negative sign indicates the percentage of reduction

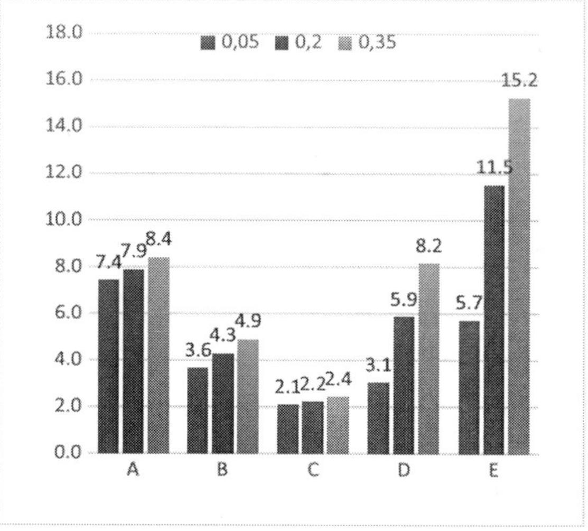

Fig. 9. Graphical dependence of liquid flow rate at different pressures.

Droppers type A and C (with pressure compensation) achieved significant uniformity of water flow and the smallest deviations from the set nominal flow rate at an operating pressure of 0.2 MPa. In type B droppers (also with pressure compensation), the uniformity of water flow decreased with increasing pressure.

Table 3 shows the coefficient of variation C_v, which expresses the uniformity of water flow through the drip trays.

The values of the coefficient of variation C_v of type D and E droppers (without pressure compensation) increased with increasing operating pressure in the system, indicating the dependence of the uniformity of flow on system pressure (Table 3).

Table 3. Coefficient of variation C_v of drip flow rate from nominal pressure.

Pressure, P MPa	C_v				
	A	B	C	D	E
0.05	0.10	0.07	0.05	0.25	0.30
0.2	0.05	0.12	0.03	0.49	0.46

0.35	0.06	0.32	0.11	1.00	0.95

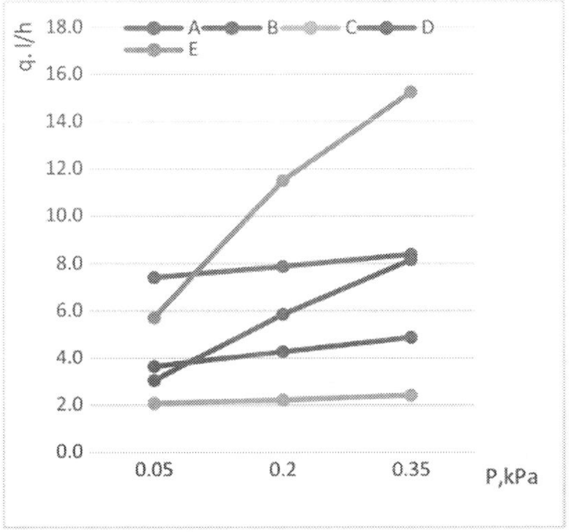

Fig. 10. Dependence of volume flow of drip trays on pressure.

We should note that for all dropper types the coefficient of variation C_v and the average deviations of the water flow rate from the specified nominal flow rate were achieved at low operating pressure in the system.

4 Conclusion

The results of these studies show that pressure-compensated drip trays are less sensitive to changes in pressure with an operating pressure of 0.35 MPa. The water consumption (q) of type A droppers increased by 5.27%, B by 27.3% and C by 9.1%. Type D and E drip trays have the lowest flow uniformity at different pressure levels. Under actual irrigation conditions, we recommend installing pressure gauges not only in the collector line, but also on the main pipes (preferably at the end of each main pipe) to determine the pressure drop and pressure loss in the system.

References

1. D. Khaliullin, I. Badretdinov, I. Naficov, R. Lukmanov, Engineering for Rural Development (Jelgava, 26–28 May 2021) **20,** 1501–1506, doi: 10.22616/ERDev.2021.20.TF321 (2021)
2. D. T. Khaliullin, et al, BIO Web of Conf. **27,** 000103, doi: 10.1051/bioconf/20202700103 (2020)
3. R. Galimov, I. G. Galiev, A. T. Kulakov, E. R. Galimov, BIO Web of Conf. **37,** 00052 (2021)
4. M. Kh. Faskhutdinov, BIO Web of Conf. **27,** 00056, doi: 10.1051/bioconf/20201700056 (2020)
5. N. M. Egorov, F. Kh. Khaliullin, S. M. Yakhin, BIO Web of Conf. **27,** 00124
6. I. Nafikov, R. Khusainov, R. Lukmanov, I. Galiyev Engineering for Rural Development (Jelgava, 26–28 May 2021) **20,** 1416–1421, doi: 10.22616/ERDev.2021.20.TF304 (2021)
7. E. R. Dalaleev, I. N. Gayaziev, B. G. Ziganshin, et al., Rural mechanic **6,** 28–29 (2017)
8. I. M. Gomaa, I. I. Kashapov, R. R. Khaidarov, et al, BIO Web of Conf. **27,** 00233, doi: 10.1051/bioconf/20201700233 (2020)
9. I. I. Kashapov, B.G. Ziganshin, Yu. A. Tsoi, et al., Bulletin of Kazan State Agrarian University **15 (3 (59),** 84–87, doi: 10.12737/2073-0462-2020-84-87 (2020)
10. B. G. Ziganshin, I. G. Galiev, R. K. Khusainov, A. Abdelfattah, BIO Web of Conf. **27,** 00119 (2020)
11. A. H. Abdelfattah, R. F. Sabirov, et al., BIO Web of Conf. **27,** 00249, doi: 10.1051/bioconf/20201700249 (2020)
12. B. Ivanov, B. Ziganshin, A. Dmitriev, et al., BIO Web of Conf. **27,** 00109
13. B. L. Ivanov, B. G. Ziganshin, et al., Bulletin of the Kazan State Agrarian University **14 (3 (54),** 103–107, doi 10.12737/article_5db969d80165a4.44685655 (2019)

Vermiculax drug effect on blood parameters, immune response, calcium and phosphorus, and weight of laying hens

Matvey Orlov[1], *Vladimir* Zaycev[1], and *Marat* Seitov[2,*]

[1] Department of Bioecology and Physiology of Farm Animals, Samara State Agrarian University", 2, Uchebnaya St., Kinel, urban-type settlement Ust-Kinelsky, Samara Region, 446442, Russian Federation
[2] Department of Non-Infectious Animal Diseases, Orenburg State Agrarian University, 18, Chelyuskintsev St., Orenburg, 460014, Russian Federation

Abstract. This paper presents the experiment on 160 laying hens of the white Leghorn breed divided into 4 groups. Group I (taken as a control group) was given a basic diet (OR). Group II had OR + Vermiculax at a dosage of 30 mg per 1 kg of the laying hen weight. Group III had OR + Vermiculax at a dosage of 37 mg per 1 kg of the laying hen weight. Group IV consumed OR + Vermiculax at a dosage of 44 mg per 1 kg of the laying hen weight. According to the results of the study, we came to the following conclusions. The drug Vermiculax increases the content of erythrocytes (by 21.45%) and hemoglobin (by 4.94%), which indicates the prevention of anemia, increases the index of total protein by 30.57%, calcium by 0.7% and inorganic phosphorus in the blood by 4.16%. Thus, it affects phagocytic processes, increases the immune response and absorption processes; increases the activity of macrophages and affects the process of degranulation of polymorphonuclear neutrophils, which in turn increases lysozyme activity by 3.9%. This is a sign of strengthening the natural antibacterial barrier. There is an increase in the retention of calcium in the body by 115.32% and phosphorus by 107.94% (on the 30th day of the study). In addition, increased retention of calcium by 8.3% and phosphorus by 8.5% (on the 180th day of the study) was observed. There is an increase in the deposition of calcium in the body by 132.8% and phosphorus by 133.45% (on the 30th day of the study), and increased retention of calcium by 66.36% and phosphorus by 66.4% (on the 180th day of the study). There is a high dynamic in weight gain by the second month of the study (increased by 21.88%), as well as a decrease in the level of feed consumption per weight gain of 1 kilogram by 13.33%.

1 Introduction

To date, the question of finding the most effective schemes for raising laying hens is relevant in poultry farming. One of the most promising and science-intensive directions is the introduction of mineral supplements into the diet. In addition, this direction is relevant because now the search is underway for more effective drugs able to replace antibiotics in the production process. This issue is actively discussed in the scientific field.

The work of Kovalevsky V.V. and Kislyakova E.M. (2013) describes the use of Calcium-MCG supplements in the diet of laying hens who have reached the age of one year. This additive was a modified form of calcium gluconate and was added in an amount of 770 grams per 1 ton of feed. In this experiment, the intensity of egg production improved (increased by 5.5%), the weight of the shell increased by 7.4%, feed costs per 10 eggs decreased by 7%, the yield of eggs suitable for incubation increased to 82.6% and the digestibility of feed increased by 2% [1].

The invention of Hongxin W. (2013) reveals a feed additive for broilers. The prepared feed additive for broilers is formed by mixing a mixed extract of Daurian lespedeza and quail lespedeza and inulin in a mass ratio, while the content of the mixed extract of Daurian lespedeza and quail lespedeza is 80%, and the inulin content is 20%. In accordance with the broiler feed additive described in the invention, 3% of the broiler feed additive is added to the main daily feed in order to have an obvious effect on the growth of broilers. The average daily increase increased by 2.62%, antioxidant protection, (serum index (U.mL -1) increased by 1.42%. This component was determined by T-SOD; this indicator increased in the heart (by 6.39%), liver (by 4.36%) and kidneys (by 2.99%), increased immunity (mortality decreased by 50.32%), etc. And the effect of

* Corresponding author: meod.adir@yandex.ru

the feed additive for broilers is better than that of the market feed additive [2].

The S. Ganjigohari experiment (2017) was conducted to assess the effect of calcium nanocarbonate (HC) instead of calcium carbonate (CC) on egg production, egg weight, egg mass, FCR, calcium in the blood and egg quality characteristics in laying hens. A total of 120 laying hens aged 23 to 33 weeks were used in the 10-week trial. Laying hens were randomly distributed into six treatments with four repetitions, five chickens each. Experimental treatments included the replacement of 50% CC in the diet with a decrease in the amount of HC and were a group 1 Basic Diet (BD) with 8.06% CC. Group 2 had 6.045% CC as negative control; group 3 had 4.03% CC replaced by 2.015% NC. Group 4 had 4.03% CC replaced by 1.01% NC; group 5 had 4.03% CC replaced by 0.252% NC, and group 6 amounted to 4.03% CC replaced by 0.126% NC. The egg weight did not affect the dietary treatment ($p > .05$). However, the percentage of egg production and egg weight in group 6 was lower than that with other treatment options ($p < 0.05$). Laying hens in the control group had the best average feed conversion rate ($p < 0.05$). Also, the lowest concentration of calcium in the blood of chickens was recorded in birds treated with group 6 ($p < 0.05$). The best eggshell quality (relative eggshell weight and eggshell weight/surface) was observed in group 1 ($p < 0.05$). Taken together, our results showed that NCC can replace CC at a lower inclusion level, but a sharp decrease in the concentration of calcium in the diet (up to 1.43% Ca in group 6) reduces productivity, egg quality characteristics, tibial thickness and calcium content in the blood, laying hens [3].

The nutritional value of limestone, oyster shells and extruded eggshells was evaluated by Mohammad Aminul Islam and Masahide Nishibori (2021) using various methods. A total of 120 18-week-old laying hens ISA Brown were divided into six dietary groups, namely D 1 (4% limestone)-calcium content 3.03%, D 2 (8% limestone))-calcium content of 3.39%, D 3 (4% of oyster shell). -calcium content of 2.95%, D 4 (8% oyster shell) -calcium content of 3.25%, D 5 (4% eggshell) -calcium content of 2.97%, and D 6 (8% eggshell) - calcium content of 2.97% to assess the effect of calcium sources on egg production, egg quality, dry matter content and cholesterol in the egg. Egg extruded shells contained 98.52, 4.24, 29.75 and 14.82% DM, CP, Ca and P, while egg extruded shells contained 99.20, 13.80, 25.53 and 13.87% DM, CP, Ca and P, respectively. Limestone and oyster shells contained 99.60, 99.51% CB, 37.12 and 35.20% Ca, respectively. Live weight (increased by 6.75% in the group with a calcium content of 3.03%), egg production (increased by 8.19% in the group with a calcium content of 2.97%), productivity per day of laying hens and egg mass, as well as FCR did not differ between diets ($p > 0.05$). Egg production tended to increase with the increase in egg production from laying hens ($p < 0.001$) in D 6, followed by D 2, D 5, D 3, D 4 and D.1, respectively. The lowest production costs and the highest net profit were noted in D 6, followed by D 2, D 4, D 5, D 1 and D 3, respectively [4].

In his work, Nikolaev S. I. (2017) draws attention to the fact that most of the feeds used in poultry farming suffer from a deficiency of mineral, energy and amino acid components. In this regard, the authors introduced several components into the diet of poultry to set up the experiment, namely: sorghum and chickpea grains with the addition of pumpkin and sunflower phosphatides with bischofite. During the study, the indicators of egg weight (which increased by 2.7, 0.9 and 1.4 g), yolk weight (increased by 15.93, 11.89 and 22.17%) were evaluated [5].

Considering the rather versatile activity of the scientific community in the creation and application of additives to the diet of laying hens. Our goal was to conduct an experiment with the introduction of the mineral-vitamin preparation "Vermiculax" into the diet of laying hens and to establish its effect on the indicators of natural resistance, haematological and biochemical. The purpose was to assess the effect on the parameters of the weight gain, feed consumption, retention and deposition of trace elements in the body of laying hens [6, 7, 8, 9].

2 Materials and methods

The experiment was conducted on 160 laying hens of the white Leghorn breed selected according to the principle of pairs of analogues. All laying hens were randomly distributed into 4 groups in equal numbers. The scheme of the experience is presented in Table 1. The main diet consisted of flour of 54%; oat flour of 11%; yeast (dry, baked) – 14.3%; Flour (fish, meat-bone, hay) -8.2%; %; rutabaga - 7.3%; fat (animal, fish)-3.8%; nettle (ground) - 1.2%; salt 0.2%. All indicators of the content of laying hens met regulatory requirements.

The air temperature was determined using a mercury thermometer TTM from 0 to 200 TU 25-2021.010-89. Humidity of the air was measured using a psychometric hygrometer VIT-2. The air velocity was measured using the Testo 440 anemometer. Illumination was determined using a luxmeter Yu-116. The concentration of NH^3 and H_2S in the air was determined by the universal gas analyser UG-2.

The water was given free rein, and was provided with a central water supply. Two valve-type micro-cup drinkers were installed in each cell. All laying hens were housed in BKM-3b cell batteries with 10 heads each.

Table 1. Scheme of experience.

Groups	Diets
I group (taken as a control group)	Basic diet
II group	The main diet + Vermiculax in a dosage of 30 mg per 1 kg of weight of laying hens
III group	The main diet + Vermiculax in a dosage of 37 mg per 1 kg of weight of laying

	hens
IV group	The main diet + Vermiculax in a dosage of 44 mg per 1 kg of weight of laying hens

Monthly records of calcium and phosphorus retained in the body were kept. Morphofunctional blood tests were carried out on the 4th and 6th month, which were carried out on an automatic hemoanalyzer BC-2800 Vet Mindray (Mindray, China). Biochemical parameters were studied using an automatic biochemical analyser Mindray BS-380 (Mindray, China) using commercial kits. From each group, 5 laying hens were sent monthly for slaughter, which was carried out externally. The amount of calcium and phosphorus was determined in the carcasses of laying hens.

Calcium was determined by a method based on the ability of calcium to precipitate under the action of ammonium oxalate according to the formula: $CaC_2O_4+H_2SO_4 \rightarrow CaSO_4+H_2C_2O_4$.

Phosphorus was determined by the Pemberton method based on the ability of phosphorous compounds with ammonium molybdenum to form a yellow phosphorous-molybdenum-ammonium salt that precipitates.

The determination took place by reaction:

$$2[(NH_4)3PO_4 \cdot 12MoO_3]+46KOH=2(NH_4)2HPO_4+23K_2MoO_4+(NH_4)2MoO_4+22H_2O.$$

Statistical processing of the received data was performed on a PC using the Microsoft Office Excel 2010 application. The obtained results were analyzed in accordance with the norms of variation statistics.

3 Results and discussion

In the course of the study of the haematology of laying hens, we found that the administration of the drug Vermiculax at a dosage of 37 mg per 1 kg of weight of laying hens favourably affects the red blood cell count. Thus, the content of erythrocytes on day 180 in group III was 21.45% higher than in group I, 21% higher than in group II and 24.74% higher than in group IV. The data obtained during the study allow us to assume that the introduction of Vermiculax into the body stimulates the immune response by increasing the number of red blood cells, which in turn adsorb various poisons on their surface and then these substances are destroyed by tissue macrophages (reticuloendothelial system).

Table 2. Haematological parameters of laying hens for the 6th and 8th month of the study.

Groups	Age, days	Hemoglobin, g/l	Erythrocytes 10^{12}/l	Hemoglobin content in erythrocyte, pg
I group	120	99.7±6.77	2.76±0.43	27.0±1.63
	180	113.3±3.95	3.03±0.52	30.6±1.32
II group	120	99.6±8.69	2.73±0.63	27.1±1.93
	180	113.3±9.21	3.04±0.62	30.6±1.46
III group	120	100.3±11.68	3.0±0.94	27.2±1.73
	180	118.9±8.62	3.68±0.37	33.2±1.54
IV group	120	98.7±14.76	2.66±0.11	26.7±1.87
	180	111.1±21.84	2.95±0.40	29.8±1.88

Also, you can see that the introduction of Vermiculax at a dosage of 37 mg per 1 kg of weight of laying hens affected the amount of hemoglobin (increased by 4.94%), which indicates the prevention of anemia with Vermiculax.

Table 3. Biochemical parameters of laying hens for the 6th and 8th month of the study.

Groups	Age, days	Total protein, g/l	Calcium mmol/l	Inorganic phosphorus, mmol/l
I group	120	57.9±1.56	3.41±1.64	1.46±1.75
	180	66.0±1.92	4.27±1.84	1.92±1.46
II group	120	58.5±0.65	3.50±1.48	1.50±1.76
	180	70.0±1.39	4.27±1.93	1.91±1.00
III group	120	58.5±1.55	3.49±1.34	1.52±1.63
	180	75.6±1.32	4.30±1.87	2.00±1.74
IV group	120	57.6±1.06	3.38±1.99	1.40±1.71
	180	65.2±1.37	4.22±1.56	1.88±1.55

During the analysis of biochemical data, we found that in comparison with the control and experimental groups, the data of group III were higher. The total protein in the blood of group III on the 180th day of the experiment was 30.57% higher than that in group I, 8% higher than that in group II and 15.95% higher than that in group IV. This phenomenon indicates a direct effect on phagocytic processes and an increase in the immune response. This indicator is consistent with high levels of calcium in the blood (the 180 day indicator of group III was 0.7% higher than the indicators of groups I and II, group IV by 1.86%) because the total protein directly affects the transfer of organic and inorganic substances in the blood and the binding of trace elements. We connect that the increase in the index of total protein, which in turn affected the immune response and absorption processes (this effect is also confirmed by an increase in the index of inorganic phosphorus by 4.16%) is due to the ingestion of laying hens of one of the components of the drug "Vermiculax" - monocalcifosphate.

Table 4. Indicators of natural resistance of laying hens for the 6th and 8th month of the study.

Groups	Age, days	Glycoproteins, mg%	Bactericidal activity, %	Lysozyme activity, %
I group	180	124.0±11.96	47.5±1.46	18.0±1.26
II group		135.8±12.39**	55.9±2.22*	21.9±1.90*
III group		159.4±11.75***	61.0±1.27***	25.3±1.48***
IV group		122.51±5.09	45.1±1.57**	17.1±0.13

The administration of the drug Vermiculax at a dosage of 37 mg per 1 kg of laying hens increases lysozyme activity by 3.9%. More precisely, it increases the activity of macrophages, cells that secrete lysozyme, and it is also possible that Vermiculax affected the process of degranulation of polymorphonuclear neutrophils, which is a sign of strengthening of the natural antibacterial barrier. An increase in lysozyme activity has a stimulating effect on bacterial activity (as confirmed by Table 4). Bacterial activity increased by 13.5%. Moreover, we found that there is also a stimulating effect on the content of glycoproteins (increased by 35.4%), which have an antiviral effect, protective function, and so on.

In the course of the study, we saw a certain relationship between the retention rates in the body of laying hens, the deposition of calcium and phosphorus in the body of laying hens and age-related changes. Thus, in group I, in comparison with the first month of studies, the retention rate of calcium in the body decreased by 18.7% on day 180. At the same time, the average monthly decline was 3.11%. In terms of phosphorus, the difference between the indicators of the first and sixth month was 21.6%, while the average monthly decline was 3.6%. In the indicators obtained after slaughter in group I, the difference between the calcium indicators at the beginning of the study and the final stage was 14%, while the average monthly decrease was 2.33%. In terms of phosphorus, the difference between the first month and the 180th day was 16.5%, while the average monthly decrease was 2.75%.

Table 5. Retention and deposition by the body of laying hens of calcium and phosphorus (without administration of the drug Vermiculax).

		Age, days					
		30 days	60 days	90 days	120 days	150 days	180 days
Retained by the body, %	Ca	35.9 ±1.73-2.99	32.5 ±0.85-0.72	27.8 ±1.65-2.72	26.0 ±1.68-2.82	17.3 ±1.74-3.02	17.2±1.67-2.79
	Ph	40.3± 1.93-2.30	36.5 ±1.64-2.69	28.8 ±1.40-1.96	27.2 ±1.14-1.3	18.7 ±1.99-3.96	18.7±1.20-1.44
Deposited in the body (based on the results of slaughter), %	Ca	25.0 ±1.13-1.27	23.25±1.10-1.21	19.3 ±1.05-1.1	19.4 ±1.00	11.2 ±1.02-1.04	11.0±1.65-2.72
	Ph	29.3 ±1.67-2.79	27.7 ±1.19-1.41	19.9 ±1.64-2.68	20.9 ±1.48-2.19	12.9 ±1.64-2.68	12.8±1.88-3.53

In group II, there was also a decrease in indicators with the age of laying hens. In comparison with the 30th day of studies, the retention rate of calcium in the body decreased by 36.2% on the 180th day. At the same time, the average monthly decline was 6.03%. In terms of phosphorus, the difference between the indicators of 30 and 180 days was 40.3%, while the average monthly decline was 6.71%. In the indicators obtained after slaughter in group II, the difference between the calcium indicators of the first month of the study and the final one was 27.3%, while the average monthly decrease was 4.55% In phosphorus indicators, the difference between the first month and the fifth was 33.1%, while the average monthly decrease was 5.51%.

Table 6. Retention and deposition by the body of laying hens of calcium and phosphorus (with the introduction of Vermiculax at a dosage of 30 mg per 1 kg of the weight of laying hens).

		Age, days					
		30 days	60 days	90 days	120 days	150 days	180 days
Retained by the body, %	Ca	60.3 ±1.95***	52.4 ±1.31***	47.5 ±1.60***	38.3 ±1.52***	24.1 ±1.66***	24.1±1.85
	Ph	65.6 ±1.74***	57.3 ±1.94***	49.3 ±1.46***	38.8 ±1.46***	25.2 ±1.45***	25.3±1.50
Deposited in the body (based on the results of slaughter), %	Ca	45.5 ±0.51***	40.6 ±1.70***	33.1 ±1.30***	30.5 ±1.03***	17.9 ±1.07***	18.2±1.03
	Ph	53.3 ±1.26***	47.7 ±1.79***	34.0 ±25***	32.7 ±1.25***	20.2 ±1.41***	20.2±1.36

Comparing the indicators of group I and II, we see that the retention rate of calcium after reaching 180 days was higher in group III by 6.9%, phosphorus by 6.6%. At the same time, based on the data of the control slaughter, the indicator of calcium deposited in the body was higher in II by 7.2%, phosphorus by 7.4%. Analysing the indicators of reduced retention of calcium and phosphorus in the body of laying hens, this indicator was higher in group II (the average monthly decrease in group II was higher by 2.92% and 3.11%, respectively). A similar pattern was observed in the indicators of calcium and phosphorus deposited in the body, the indicators of the average monthly decrease in calcium were higher in group II by 2.22% and 2.76, respectively

In group III, there was also a decrease in indicators with the age of laying hens. In comparison with the 30th day of studies, the retention rate of calcium in the body decreased by 51.8% on the 180th day. At the same time, the average monthly decline was 8.63%. In terms of phosphorus, the difference between the indicators of 30 and 180 days was 56.3%, while the average monthly decline was 9.38%. In the indicators obtained after slaughter in group III, the difference between the calcium indicators of the first month of the study and the final one was 39.4%, while the average monthly decrease was 6.65% In phosphorus indicators, the difference between the first month and the fifth was 47.1%, while the average monthly decrease was 7.85%.

Table 7. Retention and deposition by the body of laying hens of calcium and phosphorus (with the introduction of Vermiculax at a dosage of 37 mg per 1 kg of the weight of laying hens).

		Age, days					
		30 days	60 days	90 days	120 days	150 days	180 days
Retained by the body, %	Ca	77.3 ±0.67***	55.2 ±1.52***	50.1 ±0.74***	40.5 ±1.55***	25.7 ±1.13***	25.5±1.46***
	Ph	83.8 ±1.96***	60.6 ±1.45***	52.3 ±1.67***	41.0 ±2.96***	27.6 ±1.66***	27.5±1.77***
Deposited in the body (based on the results of slaughter),%	Ca	58.2 ±1.17***	42.9 ±1.29***	35.0 ±1.36***	33.3 ±1.23***	18.7 ±1.98***	18.3±1.75***
	Ph	68.4 ±1.34***	49.0 ±1.53***	36.9 ±1.74***	35.2 ±1.58***	21.6 ±1.64***	21.3±1.53***

Comparing the indicators of groups I and II, we see that the retention rate of calcium after reaching 180 days was higher in group III by 8.3%, phosphorus by 8.5%. At the same time, based on the data of the control slaughter, the indicator of calcium deposited in the body was higher in III by 7.3%, phosphorus by 8.5%. Analysing the indicators of reduced retention of calcium and phosphorus in the body of laying hens, this indicator was higher in group III (the average monthly decrease in group III was higher by 5.52% and 5.78%, respectively). A similar pattern was observed in the indicators of calcium and phosphorus deposited in the body, the indicators of the average monthly decrease in calcium were higher in group III by 4.32% and 5.1, respectively.

In group IV, there was also a decrease in indicators with the age of laying hens. In comparison with the 30th day of studies, the retention rate of calcium in the body decreased by 20.2% on the 180th day. At the same time, the average monthly decline was 3.36%. In terms of phosphorus, the difference between the indicators of 30 and 180 days was 22.4%, while the average monthly decline was 3.73%. In the indicators obtained after slaughter in group IV, the difference between the calcium indicators of the first month of the study and the final one was 12.4%, while the average monthly decrease was 2.06%. It should be noted that the decrease in calcium and phosphorus indicators was observed only from the 60th day of the study. In terms of phosphorus, the difference between the first month and the fifth was 16.5%, while the average monthly decrease was 2.75%.

Table 8. Retention and deposition by the body of laying hens of calcium and phosphorus (with the introduction of Vermiculax at a dosage of 44 mg per 1 kg of the weight of laying hens).

		Age, days					
		30 days	60 days	90 days	120 days	150 days	180 days
Retained by the body, %	Ca	40.2 ±0.32**	36.3 ±1.25***	30.3 ±1.48***	30.2 ±1.90***	20.1 ±1.36**	20.0±1.64***
	Ph	43.5 ±1.36***	41.0 ±1.43***	31.5 ±1.93	30.8 ±0.04***	21.5 ±1.47***	21.1±1.05***
Deposited in the body (based on the results of slaughter),%	Ca	24.3 ±1.91	35.0 ±1.54***	29.5 ±1.69***	28.9 ±1.66***	12.9 ±1.95	11.9±1.76
	Ph	29.0 ±1.62	40.1 ±1.45***	29.7 ±1.11***	30.2 ±1.65***	13.7 ±1.50	12.5±1.44

Table 9. Indicators of the weight of laying hens and feed consumption for the first 2 months of the study.

Groups	Weight, gr.		Feed consump-tion per 1 kg of weight gain
	30 days	60 days	
I	413 ±12	1321 ±27	2.38 ±0.76
II	424 ±31	1476 ±54**	2.29 ±0.57
III	455 ±26	1610 ±30***	2.10 ±0.39
IV	389 ±6*	1311 ±24	2.59 ±0.66

Comparing the indicators of group I and IV, we see that the retention rate of calcium after reaching 180 days was 2.8% higher in group IV, phosphorus was 0.3% lower. At the same time, based on the data of the control slaughter, the indicator of calcium deposited in the body was 8.2% higher in IV, phosphorus was 2.3% lower. Analyzing the indicators of a decrease in the retention of calcium and phosphorus in the body of laying hens, this indicator was higher in group IV (the average monthly decrease in group IV was 0.25% and 0.13% higher, respectively). A similar pattern was observed in the indicators of calcium and phosphorus deposited in the body, the indicators of the average monthly decrease in calcium were lower in group IV by 0.27% and were identical in changes in phosphorus.

Analyzing the data of mass indicators, we see that when Vermiculax was introduced into the diet at a dosage of 37 mg per 1 kg of weight of laying hens (group III) by the second month of the study. The highest results of weight gain were achieved: 1610 grams ($P < 0.001$), which is 21.88% higher than that in the control group, 9.07% higher than that in group II and 22.81% than that in group IV. At the same time, a higher rate of the weight gain between the first and second months of studies was also observed in group III (the mass index increased by 253.85% compared to the first month of the study), which is 33.99% (the mass index of group I increased by 219.85% compared to the first month of the study). Also, group III had the lowest amount of feed consumption per 1 kilogram of weight gain. This indicator was lower than that in group I by 13.33%, by 9.04% than that in group II and by 23.33% than that in group IV. These phenomena of a higher level of the weight gain are explained by the amount of deposited calcium in the body of laying hens of group III. This phenomenon has affected a large number of enzymatic processes that affect the growth and development of the body of laying hens.

Conclusion

In the course of the study, we concluded that the drug Vermiculax has the most beneficial effect on the body when introduced into the diet of laying hens at a dosage of 37 mg per 1 kg of weight of laying hens. According to the results of the study, we came to the following conclusions. The drug Vermiculax:

- increases the content of erythrocytes (by 21.45%) and hemoglobin (by 4.94%), which indicates the prevention of anemia;
- increases the index of total protein by 30.57%, calcium by 0.7% and inorganic phosphorus in the blood by 4.16%. Thus, it has an effect on phagocytic processes, an increase in the immune response and absorption processes;
- increases the activity of macrophages and affects the process of degranulation of polymorphonuclear neutrophils, which in turn increases lysozyme activity by 3.9%, which is a sign of strengthening the natural antibacterial barrier;
- increases the retention of calcium in the body by 115.32% and phosphorus by 107.94% (on the 30th day of the study), retention of calcium by 8.3% and phosphorus by 8.5% (on the 180th day of the study);
- increases the deposition of calcium in the body by 132.8% and phosphorus by 133.45% (on the 30th day of the study), retention of calcium by 66.36% and phosphorus by 66.4% (on the 180th day of the study);
- shows a high dynamics of weight gain by the second month of the study (increased by 21.88%), as well as a decrease in the level of feed consumption for weight gain of 1 kilogram by 13.33% (Kovalevsky V.V. Achievements of science and technology of agriculture, 43-48 (2013)).

References

1. W. Hongxin, Pat. CN103053843A Broiler feed additive (15.01.2013)
2. S. Ganjigohari, *Journal of Animal Physiology and Animal Nutrition*, **102 (1)**, e225–e232(2017) https://doi.org/10.1111/jpn.12731
3. M. A. Islam, Veterinary Medicine and Science, **7 (5)**, 1948–1958 (2021) doi: 10.1002/vms3.544
4. S. I. Nikolaev, *Proceedings of the Nizhnevolzhsk agro-university complex: science and higher professional education*, pp 190–197 (2017)
5. V. V. Tarabrin, Pat. RU 2738891 C1 Complex additive "Vermiculax" to increase productivity and natural resistance of poultry. Publ.18.12.2020.
6. V. V. Petryakov, *Actual problems of pork production in the Russian Federation, Materials of the All-Russian Scientific and practical conference and the XIV Interuniversity Coordinating Council "Pork"*, pp 112–114 (2005)
7. C. E. Moore, Changes in antibiotic resistance in animals, *Science*, **365,** 1251-1252 (2019)
8. N. Roth, *Poult Sci*, **98,** 1791-1804 (2019)
9. EFSA, The European Union summary report on antimicrobial resistance in zoonotic and indicator bacteria from humans, animals, and food in 2014, *EFSA J* **14,** 4380 (2016)

https://doi.org/10.1051/bioconf/20225200051

Impact of Silo Twice biopreservative on haylage quality

N. E. Zemskova[1*], *V. A.* Kornilova[1], *A. G.* Meshcheryakov[1], and *P. V.* Penkin[2]

[1] Samara State Agrarian University", 2, Uchebnaya St., Kinel, urban-type settlement Ust-Kinelsky, Samara Region, 446442, Russian Federation

[2] Agroplemconsulting LLC, urban-type settlement Ust-Kinelsky, 446442, Russian Federation

Abstract. The complex biological starter (biopreservative) "Silo Twice" is related to microbiology in feed production, in particular to the technology for obtaining complex biopreservatives - preparations used for silage of feed. In the finished biopreservative fermented Silo Twice product, all nutrients of the starting material are preserved and the relative value of the feed has increased 17 units. Compared to the classic self-preservation of haylage, the introduction of Silo Twice biopreservative provides a decrease in protein losses of 2.0–2.2 times. Feeding milk cows with haylage obtained when applying Silo Twice biopreservative provides an increase in the milk yield by 8.0%, milk fat content by 0.4%. The use in the diet of dairy cows of haylage obtained with the introduction of the biopreservative Silo Twice contributed to a reduction in the cost of exchange energy and concentrates by 2.43% and 2.95%, respectively. The use in the diet of dairy cows of haylage obtained with the introduction of biopreservative Silo Twice led to the optimization of processes in the scar, an increase in the level of volatile fatty acids and bacteria, with a decrease in the amount of ammonia by 16.3%. This fact indicates an improvement in the digestibility and absorption of nitrogen in the scar of cows. Feeding milk cows for 60 days with haylage obtained when the Silo Twice biopreservative was introduced provided a savings of 2250 rubles per cow.

1 Introduction

One of the main most priority and continuously developing areas of animal husbandry is dairy cattle breeding. The most important factor in realizing the biological potential of cows is balanced feeding, taking into account the physiological need of the animal body.

Currently, the solution of issues in the field of animal feeding is impossible without taking into account the preservation of plant nutrients in feed production. At the same time, the aspect of processing green mass into canned food is of particular relevance.

Haylage is a feed made from herbs harvested in the early phases of the growing season, boiled to a humidity of 45...55% and preserved in anaerobic conditions. Haylage canning is achieved as a result of the insufficient water content in boiled plants, when most bacteria, due to physiological dryness of the environment, cannot extract the water necessary for intensive development. As a result, acid formation in haylage is limited, putrid and oily - acidic bacteria develop weakly, which contributes to the better preservation of nutrients and, above all, sugar. Therefore, it is important for the humidity of the feed to be within 45...55%. Then the water of the feed is inaccessible to bacteria. The water-retaining force of plant cells is at a humidity of 50... 55% - 52 atm and the maximum sucking force of most bacteria is about 50 atm. But if putrid and other bacteria cannot develop under such humidity, molds that have a sucking force of more than 200 atm are intensively developed. Their development can only be prevented by careful isolation of the mass from the air. Without air access, the breathing of plant cells is also stopped and the development of thermophilic bacteria, which cause strong heating of the mass, is eliminated. The nutritional value of feed is determined by the content of exchange energy in 1 kg of natural feed or dry matter. In the norms of feeding animals, along with their need for exchange energy, energy rationing is also indicated in feed units. Sources of metabolic energy are carbohydrates, fats and proteins. To obtain 4000... 4500 kg of milk from a cow, 800... 1000 g of the average daily growth when fattening young cattle and 450... 500 g when fattening pigs, it is necessary to harvest at least 40... 45 c of food for each conditional unit per year. Units are with content of 1 feed. Units have 110... 115 g of digestible protein. At the same time, in the annual structure of feed, haylage is 7%, silage is 23%, or at least 1.5... 2 tons of haylage and 5... 6 tons of silage should be stored per 1 head of cattle. Starting with grass fermentation, traditional and new technologies of silage and haylage preparation from whole plants and partially combed stems have the same technological operations: mowing of plant raw materials with grinding, transportation of mass, its laying in storage, leveling, compaction and sealing. To compress the mass, the degree of grinding for both silage and haylage plays a crucial role. Beveled herbs on silage and boiled on haylage to optimal humidity are ground by fodder harvesting machines into particles 20... 50 mm long. The production of feed with a high nutrient content is largely determined by the harvest time of the plants. Silos should be prepared from plants that are freshly milled or tilted to a humidity of 60... 70% of the ground

* Corresponding author: zemskowa.nat@yandex.ru

mass. Quality silage is obtained from plants mowed in the following phases of growing. Corn and sorghum are from wax, milk-wax ripeness of grain, sunflower from the beginning of flowering. Sudanese grass is from sweeping panicles, perennial legumes - budding, perennial legume grasses – budding. But it should not be later than the flowering phase, perennial cereal grasses is at the end of the exit phase into the tube - the beginning of scaling. Grass mixture is from the above-mentioned growing phases of the predominant component. Annual legume grasses and legume-cereal grass mixtures are mowed in the phase of wax ripeness of legume seeds in two or three lower tiers. To obtain haylage, legume perennial herbs are mowed in the budding phase, but not later than the beginning of flowering. Legume annual plants, legume-cereal and their mixtures are mowed no later than the phase of bean formation in two or three lower tiers, perennial cereal - at the end of the pipe-forming phase until the beginning of spitting. Legumes should be boiled to a humidity of 45... 55%, cereals - 40... 55% [7; 8; 11].

Due to climatic conditions and imperfection of storage facilities, harvesting of feed in Russia often occurs in conditions of high humidity, which negatively affects the quality of feed, including silage and haylage. After depressurization of canned food, deterioration of its quality begins; putrefactive bacteria and other pathogenic microflora lead to significant losses of feed nutrients, cause a number of diseases and a decrease in animal productivity [1; 2]. This can be avoided by the use of chemical and biological preservatives containing microorganisms capable of producing high levels of lactic acid and fermenting a number of hard-to-reach plant compounds. The use of chemical preservatives has significant disadvantages: the resulting feed may contain toxic substances, which is unsafe for health [3]. Therefore, it was decided to create a preservative of biological origin that does not harm animal health and has a beneficial effect on the quality of haylage and milk productivity.

The purpose of the work is to create a new comprehensive third-generation biopreservative for silting the green mass of all plant species, flattened grain and preserving hay.

2 Material and methods

The research material was as follows.

Lactic acid-producing bacilli Enterococcus faecium had the content of enterococci in plant and animal raw materials that does not allow staphylococci, listeria and E. coli to multiply in the finished product. The main factor of antagonistic activity of enterococci is their ability to produce antimicrobial peptides - enterocins [4].

Lactic acid bacteria are Lactobacillus plantarum, Lactobacillus buchneri, Lactobacillus fermentum, producing lactic acid, which contributes to a rapid decrease in the pH of the silage mass, prevents the development of anaerobic microbes that cause food damage [5].

Propionibacterium freudenreichii subsp. Shermanii produce propionic acid and its salts (propionates), acetic and lactic acids having a pronounced bactericidal effect.

Enzyme complex contained:

- cellulase catalyzing hydrolysis of glycoside bonds in cellulose to form glucose;
- amylase catalyzing the cleavage of starch, glycogen and saccharides to simple monosaccharides which are readily digestible in the intestine.
- glucanase, which destroys beta-glucans and other non-starch polysaccharides of grain raw materials to form simple sugars;
- xylanase, which promotes the release of protein, starch, pectin from plant cells and the formation of simple sugars by decomposing arabinoxylans to a xylose molecule [6].

The method for preparation of the Silo Twice biopreservative is microbiological synthesis followed by freeze-drying and mixing of preparation components with each other and dry milk whey to normalize titer. As a result, a new bacterial substance is obtained, containing highly active strains of bacteria and a complex of enzymes for the fermentation of haylage, silage and flattened grain.

3 Results and discussion

Microbiological and physicochemical indicators of the quality of microorganisms and enzymes included in the Silo Twice biopreservative comply with the developed TS 10.89.19-012-09967133-2021 and contain the following strains and colony-forming units (CFU/g):

- Enterococcus faecium VOA-1 ECM V-28720 – 2,3×1010 CFU/g;
- Lactobacillus plantarum VKPM V-11264 - 1,2×1011 CFU/g;
- Propionibacterium freudenreichii subsp. Shermanii VKPM B-5592 - 2.2×109 CFU/g;
- Lactobacillus fermentum VKPM B-7573 - 1.2×1010 CFU/g;
- Lactobacillus buchneri VKPM V-7641 - 1.2×1011 CFU/g;
- amylases - 300 U/g;
- glucanases - 300 U/g;
- xylanases - 300 U/g;
- cellulases - 100 U/g.

Starter microorganisms provide fast preservation and enhanced effect of aerobic stability directed against putrefactive bacteria, yeast, molds and other fungi. Live Lactobacillus plantarum bacteria provide a high level of lactic acid formation in the preserved food, minimize the appearance of undesirable acids and other compounds during fermentation of the green mass, provide a high level of carotenoid preservation in the product, increasing its biological value, and have antagonistic properties with respect to undesirable microflora leading to its deterioration. It has been found that efficient processing of feed butyric acid by L. fermentum strain into hydrogen peroxide (H_2O_2) allows one to additionally fight emerging molds, fungi and pathogenic bacteria, protect the digestive tract of the animal from foodborne infections, and helps to prevent oxidative damage of feed products. The Silo Twice biopreservative enzyme complex (cellulase, amylase, glucanase, xylanase) contributes to the cleavage of non-starchy polysaccharides and the release of additional nutrients, in particular protein, thereby improving the preservation quality of difficult-to-silage plants.

The use of the Silo Twice biopreservative consists in its introduction into the embedded green mass at the rate of 1.0 g of the product (with an activity of at least 1x1011 CFU) per 1 ton of the silted raw material, diluting it in the required amount of water. The ready solution of the preparation is introduced into the silusable mass by means of spraying devices on the combine or in trench. In the absence of such devices, the drug can be introduced into the trench with a watering bottle with small holes. The silo mass must be carefully sealed after its laying and compaction for 1-2 days, since lactic acid bacteria are able to multiply strictly without air oxygen access.

There were no adverse events and complications with the use of Silo Twice biopreservative at the recommended doses. There are no contraindications to the use of the Silo Twice biopreservative. The biopreservative Silo Twice is compatible with all feed ingredients, other feed supplements and drugs. Products from animals (including birds and fish) after the use of the Silo Twice biopreservative can be used for food purposes without restrictions, as confirmed by the results of laboratory tests.

When injecting the Silo Twice biopreservative at a dose of 1.0 g/t during haylage of the green mass of Sudanese grass harvested in the panicle sweeping phase, haylage was obtained, the chemical composition of which is presented in parentheses, after the indicators of Sudanese grass.

So, the moisture content of the grass was 71.43%, including hay (76.64%), crude 3.51% protein (2.80%), 0.06% starch (1.06%), simple 0.66% sugars (0.32%), lactic acid 1.34% (1.95%), acetic acid 0.07% (0.30%), butyric acid 0.00% (0.00%), ammonia 0.03% (0.13%), the relative value of the 84 feed (101).

Therefore, in haylage, part of the protein passed into acid, which is proved by an increase in ammonia, simple sugars passed into alcohol-soluble carbohydrates and organic acids. The relative value of feed increased by 17 units compared to raw materials. Compared to the classic self-preservation of haylage, the introduction of the Silo Twice biopreservative provides a decrease in protein losses of 2.0-2.2 times.

The closest in functional properties to the Silo Twice biopreservative is a second-generation biopreservative (prototype), represented by a drug consisting of hetero-enzyme lactic acid and propionic acid bacteria with possible addition of enzymes. It contains a combination of strains of lactic acid bacteria: Lactobacillus diolivorans, Lactobacillus buchneri and Lactobacillus rhamnosus contribute to the rapid and intensive fermentation processes in the first weeks of silage of starch-rich cultures. The dosage is 1 g per ton of the silted mass [2].

Protopit has a significant drawback: a rather high cost due to foreign production.

Feeding dairy cows with haylage obtained by applying the Silo Twice biopreservative led to optimization of scar processes, an increase in the level of volatile fatty acids and bacteria with a decrease in the amount of ammonia by 16.3%, which indicates an improvement in the digestibility and absorption of nitrogen in the scar of cows. As a result, the milk yield increased by 8.0% compared to that of the control group, milk fat content increased by 0.4% (Table 1).

Table 1. Comparative analysis of the preserved food by the example of haylage from Sudanese grass (in dry substance (CB) feed).

Indicators	Control (prototype)	Experience I self-conservation	Experience II preservative-biopreservative-volume Silo Twice
Crude protein (CP), %	11.85	11.20	11.99
Rhodium-rhyme protein, % CP	56.3	61.4	57.0
Split in the scar protein, % CP	67.3	59.0	66.0
Transit protein, % CP	32.7	41.0	34.0
Starch, %	4.51	3.86	4.56
Lactic acid, %	8.28	5.30	8.33
Vinegar acid,%	1.40	0.89	1.30
Energy exchange, MJ/kg	9.45	8.68	9.93
Permafrost energy, MJ/kg	11.32	10.12	11.70

Relative fodder value	100	82	101

The amount of raw protein in all three haylage samples is normal. The optimal value of the crude protein should be < 190 g/kg CB or, translated into percentages, < 19% [9].

Soluble protein is the percentage of crude protein soluble in the buffer that is responsible for the pH of the scar. The indicator is given in % of the total raw protein. The high content of soluble raw protein suggests that a large amount of protein will be available in the scar. As a result, a high content is cleaved in the scar and a low content of protein is absorbed in the intestine. Thus, soluble crude protein is an indicator of protein quality. This parameter gives an idea of the amount of protein absorption in the scar. A high rate suggests that more protein is available to bacteria than they can absorb, so the protein is lost (not used) in the environment. Where the scar cleavage rate is low, the amounts of energy and protein are in equilibrium, whereas the negative scar cleavage rate indicates a lack of protein in the scar, which means insufficient bacterial growth. This indicator in all test groups of samples is within the limit.

Transit protein means a non-cleavable protein, this is the amount of protein that does not break down in the scar, but passes through the scar when transiting for subsequent absorption in the small intestine. The normal indicator is more than 14.0%; therefore, all test samples for this indicator correspond to the norm.

Starch levels correlate with lactic acid levels. The highest starch index (preferably) was noted in the second test group, where the Silo Twice biopreservative was used, a slightly lower amount of starch was noted in the prototype, and in the group where self-preservation was used, the lowest value was noted.

Lactic acid is one of the acids formed during the preservation process. Sugar is a fuel for lactic acid bacteria that produce lactic acid. The standard is 3-15%. The largest amount of lactic acid is contained in the second experimental group, where the biopreservative Silo Twice was used (8.33%), a slightly smaller amount in the prototype (8.28%), and the lowest value was noted (5.30%) in the group where self-preservation was used.

Acetic acid is very important, as it helps to prevent the feed from overheating. Grass silage/haylage with a low proportion of acetic acid is very sensitive to overheating. The optimal acetic acid content is 1.0-2.0% or 10-20 g/kg CB. The largest amount of acetic acid is contained in the second experimental group, where the biopreservative Silo Twice was used (1.30%), a slightly larger amount in the prototype (1.40%), and the lowest value was noted (0.89%) in the group where self-preservation was used.

The content of exchange energy in a feeding product is the amount of energy available for metabolism. Digestible (digestible) energy is gross energy minus the energy released with manure. Digestible (digestible) energy adjusted for the amount of energy lost with the release of urine and gas (mainly methane) is considered as the amount of exchange energy. This energy is available for use in the metabolic process. In the process of metabolism, most of the energy is spent on heat production. Heat-adjusted exchange energy is considered to be ultimately clean energy. Clean energy is used for growth, milk production and reproduction. The optimal level of exchange energy is > 10.5 MJ/kg CB [9]. The largest amount of exchanged and digested energy was noted in the second experimental group. The highest fodder value was noted in the second experimental group.

Calculating the cost of the diet with the introduction of haylage to each of the experimental groups, the following result was obtained: feeding milk cows for 60 days of haylage obtained when the biopreservative Silo Twice was introduced, provided a savings of 2250 rubles per cow.

4 Conclusion

After conducting a comparative assessment of the effect of preservatives on haylage quality, the following conclusions were made.

1. The extended preservation and bactericidal effect of the Silo Twice biopreservative complex of live bacteria and enzymes on the greasy plant mass led to the preservation of the raw material nutrients in the product and an increase in their digestibility compared to the self-preserving product.

2. The cost of the diet using the biopreservative Silo Twice turned out to be lower than the cost of the diet with the imported biopreservative.

References

1. AIV-2000 Plus Na is used for silo billet, Retrieved from: https://kormovit.ru/wp-content/uploads/2020/05/instrukcziya-aiv-2000-plyus-na-taminco.pdf.
2. BONSILAGE SPEED M, Retrieved from: https://schaumann.ru/product/bonsilazh-speed-m.
3. The use of biopreservatives in the preparation of feed, Retrieved from: https://agrovesti.net/lib/tech/fodder-production-tech/ispolzovanie-biokonservantov-pri-zagotovke-kormov.html.
4. Preservatives for animal husbandry, Retrieved from: https://teknofeed.org/2019/06/25/preservatives-for-livestock/
5. Feed and feed additives, Retrieved from: https://www.tsenovik.ru/articles/korma-i-kormovye-dobavki/kormovye-fermenty-rasshcheplyayushchie-nekrakhmalistye-polisakharidy/.

6. Food preservatives, Retrieved from: https://www.tsenovik.ru/articles/korma-i-kormovye-dobavki/kormovye-konservanty/.
7. S. A. Makarov, *Increasing the efficiency of the technological process of block silage extraction and haylage from trench storages and justifying the parameters of the loader working body*. PhD Dissertation, Saratov (2003).
8. S. A. Makarov, Procurement technology and methods for storing canned feed, *MNIZH* **3-3 (45)** (2016)
9. M. I. Kartashov, T. M. Voinova, N. V. Statsyuk, S. V. Rogovsky, Y. O. Grebeneva, Patent No. 2645227C1 *New biopreservative for feed silage and method of its production* (2018)
10. Soluble crude protein, Retrieved from: https://sagitov.pro/downloads/BLGG.pdf
11. I. Yu. Tyurin, Improving the technological process of additional drying of hay at the hospital: PhD Dissertation (Saratov, 2000).

Effect of biological products on spring wheat cultivation

I. A. Bobrenko, *N. V.* Goman*, *V. P.* Kormin, *V. I.* Popova, and *E. P.* Boldysheva

Omsk State Agrarian University named after P.A. Stolypin, 1, Instituskaya Square, Omsk, 644008, Russian Federation

Abstract. The studies of the meadow-chernozem soil of the Omsk region, showed that biological products are effective when used for spring wheat by seed treatment. When comparing their action, it was noted that a greater effect was obtained from the use of the variants Azotobacterin+Phosphobacterin and Azotobacterin+Phosphobacterin+Humate K. The improvement of the nutritional conditions of plants when applying biological fertilizers and biological products has a positive effect on the content of the number of amino acids in the grain, which without fertilizers increases from 9.59% up to a maximum of 9.92% with Humat K.

1 Introduction

In agricultural production, as a result of a shortage of nutrients and their constant harvest alienation, with a small use of mineral fertilizers, as well as their high cost and the economic condition of farms, there is an objective need to find alternative means and methods to preserve and increase soil fertility. At the same time, organic technologies are actively developing [1–3].

For soil enrichment microbiological and bacterial fertilizers, we use products of highly active microorganisms that improve the nutritional conditions of crops. They contain specific strains of microorganisms, using which the processes of transformation of compounds containing nutrients necessary for plants are activated in the soil. Humic acids and amino acid products have a versatile direction of action such as the activation of bioenergetic processes, stimulation of metabolism, synthetic processes, improvement of the penetration of nutrients through the plasmalemma, intensification of enzymatic systems, increase in the adaptive properties of plant organism. They are used as plant growth stimulators [4–6].

The purpose of the research is to analyze the effectiveness of the use of biological products to increase the yield and improve the quality of grain of soft spring wheat on meadow-chernozem soil.

2 Materials and methods

The experiment was carried out on the experimental field of Omsk State Agrarian University in 2021. The spring wheat of Element 22 variety was used, the soil was meadow-chernozem, thin, low-humus, heavy loamy. The objects of research were: biological fertilizers Azotobacterin and Phosphobacterin; growth stimulants Biostim and humate K. They were used by the method of seed treatment. The content in the soil of N-NO3 – 4.6, P2O5 – 135, K2O – 215 mg/kg. The location of the plots was systematic. The repetition of options was three times. The plot area was 20 m^2; the accounting area was 16 m^2. The used agricultural technology was generally accepted for the zone. In soil samples, the content of nitrate nitrogen was determined with disulfophenolic acid according to Grandval-Lage; mobile phosphorus and potassium content from one extract was determined according to Chirikov (GOST 26204-84). Quality indicators were determined by conventional methods.

The studied growing season, in general, was characterized as warm and dry, while rather anomalous conditions for plants were observed due to a sharp shortage of water content. During this period, 134 mm of precipitation fell, which amounted to only 65% of the long-term average. The average temperature for 4 months was 8.3 higher than the long-term average.

3 Results and discussion

It was found that all biological products had a positive effect on the yield of spring wheat grains on the meadow-chernozem soil of the forest-steppe zone of the Omsk region (Figure 1). With a yield in the control variant of 3.21 t/ha, the use of biological products contributed to the formation of 3.48-3.73 t/ha of spring wheat grain. In the steppe zone, the use of urea (kg a.i./ha) and zinc and copper chelates (g a.i./ha) was studied for foliar top dressing of hulled barley in the tillering phase (Table 1).

* Corresponding author: nv.goman@omgau.org

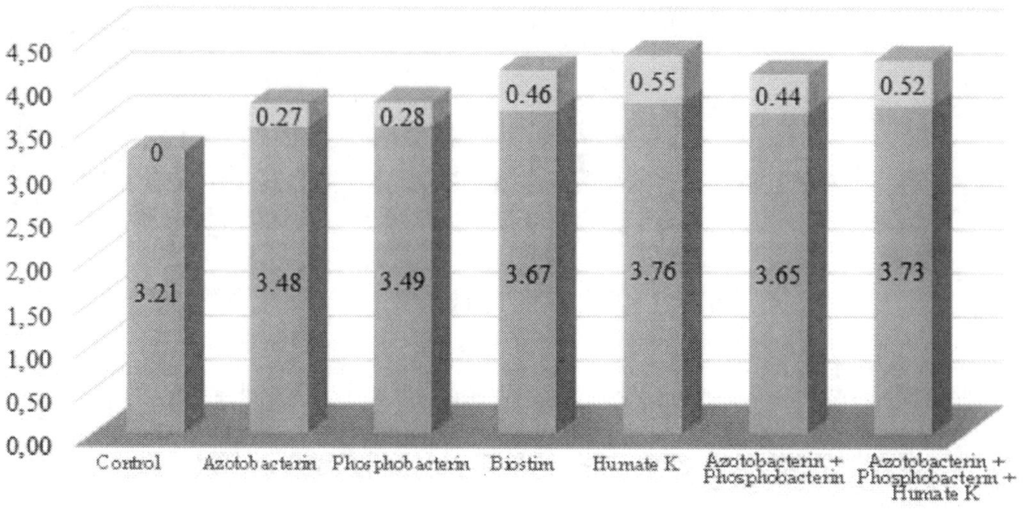

Fig. 1. Yield of spring wheat depending on biological products on the meadow-chernozem soil of the forest-steppe zone.

A noticeable increase in the yield of the variety was observed from the use of Biostim, Humate K, Azotobacterin + Phosphobacterin and Azotobacterin + Phosphobacterin + Humate K: yield increases amounted to 0.44 -0.55 t/ha. The use of Azotobacterin and Phosphobacterin separately also led to a significant increase in the yield of wheat: 0.27 and 0.28 t/ha of grain. The content of mobile forms of nutrients in the soil in Siberian chernozems was often at an insufficient level [7–9].

We studied the effect of biological products in the treatment of seeds on soil fertility (Table 1).

Table 1. Content of mobile forms of nutrients in the soil under spring wheat during the harvesting period, depending on biological products, mg/kg of soil (0–20 cm layer).

Variant	N-NO_3	P_2O_5	K_2O
Control	3.62	130	258
Azotobacterin	4.38	128	250
Phosphobacterin	3.98	138	235
Biostim	3.52	130	240
Humate K	3.62	128	255
Azotobacterin + Phosphobacterin	4.36	136	246
Azotobacterin + Phosphobacterin + Humate K	4.42	138	237

The supply of wheat plants with nitrate nitrogen during the harvesting period under the action of biological products was low (3.62-4.42 mg/kg). It is necessary to note that there is the increase in the level of nitrate nitrogen in the soil layer of 0-20 cm when using azotobacterin, both separately and in combination with other products (Phosphobacterin and Humate K): during the harvesting period, it increased by 0.74-0.80 mg /kg at the content of nitrate nitrogen in the control of 3.62 mg/kg.

The provision of wheat plants with mobile forms of phosphorus (according to Chirikov) during the harvesting period under the action of biological preparations was average (128–149 mg/kg).

The experimental data indicate that when using Phosphobacterin, the content of mobile phosphorus slightly exceeded the variants without its use. Both independently and together with other products, it increased the amount of the element by 6-8 mg/kg. Comparing the considered spring wheat cultivation technologies, it is necessary to note that, on average, according to the options, spring wheat plants cultivated using Azotobacterin, mobile phosphorus – Phosphobacterin are most provided with nitrate nitrogen.

The study of the potassium mode of the culture showed that the content of mobile potassium (according to Chirikov) in the soil during the harvesting period was very high – 412–548 mg/kg. When biological products are introduced, the content of mobile potassium in the soil does not change significantly.

Thus, the use of biologicals in the treatment of seeds has a positive effect on the content of nitrate nitrogen and mobile phosphorus in the soil. During the use of biological products in this way, their level increases. The content of mobile potassium does not increase. The use of stimulants does not significantly affect the chemical composition of the soil.

Improving the structure of the spring wheat crop depends on the type of soil and nutritional conditions, the improvement of which contributes to the mobilization of the plant's physiological resources and increased yields. The change in yield is accompanied by a change in the elements of the structure of the crop, one of the main ones is productive tillering, the mass of grains from one plant, and the weight of 1000 grains (Table 2).

It was found that the productive tillering capacity with the use of biological products generally increased - from 2.2 to 2.3-2.5 in the variants Biostim, Humate K, and Azotobacterin + Phosphobacterin + Humate K.

The mass of 1000 grains in the harvest increased by 3.1 g when using biological products (Humate K and Azotobacterin + Phosphobacterin), in general, the use of biological products increased this parameter.

Table 2. Structure of spring wheat yield depending on biological preparations.

Variant	Productive tillering capacity	Weight of grains from one plant, g	Weight of 1000 grains, g
Control	2.2	5.7	35.0
Azotobacterin	2.2	5.9	35.3
Phosphobacterin	2.2	5.9	35.9
Biostim	2.3	5.8	36.5
Humate K	2.5	5.8	38.1
Azotobacterin + phosphobacterin	2.2	6.5	38.1
Azotobacterin + phosphobacterin + Humate K	2.5	5.9	36.9

The change in nutrition has a significant impact on the content of raw gluten in the grain [10–12]. Its mass fraction during the introduction of biological products increased (Table 3) to 1.5% (Azotobacterin + Phosphobacterin + Humate K). The content of gluten in the control was 34.4.

The protein content in the grain in the control variant was 17.7%. The use of biological products provided a slight increase in the protein content of grain by 0.2–0.5%.

The analysis of qualitative indicators showed that the grown wheat grain was average natural (760–798 g/l). The highest increase in this parameter was observed in Azotobacterin + Phosphobacterin + Azotobacterin + Phosphobacterin + Humate K: they amounted to 35 and 38 g/l. Hardness was 59–62 and did not depend significantly on biological products.

Table 3. Grain quality of spring wheat depending on biological products.

Variant	Nature, g/l	Protein, %	Hardness,%	Gluten, %
Control	760	17.7	61	34.4
Azotobacterin	762	18.0	61	35.0
Phosphobacterin	764	18.2	62	35.6
Biostim	775	17.9	59	35.4
Humate K	790	18.1	61	35.4
Azotobacterin + phosphobacterin	795	18.3	59	34.6
Azotobacterin + phosphobacterin + Humate K	798	18.0	60	35.9

The amino acid composition of the protein was determined depending on the nutritional conditions in the experiment. Nutritional conditions affected the quantitative content of amino acids (Table 4). Thus, during the study of the effect of biological products on the qualitative characteristics of wheat protein, it was found that the number of amino acids increased without fertilizers from 9.59% to the highest 9.92% in the variant with Humate K (HCP_{05} 0.31). The effect of biofertilizers and biological products on the content of amino acids in grain was positive, the greatest effect was caused by Humate K and Biostim stimulants.

Table 4. Amino acid composition of spring wheat protein depending on biological products in the conditions of the Omsk region, % (2021).

Amino acid	Control	Azotobacterin	Phosphobacterin	Biostim	Humate K	Azotobacterin + Phosphobacterin	Azotobacterin + phosphobacterin + Humate K
Arginine	0.81	0.84	0.82	0.91	0.90	0.82	0.84
Lysine	0.34	0.35	0.33	0.38	0.37	0.36	0.35
Tyrosine	0.40	0.39	0.41	0.42	0.42	0.44	0.37
Phenylalanine	0.65	0.70	0.66	0.72	0.71	0.65	0.69
Histidine	0.32	0.34	0.31	0.33	0.34	0.34	0.33
Leucine+isoleucine	1.60	1.65	1.61	1.66	1.68	1.63	1.66
Methionine	0.30	0.31	0.31	0.32	0.31	0.32	0.30
Valin	0.68	0.74	0.67	0.71	0.71	0.68	0.73
Proline	2.05	2.02	2.03	2.03	2.05	2.04	2.03
Threonine	0.48	0.46	0.46	0.47	0.49	0.47	0.46
Cool	0.77	0.79	0.75	0.75	0.76	0.76	0.79
Alanine	0.51	0.52	0.50	0.53	0.54	0.51	0.52
Glycine	0.68	0.63	0.66	0.63	0.64	0.64	0.63
Amount of amino acids	9.59	9.74	9.52	9.86	9.92	9.66	9.70

We can state that the improvement of plant nutrition conditions when applying biological fertilizers and biological products have a positive effect on the content of the content of amino acids in grains.

4 Conclusion

Thus, all biological products are effective when used under spring wheat by seed treatment. With a yield in the control variant of 3.21 t/ha, the use of biological products contributed to the formation of 3.48-3.73 t/ha of spring wheat grain. When comparing the effect, it was noted that a greater effect was obtained from the use of Azotobacterin + Phosphobacterin + Azotobacterin + Phosphobacterin + Humate K. Biological fertilizers had a positive effect on crop quality indicators. It was found that the number of amino acids without fertilizers increased from 9.59% to the highest 9.92% in the variant with Humate K.

References

1. D. I. Eremin, Eurasian soil science **49 (5)**, 538–545 (2016)
2. N. A. Voronkova, I. A. N. M. Bobrenko, V. I. Nevenchannaya, IOP Conference Series: Earth and Environmental Science **659,** 022071 (2020)
3. I. A. Bobrenko, O. A. Matveychik, E. G. Bobrenko, V. I. Popova, IOP Conference Series: Earth and Environmental Science **624,** 012219 (2021)
4. N. V. Goman, V. P. Kormin, I. A. Bobrenko, V. I. Popova., E. P. Boldysheva, IOP Conference Series: Earth and Environmental Science **937,** 022126 (2021)
5. A. A. Alferov, L. S. Chemova, A. P. Kozhemyakov, Russian Agricultural Sciences **44 (1),** 53–57 (2018)
6. A. A. Alferov, L. S. Chemova, Russian Agricultural Sciences **44 (6),** 546–550 (2018)
7. A. A. Shpedt, Y. V. Aksenova, J. of Pharmaceutical Sciences and Research **10 (4),** 870–873 (2018)
8. A. A. Shpedt, Yu. V. Aksenova, M. R.

Shayakhmetov, et al., ITJEMAST **10 (3),** 309–318 (2019)

9. M. Shayakhmetov, A. Zinich, A. Gindemit ISPC **167,** 156–159 (2019)
10. Y. Aksenova, N. Nevenchannaya, V. Boiko, et al., TFTS **393,** 50–54 (2019)
11. N. V. Goman, I. A. Bobrenko, V. V. Popova, A. A. Gaidar, E. P. Boldysheva IOP: Materials Science and Engineering **659,** 012066 (2021)
12. I. A. Bobrenko, V. P. Kormin, N. V. Goman, V. I. Popova, E. P. Boldysheva, IOP Conference Series: Earth and Environmental Science **845,** 5012016 (2021)

Analysis of market prospects for bakery products based on the ranking of bakery producers in the region

Oksana Shikhova*, *Svetlana* Golubeva, *Irina* Shilova, *Marina* Selina and *Natalya* Dyakova

Vologda State Dairy Farming Academy named after N. V. Vereshchagin, 2, Str. Schmid, Molochnoe, Vologda, 160555, Russian Federation

Abstract. According to the result of the study based on a comprehensive multidimensional comparative assessment and rating of Vologda enterprises – producers of bakery products participating in the system of voluntary certification of food products "Real Vologda Product", the authors have analyzed the prospects for the development of this industry in the region. They used the following research methods: questionnaire survey methods, comparative overall assessment, and rating assessment. The study is based on the results of a specially organized marketing survey of consumer preferences. This article presents key aspects of the authors' comprehensive assessment of commodity producers based on the analysis of economic potential and the position of consumers of products as factors of sustainable development of enterprises. As a validation of the proposed methodology of multidimensional comparative rating, which includes the analysis of the financial condition of the research subjects and assessment of consumer preferences, the researchers carried out a qualitative characterization of the "Real Vologda Product" brand actors engaged in the production of bakery products. The article revealed the structural features of the producers of this commodity group in terms of their influence on the development of the regional bakery market. The study identified the main problems of the economic development of enterprises-producers of bakery products, noted the factors of demand for their products from the local consumer, identified the prospects for sustainable development and stated the ways to solve the problems identified.

1 Introduction

According to experts, the outlook for the bakery market in Russia is not promising, as the production of bakery products has dropped from 5.76 to 5.58 million tonnes. According to BusinesStat forecasts, in 2021–2025, as the economic situation in the country stabilises, sales of bread and bakery products in the Russian market will continue to decline by 0.3-1.4% per annum. In contrast, the cost of these products has been gradually increasing over the last three years, remaining in line with inflation.

Researchers point to the following reasons for this dynamic:

1. Reduction in the consumption of bread. Research by Trend Lab identified the following reasons: doctor's recommendations and health problems, e.g. autoimmune diseases that require the elimination of gluten from the diet; lack of tasty bread that one would like to consume; very caloric product; bread not liked as a product; improved well-being as a result of experimenting with no bakery consumption.

2) Competition between enterprises and their products [1]. In every Russian city, there are usually 1–2 large bakery producers and about a dozen small enterprises and bakeries, which find it difficult to compete in the price segment with the large ones. They tend to appeal to consumers in terms of taste and the range of products on offer.

It is also about improving the quality of life of Russian citizens by guaranteeing high living standards [2, 3]. On the other hand, Russia is currently in dire need of effective mechanisms for launching regional economic growth [4]. The problem of ensuring food security in Russia objectively needs to be addressed not only at the national level, but also at the regional level [5]. The most important task of state industrial policy is to prioritise the formation of a forward-looking industrial structure of the national economic complex capable of generating new sources of growth [6, 9]

Researchers, economists agree that one way out of the current situation, allowing for the support of local producers-processors of agricultural products, is branding as a way of creating their image, recognition, quality and market retention of bakery products. Therefore, many of them create their own brands or participate in already existing territorial brands [7]

The number of participants in the voluntary certification system "Real Vologda Product" is growing in the Vologda Oblast: 74 organisations that label more than 3.600 products with the trademark "Real Vologda Product" [8]. They also include bakeries, but branding has not solved the above problems.

There are various methodologies to assess simultaneously supply and demand for the whole industry in the region and each individual enterprise. However, modern science does not provide a

* Corresponding author: oksana-shikhova@yandex.ru

multidimensional assessment of producers that simultaneously considers both customer feedback and the economic situation of each enterprise. The economic situation would provide a recommendation to the market actor: what to look for in the first place to improve their situation, strengthen their economic potential and become more sustainable. This is precisely the opportunity that our methodology offers.

2 Materials and methods

The methodology of constructing any rankings is based on the technique of comparing the key characteristics of the objects being compared, and in our case the indicators of assessing the economic potential of the brand actors, producing and selling specific products. Therefore, implementing this procedure makes sense only within relatively homogeneous product groups, which was implemented in this work.

The information base for the comparative ranking of producers is a system of indicators in two categories.

I. Economic indicators — ratios of the financial condition, business activity and performance of an enterprise. The source of information for calculating the values of indicators in this category is the accounting and financial statements of enterprises published by the Federal State Statistics Service of the Russian Federation. The method of information collection is traditional document analysis (desk research). Here we consider the following indicators:

1. Financial sustainability indicators.
2. Liquidity and solvency indicators.
3. Business activity indicators.
4. Financial performance indicators.

II. Consumer Assessment Indicators were calculated from a summary of a questionnaire survey of the population. The source of information for calculating indicator values for this category is a consumer survey. The method for collecting information is qualitative (determining the proportion of respondents who gave a particular answer) and quantitative (using a scale method).

This category's indicator system included the following 8 assessment indicators:

- indicator for assessing the level of consumer loyalty to a brand actor (an indicator of commitment to the producer, “heart share”, “Product patriotism”);
- indicator for assessing the level of perceived quality of a brand actor's products (a measure of the perceived quality of the manufacturer's products;
- indicator for assessing the level of market share (market share, sales share) of a brand actor;
- indicator to assess the level of perception and recognition of the manufacturer's image as a brand actor;
- indicator for assessing the level of accessibility of a brand actor's products (an indicator of physical accessibility of products);
- indicator for assessing the level of price satisfaction with a brand actor's products;
- indicator for assessing the level of territorial prevalence of a brand actor's products (breadth of the sales market);
- indicator for assessing the level of awareness of brand actor advertising.

The purpose of the study was to assess the economic potential, prospects and factors of sustainable development of bakery products producers – actors of the brand “Real Vologda Product” by the method of multidimensional comparative rating based on the results of the marketing survey of the population.

We used a questionnaire marketing survey to investigate consumer opinion.

The list of 13 main questions in the questionnaire includes four types of questions. We applied different methods to quantify the results of respondents' answers, aggregating and normalising them to obtain a quantitative coefficient assessment of the overall result of the respondents for each question for each brand actor included in the set of compared entities to build their ranking [8]

In the next step, the resulting coefficient assessments of the respondents' overall response to each question were aggregated for two indicators: brand member loyalty (a measure of commitment to the manufacturer, “heart share”) and brand member product availability (a measure of physical product availability), for which the respondent answered several questions in the questionnaire. The other indicators, for which there was only one question in the questionnaire, did not require the aggregation of the result of the respondents' answers.

Since the aggregates are coefficients, the geometric mean methodology applied in such cases was used to summarize and determine essentially their mean.

Based on the system of indicators discussed, represented by the indicators of the economic block and the consumer assessment block, it was possible to provide an in-depth and detailed comparative analysis of brand member producers by applying the methodology of a multidimensional comparative and comprehensive assessment.

The methodology for the comparative composite assessment of brand member producers relies on a multidimensional comparison algorithm regarding the indicator system discussed and includes the following steps:

- standardising indicator values;
- constructing a square matrix for each indicator with the values of private comparative levels between each pair of brand participants in the study population;
- summarizing the derived matrices with partial levels into a final comparative assessment matrix with summary levels by determining the sum of the values of the partial levels according to the place in the matrices for the whole indicator system for each brand member;
- constructing a final assessment matrix by calculating the values of the levels of summary comparative scores for each brand actor;

- transforming the derived summative assessment levels of brand actors to produce an integrated assessment level value by converting the summative assessment level value to a non-negative form, comparing the non-negative summative assessment levels to the maximum value in their aggregate and presenting the result as an integrated summative comparative assessment level (Integrated Comparative Assessment - ICA);
- analysing and comparing the integrated assessment levels of the participating brand producers;
- determining the participation degree of the levels of indicators for each block in forming an integrated assessment level by calculating the block relative assessment levels of brand participants according to the above described methodology, based on the previously calculated private levels of comparative assessments of the indicators for the selected blocks;
- qualitative characteristics of the level of assessment received by each brand member according to the blocks of indicators based on an interval scale of values of the received integrated levels, construction of the rating and typological grouping of brand member producers.

The level of integrated comparative assessment of a brand member against one or another of the four domains outlined characterises its economic potential, consumer demand and the degree of attribute manifestation according to the highlighted blocks of indicators.

3 Results and discussion

The questionnaire survey "Real Vologda product – Bread products", conducted online via Google Form, captured the opinions of 105 respondents who wished to take part in this survey.

The questionnaire investigated consumer opinion on the products of the producers participating in the brand "Real Vologda Product", specialising in the production of bread and bakery products. The list of manufacturers included 18 brand participants from this product group.

The processing of the responses according to the methodology described above has resulted in forming a system of indicators of consumer assessment to apply in practice a multidimensional comprehensive comparative assessment of producers - participants of the brand "Real Vologda Product" and the construction of their rating for the product group "Bakery products."

Further, under the methodology of multidimensional comprehensive comparative assessment (Shikhova, Selina, Barinova, 2020), we normalised their values and calculated partial comparative assessments of the indicators for this block of consumer assessment indicators. Summarizing the values of private assessments enabled to calculate the values for the integrated level of comparative assessment of manufacturers-participants in the product group "Bakery products" of the brand "Real Vologda product" on the block of indicators of consumer assessment and to build their rating.

To calculate the values of the economic block indicators, we collected initial information on the accounting (financial) statements of enterprises from the official website of the Federal State Statistics Service of the Russian Federation.

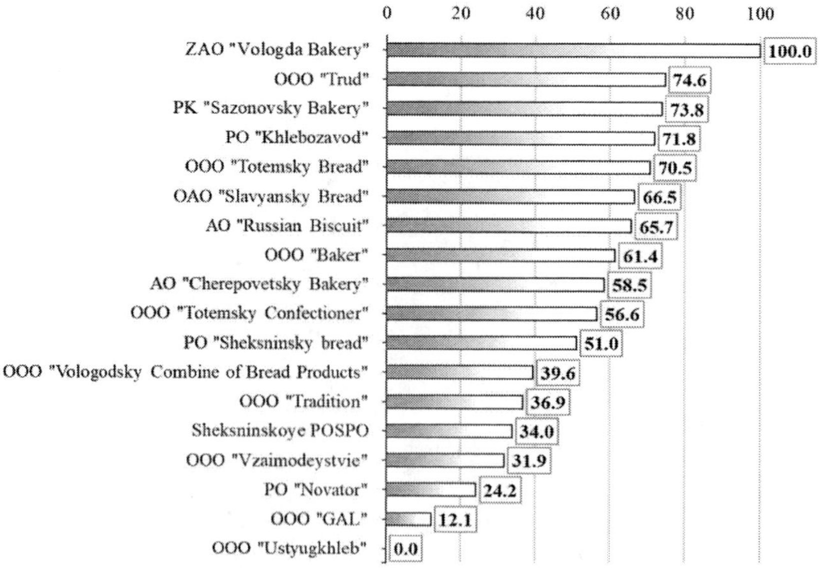

Fig. 1. Ranking of producers participating in the "Real Vologda product" brand in the product group "Bakery products" according to the value of the aggregated complex comparative assessment.

Further, under the methodology of multidimensional comprehensive comparative assessment described in [7], we normalised their values and calculated partial comparative assessments of the indicators for this block of consumer assessment indicators. Summarizing the values of private assessments, we calculated the values of the integrated level of comparative assessment of producers participating in the product group "Bakery products" of the brand "Real Vologda product" on the block of economic indicators and build their rating. In addition, the same methodology calculated sub-block integrated scores for each of the four indicator groups in this block to detail the conclusions of the overall brand assessment of the participants in a given product group.

Building the rating of producers of bakery products – participants of the brand "Real Vologda product" in the whole system of indicators, including both problematic blocks, required according to the same methodology the aggregation of assessment results by blocks. The results are shown in the figure (Figure 1).

The results of the study show that the leader of this product group in the brand "Real Vologda Product" according to the aggregate value of the integrated comparative assessment is ZAO "Vologodsky Khlebokombinat" (ICA is 100%).

According to the interval value scale of the integrated level of comparative assessment (Table 01), its value can be characterised as high. In the same interval of the scale there are values of integral estimation of four more participants of the brand – OOO "Trud" (ICA is 74.6%, 2nd place in the rating of participants), PK "Sazonovsky Khlebozavod" (ICA is 73.8%, 3rd place), PO "Khlebozavod" (ICA is 71.8%, 4th place) and OOO "Totemsky Khleb" (ICA is 70.5%, 5th place in the rating).

Table 1. Interval scale of values for the integrated level of comparative assessment of brand member manufacturers.

Level value area	Interval boundaries	Characteristics of the assessment level of the participant
1	$70 \leq ICA \leq 100$	high
2	$50 \leq ICA < 70$	above average
3	$30 \leq ICA < 50$	below average
4	$0 \leq ICA < 30$	low

We should note this high level of aggregate scores for these three branded product manufacturers is due to the rather high scores for the economic block of indicators (Figure 2).

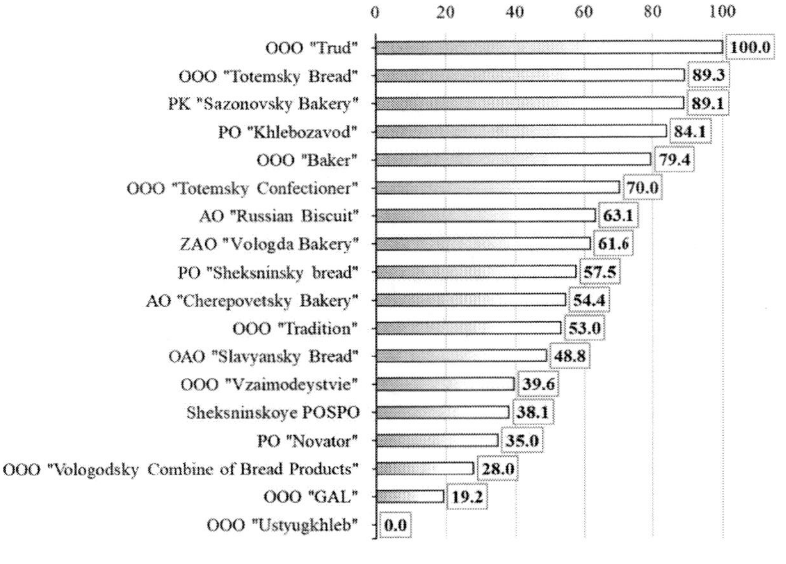

Fig. 2. Ranking of producers participating in the "Real Vologda Product" brand in the "Bakery Products" product group according to the value of the integrated comparative assessment of the economic block of indicators.

According to the results of the generalization of respondents' opinions ZAO "Vologodsky Khlebokombinat" is also the leader in the rating of consumer assessment, taking the first place by all indicators except for the indicator of product availability (score 93.6%, 2nd place in the rating after OAO "Slavyansky Khleb") (Figure 3).

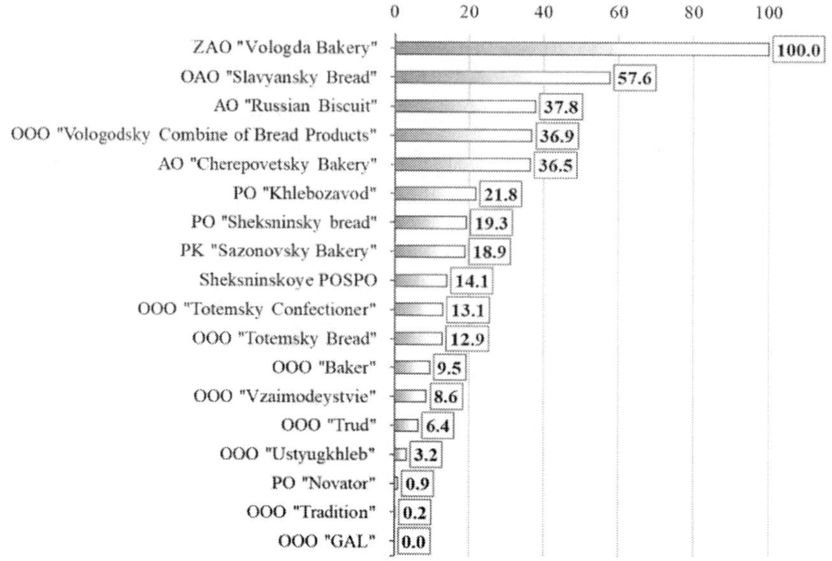

Fig. 3. Ranking of producers-participants of the brand "Real Vologda Product" in the product group "Bakery products" according to the value of the integrated comparative assessment of the block of indicators of consumer opinion.

In the economic block of indicators the level of an integrated comparative assessment of ZAO "Vologodsky Khlebokombinat" is only above average (61.6%) and 8th position in the rating for this problematic block, which is explained by the low level of ratings for indicators of liquidity and solvency (17.3%, 6th place in the rating), business activity (27.7%, 11th place of 18 in the rating of participants) (Figure 2).

The second place in the consumer rating is occupied by OAO "Slavyansky Khleb", but we should note that the value of the integrated indicator is significantly lower than that of the leader and amounts to 57.6%. Summarizing the consumers' opinions on the totality of indicators in this block, we can say that this company is second in the ranking by almost all items, except for the first place for the level of product availability and fourth place for the level of perceived product quality (62.2%) and the level of its territorial prevalence (27.2%).

The third and fourth positions in the consumer ranking belong to AO "Russky Biscuit" (37.8%) and OOO "Vologodsky Kombinat Hleboproductov" (36.9%), respectively, but their integrated score for this block is significantly lower than the average (Figure 3). According to the results of the survey, respondents were quite appreciative of the quality and availability, as well as the price level of these companies' products. However, assessments of the level of advertising visibility, image recognition of enterprises and, as a consequence, indicators of market share, product prevalence and consumer loyalty to them are low. These businesses, like all but the leaders in this product group, should pay serious attention to the effectiveness and vigour of their marketing policy.

Of course, it is difficult for small businesses in this product group to compete with the leaders in the consumer market, as they have a narrowly defined task of providing fresh bakery products and confectionery to the consumer in the immediate vicinity of the production site. This is why the products of small businesses in this product group are little known to the general public in the region and even less so outside it. If the objective is to expand the market, to reach a new customer, this will require additional financial investment to expand and modernise production, to diversify the range of products, and to build a strategy for entering new markets. This is possible through effective product branding but requires investment and support.

There is a marked difference in the alignment of the integrated assessments of the problematic blocks in this product group with the ratings for the meat and dairy producer groups. Whereas in the previous two cases of testing the comprehensive rating methodology, consumer opinion leaders mostly had high scores on the block of economic indicators, the situation in the case of the product group of bakery product producers is uncharacteristic (Figure 2).

Thus, the 2nd and 3rd positions in the rating by the level of aggregate assessment occupied OOO "Trud" (74.6%) and PK "Sazonovsky Khlebozavod" (73.8%). Such a high position was due to a high level of integrated assessment by the block of economic indicators (accordingly 100% – 1st place for OOO "Trud", and 89. 1% – 3rd place for PC "Sazonovsky Khlebozavod"), but at the same time the levels of consumer assessment for them are very low (OOO "Trud" – 6.4% and 14th position out of 18 in the consumer rating, PC "Sazonovsky Khlebozavod" – 18.9% and 8th place). However, the above leaders of the overall ranking are significantly inferior to these companies in terms of their aggregate economic capacity scores. This fact allows us to conclude that such brand participants as OOO "Totemsky Khleb", OOO "Trud", PK "Sazonovsky Khlebozavod", PO "Khlebozavod", OOO "Baker" can expect an increase in loyalty and popularity with the Vologda Oblast population when expanding the sales of manufactured products and using an active marketing policy.

4 Conclusion

The analysis of producers of bakery products participating in the brand "Real Vologda product" on the basis of a comprehensive approach allowed to identify the strengths and weaknesses of these enterprises, their economic potential and the prospects for the development of this sector of the agricultural market in the region. The identified factors of economic sustainability of this category of producers, determined by consumer demand for the products of this product group, have made it possible to identify possible areas of expansion of the sales market both within and outside the region. Products of agricultural origin made from certified organic raw materials have a clear competitive advantage over products from neighbouring regions and imported products.

The article is completed with the financial support of the Government of the Vologda region within the state scientific grant "The approaches to the evaluation of the brand power and rating of commodity producers of the Vologda region, Russia as viewed by consumers".

References

1. T. G. Yureneva, O. I Barinova, S. G. Golubeva, *Towards productive, sustainable and resilient global agriculture and food systems conference proceedings* (Nitra, Slovakia: Published by Wolters Kluwer) 1500–1514 (2018)
2. A. N. Chekavinskiy, R. Yu. Selimenkov, *Economic and Social Changes: Facts, Trends, Forecast* **4,** 226–235 (2014)
3. V. M. Selin, Economic and Social Changes: Facts, Trends, Forecast **6 (42),** 139–152 (2015)
4. E.V. Balatsky, N.A. Ekimova, Economic and Social Changes: Facts, Trends, Forecast **13 (3),** 74–92 (2020)
5. O. G. Moronova, A. P. Dorogovtsev, Economic and Social Changes: Facts, Trends, Forecast **1,** 99–107 (2010)
6. V. S. Uskov, Economic and Social Changes: Facts, Trends, Forecast **13 (6),** 134–151 (2020)
7. T. G., Butova, E. A. Demakova, S. L. Ulina, O. L. Egoshina, S. I. Mutovin, E. P Danilina, Journal of Siberian Federal University. Humanities and Social Sciences **13 (11),** 1880–1892 (2020)
8. O.A. Shikhova, M. N. Selina, O. I Barinova, BIO Web Conf. **27,** 00031 (2020)
9. V. S. Uskov, Economic and Social Changes: Facts, Trends, Forecast, **12 (2),** 128–146 (2019)

https://doi.org/10.1051/bioconf/20225200054

Influence of the microbiological fertilizer Slavol on the formation of early harvest of potatoes in the conditions of the Moscow region

M. E. Dyikanova*, *I. N.* Gasparyan, *O. N.* Ivashova, *N. F.* Deniskina, and *Sh. V.* Gasparyan

Russian State Agrarian University - Moscow Agricultural Academy named after K.A. Timiryazev, 49, Timiryazevskaya St., Moscow, 127434, Russian Federation

Abstract. The article presents the results of a study of obtaining early potato production using a microbiological fertilizer in the conditions of the Moscow region. Early potato varieties were used for the study: Udacha, Bryansk delicacy, Red Scarlett, Zhukovsky early, Meteor, Riviera. All varieties of table appointment with high palatability are recommended for cultivation in the Nonchernozem zone of the Russian Federation. The purpose of the research is to study the effect of the microbiological fertilizer Slavol on the growth, development of early potato plants and productivity. The experiment was carried out in 2020-2021. Observations and records in the experiment were carried out according to generally accepted methods when conducting field and laboratory studies on potato crops. Mass seedlings were noted in the variants with the treatment of tubers with the microbiological fertilizer Slavol in the following sequence: variety Zhukovsky, Meteor, Riviera, Udacha, Bryansk delicacy. The maximum total period from germination to harvesting on July 15 was longer in the variant with treatment in the Zhukovsky early variety (56 days), which further affected the increase in yield. The use of the microbiological fertilizer Slavol contributed to an increase in yield for all the studied varieties from 22.2 to 42.8% in relation to the control variant. Thus, in the experiment, the positive effect of microbiological fertilizer Slavol on the growth, development and yield of early potato varieties in the conditions of the Moscow region was established.

1 Introduction

In the modern world, potatoes are grown everywhere, which is very popular among people with different income levels [1]. To increase the life expectancy of the population, it is important to have a varied and safe diet, which is very difficult to provide with the constant deterioration of the environment. Breeders are constantly working on breeding varieties that meet modern requirements; however, in the production of potatoes, pests and diseases are increasingly showing themselves aggressively, due to resistance to the used pesticides [2, 3].

Under the conditions of climate change, an individual approach to the agrotechnology of early potatoes reduces the biotic and climatic load and select alternative forms of microfertilizers. At present, it is impossible to obtain high yields and safe production of early potatoes using only mineral fertilizers; therefore, scientists more often consider microbiological fertilizers as an alternative to classical technology that allows them to comprehensively influence the growth and development of plants [4, 5].

The development of plants and the formation of crops, among other things, are negatively affected by adverse climatic conditions arising from climate change around the world. There are a number of safe farming system solutions that allow agriculture to develop and increase yields and product quality. Most of the recommendations for growing environmentally friendly products are studied by college and university students in specialized disciplines, studying agronomists, vegetable growers and engineers, it is important to correctly apply modern knowledge in the production of agricultural products. To obtain environmentally friendly products, it is necessary to adhere to an integrated approach to the cultivation of early potato production. This is a scientifically based crop rotation with the inclusion of legumes, timely application of organic fertilizers in the autumn to improve soil fertility, integrated plant protection and the use of microbiological fertilizers, to develop plant resistance to environmental factors [5, 6].

* Corresponding author: dyikanova@rgau-msha.ru

 https://doi.org/10.1051/bioconf/20225200054

Intensive technologies widely used in agriculture are not always suitable for obtaining high yields of environmentally friendly potato products in the early stages [7]. The use of microbiological fertilizer allows you to accelerate the formation of yields in the early stages (on July 15). Intensive technologies widely used in agriculture are not always suitable for obtaining high yields of environmentally friendly potato products in the early stages [7]. The use of the microbiological fertilizer makes it possible to accelerate the formation of yields in the early stages (on July 15) due to the formation of a high concentration of beneficial microflora, which suppresses the development of pathogens [8]. The composition of the microbiological fertilizer Slavol includes six components, three of which are free-living nitrogen fixers that provide plants with nitrogen (Derxia spp., Azotobacter chroococcum, Azotobacter vinelandi). The following three components belong to representatives of the genus Bacillus, characterized by the ability to produce antibiotics with activity against phytopathogenic fungi. Antagonist bacteria of the genus Bacillus stimulate the formation of antimicrobial substances, vitamins, plant immunity, which is confirmed by our observations. Variants treated with the microbiological fertilizer Slavol were distinguished by faster plant development and resistance to the development of early blight, which contributed to favorable conditions during the formation of the crop [9,10].

2 Materials and methods

The research have studied the effect of the microbiological fertilizer Slavol on the growth, development and productivity of early potatoes of such varieties as Udacha, Bryansk delicacy, Red Scarlett, Zhukovsky early, Meteor, Riviera. The treatment with Slavol was carried out before planting by soaking the tubers and double foliar feeding with a frequency of 10 ... 15 days.

The studies were carried out in 2020 ... 2021 on the territory of the UNPC Vegetable Experimental Station named after V.I. Edelstein. On a plot with soddy-podzolic medium loamy soil, topsoil of 20...22 cm with the N content of 9.3 mg, P_2O_5 of 15.0 mg, K_2O of 8.3 mg per 100 grams of soil.

Research options are control and treatment with the microbiological fertilizer Slavol (before planting, soaking tubers for 3...4 hours in a solution concentration of 10 ml/l of water) + two foliar top dressings during the growing season with a frequency of 10...15 days. The first 15 days are after mass shoots, the second is in the budding phase. The studies were carried out on potato varieties: 1. Udacha, 2. Bryansk delicacy, 3. Red Scarlett, 4. Zhukovsky early, 5. Meteor, 6. Riviera. All selected varieties are early ripe, for table use and processing with good taste, adapted for the Non-Chernozem Zone.

Potato cultivation was carried out using ridge technology, agricultural technology included such operations as autumn plowing, spring milling, cutting ridges before planting, planting, inter-row cultivation, hilling, preventive treatment against the Colorado potato beetle (in 2021) by Euphoria (ks) 0.2 l /ha + Break (me) 0.3 l/ha. Soil preparation was carried out taking into account the creation of a finely cloddy soil structure, in the spring period the impact of machinery on soil compaction was minimized, the cutting of ridges was carried out 3 days before planting to warm up the soil with a height of 16 ... 18 cm. Planting tubers was carried out on May 3–4, harvesting taking into account the structure of the crop July 15, the experiment was repeated three times, randomized placement, planting pattern of 70x35 cm, planting material of the middle fraction (40 ... 80 g). Before planting, the tubers were dried for 14 days to partially lose moisture and accelerate germination in early spring. In the variant with the microbiological fertilizer, the tubers were additionally soaked for 3–4 hours in a working solution (10 ml/10 l of water). Observations and records in the experiment were carried out according to generally accepted methods when conducting field and laboratory studies on potato crops [11, 12, 13].

3 Results and discussion

When considering the similarities and differences between the variants of the experiment on the passage of the phases of plant development and the duration of the interphase periods, we started with the emergence of seedlings. The earliest mass shoots were noted in the variants with the treatment of tubers with Slavol microbiological fertilizer, in the following sequence: variety Zhukovsky early (on day 18), Meteor, Riviera, Udacha (on day 20), Bryansk delicacy (on day 22). Seedlings in the control variants appeared 2–3 days later for each variety. The beginning of budding occurred in the same sequence as the seedlings in the variants treated with Slavol, and then the control variants. The maximum total period from germination to harvesting on July 15 was longer in the variant with treatment in the Zhukovsky early variety (56 days), which further affected the increase in yield. The minimum period from germination to harvesting on July 15 was 53 days in the control variant for varieties Udacha, Bryansk delicacy, Riviera. Accordingly, the duration of the interphase periods of early potatoes depends on the biological characteristics of the variety, optimal conditions during the growing season and methods of processing planting material. The average number of stems and plant height in the control variants corresponded to the biological characteristics of the variety, and formed from 2.5 to 3.5 stems per plant. The minimum number of erect, high (up to 60 cm) stems was noted in the Meteor variety. The maximum number of stems was noted in the variety Bryansk delicacy up to 3...3.5 pieces per plant and an average height (50...55 cm). The remaining varieties in the control variants had intermediate values in terms of height and number of stems. The use of the microbiological fertilizer Slavol stimulated the emergence of more stems and increased the height of plants by an average of 32.6% and 10.8%, respectively.

The treatment of tubers and vegetative potato plants with the microbiological fertilizer Slavol in experimental variants contributed to an increase in the leaf surface area on all studied potato varieties. The minimum difference in the increase in leaf area was noted between the control and treated variants in the Red Scarlett variety, the increase was 17.9%. The maximum difference between the variants was noted in the variety Bryansk delicacy by 43% in terms of leaf area. On average, for all varieties, the increase in leaf area was 32.9% (Fig. 1).

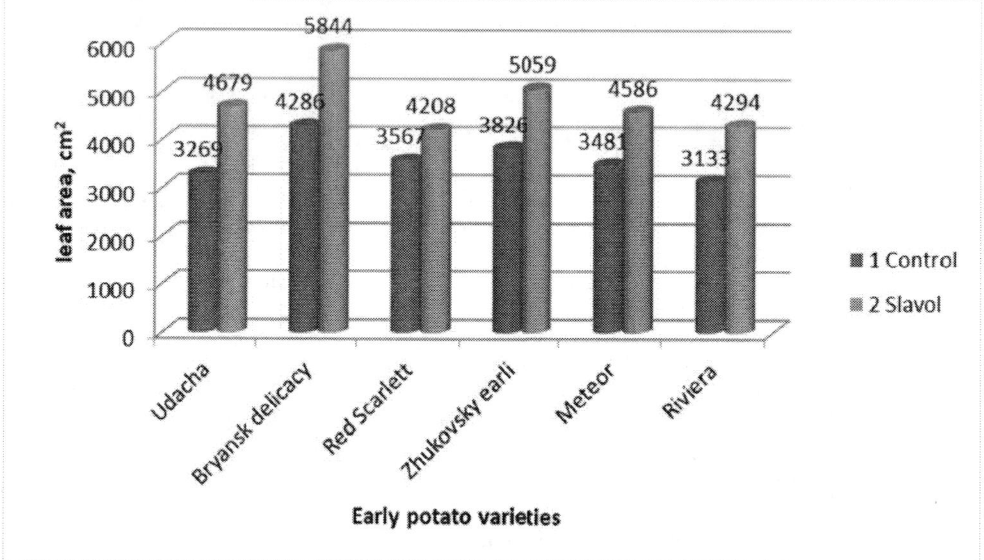

Fig. 1. Effect of the microbiological fertilizer Slavol on the leaf surface area of potato plants on average for 2020…2021 per plant, in cm^2.

In the first half of the summer of 2020-2021, favorable conditions have developed for the development of alternariosis, which develops in dry and hot weather. Alternariosis is one of the dangerous potato diseases that can drastically reduce yields. Examination of plants for damage to the leaves was carried out in the flowering phase, according to a six-point scale, and the average% of the development of Alternaria was calculated. Strong damage to individual plants was noted in the control variants in the varieties Zhukovsky early, Meteor, the average development was noted in the variety Udacha. The positive effect of the microbiological fertilizer Slavol also affected the development of alternariosis, so the listed varieties had signs of the disease, but the percentage of damage decreased 2...2.5 times. The minimum damage was noted on the varieties Bryansk delicacy and Riviera.

 https://doi.org/10.1051/bioconf/20225200054

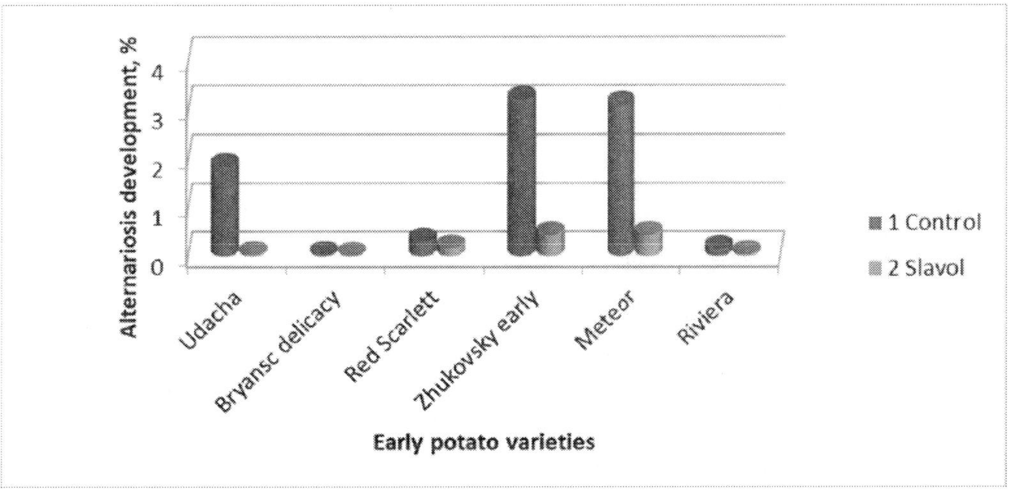

Fig. 2. Influence of microbiological fertilizer on the % development of early blight in early potato plants in the conditions of the Moscow region on average in 2020–2021.

Potato harvesting was carried out at an early date for the Moscow region, on July 15. During this period, the sale of imported potato products is mainly carried out, since the domestic potato does not have time to form a crop by this period. The rate of plant development and tuberization depends on the variety and growing conditions, especially in the first growing season. The use of the microbiological fertilizer on early potatoes contributed to the formation of the crop in the early stages and increased yields for all varieties. In the conditions of the Moscow region, early varieties of potatoes use the classical growing technology, on July 15, on average, you can get 400...500 grams of tubers from one bush. In the control variants for five varieties, the mass of tubers from one bush ranged from 409...465 grams. In both variants, the yield of the variety Bryansk Delicacy was not high, the average weight of one bush in the control variant was 208 grams. The treated plants of the variety Bryansk delicacy with microbiological fertilizer Slavol also had a low mass of tubers from one bush (320 grams); however, the increase in relation to the control variant of this variety was 53.8%. On average, for all varieties, when using the microbiological fertilizer Slavol, the yield increased from 22.2 to 42.8% compared to the control.

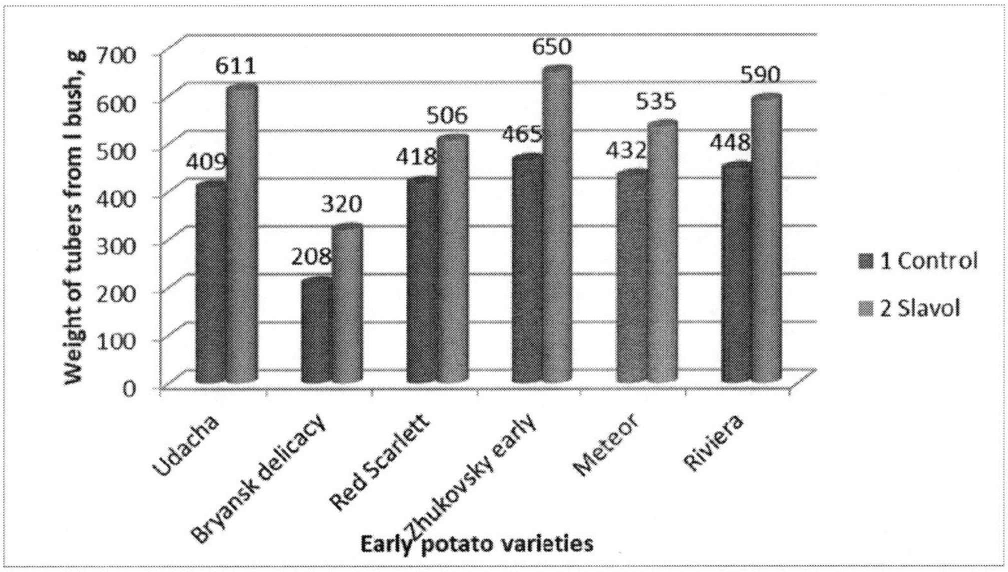

Fig. 3. Effect of the microbiological fertilizer Slavol on the average mass of tubers from one plant as of July 15 in the conditions of the Moscow Region on average in 2020–2022.

4 Conclusion

Thus, in the experiment, the positive effect of the microbiological fertilizer Slavol on the growth, development and yield of early potato varieties in the conditions of the Moscow region was established. The use of adapted varieties in combination with modern cultivation methods makes it possible to increase the yield due to the number and average weight of tubers per plant in the early stages (July 15).

Acknowledgments

The authors would like to thank their colleague for their contribution and support of the research. They are also thankful to all the reviewers who gave their valuable inputs to the manuscript and helped in completing the paper.

References

1. FAOSTAT, Retireved from: http://www.faostat.fao.org/
2. L. N. Devyatkina, Potato production: global and national discourses, *Bulletin NGIEI* **5 (84),** 122–134 (2018)
3. L. S. Fedotova., A. V. Kravchenko, In changing climatic conditions, new approaches to the cultivation of potatoes are needed, *Potatoes and Vegetables* **2,** 20–23 (2011)
4. E. Wszelaczynska, D. Wichrowska, I. Rogozinska, Effect of bioelements (Mg, N, K) and herbicides on vitamin C content in potato tubers Part II. Dy-namics of the vitamin C changes in stored potato tubers, *Journal Elementology* **10 (4),** 1117–1125 (2005)
5. L. S Fedotova., A. V. Kravchenko N. A. Timoshina, A. N. Gavrilov, The use of bacterial fertilizers in the cultivation of potatoes, *Fertility* **2 (65),** 6–9 (2012)
6. S. B. Reshnovetsky, N. V. Klimova, O. V. Balycheva, Biological preparations on potatoes, *Proceedings of the International Anniversary Scientific and Practical Conference dedicated to the 75th anniversary of the Institute of Potato Growing of the National Academy of Sciences of Belarus* (Minsk) Part 2, pp 182–185 (2003)
7. S. F. Yanni, H. H. Janzen, E. G. Gregorich, B. H. Ellert, F. J. Larney et al., Organic Carbor Convergence in Diverse Soils toward Steady State, *Soil Science Society of American Journal,* **80 (6),** 1653–1662 (2016)
8. A. M. Soloviev, *Application of Fertilizers in the Cultivation of Crops with High Technology* (Moscow: Publishing House of the Russian State Autonomy and Agricultural Academy named after K. A. Timiryazev, 2014).
9. I. Gasparyan, A. Levshin, S. Smurov, M. Dyikanova, Sh. Gasparyan, N. Deniskina, V. A. Berdyshev, Use of microbiological fertilizer in planting early potatoes (Solanum tuberosum) in the conditions of Belgorod region, Russia, *Research on Crops* **22 (4)** (2021).
10. O. Ivashova, I. Gasparyan, A. Levshin, M. Dyikanova. Justification of the possibility of cultivating in Moscov region two-crop culture of early potatoes, *Engineering for Rural Development* **19,** 399–405 (2020)
11. B. A. Dospehov, *Methods of research on potato culture* (Moscow, 1967)
12. K. V. Popkova, I. N. Kutsamanova, *Proceedings of the Scientific Conference of Young Scientists and Specialists* (Moscow: Publishing house of ICCA, 1999)

Study of the influence of various factors on the emission of carbon dioxide by the aggregate during direct sowing of grain crops

Camill A. Khafizov, *Ramil N.* Khafizov, *Azat A.* Nurmiev*, and *Olga I.* Makarova

Kazan State Agrarian University, 65, K. Marx St., Kazan, 420015, Republic of Tatarstan, Russian Federation

Abstract. The article deals with issues related to the reduction of carbon dioxide emissions into the atmosphere during the operation of sowing machine and tractor units. To identify ways to reduce CO_2 emissions, computational experiments were conducted using a mathematical model of sowing units compiled according to the efficiency indicator. This is the integral emission of carbon dioxide as the sum of CO_2 in the manufacture, maintenance, repair, control of the unit, when burning fuel and the amount of CO_2 not absorbed from the atmosphere during the formation of the grain crop due to its losses from improperly selected equipment and its parameters. The influence of a number of factors on the integral emission of CO_2 is investigated. These are the area of the treated area or field, the specific resistance of the planter, the volume of seasonal load on the unit, the annual load of the tractor in hours, the physical and mechanical properties of the soil, the pressure in the tires of the wheel propulsion engines, etc. Effective values of factors for reducing the emission of carbon dioxide by the sowing unit are revealed. The use of the totality of all the recommendations received will reduce the emission of carbon dioxide into the atmosphere by the aggregate during the technological operation of direct sowing of grain crops by 600-650 kg/ha.

1 Introduction

Among the many threats to humanity, one of the most significant is the threat of global warming, which, according to climatologists, will be accompanied by various cataclysms, such as the melting of glaciers at the poles of the Earth, rising sea levels, the intensification of seismic processes on the surface of the earth and much more [1–6].

Agricultural production makes a significant contribution to this process. In addition to freeing up part of the forest area for agricultural needs [7, 8], a rapid increase in the number of livestock, an increasingly intensive mechanical and chemical effect on the soil, accompanied by a decrease in carbon dioxide absorption and an increase in the emission of greenhouse gases such as methane [9, 10], mobile machines are used in agriculture. This is due to the high level of mechanization, performing a variety of functions and at the same time intensively burning fossil fuels [11–20]. As a result of the combustion of carbon fuels, carbon dioxide, which is a reference greenhouse gas, is emitted.

In this regard, agricultural producers face the task of reducing carbon dioxide emissions by the engines of agricultural mobile and stationary machines, maximizing its binding by cultivated plants. The article is devoted to the search for ways to reduce CO_2 emissions by machine-tractor units at the time of direct sowing of grain crops.

2 Materials and methods

To identify ways to reduce carbon dioxide emissions by machine-tractor units engaged in direct sowing of grain crops, computational experiments based on the use of a mathematical model of sowing machine-tractor units are used. The mathematical model of the sowing unit was developed according to the criterion of the efficiency of its work – the integral emission of carbon dioxide into the atmosphere. This indicator takes into account both the direct emission of CO_2 during the operation of the unit, and its indirect release at the time of manufacture, maintenance and repair of equipment. And it also takes into account CO_2 not absorbed from the atmosphere by part of the potential harvest due to its losses associated with the choice of non-optimal parameters of equipment [21, 22].

3 Results and discussion

To identify ways to reduce the emission of carbon dioxide into the atmosphere by sowing units in the course of their operation, it is necessary to study the influence of various factors on the efficiency of the unit.

* Corresponding author: azat-nurmiev@mail.ru

The main initial data for the calculation are selected as typical of the region.

Initial data:
Field area = 1÷100 ha;
Rutting length = 1 km;
Aggregate crossing distance = 3 km;
Crop seed density = 800 kg/m³;
Strength coefficient of the bearing surface of the field = 0.9;
Volume of work performed by the unit = 1000 ha;
Number of tractors performing the operation = 1;
Number of hours of operation of the unit per day = 16 h;
Planned yield of the main and by-products = 40 c/ha;
Pressure in the tractor tires = 0.16 MPa;
Number of wheels on one side of the tractor propulsion = 1;
Coefficient adhesion of the tractor wheels to the soil = 0.65;
Coefficient of resistance to rolling wheels of the tractor = 0.15;
Soil density = 1.4 g/cm³;
Specific gravity of CO_2 absorbed by the crop (wheat) = 243 kg/c;
Tractor weight = 120 kN;
Engine power = 500 hp;
Maximum working width of the unit = 18 m;
Maximum speed of the unit = 12 km/h;
Specific resistance of the seeder = 4 kN/m.

As the first factor, we select the area of the treated field, setting it by varying from 1 to 100 hectares.

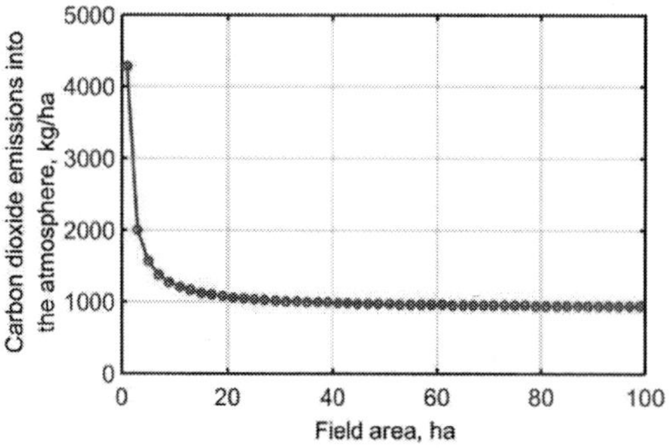

1 5 9049 Dependence of CO_2 emission on the area of the treated field.

As can be seen from Figure 1, the change in the area of the field from 1 to 100 hectares leads to a decrease in CO_2 emissions from 4300 kg/ha to 939.14 kg/ha. A particularly intense, parabolic decrease in CO_2 emissions is observed with an increase in the field area from 1 hectare to 15 hectares and this is a weighty argument for the use of plots and fields with an area of more than 15 hectares at agricultural enterprises.

The effect of the specific resistance of the planter on the release of carbon dioxide into the atmosphere is shown in Figure 2. With an increase in the resistivity of the planter, the CO_2 emission of the unit increases in stages, which is caused by a step change in the optimal width of the seeder." With an increase in the specific resistance of the planter from 3.5 kN/m to 7.5 kN/m, there is an increase in CO_2 emissions of almost 2.2 times. The resistivity of agricultural machines depends on a large number of factors, including the physical and mechanical properties of the soil (density, hardness, moisture, type of soil and its mechanical composition, etc.), on the condition and type of coulters. In the work [23] it is established that "the energetically optimal level of soil moisture when the sowing unit works with anchor coulters is 23.6%, and with paw coulters 23.4%. The direct energy consumption of a sowing unit equipped with anchor coulters is 15-20% lower than that of lancet paws". Thus, it is necessary to strive to reduce the resistivity of seeders due to organizational, structural and other measures leading to a reduction in CO_2 emissions into the atmosphere.

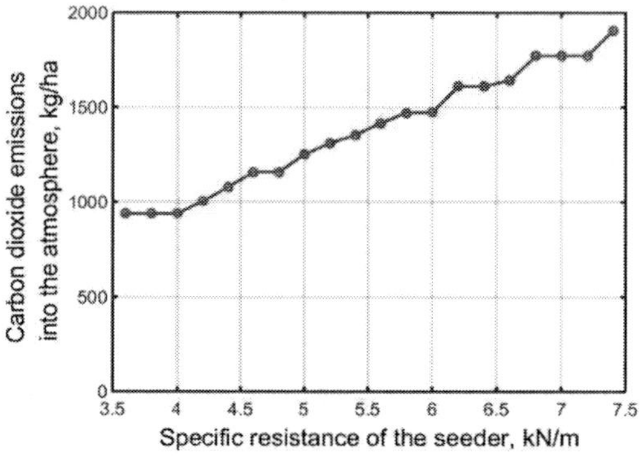

1 5 9089 Dependence of CO_2 emissions during direct sowing of grain crops on the resistivity of the planter.

Oddly enough, the release of carbon dioxide into the atmosphere is affected by the amount of seasonal load per sowing unit (Figure 3). The minimum CO_2 emission by the unit in the amount of 730 kg/ha corresponds to the load on one sowing unit equal to 200 hectares. The increase in the integral value of CO_2, as the seasonal load on the sowing unit increases above the optimal value, is explained by an increase in the amount of lost crop due to an increase in the agrotechnical period of the technological operation. That is, in this case, the cultivated crop will absorb less carbon dioxide from the atmosphere, which is equivalent to its ejection. Reducing the seasonal load less than the optimal value leads to an increase in the indirect component of the integral CO_2 emission (at the time of manufacture, maintenance and repair of equipment) due to a decrease in the annual output of the sowing unit.

The value of integral CO_2 emissions into the atmosphere is also affected by the intensity of tractor use during the year, although this influence is not so significant. Increasing the annual load of the tractor from 200 to 1600 hours leads to a reduction in carbon dioxide emissions into the atmosphere by 40 kg/ha, see Figure 4.

A significant influence on the integral amount of CO_2 emitted by the sowing aggregate is exerted by the soil density (Figure 5), which is in close correlation with the hardness of the soil. As soil density increases, potential yield losses increase due to deteriorating crop growth conditions, so increasing soil density from 1.2 g/cm^3 to 1.78 g/cm^3 leads to a tenfold increase in CO_2 emissions from 1,000 to 10,000 kg/ha.

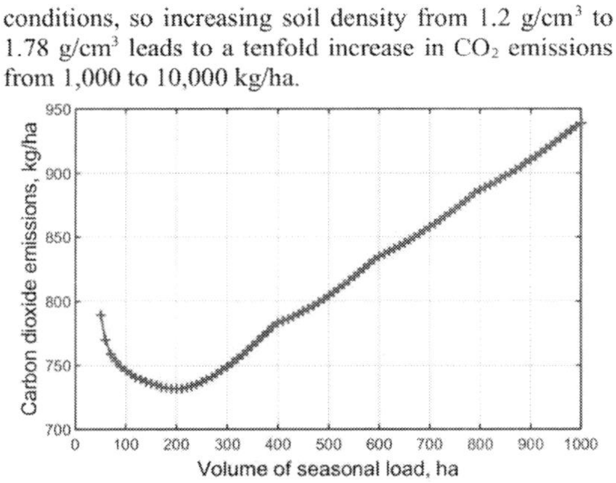

1 5 9B9 Dependence of CO_2 emissions during direct sowing of grain crops on the volume of seasonal load per hectare.

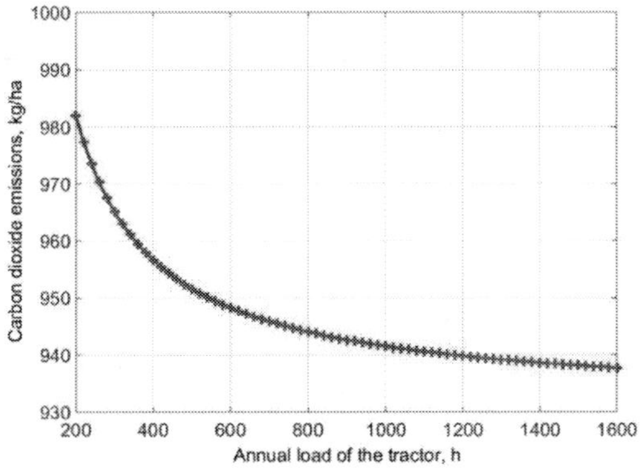

1 5 9Ø9 Dependence of CO_2 emissions during direct sowing of grain crops on the annual load of the tractor per hour.

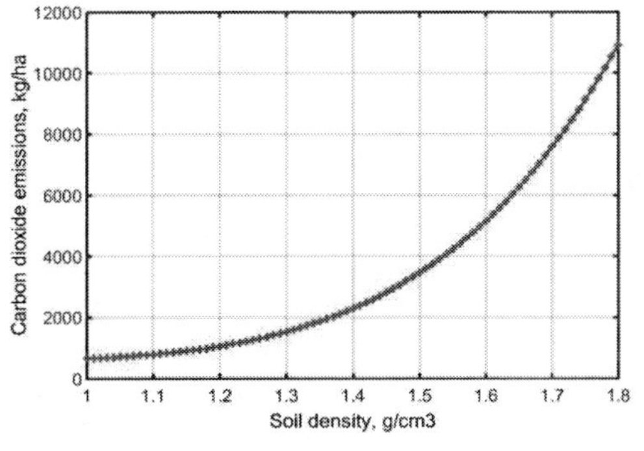

1 5 9079 Dependence of integral CO_2 emission during direct sowing of grain crops on soil density in g/cm^3.

Figures 6, 7, 8 show the dependence of carbon dioxide emissions on the number of aggregates with an annual sowing volume at an agricultural enterprise of 500 hectares in Figure 6, 1000 hectares in Figure 7 and 2000 hectares, Figure 8. As can be seen from the figures, the optimal number of units to handle the entire annual volume of work does not always grow in direct proportion to the growth in the volume of work. To process 500 hectares of area, 2 sowing units are required, 1000 hectares of area – 4 sowing units, and for sowing 2000 hectares of area it is optimal, from the point of view of minimizing the integral emission of CO_2 into the atmosphere – 7 sowing units. At the same time, the minimum value of CO_2 emitted per sowing per 1 hectare increases from 750 kg/ha (500 ha) to 830 kg/ha (2000 ha), i.e., by 80 kg/ha, which is explained by an increase in the indirect components of the integral CO_2 emission (for manufacturing, maintenance and repair).

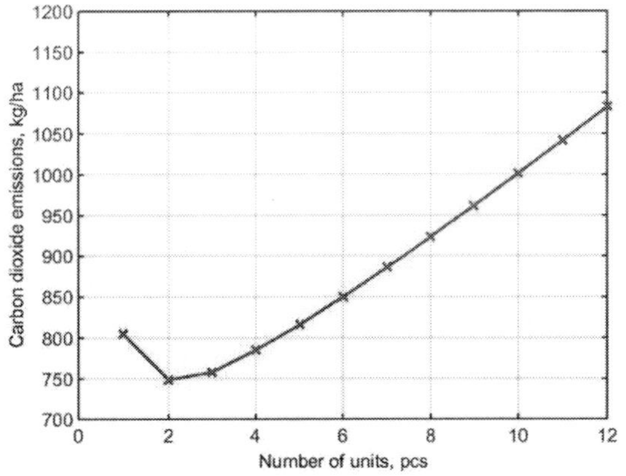

1 5 9ő9 Dependence of CO_2 emissions during direct sowing of grain crops on the number of aggregates engaged in sowing an area of 500 hectares.

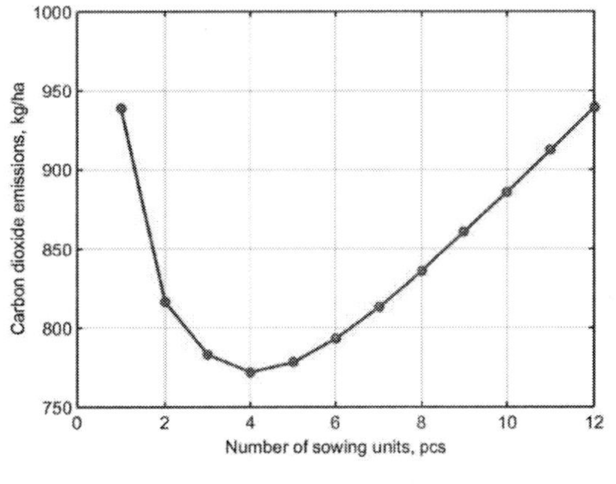

1 5 9F9 Dependence of CO_2 emissions during direct sowing of grain crops on the number of aggregates engaged in sowing an area of 1000 hectares.

An increase in the tire pressure of the wheel drive (see Figure 9) from the minimum allowable value of 0.06 MPa to the upper value of 0.2 MPa leads to an increase in CO_2 emissions from 760 to 1100 kg/ha. The difference is 340 kg / ha and this is a significant value. In this regard, within the permissible limits, it is always necessary to reduce the pressure in the tires [24–26].

Of interest is the influence of the shift coefficient of the use of the sowing unit or the total operating time of the unit during one day on the emission of carbon dioxide. From Figure 10 it can be seen that this relationship is nonlinear and with an increase in the total operating time of the unit per day, the emission of carbon dioxide into the atmosphere is reduced by hyperbolic dependence. With the single-shift use of the sowing unit for 8 hours, 1200 kg/ha of CO_2 is emitted into the atmosphere. The use of the unit in two shifts for 8 hours leads to a reduction in CO_2 emissions by 260 kg/ha. It is possible to organize the operation of the unit in three shifts.

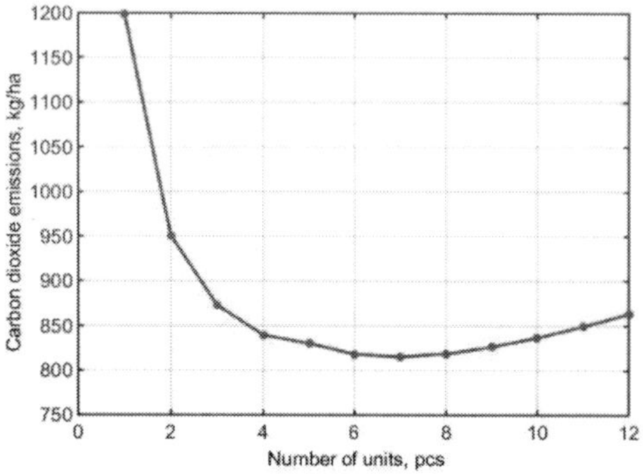

1 5 9l 9 Dependence of CO_2 emissions during direct sowing of grain crops on the number of aggregates engaged in sowing an area of 2000 hectares.

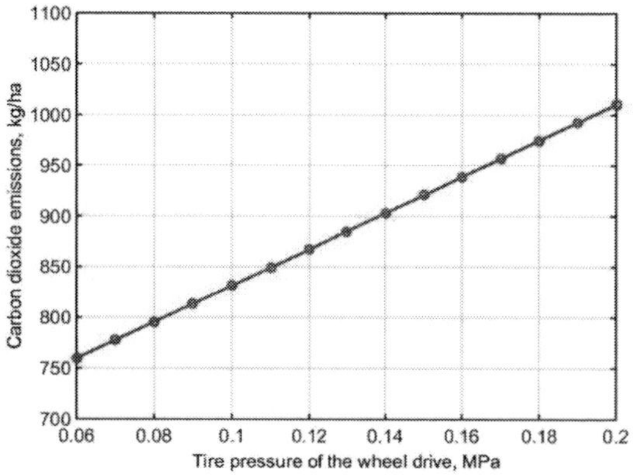

1 5 9l9 Dependence of CO_2 emissions during direct sowing of grain crops on the tire pressure of the tractor wheeled propulsion.

Let us repeat the previous calculation with the optimal values of factors in the initial data.
Updated initial data:
The field area = 100 ha;
The rutting length = 1 km;
The moving distance of the unit = 3 km;
The seed density = 800 kg/m^3;
The coefficient of strength of the bearing surface of the field = 0.89;
The amount of work performed by the unit = 200 hectares;
The number of tractors performing the operation = 1;
The number of hours of operation of the unit per day = 16 hours;
The planned yield of the main and by-products =40 c/ha;
The tractor tire pressure = 0.06 MPa;
The number of wheels on one side of the tractor propulsion = 1;
The coefficient of adhesion of the wheels of the tractor with the soil = 0.66;
The coefficient of resistance to rolling wheels of the tractor = 0.16;
The soil density = 1.2 g/cm^3;
The specific gravity of CO_2 absorbed by the crop (wheat) = 243 kg/c;
The tractor weight = 120 kN;
The engine power = 500 hp;
The maximum working width of the unit = 18 m;
The maximum speed of the unit = 12 km/h;
The specific resistance of the seeder = 4 kN/m.

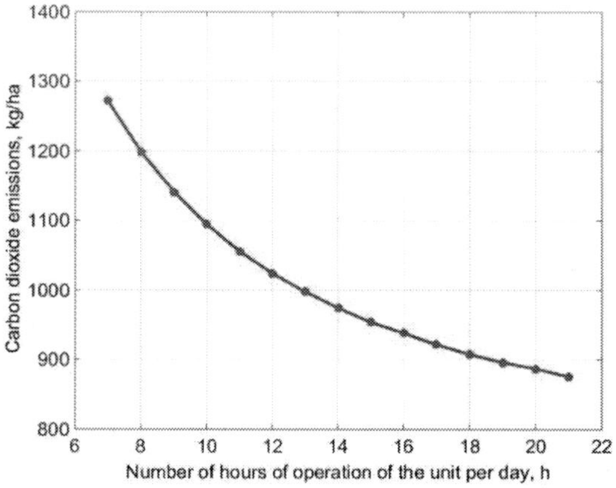

1 5 9l4 9Dependence of CO_2 emissions during direct sowing of grain crops on the number of hours of operation of the unit in the field with the optimal combination of factor values.

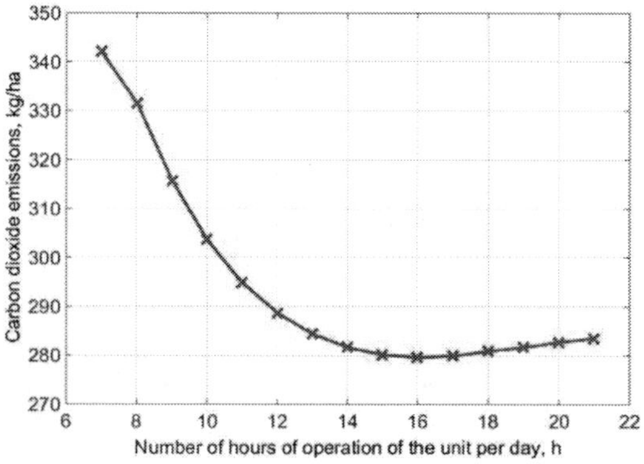

1 5 9l449 Dependence of CO_2 emissions during direct sowing of grain crops on the number of hours of operation of the unit in the field with the optimal combination of factor values.

The result of the calculation is shown in Figure 11. As you can see, comparing Figures 10 and 11, the absolute value of CO_2 emissions into the atmosphere decreased from 940 to 280 kg / ha, which is a very significant decrease. The difference is 660 kg/ha. At the same time, it is most effective to use the sowing unit in two shifts of 8 hours.

4 Conclusion

The conducted computational experiments using a mathematical model of the sowing unit, compiled according to the efficiency criterion, the integral emission of carbon dioxide by the aggregate into the atmosphere showed that in order to reduce CO_2 emissions, it is necessary:

- to take organizational, operational and structural measures to reduce the resistivity of seeders;
- to use fields with an area of more than 15 hectares for sowing grain crops;
- for each type of a sowing machine and tractor unit to determine the optimal volume of seasonal load on this technological operation in hectares;
- to maximize the annual load of the tractor in hours, using in various types of work;
- to reduce the density of the soil in the field and its hardness to reduce the compacting effect of the sowing aggregate on the soil, reduce the loss of potential yield and the amount of CO_2 not absorbed from the atmosphere;
- to set the minimum allowable pressure in the tires of the wheels of the tractor propulsion system based on the recommendations of the tire manufacturer;
- to select the optimal number of units to perform a given amount of work on a technological operation, which can be calculated using the developed mathematical model of the unit;

The use of all the recommendations in the aggregate will reduce the emission of carbon dioxide into the atmosphere during the technological operation of direct sowing of grain crops. Reducing the dissipation of CO_2 emissions in the atmosphere by means of the sowing complex with an optimal tractor weight of 120 kN, an engine power of 500 hp, a working width of 18 m at a speed of 12 km/h will be 940–280 = 660 kg/ha.

References

1. J. T. Houghton, Y. Ding, D. J. Griggs et al. (Eds.) *Climate Change 2001. The Scientific Basis. Contribution of Working Group I to the Third Assessment Report of the Intergovernmental Panel on Climate Change* (Cambridge: Cambridge Univ. Press, 2001)
2. Assessment of the risks of climate change in 2021. Retrieved from: https://www.chathamhouse.org/2021/09/climate-change-risk-assessment-2021/03-direct-climate-impacts (2021)
3. Y. A. Israel, S. M. Semenov Example of the calculation of critical boundaries of the content of greenhouse gases in the atmosphere with the help of a minimal simulation model of the greenhouse effect, *Reports of the Academy of Sciences* **2**, 533–536 (2003)
4. I. K. Rodin, Emissions of greenhouse gases and Russia's contribution to the fight against global warming. Ecological state of the natural environment and scientific and practical aspects of modern agrotechnologies, M*aterials of the V International Scientific and Practical Conference, Ryazan, Individual entrepreneur Konyakhin Alexander Viktorovich,* pp 353–355 (2021)
5. Zh. I. Sadykhova, Warming of the climate as a problem of threat to the biosphere, *Proceedings of the International Symposium "Reliability and Quality",* pp 243–245 (2020)
6. I. Maksimov, N. Adigamov, A. Mustafin, D. Khaliullin, I. Gayaziev, A. Matyashin, R. Lukmanov, Theoretical fundamentals for determining soil erosion potential, *Grupo Tchê Química* **46(34),** 540–557 (2019)
7. The World Bank proposed to reduce greenhouse gas emissions by preserving forests, *Bulletin of Environmental Education in Russia* **2,** 11 (2007)
8. V. D. Burkov, V. F. Krapivin, V. S. Shalaev, The role of forest ecosystems in the regulation of the greenhouse effect, *Bulletin of Moscow State Forest University – Forest Herald* **4,** 20–31 (2008)
9. S. I. Shiyan, K. A. Ilyinsky, I. O. Pashurenko, Processing of waste of the livestock complex as a solution to the problem of the greenhouse effect and the way to obtain a source of energy, *Science. Technique. Technologies (Polytechnic Bulletin)* **2,** 219–227 (2016)
10. E. V. Klimova, Saving agriculture, greenhouse effect, reduction of CO_2 emissions and protection of the environment: politics and prospects (Soil protection agriculture) *Ecological safety in APK, Abstract journal* **2,** 914 (2006)
11. LNG will save the world from the greenhouse effect *AvtoGasRefueling Complex + Alternative Fuel* **4 (64),**080–82 (2012)
12. O. K. Bezyukov, V. A. Zhukov, Thermal engines and greenhouse effect, *Bulletin of Mechanical Engineering* **F,** 84–88 (2013)
13. G. G. Galeev, To substantiate the needs of peasant and small farms in tractors, *Bulletin of Kazan State Agrarian University* **2(4i),** 106–107 (2010)
14. G. G. Galeev, To the calculation of transport support for harvesting units in the agro-industrial complex, *Bulletin of Kazan State Agrarian University* **3(84),** 75–77 (2011)
15. F. Kh. Khaliullin, Estimation of design parameters of the crank-connecting rod mechanism of engines for mobile agricultural machines, *BIO Web of Conferences* **4F,**00076 (2021)
16. I. G. Gainutdinov, Beet production efficiency and ways to increase it in case of negative market conditions in the commodity market, *BIO Web of Conferences* **4F,** 00108 (2020)
17. I. G. Gainutdinov, R. R. Gadelshin, Prospects for the development of production and processing of

potatoes in Russia, *Vector of economy* **6 0(2i)**, 30 (2020)

18. I. G. Galiev, A. A. Mukhametshin, I. R. Iskhakov, A. R. Shamsutdinov, Increasing the efficiency of tractors in modern conditions, *Bulletin of Kazan State Agrarian University* **4(48),0**169–172 (2009)
19. M. N. Kalimullin, R. K. Abdrakhmanov, I. G. Galiev, Perfection of potato cultivation technology *Technics and Equipment for the Village* **2,** 6–9 (2017)
20. D. Khaliullin, A. Belinsky, A. Valiev, R. Lukmanov, Optimization of plow adjustment and Gaston Bourges, *BIO Web of Conferences* **8F,** 00103 (2020).
21. K. A. Khafizov, Reduction of total energy costs on technological operations in the agro-industrial complex - the way to reduce greenhouse gas emissions in the atmosphere, *Bulletin of Kazan State Agrarian University* **3(63),** 43–47 (2021)
22. K. A. Khafizov, System mathematical model of vehicles according to the optimization criterion - minimum emission of carbon dioxide into the atmosphere, Dynamics of mechanical systems: *materials of the II International Scientific and Practical Conference dedicated to the memory of Professor A. K. Yuldashev, Kazan State Agrarian University,* pp 122–130 (2021)
23. D. A. Yakovlev, *Energy assessment of coulters during the operation of sowing units in conditions of different soil moisture of the steppe zone of Siberia*: PhD Dissertation (Barnaul, 2020)
24. R. I. Safin et al., Control of soil over compaction in resource-saving agriculture, *Methodical recommendations Kazan State Agrarian University* **2i** (2018)
25. C. Khafizov, R. Khafizov, A. Nurmiev, R. Usenkov, Optimization of main parameters of tractor and unit for deep processing of soil according to criterion - total energy costs, *Engineering for rural development*, pp 603–608 (2020)
26. Tractor Belarus 3522.5 – Operating instructions (3522.5-0000010 RE) (2011)

Development and theoretical study of the impact of the working body on the soil

Marat Kalimullin[1,*], *Marat* Salimzyanov[2], *Vladimir* Pervushin[2], *Rafkat* Latypov[3], *Nikolay* Ayugin[4], and *Renat* Sharipov[2]

[1] Kazan State Agrarian University, 65, Karl Marx St., Kazan, Republic of Tatarstan, 420015, Russian Federation
[2] Izhevsk State Agricultural Academy, 11, Studencheskaya St., Izhevsk, Udmurt republic, 426069, Russian Federation
[3] South Ural State Agrarian University, 13, Gagarin St., Troitsk, Chelyabinsk, 454080, Russian Federation
[4] Ulyanovsk State Agrarian University named after P. A. Stolypin, 1, Novy Venets Boulevard, Ulyanovsk, 432017, Russian Federation

Abstract. The article presents the theoretical results of the performance indicators of the defoliator. They depend on the parameters and modes of the defoliator operation. It operates on the basis of the description of the justification, which is provided in the work. The combined defoliator is aimed at removing potato tops with working bodies. They are multiple in number and they combine the consistency of their rotation including their vertical axis. Moreover, with the horizontal axis they are in the zone of their interaction without touching each other.

* Corresponding author: marat-kmn@yandex.ru

1 Introduction

Experience in extensive mechanized harvesting, accumulated in previous years, revealed the effectiveness of preharvest removal of potato haulm. It has the following advantages. It prevents clogging the working bodies of potato harvesters when harvesting potatoes with strongly twisted haulm, weeds. This reduces downtime of machinery and facilitates their work. Besides, preliminary removal of haulm increases the physiological maturity of tubers, their resistance to mechanical damages during harvesting, transportation, post-harvesting. It also preserves them in storages. Therefore, the effectiveness of preharvest haulm removal is proven and we should go in this direction [1–3].

In the world practice, mainly crushing and shearing types of working bodies are used. Rotary working bodies are mainly used as a working body. They are simple, reliable in operation, have high productivity. Let us consider machines in more detail with respect to the location of the axis of rotation of the working body [4–6].

Rotary crushers with the horizontal axis of rotation are a rotor (drum) with pivotally mounted elements: knives, bits, cables or chains. The length of the elements corresponds to the transverse profile of ridges or beds, which makes it possible to capture dead haulm in between the rows. The disadvantages are the high cost, energy intensity and metal consumption of the device design, which requires a special and high accuracy of balancing the rotor. Manufacturers present machines for removing or crushing potato haulm. They are domestic factories OAO Ryazselmash, BD-2M, BD-4, UMVK-1.4, LLC Agrotehmash UMVK-1.4, Kolnag-Rafale 2x75, Rafale 4x90, Belarusian POOO Techmash MBU-2.8, MBU-3.6, CJSC AgropromSelMash BTU-1.4 and foreign machines Rumptstad RSK (the Netherlands) 2000. There are also GrimmeKS 3000 (Germany); Miedema LKFA (the Netherlands); UN 3604 Kverneland (UK); Botwetcher LKB320 Baselier (Slovakia) and OBN17 MachinesPioro (Poland) [7–9].

Vertical axes of rotation scrapers include working bodies with knives, chains or other scraping elements fixed radially on the periphery and rotating in a horizontal plane. The advantage of these haulm pickers is simplicity of design, the disadvantage is relatively low percentage of haulm removal (50...60%), especially when breaking fallen haulm in inter-row areas. These are represented by the Polish 4-row RLZ-4 and its analogues, the Russian and Belarusian D-POL 2-row devices [10–12].

In order to eliminate the above-mentioned drawbacks, a combined bot-vibrator is proposed. It combines the consistency of rotation of working bodies and their number of both the vertical axis and the horizontal axis in the area of their work.

The purpose of the research is increasing the completeness of haulm removal by justification of parameters and modes of the operation of the combined haulm trimmer.

Based on the above information, and taking into account the existing real need to find the most optimal parameters of the working bodies, the main task of this work is to ensure an increase in the efficiency of the process associated with planting potatoes. For this, it is necessary to improve the existing technologies and design of the working bodies of the device for planting potatoes. In addition, the proposed design changes and operating modes are substantiated.

2 Materials and methods

Having analyzed the work of toppers in the field and works on rotary tools and machines by researchers V. P. Goryachkin, E. M. Gutyar, A. Yu Ishlinsky, V. A. Zheligovsky, I. F. Vasilenko, V. J. Kallus, E. S. Bosogo, N. E. Reznik, Matyashin Yu. I. and others, we found that the substantiation of the parameters of working tools proceeds from the consideration of the following conditions [13, 14].

The trajectory of any circular motion of the knives is expressed by parametric equations, which gives us a graphical analysis to represent them as elongated cycloids with a kinematic index λ>1:

$$x = V_M \cdot t + r_i \cdot \sin(\omega t),$$

$$y = r_i \cdot \cos(\omega t),$$

$$z = r_i \cdot \cos(\omega t),$$

where V_M is the forward speed of the machine, m/s; t is time, s; r_i is the blade radius, m; ω is the angular velocity of the rotor, s^{-1}.

Working tools or elements destroy the haulm with a non-supported cut (14...34 m/s, depending on the plant diameter), which can be provided by the critical cutting speed with the kinematic mode parameter λ>1 [1, 3–4]:

$$\lambda_i = \frac{V_{pi}}{V_M} = \frac{(\omega \cdot r_i)}{V_M}$$

where V_{pi} is the linear (circumferential) speed of knives, blades with radius (r... R), m s^{-1}.

The main indicators evaluating the quality of the haulm chopper are the completeness of haulm removal, which is characterized by the minimum height of the remaining haulm petioles h_{cr} and the minimum average cutting length l_{cr}.

Theoretically, the residual height of the haulm is the sum of the installation height of the tool h in relation to the average soil surface and the scallop height. (This is the difference between the height of the stem cut and the height of the blade in the lower position). It can be found from the absolute trajectory of the blade using the previous expressions (1-3) and after transformation we can write the following dependence:

* Corresponding author: marat-kmn@yandex.ru

$$h_{cr} = h + r - \sqrt{r^2 - \frac{V_m^2 \times \pi^2 \times r^2}{m^2 (V_m + V_n)^2}}$$

Theoretically, the length of the cut is characterized by its average value of l_{cr}. It can be determined through the coordinates of the beginning of the stem of the first knife and the next knife, taking into account the rotor rotation and translational speed using the previous expressions (1–3). And after such transformations, the following dependencies are obtained:

$$V_m = \frac{V_m(\arccos\frac{l_{cp}}{r} + \frac{2\pi}{m} - \frac{\pi}{2})}{1 - \sin(\arccos\frac{l_{cp}}{r})}.$$

where m is the number of knives rotating at a given radius, pcs.

The equation expressed through the translational velocity of the machine has the following view:

$$V_m = \frac{V_m(1 - \sin(\arccos\frac{l_{cp}}{r}))}{\arccos\frac{l_{cp}}{r} + \frac{2\pi}{m} - \frac{\pi}{2}}$$

where

$$B = \sin\alpha \cdot \sin\beta \cdot \sin(\omega t - \gamma_i) + \cos\alpha \cdot \cos(\omega t - \gamma_i).$$

Expressions (5-7) allow you to comply with the operating mode of the working body of the haulm chopper depending on the residual height of the haulm, the height of the average length of the cut stems on agricultural requirements. That is it is allowed to determine the maximum speed of the machine or the required circumferential speed of knives and their number.

Based on the above, we presented a scheme of the 2 row new combined botworms with the working organs of rotation in both the horizontal and vertical planes (Figure 1) [15].

The proposed topper has a frame 1 resting on two pneumatic support wheels 2. In the front part of the machine there is a connecting device (suspension) 3 (Figure 3) to the tractor. Behind it on the frame 1 there are three bevel gears 4, 5, 6, connected with each other by chain couplings 7. On the shafts of bevel gearboxes there are working tools with horizontal 8 and vertical 9 axes of rotation [16–18].

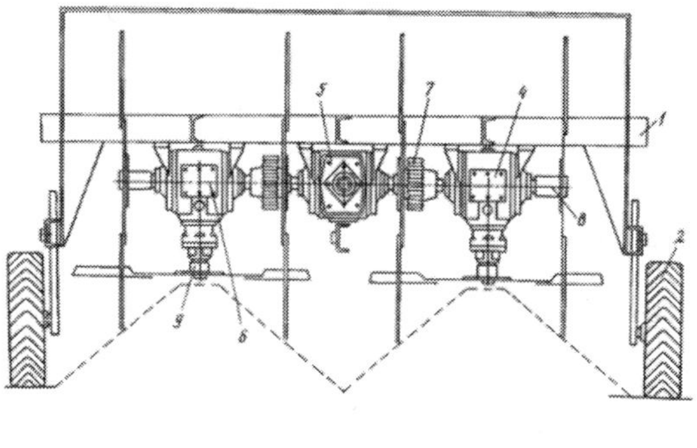

Fig. 1. Front view of the combined rotary crusher.

Working elements 8 on the horizontal axis of rotation in Figures 1, 2, 3 in the design version include a hub 10 (Figure 2) with a disc fixed to it 11, having knives 12 with additional working elements 13 in the form of elastic rods (scourers – haulm elevators).

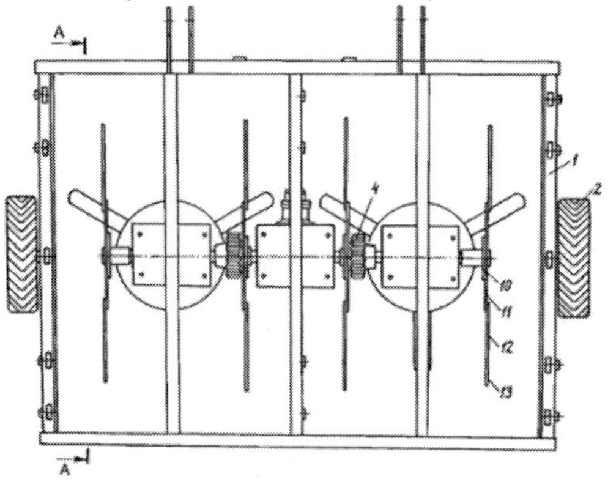

Fig. 2. Top view of the combined rotary topper.

Actuators 9 on the vertical axis of rotation are mounted on the secondary shafts 4, 6 and also include a hub 14 (Figure 3) with a disk attached to it 15 having knives 16 turned to the plane of rotation at an acute angle to reduce the energy intensity of the cut [2].

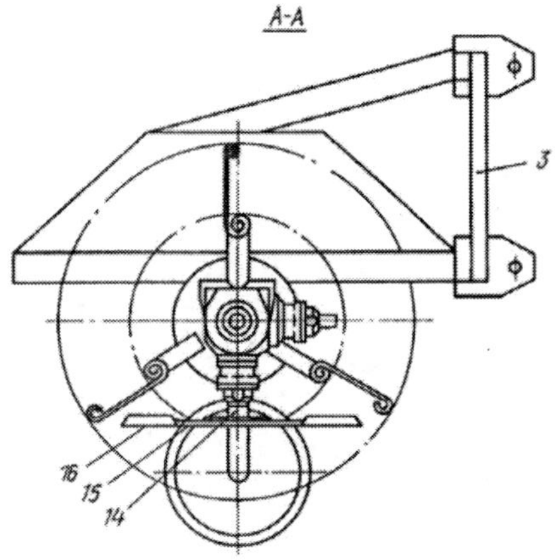

Fig. 3. Left view of the combined rotary topper.

 https://doi.org/10.1051/bioconf/20225200056

3 Results and discussion

In Figure 4 there are results of calculations of expressions (6–7) and dependences of the circular blade speed at various forward speeds of the machine within the limits of agricultural technical requirements. And there are dependences of the forward speed of the machine movement at circular blade speed V_m = 14; 20; 26 m/s on average cutting length of stems (0.05…0.1 m) in the number of knives m = 2 and 3 pieces.

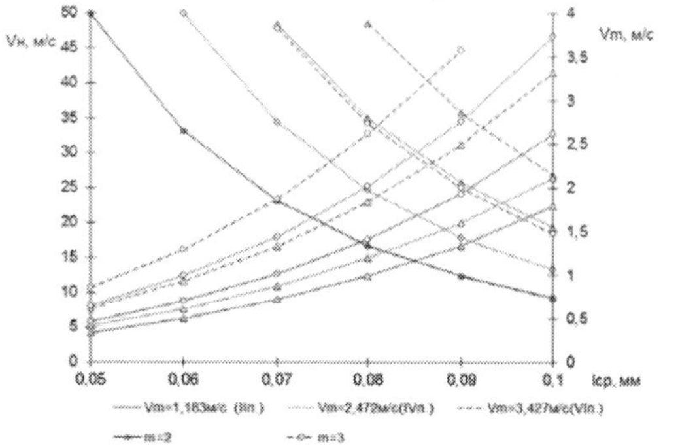

Fig. 4. Dependence of the peripheral speed of the knife and progressive speed of the machine on the average length of stalk cutting with the number of knives m = 2.

We analyzed this graph of the calculation results of expressions (6–7) for the Botvodcher and used the progressive speed of the machine v_m = 1.43...2.55 m/s and rotor revolutions n = 1365...1775 min-1. Then we solved the problems of assessing performance Botvodbiker and its parameters and modes of operation. They included the height of petioles cut h_{cr} = 0.02... 0.08 m, cutting length L_{sr} = 0.5 ... 0.10 m, the number of knives, blades m = 3, the radius of small knives r = 0.25 ...0.3 m, the radius of large beats R = 0.45...0.5 m.

Based on these calculations, we made the rotary topper with multiple working bodies and combined the consistency of their rotation, both with the vertical axis and with the horizontal axis in the area of their interaction without contacting with each other (Figure 5, 6).

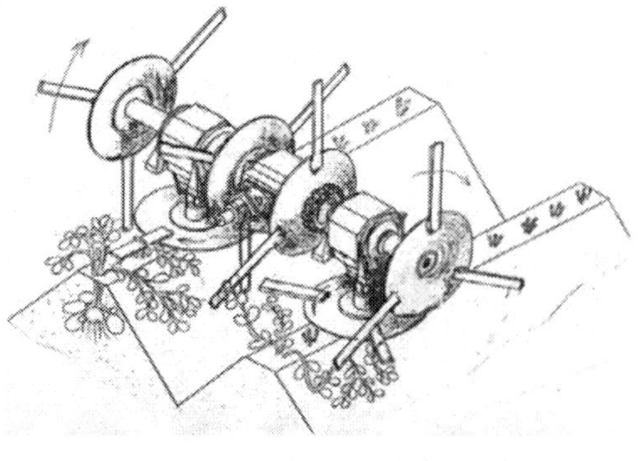

Fig. 5. Scheme of the technological process of removing the haulm (movement of the machine to us).

Fig. 6. Experimental combined rotary crusher.

During the forward movement of the machine working bodies on a horizontal axis of rotation due to elastic whips-bottom lifters, we pick up and lift the fallen haulm from the inter-row with the subsequent crushing of haulm by knives of working bodies. This includes both horizontal and vertical rotation planes, strictly coordinated with their movement. Working bodies on the vertical axis of rotation undercut the main mass of the haulm located along the ridges. Knives with a horizontal plane of rotation, turned to the latter by an acute angle, create a vacuum under their plane of rotation, thus contributing to the supply of stems to the blades of the haulm harvester [19].

4 Conclusion

As a result, the analytical dependences evaluating the work quality of the haulm crusher as residual haulm v_{cr} and average cutting length l_{cr} on the forward speed of the haulm crusher v_m. The peripheral speed of knives v_n, the radius of knives r and their number m were obtained. With the translational velocity of the machine v_m = 1.43...2.55 m/s and the rotor speed n = 1365...1775 min-1, the cutting height of petioles is equal to h_{sr} = 0.02...0.08 m, the length of cutting is l_{sr} = 0.5 ... 0.1 m.

We made a 2-row experimental combined crusher haulm with the following parameters and operating modes. The number of working elements revolutions was up to 1500 min-1, the number of knives on the rotor in the horizontal plane was 3 pieces. The number of the beech-beater was 3 pieces, the radius of knives in the horizontal plane was 0,3 m, the radius of the beech in the vertical plane was 0.5 m.

Acknowledgements

The study is supported by the grant of the President of the Russian Federation for young scientists-doctors of sciences MD-6281.2021.4.

References

1. V. Moskalevich, MOTROL, **11B**, 179–189 (2009)

2. I. Ahmadi, *Soil and Till. Res.*, **171,** 1–8 (2017)
3. O. A. Fouda, *J. Soil Sci. and Agric. Eng. Mansoura Univ.*, **7,** 929–936 (2016)
4. K. Verbist, W. M. Cornelis, W. Schiettecatte, G. Oltenfreiter, M. Van Meirvenne, D. Gabriels, *Engl. Soil Till.*, **96,** 292–302 (2007)
5. P. Borowski, M. Klimkiewicz, M. Powałka, *WEMA, Warsaw Univ. of Life Sci.*, **133** (2010)
6. A. Srivastava, E. Carroll, P. Roger, R. Dennis, *American Society of Agricultural and Biological Engineers,* pp 169–230 (2006)
7. I. Mukhametshin, A. Valiev, F. Muhamadyarov, M. Kalimullin, F. Yarullin, Kinematic analysis of conical rotary subsoil loosener for tillage, *19th Int. Sci. Conf. Engineering For Rural Development Proceedings,* (Latvia Univ. of Life Sci. and Technol., Jelgava, 20–22 May) **19,** pp 1946–1952 (2020)
8. V. P. Mazyarov, V. I. Medvedev, G. Z Gaifullin, L. P. Shershevsky, publ. Auth. St., no. SU 1373336 (1986)
9. Deep diggers give a high lift to yields, *Farmers Weekly* **115,** 42 (1991)
10. G. Sabourin, J. Richardeau, Patent FR, no. 9115937 (1993)
11. V. P. Mazyarov, V. I. Medvedev, *The sub-cover ripper significantly reduces energy consumption during tillage* (2013), retrieved from: http://www.polytech21.ru/news/1846-podpokrovnyj-rykhlitel-znachitelno-snizhaet-energozatraty-pri-pochvoobrabotke
12. A. P. Akimov, Yu. V. Konstantinov, *Tract. and Agricult. Mach.*, **12,** 21–23 (2015)
13. S. A. Sidorov, V. K. Horoshenkov, D. A. Mironov, E. S. Luzhnova, *Tract. and Agricult. Mach.*, **8,** 30–32 (2016)
14. A. Valiev, F. Muhamadyarov, *15th Int. Sci. Conf. Engineering for Rural Development Proc.*, (Latvia Univ. of Agricult.) **15,** pp 1378–1385 (2016)
15. M. N. Letoshnev, *Agricultural Machines. Agricultural Equipment. Theory, Calculation, Design and Testing* (Moscow, 1955)
16. I. S. Mukhametshin, A. R. Valiev, A. V. Aleshkin, R. I. Ibyatov, *Bull. of Ulyanovsk State Agricult. Acad.* **4,** 50–182 (2018)
17. I. S. Mukhametshin, A. R. Valiev, P. I Makarov, Patent of RF, no. RU 2522320 (2013)
18. N. N. Nikitin, *The Course of Theoretical Mechanics* (Moscow: Higher School, 1955)
19. M. Kalimullin, D. Ismagilov, R. Abdrakhmanov, M. Salimzyanov, R. Latypov, *19th Int. Sci. Conf. Engineering For Rural Development Proceedings,* (Latvia Univ. of Life Sci. and Technol., Jelgava, 20–22 May) **19,** pp 1224–1229 (2020)

Influence of woody vegetation on physico-chemical properties of sod-podzolic soils at Lesnaya Opytnaya Dacha of RSAU-MTAA named after K.A. Timiryazev

Vladimir Naumov, *Natalia* Kamennykh, *Kristina* Shmakova, *Alexey* Polyakov, and *Vasily* Rashkovich*

Russian State Agrarian University – MTAA named after K.A. Timiryazev, 49, Timiryazevskaya St., Moscow, 127434, Russian Federation

Abstract. A research of the dependence of Lesnaya Opytnaya Dacha (LOD) soil properties on the varieties of the woody vegetation grown on these soils has been conducted. Measurements of the humus content, hydrolytic soil acidity, saline soil acidity and the sum of exchange base cations were used as main soil properties. Different types of vegetation were studied, including pure deciduous and pure coniferous stands and their mixtures in different proportions. Some controversial data was acquired, according to which, it appears that deciduous stands and coniferous stands do not have a statistically significant impact on the studied soil properties.

1 Introduction

V. Dokuchaev, the founder of genetic soil science, pointed out the dependence of soil properties on the type of forest vegetation [1, 2]. In his works, he noted that in the same climatic conditions, different soils are formed under different tree species and supported this hypothesis with numerous examples. He pointed out the importance of studying the influence of individual tree species on the soil [3].

G. Morozov [4] noted that forestry is interested in the subject of plantings influence on soils. A forestry specialist has always known how to use the synergy between the forest canopy and the generated litter in its favor to preserve the forests' fertility and to facilitate the emergence of the undergrowth.

A significant number of works on the importance of the influence of individual tree species on soil properties have been published (M. Tkachenko, 1939 [5]; N. Remezov, 1953 [6]; S. Sonn, 1954 [7]; B. Zaitsev, 1964 [8]; G. Demin, 1997 [9]; L. Karpachevsky, 1997 [10]; K. Gavrilov, 2000 [11], etc.).

Currently, the forest vegetation influence on the soil is assessed in many different ways. L. Karpachevsky and M. Stroganova [12] raise the question of what is primary: changes in soil due to the influence of vegetation, or differentiation of vegetation due to the soil properties. At the same time, L. Karpachevsky [13] identifies two aspects of studying the influence of tree species on soil formation: 1) the influence of plants in pure plantings; 2) the total effect in mixed plantings.

In their works, K. Gavrilov [11] and G. Demin [9] have confirmed that there would be significant differences in soil chemistry and morphology under different types of forestry vegetation, all other factors being equal (e.g.: climate, landform, soil genesis, soil texture, etc.). It is known that each tree species has its own requirements for soil conditions and, above all, there is the moisture-to-air ratio in the soil and the amount of available nutrients.

2 Materials and methods

A. Pozdnyakov et al. [14] have noted that the soil is an inhomogeneous natural formation, characterized by both temporal and spatial variabilities. Anisotropy is common at all levels of soil organization, yet it still has to be studied further.

Through their works, N. Lukina et al. [15] indicate that trees should be regarded as ecosystem engineers, since they form the horizontal mosaic structure of forests.

According to A. Gennadiev et al. [16], M. Gerasimova and L. Isachenkov [17], certain soils correspond to each stage of the succession of woody vegetation, which differ primarily in the upper horizons.

* Corresponding author: rashkovich@rgau-msha.ru

V. Karminov [18] notes that despite soil characteristics being rather consistent in the short run, age-related changes in forestry vegetation noticeably change properties of the upper horizons of soils.

M. Gerasimova et al. [17], L.O. Karpachevsky [10], Yu. Demakov et al. [19], V.D. Naumov et al. [20] note that the current soil formation processes are affected by the woody vegetation. The dynamic soil properties are as follows: pH, cation exchange capacity, biogenic accumulation of ash constituents, etc. Alongside with these changes, heights of genetic horizons change too.

A review of available literature shows that a number of opinions regarding the effects of vegetation on soils exist. It is still debatable whether woody plants change soil properties or vice versa – soil properties determine the composition of woody vegetation. Despite a lot of conducted research, the relationship between forest vegetation and soils is still evaluated insufficiently. The obtained data is ambiguous and controversial, which has served as the reason for conducting this research.

The goal of this research is to study the influence of woody plants of various compositions on the physico-chemical properties of the upper horizons of sod-podzolic soils at the Lesnaya Opytnaya Dacha (LOD) station of the Russian State Agricultural Academy named after K.A. Timiryazev.

The study was conducted at the Lesnaya Opytnaya Dacha (Experimental Forest) station at the RSAU-MTAA n. a. K. A. Timiryazev. LOD was founded in 1862 and is used to this day as a native scientific laboratory, being the oldest of its kind in Europe. During its existence, LOD was subjected to continuous monitoring and extensive research. The forest in the area is presented by both native and artificially planted parts, where some trees are over 300 years old. Considering the age, LOD has an immense scientific and industrial significance, as it provides an opportunity to conduct long-term research and data aggregation that excludes errors of short-term research.

LOD is situated at the north-west part of Moscow, being the southmost part of RSAU university. Its shape resembles a rectangle, being approximately 1.6km wide and 2.8km long. According to soil-geographic zoning, LOD is placed at the European-North-Siberian taiga-forest continental area, in a southern-taiga sod-podzolic zone.

From the geological perspective, LOD finds itself in a Central-Moscow structural-geomorphological block, which is a gently sloping dome-like elevated area. It is worth noting that LOD has none of the upper Pleistocene cover loam, which is very commonly present through the city of Moscow [20].

The relief of the LOD territory is moraine-plain. Soil-forming rocks are represented by moraine deposits of various sizes.

3 Results and discussion

The LOD is located on a flat watershed moraine hill with a very gentle slope to the southwest (on average, 0.01 m falls per 1 linear meter) and with a steeper slope to the northeast. The highest part of this hill is 175 m above sea level and is located in the middle between the VII and XI quarters (Fig. 1). The lowest part is 160 m above sea level and is located in the northern part of the LOD on the territory of the I and III quarters. The relief is represented by a combination of Middle Pleistocene, Late Middle Pleistocene and Late Pleistocene-Holocene genetic types of surfaces that were formed as a result of the processes of accumulation of glacier deposits and flows of thawed glacial waters.

During LOD organization in 1892 by Vargas de Bardemar, a famous forester, it was split in 14 quarters, represented by roman numbers in fig. 1.

The soil cover of the LOD is represented by sod-podzolic soils, differing in the manifestation of sod, podzolic and gley processes. The taxation-forestry and soil survey of the quarters [20] showed that all the quarters have a different composition of stands.

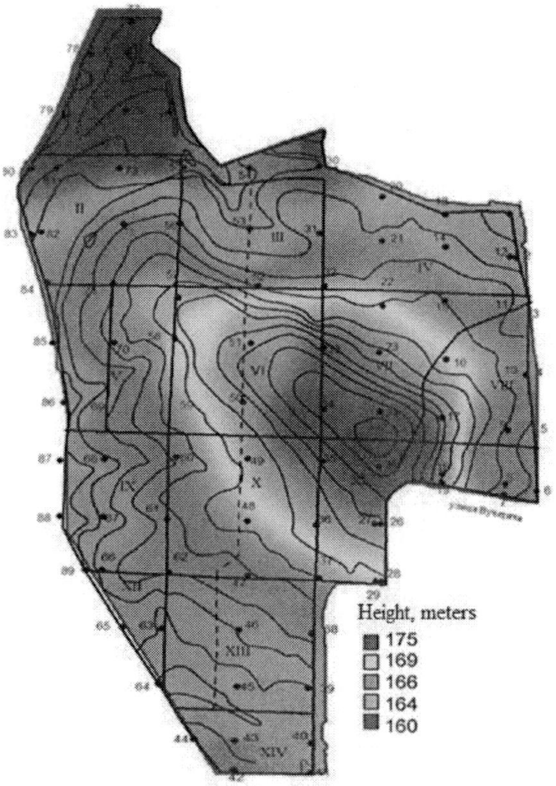

Fig. 1. LOD elevation map.

The relief of the III quarter is represented in the north and east of the site by the subhorizontal surface of the water-glacial plain, the central part is occupied by inclined surfaces of valley-like depressions, in the northwest there is a cirque–shaped depression, in the west - the slope of a moraine hill. The soil-forming rock is represented by moraine deposits of various textures (mainly medium loamy).

The soils are sod-podzolic, medium-fine, shallow, low-power in terms of the thickness of the humus horizon; in terms of the intensity of the manifestation of the podzolic process – strongly and medium-podzolic; in terms of the depth of the podzolization – deep and ultra-deep podzolic. The texture is light and medium loamy. Signs of gley-building were detected on 4 test areas: E, 6, 1, and 2. According to the depth of gleying, profile-gleyed soils are marked under mixed plantings with a predominance of conifers located on the inclined surfaces of the circus depression, superficially–gleyed soils are confined to the area of the valley-like depression.

Vegetation is mainly represented by mixed stands. For sample areas 6 and 1, vegetation is mixed with a predominance of coniferous species mixed vegetation with a predominance of deciduous species for areas 5 and 2; for area E – pure coniferous stands and area 4 – pure deciduous stands.

The relief of the IV quarter is represented by the Middle Pleistocene surfaces, mainly the subhorizontal surfaces of the water-glacial plain. Accumulative-erosive inclined surfaces of valley-like depressions are present in the southwest and northwest. The soil-forming rock is represented by moraine deposits of various texture (mainly light loamy).

The soils are sod-podzolic, medium-fine, shallow and extremely shallow in terms of the thickness of the humus horizon; in terms of the intensity of the podzolic process – strongly podzolic; in terms of the depth of podzolization – deep and ultra-deep podzolic. According to the texture, they are light and medium loamy. Signs of gley forming were revealed on 4 test areas: 2*, III, E and B. According to the depth of gley forming, the soils are mainly profiled, deeply gleyed and confined to the area of valley-like depression.

Vegetation is mainly represented by mixed stands with a predominance of coniferous species on test areas 2* and Ъ. Test areas E and B are represented by pure coniferous stands.

The relief of the V quarter is represented by the subhorizontal top of a moraine hill in the northeast of the site, turning into a northwestern slope, to the south, accumulative-erosive inclined surfaces of the valley-like depressions stretch from west to east. The soil-forming rock is represented by moraine deposits of various textures (mainly light loamy).

The soils are sod-podzolic, medium–fine and shallow in terms of the thickness of the humus horizon; strongly and medium-podzolic in terms of the intensity of the manifestation of the podzolic process; deep and ultra–deep podzolic in terms of the depth of podzolization. Texture wise, they are light loamy. Signs of gleybuilding were detected on all test areas. According to the depth of the gleying, the soils are profile gleyed.

Vegetation is mainly represented by mixed stands with a predominance of coniferous species on trial areas M4 and У. For the trial area Ж and section 85, pure coniferous stands of origin artificial were present.

The main territory of the VI quarter is situated on a subhorizontal peak and the slopes of a moraine hill. The soil-forming rock is represented by moraine deposits of various texture (mainly medium loamy).

The soils are sod-podzolic, mainly medium–shallow and partially shallow according to the thickness of the humus horizon; strongly and medium-podzolic according to the intensity of the manifestation of the podzolic process; deep and ultra–deep podzolic according to the depth of the podzolization. Texture wise soils are light loamy. Signs of gleying were detected on all test areas, except 13. According to the depth of gleying, the soils are mainly profiled.

Vegetation is mainly represented by mixed stands with a predominance of coniferous species on test areas П, B, 15. Pure deciduous stands are represented on areas Ъ and 13. Single pure coniferous stands and mixed stands, with a predominance of deciduous, are on areas E and C, respectively. Everything is of artificial origin.

The territory of the VII quarter is represented by subhorizontal peaks and a moraine hill slope. A suffusion depression is located in the central part of the site. A narrow sloping surface of a valley-like depression stretches from the north-eastern corner. The soil-forming rock is represented by moraine deposits of various textures (predominantly light loamy).

The soils are sod-podzolic, mainly medium–shallow in terms of the thickness of the humus horizon, mainly medium-podzolic in terms of the intensity of the manifestation of the podzolic process, deep- and ultra-deep podzolic in terms of the depth of the podzolization. According to the texture, light loamy soils predominate. Signs of gley building were revealed in the trial areas E and K. According to the depth of gleying, the soils are deeply and profile gleyed.

Vegetation is mainly represented by mixed plantings with a predominance of conifers on test areas E, C, K. Pure coniferous stands dominate on the areas П and P. All stands of artificial origin.

The territory of the VIII quarter consists of subhorizontal surfaces turning into the slopes of a moraine hill, with inclined and horizontal surfaces of valley-like depression stretching from the north-eastern part of the quarter. The soil-forming rock is represented by moraine deposits of various textures (mainly light loamy).

The soils are sod-podzolic, medium–fine and shallow in terms of the thickness of the humus horizon; mainly medium-podzolic in terms of the intensity of the podzolic process; deep- and ultra-deep podzolic in terms of the depth of podzolization. Soil texture is light loamy. Signs of gley building were revealed on the test areas T1, П and O. According to the depth of the soil gleying, soils on test areas T1 and П are deeply gleyed and profile gleyed on the test area O.

Vegetation is mainly represented by mixed stands with a predominance of deciduous on test areas T1 and H. Pure deciduous stands are found on areas 3e, 3 and O. The test area П is characterized by mixed stands with predominance of conifers. Stands on areas 3, O and H are of natural origin, whereas П is artificial.

The relief of the X quarter is represented by the horizontal surface of the stone terraces bordered from the northeast by the slope of the moraine hill. From west to east, the inclined and horizontal surfaces of valley-like depressions extend, in the south there is a suffusion depression. The soil-forming rock is represented by moraine deposits of various textures (mainly heavy loamy).

The soils are sod-podzolic, medium–shallow in terms of the thickness of the humus horizon; medium–podzolic in terms of the intensity of the podzolic process; deep- and ultra-deep-podzolic in terms of the depth of the podzolization. The texture is light loamy and sandy loam. The signs of gleying are found in the test area T soil profiles – deeply gleyed soils.

Vegetation is represented by mixed stands with a predominance of coniferous species on test areas 3 and 1. Single pure coniferous stands are represented on the test area T, mixed with deciduous on test area 4. Through the quarter, except for the test area T, stands have artificial origin.

The territory of the XI quarter is confined to a subhorizontal top of the moraine hill, passing into the slopes to the west, in the westernmost part of the site there is a subhorizontal surface of the water-glacial plain. In the southern part of the quarter, there is a suffusion depression and a horizontal surface of the stone terraces. The soil-forming rock is represented by moraine deposits of various textures (mainly light loamy).

The soils are sod-podzolic, medium-small and low-power in terms of the thickness of the humus horizon, occasionally extremely small; mainly medium-podzolic, partially - strongly podzolic and shallow-podzolic in depth in terms of intensity of the manifestation of the podzolic process. Texture wise soils are light loamy and sandy loam. Signs of gleying were noted in all test areas, except for K. According to the depth of gleying, the soils on the test areas E and M are deeply gleyed, the areas 3 and Г* are superficially gleyed, the areas B2 and G are profile gleyed.

Vegetation is mainly represented by pure coniferous stands (trial areas K, Г and Г*), mixed with deciduous on test areas 3 and M. Single pure deciduous stands are represented on the test area E, mixed with coniferous species on the test area B2. The origin of stands in most cases is artificial (B2, K, Г*, Г); stands on areas E and M are of natural origin.

The territory of the XII quarter is a subhorizontal surface of the water-glacial plain in the west, passing into the north-western slope of the moraine hill. In the north-eastern part of the site, a part of a horizontal surface of the stone terraces is marked. In the east there is a suffusion depression. The southernmost edge of the quarter is represented by an inclined surface of a valley-like depression. The soil-forming rock is represented by heavy loamy moraine deposits.

The soils are sod-podzolic, medium-shallow and low-power in terms of the thickness of the humus horizon; strongly podzolic in terms of the intensity of the manifestation of the podzolic process; deep podzolic in terms of the depth of the podzolization. The texture of soils is light and medium loamy. Signs of gleying were detected on test area 11. According to the depth of the gleying, the soils are deeply gleyed.

Vegetation is represented by mixed stands with a predominance of deciduous plants. The origin of the stands is not specified.

The main area of the XIII quarter is occupied by the subhorizontal surface of the glacial plain, located in the central part of the site. In the south-western part, the presence of inclined and horizontal surfaces of valley-like depression was noted. The north-eastern part is distinguished by the slope of the moraine hill passing into the horizontal surface of the kama terraces. The soil-forming rock is represented by moraine deposits of various textures (mainly medium loamy).

The soils are sod-podzolic, medium–small in terms of the thickness of the humus horizon; strongly podzolic in terms of the intensity of the manifestation of the podzolic process; mainly ultra-deep podzolic in terms of the depth of the podzolization. The texture of the soil is light loamy. Signs of gley building were noted in all soils on the test areas. According to the depth of the gleying, the soils are deeply gleyed.

Vegetation is mainly represented by mixed stands with predominance of coniferous species (test areas 9, 7, 6). Single pure deciduous stands are represented on the test area 8, mixed with deciduous woods on the test area 5. The origin of the stands is not specified.

The territory of the XIV quarter is a subhorizontal surface of the water-glacial plain in the west. An inclined surface of a valley-like depression stretches from the south-east to the north. Moraine deposits of various granulometric compositions represent the soil-forming rock.

The soils are sod-podzolic, shallow in terms of the thickness of the humus horizon; strongly podzolic in terms of the intensity of the podzolic process; deep and ultra–deep podzolic in terms of podzolization. Soil texture is light and medium loamy. Signs of gleying were revealed on test areas 1 and 3. According to the depth of gleying, the soils are deeply glued.

Vegetation is represented by mixed stands with a predominance of deciduous woods. The origin of the stands is not specified.

To establish links between the physico-chemical properties of the upper horizons of sod-podzolic soils and stands of various species composition, plants were grouped. The composition of stands was chosen as a feature. 4 groups of stands were identified (pure coniferous, pure deciduous, mixed stands with a predominance of coniferous and mixed stands with a predominance of deciduous) (Tables 1 and 2).

Table 1. Test areas of pure stands grouped by wood composition.

Coniferous			Deciduous		
Quarter	Test area	Stands composition	Quarter	Test area	Stands composition
III	E	I-10C ed.B, Lp			
		II-7D3Kl ed.Lp,B			
IV	E	9C1Lp,Kl ed.B,D,B			9D1Lp
IV	B	I-9C1Lp ed. Kl	III	4	9B1Lp,D
		II-6Kl2Lp2D	VI	Ъ	9B1Kl
V	85	10L+Lp	VI	13	10Lp
V	Ж	10L ed.D	VIII	3	I-10D ed.Lp
VI	E	9C1Kl ed.Lp	VIII	O	II-10Lp ed.Kl
VII	П	I-10L+B ed.B			9D1Lp ed.C,B
		II-6B3Kl1Lp+B	XI	E	10IVD
VII	P	10L+Lp	XIII	8	I-9Lp1D ed.C
X	T	10L ed.D,B	XIII	3e	II-6Lp3B1Kl
XI	Г*	10L+Lp ed.D			9Iv1D ed.
XI	K	10L+Lp	XIV	1	B,B,Kl,O
XI	Г	10L+Lp ed.D			

Pure coniferous stands are confined to the territory of the III quarter (E), IV quarter (E, B) V quarter (85, Ж), VI quarter (E), VII quarter (П, P), X quarter (T) and XI quarter (Г*, K, Г). The III, IV and VI quarters are characterized by the predominance of pine in stands with single inclusions of linden, maple and birch. The stands of other sites are dominated by larch, linden, oak with birch being singly marked. For test areas E of the III quarter, B of the IV quarter, П of the VII quarter, a distinctive feature is the presence of 2 tiers of mixed stands with a predominance of deciduous woods.

Pure deciduous stands are represented in the III quarter (4), VI quarter (Ъ, 13), VIII quarter (3, O), XII quarter (E), XIII quarter (8, 3e) and XIV quarter (1). In the VI quarter, the predominance of birch is in the composition of plantings with 10% admixture of linden, maple and oak. Oak is predominant on the III, VIII (O) and XI quarters. The VIII (3) and XIII (3e) quarters are distinguished by the predominance of linden in the composition of stands. Willow prevails in the stands of the XIII (8) and XIV quarters with an admixture of elm and oak. The second tier of stands, represented by mixed deciduous plantings, was identified in the VIII and XIII quarters.

Table 2. Test areas of mixed stands grouped by wood composition.

Mixed with coniferous predominance			Mixed with deciduous predominance		
Quarter	Test area	Stands composition	Quarter	Test area	Stands composition
			II	1	8B2Ded.Lp
			II	2	6B3D1Lp ed.OC
III	6	5C4B1B ed. Lp,K, O	II	3	5Lp4C1B ed.OC
III	1	6C4D ed.Lp, Kl, O	II	4	8D2C ed.Lp
IV	2*	5C3D2L+B,Lp, ed. E	III	5	7D3C
IV	Ъ	8C1Lp1D ed.B,Kl			ed. Lp,Kl,O
V	M4	5C5B ed.D,Lp	III	2	6D3Lp1C
V	У	6L3B1Kl ed.C,D			ed.Kl,O
VI	П	5C2B1Lp1B1D,Kl,E	IV	Ш	4C6B ed.B,Kl,D
VI	B	8C1Lp1D,Kl ed. B	V	B	4C6Lp
VI	15	5C5Lp ed.D,B	VI	C	5B3C1B1D ed.Kl
VII	E	I- 6L2Kl1C1Lp ed.B	VIII	T1	4Kl2L2B2D
		II- 8Kl1Lp1B	VIII	H	8D2Lp ed.C
VII	C	ed.B 8L2Lp	X	4	5B2Kl2B1D
VII	K	I- 8C2Kl ed.B			ed.C
		II- 10Kl ed. B,D	XI	B2	4C6Lp ed.B
VIII	П	5L5Bed.D,Kl,B,C,Lp	XI	3	5Kl4D1P
X	3	I- 7L3C ed.Lp,E,B	XI	M	4D5Lp1B
		II- 10Kl ed.Lp,B			ed.B,Kl
X	1	I- 8C1B1Lp ed.E,D	XII	10	7D3Lp ed. Kl,O
		II- 8Kl2Lp +B	XII	11	5B3D1Lp1Kl
XIII	9	5L4B1Lp ed. Kl, O			ed. O
XIII	7	6C3B1Kl,O	XIII	5	8B1Lp1KlO
XII	6	6L2C1Lp1Kl,O	XIV	3	8D2B ed. Kl,OB
			XIV	4	8D1C1Iv ed.DB

Mixed stands with a predominance of coniferous species are widespread throughout the territory of the LOD in VI, VII and XIII quarters. The predominant species is pine, less often larch. The admixture of deciduous plants is mainly represented by birch, oak and maple.

Mixed stands with a predominance of deciduous woods are also evenly distributed throughout the LOD. A greater number of such stands were noted on the II and XI quarters. The predominant species are oak and birch with linden and maple being less common.

For each group of stands, statistical processing of data from analyses of the upper horizons of sod-podzolic soils was performed, representing the following physico-chemical properties: humus content, hydrolytic acidity, sum of exchange bases and exchange acidity.

The humus content in the sod-podzolic soils of LOD ranges from 1.74 to 7.04%. For statistical processing of the obtained results, they were divided into 5 group soils according to the humus content (Fig. 2).

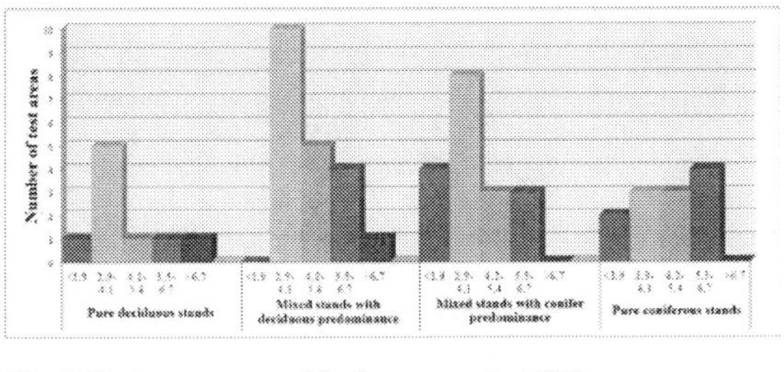

Fig. 2. Test areas grouped by humus content (%).

The first group included soils with a humus content of less than 2.9%, the second – 2.9-4.1%, the third – 4.2–5.4%, the fourth – 5.5–6.7%, the fifth – more than 6.7%. From Fig. 2 it can be seen that the maximum number of stands is in 3 groups: pure deciduous, mixed with a predominance of deciduous and mixed with a predominance of coniferous are confined to soil areas with a humus content of 2.9–4.1%. The humus content in sod-podzolic soils varies significantly, including soils under stands belonging to the same group by rock composition. The calculation of humus reserves in soils showed the following (Fig. 3). The least humus reserves were found in soils under purely coniferous and purely deciduous stands. The maximum reserves are under mixed stands with a predominance of deciduous species.

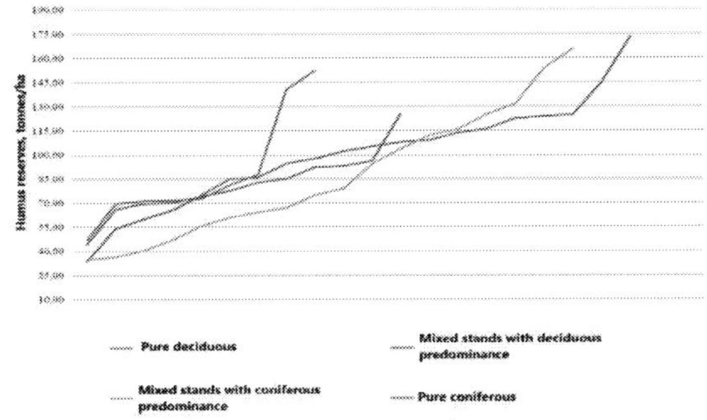

Fig. 3. Humus reserves in sod-podzolic soils of LOD.

The value of hydrolytic acidity (Ha) in sod-podzolic soils of LOD ranges from 1.50 to 17.06 mg-eq /100 g of soil. For statistical processing of the obtained data, five groups of soils were identified: the first group with an Ha value of less than 4.9 mg-eq/100 g of soil, the second – 4.9–8.3 mg-eq/100, the third – 8.4–11.8 mg-eq/100, the fourth – 11.9-15.3 mg-eq/100, the fifth is more than 15.3 mg-eq per 100 g of soil. Figure 4 shows fluctuations in the value of hydrolytic acidity in soils under stands of various compositions.

Soils under deciduous, mixed with a predominance of coniferous and pure coniferous, are mostly presented by group four, where the Ha value varies from 8.4 to 11.8 mg-eq/100g soil. Peak hydrolytic acidity was determined in soils under mixed stands with a predominance of conifers.

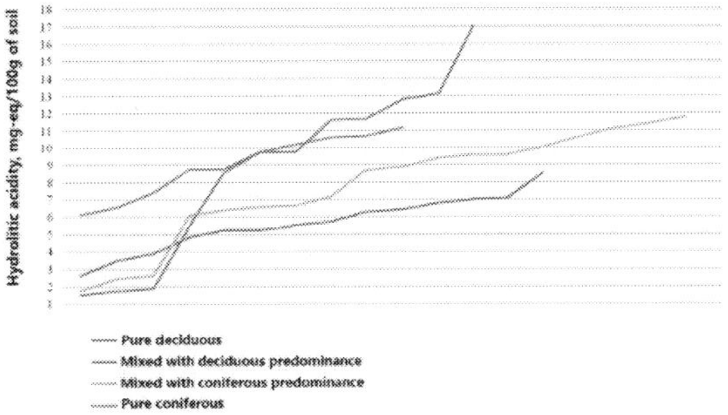

Fig. 4. Hydrolytic acidity of LOD sod-podzolic soils.

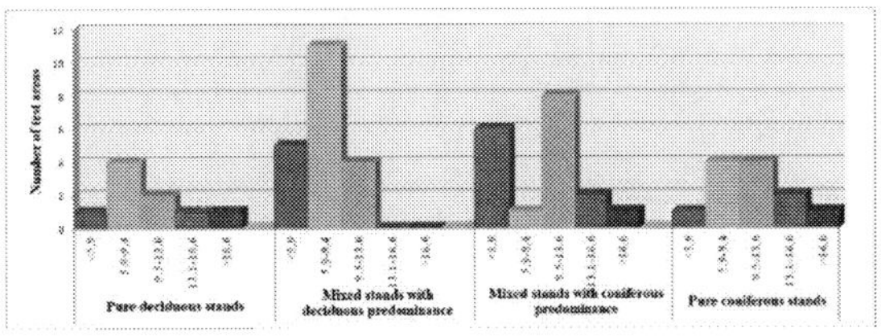

Fig. 5. Test areas grouped by sums of exchangeable base cations (mg-eq/100g).

The sum of exchange bases in sod-podzolic soils ranges from 2.40 to 18.21 mg-eq/100 of soil. For statistical processing of the obtained data, soils were split in 5 groups by the amount of exchangeable bases: less than 5.9; 5.9–9.4; 9.5–13.0; 13.1–16.6; more than 16.6 mg-eq/100 of soil (Fig. 5).

In soils under purely deciduous stands and mixed stands with a predominance of deciduous species are the maximum number of stands grows on soils with intervals of values of the sum of exchange bases of 5.9–9.4 mg-eq/100g soils. According to these values, a group of mixed with a predominance of deciduous stands is distinguished. A more uniform distribution of stands across different soil groups according to the content of the sum of exchange bases was revealed in mixed with a predominance of deciduous and mixed with a predominance of coniferous stands. The distribution of soils in the selected groups according to the sum of exchange bases under pure coniferous stands is close in nature to pure deciduous stands.

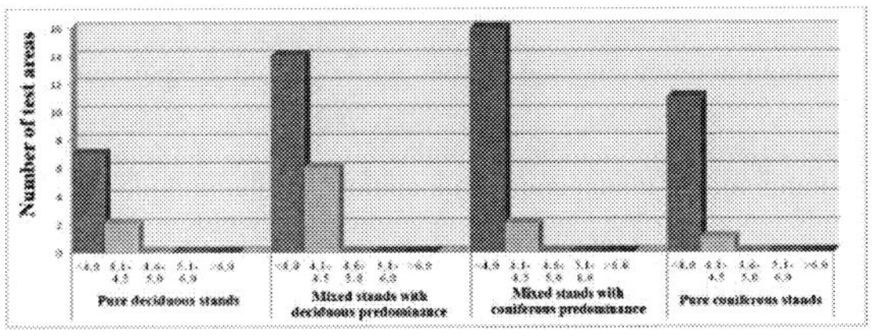

Fig. 6. Test areas grouped by pH_{KCl} of sod-podzolic soils.

The pH_{KCl} value in sod-podzolic soils ranges from 3.29 to 4.45. The following groups of soils were identified by this pH value: less than 4.0; 4.1–4.5; 4.6–5.0; 5.1–6.0; more than 6.0 as represented in fig. 6.

As seen from fig. 6, most of the soils under stands of various species composition have a pH_{KCl} value under 4.0. Significantly fewer stands grow on soils where the pH range is 4.1–4.5. A small number of stands grow on soils with a pH value of more than 4.6. Thus, on the territory of the LOD, most of the stands of various species composition grow on strongly acidic soils.

In this work, the methods of descriptive statistics and ANOVA analysis were used. The statistical evaluation of the data showed that the influence of different vegetation groups on soils in terms of the humus content, hydrolytic acidity, the amount of exchange bases and exchange acidity is similar. Comparison of the average values of the studied physico-chemical properties of soils showed that, according to the Fisher criterion, the level of significance of the differences between the compared groups in the composition of the stand is very low, which gives reason to conclude that there are no statistically significant differences (Table 1).

Table 2. ANOVA results.

Stands type		Humus, %	Ha, mg-eq/ 100g	Sum of base cations, mg-eq/ 100g.	pH_{KCl}
Pure deciduous stands	Average	4.21	8.81	10.18	3.92
	N	9	9	9	9
	St.dev.	1.53	1.80	3.90	0.24
Mixed stands with deciduous predominance	Average	4.49	7.08	7.46	3.84
	N	20	20	20	20
	St.dev.	1.08	2.79	2.35	0.26
Mixed stands with coniferous predominance	Average	3.86	7.83	9.35	3.77
	N	18	18	18	18
	St.dev.	1.23	310	4,71	0,20
Pure coniferous stands	Average	4,40	8,74	10,30	3,73
	N	12	12	12	12
	St.dev.	1.35	5.07	4.32	0.22
F-criterion		0.895	0.870	1.871	1.430
Significance		0.450	0.462	0.145	0.244

4 Conclusion

The studied groups of stands are located on the territory of the LOD and differ in: 1) genesis of geomorphological surfaces; 2) thickness of the humus horizon in the soils; 3) the intensity of the manifestation of the podzolic process; 4) depth of the appearance of the podzolic horizon; 5) depth of gley appearance.

The humus content in the sod-podzolic soils of the LOD ranges from 1.74 to 7.04%. Most of the purely deciduous stands and mixed stands with a predominance of deciduous and coniferous, grow on soils with a humus content of 2.9-4.1%. It was found that mixed stands and pure coniferous stands are more evenly distributed through the test areas on soils with both higher and lower humus contents in comparison with pure deciduous stands. The minimum reserves of humus were determined in soils under purely coniferous and purely deciduous plantings. The maximum reserves are found under mixed stands with a predominance of deciduous.

The value of hydrolytic acidity in sod-podzolic soils of LOD ranges from 1.5 to 17.06 mg-eq /100 g of soil. The maximum values of the hydrolytic acidity are determined in soils under mixed stands with a predominance of conifers.

The sum of exchange bases in sod-podzolic soils ranges from 2.40 to 18.21 mg-eq/100 of soil. Most of the plants in purely deciduous stands and mixed stands with predominance of deciduous, grow on soils with the exchange base sum in the range of 5.9 to 9.4 mg-eq/100 of soil, and in the group of mixed stands with a predominance of coniferous in the range of 9.5-13.0 mg-eq/100 of soil.

The pH_{KCl} value in sod-podzolic soils ranges from 3.29 to 4.45. In all groups of the studied stands, most of the plants grow on soils with a pH value under 4.0.

The influence of different groups of stands on the properties of the upper horizons of sod-podzolic soils of LOD has not been established.

References

1. V. V. Dokuchaev. *Lectures by Professors V.V. Dokuchaev and A. F. Fortunatov* (Byuro Polt. gub. Zemstvo, 1901)
2. V. V. Dokuchaev. *On soil science (lectures given to the statistical staff of the Poltava Provincial Zemstvo in the summer of 1900) (*Khutoryanin, 1900)
3. N. G. Remezov. *Forest soil science (*Lesnaya promyshlennost', 1965)
4. G. F. Morozov, *The significance of V. V. Dokuchaev's works for forestry* (Pochvovedenie, 1904)
5. M. E. Tkachenko. *General forestry* (Goslestekhizdat, 1939)
6. N. G. Remezov, *On the role of forests in soil formation,* (Pochvovedenie, 1953)
7. S. V. Zonn, *The influence of forest on the soils* (Akad. nauk SSSR, 1954)
8. B. D. Zaitsev, *Forest and soils,* (Lesnaya promyshlennost', 1964)

9. G. V. Demin, *Influence of different types of forests on the content and qualitative composition of humus of sod-podzol soils (*Lesa Bashkortostana: sovremennoe sostoyanie i perspektivy, 1997)
10. L. O. Karpachevskii, *Forest and forest soils,* (Lesnaya promyshlennost', 1981)
11. K. A. Gavrilov, The influence of various forest species on the soil *Lesnoe khozyaistvo* **3,** 30–37, (2000)
12. L. O. Karpachevskii, M.H. Stroganova, General patterns of soil formation in the forest zone *Pochvoobrazovanie v lesnykh biogeotsenozakh* (1989)
13. L.O. Karpachevskii, V.A. Rozhkov, Forest, soil and forest soil science, *Pochvovedenie* **5,** 586–598, (1996)
14. A. I. Pozdnyakov, *A*nisotropy of some soils of the zonal series, *Pochvovedenie* **11,** 1308–1319 (2009)
15. N. V. Lukina, Fertility of forest soils as the basis of the soil–vegetation relationship, *Lesovedenie* **5,** 45–56 (2010)
16. A. N. Gennadiev, N. P. Solntseva, M. I. Gerasimova, On the principles of grouping and nomenclature of technogenically altered soils), *Pochvovedenie* **2,** 49–61 (1992)
17. M. I. Gerasimova, L. B. Isachenkova, *"Short memory" of sod-podzolic soils in reforestation successions* (Moscow: LKI, 2008)
18. V. N. Karminov, O. V. Martynenko, Dynamics of soil properties in connection with the age of pine forests, *Lesnoi vestnik,* **7,** 50–62 (2007)
19. Yu. P. Demakov, A. V. Isaev, Dynamics of productivity and composition of stands in various ecotopes of the Bolshaya Kokshaga Nature Reserve, *"Bol'shaya Kokshaga"* **4,** 64–67 (2009)
20. V. D. Naumov, A. N. Polyakov, *145 years of Forest experimental dacha of RSAU-MTAA named after K. A. Timiryazev,* (Moscow, 2009)

The influence of supplementary lighting sources on agro-biological performance in greenhouse-grown cucumbers

V. I. Terekhova*, *M. A.* Bocharova, and *E. Yu.* Yembaturova

Russian State Agrarian University – Moscow Timiryazev Agricultural Academy, 49, Timiryazevskaya St., Moscow, 127550, Russian Federation

Abstract. The paper presents the research of the influence of supplementary lighting sources on growth and yielding capacity of cucumber hybrids. The experiment was carried out in 2020-2021 in commercial greenhouses. Two sources of supplementary lighting were studied (option 1 – high-pressure sodium arc reflector lamps (SRL) "super" (Reflux), option 2 – LED-lamps (Phillips) on two cucumber hybrids: F1 Meva and F1 Svyatogor. The present research revealed statistically significant influence of factor B (source of supplementary lighting) on the overall crop productivity. To obtain the maximal yield (Meva F1 – 20,86 kg/m^2, Svyatogor F1 – 18,21 kg/m^2), we recommend using LED-lamps as supplementary lighting sources.

1 Introduction

Greenhouse facilities have been developed successfully worldwide, providing cutting-edge solutions for technical, economical and biological problems in vegetable crops cultivation [1]. To provide year-round supply of fresh vegetables to the population according to medical requirements [2] and to ensure sustainable off-season supply of greenhouse-grown vegetables, better pricing in retail, supplementary lighting is used in addition to the natural one in order to optimize the light environment [3]. Light is one of the key factors for plant growth and development [4]. For green plants as primary producers, light is the source of energy, eventually to be consumed by humans at the end of the food chain. It also serves for regulatory purposes, influencing cell elongation and plant growth, absorption of plant food compounds, stomatal opening and transpiration [5]. Several literary sources have been dedicated to the study of light quality and its influence on the growth and development of plants [6, 7]. High-Pressure Sodium (HPS) lamps are the predominant lighting source in industrial greenhouses due to their high efficiency in converting energy into active photosynthetic emissions [8]. However, HPS lamps are spectrally and energetically non-optimal, since most of the emissions are in the yellow and orange ranges with a small amount of red in the 550-650 nm range, and about 5% fall in the blue range of 400-500 nm. The insufficient range of blue light and other wavelengths that affect photosynthesis causes this light source's low efficiency [9; 10]. A published study assessed the effects of blue light on leaves and whole plants [11]. The blue spectrum of light affects many plant processes such as photomorphogenesis, phototropism, photosynthetic functioning of leaves [12].

In commercial greenhouses, to obtain high crop productivity and high quality when natural illumination is low, supplementary lighting is used up to 18-120 hours a day with the intensity from 200 to 500 µmol · m^{-2} · c^{-1} [13; 14]. High pressure sodium arc (reflector) lamps (hereinafter referred to as SRL) appear to be main light sources in greenhouses, as well as LED bulbs, which have some additional advantages, such as reduction of energy consumption [15; 16]. Moreover, rapidly developing LED technologies form an alternative light source [17]. LEDs can provide a light spectrum from ultraviolet to infrared. The LED lamp service life is longer than that of HPS and can last up to 100,000 hours [18]. Previously, authors noted [19] that LED light sources can provide energy savings during greenhouse operations. A noteworthy LED lamp characteristic is the emission of less heat during operation compared to that of sodium lamps. As a result, it is necessary to regulate one of the microclimate parameters, air temperature, in order to maintain leaf temperature at the same level and eliminate fluctuations in thermal emissions when switching on and off the lamps [20]. –

The current research is aimed at estimating the influence of a supplementary lighting source on the growth, development and productivity of parthenocarpic (seedless) cucumber hybrids.

2 Materials and methods

The experiment was carried out in commercial greenhouses in 2020-2021 in light zone III during the winter-spring period. Parthenocarpic cucumber hybrids Svyatogor F1 and Meva F1 were used (originator is

* Corresponding author: v_terekhova@rgau-msha.ru

Rijk Zwaan, the Netherlands) in tall Venlo greenhouses.

The present two-factor experiment involved Factor A: cucumber "hybrid genotype" (Meva F1, Svyatogor F1), and Factor B: "supplementary lighting source" (option I: high-pressure sodium arc reflector lamps (SRL) "super" (Reflux); option II: light emitting diode (LED) bulbs (Phillips)). SRLs 600 W were hung over each row spaced 2.5 m between them. The microclimatic parameters were controlled by the Hoogendoorn climate computer; the CO_2 content was maintained at 0,05%-0,09%.

20-day seedlings were planted in Grodan slabs with the density for hybrids; under LED bulbs the crop crowdedness was maintained at 2,5 plants/m^2, under sodium reflector lamps – 2,8 plants / m^2.

Photoculture technology involves switching lamps on and off. The dark period lasted from 5 pm till 9 pm. The illumination of leaves at different levels was measured using LI-190 SA meter (Licor Inc., UK). For measurements, fully developed leaves from 10 model plants were used. To measure the leaf weight and area, 10 plants were divided into organs and weighted. Leaf samples were then dried at 70°C. The specific leaf surface area was calculated as their weight / area ratio. The leaf area index (LAI, m^2/m^2) was calculated by multiplying the leaf area of a plant by the number of plants per 1 m^2.

The yield was accounted dynamically by weighing cucumbers harvested from a plot upon each harvesting [21].

When statistically processing the obtained results, the significance of differences was estimated by means of Student's test and considered statistically significant at $p \leq 0,05$. For statistical purposes, Microsoft Excel 7.0 and STATISTICA software were used.

3 Results and discussion

Table 1 contains the experimental data (averaged values per month for the entire growing period) on the formation of the leaf surface in the studied cucumber hybrids under different supplementary lighting sources as a factor of their adaptation capacities to evaluate their flexibility.

Table 1. Leaf surface formation factors in cucumber hybrids with different supplementary lighting.

Hybrid genotype (A)	Supplementary lighting source (B)	Number of leaves	Leaf length, cm	Leaf width, cm	Leaf area, dm^2	Leaf area index (LAI), m^2/m^2
Meva F_1	LED	19.6	26.8	35.2	953	4.9
	Sodium reflector lamp	20.2	25.7	31.7	81.8	2.0
Svyatogor F_1	LED	20.4	25.8	35.1	91.1	4.7
	Sodium reflector lamp	20.8	25.1	32.0	80.6	1.8
$HCP_{05}B$=		–	–	–	6.3	0.2

Significant differences in the leaf surface area using different supplementary lighting sources (LED vs sodium reflector lamps) in hybrids were revealed ($HCP_{05}B$=6,3 dm^2). The contribution of the influence of Factor B (supplementary lighting source) on the leaf surface area is significant, comprising 14%. Averages for the leaf surface area in hybrids did not differ significantly in the experiment. Factor A (hybrid genotype) was not found to have a significant effect on the leaf surface. The highest contribution in terms of influence on the leaf surface area (71%) appeared to be due to random factors.

Only Factor B was found to significantly influence the LAI ($HCP_{05}B$=0,2 m^2/m^2), the contribution of the factor's influence accounted for 32%; the highest effect on LAI (73%) was due to random factors.

Growth dynamics in partenocarpic cucumber hybrids with different supplementary lighting sources is presented in Table 2.

The height of the main stem reflects photosynthetic potency of plants and appears to change in ontogeny. In the propagation department, all hybrids had the same source of light, SRL-600 lamps; in this relation, the difference in the plant heights on the 23rd day was influenced by the hybrid genotype. According to the two-factor experiment results, having evaluated the effect of Factor A (hybrid genotype) and Factor B (supplementary lighting source), we must indicate that both factors had statistically significant effects on the growth dynamics in cucumber plants. According to the biometric observations, the greatest daily increase in the plant length was found on the 102nd day and further, which can be explained by the increase of natural lighting.

Yield is the key criterion for each variety or hybrid, reflecting the effectiveness of these or those factors, techniques or methods. The effect of the supplementary lighting source on yields in the winter-spring period was studied. According to the obtained experimental data it was revealed that the overall crop productivity of the Meva F1 hybrid under SRL lamps was 19,26 kg/m^2, whereas of the Svyatogor F1 hybrid it was 17,65 kg/m^2. The highest crop productivity appeared to be in Option II – when using LED bulbs (Meva F1 was 20,86 kg/m^2, Svyatogor F1 was 18, 21 kg/m^2).

Factor B (supplementary lighting source) had a significant effect on the overall crop productivity ($HCP_{05}B$=2,5 kg/m^2), contribution of influence was 28%. Factor A (hybrid genotype) had an insignificant effect (contribution was 6%, $HCP_{05}A$ = 1,2 kg/m^2).

Table 2. The influence of hybrid genotype and supplementary lighting sources on growth dynamics in cucumbers.

Hybrid genotype (A)	Supplementary lighting source (B)	Plant age (days after mass sprouting)						Average daily growth during cultivation period, cm
		23		68		102		
		11.01		26.02		02.04		
		1	2	1	2	1	2	
Meva F_1	LED	39.4	–	370.0	52.4	670.2	60.0	59.1±7.2
	SRL	46.5	–	355.4	52.1	629.2	54.8	55.2±4.1
Svyatogor F_1	LED	33.1	–	406.2	60.6	752.5	62.3	64.5±6.2

	SRL	40.1	–	394.5	57.7	723.3	63.8	60.0±5.2
HCP$_{05}$A								5.4
HCP$_{05}$B								6.1

Note: 1 is the plant height, cm; 2 is the weekly growth, cm.

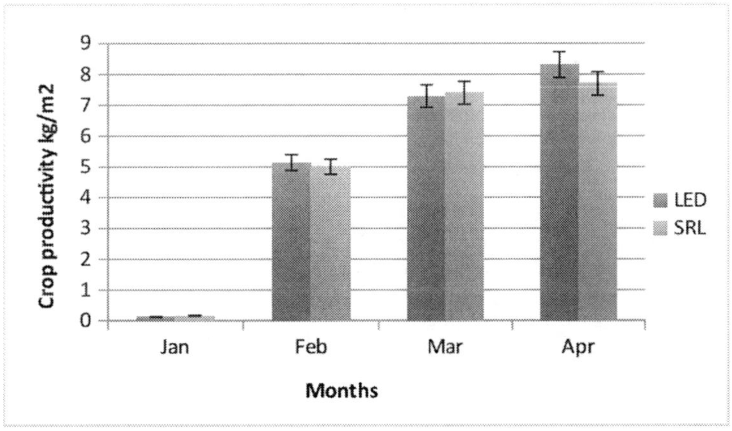

Fig. 1. Crop productivity dynamics in Meva F1 using different supplementary lighting sources.

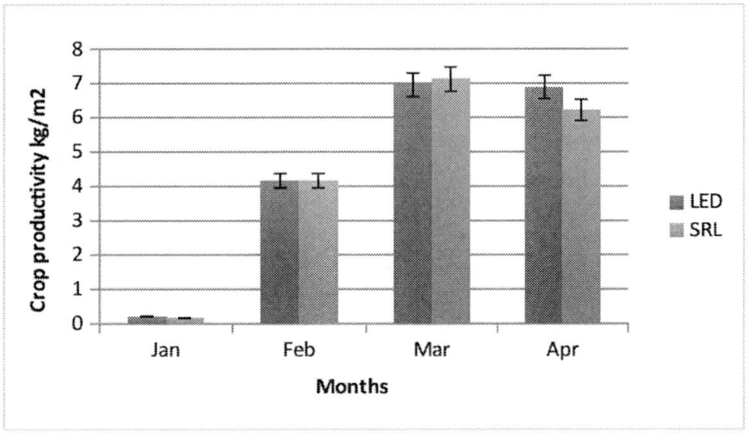

Fig. 2. Crop productivity dynamics in Svyatogor F1 using different supplementary lighting sources.

4 Conclusion

In our studies, the sources of supplementary illumination influenced the agrobiological indicators of the cucumber greenhouse culture. Investigation of measuring the distribution of light in the cenosis and the efficiency of using the thermal culture of cucumber in the winter-spring circulation of scattering coming from LED lamps. Cases of revealing differences in leaf surface areas during the collection of various sources of supplementary illumination (LED lamps and DNaZ lamps) of the studied cucumber plants were established.

The "test genotype" factor is no exception. In our studies, the largest share of the analysis of leaf area and leaf area index (FLI) falls on random factors - microclimate parameters. The conducted studies have revealed the prospects of the LED lamp as artificial lighting, which is based on objective measurements of completeness. The lighting mode used in the experiment makes it possible to obtain a positive dynamics of the oral processes of the studied cucumber hybrids Meva F1 and Svyatogor F1 and to consistently obtain yields in the winter-spring turnover.

To obtain maximal productivity, we recommend using LED bulbs as sources of supplementary lighting, so crop productivity of Meva F1 is 20,86 kg/m^2; that of Svyatogor F1 is 18,kg/m^2.

References

1. I. V. Dal'ke, E. E. Grigorai, T. K. Golovko,

Photosynthetic productivity and light energy efficiency of cucumber greenhouse culture with additional illumination inside the cenosis. *Izvestia TSHA* **5,** 13–23 (2014)

2. O. Oyebode, V. Gordon-Dseagu, A. Walker, J. S. Mindell, Fruit and vegetable consumption and all-cause, cancer and CVD mortality: Analysis of health survey for england data, *Journal of Epidemiology and Community Health*, **68,** 856–862 (2014)
3. I. G. Tarakanov, Current state and prospects for the development of plant light culture, *Materials of the V International Symposium on Plant Light Culture in Lillehammer* (Norway, June 21–24), **6.34–3** (2005)
4. K. M. Folta, M. Kevin, K. S. Childers, Light as a growth regulator: controlling plant biology with narrow-bandwidth solid-state lighting systems? *HortScience* **43.7,** 1957–1964 (2008)
5. C. Fankhauser, J. Chory Light control of plant development, *Annual review of cell and developmental biology* **13.1** 203–229 (1997)
6. Y. A. Berkovich, I. O. Konovalova, S. O. Smolyanina, A. N. Erokhin, O. V. Avercheva, E. M. Bassarskaya, G. V. Kochetova, T. V. Zhigalova, O. S. Yakovleva, I. G. Tarakanov, LED crop illumination inside space greenhouses, *Reach. Rev. Hum. Space Explor.* **6,** 11–24 (2017)
7. T. Ouzounis, E. Rosenqvist, C. O. Ottosen, Spectral effects of artificial light on plant physiology and secondary metabolism: A review, *Hort. Sci.* **50** (2015)
8. W. Van Ieperen, G. Trouwborst, The application of LEDs as assimilation light source in greenhouse horticulture: A simulation study, *Acta Hort* (2008)
9. J. C. Sager, J. C. McFarlane, Plant growth chamber handbook, *Radiation* pp 1–29 In: R. W. Langhans, T.W. Tibbits (eds.). *Iowa Agriculture and Home Economics Experimental Station Special Report no. 99. North Central Region Research Publication* **340** (Iowa State University Press, Ames, IA, 1997)
10. L. F. M. Marcelis, A. Elings, M. Bakker, E. Brajeul, J. A. Dieleman, P. H. B. de Visser, E. Heuvelink, Modelling dry matter production and partitioning in sweet pepper *Acta Hort* (2006)
11. R. Matsuda, K. Ohashi-Kaneko, K. Fujiwara, K. Kurata, Effects of blue light deficiency on acclimation of light energy partitioning in PSII and CO2 assimilation capacity to high irradiance in spinach leaves, *Plant Cell Physiol.* **49** (2008)
12. G. Whitelam, K. Halliday, *Light and Plant Development* (Blackwell: Oxford, UK, 2007)
13. R. Paradiso, E. Meinen, J. F. Snel, P. De Visser, W. Van Ieperen, S. W. Hogewoning, L. F. Marcelis, Spectral dependence of photosynthesis and light absorptance in single leaves and canopy in rose, *Scientia Horticulturae* **127 (4),** 548–554 (2011)
14. A. Savvides, F. Dimitrios, W. van Ieperen, Coordination of hydraulic and stomatal conductances across light qualities in cucumber leaves, *Journal of Experimental Botany* **63.3,** 1135–1143 (2012)
15. Z. Bian, N. Jiang, S. Grundy C. Lu, Uncovering LED light effects on plant growth: new angles and perspectives-LED light for improving plant growth, nutrition and energy-use efficiency. In International Symposium on New Technologies for Environment Control, *Energy-Saving and Crop Production in Greenhouse and Plant* **1227,** 491–498 (2017)
16. W. Van Ieperen, G. Trouwborst, The application of LEDs as assimilation light source in greenhouse horticulture: a simulation study. *International Symposium on High Technology for Greenhouse System Management* (Greensys) **801,** 1407–1414 (2007)
17. G. D. Massa, H. H. Kim, R. M. Wheeler, C. A. Mitchell, Plant productivity in response to LED lighting, *HortScience* **43 (7)** (2008)
18. M. C. Bourget, An introduction to light-emitting diodes, *HortScience* **43** (2008)
19. R. C. Morrow, LED lighting in horticulture, *HortScience* **43** (2008)
20. K. M. Folta, K. S. Childers, Light as a growth regulator: Controlling plant biology with narrow-bandwidth solid-state lighting systems, *HortScience* **43** (2008)
21. B. A. Dospekhov, *Field Experiment Methodology* (Moscow: Agropromizdat, 1985)

The effect of growth stimulators on the vegetative reproduction of North American tree species in the conditions of the Omsk region

A. I. Degtyarev*, *G. V.* Barayshchuk and *A. A* Gayvas

Department of Horticulture, Forestry and Plant Protection, Omsk State Agrarian University named after P. A. Stolypin, 1, Institutskaya ploshchad St., Omsk, 644008, Russian Federation

Abstract. The relevance of studying the growth and development of North American cypress tree species in the Omsk region is due to the short period of their use in urban landscaping. The issues of effective production of planting material of these rocks in the West Siberian region have not been fully studied. The use of North American species in the landscaping of the city will expand the range of conifers and reduce the anthropogenic load. The research is devoted to the study of the effect of growth stimulants on vegetative reproduction by green cuttings of representatives of the Cypress family in 2018-2021 in the conditions of the southern forest-steppe of the Omsk region. It has been established that all the studied preparations have a positive effect on the rooting of green cuttings of different forms of Western thuja. The green cuttings of the studied junipers took root better under the influence of heteroauxin, kornevin and ribav extra. An increase in the diameter of the root neck in the studied rocks was observed under the influence of kornevin.

1 Introduction

Green spaces in urban ecosystems perform health, structural planning and decorative functions, enrich the atmosphere with oxygen. At the same time, they experience a high anthropogenic load, are exposed to chemical, physical, and biological pollution. Performing the role of light filters, plants are often influenced by unfavorable conditions of the urban environment: their physiological and biochemical processes change markedly, structures responsible for photosynthesis, respiration, and water exchange are destroyed. The development of an adaptive response under mild stress contributes to the better survival of plants in changing climatic conditions and of technogenic pollution.

Currently, an urgent task is to expand the species composition and introduce new tree and shrub species used in landscaping settlements in Russia. In comparison with European countries, the total number of coniferous species used in the practice of landscape design is small. This is especially true for the territory of Siberia and the Far East [1]. Due to the short period of use of species of North American origin in the Omsk region, the nature of the growth of these species is still insufficiently studied, the issues of their sustainability in the urban environment are practically not covered. All this underlines the need to study the ecological and biological features of the introducers [2].

Recently, North American tree species of the Cypress family have become interesting for landscaping the city of Omsk, the reproduction of which is of great practical importance. The mother plants of this family grow in the arboretum parks of the city [3]. Cypress (*Cupressaceae* Neger) is the largest coniferous family by the number of genera and the third by the number of species. One of the values of these plants is a huge amount of essential oils evaporating from the leaves, which purify the air from microbes. Cypress – evergreen long-lived (90–800 years) shrubs and trees, which are valuable material for park construction. The advantages of these plants are well known: longevity, color of needles that persists in winter and summer, high decorative qualities, health-improving properties. Since the flora of the Omsk region is not rich in species diversity of coniferous species, studies on the technology of growing introduced species in the conditions of the southern forest-steppe of the Omsk region are of undeniable practical significance [4, 5]. Due to the large number of highly decorative artificially bred forms, winter hardiness, durability and resistance to urban conditions, North American tree species are

* Corresponding author: a.degtyareff2014@yandex.ru

widely distributed in decorative gardening on all continents in many climatic zones. Numerous garden centers strive to meet the demand for planting material, but often this material is not viable for our conditions (other climatic factors). Plants begin to develop various diseases, most often there is drying of branches and sunburn. Therefore, it is important to obtain planting material from acclimatized plants by vegetative propagation.

Previously, vegetative reproduction was used in garden practice, but the rooting of most coniferous plants is extremely slow. There are very few works on vegetative reproduction of conifers. Most of them have been released only in the last 25 years. This phenomenon is not accidental, it is explained by the difficulty of root formation of stem cuttings of coniferous species and especially the duration of rooting of cuttings. The manifestation of interest in cuttings over the past 25 years coincides with the discovery of special substances (auxins) in plants that promote root formation, and simultaneously with the production of some organic acids, which in their action turned out to be equivalent to auxins [6].

To accelerate and enhance shoot formation in the cuttings of woody conifers, which take root especially difficult, they are pre-treated with organic and mineral acids. The action of acids on cuttings causes increased metabolism and accumulation of organic substances in the lower part of the cuttings that promote root formation [7].

The most common method of reproduction is by green cuttings. The plants used in the landscaping of the urban environment grow in conditions of various technogenic loads. Therefore, they must be adapted to the stress caused by both technogenic pollution and sudden fluctuations in weather conditions, which has been noted in recent years. It is possible to increase the stability of introducers in various ways: hardening, exposure to variable temperature, as well as growth stimulants [8].

The use of growth stimulants reduces the period required for root formation under normal conditions and provides significant time savings. The effectiveness of growth stimulants depends on many factors: concentration, duration of treatment, condition of cuttings and mother plants, and of course, the type of stimulant.

The effect of growth stimulators can be species- and non-species-specific, therefore, when studying their effect on some cultures, it is impossible to accurately predict the result with others. However, the ultimate goal of using growth stimulants is to increase the percentage of rooting, increase growth, which is achieved by the correct selection of concentration and exposure when processing the material [9, 10].

The use of vegetative reproduction of conifers is of practical importance, so it allows you to obtain plants with the necessary economically valuable characteristics. Cuttings of most coniferous species do not take root in the open ground, since constant temperature and humidity are needed for their root formation. Different breeds require different conditions for rooting cuttings, so different types of greenhouses are used, in which the necessary regime is artificially created. One of the important factors during rooting is that the substrate temperature should be 3-5 °C above the air temperature [11].

In our experience, we studied the effect of various growth stimulants on the rooting of green cuttings in the conditions of the city of Omsk. Plant growth regulator ribav-extra contains a natural complex of biologically active substances produced by mycorrhizal fungi isolated from ginseng root: L-glutamic acid (0.00196 g/l), L-alanine (0.00152 g/l). Heteroauxin contains the active substance β-indolylacetic acid, a chemical of high physiological activity that affects growth processes. Zircon contains a mixture of hydroxycoric acids (chlorogenic, chicory and kaftaric). Kornevin is a biostimulating preperation for plants, which includes indolyl butyric acid (5 g/kg), which, when it gets on the plant, irritates its integumentary tissues, stimulating the appearance of callus and roots.

The purpose of the experiment is to study the influence of growth regulators on the process of root formation and biometric indicators of the studied North American species.

2 Materials and methods

The experiments were conducted in 2018–2021 in the educational and scientific laboratory of perennial crops "Garden named A. D. Kizyurin" of the educational and experimental farm: propagation by green cuttings - in a film greenhouse. In the experiment, the effect of growth regulators on the regenerative ability of green cuttings of representatives of the Cypress family, as well as on biometric indicators (height, length and number of roots, diameter of the root neck), was studied.

Stem cuttings of 10–15 cm representatives of the Cypress family (Western thuja, Western thuja Woodwardi variety, Cossack juniper, common juniper) were used to lay the experiment. Before planting, green cuttings were treated with growth regulators by immersing cuttings: for 8 hours in a 1% solution of kornevin, for 14 hours – in a 1% solution of heteroauxin, for 16 hours – in a 1% solution of ribav extra, for 12 hours – in a 1% solution of zircon. Cuttings without treatment with growth stimulants were used as a control. All experiments were carried out in three-fold repetition, in each repetition of 25 cuttings. The cuttings were planted in a stationary greenhouse with fine dispersion irrigation. Sand was used as a substrate. 5 days after planting, by carefully pulling out the cuttings, systematic monitoring of the development of the callus and the formation of roots was carried out.

3 Results and discussion

Studies have shown that different forms of the same species can be cut with more or less success.

The appearance of roots on cuttings is preceded by the process of formation of new healing tissues, which grow and form an outgrowth - a callus. The appearance of a callus does not always lead to the formation of roots. According to the results of our research, it was found that the callus began to appear on the 20th day after planting in cuttings of the genus thuja and on the 25th day in junipers. Callus formation in green cuttings of Western thuja was intensively carried out when treated with growth stimulators heteroauxin (95.4%) zircon (96.4%), ribav extra (97.8%) under control (80.8%). The same trend was observed in thuja of the Western Woodward variety: heteroauxin (89.2%), ribav extra (86.6%), zircon (84.6%) under control (76.8%). In Cossack and common juniper, the best indicators of callus formation were in variants with heteroauxin, ribav extra and kornevin.

In the control variant of the western thuja planted without treatment with growth stimulants, the root ability was: in 2018 – 78.2 %, in 2019 – 79.2 %, in 2020 – 80.8 %, in 2021 – 70.7 %. In general, the highest rooting of green cuttings of Western thuja was in experiments with treatment with heteroauxin, ribav extra, zircon (Fig. 1).

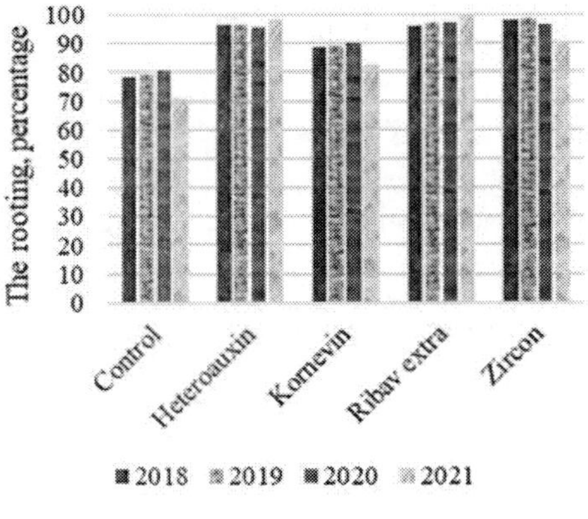

Fig. 1. Rooting of green cuttings of western thuja, LSD_{05} = 1.09.

The control variant of the Western thuja has the Woodward variety: in 2018 – 73.3 %, in 2019 – 74.2 %, and in 2020 – 76.8 %, in 2021 – 70.6 %. The highest percentage of rooting in the Western thuja is spherical in variants using heteroauxin and ribav extra (Fig. 2).

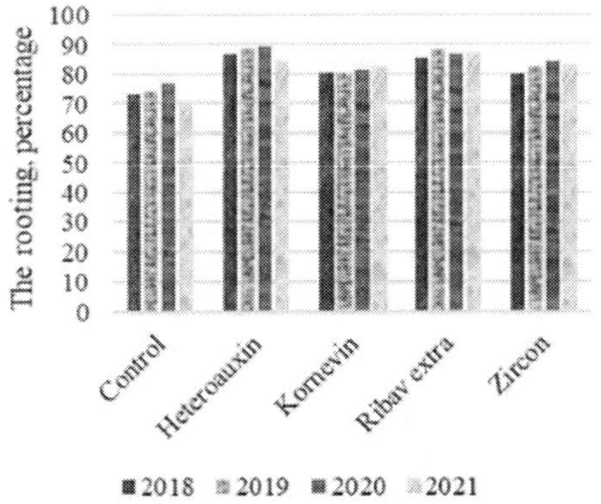

Fig. 2. Rooting of green cuttings of thuja of the western Woodward variety, LSD_{05} = 1.12

When rooting green cuttings of Cossack juniper in the experiment with the use of zircon, the percentage of rooting was lower than control. Treatment of green cuttings of Cossack juniper with heteroauxin, kornevin, ribav extra positively affected rooting (Fig. 3).

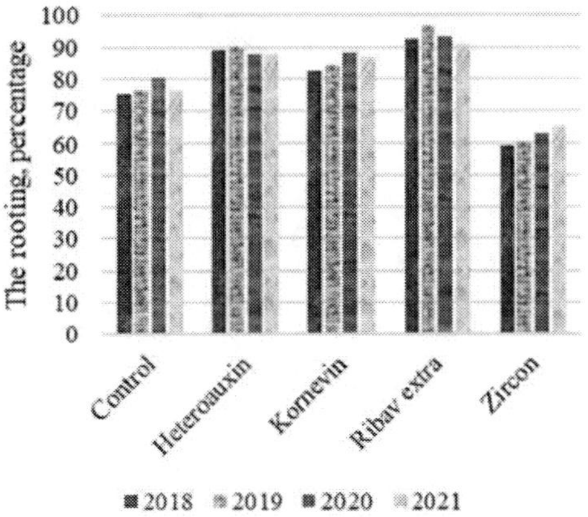

Fig. 3. Rooting of green cuttings of Cossack juniper, LSD_{05} = 1.08.

In the control variant of the common juniper, planted without treatment with growth stimulants, the rooting rate was 79.9 % in 2018, 80.6 % in 2019, 80.1 % in 2020, and 77.3 % in 2021. Whereas in the variants with heteroauxin, kornevin, ribav-extra, the rooting of green cuttings of ordinary juniper was more successful. At the same time, the percentages of rooting with the use of zircon were lower than the control (Fig. 4).

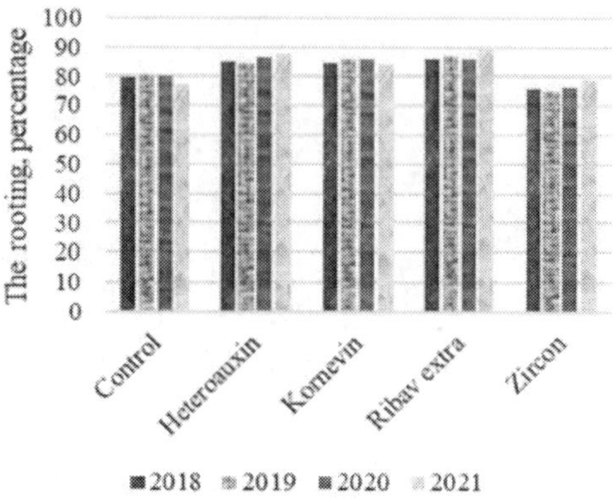

Fig. 4. Rooting of green cuttings of common juniper, LSD_{05} = 1.16

Thus, during vegetative reproduction of representatives of the genus Thuja, growth stimulators heteroauxin, kornevin, zircon, ribav extra increased the rooting of cuttings, and during the reproduction of junipers, an increase in the percentage of rooting was observed when using heteroauxin, kornevin, ribav extra.

The data obtained allow us to state that an increased increase in relation to the control was observed when treated with growth stimulators ribav extra (Fig. 5). The growth of rooted cuttings in the treatment variant with kornevin was at the control level. The growth of green cuttings of Western thuja treated with the stimulant heteroauxin was lower or at the control level.

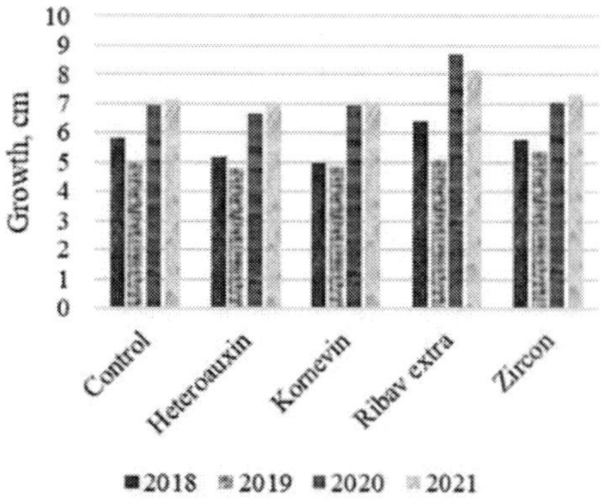

Fig. 5. Growth of rooted cuttings of Western thuja, LSD_{05} = 0.12

When rooting green cuttings of Western thuja in comparison with the control, the stimulator of root formation of the roots contributed to the formation of a larger diameter of the root neck. The diameter of the root neck in green cuttings treated with ribav extra, heteroauxin, zircon compared with the control was smaller (Fig. 6).

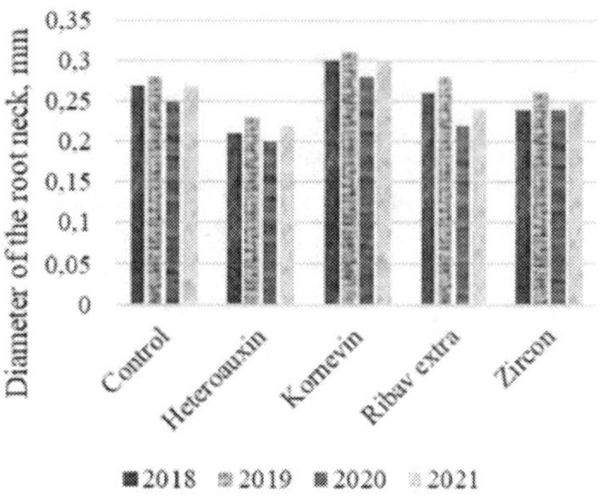

Fig. 6. Diameter of the root neck of rooted cuttings of Western thuja, LSD_{05} = 0.01

Thus, when propagated by green cuttings of Western thuja, kornevin contributed to the production of stronger cuttings seedlings.

Observations of rooted cuttings of thuja of the Western Woodward variety showed an increased increase in comparison with the control when treated with growth stimulators ribav extra and heteroauxin. The green cuttings of the Western thuja are spherical, treated with the stimulants zircon and kornevin, had an increase below control (Fig. 7).

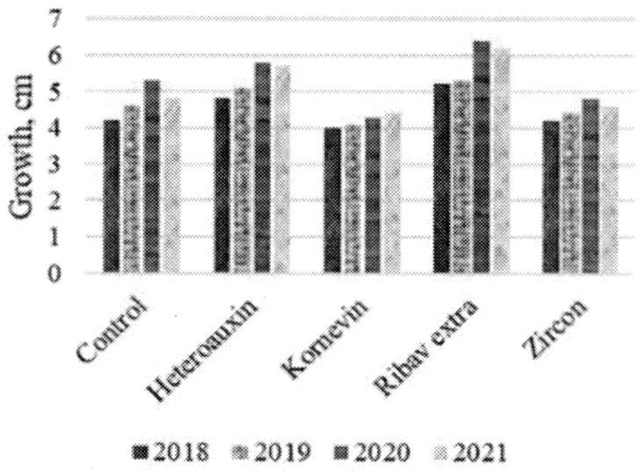

Fig. 7. Growth of rooted cuttings of thuja of the western Woodward variety, LSD_{05} = 0.13.

The root formation stimulators heteroauxin and kornevin contributed to the formation of a larger diameter of the root neck in the cuttings of the Western thuja, the shape is spherical in comparison with the control. Green cuttings treated with ribav extra and zircon, in comparison with the control, had a smaller diameter of the root neck (Fig. 8).

 https://doi.org/10.1051/bioconf/20225200059

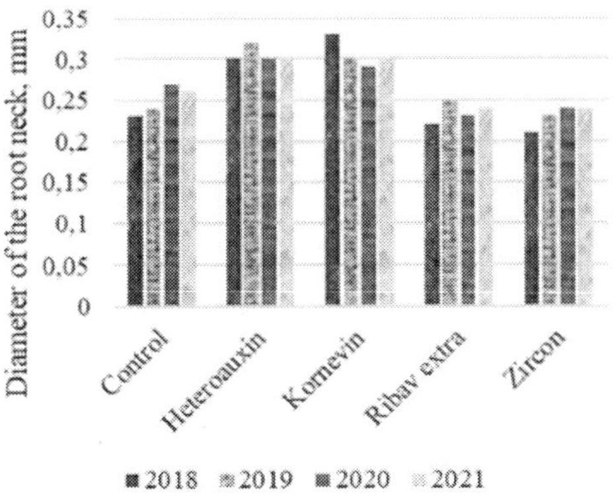

Fig. 8. Diameter of the root neck of rooted cuttings of thuja of the Western Woodward variety, $LSD_{05} = 0.01$.

In the rooted cuttings of Cossack juniper in the variant with ribav extra and kornevin, the growth and diameter of the root neck were higher in comparison with the control (Fig. 9, 10).

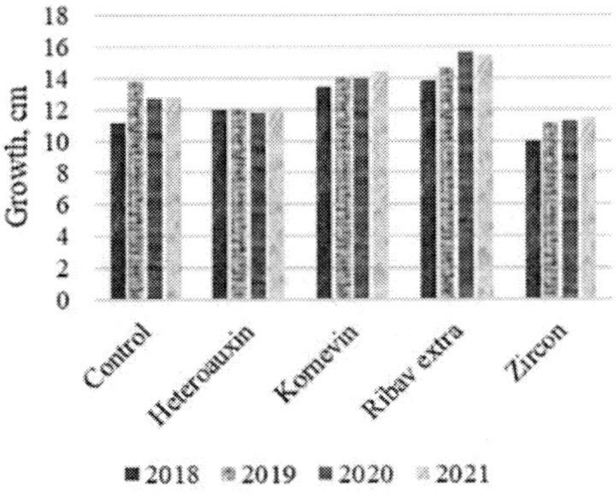

Fig. 9. Growth of rooted cuttings of Cossack juniper, $LSD_{05} = 0.15$.

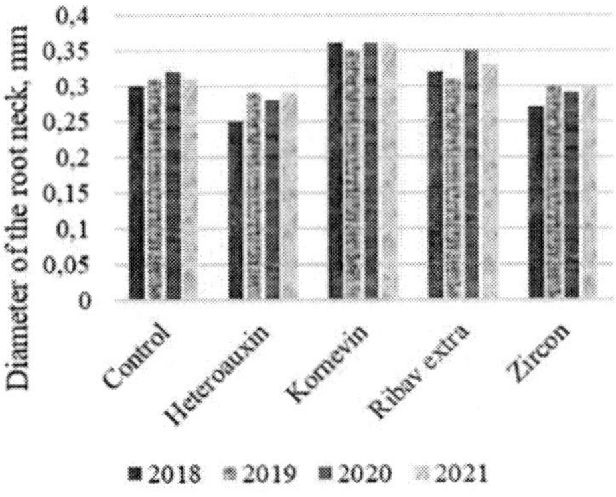

Fig. 10. Diameter of the root neck of rooted cuttings of Cossack juniper, $LSD_{05} = 0.01$.

The growth of green cuttings in common juniper when treated with growth stimulants exceeded the control variant in all experimental variants, except for the experiment with the use of heteroauxin (Fig. 11).

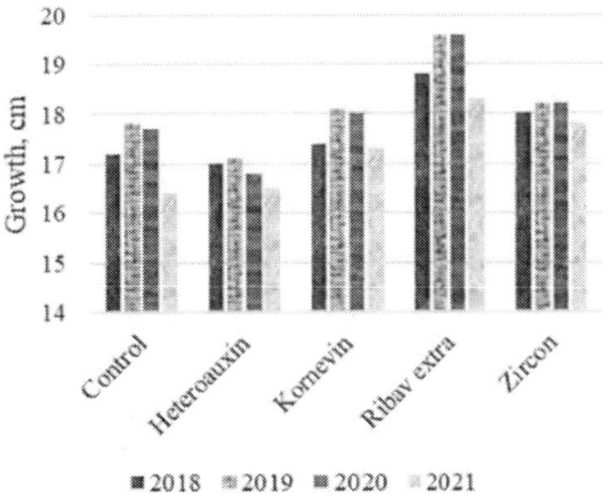

Fig. 11. Growth of rooted cuttings of common juniper, $LSD_{05} = 0.20$

The diameter of the root neck of the rooted cuttings of ordinary juniper was higher than the control only in the variant with the root (Fig. 12).

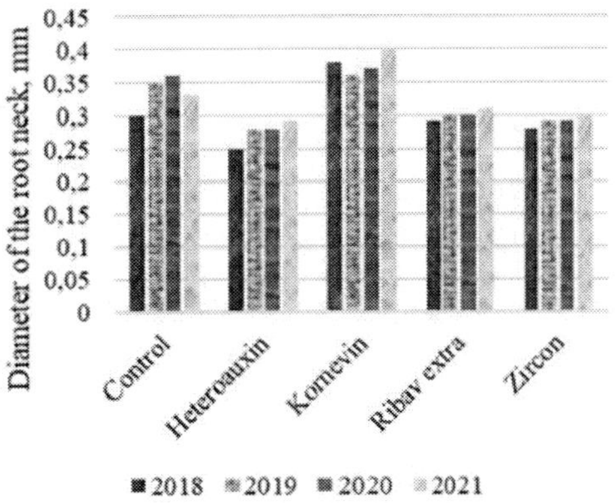

Fig. 12. The diameter of the root neck of the rooted cuttings of the common juniper, $LSD_{05} = 0.01$.

The preparations heteroauxin, ribav extra, zircon had a significant stimulating effect on root growth in Western thuja (Fig. 13, 14). Compared with the control, the number of roots in the Western thuja increased by 29.7–162.5 %, while the length of the roots increased from 3 to 32.4 %.

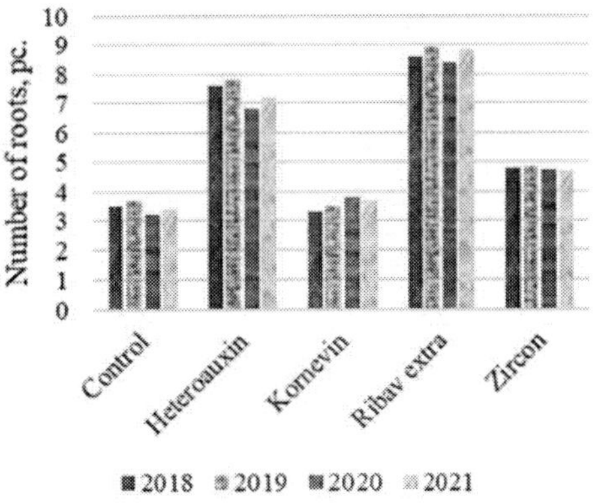

Fig. 13. The number of roots in rooted cuttings of Western thuja, $LSD_{05} = 0.32$.

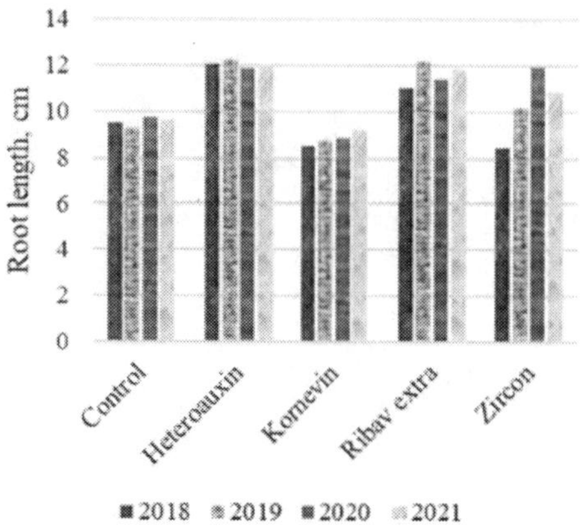

Fig. 14. The length of the roots of the rooted cuttings of the Western thuja, $LSD_{05} = 0.64$

The Western thuja has a spherical shape, the number of roots increased by 60.5-147 % with the use of heteroauxin, ribav extra, zircon (Fig. 15). At the same time, the root length of this form of Western thuja increased from 4.8 to 34.2 % under the action of heteroauxin (Fig. 16). Observation of the growth of the root system of the green cuttings of the Western thuja of the two studied forms showed that the growth stimulator kornevin inhibited the process of root formation. The root-forming ability of cuttings of different forms of thuja treated with this stimulant is significantly lower than that in the control. It is possible that the concentrations and consumption rates of the preparation recommended for all fruit and berry and ornamental crops require further adjustments for specific tree species. The growth regulators heteroauxin, ribav extra and zircon consistently exceeded the control variant in the number of roots, but had a lower value in the length of the roots (Fig. 13, 14, 15, 16).

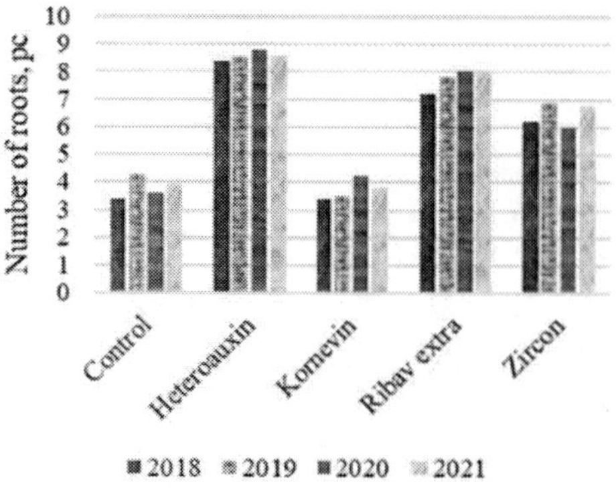

Fig. 15. The number of roots in rooted cuttings of Western thuja Woodwardi variety, $LSD_{05} = 0.22$.

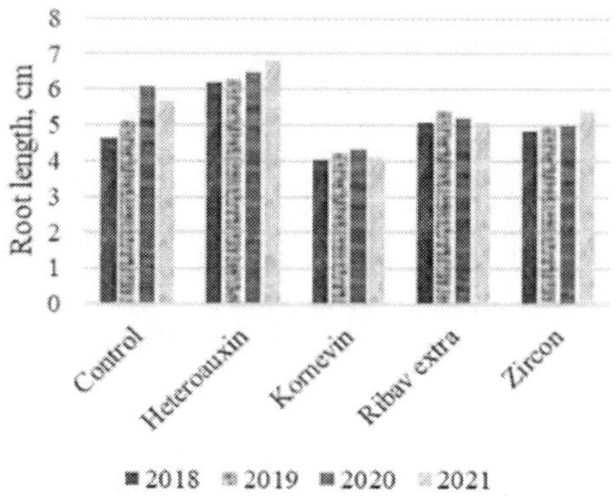

Fig. 16. Root length of rooted cuttings of Western thuja Woodwardi variety, $LSD_{05} = 0.56$.

In the studied junipers, the number of roots and the length of the roots exceeded the control variant under the influence of all the growth regulators tested (Fig. 17, 18, 19, 20). In all experimental variants, the number of roots and the length of the root system were higher than the control at a significance level of 95 %.

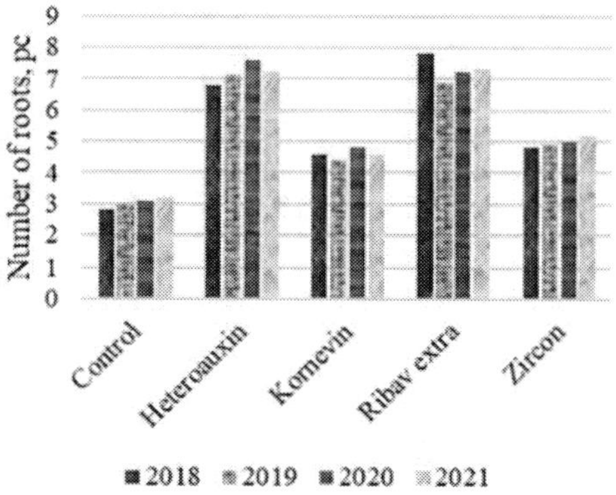

Fig. 17. The number of roots in rooted cuttings of Cossack juniper, $LSD_{05} = 0.28$.

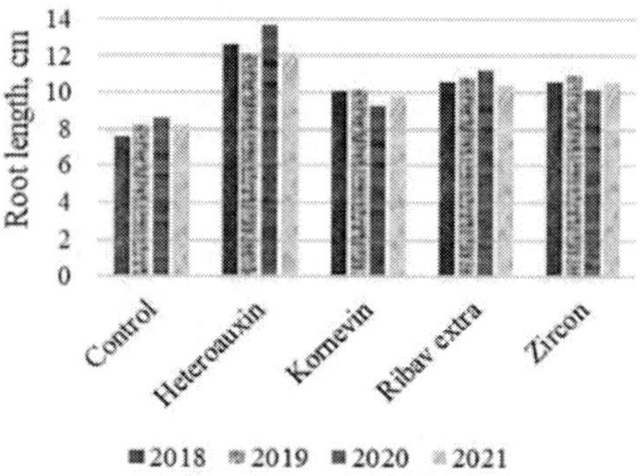

Fig. 18. The length of the roots of the rooted cuttings of the Cossack juniper, $LSD_{05} = 0.66$.

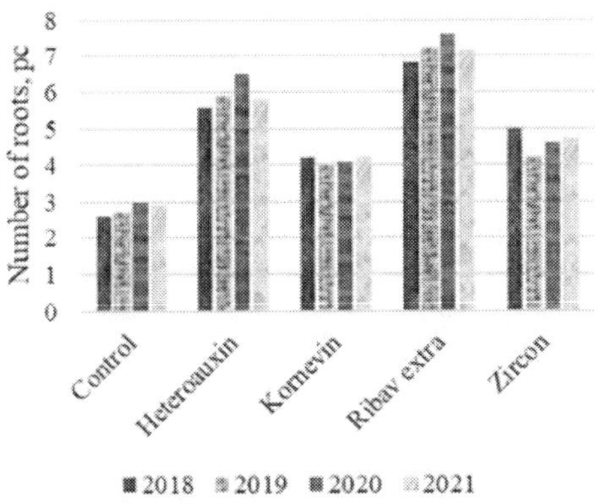

Fig. 19. The number of roots in the rooted cuttings of the common juniper, $LSD_{05} = 0.33$.

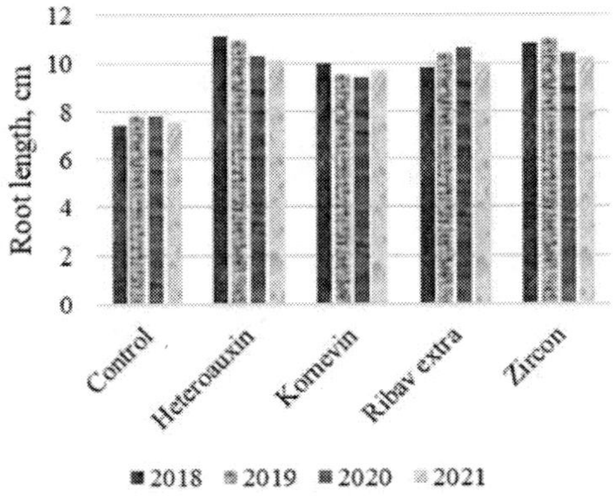

Fig. 20. The length of the roots of rooted cuttings of common juniper, $LSD_{05} = 0.62$.

4 Conclusion

Cuttings are a fast and cost-effective way to propagate decorative and valuable, but slow-growing conifers. The use of chemical growth stimulants increases the root ability of cuttings, increases the efficiency of reproduction and the practical yield of finished products, contributing to the strengthening and adaptation of young plants to environmental conditions.

During vegetative reproduction of representatives of the genus Thuja, active callus formation and high rooting were influenced by growth stimulants: heteroauxin, kornevin, zircon, ribav extra, and during reproduction of representatives of the genus Juniper by heteroauxin, kornevin, ribav extra.

The greatest stable increase in growth was found for thuja and juniper when they were treated with ribav extra.

The greatest increase in the diameter of the root neck was found in all the studied species of thuja and juniper when they were treated with kornevin.

All the growth regulators studied had a positive effect on increasing the number and length of thuja and juniper roots, except for the recorded inhibitory effect of kornevin on cuttings of western thuja seedlings.

References

1. N. A. Kolyada, About the use of North American coniferous plants in the landscaping of Russian cities, *Bulletin of Krasnoyarsk State Agrarian University* **9,** 73–76 (2010)
2. M. Yu. Vyazankin, G.V. Barayshchuk, Cultivation of adaptive planting material of Western thuja in the conditions of the city of Omsk, *Bulletin of Omsk State Agrarian University* **1 (1),** 23–26 (2011)
3. E. A. Gorb, G. V. Barayshchuk, Evaluation of microbiological preparations effect on introduced species of trees, in the Fifth Technological Order, *Prospects for the Development and Modernization of the Russian Agro-Industrial Sector* TFTS 119–123 (Atlantis Press) (2019)
4. A. I. Degtyarev, G. V. Barayshchuk, Representatives of the Cypress family in an urban environment, *The state and prospects of development of the agro-industrial complex. Rostov-on-Don*, pp 496–499 (2020)
5. E. S. Zalesova, S. V. Zalesov, N. P. Bunkova, N. P. Kletsko, M. V. Solovyova, Ya. A. Krekova, Promising coniferous introducers for landscaping and expanding biological diversity in the Middle Urals, *Forest science in the implementation of the concept of the Ural Engineering School: Socio-economic and environmental problems of the forest sector of the economy* (Yekaterinburg), pp 169–172 (2019)
6. B. S. Ermakov, Reproduction of woody and shrubby plants by green cuttings, *Mold. scient.-research. in-t fruit growing NGO "Kodru"* (Chisinau: Stiinza, 1981)
7. Z. Ya. Ivanova, *Biological bases and techniques of vegetative reproduction of woody plants by stem cuttings* (Kiev: Nauk. Dumka, 1982)
8. T. V. Baranova, Accelerated production of plants resistant to urban conditions, *Ecology and industry of Russia,* **4,** 65–67 (2013)
9. A. I. Degtyarev et al., The influence of growth stimulators on the rooting of green cuttings of the Cypress family in the conditions of the city of Omsk, *IOP Conference Series: Earth Environmental Science* **954,** 012019 (2022)
10. S. V. Rezvyakova, A. G. Gurin, E. S. Rezvyakova, Reproduction of coniferous rocks with green cuttings using new biological products, *Bulletin of Agrarian Science* **2 (65),** 9–14 (2017)
11. E. A. Gorb, G. V. Barayshchuk, Greening coniferous breeding technology for urban landscaping, *The European Proceedings of Social and Behavioural Sciences,* pp 156–162 (2021)

Modeling the destruction of the grain shell of cereal crops

Elmas Nurullin[1,*], *Andrey* Dmitriev[1], *Damir* Khaliullin[1], and *Igor* Malanichev[2]

[1] Kazan State Agrarian University, 65, K. Marx Str., Kazan, 420015, Republic of Tatarstan, Russian Federation
[2] Kazan State University of Architecture and Engineering, 1, Zelenaya St., Kazan, 420043, Republic of Tatarstan, Russian Federation

Abstract. When processing grain cereals, hulling is an important part being the process of destruction of shell and separation of the fruit, from which cereals are prepared, which are a valuable food for people. As a result of theoretical research, a physicomathematical model of grain of cereal crops has been developed as a hulling object consisting of two separate structural elements (spherical shell and spherical kernel), each of which has its own geometric parameters and rheological properties. Mathematical dependences of the destructive stresses of the shell and kernel during static and dynamic interactions with working surfaces on their strength characteristics and geometric parameters are established. The obtained theoretical dependencies, together, represent a mathematical model describing the process of destruction of the grain shell of cereal crops under static and dynamic influences. It can be used to simulate sunflower seed shell hulling in vegetable oil production. The developed model provides quick and qualitative computational experiments with the use of computers and software, with the required set of given values of physical-mechanical and technological properties of the processed grain and structural and technological parameters of machines for destruction and separation of the shell. When using the appropriate software, it is possible to visualize the process of shell destruction with a full frame-by-frame picture of the distribution, magnitude, direction of loads and arising stresses. When using the appropriate software, it is possible to visualize the process of shell destruction with a full frame-by-frame picture of the distribution, magnitude, direction of loads and arising stresses. The possibility of solving this model using computer technology and modern software allows you to significantly reduce the labor intensity of natural experiments, while excluding random and systematic errors associated with the experimental equipment and the researcher.

1 Introduction

The hulling as the process of destructing the shell and separating the kernel (fruit) plays an important role in the technology of cereal grains and sunflower seeds processing [1-3].

The hulling quality is determined by the amount of maximum extraction of whole kernels in a single pass through the hulling machine [4]. The main factors affecting the quality are the anatomical structure and physical and mechanical properties of the processed grain, the method of hulling and design of the hulling machine. Experimental and theoretical material available in the literature shows a wide range of technologies based on the use of different methods and machines for hulling [3, 5-13].

When creating hulling machines, it is necessary to take into account the anatomical structure, physical and mechanical properties of the shell and kernel of the processed grains. Currently, this problem is solved experimentally individually for each crop and there are no theoretical provisions to justify the most effective method and design of the hulling machine.

The purpose of the study is to simulate the process of destructing the shell of cereal crops and the interaction with the working surfaces of hulling machines.

2 Materials and methods

Hulling is considered as a process of deformation of the shell beyond the elastic limit state and its destruction with separation from the kernel with the maximum preservation of the integrity of the latter. Before modeling the process of destruction and separation of the shell from the kernel, it makes sense to consider the physical and mathematical model of the grain, consisting of two elements - the shell and the kernel. Moreover, the shell is an object with different mechanical properties from the kernel, a separate object spatially separated from it by an air layer. The geometrical dimensions that characterize the shell as a hulling object are the diameter of the sphere D_s, shell thickness δ, air gap thickness δ', surface area S_s, volume V_s. The geometric parameters of the spherical body (kernel) are: ball diameter D_k, surface area S_k, volume V_k.

For mathematical recording of the equations establishing the relationship between these parameters, let's connect a spherical structural body modeling the hulling object with a moving coordinate system. And the

*Corresponding author: nureg@mail.ru

coordinate system is chosen so that the origin of the coordinate system lies in the center of the spherical body. Since in this case we consider the model of a whole unshelled grain with undamaged structure, it is reasonable to express some parameters through those that can be determined experimentally.

Mathematical equations establishing the relationship between the listed geometrical parameters:

$$S_s = \pi D_s^2 ; S_k = \pi D_k^2, \quad (1)$$

$$V_k = \frac{\pi D_k^3}{6} ; V_s = \frac{\pi}{6}\left[D_s^3 - (D_s - 2\delta)^3 \right], \quad (2)$$

$$D_s = 2R_s ; D_k = 2R_k. \quad (3)$$

The surface bounding the simulated body will be described by the following expression:

$$X^2 + Y^2 + Z^2 = R^2. \quad (4)$$

In addition to geometric parameters, structural elements have physical and mechanical properties such as mass, density, humidity, which determine the physical state of the hulling object. Here it should be noted that the moisture content of the hulling object and mainly the difference in the moisture content of the W_k and seedcoat W_s have a significant impact on the process of hulling a single grain and on the indicators of technological efficiency of the working process in the flow.

From the point of view of fracture mechanics, a spherical shell is a solid elastic deformable body with certain strength characteristics. In addition, the shell has a number of properties, among which we will highlight its cleavability, fractibility under static and dynamic (shock) influences, which directly depend on the strength characteristics. The ball-shaped body (kernel) is an elastic and plastic body. Here, too, there are a number of properties reflecting crushability, destructibility, which are closely related to the moisture content of the kernel. Grain of cereal crops as a subject of hulling has a number of strength properties, which characterize the rigidity of structural elements under external force action. The most important ones, in our opinion, are the elastic modules of the shell E_s and kernel E_k, shell Poisson's coefficients μ_s and kernel μ_k.

We note that these characteristics depend on the physical state of the hulling subject and mainly on the moisture content. The listed properties determine the values of the limiting forces leading to the destruction of the shell, with the maximum preservation of the original geometrical parameters of the kernel.

The research used general scientific methods of mathematics, physics, theory of elasticity, theory of shells, physical and mathematical modeling, as well as private and specific methods of research cereal grains and technological processes of their processing [3,14-19].

3 Results and discussion

Based on the above materials, the model of the subject of hulling (grains) is presented in the following form (fig. 1).

The presented model of the subject is the basis for modeling the process of destruction of the grain shell of cereal crops. The presented model is universal, which allows one to describe the geometry and condition of almost all cereal seeds with different anatomical structure and physical and mechanical properties. The destruction of the shell of grain in the hulling machine is a consequence of their interaction with the working surfaces, which have certain mechanical characteristics. Therefore, in this case it makes sense to consider simplified models of a single-grain shell fracture during interaction with the working surface. Let us consider the process of interaction of the grain with the working surface on the assumption that it behaves as an elastic shell of the finite thickness. The stiffness of the working surface is an order of magnitude higher than the stiffness of the shell, so only its deformation occurs. The elastic material properties of the working surface are taken into account in the calculations through the modulus of elasticity (Ep) and Poisson's ratio (μp).

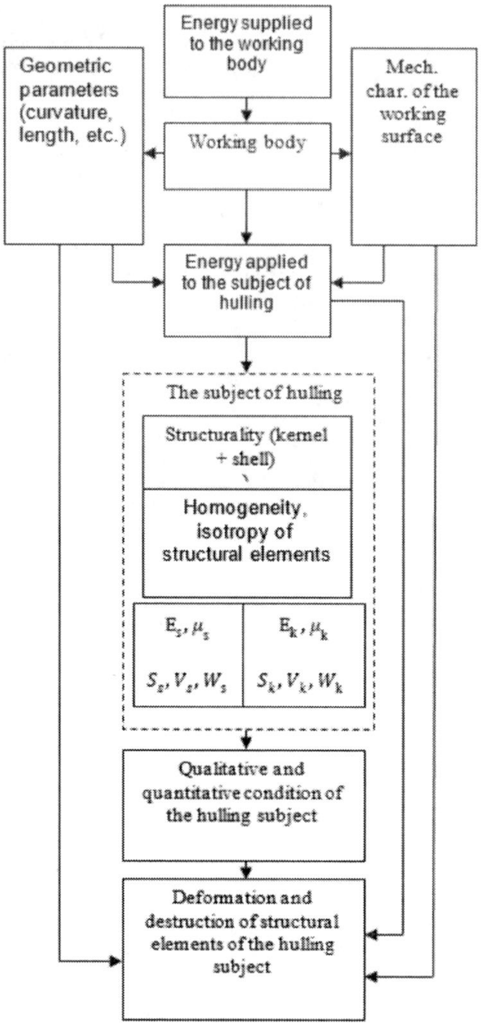

Fig. 1. Structure diagram of the hulling subject model

The destruction of the shell in the simulation is considered as its deformation to the supercritical stage and its collapse with the formation of cracks along the meridians of the sphere.

Based on the theory of shells, the potential energy of the grain shell of cereal crops, depending on the deformation, in the interaction with the working surface is determined by the following dependence:

$$U(h) = k_s h_s^{\frac{3}{2}}, \qquad (5)$$

where $U(h)$ is the potential energy of deformation of the fruit shell of the hulling subject as a function of deformation, $\frac{kg \cdot m^2}{s^2}$; k_s is the factor of proportionality, $\frac{H}{\sqrt{m}}$; h_s is the shell deformation, m.

The coefficient k_s depends on the strength characteristics of the hulling object shell, and the material of the working surface, as well as the curvature of their surface in the contact zone, is equal to:

$$k_s = 2{,}4\,\delta^{\frac{5}{2}} \frac{1}{D_s} \text{¿}, \qquad (6)$$

where D_s is the shell diameter, m; δ is the shell thickness, m; E_s, E_{ws} are elastic modules of the shell and the working surface, N/m^2; μ_s, μ_{ws} are shell and work surface Poisson's coefficients.

During the interaction of the grain with the working surface, its total energy is equal to the sum of the kinetic energy of the deforming part of the structural elements $\left(\frac{m_g d h^2}{2 d t^2}\right)$ and potential energy of elastic deformations $\left(k_s h^{\frac{3}{2}}\right)$. According to the law of energy conservation, we can write:

$$\frac{m_g \vartheta_a^2}{2} = \frac{m_g}{2} \cdot \frac{d h_s^2}{d t^2} + k_s h_s^{\frac{3}{2}}, \qquad (7)$$

where ϑ_a is the velocity of the hulling object at the beginning of interaction, m/s; m_g is the mass of the hulling object, kg.

Maximum strain value h_{sm} will take place when the relative strain rate turns to zero; hence we can write:

$$h_s = \left(\frac{m_g}{2 k_s}\right)^{\frac{2}{3}} \cdot \vartheta_a^{\frac{4}{3}}. \qquad (8)$$

Let us determine the maximum value of the deforming force. To do this, we differentiate equation (5) by strain:

$$F_m = \frac{du}{dh} = 1{,}5 k_s h_s^{\frac{1}{2}}. \qquad (9)$$

Then let us substitute formula (8) into the obtained equation. The value of the deforming force will be determined by the following equation:

$$F_m = 1{,}2 k_s^{\frac{2}{3}} m_g^{\frac{1}{3}} \vartheta_a^{\frac{2}{3}}. \qquad (10)$$

During the theoretical study of the process of deformation and destruction of the shell we apply the first theory of the local limit state, the theory of maximum normal stresses. It is based on the following hypothesis: the limit state of the shell will occur when the normal stresses in the shell section in the area of contact with the surface of the working bodies reach the limit value σ_{max}. The force, directed along the centerline of the grain, acts on part of the shell for a time t and deforms the shell by the amount of h. This creates a contact area in the form of a circle and creates maximum stresses in the cross-section of the shell, which is determined by the following expression:

$$S_d = \pi d \delta, \qquad (11)$$

where d is the diameter of the contact area, m, ($d = 2r$); δ is the shell thickness, m.

These stresses are directed tangentially to the surface of the spherical shell along the perimeter of the contact site diameter.

The shell stability condition is written in the following form:

$$F_m^s = R_z(\sigma), \qquad (12)$$

where $R_z(\sigma)$ is the projection on the axial line of the hulling object of the equipotential forces of internal stresses acting on the shell in the section with the maximum diameter d;

F_m is the maximum value of the external shock load; at this:

$$R_z(\sigma) = \pi d \delta \sigma \sin\theta, \qquad (13)$$

where θ is the angle indicating the direction of normal stresses; σ is the value of normal stresses arising in the section of the shell, N/m^2.

Keeping in mind that $\sin\theta = \frac{d}{D_s}$, we write equality (9) in the following form:

$$R_z(\sigma) = \frac{\pi d^2 \delta\sigma}{D_s}. \qquad (14)$$

To determine the diameter of the contact area d, we use the known approximate formula for determining the surface area of the ball segment $Dh = \frac{d^2}{4} + h^2$.

For the considered section of the shell of the hulling object when deformed by the value of h, the value of d^2 will be determined as:

$$d^2 = 4h(D_s - h). \qquad (15)$$

Substituting equality (15) into equation (14), we obtain an expression to determine the projection of internal maximum stresses acting on the shell element in the considered section:

$$R_z(\sigma) = \frac{4\pi\delta h \cdot \sigma_e^s (D_s - h)}{D_s}. \qquad (16)$$

The value σ_e^s in the expression can be considered as the sum of some stresses in the uniaxial stress state of the shell in the considered section, equivalent in effect to the stresses in the complex stress state of the whole shell. The effect here is understood as the occurrence of the limit state of the shell during deformation. Therefore, in further considerations we will use equivalent stresses instead of maximum stresses.

Condition (12), taking into account equality (9) and equation (16), will be written in the following form:

$$1{,}5 k_s h_s^{\frac{1}{2}} = \frac{4\pi\delta h \sigma_e^s (D_s - h_s)}{D_s}. \qquad (17)$$

From here, let us express the equivalent stresses that may occur in the fruit shell during the interaction of cereal grains with the working surfaces:

$$\sigma_e^s = 0{,}12 \frac{k_s D_s}{\delta h^{\frac{1}{2}} (D_s - h_s)}. \qquad (18)$$

The right part of the resulting equation includes the coefficient k_s, which characterizes the mechanical and geometrical parameters of the hulling object and the working surface. For a more complete analysis, we substitute expression (6) into equation (18) and after some transformations we obtain:

$$\sigma_e^s = 2{,}88 \frac{\delta^{\frac{1}{2}} \left(\frac{E_s}{(1-\mu_s^2)^{\frac{3}{4}}} + \frac{E_{ws}}{(1-\mu_{ws}^2)^{\frac{3}{4}}} \right)}{h_s^{\frac{1}{2}} (D_s - h)}. \qquad (19)$$

Analysis of expression (19) shows that the stresses arising from deformation of the shell during interaction with the working surfaces depends on the physical and mechanical properties of the grain and the type of working surface. Consequently, by changing the physical and mechanical properties of seedcoat, flowering films of cereal seeds, as well as by changing the type of working surface of working bodies of hulling machines, we can determine the specific values of limiting stresses, under which the destruction of the shell occurs. Let's call these stresses critical stresses and denote by $[\sigma_s]$. Then the condition under which the shell breaks down will be written in the following form:

$$\sigma_e^s > [\sigma_s]. \qquad (20)$$

Using expression (19) and condition (20), the allowable stresses under static and dynamic (impact interaction) loads for different conditions can be determined experimentally. However, it should be noted that during the impact interaction of grain with the working surfaces of hulling machines there is an effect of "delayed fracture". This effect is explained by the insufficient duration of time, during which the stresses resulting from the interaction act. The value of critical shell stresses is determined by static loading, where the loading time is long enough. Since in shock interaction, the load time span and, hence, the stresses are not large enough, it is necessary to increase the value of the stress in order to destroy the shell. Then the dynamic destructive stresses will be calculated by the following formula:

$$[\sigma_s]_{ws} = k_t^s [\sigma_s], \qquad (21)$$

where $[\sigma_s]_{ws}$ is critical ultimate stresses of shell fracture during impact interaction with the working surface, N/m^2; k_t^s is the coefficient, which takes into account the duration of the stress-strain state of the shell during the dynamic action on the grain.

In an impact interaction, the stresses must reach a critical value if the shell is to collapse $[\sigma_s]_{ws}$. To determine the limiting impact velocity, we substitute expression (10) and (16) in condition (12). Then the condition of the limit state under dynamic influences taking into account (19) is defined by the dependence:

$$1{,}2 k_s^{\frac{2}{3}} m_g^{\frac{1}{3}} \vartheta_a^{\frac{2}{3}} = \frac{4\pi\delta h [\sigma_s]_{ws} (D_s - h)}{D_s}. \qquad (22)$$

After transforming equality (22), it can be written in the following form:

$$\upsilon_p^0 = 19,47 \frac{\left[\left[\sigma_s\right]_{ws} h\left(D_s - h\right)\right]^{\frac{3}{2}}}{s\, D_s^2 \delta \rho_g^{\frac{1}{2}} \left[\frac{E_s}{\left(1-\mu_s^2\right)^{\frac{3}{4}}} + \frac{E_{ws}}{\left(1-\mu_{ws}^2\right)\frac{3}{4}} \right]} \quad (23)$$

The obtained expression allows us to calculate the impact velocity at which the critical stresses arise in the shell. To destroy the shell in dynamic hulling methods, the impact velocity must be greater than the value of the limiting velocity. Then expression (23) can finally be written in the form of the following inequality:

$$\upsilon_a^0 > 19,47 \frac{\left[\left[\sigma_s\right]_g h_s\left(D_s - h_s\right)\right]^{\frac{3}{2}}}{D_s^2 \delta \rho_g^{\frac{1}{2}} \left[\frac{E_s}{\left(1-\mu_s^2\right)^{\frac{3}{4}}} + \frac{E_p}{\left(1-\mu_p^2\right)^{\frac{3}{4}}} \right]}. \quad (24)$$

The obtained expression makes it possible to determine the region of impact velocity values at which the destruction of the shell occurs. However, it should be noted that the maximum velocity value should be limited to the impact velocity values at which the shell destruction begins.

As can be seen from the obtained inequality, the impact velocity in hulling methods should be greater than the right part of expression (24), which includes the limiting stress under which the shell destruction occurs, the elastic modulus, including geometrical and physical parameters of the shell. To determine specific values of velocity, it is necessary to conduct a number of experiments to determine geometrical, physical and strength characteristics of the fruit grain shell of cereal crops.

4 Conclusion

As a result of the research, a physical and mathematical model of cereal grains as the subject of hulling, consisting of two separate structural elements (spherical shell and spherical kernel), each of which has its own geometric parameters and rheological properties, mathematical dependences of destructive stresses of the shell and kernel during static and dynamic interactions with the working surfaces on their strength characteristics and geometric parameters have been developed.

The obtained theoretical dependences, taken together, constitute a mathematical model describing the process of destruction of the shell of cereal grains under static and dynamic influences. It can be used to simulate the collapse of sunflower seed shells in the production of seed oil.

The developed model enables to carry out computational experiments with the use of computers and software with a required set of given values of physical-mechanical and technological properties of the processed grains and construction-technological parameters of machines intended for destruction and separation of the shell. When using the appropriate software it is possible to visualize the process of shell fracture with obtaining a full frame-by-frame picture of distribution, magnitude, direction of loads and arising stress

In the future, to expand the practical usefulness of the obtained model, it is necessary to develop a model of destruction of the kernel of cereal crops (the most valuable part used as human food), which will allow one to identify the conditions of stresses, under which the destruction of the shell with the maximum preservation of its integrity occurs.

References

1. E. G. Nurullin, The main results of the scientific direction to create new technical means of pneumomechanical type to implement energy-saving environmentally safe technologies for grain production and processing, *Modern achievements of agrarian science: Proceedings of the All-Russian (national) scientific and practical conference dedicated to the memory of honored worker of science and technology of Russia, professor, academician of Academy of Agricultural Education, Laureate of State Prize of Russia in Science and Technology, Honored Scientist* p. 109-116. Kazan: Kazan State Agrarian University, 2021.
2. S. N. Brasalin Assessment of technological efficiency of grain husking in cereal production, *Khleboprodukty*, **9**, 60-62, (2014).
3. E.G. Nurullin Pneumomechanical grain huskers (theory, design, calculation) (Kazan: Kazan University, 2011).
4. A. G. Rodina, Main qualitative indicators for assessing the work of caving machines, *Agriculture development strategy in modern conditions - continuation of the scientific heritage Listopad G.E., Academician of VASKhNIL (RAAS), Doctor of Technical Sciences, Professor*, p. 122-127, Volgograd, 06-07 November 2018.
5. S. S. Ravshanov Improving the design of the device for hulling grain *Universum: Technical Sciences*, **5-2(74)**, 37-40, (2020).
6. R. Ibyatov, A. Dmitriev, B. Ziganshin, D Khaliullin and A. Zinnatullina, Mathematical modeling of the grain trajectory in the workspace of the sheller with rotating decks. *BIO Web Conf.* **17**, 00093, (2020). DOI: 10.1051/bioconf/20201700093.
7. V. A. Mar'in, A. L. Vereshchagin, A. A. Ivanov, Feasibility of using a deck of visco-elastic material in husking buckwheat grain, *Technique and technology of food production.* **50(1)** 87-95, (2020).
8. Y.F. Lachuga, R.I. Ibyatov, B.G. Ziganshin, et al. Simulation of the Grain Trajectory along Working Bodies of the Pneumatic Mechanical

Dehuller. *Russian Agricultural Sciences.* **46,** 534-538 (2020).

9. A. V. Dmitriev, E. G. Nurullin Theoretical determination of the energy of hulling on pneumomechanical grain peelers, *Bulletin of the Kazan State Agrarian University,* **6,1(19),** 101-102 (2011).
10. I. V. Malanichev, E. G. Nurullin, Computer model of pneumomechanical husking process of buckwheat grain, *Bulletin of Kazan State Agrarian University* **3(1(7))** 169-171
11. A. Dmitriev, B. Ziganshin, D.Khaliullin, A. Aleshkinб Study of efficiency of hulling machine with variable deck. *19th International Scientific Conference Engineering for Rural Development*, Volume 19, p. 1053 – 1058, May 20-22, Latvia University of Life Sciences and Technologies Faculty of Engineering, Jelgava, 2020.
12. V. I. Lobanov, S. Yu. Buzoverov, The dependence of sunflower seed capping efficiency on the number of revolutions of the working body of the centrifugal husker *Bulletin of Altai State Agrarian University* **9(179)** 145-149 (2021).
13. D. T. Khaliullin, E. G. Nurullin, Study of grain movement in the confuser of pneumomechanical sunflower seed breaker, *Bulletin of Kazan State Agrarian University,* **5(4(18)),** 122-124 (2010).
14. E. Nurullin, R. Ibyatov, A. Dmitriev, D. Khaliullin, Modeling of grain processing in a pneumomechanical dresser. *BIO Web Conf.* **27,** 00077, (2020.) DOI: 10.1051/bioconf/20202700077
15. L.D. Landau, E.M. Lifischtz, *Theoretical Physics. Theory of elasticity,* (Moscow: Nauka, 1987)
16. R.I. Nigmatullin, *Fundamentals of mechanics of heterogeneous media.* (Moscow: Nauka, 1979).
17. S. G. Mudarisov, I. D. Badretdinov, Z. S. Rakhimov[et al. Numerical simulation of two-phase "Air-Seed" flow in the distribution system of the grain seeder, *Computers and Electronics in Agriculture*, **168,** 105151,(2020).
18. V.Z. Parton. *Mechanics of fracture* (Moscow: Nauka, 1990).
19. G.P. Cherepanov, *Mechanics of Brittle Fracture* (M.: Publishing House "Nauka" 1974).
20. A.V. Pogorelov *To the Theory of Convex Elastic Shells in the Zacritical Stage* (Kharkov: Publishing House of Kharkov University, 1960).

Productivity of various types of spring wheat under the influence of the area and nutrition status in the Pre-Kama of the Republic of Tatarstan

Igor Serzhanov*, *Farit* Shaikhutdinov, *Albina* Serzhanova , *Razil* Garaev, and *Adilya* Khamitova

Kazan State Agrarian University, 65, K. Marx Str., Kazan, 420015, Republic of Tatarstan, Russian Federation

Abstract. This paper summarizes the experimental data on the results of two years of experiments on the study the effect on the yield and grain quality of spring soft wheat of the Ulyanovskaya 105 variety and emmer wheat (Triticum Dicoccum Schvebl, spelt) of Runo variety and when applying a full dose of mineral fertilizers (N30-34P49-52K33-35) for pre-sowing cultivation. Natural background without fertilizers was in the control. Four seeding rates were tested at each nutritional level: 4; 5; 6 and 7 million pieces viable seeds per 1 ha. Meteorological conditions during the research years were different. The hydrothermal index (HTI) for the growing season of spring wheat in 2020 was 1.38, growth and development took place in favorable conditions, and in 2021 the HTI was 0.29, which had a very negative impact on the yield of the tested wheat varieties. The highest yield in the research years on both statuses of nutrition in the Ulyanovskaya105 variety was provided when sowing 6 million grains per hectare, in wheat variety Runo spelt (1.95–2.6...2.63–2.75), respectively.

1 Introduction

Grain production remains the main direction in crop production in the state program for the development of agriculture in the Russian Federation. At the same time, the implementation of the predicted parameters is associated with the need to solve a number of scientific and production problems. The most important of them are:

- the increase in the share of products with the highest marks in terms of indicators characterizing the technological, dietary and environmental properties of grain;
- the increase in the diversity of grain crops, including the expansion of the share of cereals and cereal crops unique in the direction of the use of spring wheat (spelt) [4].

The grain of spring soft wheat has a great nutritional value. It contains 12-16% of protein, 63-65% of starch, 1.5% of fiber, 1.7% of ash, as well as various vitamins, enzymes and other substances [2, 3].

Emmer wheat (spelt) occupies a special position among other types of this crop due to the highest quality of grain.

Spelt wheat grain has a high Hardness, which reaches 95% and more. Increased Hardness is explained by a higher content of protein and a mass fraction of gluten. The protein content in spelt grain is 4–6% higher compared to common wheat, and the gluten protein of this culture includes 18 essential amino acids that cannot be obtained from animal food [8, 10].

Emmer wheat (spelt) is an independent botanical variety of Triticum Dicoccum. It differs from the soft one in a smaller distribution area. In the middle of the twentieth century, it was widespread in the middle Volga region and in the North Caucasus. With the change in the structure of sown areas, the expansion of spring wheat and tilled crops, the sowing of spelt wheat has sharply decreased [7, 11].

The yield value largely depends on the density of the productive stem before harvesting, which, in turn, is mainly determined by the seeding rate, the completeness of seedlings and the survival of plants during the growing season [1; 4; 9]. A significant part of conditioned seeds do not form shoots after sowing. According to a number of authors, the completeness of seedlings depends on meteorological conditions, soil properties and moisture, as well as agricultural practices [12, 13].

Among the agrotechnical measures that affect the yield and properties of grain, fertilizers occupy an important place. Types, doses and timing of their introduction should be recommended based on their specific conditions of the area and farms, taking into account the requirements of the culture, variety [5, 6].

Under the conditions of Tatarstan, spelt has been little studied. The lack of theoretical data and practical recommendations makes it difficult to develop a unified technology for cultivating this crop in the region [7, 8]. The purpose of this research is to study the influence of various seeding rates and nutritional status on the formation of grain yield and its quality of spring soft and emmer wheat (spelt) in the conditions of the Pre-Kama region of the Republic of Tatarstan.

2 Materials and methods

The field studies were carried out in 2020–2021 on the territory of "Agrobiotechnopark" of Kazan State

* Corresponding author: igor.serzhanov@mail.ru

Agrarian University. The soil was gray forest medium loamy with the following agrochemical indicators in the soil layer of 0–20 cm: the content of organic matter (humus) was 3.1–3.4% (according to Tyurin), P_2O_5 – 229–240, K_2O – 84–117 mg/1000 g of soil (according to Kirsanov), the number of absorbed bases was 26.4–32.0 mmol (100 g of soil, pH_{salt} was 6.0–6.3 (acidity was close to neutral). Winter wheat was used as a precursor rye after pure fallow. The area of each plot was 27.5 m^2 (25 × 1.1), the repetition was 4-fold. The placement of the plots was consistent. The object of the study was the zoned varieties of spring soft wheat Ulyanovskaya 105 and spelt wheat Runo with the generally accepted technology of spring cultivation wheat in the region.

Research scheme. To study the purpose of the research, the experiment was carried out in the crop rotation link: bare fallow - winter rye - spring wheat according to the following nutritional statuses. Factor A (fertilizers):

1. Natural background (control);
2. NRK calculated per 3 tons of grain per hectare of Factor B (seeding rate). Four seeding rates were tested on each nutritional status: 4 million germinating seeds per ha; 5 million; 6 million and 7 million pieces per ha.

Fertilizers were applied in the form of ammonium nitrate, Azofoska N_{16} R_{16} K_{16} and potassium chloride. In both years of research, after pre-sowing cultivation, sowing was carried out at the optimally early time (the 1st decade of May) with seeds of the 1st class. Sowing was carried out with a seeder Wintersteiger to a depth of 4–5 cm. The calculated sowing rates deviated from those actually sown by 1.5–3.0 percent.

Crop care was performed in accordance with the requirements of advanced technology. The yield was determined by plots after threshing in the phase of full ripeness with a SAMPO-500 combine.

3. Results and discussion

During the study, the weather conditions in the growing season were not the same, both in terms of the amount of precipitation and the temperature regime. In 2020, soil moisture under the crops of the studied objects was quite high at the optimal temperature regime (HTI-1.38), which had a positive effect on the future yield of spring wheat agrocenosis.

The meteorological indicators of the growing season in 2021 were characterized as extremely unfavorable for the formation of spring wheat productivity. The months of May and June were extremely dry, HTI was 0.29. Precipitation in July (57% long-term norm) did not have a significant impact on the yield.

Our observations in both years of research showed that seedlings of spring soft wheat and emmer wheat (spelt) appeared in all variants of the experiment together on the 10th and 11th days after sowing. On a fertilized background, seedlings had a more intense green color compared to plants on a natural background (control). The density and completeness of seedlings in both types of wheat were significantly affected by seeding rates, regardless of the status of nutrition. Thus, on average, for two years, spring soft wheat on both statuses of nutrition at a rate of 4 million per 1 ha, or 400 pcs. per 1 m^2, on an area of 1 m^2 there were 331–334 plants, with an increase in the seeding rate to 7 million, or 700 pcs. per 1 m^2, the number of seedlings increased by 217–222 pcs and amounted to 548–556 plants per 1 m^2. In wheat, spelt, respectively: 338–353; 225–234 pcs and 563–587 plants per 1 m^2. If with the increase in the sowing rate in both types of wheat, the number of plants increased, then the completeness of seedlings decreased to some extent (by 4.1–4.5%). In general, it remained at a high level (78.3–82.8; 79.4–83.5%) in soft wheat and 80.4–84.5; 83.8–88.3% in emmer wheat (spelt).

At all levels of nutrition in both types of wheat, as the seeding rate increased, the fall of plants to full ripeness from the number of seedlings increased by 7.9–10.5 percent. Consequently, the sowing rate is a very significant factor in the regulation of the density of the stem and one of the main elements of the cultivation technology, both spring soft wheat and spelt.

The weather conditions during the years of research on various levels of nutrition and seeding rates determined the characteristics of plant growth and development, the yield and grain quality of spring soft wheat and spelt wheat (Diccocum).

At all the levels of nutrition, with the increase in the seeding rate, the yield increased. However, the proportionality of the increase in seeding rates was noted only up to a certain limit (Table 1).

Table 1. Productivity of various types of spring wheat depending on the level of nutrition and seeding rates, t/ha.

Type and variety of spring wheat (A)	Status power (V)	Seeding rate, mln.pcs/ha (C)	Year		Average for 2 years	Increase	
			2020	2021		from status power	from seeding rates
Soft wheat – Ulyanovskaya 105	Natural - background (control)	4	2.16	1.44	1.80	–	–
		5	2.79	1.59	2.19	–	0.39
		6	2.38	1.79	2.09	–	0.29
		7	2.39	1.82	2.07	–	0.29
	NRK per 3 tons of grain per hectare	4	3.05	2.0	2.53	0.73	–
		5	3.14	2.2	2.67	0.48	0.14
		6	3.67	2.4	3.04	0.95	0.51
		7	3.62	2.5	2.91	0.84	0.38

Emmer wheat (spelt) Runo	Natural background (control)	4	2.13	1.77	1.95	–	–
		5	2.35	1.89	2.02	–	0.07
		6	2.30	1.95	2.13	–	0.18
		7	2.31	1.96	2.07	–	0.12
	NRK per 3 tons of grain per hectare	4	2.49	2.17	2.33	0.38	–
		5	2.79	2.32	2.56	0.44	0.23
		6	2.95	2.63	2.79	0.66	0.46
		7	2.96	2.69	2.68	0.61	0.35

The maximum reliable yield in 2020, favorable for weather conditions for both types of wheat was obtained when sowing 5 million germinating seeds per hectare on a natural background, and 6 million on fertilized background.

In the extremely dry year 2021, on both backgrounds of nutrition, soft wheat and spelt had higher yields (2.09–3.04 t/ha and 2.02–2.79 t/ha) at a sowing rate of 6 million seedlings. seeds per 1 ha.

Fertilizers applied at a yield of 3.0 t/ha actually provided 3.05–3.62 t/ha in 2020 and 2.0–2.2 t/ha in the extremely dry 2021. In wheat, spelt, respectively: 2.49–2.86 and 2.17–2.49 tons per hectare.

The data in the table show that fertilizers had a significant effect on yields on both types of wheat. In average for 2 years, depending on the seeding rates, with the application of fertilizers, the excess yield over the control in soft wheat was 0.73–0.84 t/ha, and spelt wheat was 0.38–0.61 t/ha.

The content of protein in the grain, the mass fraction of gluten, and the weight of 1000 grains decreased in both statuses of nutrition in the studied varieties of spring wheat, with the increase in the sowing rate from 4 to 7 million germinating seeds per hectare (Table 2).

Table 2. Technological qualities of spring wheat and emmer wheat (spelt) depending on the main elements of cultivation technology (cf. 2020-2021).

Type and variety	Status of nutrition	Seeding rate, million/ha	Weight of 1000 grains, g	Nature, g/l	Hardness, %	Protein content, %	Content of the mass fraction of gluten, %	Quality of the mass fraction of gluten according to IDK
Soft wheat Ulyanovsk 105	Natural background (control)	4	33.0	742	44.7	12.8	27.1	69
		5	32.4	742	45.3	12.7	27.0	68.7
		6	32.1	744	45.2	12.7	26.3	68.0
		7	32.0	747	45.0	12.4	26.3	67.6
	NRK per 3 tons of grain per hectare	4	33.4	748	46.9	14.2	31.2	75.4
		5	33.1	750	47.3	14.0	30.7	75.4
		6	32.9	753	48.0	13.1	30.4	75.0
		7	32.6	755	48.0	13.0	30.0	75.0
Emmer wheat (spelt) Fleece	Natural background (control)	4	28.4	–	68.0	13.9	28.6	90.0
		5	28.2	–	68.0	13.6	28.4	90.0
		6	28.0	–	68.0	13.5	28.4	90.0
		7	27.9	–	68.0	13.3	28.4	90.0

	NRK per 3 tons of grain per hectare	4 5 6 7	29. 3 29. 0 28. 7 28. 4	– – – –	81. 0 81. 0 81. 0 81. 0	15. 8 15. 3 15. 0 15. 0	33.0 32.3 31.7 31.4	85. 0 85. 0 85. 0 85. 0

The changes in the level of nutrition with the introduction of calculated doses of mineral fertilizers had a positive effect on some indicators of grain quality. The protein content in the grain of spring soft wheat increased by 0.4% at a sowing rate of 6 million pieces seeds per hectare, weight of 1000 grains per 0.8 g and grain Hardness increased by 2.8%. In emmer wheat (spelt) these indicators were 1.5%; 0.7 g and 13% respectively.

In modern conditions, along with agronomic efficiency, it is even more important to assess the impact of the area and status of nutrition of various types of spring wheat in the Pre-Kama region of the Republic of Tatarstan in terms of such indicators as the yield of additional products in value terms, the amount of profit per unit area and the level of profitability [15].

The calculations of the economic efficiency of the use of the tested rates of fertilizer application, at different seeding rates of spring soft wheat of the Ulyanovskaya 105 variety in conditions of gray forest soil are shown in Table 3.

Table 3. Economic efficiency of the use of various rates of application of mineral fertilizers, sowing, in crops of spring soft wheat (the average number for 2020–2022).

Indicators	Status of nutrition and seeding rates, mln/ha							
	Natural, without fertilizers (control)				Average fertilizer rates (for 3 tons of grain/ha)			
	4 million	5 million	6 million	7 million	4 million	5 million	6 million	7 million
Productivity, t/ha	1.80	2.19	2.09	2.07	2.53	2.67	3.04	2.91
Yield increase, t/ha	–	–	–	–	0.73	0.48	0.95	0.84
Additional products, rub.	–	–	–	–	9673	6360	12587	11130
Additional costs - total, rub.	–	–	–	–	7013	7094	7386	7212
Including:								
Fertilizer, rub.	–	–	–	–	6856	6856	6856	6856
Depreciation, rub.	–	–	–	–	38	38	38	38
Salary, rub.	–	–	–	–	19	19	19	19
Fuel, rub.	–	–	–	–	25	25	25	25
Motor transport works, rub.	–	–	–	–	75	75	75	75
Additional profit, rub.	–	–	–	–	2660	2660	5201	3918
Profitability level of additional costs, %	–	–	–	–	37.9	37.9	70.4	54.3

* Legend: At the beginning of 2020, the average selling price of wheat grain of the third class was 13.250 rubles.

The cost of additional production was calculated according to the size of the increase in yield and the average selling price of 1 ton of grain of the third class at the beginning of 2020 (13,250 rubles). Mineral fertilizers at average rates at the optimal sowing rate (6 million germinating seeds per 1 ha) gave a very significant increase in marketable products: the cost of additional products from which they increased by 12,587 rubles/ha, respectively.

Additional costs for the use of fertilizers, including harvesting and finalizing the increase in grain, were calculated according to the flow charts for the cultivation of spring wheat, developed by scientists of Kazan State Agrarian University on a self-supporting assignment of the Ministry of Agriculture and Food of the Republic of Tatarstan. The use of mineral fertilizers at average rates during sowing with a seeding rate of 6 million seeds per hectare increased the cost of the cultivation of spring wheat by 7,386 rubles per hectare, respectively. Despite this, the use of mineral fertilizers at rates calculated in order to obtain 3.0 t/ha of spring wheat at an optimal seeding rate turned out to be economically justified. As we can see, the additional profit from each hectare from the introduction of average fertilizer rates at a seeding rate of 6 million amounted to 5201 rubles. This is also proved by the relative indicator of the economic efficiency of the use of fertilizers – the level of profitability: the use of average fertilizer rates at a sowing rate of 6 million seeds per hectare provided profitability at the level of 70.4%.

Thus, the use of mineral fertilizers in the norms established by the calculation-balance method to obtain

3.0 t/ha of spring wheat grain during sowing 6 million germinating seeds per 1 ha, taking into account local delivery coefficients and use of NRK from soil and fertilizers in conditions of gray forest soil was agronomically and economically beneficial and provided additional profit per hectare in the amount of 5201 rubles.

4 Conclusion

According to the research results, we can draw the following conclusions:
Mineral fertilizers applied according to the calculation method is an effective method for the increase in the yield of both types of wheat in the conditions of the Pre-Kama zone of the Republic of Tatarstan.

Despite the fact that the yield of the studied wheat varieties in both years of research increases to a seeding rate of 7 million pcs. seeds per hectare, depending on the status of nutrition, the norms are more effective - 5-6 million viable seeds per hectare.

Fertilizers and optimal seeding rates provide high quality food grains.

The use of mineral fertilizers at the optimal seeding rate established by the calculation-balance method to obtain 3.0 t/ha of spring wheat grain, in conditions of gray forest soil, was agronomically and economically beneficial and provided additional profit per hectare in the amount of 5201 rubles.

References

1. M. F. Amirov, L. G. Sagitov, R. N. Salavatullin, Grain economy of Russia **2 (50),** 6–9 (2017)
2. M. F. Amirov, Bulletin of the Kazan State Agrarian University **12 (4–2 (47),** 5–8
3. V. G. Vasin, N. A. Prosandeev, *Proceedings of the Orenburg State Agrarian University* **3,** 53–56 (2012)
4. R. I. Ibyatov F. Sh. Shaikhutdinov, A. A. Valiev, Grain economy of Russia **2 (50),** 17–22 (2017)
5. A. N. Kshnikatkina, I. G. Rusyaev, Agrochemical Bulletin **3,** 48–50 (2018)
6. I. G. Rusyaev, Agrochemical Bulletin **3,** 48–50 (2018)
7. F. N. Safiollin, G. S. Minnullin, M. M. Khismatullin, S. V. Sochneva, Grain economy of Russia **2 (50),** 29–33 (2017)
8. I. M. Serzhanov, F. Sh. Shaikhutdinov, R. I. Ibyatov, et al., Bulletin of the Kazan State Agrarian University **12 (4–2 (47),** 62–66
9. F. Sh. Shaikhutdinov, I. M. Serzhanov, R. I. Ibyatov, et al., Bulletin of the Kazan State Agrarian University **13 (4 (51),** 103–108 (2018)
10. M. Kh. Sharafutdinov , L. S. Nizhegorodtsev, R. I. Safin, Grain economy of Russia **2 (50),** 69–72 (2017)
11. V. Petrenko, T. Sheiko, R. Spychaj, O. Prysiazhniuk, L. Khudolii, Romanian Agricultural Research **35,** 255–264 (2018)
12. N. Daskalova, S. Doneva, P. Spetsov, Cereal Research Communications **44 (3),** 453–460 (2016)
13. R. I. Ibyatov, F. Sh. Shaikhutdinov, A. A. Valiev, Grain economy of Russia **2 (50),** 17–22 (2017)
14. V. A. Kolesar, A. A. Ziganshin, R. I. Safin Grain economy of Russia **2 (50),** 45–47 (2017)
15. A. R. Valiev, I. Kh. Gabdrakhmanov, R. I. Safin, B. G. Ziganshin, *Agriculture system of the Republic of Tatarstan* (Kazan: Center for Innovative Technologies LLC) (2014)
16. M. Yu. Gilyazov, A. A. Lukmanov, M. R. Muratov, *Long-term use of fertilizers and arable land productivity* (Kazan: Publishing House of Kazan University) (2016)

Productivity of spring spiked crops when using biological products based on *Bacillus subtilis* in the conditions of the Republic of Tatarstan

Farit Shaikhutdinov, *Igor* Serzhanov, Albina *Serzhanova*, *Razil* Garaev, and *Pavel* Semenov

Kazan State Agrarian University, 65, K. Marx Str., Kazan, 420015, Republic of Tatarstan, Russian Federation

Abstract. The effectiveness of the use of an experimental strain of Bacillus subtilis RECB-95B in the treatment of seeds of barley and spring wheat against diseases was studied. The growing season during the years of research was favorable for the growth and development of the tested crops. Barley and spring wheat seeds were treated with various doses of the experimental preparation. The scheme of the experiment included: control without seed fertilizing; the industrial biological product Rhizoplane was used as the standard. The smallest development of striped and brown spots on plants was noted on the variant, the seeds of which were treated with the biological product Rhizoplane. Plant damage by net blotch and root rot was minimal in the variant with Bacillus Subtilis RECB-95B 2.0l/t seed treatment. The highest yield of barley and spring wheat in the experiment was achieved with the treatment of seeds with Bacillus Subtilis RECB-95B 2 l/t (3.23 t/ha and 3.47 t/ha).

1 Introduction

Spring wheat is the main grain fodder crop of the Republic of Tatarstan. Among the most important agroecological features of barley is its higher potential for genotypic adaptation compared to spring wheat [7]. The yield and sowing qualities of spring wheat and barley seeds are determined by many factors, among which the correct high agricultural technology, meteorological conditions, the use of various groups of biological products and the adapted variety mainly play a determinant role. In modern zonal farming systems, the quality of seeds is of paramount importance, because without the full provision of farms with conditioned seeds of the best varieties, the efficiency of all other parts of the agrotechnical complex decreases [2, 3]. The formation of the crop of spring wheat and barley is determined under the influence of a complex set of conditions, each of which affects its quantity and quality [5, 6]. One of the important aspects of the influence of the environment surrounding the mother plant is the change in the viability of seeds as a result of infection with fungi, bacteria and viruses that cause various diseases of germinating seeds, seedlings and adult plants [1]. Resource saving in the field of crop production involves the widespread use of the achievements of modern biotechnology, including the use of various groups of biological products in the technology of growing crops [4].

2 Materials and Methods

Microfield experiments were performed on the experimental field of Kazan State Agrarian University in 2018. The soil of the experimental plot was gray forest. The content of humus was 4.1%, the pH of the salt extract was 5.5, easily hydrolysable nitrogen was 98–112, mobile phosphorus (according to Kirsanov) was 206-232, exchangeable potassium (according to Kirsanov) was 89-93 mg/kg of soil. The plot area was 1.0 m^2. The experiments were performed six times. The predecessor was winter rye. Fall plowing was carried out in August with preliminary stubble peeling. Fertilizers were applied under pre-sowing cultivation. The fallow harrowing was carried out on April 30, pre-sowing cultivation, respectively, on May 6. Sowing was carried out with a SN-16 seeder and an MTZ-82 tractor. The sowing rate for wheat was 5, and for spring wheat 6 million germinating seeds per 1 ha.

The object of the study was spring wheat and barley of Raushan and Ulyanovskaya 100 varieties.

The scheme of the experiment included the study of the following options:

1. No treatment (control);
2. *Chemical fungicide Armor* (seed treatment, standard KS 1.5 l/t);
3. *Pseudomonas fluorescens* Rhizoplane (seed treatment) + Rhizoplane, 1 l/ha (plant spraying);
4. *Bacillus subtilis* RECB – 95 B (1.0 l/t) (seed treatment) + RECB-95 B, 1.0 l/t (plant spraying);
5. *Bacillus subtilis* RECB – 95 B (2.0 l/t) (seed treatment) + RECB-95 B, 2.0 l/t (plant spraying);
6. *Trichoderma viride* RECB – 74 V (2.0 l/t) (seed treatment) + RECB-74 V, 2.0 l/t (plant spraying).

In 2019, the research was continued with spring wheat on the experimental field of Kazan State Agrarian University in the Laishensky district and in three state variety plots of the Republic of Tatarstan (Chistopolsky, Zainsky and Buinsky).

Table 1 shows the grain yield of spring wheat depending on presowing seed treatment and spraying (microfield experience) in 2018. In this experiment, biological agents were used, but the agent *Bacillus subtilis* RECB – 95 B (2.0 l/t) gave the highest increase to the control.

3 Results and Discussion

Seed treatment and the use of spraying in the experiments in 2018 contributed to the formation of higher yields of both wheat and spring wheat (Figure 1).

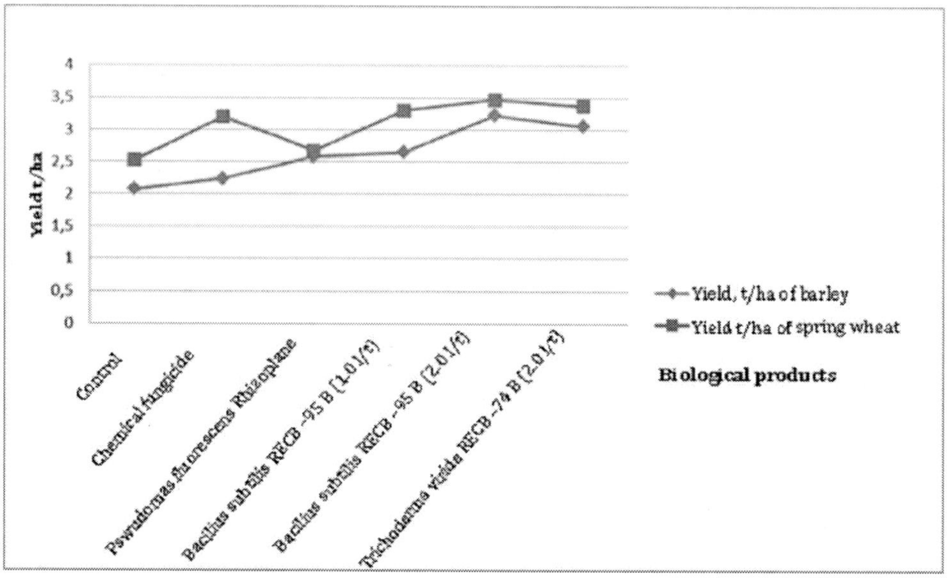

Fig. 1. Grain yield (t/ha) of spring wheat and barley depending on pre-sowing seed treatment and spraying (microfield experience), 2018.

The largest increase in grain yield of spring wheat - 0.94 t/ha and barley – 1.15 t/ha were obtained using *Bacillus subtilis* RECB – 95 B (2.0 l/t). The use of *Trichoderma viride* RECB – 74 V (2.0 l/t) also provided a significant increase of 0.84 t/ha and 0.98 t/ha in grain yield. The great importance of increasing the protein content in grain is generally recognized. Protein content is a quantitative trait with polygenic inheritance and high sensitivity to environmental conditions. In order to obtain high quality grain, nutrients and products that contribute to their better use are of great importance. The maximum protein content in the grain of spring wheat and wheat was 13.7–16.7% and nature 703–784 g/l was in the variant with drug treatment *Bacillus subtilis* RECB – 95 B (2.0 l/t) (Table 1).

Table 1. Grain quality indicators of wheat and spring wheat depending on presowing seed treatment and spraying (microfield experience), 2018.

Variant	Grain nature, g/l		Protein content, %	
	Barley	Spring wheat	Barley	Spring wheat
Control	672	763	12.5	10.8
Chemical fungicide	679	783	12.9	10.0
Pseudomonas fluorescens Rhizoplane	684	774	13.2	13.0
Bacillus subtilis RECB – 95 B (1.0 l/t)	677	778	13.3	14.9
Bacillus subtilis RECB – 95 B (2.0 l/t)	703	784	13.7	16.7
Trichoderma viride RECB – 74 V (2.0 l/t)	671	778	12.6	13.5

In 2019, in field experiments in the conditions of the Laishevsky district (the Pre-Kama) on gray forest soils, a significant increase in the yield of spring wheat was obtained using *Trichoderma viride* RECB – 74 V (2.0 l/t) and *Pseudomonas fluorescens Rhizoplane* 0.26 and 0.25 t/ha, respectively (Figure 2).

In field experiments conducted in the same year according to the same scheme, but in the conditions of the Western the Pre-Kama (Chistopol state variety testing plots) on leached chernozems, the maximum spring wheat yield of 4.53 and 4.50 t/ha was obtained using *Bacillus subtilis* RECB – 95 B (1.0 l/t) and *Trichoderma viride* RECB – 74 V (2.0 l/t) (Fig. 3).

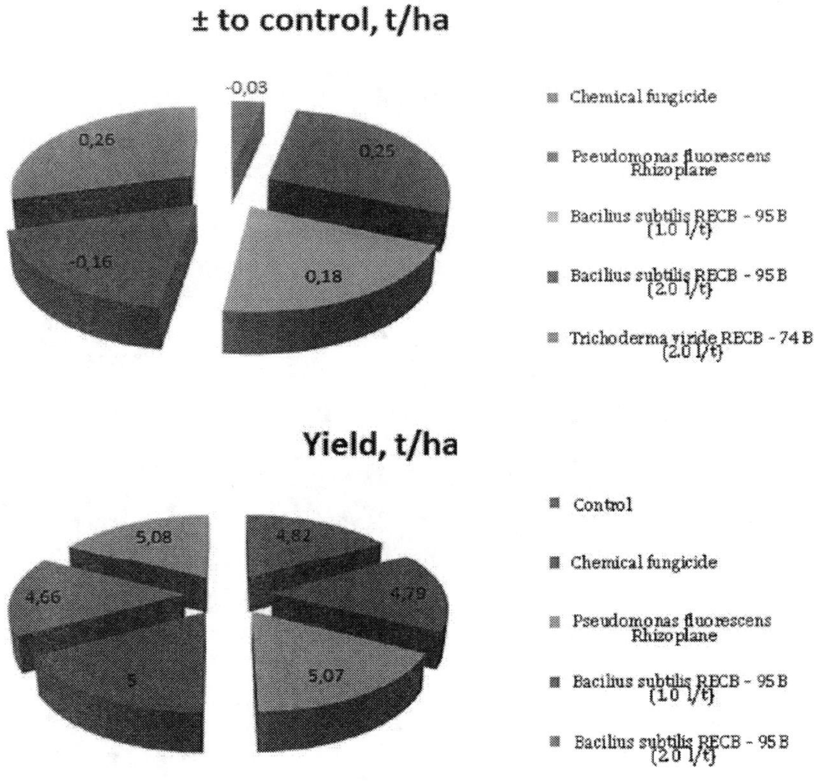

Fig. 2. Grain yield (t/ha) of spring wheat on gray forest soils of the Pre-Kama (Laishevsky district) depending on presowing seed treatment and spraying (field experience), 2019.

The results of studies in the conditions of the Eastern Trans-Kama region (Zainsky state variety testing plots) are shown (Fig. 4) A significant increase in the yield of spring wheat was obtained using *Bacillus subtilis* RECB – 95 B (1.0 l/t) and (2.0 l/t) – 0.50 and 0.35 t/ha. The use of *Trichoderma viride* RECB – 74 V (2.0 l/t) provided the increase in spring wheat grain yield of 0.37 t/ha compared to the control one.

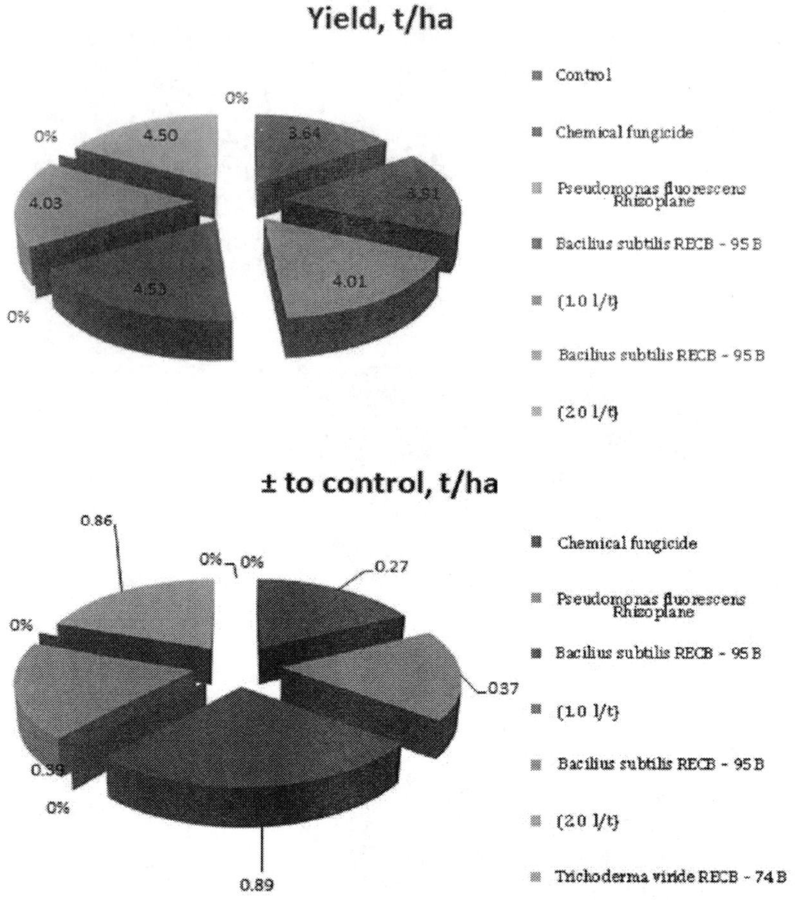

Figure 3. Grain yield (t/ha) of spring wheat on the chernozems of the Western Trans – Kama region, depending on pre-sowing seed treatment and spraying (field experience), 2019.

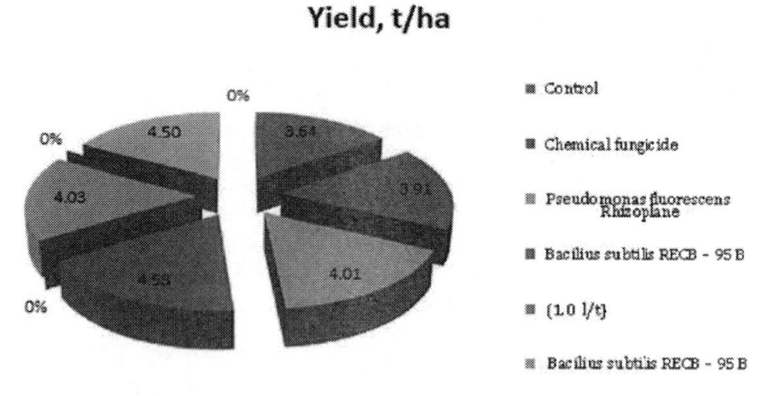

https://doi.org/10.1051/bioconf/20225200062

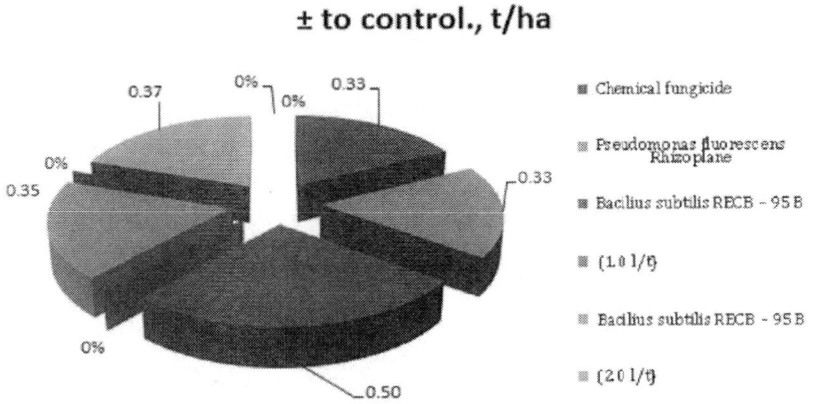

Fig. 4. Grain yield (t/ha) of spring wheat on chernozems of the Eastern Trans-Kama, depending on the pre-sowing treatment of seeds and spraying (field experience), 2019.

4 Conclusion

In the conditions of the Pre-Kama region of the Republic of Tatarstan on gray forest soils in 2018, the highest yield of barley and spring wheat was obtained when using *Bacillus subtilis* RECB – 95 B (2.0 l/t) and *Trichoderma viride* RECB – 74 B (2.0 l/t), and in 2019, only *Trichoderma viride* RECB – 74 V (2.0 l/t) gave a significant increase.

In the conditions of the Trans-Kama region of the Republic of Tatarstan on leached chernozems, the highest yield of spring wheat was obtained using *Bacillus subtilis* RECB – 95 B (1.0 l/t) and *Trichoderma viride* RECB – 74 B (2.0 l/t).

In order to increase the protein content in barley grain and to stabilize high yields of the main grain fodder crop in the Republic of Tatarstan, it is advisable to use the biological product *Bacillus subtilis* RECB – 95 B for seed treatment before sowing and spraying plants during the growing season.

References

1. M. F. Amirov, A. M. Amirov, Bulletin of Kazan State Agrarian University **1 (35),** 98–102 (2015)
2. M. F. Amirov, Bulletin of Kazan State Agrarian University **2 (44),** 5–8 (2017)
3. M. F. Amirov, V. P. Vladimirov, I. M. Serzhanov, F. Sh. Shaikhutdinov, *Adaptive technologies for the cultivation of field crops* (Kazan: Brig Publishing House) (2018)
4. I. M. Serzhanov, F. Sh. Shaikhutdinov, A. R. Serzhanova, R. I. Garaev, Bulletin of Kazan State Agrarian University **2 (53),** 52–57 (2019)
5. L. V. Karpova, *Proceedings of the Orenburg State Agrarian University* **1,** 13–15 (2009)
6. A. M. Ganiev, I. M. Serzhanov, F. Sh. Shaikhutdinov, Grain economy of Russia **2 (50)**, 12–17 (2017)
7. L. Z. Karimova, L. S. Nizhegorodtseva, R. I. Safin, Bulletin of Kazan State Agrarian University **2 (36),** 161–163 (2015)
8. F. Sh. Shaikhutdinov, I. M. Serzhanov, A. R. Serzhanova, R. I. Garaev, *Scientific works of the All-Russian (national) scientific and practical conference dedicated to the memory of Kh. S. Gainanov* (Kazan, February 26, 2021) (Kazan: Kazan State Agrarian University) pp 357–361 (2021)
9. F. Shaikhutdinov, M. Amirov, I. Serzhanov, et al., BIO Web of Conf. **17,** 00076 (2020)
10. G. Kh. Khusainova, R. I. Safin, *Reproduction of soil fertility and food security in modern conditions: Proceedings of the international scientific and practical conference* (Kazan, March 17, 2021) (Kazan: Kazan State Agrarian University) pp 294–299 (2021)

Features of the granulometric composition and properties of weakly and low-humus agro-gray soils in the Vyatka strip of the forest-steppe zone

Lilia Gaffarova*, *Ragat* Minicaev, *Anastasia* Akhrarova, and *Lilia* Nikifirova

Kazan State Agrarian University, 65, K. Marx Str., Kazan, 420015, Republic of Tatarstan, Russian Federation

Abstract. Statistical parameters of agro-gray soil varieties obtained on the basis of generalization of information are the zero-monitoring cycle in the studied region and serve as reliable criteria for the diagnosis of these soils. Numerical indicators reflect the results of past and current soil processes. In agricultural production, the role of the arable layer with all the properties that determine the level of grain yield is especially important. Thus, in the studied varieties, there is a stable relationship between the amount of silt fraction, food elements, and humus. Thus, the clay variety is characterized by the highest fertility of the arable layer - an average content of mobile potassium (10.5 mg/100 g), an increased content of phosphorus (12.4 mg/100 g), low-concentration of humus (2.9%), close to neutral reaction of the medium (5.8 pH). Comparative analysis of soil formation products in the analyzed soil varieties shows their regular change in morphometric characteristics and properties of soils, reflects the result of soil formation processes, partly-economic activity.

* Corresponding author: gaffarovalylya@mail.ru

1 Introduction

The analyzed zonal soil varieties are at the equilibrium stage, with minor changes under the influence of natural factors of soil formation [1]. Current transformations mainly occur under the influence of economic activities, which include plowing the soil with a violation of natural composition, its annual processing for growing a whole set of crops – from grain to row crops, alienation of crops from the place of cultivation, application of organic and mineral fertilizers, protection of cultivated plants with modern means, etc. In modern conditions, intensive farming is carried out by agricultural machines, which also negatively affects the soil cover of arable land.

Despite the existing changes in the morphological structure and chemical composition of soils, the main patterns of distribution of properties are still preserved. At the same time, human economic activity and its intensity have a general tendency to increase over time. It is aimed at increasing soil fertility by improving their humus state, providing basic nutrients and neutralizing soil solution through liming and phosphorization of acidic soils.

For the purpose of directed regulation of soil changes under the influence of economic activity, it is necessary to take into account natural patterns of spatial changes in soil properties.

2 Materials and methods

We have considered the regularities of changes in soil parameters depending on the granulometric composition. For processing data of soil sections, clay (c), heavy loam (h), medium loam (m), light loam (l) and sandy loam (s) varieties of agro-gray soils of the Republic of Tatarstan in the North-Eastern part of the forest-steppe were selected [2, 3]. The region occupies the Northern strip of the forest-steppe zone, occupies one of the leading places in modern agricultural production, crop production and animal husbandry. The obtained samples of soil indicators of multiple sections of agro-gray soils were processed by the method of mathematical statistics [4].

3 Results and discussion

Each soil property and morphological indicators have their own significance in the formation of agricultural products. At the same time, statistical indicators of the morphological structure and properties of soils (arithmetic mean, average error of the arithmetic mean, standard deviation, coefficient of variation, sample size, limit values) are more informative than information obtained from typical soil sections. The significance of statistical characteristics increases at the regional level, and the uniqueness of these soil varieties is determined by the same set of soil formation factors.

The thickness of the arable horizon in the varieties under consideration varies in the following order (tables 1–5): 27(c) – 24.8 (h) – 24.7 (m) – 24.2 (l) – 23.8 cm (s). The Maximum difference between them is 3.2 cm, which confirms the anthropogenic nature of this soil layer and the depth of humus-accumulative and transition horizons in the varieties.

The transition horizon AEl lies below the arable layer, in many cases it is part of it, so information about its lower boundary is fragmentary.

The lower boundary of the BEL horizon has the following range: 32.7 (c) – 28.0 (m) – 34.2 (l) cm – 32.7 cm (s); the differences between them are also insignificant.

The lower boundary of the horizon B lies at a depth of 80.8 (c) – 96.7 (h) – 84.9 (m) – 99.6 cm (l) – 73.3 (s). Significant difference between them observed only is in the sandy loam variety. The formation of a powerful illuvial horizon should be explained by a periodically flushing type of water regime.

The distribution of the humus content in the arable horizon from the granulometric composition is more clearly reflected, it is as follows: 2.9 (c) – 2.8 (h) – 2.4 (m) – 2.0 (l) - 1.2 (s) %. The feature is a decrease in the humus content by 1.7% between clay and sandy loam varieties, at a depth of more than 30 cm, the differences are almost erased. All this is evidence of improved conditions for the accumulation of humus of clay and heavy loam varieties.

The sum of absorbed bases represents a positive function of the humus content and fine fraction of the granulometric composition and has the following series for these representatives of the arable horizon soils: 25.7 (c) – 19.8 (h) – 17.7 (m) – 16.1 (l) – 14.5 (s) mmol/100 g. In this series, the difference in extreme values is 11.2 mmol/100 g of soil. The hydrolytic acidity of the arable layer has the following values: 2.2 (c)–3.1(h)–2.6 (m)–2.3 (l)-2.0 (s) mmol/100 g of soil. The maximum difference between them is 1.1 mmol/100 g of soil, which indicates a uniform distribution. The profile distribution of the sum of absorbed bases and hydrolytic acidity is consistent with changes in both the humus content and the fine fraction, as well as with a local increase in the podzolic process in the upper part of the profile.

The pH value of the salt suspension of soils in the arable layer has the following range: 5.8 (c)–5.5 (h) – 5.6 (m) – 5.8 (l) – 5.4 (s).

According to the pH value of the illuvial horizon, the group of soils has the following range: 5.3 (c) – 4.9 (h) – 5.0 (m) – 5.4 (l) - 5.6 (s). It reflects the gradual neutralization of pH as it moves from the upper horizons to the soil-forming rock. In this direction, there is an improvement in the conditions for the absorption of trace elements for the root system, along with better migration in light soils.

Mobile elements of nutrition-phosphorus and potassium are mainly inherited by the breed in the form of gross forms. In the process of weathering rocks and minerals, they pass into water-soluble forms and become available to the root system of plants. In extensive

* Corresponding author: gaffarovalylya@mail.ru

agriculture, the main source of phosphorus and potassium is their gross forms, and in intensive agriculture, a new source appears in the form of mineral and organic fertilizers. Accordingly, the phosphate and potash regimes of the arable horizon are the result of reflecting the influence of these two components.

The content of mobile phosphorus of the studied soils in the arable horizon has the following range in the direction from average to high security: 12.4 (c)–10.1 (h)–11.1 (m)–11.9 (l)–8.0 (s) mg/kg of soil. Of course, such a change in the phosphorus content in arable soils is the result of the use of phosphorus-containing mineral fertilizers and the possibilities of its accumulation. The criterion for effective use of fertilizers is a positive balance of phosphorus, which is confirmed by calculations in recent years [5, 6].

A similar trend is also observed with respect to mobile potassium. The studied soils in the arable horizon have a corresponding range: 10.5 (c)–9.1 (h)–7.6 (m)–9.7 (l)–3.6 (s) mg/kg of soil. The content of mobile potassium gradually increases from a very low degree of security in sandy loam varieties to an average security in others. Such a relatively high level of potassium supply is an integral process of the following phenomena: decomposition of rock containing potassium in the composition, alienation of potassium from the soil with the crop yield, loss of mobile potassium by water flows during intense precipitation, negative potassium balance in recent years [7, 8].

The profile distribution of granulometric particles < 0.01 and < 0.001 mm of the studied soils has an eluvial-illuvial type according to B. G. Rozanov (2004), which is the result of elementary soil processes, primarily due to the movement of a fine soil fraction [9]. This requires a wash or periodically wash type of water regime and an acid reaction of the soil solution to disperse soil particles. These conditions are most optimal in the sandy loam variety.

The fundamental components of soils - particles of physical clay (<0.01 mm) and silt fraction (<0.001 mm) are less subject to change over time. As is known, the granulometric composition and the ratio of granulometric particles in the soil is inherited from the soil-forming (parent) rock and changes slightly in the process of soil formation [10]. The silty fraction of the granulometric composition in the middle part of the soil is more mobile and variable than the fraction of physical clay, this is evident from the corresponding series: 23.2 (c) – 16.5 (h) – 13.8 (m) – 7.2 (l) - 5.1 (s) % of the soil. Thus, the arable layer of agro-gray forest soils contains 16.5 % silt, in the horizon B-34.2-37.2 %, in the soil-forming rock-32.8 %. This difference in the values of this fraction occurred in the process of soil formation. Thus, the upper part of the agro-gray soil undergoes acid hydrolysis as a result of the podzolic process. Therefore, the upper horizons are impoverished by the silty fraction, and the products of destruction are moved by the soil solution to the illuvial horizon, where they accumulate. Some of the silty particles migrate without significant destruction and these two processes go together.

Table 1. Statistical averages of indicators of signs and properties of light gray clay arable soils in the Vyatka strip of the pre-Kama region of the Republic of Tatarstan.

Horizons	lower bound, cm	Particles, %		Humus, %	mg-EQ/100g		PH, pH*	Mobile, mg/100g	
		<0.001	<0.01		Ca+Mg	Hr		P_2O_5	K_2O
P	27.0	23.2	52.1	2.9	25.7	2.2	5.8	12.4	10.5
Bt_1	44.4	37.8	59.3	0.7	26.7	2.0	5.3	7.3	8.8
Bt_2	80.8	41.2	58.9	0.6	28.9	–	5.0	–	–
BC	107.5	37.9	61.6	–	–	–	7.0	–	–
C	136.2	34.1	58.9	–	–	–	7.0	–	–

Table 2. Statistical average indicators of characteristics and properties of light gray forest heavy loam arable soils (n = 111–144).

Horizon	lower bound	Particles, %		Humus, %	mmol/100 g		pH_{KCl}	Mobile forms, mg / kg	
		<0.001 mm	<0.01 mm		Ca+Mg	H_r		P_2O_5	K_2O
P	24.8	16.5	44.1	2.8	19.8	3.1	5.5	101	91
AEl	28.5	16.7	43.0	1.7	16.4	2.8	5.3	96	75
BEL	32.7	22.6	47.1	1.1	19.8	3.4	4.8	69	85
Bt_1	48.3	34.2	53.1	0.7	22.8	3.1	4.9	106	90
Bt_2	96.7	37.2	56.1	0.5	25.8	2.8	4.7	–	–
BC	122.7	34.3	55.5	–	–	–	5.9	–	–
C	160.0	32.8	54.7	–	–	–	6.3	–	–

Table 3. The statistical average of the indicators characteristics and properties light grey medium loam arable soil Privatsky strip of pre-Kama region of the Republic of Tatarstan.

Horizons	lower bound, cm	Particles, %		Humus, %	mg-EQ/100g		PH, pH*	Mobile, mg/100g	
		<0.001	<0.01		Ca+Mg	Hr		P_2O_5	K_2O
P	24.7	13.8	36.6	2.4	17.7	2.6	5.6	11.1	7.6
AEl	31.3	11.3	35.8	1.4	14.5	2.2	5.3	11.1	4.5
BEL	28.0	17.4	41.0	0.7	16.0	2.8	5.2	14.0	7.8
Bt_1	45.0	28.0	44.8	0.7	20.4	2.3	5.0	15.2	9.2
Bt_2	84.9	32.4	47.4	0.4	23.4	2.6	4.8	–	–
BC	120.1	30.6	44.2	–	–	–	5.6	–	–
C	149.2	28.0	45.0	–	–	–	5.5	–	–

Table 4. The statistical average of the indicators characteristics and properties of light-gray, light loamy arable soils Privatsky strip of pre-Kama region of the Republic of Tatarstan.

Horizons	lower bound, cm	Particles, %		Humus, %	mg-EQ/100g		PH, pH*	Mobile, mg/100g	
		<0.001	<0.01		Ca+Mg	Hr		P_2O_5	K_2O
P	24.2	7.2	25.2	2.0	16.1	2.3	5.8	11.9	9.7
AEl	39.5	-	-	1.5	14.3	2.3	5.4	8.3	7.9
BEL	34.2	11.8	26.3	-	15.3	1.8	6.1	16.7	4.0
Bt_1	51.1	19.4	30.8	0.6	19.5	1.9	5.4	21.1	10.0
Bt_2	99.6	21.9	33.6	0.4	21.3	-	4.8	-	-
BC	125.0	28.6	40.4	-	-	-	3.4	-	-
C	155.6	18.8	28.5	-	-	-	5.0	-	-

Table 5. Statistical averages of indicators of signs and properties of light gray sandy loam arable soils in the Vyatka strip of the pre-Kama region of the Republic of Tatarstan.

Horizons	lower bound, cm	Particles, %		Humus, %	mg-EQ/100g		PH, pH*	Mobile, mg/100g	
		<0.001	<0.01		Ca+Mg	Hr		P_2O_5	K_2O
P	23.8	5.1	17.9	1.2	14.5	2.0	5.4	8.0	3.6
BEL	31.0	–	–	0.5	10.0	0.9	5.8	15.0	2.7
Bt_1	43.8	–	–	0.3	11.3	0.9	5.6	17.5	–
Bt_2	73.3	–	–	–	–	–	5.2	–	–

Table 6. Parameters of correlation of particle content <0.01 (x) and <0.001 (y) m, % of light gray heavy loam arable soil.

Properties		τ	Regression equation	
<0.01(x)	0.001(y)		x = a × y + b	y = a × x + b
P	P	0.429	x = 0.3577 × y + 38.20	y = 0.5133 × x − 6.15
AEl	AEl	0.846	x = 1.0342 × y + 25.70	y = 0.6914 × x −13.01
BEL	BEL	0.584	x = 0.5648 × y + 34.36	y = 0.6039 × x – 5.89
Bt_1	Bt_1	0.637	x = 0.7622 × y + 27.03	y = 0.5325 × x + 5.95
Bt_2	Bt_2	0.740	x = 0.7548 × y + 28.00	y = 0.7264 × x – 3.5389
BC	BC	0.525	x = 0.903 × y + 24.48	y = 0.30551 × x + 17.37
C	C	0.592	x = 0.7713 × y + 29.47	y = 0.454 × x + 7.96

Table 7. Parameters of particle content <0.01(x) and <0.001 (y) mm, % of light gray clay arable soils.

Properties		τ	Regression equation	
x	y		x = a × y + b	y = a × x + b
P	P	0.165	x = 0.3036455 × y + 45.09	y = 1.247229 × x - 41.81
Bt_1	Bt_1	0.098	x = 0.04300493 × y + 57.71	y = 0.2244788 × x + 24.50
Bt_2	Bt_2	0.474	x = 0.2597691 × y + 48.19	y = 0.8633741 × x - 9.65
BC	BC	0.744	x = 0.6039882 × y + 38.68	y = 0.9162761 × x - 18.49
C	C	0.986	x = 1.14195 × y + 19.98	y = 0.8514854 × x - 16.07

Table 8. Parameters of particle content <0.01 (x) and <0.001 (y) m, % of light gray medium loamy arable soils.

Properties		τ	Regression equation	
x	y		x = a × y + b	y = a × x + b
P	P	0.382	x = 0.4007527 × y + 31.08	y = 0.3648497 × x + 0.46
AEl	AEl	0.495	x = 0.9151309 × y + 25.51	y = 0.2677929 × x + 1.67
BEL	BEL	0.957	x = 0.910478 × y + 25.17	y = 1.004879 × x − 23.82
Bt_1	Bt_1	0.836	x = 0.8492653 × y + 21.01	y = 0.8235409 × x − 8.87
Bt_2	Bt_2	0.966	x = 1.315886 × y + 4.74	y = 0.7092959 × x − 1.20
BC	BC	0.960	x = 1.434731 × y + 0.25	y = 0.6429685 × x + 2.22
C	C	0.834	x = 0.9862598 × y + 17.43	y = 0.7056744 × x − 3.80

Table 9. Parameters of particle content <0.01 (x) and <0.001 (y) m, % of light gray light loamy arable soils.

Properties		τ	Regression equation	
x	y		x = a × y + b	y = a × x + b
P	P	0.731	x = 1.120957 y + 17.18	y = 0.4762433 x − 4.83
BEL	BEL	0.946	x = 1.3541 y + 10.25	y = 0.6602341 x − 5.51
Bt_1	Bt_1	0.977	x = 1.351128 y + 4.60	y = 0.7059461 x − 2.35
BC	BC	0.704	x = 0.9199553 y + 13.38	y = 0.5387522 x + 3.85
C	C	0.979	x = 1.444575 y + 1.38	y = 0.6632491 x − 0.13

Table 10. Correlation parameters of particle content <0.01(x) and <0.001 (y) mm, % of light gray sandy loam arable soil.

Properties		τ	Regression equation	
<0.01(x)	<0.001(y)		x = a × y + b	y = a × x + b
P	P	0.806	x = 0.7497177 × y + 14.13	y = 0.8666015 × x − 10.47

The content of physical clay in the profile confirms the eluvial-illuvial type [9]. The content of physical clay in the studied varieties corresponds to the following series: 52.1 (c) – 44.1 (h) – 36.6 (m) – 25.2 (l) - 17.9 (s) % of the soil. The data obtained correspond to the granulometric composition of the upper horizon of the varieties.

The silty fraction (particles <0.001 mm) is part of the physical clay (particles <0.01 mm), which allows us to consider the correlation between them by genetic horizons (tables 6-10). At the same time, the compared indicators characterize different soil sections laid down throughout the territory of the studied region, that is, these coefficients are obtained in a spatial series of soils, even soil horizons.

The obtained coefficients are reliable at a significance level of 0.05, and they characterize a tendency to strengthen the connection in the upper part of the soil profile than in the lower part. Exceptions are indicators of the clay variety and between the content of particles in the arable horizon. Apparently, the decrease in the tightness of communication is due to the processes of water erosion of soils and maximum migration of silt particles.

4 Conclusion

The genetic profile and location of horizons are fundamental soil features that carry information about past and current processes that form the main soil taxa. The obtained morphometric data can be used to diagnose aggressive soil types in this region.

The soil is a natural body formed as a result of a combination of natural factors, properties of the soil itself, and indicators of the profile structure. This relationship is more pronounced between soil indicators. The coefficients of pair correlation between the content of physical clay and silt fraction within the same genetic horizon and the sample are carriers of quantitative information about the intensity of their movement and the formation of genetic horizons.

References

1. V. V. Dokuchaev, *Russian Chernozem,* pp 25–435 (Moscow: Selkhozgiz, 1948)
2. M. I. Gerasimov, Classification of soils in Russia: The way to the next version by, *Soil Science* **1,** 32–42 (2019)
3. L. L. Shishov, V. D. Tonkonogov, I. I. Lebedeva, M. I. Gerasimova *Classification and Diagnostics of Doils in Russia* (Smolensk: Oikumena, 2004)
4. E. A. Dmitriev, *Mathematical Statistics in Soil Science* (Moscow: Book house "LIBROKOM", 2019)
5. I. D. Davlyatshin, et al., *Handbook of Agrochemists* (Kazan: MedDoc Ltd., 2013)
6. P. A. Chekmarev, et al., *Directory of the Agrochemist of the Republic of Tatarstan* (Kazan: IE Sheikhutdinova A. I, 2015)
7. I. D. Davlyatshin, A. A. Lukmanov, A. M. Badikov, Potassium in arable soils of the forest-steppe, *Fertility* **2,** 27–28 (2013)

8. I. D. Davlyatshin and L. G. Gaffarova Agrochemical properties of light gray forest soils and productivity of winter rye *Agrochemical Gazette* **6**, 7–9 (2016)
9. B. G. Rozanov, *Soil Morphology* (Moscow: Academic project, 2004)
10. L. G. Bogatyrev, V. D. Vasilevskaya, A. S. Vladychensky, Types of soils, their geography and use in *Soil* Science (Moscow: Vysshaya shkola, 1988)

Digital transformation of management system of program apparatus complex based on correlation-regression analysis

N. A. Safiullin[1], *F. T.* Nezhmetdinova[1,*], *R. I.* Zinurova[2], *F. F.* Gatina[1], and *A. R.* Tuzikov[2]

[1] Kazan State Agrarian University, 65, K. Marx St., Kazan, 420015, Republic of Tatarstan, Russian Federation
[2] Kazan National Research Technological University, 68, K. Marx St., Kazan, 420015, Republic of Tatarstan, Russian Federation

Abstract. The paper is devoted to the study of the modern state and development of the system for the provision of electronic public services in the Volga Federal District. The data on the dynamics of the number of users on the portal of public services are given. The rating of the regions of the studied district in terms of the quality of the provision of electronic public services is presented. The factors affecting the score-rating system are outlined. A correlation-regression analysis is carried out, which reveals the correlation between the resultant and factor characteristics. It is proved that the quality of the electronic provided services can be brought to the level of 67.7 points by changing the degree of influence of these factors. The experience of digital transformation of state and municipal government in the Udmurt Republic, as the leader of the Volga Federal District in terms of the quality of the provision of electronic public services, is described.

1 Introduction

The process of digital transformation of state and municipal government includes deep transformations in the provision of public services to the population and organizations based on information and communication technologies. The virtual interaction of citizens with the authorities allows improving the quality of life in the country and the efficiency of managerial decision-making.

The scientific community supports the priority of the improvement of the quality of services in general, but the authors especially emphasize the role of public sector services as special forms of service that ensure the socio-economic stability of national economy [1].

Nowadays, public services in electronic form are provided through the federal and regional portals of public services, as well as on the basis of multifunctional centers. Citizens can receive about 80% of basic services and services remotely. By the beginning of 2022, more than 93 million Russians had registered a verified account on the Unified Portal of Public Services [2]. The total number of registered users reached 137 million (Figure 1).

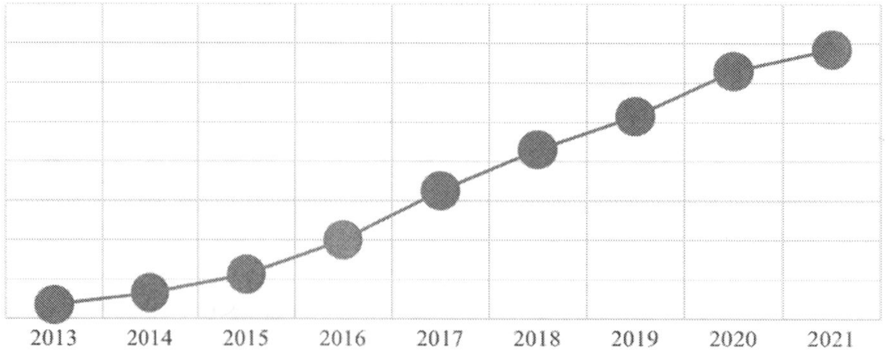

Fig. 1. Dynamics of the number of registered users of Unified portal of state and municipal services of the Russian Federation, million users.

According to the Decree of the Government of the Russian Federation since 2022, the list of state services,which were previously provided either on regional portals or in the traditional form, is significantly expanding [3]. The resolution approved 212 services, including a number of benefits for children, technical inspection of self-propelled vehicles and other types of equipment, registration of rights to real estate, acceptance of tax returns on income of individuals,

* Corresponding author: nadgmi@mail.ru

extracts from state forest register, various licenses and permits, etc.

Thus, the importance of the quality of the provision of public services increases, since the level of satisfaction of citizens and organizations of state and municipal government, as well as the level of trust in the authorities, depends on this.

The assessment of the quality of electronic public services can also be carried out from the point of view of the internal environment, that is, a performer in the form of a set of quality of internal processes for the provision of services [4].

The main target indicators of the quality of public and municipal services are enshrined in the Decree of the President of the Russian Federation, including the satisfaction of citizens with the quality of public services [5]. By 2018, it was planned to reach the indicator value of at least 90%. In accordance with this decree, the Ministry of Economic Development of the Russian Federation annually monitors the quality of public services. In 2019 and 2020, monitoring began to be carried out on the basis of a score, the maximum value of which was 120 points.

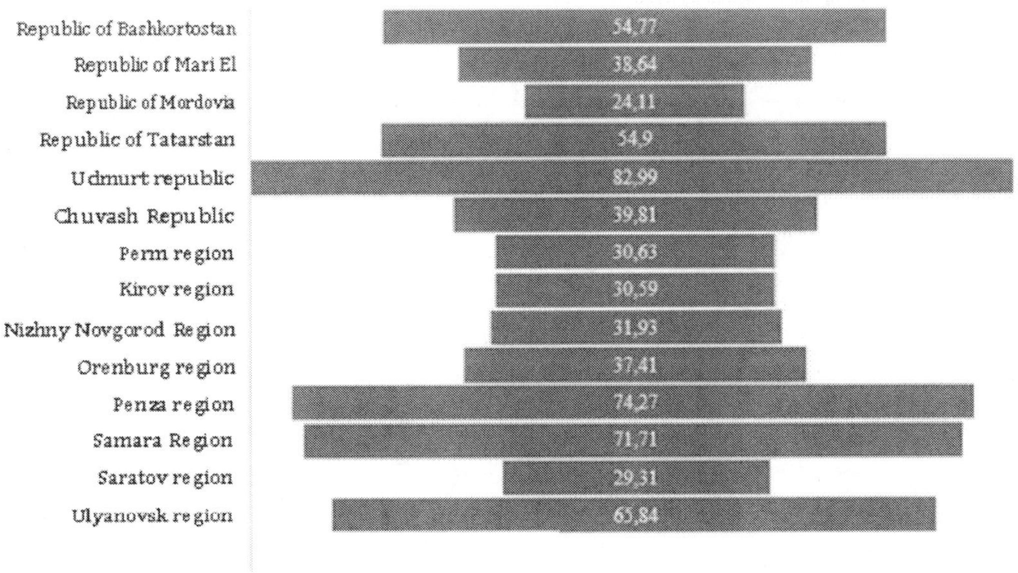

Fig. 2. Rating of the regions of the Volga Federal District in terms of the quality of the provision of electronic public services in 2019, points.

As we see from the presented data, the leaders of the Volga Federal District were the Udmurt Republic, the Penza and Samara regions. The quality of the provision of public services is reasoned by many factors, especially the socio-economic development of individual regions, the technical condition of the infrastructure and the attitude of citizens to regional authorities.

2 Materials and Methods

The adoption of rational managerial decisions is based primarily on the results of a factor analysis of socio-economic phenomena. In economic practice, the ability to manage the factors of socio-economic phenomena makes it necessary to measure their interrelations. The quality of the provision of electronic public services and services depends on many factors.

In order to assess the degree and nature of the relationship between the rating of regions in terms of the quality of electronic services and the factors that determine it, we conducted a correlation-regression analysis solving problems of multiple correlation. For this purpose, two multifactorial correlation-regression models of the state and development of the state system for the provision of electronic services to the population in the regions of the Russian Federation for 2019 were created. It covers the totality of the subjects of the Volga Federal District (14 units) [6].

In these models, effective (y) and factor (x) features are distinguished.

In socio-economic processes, performance factors include indicators that directly or indirectly reflect the results of socio-economic activities. In this regard, an indicator that characterizes the score-rating assessment of the regions in terms of the quality of the provision of electronic services acts as an effective factor.

In turn, factor indicators should reflect the level of their impact on the score-rating system:

y – rating of regions in terms of the quality of electronic services, points;

$x1$ – the average monthly subscription fee for Internet access, rub.;

$x2$ – the share of people employed in the ICT sector in the total number of employed population, %;

$x3$ – the share of public authority (PA) and local public authority (LPA) that had a data transfer rate over the Internet of less than 2 Mbps, in the total number of surveyed organizations of PA and LPA, %;

x4 – the share of healthcare institutions using the Internet in the total number of healthcare institutions, %;

x5 – the average number of personal computers per 100 households, pcs.

Multivariate correlation-regression analysis allows estimating the degree of influence on the studied performance indicator of each of the factors included in the model with a fixed position of the remaining factors, as well as with any possible combinations of factors with a certain degree of accuracy. In this case, the important condition is the absence of a functional relationship between the factors.

The assessment of the parameters of the solution results is reduced to the comparison of the obtained parameters with the norms of mathematical statistics. If the parameters meet the standards, then they proceed to the interpretation of the results of the solution. Otherwise, the correlation-regression model is corrected by screening out observations that have gross (anomalous) errors and excluding or replacing factor features.

A feature with insufficient variation range is excluded from the model. If the coefficient of variation is higher than 30%, the model can be corrected by dropping observations with critical (highest and lowest) values of the feature. However, economic studies are characterized by a high level of variation range. Therefore, in practice, a variation range level of 50–60% is allowed. In our case, the values of the coefficients of variation reflect the normal spread (the maximum value of the coefficient of variation is *vx2* = 33.048%), therefore, they do not require adjustment of the correlation-regression model.

Table 1. Results of correlation and regression analysis for the Volga Federal District (compiled by the author).

Result and factors	*Y*	*X1*	*X2*	*X3*	*X4*	*X5*
Errors of correlation ratios	0.070	0.214	0.180	0.222	0.242	0.212
Reliability coefficient ratio	12.235	-2.072	3.167	-1.854	-1.266	2.155
Arithmetic averages	47.636	495.417	1,490	64.336	98.121	117,000
Standard deviations	18.778	56.143	0.492	4.277	1.406	11.778
Variation ratio, %	39.419	11.333	33.048	6.649	1.433	10.066
Regression ratio	277.074	-0.038	18.502	-1.669	-2.178	0.710
Beta ratio		-0.114	0.485	-0.380	-0.163	0.445
Determination ratios,%	73.739	19.747	32.568	16.935	9.392	20.804
Individual Definitions ratio	73.739	5.086	27.693	15,650	4.997	20.312
Multiplier Correlations ratio	0.859					
Paired correlation ratio	*Y*	-0.444	0.571	-0.412	-0.306	0.456
	X1		-0.548	-0.002	-0.081	-0.175
	X2			-0.075	-0.027	-0.023
	X3				0.162	0.070
	X4					-0.175

According to the results of the solution of the five-factor model, the correlation ratio between the resultant and factor characteristics selected in the correlation model is very strong in the Volga Federal District (R = 0.859), as evidenced by the numerical value of the multiple correlation coefficient. At the same time, the error of the correlation ratio (*OR1* = 0.070) and the reliability of the correlation ratio (*TR* =12.235) confirm the existence of a connection between the effective and factor signs and indicate sufficient reliability of the multiple correlation coefficient. If the value of the multiple correlation ratio, as well as the pair correlation, is below 0.2 and the absolute value of the reliability of the correlation ratio is below t = 2.0 (with a probability value of P = 0.95450), then this indicates the absence of a significant and reliable relationship between the effective and factor traits. In this case, further analysis of the results of the solution of the problem is not important.

The assessment of the problem solution parameters shows that two factor signs *x3* and *x4* (*t3* = -1.854, *t4* = -1.266) should be excluded from the correlation – regression model. At the same time, the composition of indicators in the new, corrected model will be as follows:

y – the rating of regions in terms of the quality of electronic services, points;

x1 – the average monthly subscription fee for Internet access, rub.;

x2 – the share of people employed in the ICT sector in the total number of employed population, %;

x3 – the average number of personal computers per 100 households, pcs.

Table 2. Results of a three-factor correlation-regression analysis for the Volga Federal District (compiled by the author).

Result and factors	*Y*	*X1*	*X2*	*X3*
Errors of correlation ratios	0.121	0.214	0.180	0.212
Reliability coefficient ratio	6.143	-2.072	3.167	2.155
Arithmetic averages	47.636	495.417	1,490	117,000
Standard deviations	18.778	56.143	0.492	11.778
Variation ratio, %	39.419	11.333	33.048	10.066
Regression ratio	-57.716	-0.022	20.787	0.729
Beta ratio		-0.066	0.545	0.457
Determination ratio, %	54.879	19.747	32.568	20.804
Individual Definitions ratio	54.879	2.919	31.112	20.847
Multiplier Correlations ratio	0.741			
Paired correlation ratio	*Y*	-0.444	0.571	0.456
	X1		-0.548	-0.175
	X2			-0.023

The assessment of the parameters of the repeated solution is satisfactory.

According to the results of the solution of the three-factor model, the correlation ratio between the resultant and factor features selected in the second correlation-regression model remains quite strong ($R = 0.741$, $TR = 6.143$).

Paired correlation ratio – r_{YXj} – a parameter that reflects the correlation ratio (strength) between a separate factor and resultant feature. The coefficient takes values ranging from -1 to +1. A positive value of the coefficient means a direct relationship between features, a negative value means feedback.

Absolute values r_{YXj} reflect the following bond strength:

$r_{YXj} = 0$ – no connection;

$0.00 < r_{YXj} < 0.20$ - insignificant, unreliable correlation;

$0.20 < r_{YXj} < 0.40$ – weak correlation;

$0.40 <= r_{YXj} < 0.60$ – average correlation;

$0.60 <= r_{YXj} < 0.80$ – strong correlation;

$0.80 <= r_{YXj} < 1.00$ – very strong correlation;

$r_{YXj} = r$,00 – functional linear correlation

Thus, the obtained values of the paired correlation coefficients allow concluding that in the Volga Federal District the quality of electronic services is most influenced by the factor *x2* – the share of those employed in the ICT sector in the total number of employed population ($ryx2 = 0.571$ at $t = 3.167$). At the same time, the factors *x1* (subscription fee for accessing the Internet) and *x3* (the number of personal computers per 100 households) are almost equally dependent on the effective feature ($r\ ux1 = -0.444$, $r\ ux3 = 0.456$). It is necessary to note that the correlation between *x1* and *y* is average, inverse; and the correlation between *x3* and *y* is medium, straight line.

The interpretation of the solution results involves the study of regression coefficients, determination coefficients and coefficients of a separate determination.

The ratio of the regression equation are the coefficients of the regression dependence equation (VFD):

Y1\u003d – -57.716 – 0.022x1 + 20.787x2 + 0.729x3

If we substitute the arithmetic mean values of factor features – *x1, x2, ..., x5* into the equation, then the calculated value of the effective feature – *Y*, will also be the arithmetic mean.

The parameters of the linear regression equation with a factor feature show what change in the average value of the resultant values leads to a change in the factor value by one unit.

That is, if the share of people employed in the ICT sector in the total number of employed population is increased by 1%, then the quality of electronic services to the population of the Volga Federal District will increase by 20.787 points in the rating of regions. If the number of personal computers per 100 households is increased by 1 unit, the rating will increase by 0.729 points. However, the growth of such a factor as the average monthly subscription fee for accessing the Internet causes a decrease in the quality of electronic services, that is, if the subscription fee is increased by 100 rubles, then the rating will decrease by 2.2 points.

 https://doi.org/10.1051/bioconf/20225200064

Thus, changing the values of factor attributes, it is possible to achieve the maximum (or minimum) possible value of the effective attribute. However, it is legitimate to change factor values in the direction of decrease or increase only within the variation range or coefficient of variation from their arithmetic mean:

$$Ymax = -57.716 - 0.022x1 - min + 20.787x2 - max + 0.729x3 - max = -57.716 - 0.022 * 433.14 + 20.787 * 1.98 + 0.729 * 128.78 = 67.7 \text{ points}$$

The results of the solution show that the value of the effective feature - Y grows with the increase in the values of the factor features x2 and x3 and the decrease in the value of the factor feature x1. Therefore, in order to determine the maximum possible value of Ymax, the maximum allowable values of factor signs x2 and x3 and the minimum allowable value of factor value x1 should be substituted into a certain regression equation.

This means that changing the degree of impact of these factors, the quality of the provided electronic services can be brought to the level of 67.7 points on the rating scale of the regions of the Volga Federal District, which is 20.1 points or 42.2% higher than the average regional value.

Thus, the parameters of the developed regression equations can be used to predict the results of the activities of state authorities in the field of the provision of electronic services to the population of the Russian Federation.

According to Figure 2, the Udmurt Republic occupies a leading position in the ranking, so it makes sense to consider the experience of this region in the digitalization of state and municipal government.

3 Results and Discussion

The Udmurt Republic has a fairly developed digital infrastructure. The compact location of settlements allowed covering 95.4% of the population with 3G network and 93.1% with 4G network. More than 400 settlements in the region are covered by fiber-optic networks, the length of which exceeds 7.800 km. In 2021, it is planned to provide high-speed Internet to all schools and feldsher-obstetric stations in the region. As a part of the elimination of the digital gap, it is planned to build 47 mobile base stations. This fact allowed decreasing the subscription fee for Internet access to a minimum value in the Volga Federal District (436.95 rubles) which is a key factor in the conducted correlation-regression analysis.

In 2020, the Udmurt Republic entered the top ten constituent entities of the Russian Federation in terms of the quality of the provision of electronic public services. The share of services received electronically by the population has not fallen below 75% for several years.

The regional portal of state and municipal services uslugi.udmurt.ru includes about 200 services [8]. In 2020, more than 8 million services were provided. During the coronavirus pandemic, electronic services for registering self-isolation and returning to work were added to the portal. In 2021, it is planned to add regional super services Healthy Citizen, Tourist, Driver and others.

There are 33 offices of Multifunctional centers (MFC) for the provision of state and municipal services in the republic, in which 238 regional and 3545 municipal services are provided. In 2020, 934,307 services were provided to individuals, 8,200 of them were provided to legal entities. A single call center was created on the basis of the MFC, in which more than 2,600 calls are processed daily. In this case, the response of the operator is 4 seconds. Due to the callback service and 52 operators, the call center covers 100% of applicants. In 2021, it is planned to expand the network of windows, introduce Lean MFC tools and offer the population a mobile application, as well as expand the range of services.

By 2030, the Republic of Udmurtia plans to bring the indicator of digital maturity of economic and social sectors to 100%. In 2020, the value of this indicator was 28 % (the planned one was 4%).

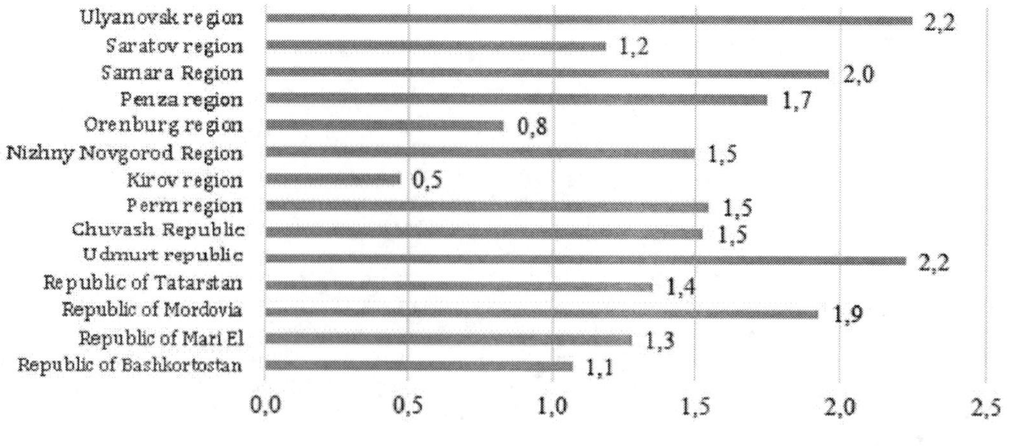

Fig. 3. Share of people employed in the ICT sector in the total number of employed people in the Volga Federal District in 2019, % (compiled by the author).

According to the conducted correlation-regression analysis, one of the factors affecting the quality of the public services in electronic form is the share of those employed in the ICT sector in the total number of employed people. The Udmurt Republic has the highest value of this indicator.

In 2020, the annual turnover of organizations engaged in the field of information and communication technologies in the republic exceeded 58 billion rubles, which was 19.5% higher than last year level. The average level of wages in the industry amounted to 46,589 rubles, which was 14.1% higher than in 2019.

The leadership of the region actively supports the enterprises of the ICT sector. They have the opportunity to receive subsidies for loans, as well as to arrange a mortgage for young professionals with a preferential rate of 2.9%. The participation of organizations involved in the field of information and communication technologies in competitions and grants attracted 232 million rubles to the Republic of Udmurtia, which was 58.9% more than last year.

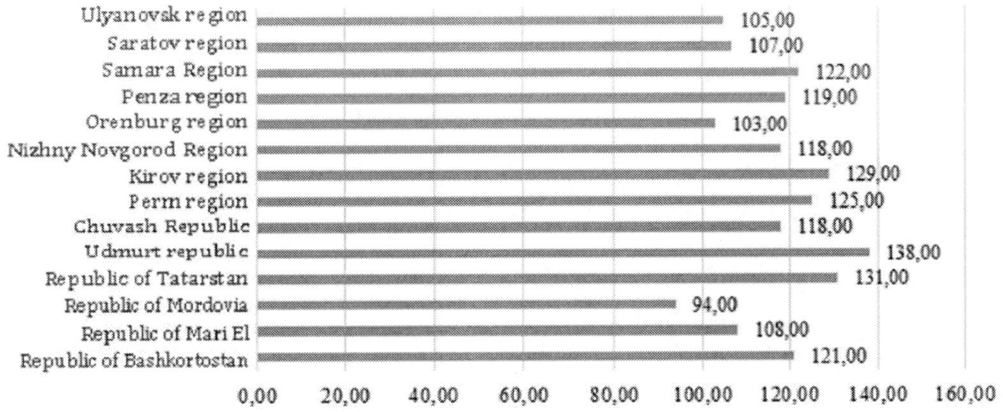

Figure 4. Number of personal computers per 100 households (compiled by the author).

The level of satisfaction with the quality of the provision of electronic public services based on the results of the correlation and regression analysis is influenced by the provision of the population with technical means, including the availability of personal computers. The Udmurt Republic ranks first in the Volga Federal District in terms of the number of personal computers per 100 households (138). Due to this, in 2020, about 70% of citizens over the age of 14 of this Republic had a verified ESIA account [9. 10.11].

At the same time, the digital transformation of the agro-industrial complex is facing many problems, which include not only the backwardness in the digital competencies of a significant part of the personnel of the agricultural sector, but also the risks posed by digital technologies related to living systems [12, 13,14].

4 Conclusion

The results of the study indicate the existing reserves for the improvement of the quality of the provision of electronic public services. Carrying out digital transformation, executive authorities need to use end-to-end technologies such as artificial intelligence, big data and the Internet of things. The Internet of Things technology is recognized as a priority in the development in terms of the potential to change business models. The development of such technologies in state and municipal administration will significantly expand the capabilities of state bodies and increase the level of confidence of citizens and organizations in the system of public services.

The experience of the Udmurt Republic in the field of informatization and communications can be used in the digital transformation of public services in other regions of the country. It will also improve the quality-of-service provision and rise in the corresponding rating.

References

1. M. V. Vasilyeva, V. A. Budasova, Perspective directions for assessing the quality of public services implemented in electronic form, *Intellect. Innovation. Investments* **6** 24–36 (2021)
2. Gosuslugi, *Interesting and useful*, retrieved from: https://www.gosuslugi.ru/help/news (2022)
3. The Decree of the Government of the Russian Federation of September 1, 2021 N 1458 "On Amendments to the Regulations on the Federal State Information System" Unified Portal of State and Municipal Services (Functions) (2021)
4. N. A. Safiullin, S. S. Kudryavtseva, Management of digital transformation of state and municipal services at the mesolevel, *Economic Bulletin of the Republic of Tatarstan* **1**, 19–24 (2021)
5. The Decree of the President of the Russian Federation of May 7, 2012 No. 601 "On the main directions for improving the system of public administration" (2021)
6. Rosstat, *Russian statistical yearbook 2020* (Moscow, 2020)

7. A. K. Subaeva, N. R. Aleksandrova, State support for the digitalization of agriculture, *Bulletin of the Kazan State Agrarian University* **16-4 (64)** 130–135 doi: 10.12737/2073-0462-2022-130-135 (2021)
8. Public Services of the Udmurt, *On the Regional Portal of State and Municipal Services of the Udmurt Republic*, retrieved from: https://uslugi.udmurt.ru/about_portal (2021)
9. E. F. Amirova, O.V. Kirillova, M. G. Kuznetsov, Sh. M. Gazetdinov Internet of things as a digital tool for the development of agricultural economy, *BIO Web of Conferences* **17,** 00050 doi: 10.1051/ bioconf /20201700050 (2020)
10. E. F. Amirova, M. G. Kuznetsov, E. Khakimova, A. Tolmacheva, Integrated development of digital agribusiness platform to support import substitution of food products, *BIO Web of Conferences*, **27,** 00055 doi: 10.1051/bioconf/20202700055 (2021)
11. G. Krupina, et al. Main directions of popularization of electronic state and municipal services among rural populationб *BIO Web of Conferences* **27** (2020)
12. F. T. Nezhmetdinova, N. S. Barabash, Transformation of education in the context of the formation of the digital economy *Innovation and expertise: scientific works* **2 (23),** 120–131 (2018)
13. F. T. Nezhmetdinova, M. E. Guryleva, N. Kh. Sharypova, R. I. Zinurova, A. R. Tuzikov, Risks of modern biotechnologies and legal aspects of their implementation in agriculture, *BIO Web of Conferences* **17,** 00227 (2020)
14. F. T. Nezhmetdinova G. R. Fassakhova, L. R. Shagivaliev, N. Kh. Sharypova, R. I. Zinurova, Transformation of personnel training for the agro-industrial complex in digital economy. In the collection: *Agriculture and food security: technologies, innovations, markets, personnel. Scientific works of the international scientific-practical conference dedicated to the 100th anniversary of agricultural science, education and enlightenment in the Middle Volga region* pp 721–725 (2019)

Stimulation factors of innovation activity as an important condition for the effective development of agriculture

D. A. Pinaeva [1], *F. T.* Nezhmetdinova[1*], *R. I.* Zinurova[2], and *A. R.* Tuzikov[2]

[1] Kazan State Agrarian University, 65, K. Marx St., Kazan, 420015, Republic of Tatarstan, Russian Federation
[2] Kazan National Research Technological University, 68, K. Marx St., Kazan, 420015, Republic of Tatarstan, Russian Federation

Abstract. The paper is devoted to the study of urgent problems of innovation activity stimulation on the example of agricultural universities as an important condition for the improvement of the efficiency of the agro-industrial complex and the development of the agricultural industry in general. The ambitious tasks for a significant increase in labor productivity in the agricultural sector, as well as the creation of high-tech jobs are impossible without proper staffing and scientific support. The effective import substitution and maintaining food independence is impossible without the introduction of new technologies, a significant increase in labor productivity in agriculture and the increase in the attractiveness of rural life. These challenges and priorities set the agenda for the formation of requirements for the system of agricultural education and new competencies of youth and teachers. In this regard, the problem of innovative activity of agricultural universities is being updated not only as a source of highly qualified personnel for agriculture, but also as centers for the development of innovations in this area. In paper, it is shown that systemic problems in the field of education are organically linked with the transformations of university science in the second half of the 20th century, when the basic principles and approaches to the organization of scientific research in universities were laid down. It is pointed out that despite numerous attempts to intensify science in agricultural universities, there are a number of constraining factors that affect their potential in the development of innovations in the agricultural sector.

1 Introduction

It is well known that the development of innovations is an important factor in the economic and political independence of the state. In modern conditions, the agricultural sector of economy is of strategic importance for the provision of food independence and sustainable development of society. The exceptional social significance of agriculture is confirmed by current events and indeed by the whole history, when, in any crisis, the demand for agricultural products grows first of all. The effective import substitution and maintaining food independence is impossible without the introduction of new technologies, a significant increase in labor productivity in agriculture and the increase in the attractiveness of rural life. These challenges and priorities set the agenda for the formation of requirements for the system of agricultural education and new competencies of youth and teachers.

Continuous deepening and updating of knowledge and improvement of professional competence are becoming the most important areas for the formation of personnel in the agro-industrial complex. A modern agrarian specialist is a person with broad general and specialized knowledge, able to quickly respond to changes in machinery and production technologies. He needs basic knowledge, analytical thinking, information literacy, socio-psychological competence and intellectual culture.

Science-intensive technologies and active innovative activity are the initial driving force of the entire economic life of the republic. The predominant increase in agricultural production and the accelerated development of the agro-industrial complex are ensured through the reproduction of new knowledge and the implementation of scientific and technological achievements.

At the same time, it is necessary to understand that modern society is at the stage of development of the post-industrial economy, i.e. requiring the increase in the processability and science intensity of all industries, including the agro-industrial complex (focusing on advanced development in terms of precision farming, production robotization, genetic engineering, genomic technologies, digitalization, artificial intelligence, etc.).

Entering a new stage of the scientific and technological revolution implies a significant transformation in agriculture, including a focus on the so-called precision farming, robotization of production, the introduction of genetic engineering achievements and

* Corresponding author: nadgmi@mail.ru

a high level of digitalization of all industries [1].

2 Materials and methods

The purpose of this paper is to determine systemic problems of agricultural education that reduce the effectiveness of the development and implementation of innovations in the agricultural sector of economy. The research is based on the modernization approach, which allows considering the role of agricultural science in innovation processes in the context of transformational changes in Russian society through the prism of modernization. Policy in this area largely determines the scale and speed of innovation, the ability of economy and society to perceive the achievements of science and innovation. In order to achieve this purpose, various research methods were used, the main of which was the analysis of regulatory documents: the Decrees of the Government of the Russian Federation and regions, various state programs for the development of the agricultural sector, as well as reports of scientific and educational organizations. This method allows comparing plans, ways to implement them, as well as correlate the planned indicators with real figures and facts.

3 Results and discussion

One of the acute problems in the implementation of the innovative path of development of the Russian agro-industrial complex is the reduction in the human resources potential of agricultural science, which is expressed, first of all, in the decrease in the number of researchers, as well as the aging of scientific personnel. Thus, according to the data for 2020, the proportion of scientists under the age of 40 is about 40%, and about 30% are scientists over 60 years old. The total number of researchers in the field of agricultural sciences for 2015-2018 decreased by almost 25%, and in the total number of researchers in 2018 it was only 2.8% [2].

According to the report of the Russian Science Foundation (RSF) 2020, in the total volume of projects supported by the Foundation according to the priorities of the Strategy for Scientific and Technological Development of Russia, highly productive agriculture and aquaculture accounted for only 6.3% [3]. In the distribution of applications by branches of knowledge, agricultural sciences accounted for only 2.9% [3]. Among the supported projects, agricultural projects accounted for 3% [3]. According to our calculations (there is no official data from the RSF report for 2021 yet), the share of supported research in the field of agriculture has decreased by at least 1%. It is necessary to note that in the most supported priority area of research - advanced digital technologies, which accounts for 36.3% of all projects supported by the Fund according to the priorities of the Strategy of Scientific and Technological Development of Russia, agricultural research is not presented at all. That is, in fact, the digitalization of agriculture, which is one of the priorities of the state agrarian policy, is not represented by scientific research [3]. There is a significant gap between the needs of the agricultural sector of economy and the real possibilities of modern agricultural science. At the same time, one of the key problems of the agricultural sector remains low labor productivity, which is largely due to the slow introduction of innovations and advanced technologies in this sector of economy. It is necessary to note that this problem is recognized at the state level. Thus, for example, according to the State Program "Development of agriculture and regulation of agricultural products, raw materials and food markets in the Republic of Tatarstan for 2013-2025" by 2025, labor productivity in agriculture should increase by 1.8 times compared to 2013 and by 20215, 4944 units of highly productive jobs should be provided [4]. It is shown that the possibility of the realization of these goals is closely related to the staffing of the agro-industrial complex [5].

The difficult situation in the development of agrarian science is also shown by the low level of the contribution of Russian scientists to the total volume of publications indexed in Scopus and Web of Science. Thus, at the beginning of 2022, Russia ranked the 27th position in terms of the number of scientific publications in the field of agricultural sciences according to Web of Science. It is also indicative that the financing of the plan for the development of science and scientific research in the Republic of Tatarstan for 2014–2025 was completed only for 74% in 2020 [6].

In fact, the problem of the intensification of scientific research and improvement of the organizational forms of science development is not fundamentally new. Russian science faced the same problems in the second half of the 20th century. Thus, already in the decisions of the 20th Congress of the CPSU, it was indicated that higher education should radically improve research work, concentrate scientific forces on the solution of the most urgent scientific and technical problems and ensure the introduction of the achievements of science and technology into national economy. To solve this problem, a set of measures was developed, which also included, structural adjustments in universities. Thus, two types of laboratories were to be opened in higher educational institutions: industry-specific, financed from the economic activities of a university itself, and fundamental research laboratories, the financing of which was directly carried out from the budget [7, p. 218-226]. The relevant ministries and departments received permission to allocate funds to university laboratories in order to provide them with equipment, materials and instruments. The laboratories created at the expense of budget financing were on the balance of a university, but they could only work on topics agreed with the relevant ministries and departments.

The fundamental research laboratories were opened at the departments. Their work was attended mainly by the teaching staff, but they could also have their full-time scientific, engineering and technical staff and service personnel not related to the work of the departments. Although the activities of fundamental research laboratories were financed from the state budget, within the limits of approved plans and estimates, they had operational independence.

Industry laboratories were created at universities to solve applied problems primarily at the expense of industrial enterprises and worked under contracts concluded for research work with ministries and departments in close cooperation with enterprises and industry research institutes.

The creation of industry laboratories was preferred in cases where there were no research institutes in the region, or design and technology and design organizations of the industry, but there were universities that had personnel capable of solving certain applied problems [8, p. 116].

Fundamental research laboratories of universities were called upon to conduct theoretical and experimental research in the field of problem areas approved for them, provide assistance to the relevant departments and organizations in design development, participate in pilot tests of new models of machines, equipment, instruments, products, materials and technological processes developed in laboratories, as well as in their implementation in production. By 1965, 41 research and 20 branch laboratories were formed at 19 universities of the RSFSR [9, p. 190]. The problem with this transformation was that the reform increased the number of bureaucratic procedures. For example, a university that had industry-specific or fundamental research laboratories had to coordinate the plans for the work of these laboratories with the Ministry of Higher Education or another ministry in a department of which a university was located. The ministries, in turn, had to coordinate the work plans of the universities' laboratories with the State Engineering Committee of the USSR, the State Construction Committee of the USSR and the State Economic Commission. Such a multi-stage structure of approvals resulted in the increase in the number of various plans, justifications, etc., but did not result in the real innovative activity of universities [10, p. 77–78]. For comparison, this problem remains relevant today. Thus, for example, agrarian universities in Russia, on the one hand, are subordinate to the Ministry of Agriculture of the Russian Federation, and on the other hand, they are obliged to comply with the general educational standards of the Ministry of Science and Higher Education of the Russian Federation. Accordingly, on the one hand, this is a double reporting, and on the other hand, agricultural universities, being departmental, do not have the opportunity to use tools to support research activities that can be used by universities directly subordinate to the Ministry of Science and Higher Education. Namely, for example, agricultural universities, can not form scientific and educational centers, since this is not provided for by the budget of the Ministry of Agriculture.

It is also fair to say that in the summer 2019 it was widely announced that within the framework of the national project “Science and Universities”, 14 agricultural research and educational centers would be created in Russia [11]. However, in this case, we are talking not only about supporting agricultural universities. In addition, information on how this initiative was eventually implemented could not be found. Agrarian universities in order to work in the system of scientific and educational centers are forced to cooperate with other universities subordinate to the Ministry of Science and Higher Education. On the other hand, an agricultural university can not deviate from the requirements of the Ministry of Science and Higher Education, for example, in order to extend practice at agricultural enterprises. All this creates significant difficulties, deprives universities of independence and flexibility and also diverts significant resources to perform purely administrative work in order to create all kinds of plans, roadmaps, monitoring and other reporting [12].

This situation is aggravated by the fact that the main scientific foundations, the purpose of which is to support promising scientific research in various fields, indicate in the requirements of the competitive documentation a number of requirements that are not directly related to science. Thus, the RSF, the main Foundation through which scientific research is financed, in the application documents (clause 2.17 of the Competitive documentation for fundamental scientific research and exploratory scientific research by small separate scientific groups) asks to indicate the experience of educational activities over the past 5 years (information about the management of graduate students, adjuncts, interns, residents, development and implementation of new educational courses in Russian and foreign universities) [13]. Taking into account that a project manager during the implementation of the project must be in labor relations with the organization through which the funding is provided (clause 2.16), all scientists who work in non-core universities for their qualification are almost automatically cut off from financial support. For example, a chemist or mathematician under such conditions will not be able to receive a research grant while working in an agricultural university, since, at a minimum, he will not have the necessary experience in educational activities, since in a non-core university he can not manage postgraduate students, etc. It is problematic in such a situation to attract specialists-scientists.

At the same time, it is important to understand that the training of a high-quality specialist is a complex multifactorial process. A modern specialist must have good knowledge in various fields of science, as well as have developed analytical thinking, which is formed not only and not so much by narrow professional disciplines, but by disciplines that are non-core for a given university, for example, mathematics, philosophy, etc. However, teachers and scientists in such a situation tend to affiliate with universities specialized in their field of study, since this increases the chances of to get grant funding, while departmental universities continue to experience difficulties in attracting high-quality specialists.

If we turn again to the experience of the USSR, we can state that the multi-level system of the management of the research work of universities and the division of laboratories into research and industry ones did not give a systemic positive result. In the work of fundamental research laboratories, formalism often prevailed, expressed in the substitution of major scientific tasks for one-time assignments, which led to the dispersion of forces, concentration on narrow subjects and work without a long-term perspective. On the other hand, industry laboratories were sometimes limited to the solution of utilitarian and narrow tasks performed in order to obtain funds from a customer for the purchase of equipment. To improve the efficiency of university science, the Central Committee of the CPSU and the Council of Ministers of the USSR on February 20, 1964 adopted a resolution "On the further development of research work in higher educational institutions" [14, p. 450], according to which university science should be included in national plans for the development of science and technology, which increased the importance of university research work. The resolution stipulated that during the development of plans for scientific research and introduction of the achievements of science and technology into national economy, universities should take a direct part in the solution of promising scientific and technical problems. A network of university research institutes also developed. The Research Institute of Nuclear Physics of Moscow State University, the Research Institute of Mathematics and Mechanics of Kazan University, the Research Institute of Physics of Leningrad University, etc. became well-known scientific institutions in the country. In total, in higher educational institutions of the USSR, by the end of 1977, there were 58 research institutes, more than 540 problematic and 770 industrial laboratories [15, p. 158].

In the 1970s the measures were taken to establish a real connection between the beginning of university research and their completion, followed by the use of research results in social production [13, p. 80–81]. To this end, from 1971-1972 all types of higher education institutions of the USSR, research and development sectors (RDS) began to be created. At the same time, two types of RDS were most widely used – centralized and decentralized.

The centralized RDS of the university had its own accounting department, personnel and supply services, a planning and financial department, etc. The presence of these services allowed collecting in a timely manner the information necessary to make decisions on the advisability of concluding certain economic contracts for research by the scientific divisions of the university in order to monitor performing R&D more carefully.

The decentralized RDS was the coordinating and controlling body in the conduct of research work at the university. Its services (accounting, personnel, etc.) were merged with the corresponding services of the university. This form of RDS prevailed in universities with a small amount of research work.

The above-mentioned measures on the improvement of the organization of university science contributed to strengthening its connection with practice and created more opportunities for the solution of topical theoretical and, mainly, scientific, technical, and applied problems.

However, in most universities, not everything was so good. It was precisely this circumstance that caused the adoption of the next resolution of the Central Committee of the CPSU and the Council of Ministers of the USSR "On the further development of higher education, improving the quality of training" on July 12, 1979. The main thing that the Resolution drew attention to was the discrepancy in the learning process with the realities of production and the latest achievements. In this regard, it was proposed to improve the professional training of university graduates through the involvement of academicians and leading scientists in teaching, improvement of scientific and technical information and the promotion of advanced national and foreign production experience among students, provision of the predominant development of full-time postgraduate studies, etc.

It is necessary to emphasize that the party and state leadership of the country in the late 1970s no longer had any extraordinary opportunities to increase the role of the department. Therefore, in the mentioned resolution of the Central Committee of the CPSU and the Council of Ministers of the USSR, the increase in the role of the department was conceived primarily by assisting young teachers in mastering pedagogical skills and implementing measures to improve the moral and material incentives for pedagogical work. However, these attitudes could no longer significantly enhance the role of the department. First, it remained unclear what the assistance to young teachers in the improvement of their pedagogical skills should consist of and who and how should provide this assistance. Secondly, the measures of personal stimulation (gratitude, Certificate of Honor, Board of Honor) were outmoded. Thirdly, financial incentives for university teachers in the conditions of economic stagnation, total shortages were just a wish.

The state did not have the opportunity to radically improve the material and technical situation of most departments. For decades, there was a lack of premises, equipment, instruments and reagents in departments. Therefore, the resolution stated that ministries and departments should participate in the intensification of the material base of universities by the provision of educational institutions with the opportunity to use the premises and equipment of relevant enterprises and organizations, creating on their basis, the branches of specialized departments if necessary.

However, not all of 883 universities in the country were able to use the provisions of this resolution. The main reason was that most organizations and enterprises were not in hurry and more often they were not going to give universities the opportunity to organize branches of departments on their premises, since it was easier for them to turn to applied research and development centers to develop important scientific and technical problems for them it was easier for them to apply to applied research institutes in their industry.

In addition, scientific work in universities did not develop rapidly enough due to the fact that the teaching staff had almost no time for scientific studies. Teaching, methodological and educational work were in the first place in their activities. According to an individual plan, a university teacher was assigned from 1 to 4 hours a day for scientific work. The situation was aggravated by the established practice of financing scientific work in higher education: if the USSR Academy of Sciences received 8 out of every 10 rubles from the state budget and only 2 earned itself under economic contracts, then the university from the 8 rubles it earned itself had the right to spend only about 40 kopecks for the development of its material and technical base, its own experimental and research production [16, p. 349].

The state order for scientific research was considered as a measure that would increase the effectiveness of scientific research in universities in the late 1980s. In 1988, it amounted to 7 million rubles and according to the calculations of the USSR State Committee for Public Education its implementation allowed increasing the volume of research in higher education by 1995 to 1 billion rubles. However, the universities themselves were not ready for a serious restructuring of the entire mechanism that had developed over decades. In the system of higher education during the years of Soviet power there was a practice when a plan for recruiting students and graduating specialists was initiated by state. In this case the state regulated everything that was possible: the ratio of students and teachers, staffing, workload, etc. The race of assembly line for the training and graduation of specialists not only did not contribute to progress, but also led to the depreciation of higher education itself. This trend continues to a large extent to this day. The problem of formalism, the pursuit of quantitative indicators instead of qualitative ones, has not been eliminated either at the level of university reporting or at the level of research.

It is well known that if in order to receive funding in the form of grant support, it is necessary, at least, to fulfill a number of quantitative requirements for publications, the age of performers etc. In the form of a report, quantitative indicators are the main ones (the number of articles published in various citation bases). Then scientists are forced by any means to try to increase the number of publications. Usually this comes at the expense of quality. The result of this approach is the impossibility of a qualitative assessment of the projects themselves. Thus, with such numerous and rather high requirements for project managers, the RSF in March 2022 was forced to publish an appeal from the RSF expert councils on the quality of reporting materials. This appeal pointed out that often reports on supported projects are essentially formal replies that do not contain specific and verifiable results, characterized by lexical mistakes, as well as excessive brevity. It is shown that the reports contain an erroneous indication of the quartile of the journal or incorrect information about the indexing of the journal in which the work is published, duplication of publications, including in different languages, etc. [17].

At the same time, important documents have already been adopted. They determine the relevance of this area of work, taking into account global challenges [18, 19, 20]. A unified global economic, legal and information system is being purposefully formed. The intensification of global competition covers not only the traditional markets for goods, capital, technology and labor, but also the systems of national governance, innovation support and human development. If we formulate the main challenges of a global nature, then they include:

- The transition of the world economy to a new technological order "Industrialization 4.0" and digital economy.
- The acceleration of technological development of the world economy. The real competitors of Russia are not only the leading countries in the field of innovation, but also many developing countries, the CIS countries.
- The intensification of global competition for the factors that determine the competitiveness of innovation systems, primarily for a highly skilled workforce and smart money (the investments that attract new knowledge, technologies and competencies to projects), education and a sharp increase in the mobility of these factors.
- Challenges faced not only by Russia, but also by humanity in general, which determine the priority areas of scientific and technological development of the Russian Federation - climate change, the problem of ecology and sustainable development, population aging and health challenges, food security on a global context.

According to the Strategy for Scientific and Technological Development of the Russian Federation, the following areas should be determined as priorities:

1. Food security, understood as the need to ensure food security and food independence of Russia, the competitiveness of domestic products in the world food markets and the reduction of technological risks in the agro-industrial complex.
2. Rational agriculture and aquaculture, environmental protection, safe food in the sense of transition to a highly productive and environmentally friendly agriculture and aquaculture, development and implementation of systems for the rational use of chemical and biological protection of agricultural plants and animals, storage and efficient processing of agricultural products, creation of safe and high-quality, including functional, food products.
3. Effective interaction between man, nature and technology, providing the possibility of an effective response of Russian society to great challenges, taking into account the interaction of man and nature, man and technology, social institutions at the present stage of global development, including the application of the methods of humanities and social sciences.

The national development goals of the Russian Federation, indicated in the Decree of the President of the Russian Federation of July 21, 2020 are of particular importance for the development of the goals and objectives of the development program of Kazan State Agrarian University. First of all, we are talking about the opportunity for self-realization and development of talents, where the positions that universities should achieve by 2030 are clearly defined:

– the formation of an effective system for the identification, support and development of abilities and talents in children and youth, based on the principles of justice, universality and aimed at self-determination and professional orientation of all students;
– the provision of the presence of the Russian Federation among the ten leading countries in the world in terms of research and development, including through the creation of an effective system of higher education;
– the creation of conditions for the formation of a harmoniously developed and socially responsible personality based on the spiritual and moral values of

the peoples of the Russian Federation, historical and national cultural traditions.

During the formation of the priorities for the development of the University, the Strategic Initiatives formulated in the Decree of the Government of the Russian Federation dated October 6, 2021 No. 2816-r "List of initiatives for the socio-economic development of the Russian Federation until 2030": Circular Economy and Agricultural Science – a step in the future development of the agro-industrial complex were taken into account.

4 Conclusion

There is an urgent need to learn from the past experience. Following the path of the division of universities, increasing reporting and giving priority to quantitative indicators, it is hard to speak of a systemic restructuring of universities for innovation processes. The introduction of innovative mechanisms, including the digitalization of education, should also not occur at the expense of a sharp increase in extracurricular workload on teachers. The reforms in this area require a reasonable approach and a qualitative analysis of possible problems and risks. The positive moment is that there is a clear understanding that universities, where scientific work is not conducted or poorly organized, are not able to train high-quality specialists corresponding to the current level of development of science, technology and production organization.

References

1. Strategy of agricultural education in the Russian Federation until 2030, retrieved from: https://bsaa.edu.ru/sveden/files/Strategiya_AO.pdf
2. Innovative development of the agro-industrial complex in Russia Agriculture 4.0, *Reports to the XXI International Scientific Conference on the Development of the Economy and Society* (Moscow: Publishing House of the Higher School of Economics) pp 70–71 (2020)
3. Report of the Russian Science Foundation for 2020. retrieved from: https://rscf.ru/upload/iblock/235/235865f821df825152d6c35f5d853bd8.pdf
4. On approval of the State program "Development of agriculture and regulation of agricultural products, raw materials and food markets in the Republic of Tatarstan for 2013-2025". The decree of the Cabinet of Ministers of the Republic of Tatarstan dated April 8, 2013 No 235, retrieved from: https://agro.tatarstan.ru/documents/post.htm?pub_id=586749
5. G. R. Fassakhova, F. T. Nezhmetdinova, N. Kh. Sharypova, N. N. Hamidullin, BIO Web of Conf. **27,** 00146 (2020) doi: 10.1051/bioconf/20202700146
6. Financing of state programs of the Republic of Tatarstan for 2020, retrieved from: https://mert.tatarstan.ru/rus/file/pub/pub_2913710.xlsx
7. *A Collection of Basic Regulations, Orders and instructions*: Part 1 (Moscow: Vysshaya shkola, 1965)
8. Vysshaya shkola of the USSR for 50 years (1917–1967) (Moscow: Vysshaya shkola, 1967)
9. E.V. Bodrova, V. N. Krasivskaya, Russian technological journal **4 (5–14),** 189–201 (2016)
10. D. A. Pinaeva, R. R. Yunusova, European Social Science Journal **12–1,** 75–83 (2018)
11. *News of the Ministry of Science and Higher Education*, retrieved from: https://minobrnauki.gov.ru/press-center/news/?ELEMENT_ID=23370
12. F. T. Nezhmetdinova, G. R. Fassakhova, N. Kh. Sharypova, Faizrakhmanov. BIO Web of Conf. **27** 00145 (2020) doi: 10.1051/bioconf/20202700145
13. Competitive documentation for an open public competition for grants from the Russian Science Foundation in the priority area of activity of the Russian Science Foundation "Conducting fundamental scientific research and exploratory scientific research by small individual scientific groups", retrieved from: https://rscf.ru/upload/iblock/c65/0rgd1myftqp3qkct0zdl6nbq4pi692cd.pdf
14. K. U. Chernenko, M. S. Smirtyukov, *Decisions of the party and government on economic issues: a collection of documents* **5** (Moscow: Politizdat, 1968)
15. E. A. Belyaev. *Formation and development of a network of scientific institutions of the USSR* (Moscow: Nauka, 1979)
16. S. P. Strekopytov, History of scientific and technical institutions in Russia (the second half of the XIX – XX centuries) (Moscow, Russian SHU, 2002)
17. Appeal of the expert councils of the Russian Science Foundation on the quality of reporting materials, retrieved from: https://rscf.ru/news/found/obrashchenie-ekspertnykh-sovetov-rnf-o-kachestve-otchetnykh-materialov/
18. F. T. Nezhmetdinova, N. S. Barabash, Innovation and expertise: scientific works **2 (23),** 120–131 (2018)
19. F. T. Nezhmetdinova M. E. Guryleva, N. Kh. Sharypova, R. I. Zinurova, A. R. Tuzikov, BIO Web of Conf. **17,** (2020) 00227
20. F. T. Nezhmetdinova, G. R. Fassakhova, L. R. Shagivaliev, N. Kh. Sharypova, R. I. Zinurova. BIO Web of Conf. **17,** (2020) 00028

Lysozyme, anti-lysozyme, hemolytic and adhesive activity in serration cultures isolated from dairy products

R. Kh. Gairabekov[1*], *L. G.* Molochaeva[1], and *N. M.* Mirzoeva[2]

[1] Chechen State UniversityNo. 31, L. Yashin Str., Grozny, 366300, Russian Federation
[2] Kabardino-Balkarian University named after H. M. Berbekov, 73, Chernyshevsky Str., Nalchik, 360004, Russian Federation

Abstract. Isolation of pathogenic or opportunistic bacteria from surrounding environment, food products, animals and humans requires establishing their potential pathogenic potential. In this work, the authors have studied 109 cultures of bacteria belonging to the species Serratia marcescens of which: 61 museum cultures, 29 cultures isolated from small ruminants and environmental objects, 19 cultures isolated from dairy products produced on private farmsteads in the Chechen Republic for the presence of lysozyme, antilysozyme, hemolytic and adhesive activity in them.

1 Introduction

Provision of population of the country with healthy food is an important task afflicting state institutions responsible for national health. An especially important problem is prevention of development of pathogenic and opportunistic microorganisms in food products. It includes prevention of possible contamination of products, prevention of circulation of pathogenic and opportunistic microorganisms in population of a given territory and in the natural ecosystem as a whole, as population is an element of this ecosystem. This problem has become especially relevant to population now, as the population suffers from a pandemic of SARS-COV-2 virus (coronavirus), the worst pandemic for the last 100 years, resulting in death of over a million of people and deep decline in the global economy.

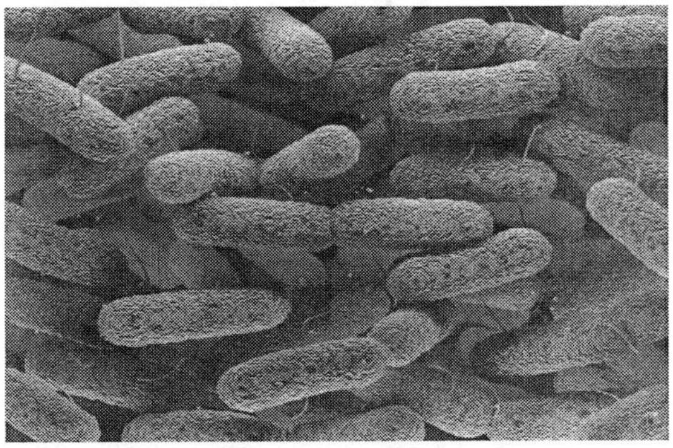

Fig. 1. *Enterobacteria*: a large family of *Gram-negative* bacteria that includes many of the more familiar pathogens, such as *Salmonella pestis, Escherichia Coli, Yersinia, Klebsiella and Shigella, Proteus, Enterobacter, Serratia,* and *Citrobacter*.

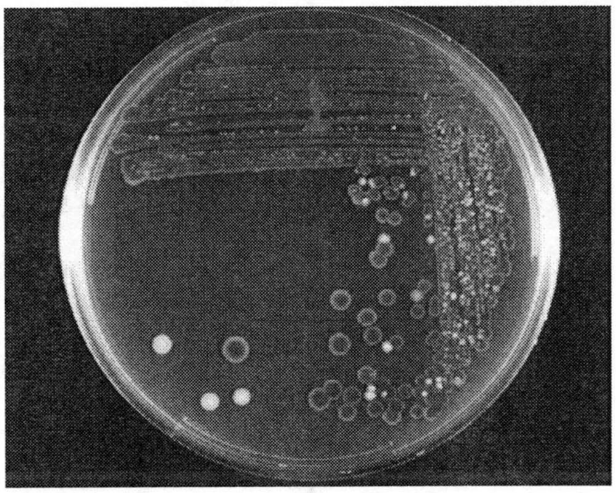

Fig. 2. *Serratia marcescens* under the microscope.

29 bacterial cultures identified as *Serratia marcescens* [9–13] were successfully isolated from sick animals (goats and sheep) and environmental object in the territory of the Chechen Republic.

The authors conducted studies aimed at identification of pathogenic and opportunistic microorganisms in dairy products produced for sale in private farms of the Republic.

The study of dairy products was finalized by isolation of 19 bacterial cultures of *Enterobacteriaceae* (Figure 1) bacterial family, which, after morphological, culture-based and biochemical assay, were identified as bacteria of genus *Serratia*, species *Serratia marcescens* (Figure 2) [7, 8].

2 Materials and methods

This research considered 19 cultures identified as *Serratia marcescens*, isolated from dairy products form private farms, 61 museum cultures and 29 cultures

* Corresponding author: ruslan.sarmat@mail.ru

isolated from goats and sheep and environmental objects in the territory of the Chechen Republic [7–11].

Lysozyme activity was determined by the method of delayed antagonism according to Bukharin et al [2, 4, 9, 13] on 1.5% nutrient agar. Plate cultures were incubated at 37^{0}C for 18-24 hours, then grown colonies were exposed to chloroform vapors, after that they were subjected to being couched together using two-layer method with washout of one-day culture of *Micrococcus luteus var. Lysodeikticus* (Figure 3) Presence of lysis area around inoculated strains of tested cultures indicates their lysozyme activity.

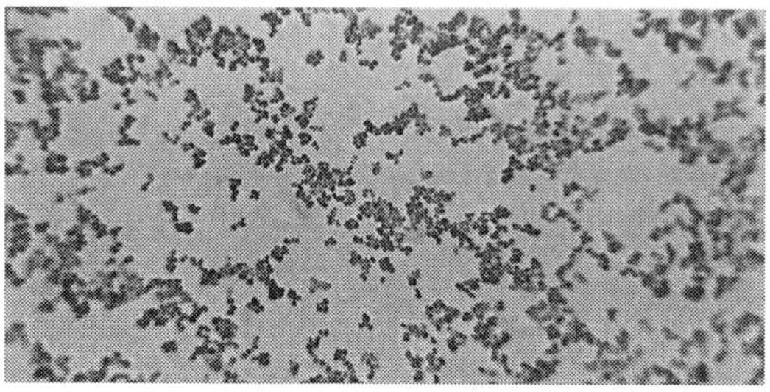

Fig. 3. *Micrococcus luteus var. Lysodeikticus.*

Antilysozyme activity of isolated *Serratiae* was determined according to Bukharin et al [3, 4, 9, 10, 13] by growth of micrococcus on Petri dishes with various content of lysozyme. This method allows determining qualitative and quantitative characteristic of antilysozyme activity in studied cultures.

Adhesive properties of tested cultures were identified by their ability to hemagglutinate sheep erythrocytes in a reaction with 3% of fresh sheep erythrocytes. D-mannose resistant hemagglutinating activity was determined by ability of serrated cultures to hemagglutinate sheep erythrocytes in presence of 1.5% of mannose [1, 9, 12].

Presence of hemolysine in tested *Serratia* cultures were determined by cultivation on 1.5% nutrient agar with addition of 3-5% of rabbit erythrocytes. In order to identify hemolysines, production of which depends on presence of thiol in blood agar, the blood agar was doped with 0.002% L-cysteine (tiol-dependent hemolysines) [5, 6, 9, 12, 13].

3 Results and discussion

Ability to lyse rabbit erythrocytes was recorded in 11 out of 19 cultures sourced form dairy products (amounting to 57.9%), in 15 strains isolated from goat, sheep and environmental samples and in 27 museum *Serratia* strains. Ability to hemagglutination in the presence of L-cysteine (thiol-dependent hemolytic activity) was found in 4 museum cultures of *Serratia* (6.6%), while no bacteria isolated from dairy products, sheep, goats or environmental objects demonstrated thiol-dependent hemolytic activity. (Table 1).

Table 1. Hemolytic activity in tested cultures of *Serratia marcescens.*

Property	Museum strains positive for the property		Strains sourced from sick animals and environmental objects, positive for the property		Strains sourced from dairy products, positive for the property		Total	
	Number of cultures	in %	Number of cultures	in %	Number of cultures	in %	Number of cultures	in %
Production of hemolysines	27	44.3	15	51.7	11	57.9	53	48.6
Production of thiol-dependent hemolysines	4	6.6	–				4	3.7

From studied 19 cultures of *Serratia* isolated from dairy products, ability for hemagglutination was shown for 14 strains, amounting to 73.7%; out of 61 museum strains, adhesive activity was identified for 45 cultures, i.e., 73.8%, while out of 29 cultures sourced from goats, sheep and environmental objects, there were 24 cultures positive for this property, amounting to 82.8%. Ability to agglutinate sheep erythrocytes independent of the presence of D-mannose was recorded for 8 strains out of 19 tested cultures sourced from dairy products (42,.1%), in 16 out of 29 cultures sourced from goats, sheep and environmental objects (55.2%) and in 35 out of 61 museum cultures (57.4) (Table 2).

No lysozyme-producing strains were produced out of *Serratia* cultures isolated from dairy products, while among 61 museum cultures 3 strains out of 29 produced lysozymes, amounting to 4.91%;

Table 2. Adhesive activity in tested cultures of *Serratia marcescens*.

Property	Museum strains positive for the property		Strains sourced from sick animals and environmental objects, positive for the property		Strains sourced from dairy products, positive for the property		Total	
	Number of cultures	%%	Number of cultures	in %	Number of cultures	in %	Number of cultures	in %
Adhesive activity in the absence of D-mannose	45	73.8	24	82.8	14	73.7	83	76.2
D-mannose-resistant adhesive activity	35	57.4	16	55.2	8	42.1	59	54.1

Out of samples sourced from goats, sheep and environmental objects, there were 2 strains positive for this property, amounting to 6.89%. (Table 3).

Table 3. Lysozyme and antilysozyme activity in tested cultures of *Serratia marcescens*.

Property	Museum strains positive for the property		Strains sourced from sick animals and environmental objects, positive for the property		Strains sourced from dairy products, positive for the property		Total	
	Number of cultures	in %	Number of cultures	in %	Number of cultures	in %	Number of cultures	in %
Ability to produce lysozyme	3	4.91	2	6.89	–	–	5	5.56
Antilysozyme activity	52	85.2	25	86.2	15	78.8	59	84.4

Ability to produce antilysozyme (Figure 4) was discovered in 15 out of 19 strains sourced from dairy products, amounting to 78.8%; among 61 museum strains, 52 strains demonstrated antilysozyme activity, amounting to 85.2%; out of 29 cultures sourced from goat, sheep and environmental objects, ability to produce antilysozyme was identified for 25 strains, that is, 86,2%. (Table 3). Quantitatively, by concentration of produced antilysozyme, the tests conducted on *Serratia* cultures showed that their antilysozyme activity varies in a range from 1 to 6 µg/ml.

4 Conclusion

Studies of lysozyme, antilysozyme, hemolytic and adhesive activity of Serratia on 61 museum strains, 19 strains isolated from home-made dairy products and 29 cultures sourced from goats, sheep and environmental objects allowed establishing that the stated factors are common among these bacteria. So, there were no strains found capable of producing lysozyme among 61 strains sourced from dairy products; out of 61 museum cultures, only 3 cultures produced lysozyme, that is, 4.91%, while out of 29 cultures isolated from goats,

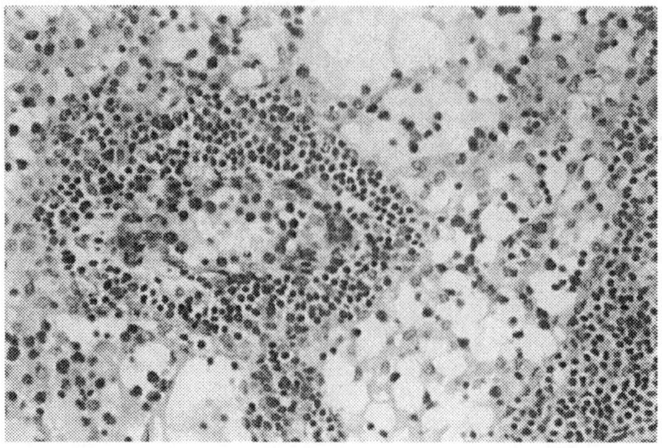

Fig. 4. Produced *antilysozymes.*

sheep and environmental samples, there were 2 strains positive for that property, amounting to 6.89%. Antilysozyme was produced by 15 cultures out of 19 isolated from dairy products (78.8%); some antilysozyme activity was demonstrated by 52 strains out of those 61 sourced from museum cultures; 25 out of 29 cultures sourced from goats, sheep and environmental samples were found capable of producing antilysozyme. Concentration of antilysozyme produced by various strains varied between 1 and 6 µg'ml.

Hemagglutination (adhesive activity) was found typical of 14 cultures isolated from dairy products (73.7%), 45 museum cultures (73.8) and 24 cultures isolated from goats, sheep and environmental objects (82,8%). D-mannose-resistant hemagglutination was detected for 8 strains (42.1%), isolated from dairy products, 16 strains (55,2%) sourced from goats, sheep and environmental objects and 35 museum strains (57.4).

Hemolytic activity with respect to rabbit erythrocytes were typical of 11 cultures (57.9%) sourced from dairy products, 15 cultures (51.7%) those sourced from goats, sheep and environmental objects and 27 museum strains (44.3%).

Thiol-dependent hemolytic activity was not detected for any tested strain of *Serratia* from dairy products or from goats, sheep and environmental objects. Among museum cultures, thiol-dependent hemolytic activity was typical of 4 strains, amounting to 6.6%.

References

1. R. A. Acharv, R. Kh. Gairabekov, F. S. Turlova, Z. E. Umieva, Adhesive ability of some opportunistic enterobacteria isolated from sheep with associated helminthosis and bacterial infection, *Natural Sciences* **2 (35),** 113–118 (2011)
2. O.V. Bukharin, N. V. Vasilyeva, B. Ia. Usviatsov, *Lysozyme of Microorganisms* (Tomsk, 1985)
3. O. V. Bukharin, B. Ia. Usvatsov, N. V. Nemtsova, Method for determination of antilysozyme activity of microorganisms, *Animal Microbiology* **2,** 27–29 (1984)
4. I. M. Gabrilovich, R. Kh. Gairabekov, T. Ch. Zhugova, K. Iu. Lotiev, O. S. Yakushenko, Lysozyme and antilysozyme activity of opportunistic enterobacteria, *Abstracts of the 18th All-Union Society of Epidemiologists, Microbiologists and Parasitologists named after I.I. Mechnikov* (Alma-Ata) pp 163–164 (1989)
5. I. M. Gabrilovich, R. Kh. Gairabekov, V. B. Boziev, M. V. Zarochentsev, Thiol-dependent hemolysines in opportunistic enterobacteria, *2nd All-Union Conference on Bacterial Toxins* (Yurmala) p 24 (1989)
6. I. M. Gabrilovich, R. Kh. Gairabekov, M. A. Sheozhev Thiol-dependent hemolysines as a factor of pathogenic power in *Citrobacter* and *Serratia, Materials of Bacterial Persistence Conference* (Kuibyshev) pp 69–71 (1990)
7. R. Kh. Gairabekova, M. R. Dakaeva, Pathogenic and opportunistic bacteria in milk and home-made dairy products, *Reflection* **1,** 19–23 (2019)
8. R. Kh. Gairabekova, M. R. Dakaeva, Pathogenic streprococci, staphylococci and opportunistic bacteria in milk and home-made dairy products, *Annals of Medical Institute* **15 (1),** 52–58 (2019)
9. R. Kh. Gairabekov, R. Kh. Gairabekova, Ia. A. Gaisumov, L. G. Molochaeva, F. S. Turlova Some peculiarities of epizootic processes accompanying enterobacterioses of goats and sheep in the territory of the Chechen Republic, *Reflection* **4,** 6–12 (2019)
10. R. Kh. Gairabekov, R. Kh. Gairabekova, S. A. Gubkhanova, F. S. Turlova, Z. E. Umieva Antilysozyme, anti-interferon and anti-complementary activity of some bacteria of family *Enterobacteriaceae, Modern problems of science and education* **5 (95),** 329 (2016)
11. R. Kh. Gairabekova Biological properties of *Serratia marcescens*, isolated from sheep in cases of post-dehelmintation disbacteriosis in the territory of the Chechen Republic, *Annals of Saratov State Agricultural University named after N.I. Vavilov* **4** 3–5 (2010)
12. 11 F. S. Turlova R. Kh. Gairabekov, Production of hemolysines by bacteria of genus *Serratia*, isolated from various sources in the territory of the Chechen Republic, *Annals of the Chechen State University* **1 (17),** 126–128 (2015)
13. 12 O. S. Yakushenko, A. B. Azamatova, V. V. Bugaeva, V. B. Boziev, K. Iu. Lotiev, R. Kh. Gairabekov, I. M. Gabrilovich, Factors of patogenic power of some opportunistic enterobacteria, *Materials of Conference Diagnostics, treatment and prevention of invasive and infectious diseases of agricultural animals* (Stavropol) pp 55–59 (1991)

BIO Web of Conferences 52, 00067 (2022)
FIES 2022
https://doi.org/10.1051/bioconf/20225200067

Circular economy in agriculture as a vector for energy security strengthening

Tatsiana Zoryna[1], *Julia* Valeeva[2], *Volha* Liubchyk[3], *Irina* Morozova[4], and *Elmira* Uteeva[5]

[1] Energy Economy Department, Institute of Power Engineering of National Academy of Science of Belarus, 15/2, Academicheskay Str., Minsk, 220072, Belarus
[2] Kazan State Power Engineering University, 51, Krasnoselskaya Str., Kazan, Republic of Tatarstan, 420066, Russian Federation
[3] Belarusian National Technical University, UNESCO Chair "Energy Conservation and Renewable Energy Sources", 65, Nezavisimosti Ave., Minsk, 220013, Belarus
[4] Department of Marketing Management, Kazan Innovative University named after V. G. Timiryasov (IEML), 42, Moskovskaya Str., Kazan, Republic of Tatarstan, 420111, Russian Federation
[5] Kazan State Medical University, 49, Butlerova Str., Kazan, 420012, Republic of Tatarstan, Russian Federation

Abstract. The article examines the principles of circular economy, its influence on energy security and sustainable development of the country. The possibilities of circular economy implementation in the sector of agriculture are described. The results of circular economy survey are analyzed, main barriers of waste use in agriculture of the Republic of Belarus are detected. The dynamics of livestock and poultry heads number was analyzed and its main trends are found, the forecast for the period of 2022-2030 was made. The assessment of current theoretical and technically available biogas production potential in the Republic of Belarus, as well as the forecast till 2030, is made. The authors evaluated the effect from biogas from agricultural waste, such as waste from livestock and poultry vital activity usage on fuel and energy resource consumption and the indicators of energy security of the Republic of Belarus and Tatarstan.

1 Introduction

Green economy requires a shift towards green energy production based on renewable energy sources to replace fossil fuels, as well as energy conservation and energy efficiency (Figure 1). Obviously, the share of the green economy in world market capitalization is growing, while the fossil fuel sector is shrinking. In particular, as of 2020, green economy accounted for 6% of the market capitalisation of global energy companies by 2020 [1].

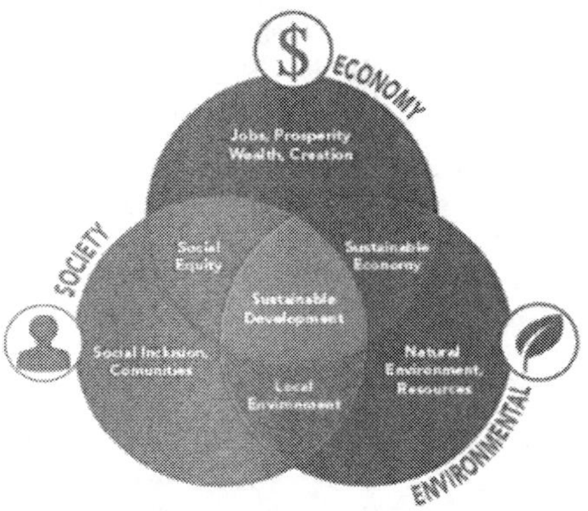

Fig. 1. Components of sustainable development and green economy [2].

Circular economy, which is also referred to as the "circular economy" [3], is "a model of production and consumption, including common use (leasing), reuse, improvement, recycling, and disposal of resources and products" (Figure 2) [4]. The theoretically identified model just promotes, it can be found that it is aimed at solving the most serious problems, in particular climate change, biodiversity loss, waste and pollution, probably beyond the thermal pollution that within the circular economy is not discussed explicitly. This is the fundamental difference between the circular economy model and the linear (it can be defined as a standard) model of the economy, which will be considered as a green economy.

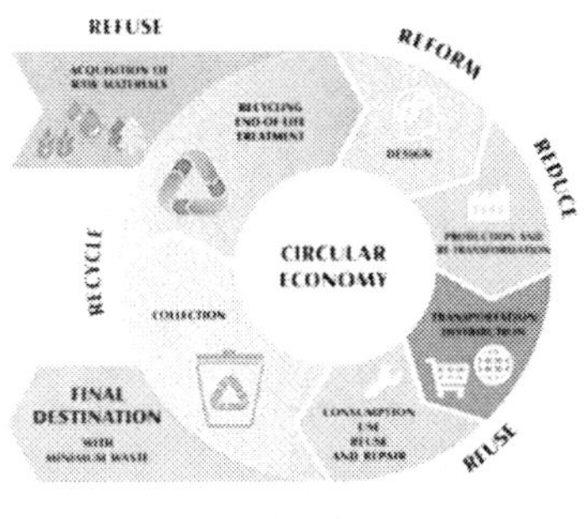

Fig. 2. Circular economy model [5].

The main purpose of this study is to consider the model of the circular economy of its main elements from the standpoint of organizing a closed cycle of food waste processing in agriculture and the possibility of obtaining biogas, as well as forecasting the consumption of electrical energy for all agriculture in general, including energy from biogas.

The scientific significance of this article is the formation of a forecast for the increase in the livestock of farm animals in the Republic of Belarus and the Republic of Tatarstan. Also, it has been established that biogas production is one of the tools for ensuring the country's energy security and implementing mechanisms in the Green Economy.

The practical orientation of a circular economy is to preserve products, materials, equipment and infrastructure for as long as possible [6]. Hence a possible increase in the productivity of energy resources. Energy should trigger the process of increasing the value of waste: either as a new component or as a regenerated resource for continued production. Biogas is an additional source of energy for agriculture.

As a result, natural resources (e.g. compost) should be regenerated. The level of environmental friendliness is increased, additional energy, the level of waste from animal rearing is reduced, and the recycling cycle is increased by converting animal life into energy.

Translated with www.DeepL.com/Translator (free version)The Ellen MacArthur Foundation (EMF) defines circular economy as an industrial economy that is restorative or regenerative in design and purpose [7, 8], which cannot be disagreed with. The circular economy is based on three main components of the economic system – resource providers, waste processors, sources of utility – and promotes the cyclical use of resources to create, supply and consume manufactured products. It is about extending the shelf life of products that are produced within a production system, making them more profitable and sustainable over time. At the moment, the global initiative to promote the circular economy paradigm belongs to the Ellen MacArthur Foundation, which goal is to facilitate the transition to it as an established practice [9].

The theoretical and methodological approaches that form the basis of the circular economy are not fundamentally new, but accumulate the results of scientists' research in the field of industrial ecology, eco-efficiency, resource efficiency, etc. As a concept, it is based on the principles of 9 R[1] and has three key implementation features:

– reserves of natural resources and maintaining a sustainable balance of renewable resources to preserve and maintain an inexhaustible level of natural capital;
– optimization of consumption processes by developing and distributing products, components and materials that meet the highest level of their reuse;
– identification and prevention of negative external effects of current production activities in order to increase the efficiency of economic and ecological systems[2].

Relating to agriculture, this direction is increasingly seen as a promising option for sustainable development, which can be achieved if innovative methods and technologies that minimize the cost of limited resources, stimulate their replacement with renewable ones, prevent losses and stimulate reuse and recycling, which together allows to ensure a qualitative transition from production efficiency to resource use efficiency.

It should be emphasized that the concept of circular economy does not contradict the basic principles of agriculture; moreover, due to the specifics of the industry, it has favorable conditions for implementation. Thus, traditionally, animal waste rich in valuable substances (nitrogen, phosphorus, potassium) was used as a source of fertilizers to improve soil quality; and organic waste from crop and livestock can be used to obtain environmentally friendly energy sources [10].

The problem of food waste management is an important direction in the environmental policy of the country. the stage-by-stage process of production, consumption and processing of food waste by 5 indicators and 25 indicators was studied. found that the level of intensity increases the amount of food waste and increase the duration of the closed cycle of the green economy. Effective management decisions aimed at the formation of a closed cycle will reduce the amount of food waste [34].

A good case in the formation of a closed cycle chain is the production of broilers and neighboring cultivation of feed to feed them. In general, broiler waste at all stages of the ah-ah life cycle is harmful in terms of nitrogen and carbohydrate content. to important decisions is the production of compass protein and the formation of energy [35].

It is necessary to highlight a position regarding the implementation of circular economy projects aimed at decentralizing the processing of food waste and localizing operations within a closed cycle. to This allows, as practice has shown, to mobilely process food waste and receive enough energy in the face of covid restrictions. the limited geographical location and time interval allows optimal timing to receive energy and process food waste [36].

The summation of the closed cycle makes it possible to replace the necessary minerals through the processing of food waste. At the same time, it is

[1] "Refuse" – rejection of excessive use of raw materials; "Recover" – production of energy from materials; "Recycle" – processing and recycling of materials; "Repurpose" – use of the product for other purposes; "Remanufacture" – production of new products from elements of the old; "Refurbish" – updating or restoring old products while keeping most of them unchanged; "Repair" – maintenance and repair of a defective product in order to extend its service life; "Reuse" – reuse of a product that has lost its value for one user, but is necessary for another; "Reduce" – reducing the use of raw materials, aimed at the total elimination of waste production.

[2] Delivering the circular economy: a toolkit for policymakers Delivering the circular economy: a toolkit for policymakers [Electronic resource] / Ellen MacArthur Foundation. [Cowes], 2015. Mode of access: https://www.ellenmacarthurfoundation.org/publications/delivering–the–circular–economy–a–toolkit–for–policymakers. Date of access : 07/08/2020.

https://doi.org/10.1051/bioconf/20225200067

important to note the involvement of not only large agricultural organizations, but also personal subsidiary plots that also have food waste, and can also participate in the formation of green economy processes. [37]

Materials and methods. The pathology of this study is based on the data of statistical reporting of livestock production by the Regional State Statistics Committee of the Republic of Tatarstan and the Federal State Statistics Service of the Republic of Belarus. The main parameters for assessing the potential without gas are based on a calculation formula that allows you to estimate the number of livestock and the possible production of biogas. Using the trend method, a biogas production forecast was formed, which in the future will form tools to ensure the energy security of the country as a whole. One of the important directions in the study of the potential for the development of biogas as a tool for obtaining energy is the use of the questionnaire method, which made it possible to identify factors that hinder the implementation of this project. In general, the results of the presented study are of practical importance for the implementation of the biogas production project both in the Republic of Tatarstan and in the Republic of Belarus.

As a result of analising data of circular economy survey conducted by the State Scientific Institution “Research Economic Institute of the Ministry of Economy of the Republic of Belarus” in the second quarter of 2019 by order of the Center for Economic Research Beroc received information about the attitude of agricultural organizations and farms to involvement of waste in reuse [11].

2 Materials and methods

The most common way to handle waste is to use it in their own activities, this was indicated by about 45% of farmers, while burial is acceptable only for a fifth of the respondents (Table 1).

The methods of waste management change significantly depending on the volume of their generation. Thus, 41.2% of farmers whose waste volume does not exceed 10 tons use them in their own activities, and only 5.9% sell them. With an increase in waste generation volume (over 10 tons), the share of waste used in own activities decreases to 30.0%, and the share of waste sold increases to 20.0%. At the same time, the share of buried waste is at the level of 30% and does not depend on the volume of their generation.

The totality of barriers identified by the surveyed agricultural enterprises and their classification according to the rate of waste use is presented in Table 2. The survey data show that agricultural business entities, regardless of their form of ownership and size, identify the same barriers that slows down the use of waste in agriculture. So, technological (lack of processing technologies) and economic (economic inexpediency and lack of resources to focus the production process on the use of secondary raw materials) barriers are the most significant both for large-scale agricultural organizations and for farms (1-3 place in the rating). Lack of sales markets for products from recycled materials and a decrease in the quality of manufactured products when using secondary raw materials have 5-6th place in the rating. At the same time, there are enough examples in the world of the successful use of the principles of the circular economy in agriculture. In particular, in the state of North Carolina (USA), which is a leading producer of poultry and eggs, the problem of poultry farm waste has been solved in a comprehensive manner since 2007. During this period, Georgia Renewable Power (GRP) created several biomass power plants that process poultry waste by mixing it with wood chips to generate electricity. In May 2017, Veolia, a company with extensive experience in implementing circular projects, was engaged to optimize technological solutions, which significantly increased the economic efficiency of waste processing. The plant currently processes up to 285,000 metric tons of poultry waste annually and has a capacity of 25 MW. Ambitious goals have been set to ensure 100% recycling of poultry waste [12].

Based on current trends analysis in order to assess the national economy sectors future needs in energy resources, the State Scientific and Production Institution “Institute of Energy of the National Academy of Sciences of Belarus” developed optimistic and pessimistic scenarios for electricity consumption in the agricultural sector of the Republic of Belarus. Forecasting long-term trends in electricity consumption in the agricultural sector of the Republic of Belarus is determined by the influence of a number of factors, the main of which include: the structure of the economy and its structural transformation; economic growth rates; rate of technological progress. Thus, the main indicators in forecasts preparation were: GDP growth rate; the rate of change in electrical and heat capacity.

According to scenario assumptions, it is assumed that the structure of the Belarusian economy will remain unchanged starting from 2030. The share of agricultural, forestry and fisheries sector in GDP from 2030 will be 6% (Fig. 3).

Analyzing the results of the forecasts presented in Fig. 3, we can conclude that according to the pessimistic scenario, electricity consumption in agriculture will increase by 1.36 times in 2050 compared to 2020, and by 1.53 times according to the optimistic scenario. The difference in electricity consumption according to the scenarios in 2050 will reach 314 million kWh and it can be covered through the introduction of circular economy principles in this sector, in part, the use of agricultural waste for the production of electricity.

Table 1. Attitudes towards the waste reuse in farms, depending on the amount of waste generated.

Name of indicator	All respondents	volume of generated waste		
		less than 1 ton	1–10 tons	more than 1 ton
General characteristics of a typical farm participating in the survey				
Average number of employees for 2018, people	6	5	6	10
Annual volume of generated waste, tons	4.4	less than 1 ton	1–10 tons	more than 1 ton
Distribution of answers to the question "How do you handle industrial waste?", %				
– use in our own activities	44.8	69.2	41.2	30.0
– transfer for processing	10.3	0.0	14.7	0.0
– sell	12.1	23.1	5.9	20.0
– bury / transfer for burial (neutralization)	22.4	0.0	29.4	30.0
– transfer to storage objects	10.4	7.7	8.8	20.0
Distribution of answers to the question "Does your company use third-party secondary raw materials?", %				
yes, for energy production	1.6	3.4	0.0	0.0
yes, in production	6.5	6.9	8.3	0.0
yes, in servicing the main activities of the economy	9.7	3.4	16.7	20.0
no, do not use	82.2	86.3	75.0	80.0

Note: The total selection includes 64 respondents.

Table 2. Rating of barriers to the use of waste in agriculture.

Barrier	Wastevolume								
	Agricultural organizations					Farms			
	All respon dents	Volume of generated waste				All respondents	Volume of generated waste		
		less than 1 ton	1–20 tons	21–50 tons	more than 50 tons		less than 1 ton	1–10 tons	more than 10 tons
Economic inexpediency	2	2	3	3	2	2	1	2	1
Lack of processing technologies	1	1	1	1	1	1	2	1	1
Lack of resources to orient production process for the use of secondary raw materials	3	n. b.	2	2	3	3	3	4	n. b.
Decrease in quality of products when using secondary raw materials	6	Nb	4	n. b.	Nb	6	n. b.	n. b.	1
Lack of markets for recycled products	5	n. b.	n. b.	3	5	5	n. b.	5	1
Lack of information on the volumes of available third-party secondary raw materials	8	n. b.	n. b.	n. b.	6	4	4	5	1
Lack of access to third-party recycled materials	6	Nb	5	n. b.	6	5	4	5	n. b.
Lack of necessary legislative framework regulating relations in the field of waste	4	n. b.	4	n. b.	4	3	3	3	n. b.
Other	7	Nb	4	n. b.	Nb	Nb	n. b.	n. b.	n. b.

Note: the barriers are ranked according to the "traffic light" principle: the red zone is the most important (the most popular answers, 1-2 places in the rating); yellow zone - quite significant (located in the upper part of the rating, 3-4 positions in terms of popularity); green zone - not very significant (5th-9th place in the rating), nb - not identified by respondents as a barrier.

1	2	3	4	5	6	7	8	9
great influence								least influence

Source: own calculations based on survey data of the State Scientific Institution "Research Economic Institute of the Ministry of Economy of the Republic of Belarus".

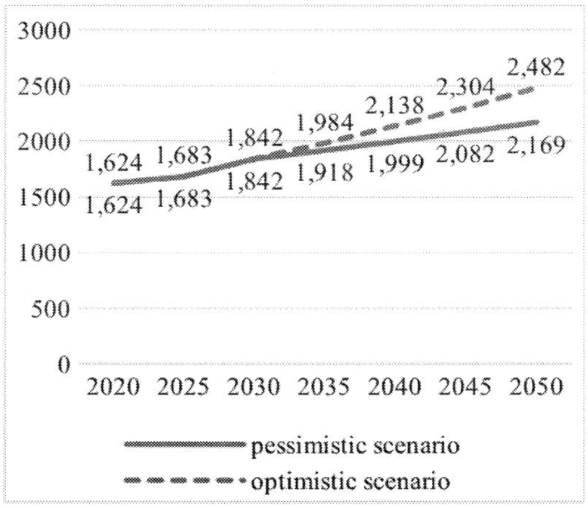

Fig. 3. Forecast of electricity consumption in the agricultural sector of the Republic of Belarus until 2050, million kWh.

3 Results and discussion

Electricity production from biogas in the world has doubled over ten years: from 46.3 thousand GWh in 2010 to 91.8 thousand GWh in 2019, while global electricity consumption increased over the same period by a quarter and by 2019 amounted to 25.0 million GWh [13, 14]. 67% of all electricity produced from biogas was obtained in Europe, the leader in production is Germany, where more than a third of the world's energy production from biomass is generated. Among other countries actively implementing biogas technologies are Italy, China, France, Czech Republic, Thailand, Turkey. The cost of energy production from biogas in Germany is 0.07-0.11 EUR/kWh, while the cost of energy from natural gas it ranges from 0.09 to 0.20 EUR/kWh [15]. It should be noted that in both cases the fuel component is excluded from the cost due to free substrate in many cases for the operation of a biogas plant and changing price of natural gas. The payback period for biogas plants, according to studies [16-19], does not exceed 10 years.

On the territory of the Republic of Belarus there is a large number of livestock complexes, which allows us to consider the waste products of livestock and poultry kept there as a reliable raw material base for biogas production.

In the Energy Potential Development Strategy of the Republic of Belarus, approved by the Decree of the Council of Ministers of the Republic of Belarus on August 9, 2010 No. 1180, the volume of biogas production is estimated at 503.7 million m3, which is equivalent to 433.2 thousand tons of coal equivalent [20].

According to the National Program for the Development of Local and Renewable Energy Sources for 2011–2015, approved by the Decree of the Council of Ministers of the Republic of Belarus on May 10, 2011 No. 586, the total biogas production potential only at cattle, pigs and poultry farms and complexes is 4158.5 million m3, which is equivalent to 3.18 million tce ; however, use of installations with electric power of more than 150 kW, which is feasible only at large farms, the effect of biogas is reduced to the replacement of 635.5 thousand tce [21]. In a study of Belarusian scientists [22], the theoretical potential of biogas in 2011 is estimated at 1603 million m3, which is equivalent to 1259 thousand tce .

Considering the significant differences in data on the existing potential of biogas, as well as the livestock and poultry, it is advisable to assess the biogas potential in the Republic of Belarus at the present time, as well as build a forecast for the coming years.

3.1 Dynamics and forecast of the livestock and poultry number at farm in the Republic of Belarus

As of the beginning of 2021, the number of cattle in farms of all categories in the Republic of Belarus reached 4292 thousand heads, pigs – 2872 thousand heads, sheep and goats - 148 thousand heads, horses – 34 thousand heads, poultry – 48 million heads [23]. The dynamics of the most numerous types of livestock and poultry in the farms of the Republic of Belarus is shown on Figure 4.

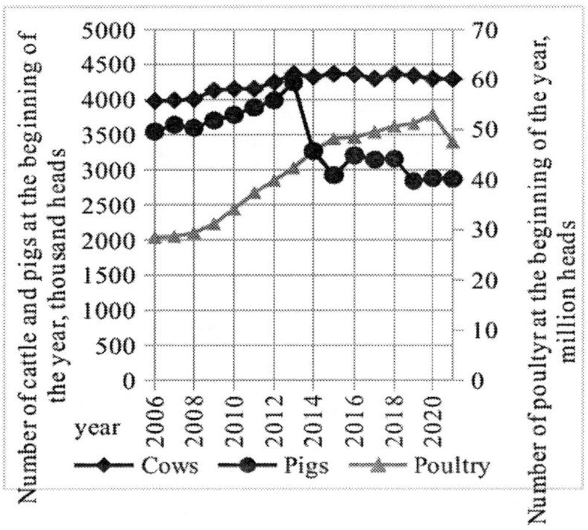

Fig.4. Dynamics of cows, pigs and poultry number in the Republic of Belarus. Source: Own development based on [24].

The increase in the number of cows in the period from 2007 to 2013 is associated with the development of not only meat, but also the dairy industry. During the specified period, against the background of a general decrease in the share of household spending on food, spending on dairy products has the highest growth compared to previous years (spending growth by 22%

from 2006 to 2013) compared to other categories of food products with a similar consumer price index [25, 26]. From 2013 to 2021, the number of cows remains relatively stable with fluctuations of no more than 1% of the average value for the specified period. This fact can be interpreted as finding a certain balance point between production, consumption, imports and exports, as well as making an assumption about stability in the coming years in the absence of significant external or internal unpredictable disturbances.

Until 2013, there was a clear increase in the number of pigs in the Republic of Belarus. The sharp decline in the number of pigs that took place in 2014 was caused with the aggravation of the epidemiological situation associated with recommendations African swine fever and massive for pigs' slaughter [27, 28]. Thereafter, the population stabilized. The lack of indicator tendency to the values of the previous period can be explained by the partial compensation of pork by meat of other livestock and poultry, as well as by the reduction in the share of private entrepreneurship in this industry and the share of exports. Since 2014, some stabilization in the number of pigs can be noted, despite the existing fluctuations.

The increase in the number of poultry in the Republic of Belarus has been observed since 2008. The production of poultry meat is more profitable than the production of beef or pork due to the lower cost and a shorter technological cycle, which is reflected in the cost of products for sale. For this reason, even during periods of crisis, the demand for poultry meat does not fall [25]. Over the period from 2015 to 2020, the growth in the number of poultries decreased slightly, but still took place. At the same time, a sharp drop to 48 million head can be observed in 2021.

If we consider in comparison with the Republic of Tatarstan, Russia (Table 3).

Table 3. Number of cattle in dynamics in the Republic of Tatarstan, thousand units.

Animal species	2010	2012	2015	2019	2020
Number of cattle	1124.6	1076	1033.6	707.8	689
Cows	420	403.2	343.60	235.3	229.05
Pigs	709.6	658.9	670.56	459.2	447.0
Sheep, goats	403	372.1	83.67	57.3	55.77
Poultry, million heads	34	31	24.53	16.8	16.35
Horses	36	32	27.59	18.9	18.39

In the agriculture of the Republic of Tatarstan there is no clearly defined specialization in the production of any product. It is approximately equally distributed between crop production (49.0% of all agricultural production in the region in value terms) and animal husbandry (51.0%). The animal husbandry of Tatarstan is characterized by high volumes of production of beef, milk, eggs, poultry meat, lamb and goat meat. Pork production, on the contrary, has been declining over the past few years, although it still exceeds the figures of 10, 15 years ago. The number of pigs in all categories of farms in the Republic of Tatarstan at the end of 2015 amounted to 481.2 thousand heads. This is 2.2% of the total number of pigs in Russia (11th place in the ranking of regions). Pork production in the Republic of Tatarstan in all categories of farms in 2015 amounted to 94.4 thousand tons in live weight (73.4 thousand tons in terms of slaughter weight). For comparison, in 2010 pork production in Tatarstan amounted to 100.2 thousand tons in live weight (75.7 thousand tons in terms of slaughter weight). At the end of 2015, the republic is in 11th place in terms of pork production (2.4% of the total pork production in the Russian Federation).

The agriculture of Tatarstan is largely based on cattle breeding. The cattle breeding of Tatarstan provides the region with high positions in the production of beef and milk in the Russian Federation. In contrast to the all-Russian trends, characterized by a significant decrease in beef production, in Tatarstan, the production of this type of meat is relatively stable - in 2015 it was at the level of a decade ago. In general, in the Russian Federation over 10 years it decreased by 10.1%.

As of the end of 2015, in terms of the total number of cattle, Tatarstan is in 2nd place in Russia (second only to the Republic of Bashkortostan) - 1,033.7 thousand heads (5.5% of the total number of cattle in the Russian Federation). Including the number of cows, there were 366.7 thousand heads (4.4% in the all-Russian number of cows, 3rd place in the Russian Federation).

Beef production in Tatarstan in 2015 was at the level of 150.2 thousand tons in live weight (85.4 thousand tons in terms of slaughter weight). For 5 years, production decreased by 9.4% or by 15.6 thousand tons in live weight. At the end of 2015, the republic took 2nd place in the ranking of beef producing regions. share in general volume production beef in country amounted to 5.2%.

To determine the number of livestock and poultry, as well as their percentage grouping across farms of various types in the coming years, forecasting methods based on the trend model construction were used: regression model and forecasting method based on an average value.

When choosing a method for livestock and poultry number forecasting, the number of which fluctuated in recent years, but remained close to the average value in these years, the average value method was applied (Table 4). The regression model was built for those kinds where an increase or decrease was observed in previous years (Table 5).

Table 4. Forecast of livestock number in the Republic of Belarus (cows, pigs, sheep and goats).

Kind of livestock	Number, thousand heads in a year	
	2021 facts	2022–2030
Cows	1485	1506
Calves	2807	2822

Pigs	2872	3036
Sheep, goats	148	148

Also, a regression model was built to forecast the share of livestock and poultry at farms of agricultural organizations (Table 6, 7).

3.2 Biogas production potential determination and forecasting

Based on the data on the biogas production volume from one animal or bird, proposed in [22, 29, 30], we will evaluate the theoretical and technically available potential [31] of biogas production in the Republic of Belarus, obtained from the waste products of livestock and poultry.

When calculating, we take into account a number of features.

1. As known, for the operation of a biogas plant, a certain amount of thermal energy is required to heat the raw materials and maintain the set temperature in the digesters. According to the set out in [22], one fifth of the energy produced recommendations is spent on the own needs of a biogas plant.

Table 5. Forecast of the livestock and poultry number in the Republic of Belarus (horses, poultry).

Kind of livestock or poultry	number									
	2021 facts	2022	2023	2024	2025	2026	2027	2028	2029	2030
Horses, thousand heads	34	29	26	23	twenty	eighteen	16	fourteen	12	eleven
Poultry, million heads	48	51	52	52	52	52	52	52	53	53

Table 6. Forecast of livestock and poultry share kept in farms of large agricultural organizations in the Republic of Belarus.

Kind of livestock or poultry	Share of livestock and poultry kept in farms of large agricultural organizations, %									
	2021 facts	2022	2023	2024	2025	2026	2027	2028	2029	2030
Cows	97.9	98.0	98.0	98.1	98.2	98.2	98.3	98.3	98.4	98.4
Calves	96.1	96.2	96.3	96.5	96.7	96.8	96.9	97.1	97.2	97.3
Pigs	89.1	89.3	90.0	90.6	91.3	91.9	92.6	93.3	93.9	94.6
Sheep, goats	8.1	8.8	9.1	9.4	9.7	10.0	10.3	10.5	10.8	11.1
Horses	40.0	40.3	40.5	40.7	40.9	41.1	41.3	41.5	41.6	41.8
Poultry	90.8	91.1	91.2	91.4	91.6	91.7	91.8	91.9	92.0	92.1

Table 7. Forecast of the number of farm animals, in the Republic of Tatarstan.

Kind of livestock or poultry	Share of livestock and poultry kept in farms of large agricultural organizations									
	2021 facts	2022	2023	2024	2025	2026	2027	2028	2029	2030
Cows	198.3	200.28	202.29	204.31	206.35	208.42	210.50	212.60	214.73	216.88
Pigs	422.2	426.42	430.69	434.99	439.34	443.74	448.17	452.66	457.18	461.75
Sheep, goats	53	53.53	54.07	54.61	55.15	55.70	56.26	56.82	57.39	57.97
Bird, million units	fifteen	15.15	15.30	15.45	15.61	15.77	15.92	16.08	16.24	16.41
Horses	17	17.17	17.34	17.52	17.69	17.87	18.05	18.23	18.41	18.59

Let's make a forecast for the number of cattle in the Republic of Tatarstan until 2030. At the same time, it is important to take into account that there has been a downward trend in the number of livestock since 2010. At the same time, a number of programs are being implemented in the Republic aimed at supporting agricultural producers to increase the number of livestock. Therefore, within the framework of an optimistic forecast, this figure is planned not to decrease, to have a small increase or remain on the same level .

2. Despite the presence of low-capacity biogas plants designed to process the waste products of several heads of livestock [32, 33], the most stable and efficient operation of plants takes place at complexes with a large number of livestock kept. In this regard, when determining the theoretical potential for obtaining biogas in the country, the entire number of livestock and poultry is taken into account, when determining the technically available - only the part kept at large agricultural organizations.

3. Manure removal systems do not always allow the use of the entire generated waste products volume for biogas production. In this connection, it becomes necessary to use corrective coefficients:

k_1 = 0.7 – coefficient that takes into account the impossibility of collecting and using cows manure for biogas production (a similar coefficient will be applied to horses, sheep and goats);

k_2 = 0.8 – coefficient taking into account the specificity of using a pig manure water flushing system;

k_3 = 1.0 – coefficient that takes into account the specificity of the use of the poultry manure collection system [20].

To calculate the theoretical potential of biogas production, it is proposed to use the formula obtained on the basis of the study [22]:

$$P_{theoretical} = (1-0.2)\cdot N \cdot g \quad (1)$$

where 0.2 – share of energy for own needs; N – number of livestock or poultry of one kind; g – estimated biogas production per head of livestock or poultry per year.

The technically available potential requires taking into account the specifics of the disposal of waste and the proportion of livestock and poultry kept at the large farms of agricultural organizations, and is calculated with the formula:

$$P_{technical} = \frac{P_{theoretical} \cdot k_x \cdot \varphi}{100} \quad (2)$$

where k_x – coefficient that takes into account the removal of waste and is specific to x animal kind; φ – share of livestock and poultry kept at large agricultural organizations.

The potential of biogas production from livestock and poultry waste products, taking into account their actual number at the beginning of 2021, is presented in Table 8.

Tables 9 and 10 show the forecast of the theoretical and technically available potential of biogas production from livestock and poultry waste product for the period of 2022-2030.

Table 8. Potential of biogas production from livestock and poultry waste products at the beginning of 2021.

Type of livestock or poultry	Biogas output per head	Biogas output		Theoretical potential		Technically available potential	
	m³/year	million m³/year	thousand tce/year	million m3/year	thousand tce/year	million m³/year	thousand tce/year
Cows	450–548	668–813	528–642	535–650	422–514	359–437	284–345
Calves, heifers	315–330	884–926	699–732	707–741	559–585	484–507	383–401
Pigs	73–1010	210–314	166–248	168–252	133–199	118–178	94–140
Sheep, goats	120–150	18–22	14–17.5	14–18	11.2–14	0.8–1	0.6–0.8
Horses	300–350	10–12	8.1–9.4	8.2–9.5	6.4–7.5	2.3–2.7	1.8–2.1
Poultry	5.5–6.2	260–295	205–233	208–236	164–186	189–215	149–170
Total:	–	2050–2383	1620–1883	1640–1906	1296–1506	1154–1340	917–1059

Table 9. Forecast of theoretically and technically available biogas production from livestock and poultry waste products potential in Republic Belarus.

Year	2022	2023	2024	2025	2026	2027	2028	2029	2030
theoretical potential (min)	1327	1327	1327	1328	1328	1328	1328	1328	1329
theoretical potential (max)	1545	1545	1545	1546	1546	1546	1546	1546	1547
technically available potential (min)	938	941	943	945	947	949	951	953	955
technically available potential (max)	1092	1095	1098	1101	1103	1106	1108	1111	1113

Table 10. The potential for obtaining biogas from the waste of agricultural animals, taking into account their actual livestock at the beginning of 2021 in the Republic of Tatarstan.

Kind of animal	Biogas output per head	Biogas output, total		Theoretical potential		Technically possible potential	
	m³/year	million m³/year	thousand tce/ year	million m³/year	thousand tce/ year	million m³/year	thousand tce/year
Cows	450–547.5	56–87	54–82	56.2–87.3	54.56–82	359–437	283.5–344.9
Pigs	315–330	79–82	77–79	79.3–82.4	77–79.3	484–507	382.6–400.8
Sheep and goats	73–109.5	19–29	17–27	19.5–29.2	17–27.2	118–178	93.6–140.4
Bird , million heads	120–150	9–11	7–9	9.67–11.4	7–9.3	0.8–1	0.6–0.8
Horses	300–350	2–4	0.9–1.2	2.67–4.1	0.9–1.2	2.3–2.7	1.8–2.1
Total:	–	887–992	778–889	887.45–992.5	778–889.4	1154–1340	911.6–1058.5

4 Conclusion

Based on the study, the following conclusions can be drawn in Republic Belarus:

– the theoretical potential at the beginning of 2021 is 1640-1906 million m3/year, which is equivalent to 1296–1506 thousand toe/year and allows to replace 1126–1309 million m3 of natural gas annually; the technically available potential is estimated at 1154–1340 million m3/year, which is equivalent to 912–1059 thousand tce /year and allows to replace 794-922 million m^3 of natural gas annually;

– by 2030, the theoretical potential for biogas production will increase slightly: up to 1682–1958 million m3/year, which is equivalent to 1329–1547 thousand tce /year or the replacement of 1155–1345 million m3 of natural gas; technically available potential increases to 1209–1409 million m^3/year, which is equal to replacement of 955–1113 thousand tce /year or 830–968 million m3 of natural gas.

– full use of technically available potential of biogas from livestock and poultry waste products will replace 3–4% of fuel and energy resources consumed in the Republic of Belarus or 5–6% of the fuel and energy resources spent in the country to produce heat and electricity.

– in addition, there will be an increase in the energy security of the country, which can be numerically expressed in values increase of the indicators of the block "Energy independence": No. 1 "The ratio of primary energy production to the gross consumption of fuel and energy resources" and No. 2 "The ratio of primary energy production from renewable energy sources to gross consumption of fuel and energy resources", each by 3–4%, and a decrease in the indicator value of the block "Diversification of suppliers and types of energy resources" No 4 "Share of the dominant type of fuel in gross consumption of fuel and energy resources" also by 3–4% by 2030.

As for Tatarstan (Table 10), there is a positive trend here, and the theoretical potential of biogas from cows is 54 million cubic meters a year, from pigs – 77, sheep and goats – 17, birds – 9 horses – 1.2. This is a good tool for developing other energy resources in rural areas. In addition, it is a good tool for reducing environmental friendliness and organizing a closed-loop green economy. This resource of the Republic can be fully used. Here it is important to take into account the increase in livestock. And the increase in energy consumption due to mechanization, automation of machinery and equipment in rural areas. in general, within the framework of the study, all the objectives were achieved; in the framework of subsequent studies, other sources of renewable energy will be considered.

References

1. A. Atasu, C. Dumas, L. N. Van Wassenhove, *Harvard business review from the Magazine, July–August* (2021)
2. Annual North Central Washington Regional Economic Forum, retrieved from: https://www.lakechelan.com/event/annual-north-central-washington-regional-economic-forum/
3. G. G. Potapov Center for Environmental Industrial Policy "Ministry of Industry and Trade of the Russian Federation"
4. O. A. Filchenkova, *Topical Issues of Economics and Management: Proceedings of the VII International Scientific Conference* (Sankt- Petersburg, April) **11** (2019)
5. Circular economy, retrieved from: https://www.portoprotocol.com/category/circular-economy/
6. R. Clift, J. Allwood, *Chem. Eng.*, **837**, 30 (2011)
7. D. Reike , J. Kirchherr , M. Hekkert *Resources Conservation and Recycling* **127,** 221 (2017)
8. S. Jabeen, S. Malik, S. Khan, N. Khan, M. I. Qureshi, M. Sh. Md Saad, *International Journal of Energy Economics and Policy* **11 (1),** 270 (2020)
9. M. I. Lisitca, E. S. Khutieva, O. A. Doroshenko, A. A. Konariova, Circular economy as an alternative to the green economy, *International Scientific and Practical Conference "Developing the Energy Agenda of the Future" for representatives of the community of young fuel and energy engineers* (Saint-Petersburg) 255 (2021).
10. N. N. Batova, I. E. Tochiczkaya, P. V. Sachek. Circular economy in agriculture: conceptual features and opportunities for implementation in the Republic of Belarus, *Proceedings of the National Academy of Sciences of Belarus: Agrarian Sciences Series* **59 (3),** 277 (2021)
11. N. N. Batova, P. V. Sachek, E. S. Shershunovich, I. E. Tochiczkaya, Circular economy in agriculture in Belarus, retrieved from: https://www.beroc.org/upload/iblock/a88/a8809b421bea7f02e8ed7cdaff5b41b7.pdf
12. Veolia, *From Chickens to Light Bulbs... What if Our Poultry Could Produce Energy*? Retrieved from: https://www.planet.veolia.com/en/biomass-green-energy-chicken-droppings-lumberton-usa (2019)
13. Bioenergy, retrieved from: https://www.irena.org/bioenergy
14. Data and Statistics, retrieved from: https://www.iea.org/data-and-statistics?country=WORLD&fuel=Electricity%20and%20heat&indicator=TotElecCons
15. Levelized Cost of Electricity, retrieved from: https://www.ise.fraunhofer.de/content/dam/ise/en/documents/publications/studies/EN2018_Fraunhofer-ISE_LCOE_Renewable_Energy_Technologies.pdf
16. S. Karellas, I. Boukis, G. Kontopoulos. *Renewable and Sustainable Energy Reviews* **14 (4),** 1273 (2010)
17. Biogas from fecal sludge at Kibera communities at Nairobi, retrieved from:

https://www.iwmi.cgiar.org/Publications/Books/PDF/resource_recovery_from_waste-114-123.pdf

18. A. Menind, Boigas *Plant Investment Analysis, Cost Benefit and Main Factors* (Estonian University of Life Sciences, 2009), retrieved from: https://www.researchgate.net/publication/242547266_BIOGAS_PLANT_INVESTMENT_ANALYSIS_COST_BENEFIT_AND_MAIN_FACTORS
19. V. F. Klinczova. *Advanced technologies and materials of the future: collection of articles of the IV International Scientific and Technical Conference "Minsk Scientific Readings 2021"* (Minsk, December 9) **3 (1),** 140 (2021)
20. Strategic Development of the Energy Potential of the Republic of Belarus, retrieved from: https://minenergo.gov.by/law/postanovleniya-soveta-ministrov-respubliki-belarus/
21. National Program for the Development of Local Renewable Energy Sources for 2011–2015, retrieved from: http://www.government.by/upload/docs/file663fb27db70962e8.pdf
22. S. P. Kundas, V. A. Pashinskij, A. A. But`ko, *Energy Efficiency*, **4 (32)** (2013)
23. National Statistical Committee of the Republic Belarus, retrieved from: https://www.belstat.gov.by/
24. Agriculture of the Republic of Belarus: Collection of Statistical Data, retrieved from: https://www.belstat.gov.by/upload/iblock/3a9/3a9942589996c1bd248d5b05512fd7d7.pdf 8.01.2021
25. Household income and expenditure, retrieved from: https://www.belstat.gov.by/ofitsialnaya-statistika/solialnaya-sfera/uroven-zhizni-naseleniya/dokhody-i-potreblenie-domashnikh-khozyaystv/godovye-dannye/
26. Consumer price indices for basic products, retrieved from: https://www.belstat.gov.by/ofitsialnaya-statistika/realny-sector-ekonomiki/tseny/potrebitelskie-tseny/godovye-dannye/
27. Due to the suspicion of a new outbreak of African plague, private farms in the Vetkovskogo area were recommended to slaughter pigs, retrieved from: https://people.onliner.by/2014/07/12/pigs
28. African swine fever in Belarus in the summer of 2014: Is there or not? Retrieved from: https://doktora.by/novosti-mediciny/afrikanskaya-chuma-sviney-v-belarusi-letom-2014-goda-est-ili-net
29. 29. Stroitel`stvo biogazovy`x ustanovok: Quick guide, retrieved from: https://fermer.ru/files/v2/advice/19087/broshuraa5rus.pdf
30. O. X. Kil`chukova. Improving the design and modes of work of a biogas plant for small agricultural (2020)
31. A. A. Mixalevich, Renewable energy source: Potential, achievements, prospects, *Proceedings of the International Seminar of its Experts* **7** (2013)
32. Biogas plant small and medium size, retrieved from: http://www.belpg.com/oborudovanie/energosberegayuschie-tehnologii/biogazovye-stancii/malyh-i-srednih-razmerov
33. 33. Bioreactor BUG for processing waste into biogas and fertilizers, retrieved from:
34. https://bmpa.ru/reaktory-dlya-pererabotki-biomassy.
35. Exploring the green waste management problem in food supply chains: A circular economy context,
36. Sh. Kharolaa, M. Ramab, S. K. Manglac, N. Goyala, O. P. Nautiyald, D. Pantd, Y. Kazancoglu, *Journal of Cleaner Production* **351,** 131355 doi: 10.1016/j.jclepro.2022.131355 (2022)
37. M. Chiarelotto, J. C. P. S. Restrepo, H. E. F. Lorin, F. M. Damaceno, Composting organic waste from the broiler production chain: A perspective for the circular economy, *Journal of Cleaner Production* **329,** 129717 doi: 10.1016/j.jclepro.2021.129717 (2021)
38. H. B. Sharma, K. R. Vanapalli, B. Samal, V. R. Sankar Cheela, B. K. Dubey, J. Bhattacharya, Circular economy approach in solid waste management system to achieve UN-SDGs: Solutions for post-COVID recovery, *Science of The Total Environment* **800,** 149605 doi: 10.1016/j .scitotenv.2021.149605 (2021)
39. Yumei Wang, Zengwei Yuan, Ya Tang, Enhancing food security and environmental sustainability: A critical review of food loss and waste management, Resources, Environment and Sustainability, **4,** 100023 (2021) doi: 10.1016/j.resenv.2021.100023

Biohumus "Sis" for the ecologically pure agricultural production

Tatevik A. Jhangiryan[1], *Albert O.* Markosyan[1], *Hovhannes* K. Yeritsyan[2], *Julia* Valeeva[3], *Margarit* T. Petrosyan[4*], and *Gulnara* Valeeva[5]

[1] H. Petrosyan Scientific Center of Soil Science, Melioration and Agrochemistry, Branch of Armenian National Agrarian University, 74, Teryan St., Yerevan, 0009, Armenia
[2] Armenian National Agrarian University, 74, Teryan St., Yerevan, 0009, Armenia
[3] Kazan State Power Engineering University, 51, Krasnoselskaya St., 51, Kazan, Republic of Tatarstan, 420066, Russian Federation
[4] Department of Biochemistry, Microbiology & Biotechnology, Biology Faculty, Yerevan State University, 1A, Manoogian St., Yerevan, 0025, Armenia
[5] Kazan State Agrarian University, 65, Karl Marx St., Kazan, Republic of Tatarstan, 420015, Russian Federation

Abstract. This study aimed to investigate the influence of the microbial biohumus obtained from organic non-toxic wastes on the yield of three varieties of peppers, identification of the effect of biohumus on soil physicochemical parameters, to determine the ecological significance and economic feasibility of biohumus application. The photosynthetic activity of plants and accumulation of ascorbic acid in the tissues of peppers were investigated. The experiments in the vegetation cabin condition have been carried out according to the following variants: 1. Control, 2. Biohumus, 3. $N_{60}P_{60}K_{60}$, 4. Biohumus + $N_{60}P_{60}K_{60}$. The yield of a variety of Jermatnayin hska pepper in the vegetation cabin conditions in the Biohumus (237 c/ha) variant was lower than in the "Biohumus + $N_{60}P_{60}K_{60}$" variant and was higher from "$N_{60}P_{60}K_{60}$" and "Control" variants (277c/ha, 229 c/ha, and 191c/ha respectively). The yield of a variety of Arajnek pepper in the Biohumus (286 c/ha) variant was lower than in the "Biohumus + $N_{60}P_{60}K_{60}$" variant and was higher "$N_{60}P_{60}K_{60}$" and "Control" variants (320c/ha, 265c/ha, and 228c/ha respectively). The yield of a variety of Loshtak pepper in the "Biohumus" (335.7 c/ha) variant was lower than in the "Biohumus + $N_{60}P_{60}K_{60}$" variant and was higher in "$N_{60}P_{60}K_{60}$" and "Control" variants (391.9c/ha, 314c/ha, and 239.8c/ha respectively).

1 Introduction

The required amount of organic fertilizers (manure, poultry manure) in the Republic of Armenia (RA) is currently higher than the quantity that is being produced. To get the desired outcome development and application of alternative biofertilizers are of great importance.

In recent years, the increasing demand for organic agriculture calls for a search for new sources of biofertilizers used in a broad spectrum to enhance plant growth and productivity [19]. There are different approaches to generating biofertilizers. Composting is a biochemical and microbiological process by which organic matter is hydrolyzed to a stable and disinfected humus, which can be obtained from both household waste and various organic wastes [1, 10].

Biofertilizers contain living microorganisms that are agriculturally useful in terms of nitrogen fixation, phosphorus solubilization, or nutrient mobilization to increase the productivity of the soil and crops [2, 5]. Bacteria break down organic substrates as a source of energy, using carbon and nitrogen, as well as oxygen, to produce water, carbon dioxide, and heat, enriching the soil with compost. The formed compost contains biologically stable significant humus, which acts as an additional additive to the soil [3]. Next, the spontaneous increase in temperature helps eliminate pathogens and makes compost safe to use [4].

Fertilization with beneficial microorganisms increased tomato plant growth and decreased tomato defenses against insect pests, in comparison with chemical fertilizer (NPK) [16, 17]. In addition, during the application of microbial fertilizers, microbes enhance plant growth by various direct and indirect promotion mechanisms production of enzymes, as well as solubilization of potassium, zinc, and phosphorus [6]. According to Glick (2005), the enzyme 1-aminocyclopropane-1-carboxylate (ACC) deaminase, produced by soil microorganisms, decreases ethylene levels and allows the plant to be more resistant to a wide variety of environmental stresses [7]. It was reported that inoculation with the combination of three bacterial species (*Bacillus lentus, Azospirillum brasilens, Pseudomonades sp*) improves growth, significantly improves CAT enzyme activity (GPX and APX activity) and chlorophyll content in leaves in basil plants under water stress [8]. The plant-associated microbes have the ability to promote the growth of plants under natural as well as extreme conditions. According to Marulanda (2007), inoculation of plants with native beneficial microorganisms may increase the drought tolerance of plants growing in arid or semi-arid areas [7].

* Corresponding author: margaritpetrosyan@ysu.am

The utilization of agricultural waste using beneficial microorganisms and its conversion to biofertilizer can be one of the successful alternative ways for optimizing resource use and generating income [11].

RA is one of the countries where abiotic stresses, such as drought, salinity, heat, and cold, decrease yields, destroy soil fertility, and in some cases, make continued farming impossible. In RA, most of the cultivated lands are dry lands and are not provided with proper irrigation systems. The application of commercial chemical inorganic fertilizer (NPK) is, however, associated with some disadvantages including high cost and harmful impact on the environment and human health. Therefore, biofertilizers as alternative organic fertilizers are of great importance for sustainable agriculture. In this study, we aimed to evaluate the combined and individual effects of Biohumus and inorganic fertilizers on crops.

2 Materials and methods

2.1 Plant material

Seedlings of peppers of crops belonging to the *Solanaceae* family: Loshtak, Arajnek, Jermatnayin hska varieties were used as objects of investigation. Seedlings were provided by the staff of the "Vegetable and Technical Crops Scientific Center" CJSC of the RA Ministry of Economy. Seedlings of three pepper varieties were grown in vegetative containers with 85 kg of soil, intended for the growth of plants in the scientific center of "Soil Science, Melioration and Agrochemistry Scientific Center after H. Petrosyan" Branch of ANAU. During the experiments were measured growth and development parameters.

Biohumus was obtained from various organic wastes (weeds, crop residues, leaf mass of trees and bushes, sawdust, and others) on the basis of "Sis" bioproduct a complex of mixed culture of soil bacteria (lactic acid, photosynthetic, nitrogen-fixing bacteria, aerobic cellulose-degrading bacteria, actinomycetes, yeast, and fungal strains). Jam juice was used as a carbon source for the growth of microorganisms. After the growth of bacteria, the bacterial solution was added to the accumulated mass of organic waste, moistened with a solution of "Sis" (1:100 ratio, in water), and covered with polyethylene film. To speed up this operation and increase the nutritional value of compost, manure or poultry manure was used, in the amount of more than 20% of the total mass. Then soil was added in the proportion of 10% of the accumulated mass (which promotes long-term moisture retention in the accumulated mass and absorption of nitrogen compounds). Periodically, once every 8-10 days, containers were opened, watered, mixed, and covered again. The preparation process of "Sis" Biohumus by this method takes two months [1].

2.2 The soil's physical and chemical properties and agrochemical indicators

The impact of "Sis" biohumus on the physicochemical parameters of the soils [14], was investigated before and during the experiments. Humus content was investigated according to I. V. Tyurin, pH-potentiometrically method, mechanical composition-according to N. A. Kachinskiy, available nutrients (nitrogen (N)-according to I. V. Tyurin and M. M. Kononova, phosphorus (P_2O_5)-according to B. M. Machigin, potassium (K_2O)-according to A. L. Maslova), Sand weight- according to N. I. Savinov [12,13,14].

2.3 Determination of photosynthetic pigments

Chlorophylls a and b, as well as total carotenoid content, were determined using a ceton and ethanol solvents by Spectrophotometric methods [15].

2.4 Determination of ascorbic acid content

The ascorbic acid content was investigated by iodometric titration [18].

The experimental results were expressed as mean ± standard deviation (SD) of three replicates. The smallest significant difference ($SSD_{0.95}$c/ha) was calculated. Mathematical analysis of the results of the vegetation experiments was performed by the method of dispersion analysis. The relative error of the experiment (S_x%) was calculated.

3 Results and discussion

The experiments were carried out in 2021in the vegetation cabin of the scientific center of "Soil Science, Amelioration and Agrochemistry Scientific Center after H. Petrosyan" Branch of ANAU. To achieve higher productivity and soil sustainability microbial biohumus was used. Experiments have been carried out in the vegetation cabin conditions according to the following variants: 1. Control, 2. Biohumus, 3.$N_{60}P_{60}K_{60}$, 4. Biohumus + $N_{60}P_{60}K_{60}$.

In the 2021 experimental year, the vegetation period was characterized by high temperature and dryness, which certainly affected plant growth, development, and productivity. According to data received, the soils of the vegetation cabin had the following content: 3.1% humus, 4.16, 35.00, 117.82 mg/100g soil of available nutrients (N, P_2O_5, K_2O), respectively. The content of available nutrients (N, P_2O_5, K_2O) for plants in the obtained biohumus was 37.5; 27.5, and 13.5 mg/100g, respectively.

Table 1. The effect of "Sis" Biohumus and chemical fertilizer on the growth and development of different pepper varieties.

Plant varieties	Variant	Plant height / average cm /26.04.2021	Plant height /average cm/ 14.05.2021
Loshtak	Control	11± 0,53	21±0,3
	Biohumus	13±0,1	24±0,23
	$N_{60}P_{60}K_{60}$	12,5±0,36	21±0,12
	Biohumus + $N_{60}P_{60}K_{60}$	14±0,2	23±0,12
Arajnek	Control	24±0,26	27±0,2
	Biohumus	21±0,36	24±0,31

	$N_{60}P_{60}K_{60}$	23 ±0,26	25±0,2
	Biohumus +$N_{60}P_{60}K_{60}$	19±0,17	22±0,12
Jermatnayin hska	Control	30±0,36	34±0,31
	Biohumus	29±0,53	33±0,15
	$N_{60}P_{60}K_{60}$	33,5±0,96	38±0,26
	Biohumus +$N_{60}P_{60}K_{60}$	29±0,6	34±0,21

At the beginning of flowering of Loshtak plants, 1-4 variants were registered on the same day. Among the plants of a variety of Arajnek and Jermatnayin hska growing in "Biohumus" and "Biohumus+$N_{60}P_{60}K_{60}$", variants are flowering 3 days earlier than plants growing in "Control" and "Biohumus" variants.

Table 1.1. The effect of Sis biohumus and chemical fertilizer on growth and development of different pepper varieties, 2021.

Plant varieties	Variant	Plant height/ average cm/04.06	Beginning of flowering	Beginning of fruiting
Loshtak	Control	52±0,17	17.05	07.06
	Biohumus	52±0,26	17.05	01.06
	$N_{60}P_{60}K_{60}$	54,3±0,44	17.05	01.06
	Biohumus +$N_{60}P_{60}K_{60}$	57,7±0,26	17.05	07.06
Arajnek	Control	48±0,1	17.05	01.06
	Biohumus	47±0,2	17.05	01.06
	$N_{60}P_{60}K_{60}$	50±0,2	14.05	07.06
	Biohumus +$N_{60}P_{60}K_{60}$	53±0,26	14.05	14.06
Jermatnayin hska	Control	63±0,35	17.05	04.06
	Biohumus	65±0,26	17.05	13.06
	$N_{60}P_{60}K_{60}$	64±0,46	14.05	14.06
	Biohumus +$N_{60}P_{60}K_{60}$	71±0,26	14.05	02.06

Plants of Loshtak variety growing in "Biohumus" and "$N_{60}P_{60}K_{60}$", and plants of Arajnek variety were different in terms of fruiting. In the "Biohumus" and "$N_{60}P_{60}K_{60}$" variants, fruiting is 6 days ahead compared to the "Control" and 4th variants. In the 1-4 variants of Loshtak plants, from seedling to fruiting, on average 48.0, 39.0, 48.0, and 43.0cm of plant stem lengths were registered, compared to the initial length of seedlings, 4.0, 5.0, 4.0 and 6.0 cm, respectively. Plant stem lengths of Arajnek 1-4 variants, from seedling to fruiting, on average were 43.0, 41.0, 44.0, 47.0 cm, respectively, compared to the initial length of seedlings 5.0, 6.0, 6.0, and 6.0 cm, respectively (Table 1).

Early fruiting was observed among the plants of Jermatnayin hska in "Biohumus+ $N_{60}P_{60}K_{60}$", and control variants. They are almost 10 and 11 days ahead of the "Control" and "Biohumus" variants. An average of 55.0, 56.0, 56.0, and 64.0 cm of plant stem length growth from seedling to flowering was registered in 1-4th variants, compared to the initial length of seedlings, 8.0 cm, 6.0 cm, 8.0 cm, and 7.0 cm, respectively. The Jermatnayin hska plants are distinguished by their high growth rate. The highest result was registered in the 4th variant (64 cm length) compared to the initial plant length of the seedling (7.0 cm) (Table 1,1.1).

Investigation of the influence of growth conditions on the yield of plant varieties of Arajnek, Jermatnayin hska, and Loshtak have shown that plants are distinguished by yield. According to the results "Biohumus" promotes the production of the yield of investigated plants. Particularly, the yield of pepper in the plants of the Arajnek variety was 228; 286; 265; 320 c/ha, respectively; and the yield increase was 25.4, 16.2, and 40.35%, respectively, compared to the Control variants. In the case of Jermatnayin hska plants, the recorded yield was 191, 237, 229, and 277 c/ha, respectively. The yield increase was 24.08,19.89 and 45.02%, respectively, compared to the Control variant. The yield of the Loshtak variety, especially for the crop supplement is different from the previous two varieties. Specifically, the "Biohumus" and "Biohumus+ $N_{60}P_{60}K_{60}$", version crop supplement is 40% and 63.4%, respectively, and the "$N_{60}P_{60}K_{60}$", version is 31.3% (Table 2).

Table 2. Biohumus and chemical fertilizer effects on the yield and crop supplement of pepper plant varieties.

Plant varieties	Variant	Yield, c/ha	Crop supplement	
			g/ha	%
Arajnek	Control	228		
	Biohumus	286	58.0	25.4
	$N_{60}P_{60}K_{60}$	265	37.0	16.2
	Biohumus+ $N_{60}P_{60}K_{60}$	320	92.0	40.35
S_x%	1.76			
$SSD_{0.95}$c/ha	34.23			
Jermatnayin hska	Control	191		
	Biohumus	237	46.0	24.08
	$N_{60}P_{60}K_{60}$	229	38.0	19.89
	Biohumus+ $N_{60}P_{60}K_{60}$	277	86.0	45.02
S_x%	1.28			
$SSD_{0.95}$c/ha	39.7			
Loshtak	Control	239.8		
	Biohumus	335.7	95.9	40.0
	$N_{60}P_{60}K_{60}$	314.8	75.0	31.3
	Biohumus+ $N_{60}P_{60}K_{60}$	391.9	152.1	63.4
S_x%	2.06			
$SSD_{0.95}$c/ha	23.9			

High temperature and dryness certainly affected the photosynthetic activity of the plants. Under these conditions, the activity of chlorophyllase, particularly, increases, and the lipoprotein complex of chlorophyll is degraded. By detecting these changes, it is possible to assess the effect of nutritional conditions and relative drought on the stability of plants. Inoculation with rhizobacteria could be efficiently used to improve growth, antioxidant status, and photosynthetic pigments in basil under water stress [8].

The results of spectrophotometric determination of pigments in acetone extracts of experimental plants

 https://doi.org/10.1051/bioconf/20225200068

showed that the content of bound chlorophyll (a + b) (18.88 mg/l) and chlorophyll "a" (9.353 mg/l) compared to control plants is higher (15.1mg/land 7.876mg/l respectively). The content of carotenoids also increased in comparison with the pigments of other plant variants. The ratio of chlorophyll Chl "a" and "b" is higher in all the studied variants. Which indicates the relative resistance of plants to high temperatures and drought 1. "Biohumus", 2. "$N_{60}P_{60}K_{60}$", 3. "Biohumus+$N_{60}P_{60}K_{60}$" conditions [Figure 1] [8, 9]. Under water stress, chlorophyll was increased significantly.

Table 3. Physico-chemical parameters of soils in vegetation pots after the end of the vegetation period.

Variant	Humus content, %	p^H	Mechanical composition:, <0,01mm. %
1. Control	3.1	8.0	22.27
2.Biohumus	5.0	8.1	24.83
3.$N_{60}P_{60}K_{60}$	4.3	8.0	25.55
4.Biohumus+ $N_{60}P_{60}K_{60}$	5.4	8.0	31.27

Table 3.1. Physico-chemical parameters of soils in vegetation pots after the end of the vegetation period.

Variant	Sand weight	Available nutrients, mg/100g in the soil		
		N	P_2O_5	K_2O
1. Control	3.1	4.16	35.00	117.82
2.Biohumus	26.61	6.24	39.12	119.39
3.$N_{60}P_{60}K_{60}$	27.35	5.20	35.41	102.11
4.Biohumus + $N_{60}P_{60}K_{60}$	22.93	7.28	37.06	107.45

Factors that may affect the ascorbic acid levels in fruits include climate, temperature, and soil nutrients. The composition of plant tissues during growth and development is also affected by temperature which varies from region to region. Determination of ascorbic acid content was done by iodometric titration. The amount of ascorbic acid in the stage of technical maturation is higher than the content of ascorbic acid in the fruits in the stage of biological maturation. It is higher in Jermatnayin hska fruits grown in the presence of "$N_{60}P_{60}K_{60}$" [Fig. 2].

The results of the analysis of the experimental soil samples in the vegetation cabin of the 20t/ha "Biohumus" variants after the end of the vegetation period show that the content of humus (in the layer of 0–20 cm) increased 1.54 times compared to the control. In these conditions, according to the obtained results, the available nitrogen, phosphorus, and potassium for plants increased 1.2, 0.02, and 1.01 times, respectively the soil is strongly carbonated (Table 3, 3.1).

In the variants of "$N_{60}P_{60}K_{60}$" (in the layer of 0–20cm) the content of humus increased 1.4 times compared to the control, and the available nitrogen, phosphorus, and potassium increased 1.25, 1.01, and 0.86 times, respectively the soil is strongly carbonated.

In "Biohumus" 20t/ha and "$N_{60}P_{60}K_{60}$" kg/ha combined variants the content of humus (in the layer of 0–20cm) increased 1.7 times compared to the control, and the available nitrogen, phosphorus, and potassium increased 1.75, 1.05 and 0.9 times, respectively. No significant change was observed in the reaction of soil solution, physical clay – water-resistant aggregates, the soil is strongly carbonated. One of the ways to improve the agrochemical parameters of soils is the combined use of biohumus and mineral fertilizers.

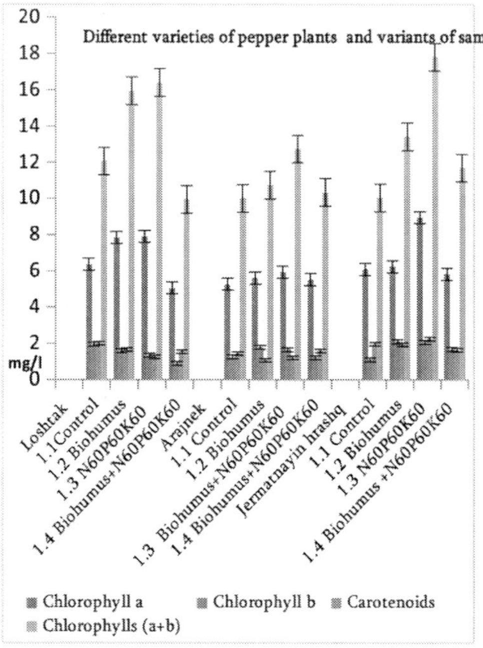

Fig.1. Organic and chemical fertilizer effect on the accumulation of photosynthetic pigments; chlorophyll "a", "b" and carotenoids in the tissues of the leaf of Loshtak, Arajnek, Jermatnayin hska varieties of pepper.

4 Conclusion

It was shown that the application of "Sis" microbial biohumus as a biofertilizer increased the bulk yield of three varieties of pepper plants and contributed to the recycling of farm wastes. It also allowed us to achieve the specific yield (per plant) level provided by traditional fertilizers. As the difference between the results of the pepper plant variants and the control is higher than the SSD, the difference compared to the control is real and there is a crop supplement. Our study demonstrated the multiple benefits of the combined use of biohumus with inorganic fertilizer (NKP). The results indicate the possibility of partial replacement of chemical fertilizers with biofertilizers (microbial biohumus) in vegetable crop culture farming.

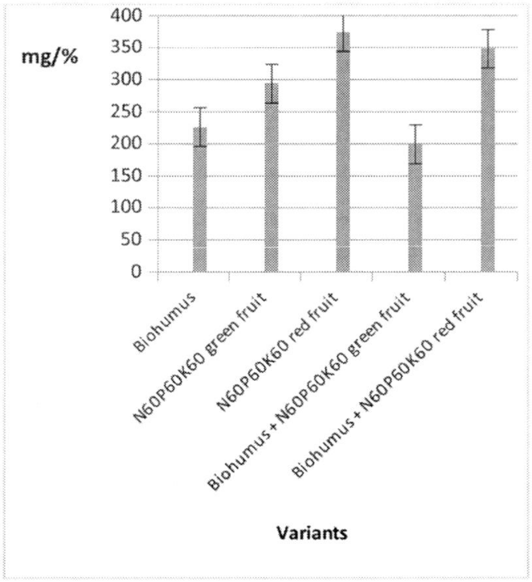

Fig.2. Organic and chemical fertilizer effect on the accumulation of ascorbic acid in the tissues of pepper.

Acknowledgements

The work was supported by the Science Committee of MESCS RA in the frames of the research project № 20APP-4C011.

References

1. Y. Wei, D. Shi, G. Liu, Y. Zhao, T. Shimaoka, Environmental challenges impeding the composting of biodegradable municipal solid waste: A critical review, *Resour. Conserv. Recycl.* **122,** 51–65 (2017)
2. E. Malusa, N. Vassilev, A contribution to set a legal framework for bio-fertilizers, *Appl. Microbiol. Biotechnol.* **98(15),** 6599–6607 (2014)
3. M. Rastogi, M. Nandal, B. Khosla, Microbes as vital additives for solid waste composting, *Helyon* **6 (2),** doi: 10.1016/j.heliyon. 2020.e03343 (2020)
4. I. Białobrzewski, M. Mikš-Krajnik, J. Dach, M. Markowski, W. Czekała, K. Głuchowska, Model of the sewagesludge-strawcomposting process integrating differenth eatgeneration capacities of mesophilic and thermophilic microorganisms, *Waste Manag.* **43,** 72–83 (2015)
5. G. Yu. Rabinovich, N. G. Kovalev, Yu. D. Smirnova, Application of new biofertilizers and biological products in the cultivation of spring wheat (*Triticum Aestivum l.*) And potato (*solanum tuberosum l.*). *Agricultural Biology* **50 (5),** 665–672 (2015)
6. K. Divjot, L. R. Kusam, A. N. Yadav, NeelamYadav, M. Kumar, V. Kumar, P. Vyas, H. S. Dhaliwal, A. K. Saxena, Microbial biofertilizers: Bioresources and eco-friendly technologies for agricultural and environmental sustainability. *Biocatalysis and Agricultural Biotechnology* **23,** 101487 (2020)
7. A. Marulanda, R. Porcel, J. M. Barea, R Azco, Drought tolerance, and antioxidant activities in lavender plants colonized by native drought-tolerant or drought-sensitive Glomus species, *Microb. Ecol.* **54,** 543–552 (2007)
8. M. Heidari, A. Golpayegani, Effects of water stress and inoculation with plant growth promoting rhizobacteria (PGPR) on antioxidant status and photosynthetic pigments in basil *(Ocimumbasilicum L.), Journal of the Saudi Society of Agricultural Sciences* **11,** 57–61 (2012)
9. R. B. Glick, Modulation of plant ethylene levels by the bacterial enzyme ACC deaminase, *FEMS Microbiology Letters* **25 (1),** 11–17 (2005)
10. M. Hussein Sabkia, Ch. Tin Leea, C. Phun Cien Bonga, Zh. Zhangb, L. Chunjie, J. Jaromír Klemešc, Sustainable Organic Waste Management Framework: A Case Study in Minhang District, Shanghai, China. *Chemical Engineering Transaction* **72,** doi: 10.3303/CET1972002 (2019)
11. S. Zulaiha Hanapi, M. Hassan Awad, Sh. Imranudin Sheikh Ali, S. Hajar Mat Sarip, M. Roji Sarmidi, R. Aziz, Agriculture wastes conversion for biofertilizer production using beneficial microorganisms for sustainable agriculture applications, *Malaysian Journal of Microbiology* **9 (1)** 60–67 (2013)
12. N. A. Kachinsky, *Mechanical and Micro Aggregate Composition and Methods of Its Study* (Moscow: Academy of Sciences of the USSR, 1958)
13. B. A. Dospekhov, *Methodology of Field Experiment* (Moscow: Kolos, 1973)
14. E. V. Arinuskina, *Manual for Soils Chemistry Analysis* (Moscow: Moscow State University, 1962)
15. A. R. Wellburn, The spectral determination of Chlorophylls "a" and "b", as well as total Carotenoids, using various solvents with spectrophotometers of different resolution, *Journal Plant Physiology* **144,** 307–313, doi: 10.1016/S0176-1617(11)81192-2 (1994)
16. L. Megali, G. Glauser, S. Rasmann, Fertilization with beneficial microorganisms decreases tomato defenses against insect pests, *Agron Sustain Dev.* **34,** 649–656, doi: 10.1007/s13593-013-0187-0 (2014)
17. A. Olasekan Adekiya, W. Sunny Ejue, A. Olayanju, Oluwagbenga Dunsin, Ch. Muyiwa Aboyeji, Ch. Aremu, K. Adegbite, O. Akinpelu, Different organic manure sources and NPK fertilizer on soil chemical properties, growth, yield, and quality of okra. *SciRep.* 1016083 doi: 10.1038/s41598-020-73291-x (2020)
18. I. J. Dioha, O. Olugbemi, T. U. Onuegbu, Z. Shahru, Determination of ascorbic acid content of some tropical fruits by iodometric titration, *Int. J. Biol. Chem. Sci.*, **5 (5),** 2180–84 (2011)
19. F. T. Nezhmetdinova, M. E. Guryleva, N. Kh. Sharypova, R. I. Zinurova, A. R. Tuzikov, Risks of modern biotechnologies and legal aspects of their implementation in agriculture, *BIO Web of Conferences* **17,** doi: 10.1051/bioconf/20201700227 (2020)

The efficiency of using synthesized fertilizers via foliar feeding and drip irrigation in various soil and climatic zones of the Republic of Armenia

Gayane H. Gasparyan[1], *Lousine S.* Yeritsyan[1,2], *B. T.* Gali[3], *N. Kh.* Sharypova[3], and *Seryozha K.* Yeritsyan[1,*]

[1] H. Petrosyan Scientific Center of Soil Science, Melioration and Agrochemistry, Branch of Armenian National Agrarian University, 74, Teryan St., Yerevan, 0009, Armenia
[2] Armenian National Agrarian University, 74, Teryan St., Yerevan, 0009, Armenia
[3] Kazan State Agrarian University, 65, Karl Marx St., Kazan, Republic of Tatarstan, 420015, Russian Federation

Abstract. In the current research paper, a new water-soluble combined fertilizer (WCF) is introduced, the composition of which is determined via the agrochemical composition of the soils of Armenia, and besides, its effectiveness has been further investigated. It has been established that when applying the recommended fertilizer through foliar feeding or soaking the seeds, the yield is significantly increased and its quality is improved. Under the impact of WCF, the yield increase compared to the NPK variant was as follows: for winter wheat under non-irrigated conditions – 5.1 c/ha, for tomato – 49 c/ha, for grapevine – 20 c/ha, while when using 20:20:20 the increase in yield capacity was 10.0 c/ha, 3.7 c/ha, and 8.0 c/ha, respectively.

1 Introduction

The use of fertilizers plays an important role in maintaining soil fertility and increasing crop yield capacity [1,2]. Since 1991, mineral fertilizers have not been produced in Armenia, but they have been imported, and among them, fertilizers used for foliar feeding and drip irrigation are of high priority [3]. However, it should be noted that their composition does not stem from the chemical composition of the soils of the Republic of Armenia. Taking into account the importance of fertilizers applied in the agriculture of the Republic of Armenia for foliar feeding, similar fertilizers have been synthesized in the Scientific Center of Soil Science, Agrochemistry and Melioration named after H. Petrosyan (branch of the Armenian National Agrarian University). At the same time, the chemical composition of fertilizers has been identified taking into account the agrochemical composition of the soils of the republic, which enabled it to ensure much higher efficiency as compared to that recorded in the case of imported fertilizers [4]. Meanwhile, it is worth mentioning that the application of the mentioned fertilizer will grow up since the prices of the soil-introduced fertilizers have grown by 2.5-3 times, while the financial capacities of many farmers in the republic are rather low.

Numerous research works designate the significance of applying fertilizers through the method of foliar nutrition [5,6]. Such studies were carried out still in the 30s of the twentieth century in the United States [7]. However, very few studies have been carried out in the Republic of Armenia so far.

It is known that the method of applying foliar fertilizers makes it possible to faster affect plant nutrition, hence, accelerating the formation of fruit buds, reducing fruit drop, and increasing fruit size and yield capacity [8]. For instance, it has been found, that using applying foliar nutrition to the cereal crops in

the tillering stage or even later, the plant growth improves, the yield capacity, weight of 1000 grains, and protein content grow up, while seed germination capacity and energy are increased [9].

For vegetable crops, it is possible to reduce (prevent) the drop of the flowers and fruits through foliar feeding, which also promotes the increase of the fruit size and quality [10]. When used in vineyards, along with the noticeable increase in the crop's yield capacity, it is also possible to change the ratio of sugar and acids in favor of sugar [11].

2 Materials and methods

Water-soluble combined fertilizer (WCF) is a complex organo-mineral fertilizer that contains macro (N, P, K, Fe) and micro (B, Mo, Mn, Zn, Cu, Co) elements and is highly soluble in water. It is used for foliar nutrition, by soaking seeds before sowing or earlier, as well as through drip irrigation. It is also allowed for use in organic agriculture.

* Corresponding author: s_eritsyan@yahoo.com

The effectiveness of the WCF was studied upon field experiments and chemical analysis of soils and plants (fruits) for 2020-2021. Experiments on winter wheat were carried out in non-irrigated conditions in the dry steppe zone, on chestnut soils. Here, the amount of annual precipitation is 410-450 mm and the average air temperature is 8.5-9.5^{0}C. The experiments were carried out according to the scheme indicated in diagram 1; the area of the experimental plot is 98 m^2.

Experiments on tomatoes were carried out in the semi-desert zone of the Ararat Valley on irrigated meadow-brown soils in irrigated conditions. Here the amount of annual precipitation is only 250-320 mm, while the average annual air temperature is 11.5-12.5^{0}C. The experiments were carried out according to the scheme indicated in diagram 2. The experiment on the grapevine was also carried out in a semi-desert zone on meadow-brown irrigated soils under irrigation conditions according to the scheme indicated in diagram 3. All experiments were carried out in four replications.

Agrochemical analysis of soils of experimental plots and the yield chemical analysis were carried out according to the accepted methods in agrochemical studies [12, 13, 14, 15].

The main mineral fertilizers (ammonium nitrate, double superphosphate, and potassium chloride) were applied by introducing them to the soil, while water-soluble combined fertilizer (WCF) and N:P: K=20:20:20 (standard) were applied via foliar feeding. For winter wheat, this was carried out in spring in the full tillering phase of plants with an application rate of 2.5 kg/ha. In the experiment on tomatoes, water-soluble fertilizers were applied after 2 weeks of transplanting seedlings to the field, three times with an application interval of 10-12 days. The rate of application of WCF each time was 2.5 kg/ha. The application rate of 20:20:20 was 5 kg for each time.

3 Results and discussion

Field experiments were carried out in different natural and climatic zones, on soils with different agrochemical parameters.

According to the data obtained, in the upper horizons (0-30 cm) of the soils of the experimental plots, the humus content is low (2.51–3.12%), which is typical for these soil types, and the soils are calcareous (3.06–4.45%), the reaction is slightly or medium alkaline (pH = 7.9–8.3), the mechanical composition is heavy loamy (physical clay - 47.6–51.6%). According to the scale of soil provision with available nutrients adopted in Armenia, the experimental plot of winter wheat and tomato is poorly supplied with nitrogen and phosphorus and well supplied with potassium (N-3.14–3.61, P_2O_5-1.16–1.58, K_2O-39.5–51.1 mg for 100 g soil). The soil of the experimental plot of the grapevine is poorly provided with nitrogen, moderately - with phosphorus, and well - with potassium (N-3.06, P_2O_5-2.76, K_2O-49.8 mg for 100 g soil). The use of the main mineral fertilizers and WCF or N:P:K = 20:20:20 on their background has exerted a remarkably positive effect on the yield capacity and chemical composition of grain, as well as on the germination capacity and germination energy of winter wheat seeds. At the same time, it is worth noting that the relatively low yield is due to dry weather conditions in non-irrigated areas. Regardless of the weather conditions, the highest grain yield (35.7 c/ha) has been obtained where WCF against the background of $N_{90}P_{60}K_{60}$ has been applied. In this variant, the grain yield was 35.7 c/ha, which is 5.1 c/ha more than the background option, and 3.7 c/ha more than the option of background + N:P:K = 20:20:20 (Figure 1).

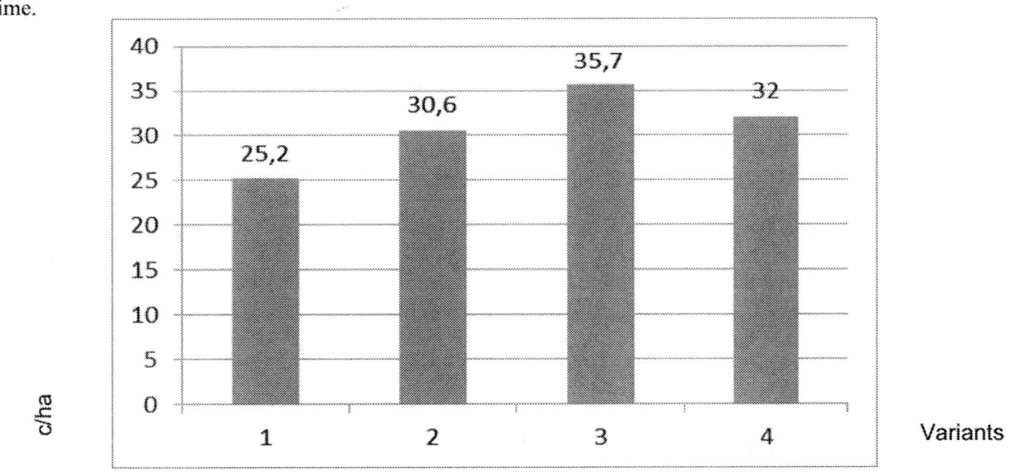

Fig. 1. The effect of water-soluble combined fertilizers on the yield of winter wheat, c/ha. **Variants**: 1. 0; 2. $N_{90}P_{60}K_{60}$, 3. $N_{90}P_{60}K_{60}$ + WCF – 2.5 kg/ha, 4. $N_{90}P_{60}K_{60}$ + 20:20:20 -5 kg/ha **Soil content**: humus – 3.12 %; pH- 7.3; mobile nitrogen (N) - 3,14 mg, phosphorus (P_2O_5) – 1.58 mg, potassium (K_2O) – 47.8 mg for 100 g soil.

The positive effect of water-soluble combined fertilizers is also perceptible in the chemical composition of the grain, the germination capacity, and the vigor of seed germination. Herewith, the highest result has been received where WCF against the background of $N_{90}P_{60}K_{60}$ has been applied. In this variant, the weight of 1000 grains is 1.2–1.6 g more than in the control and background variants, whereas the content of nitrogen exceeds by 0.34–0.38%, phosphorus - by 0.147–0.21%, and potassium – by 0.07–0.08%.

When using WCF, the germination energy is 70.4%, and the germination capacity is 95.9%, while when using 20:20:20, these indicators are 56.7% and 93.2%, respectively and in the variant of $N_{90}P_{60}K_{60}$ (background), they make 45.9-56.0 and 88.4-90.3%, respectively.

In the experiments on tomatoes, the highest yield has been obtained when, against the background of main mineral fertilizers, a water-soluble combined fertilizer or 20:20:20 (Figure 2) was applied. The highest fruit yield has been harvested (650 c/ha) when using the background ($N_{90}P_{90}K_{60}$) + WCF, which is higher than the control variant by154 c/ha (31.0%) and compared to the background variant – by 49 c/ha (8.2%).

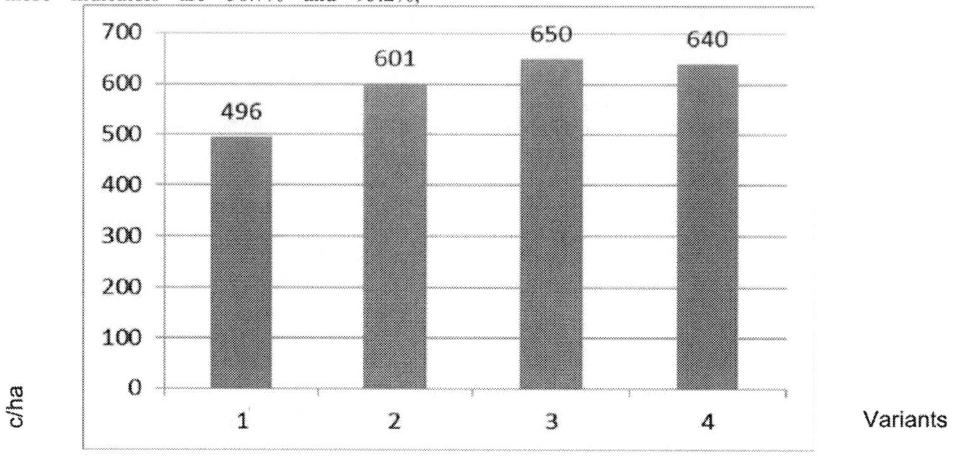

Fig. 2. The effect of foliar feeding on the yield capacity of tomatoes, c/ha. **Variants**: 1. 0; 2. $N_{90}P_{90}K_{60}$, 3. $N_{90}P_{90}K_{60}$ + WCF – 2.5 kg/ha (3 times), 4. $N_{90}P_{90}K_{60}$ + 20:20:20 – 5 kg/ha (3 times). **Soil content**: humus–2.71 %; pH-7.8; mobile nitrogen (N)–3.61 mg, phosphorus (P_2O_5)–1.12 mg, potassium (K_2O) – 51.14 mg for 100 g soil.

Regarding the efficiency of water-soluble fertilizers, similar results have been obtained in experiments with a grapevine (Figure 3), that is, the highest fruit yield has been obtained in the variant of background ($N_{90}P_{90}K_{60}$) + WCF – 144 c/ha, which is higher than the background variant by 20 c/ha (16.1%) and exceeds the background variant + 20:20:20 by 8.0 c/ha (5.9 %).

Water-soluble combined fertilizers affected the content of sugar and acids in fruits. At the same time, under the influence of WCF, the sugar content in fruits makes 19.4%, acidity - 4.1%, while in the background variant the sugar content is 17.8%, acids - 6.2%, and in the option of background + 20:20:20, these indices make 18.6% and 4.6%, respectively.

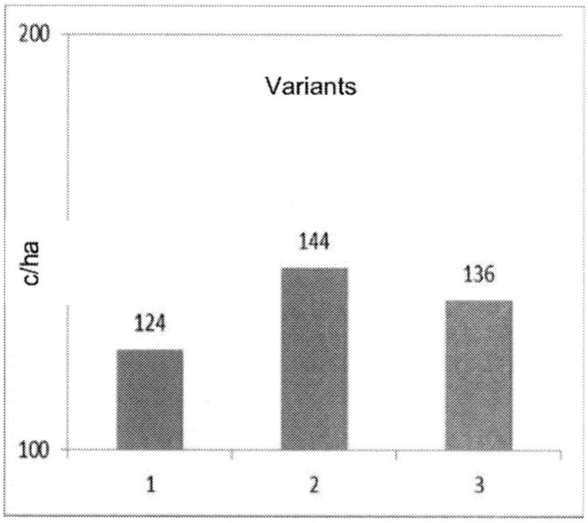

Fig. 3. The effect of foliar feeding with water-soluble fertilizer on the yield capacity of grapevine, c/ha. **Variants**: 1. 0; 2. $N_{90}P_{90}K_{60}$, 3. $N_{90}P_{90}K_{60}$ + WCF – 2.5 kg/ha (3 times), 4. $N_{90}P_{90}K_{60}$ + 20:20:20 - 5 kg/ha (3 times) **Soil content**: humus – 2.51 %; pH-7,5; mobile nitrogen (N) – 5.13 mg, phosphorusa (P_2O_5) – 4.58 mg, potassium (K_2O) – 71.61 mg for 100 g soil.

4 Conclusion

Thus, studies on the effect of water-soluble fertilizers (WCF and 20:20:20) on winter wheat, tomato, and grapevine have designated the following.

In Armenia, the yield capacity of the mentioned crops can be still increased if WCF or 20:20:20 via foliar feeding is also included in the system of fertilizers application. In winter wheat, the increase in yield capacity occurred due to the enhancement of productive tillering and weight of thousand grains, besides, the quality of seeds improved. The content of nitrogen, phosphorus, and potassium in the seeds increased, while in tomatoes the germination capacity and germination energy improved. The yield capacity increase was due to an increased number and weight of plant fruits, and the improvement of fruit quality occurred due to an increase in dry matter and sugar content in fruits. In the grapevine, the increase in yield capacity occurred due to an increased weight of bushes and fruits, as well as due to the growth of sugar content in the fruits.

When applying WCF, the yield increase compared to the control variant makes 10.5 c/ha (41.7%) for winter wheat, for tomato it makes 154 c/ha (31.0%), and for grapevine ($N_{90}P_{60}K_{60}$) 20.0 c/ha (16.1%), while as compared to the background variant, it makes 5.1 c/ha (16.7 %), 49 c/ha (8.2%) and 20c/ha (16.1%), respectively and against the option of background + 20:20:20 the mentioned indices equaled to 3.7 c/ha (11.6%), 10 c/ha (1.6 %) and 8.0 c/ha (5.9 %), respectively.

Acknowledgments

The work was supported by the Science Committee of MESCS RA, in the frames of research project № 20APP-4C015.

References

1. A. V. Volkova Market of mineral fertilizers, *Bulletin of National Research Institute. Higher School of Economics*, pp 48–50 (2015)
2. K. G. Melkonyan, Kh. Kh. Ghazaryan, R. R. Manukyan *The current state of the ecological state of agricultural lands, the level of land use, improvement of the management system, and ways to increase efficiency in the Republic of Armenia* (Yerevan: H. Petrosyan Scientific Center for Soil Science and Agrochemistry, 2004)
3. S. Yeritsyan, A. Sahakyan, A. Amyan, L. Yeritsyan The impact of various fertilizers application on the yield of potato sorts Agria and Red Scarlett and the tuber quality in conditions of Lori region, *Bulletin of ANAU* **1,** 21–26 (2017)
4. S. Yeritsyan, A. Sahakyan, V. Makyan, L. Yeritsyan The influence of the application of different fertilizers on the growth and development of potato varieties Agria and Red Scarlet in Lori marz conditions, *Bulletin of ANAU* **1,** 9–14 (2016)
5. A. I. Osipov, E. S. Shkrabak, The role of foliar nutrition in increasing the productivity of crops *Proceedings of the St. Petersburg State Agrarian University* **1 (54),** 44–52 (2019)
6. A. B. Khoroshkin, Foliar feeding of potatoes: Brief review, *Potatoes and vegetables* **11,** 25–26 (2015)
7. Z. A. Metlitsky, L. F. Blinova, *Mineral Nutrition of Fruit and Berry Crops* (Moscow: Selkhozizdat, 1960)
8. A. E. Mandonkaeva, Ya. N. Losinska, The effect of foliar micro fertilizers on agrobiological indicators and output of standard products of table grape varieties *Viticulture and winemaking* **40** 42–44 (2010)
9. O. A. Mitrokhina, The efficiency of foliar treatment with complex mineral fertilizers of winter wheat crops in the Kursk region, *Agriculture* **5,** 21–22 (2015)
10. M. Yu. Anishka, Sh. B. Bayrambekov, E. D. Garyanova, G. V. Gulyaeva. The effect of foliar feeding on the productivity of tomato in conditions of the Volga delta, *Bulletin of the Nizhnevolzhsky Agro-University Complex: Science and Higher Professional Education* **1,** 5–7 (2019)
11. E. V. Polukhina, E. N. Ivanenko, D. E. Morozov, M. V. Vlasenko, The effect of foliar fertilizers on economically valuable indicators of table grape varieties, *Bulletin of the Nizhnevolzhsky Agro-University Complex: Science and Higher Professional Education* **4 (52),** 185–191 (2018)
12. A. S. Piskunov. *Methods of Agrochemical Research* (Moscow: Kolos, 2004)
13. F. T. Nezhmetdinova, M. E .Guryleva, N. Kh. Sharypova, R. I. Zinurova, A. R. Tuzikov, *Risks of modern biotechnologies and legal aspects of their implementation in agriculture,* BIO Web of Conferences **17,** 00227 (2020)
14. F. T. Nezhmetdinova, G. R. Fassakhova, L. R. Shagivaliev, N. Kh. Sharypova, R. I. Zinurova, Digital economy and transformation of personnel training for AIC, *BIO Web of Conferences* **17,** 00228 (2020)
15. A. Z. Ibatova, F. T. Nezhmetdinova, F. F. Sitdikovб Global challenges for the agrarian sector of the Russian economy and its human resources, *Spaces* **39 (26)** (2018)

Biological parameters of juveniles of *Gadus morhua* Linnaeus, 1958 and *Pollachius virens* (Linnaeus, 1958) in the fjords affected by anthropogenic load, by the example of the Kola Bay, the Barents Sea

Oleg V. Bondarev[1,*] and *Olga S.* Tyukina[2]

[1] Murmansk Marine Biological Institute of the Russian Academy of Sciences, 17, Vladimirskaya Str., Murmansk, 183010, Russian Federation
[2] Murmansk State Technical University, 13, Sportivnaya Str., Murmansk, 183010, Russian Federation

Abstract. The paper continues the series of studies aimed on analyzing the biological processes taking place in the coastal zone of the Barents Sea. The size and weight parameters, age and sex structure, as well as the feeding of juveniles of Atlantic cod and saithe in the intertidal zone. There are pronounced seasonal changes in the studied parameters. Gammaridae play a large role in nutrition, but cannibalism is well-pronounced. Malformation of the otolith structure is recorded. The data obtained may be used for further monitoring or applied for the ecosystem modeling.

1 Introduction

Recently, the development of the exploitation of the natural resources of the Russian Arctic, accompanied by technogenic and environmental consequences, attracts much attention [1]. The Kola Bay located on the southern coast of the Barents Sea is an example of a vulnerable Arctic ecosystem under the constant anthropogenic pressure. A number of construction works are being carried out in different parts of the Kola Bay, so there is an increase in gas-oil and transport marine activities [2]. Therefore, there is an increasing need for environmental monitoring. The biodiversity of the fish community is one of the indicators of the stability of the ecosystem, the physiological state of the ichthyofauna is the other. The species diversity of the fish community of the Kola Bay is well studied [3–7]. However, data on the biology of fish are limited to several species [8, 5, 9, 7, 10], and no information for Gadidae have been found even in the Russian scientific literature. The study aims to analyze seasonal changes in the biological parameters of juvenile Atlantic cod and saithe in the central part of the Kola Bay in 2018.

2 Materials and methods

Ichthyological material was collected in 2018 in the central part of the Kola Bay (Figure 1), the study period and the sample size are described in Table 1. The fish were caught by a 15-meter fry seine (height of 1.8 m; mesh diameter: 6 mm in the wings, 5 mm in the gullet, 4 mm in the cod end) and different fixed nets (length of 10.0 m, height of 1.8 m) of 30-, 40-, and 50-mm mesh diameter. The area of one harvesting by fry seine was about 750 m^2. Fixed nets with different mesh size were installed for 24 hours perpendicular to the coastline at the depth range of 1.5–5.0 m during the period of maximum low tide.

The collected material was processed according to standard ichthyological methods [11–13]. The absolute body length (*TL*) was measured from the tip of the snout to the end of the longest rays of the caudal fin. The fish was weighted on a torsion balance with a 0.01-g accuracy. Young fish aged from 0+ to 4+ years were analyzed. The age of fish was determined by otolith method [14]. The average score of stomach filling according to Lebedev [15] was applied for assessing the feeding intensity. The food items were determined to the possibly lowest taxon [16].

Table 1. Study period and amount of material sampled and analyzed in 2018.

Parameter	May	July	October
Number of fry seine, pcs.	5	5	3
Number of fixed nets, pcs.	6	6	6
Saithe, ind.	40	19	24
Atlantic cod, ind.	9	48	6

* Corresponding author: bondarev@mmbi.info

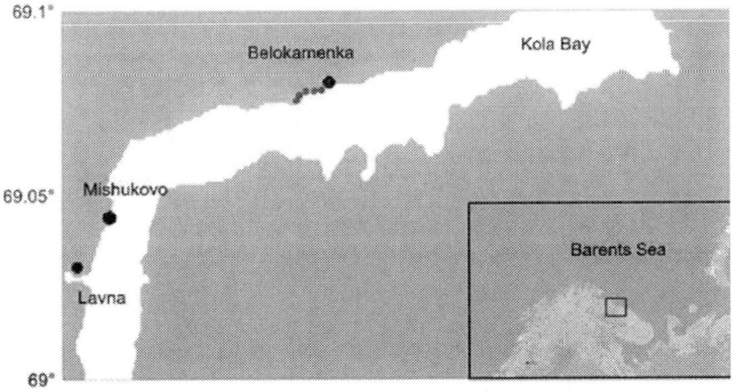

Fig. 1. Schematic map of ichthyological sampling stations.

3 Results and discussion

Since there are no published data for Gadidae in the Kola Bay, the obtained results have been compared with other bays of the Kola Peninsula. There are information on the body size of immature Atlantic cod in the Teriberskaya, Opasova, and Dolgaya bays [17], and in the Zelenetskaya and Yarnyshnaya bays [18]. Data on saithe of different ages are presented for the Kislaya Bay and the bays of the eastern Murman Coast [19–20].

The size-age composition of juvenile Atlantic cod exhibits clear dynamics (Table 2). The fish length varied from 4.2 cm in July up to 34 cm in October; the weight, from 0.2 g to 392.5 g, respectively. There is also a regular increase in the average length and body weight of fish from May to October of all age groups, except for that of 1+. Comparing these results with the published earlier, it can be noted that there were no young-of-the-year in the summer catches of 2008 [17]. Apparently, this was due to different fishing gear. In the Zelenetskaya and Yarnishnaya bays, the group 0+ was present in the catches; moreover, there were significant differences in the average size of fish [10].

In May, the main share of cod (56%) were specimens aged 1+ (Figure 2), three-year-olds and four-year-olds accounted for 22%, respectively. In July, the young-of-the-year comprised 73%, and the fish aged 4+ were recorded in the catches, too. In October, young-of-the-year were not caught at all, the dominant group was 2+. In Norwegian fjords, Atlantic cod lives up to 8 years, although most of the population is presented by 6–7-year old specimens [21].

Table 2. Size-age structure of Atlantic cod population.

Age, years	Average body length, cm			Average body weight, g		
	May	July	October	May	July	October
0+	–	5.6±0.8	–	–	1.2±0.7	–
1+	12.1±2.2	13.8±0.5	10.2±1.0	14.5±8.1	22.1±3.7	9.5±2.9
2+	15.8±0.2	18.9±1.4	25.7±3.2	36.3±0.1	61.7±13	157.7±67.3
3+	27±1.4	28.6±1.2	34	201.3±6.0	213.9±73.2	392.5
4+	–	31.9±0.6	–	–	325.5±6.4	–
Min	9.3	4.2	9.2	5.7	0.2	6.1
Max	28	32.3	34	205.6	330	392.5
Total	16.2±6.5	10±8.3	20.8±8.8	60.9±80.9	35.9±83.4	118.8±119.9

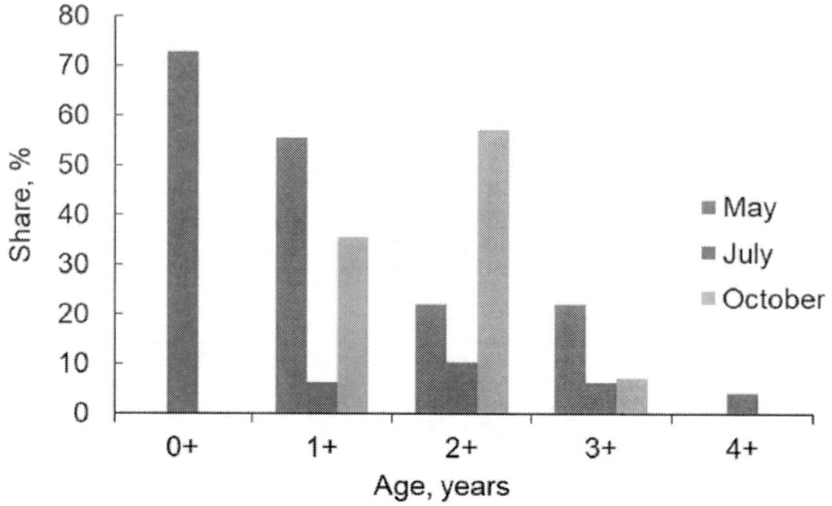

Fig. 2. Age composition of Atlantic cod in the central part of the Kola Bay.

All the females and males caught were represented by immature fish with maturing gonads at the maturity stage II. In spring, most juveniles of Atlantic cod were males (63%), followed by females (25%); immature juveniles represented only 12% of the total population (Fig. 3). In summer, due to the appearance of young-of-the-year, the proportion of juveniles increased up to 73%, the share of females and males was nearly the same (the difference did not exceed 7%). In autumn, there was a slight predominance of females (40%), and the share of males and juvenile fish was 30% each.

Analyzing the contents of the stomachs of Atlantic cod, it was found that in May the feeding intensity was low: stomach fullness score (SFC) was 1.2. The share of feeding fish was 71.4%, the food spectrum was mostly Gammaridae (86%) and juveniles of Gadidae (14%). In July, the average SFC was higher (2.1), the share of feeding fish was 72.9%. The food spectrum was much wider; young-of-the-year fed mainly on copepods, amphipods, chironomids, and, rarely, on isopods. The food spectrum of elder age groups also consisted of amphipods mostly, but juveniles of Atlantic herring and Atlantic cod, as well as polychaetes and shrimps, were also found. In October, all the fish was feeding, the average SFC was 2.2; Gammaridae were the "favorite" food; polychaetes, unidentified fish species, and capelin were also presented in the gut. Atlantic cod, feeding mainly on benthos, was reported in 1984 in the bays of the western Murman Coast, instead of mostly pelagic predating [22].

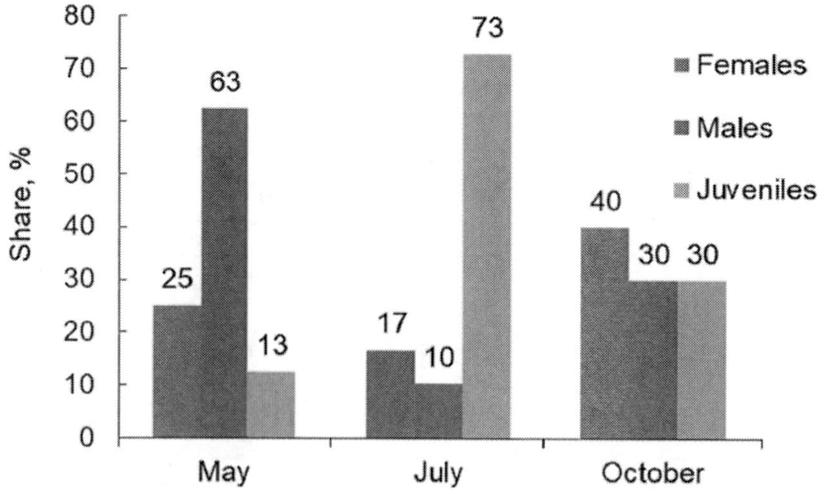

Fig. 3. Sex structure of Atlantic cod population in 2018.

Saithe. The size of saithe in the central part of the Kola Bay in 2018 varied from 4.8 to 46.6 cm, weight, from 0.84 g to 861 g (Table 3). The average length and weight of saithe aged 2+ and 3+ increased from May through October; however, in July, the parameters of three-year-olds were somewhat lower than those observed in spring.

In the bays of the western Murman Coast, the length of fish aged 1+ varied from 14 to 23 cm, and the size of the three-year-old saithe, from 24 to 31 cm [19]. According to N.V. Mironova [20], the average length of saithe at the age of 1+ is 14.5 cm, which fits well with our data; in July, the saithe of the age of 2+ is 30.4-cm long ($n = 62$). This is less that reported for saithe population from the North Sea [23]. In the North-Eastern Atlantic population of saithe, positive relationship was detected between growth performance and water temperature; in addition, body weight, length, Fulton's condition factor (K), and specific growth rate increased under positive temperature anomalies. In this population, acclimatization to local temperatures per se was likely minor, and caught specimens had low growth performance [23].

Saithe in the studied water area was represented by both young-of-the-year (0+), which were recorded only in July, and by older fish up to 4+ (Fig. 4). In May, most of the fish belonged to 2+ age group (65%); in July, half of the caught fish (50%) were at the age of 3+, the share of young-of-the-year was 39%; in October, two-year-olds appeared (36%), and the share of three-year-olds did not exceed 45%. The saithe, inhabiting Icelandic waters, lives mostly 2–3 years, rarely reaching 6+ age [24].

Table 3. Size-age structure of saithe population.

Age, years	Average body length, cm			Average body weight, g		
	May	July	October	May	July	October
0+	–	5.9±0.7	–	–	1.7±0.6	–
1+	18.5±4	–	11.7±1.7	53.7±33.3	–	12.4±5.9
2+	24.6±1.7	23±0.7	29.9±2.3	126.6±24.1	106.7±6	244.8±26.0
3+	25.2±1.7	26.8±1.7	34±0.7	135.4±30.3	170.3±37.8	372.8±37.2
4+	32.2±12.3	–	–	379.7±416.8	–	–
Min	15.5	4.8	9.5	29.2	0.84	6.15
Max	46.6	28.5	34.5	861	226.8	410.0
Total	24.9±4.3	18.2±10.3	24.0±9.7	141.9±121	97.7±85.3	183.6±141.4

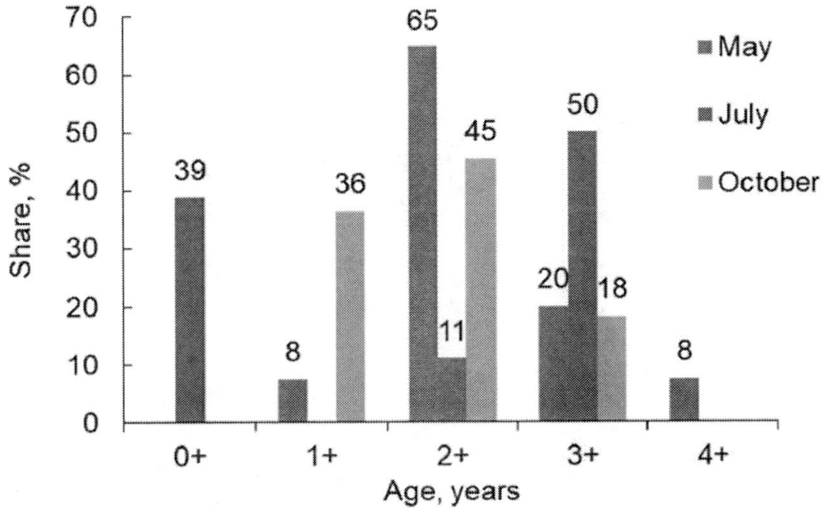

Fig. 4. Age structure of saithe population in the central part of the Kola Bay.

When determining the age of Atlantic cod and saithe, abnormal otoliths were found, which had a sevenfold spherical outgrowth in the center of the otolith (Fig. 5). At transverse fracture, there were no clear rings observed, which made it difficult to determine the age; the reasons for such formations were not found in the open sources.

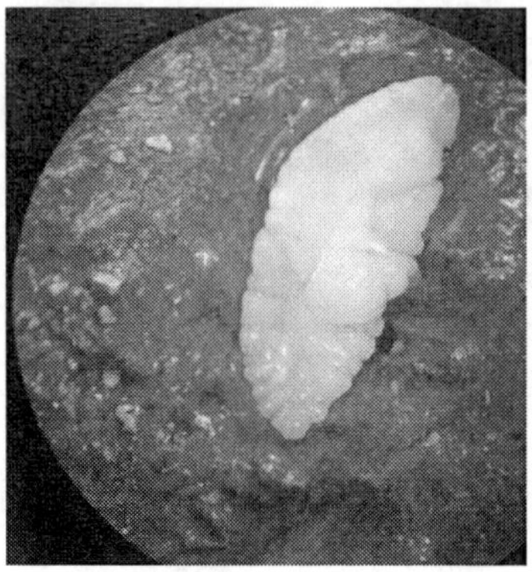

(a)

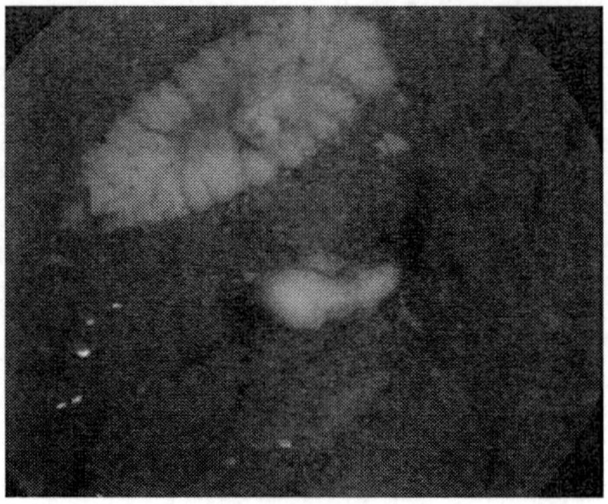

(b)

Fig. 5. (a) – Atlantic cod otolith, (b) – saithe otolith.

In 2018, the sex ratio of saithe in the central part of the Kola Bay indicates the predominance of males in spring and autumn, 73% and 81% (Figure 6). The share of females was higher only in summer (44%), the young-of-the-year were also present (39%). All females and males were with gonads at maturity stage II.

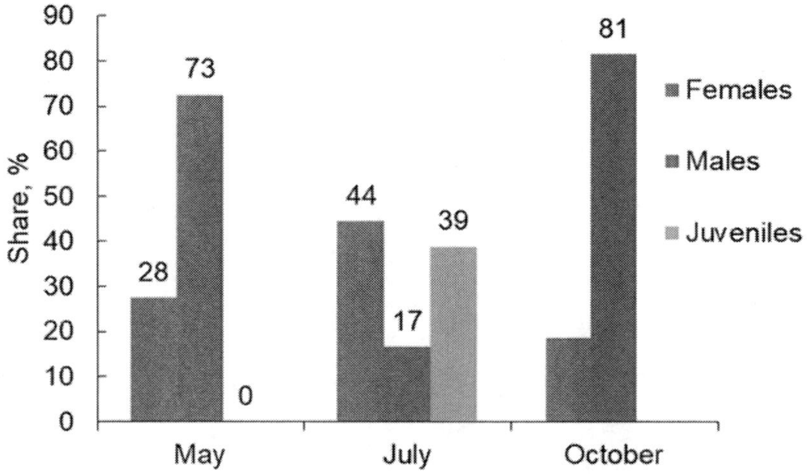

Fig. 6. Sex structure of saithe population in 2018.

The food spectrum and feeding rate of saithe differed from those of Atlantic cod. In May, SFC was 2.9; by frequency of occurrence, Gammaridae prevailed (64%), followed by polychaetes (35%); capelin up to 8.5-cm length and Atlantic cod up to 10-cm (20%) were also noted in stomachs (Figure 7). Single specimens of large saithe (23–24 cm) ate copepods. In July, three-year-olds and four-year-olds continued to feed on amphipods, polychaetes, and fish; the food spectrum of young-of-the-year was represented by copepods, small amphipods, and isopods (SFC = 2.7).

In October, the intensity of feeding decreased down to SFC = 1.8, capelin and juveniles of Gadidae were the basis of the diet; one specimen (27 cm) fed on polychaetes. It should be noted that the food spectrum of saithe in the bays of the western Murman Coast is somewhat different [19].

Fig. 7. Saithe gut filled with amphipods.

Young-of-the-year prefer to consume copepods (90% of the food bolus), and older fish feed on juveniles of their own species, as well as on bivalve mollusks, polychaetes, and gerbil, constituting a large part of the diet.

4 Conclusions

The relevance of ecological studies in the Kola Bay due to the ongoing anthropogenic pressure is beyond doubt. Our data on the biology of juvenile Gadidae supplement previous studies. The size and weight characteristics of Atlantic cod and saithe increase from May to October, the obtained mean values are consistent with the literature data. The studied fish species are represented by young-of-the-year and fish of older age groups, up to 4+ years. The sex ratio changes through the season, when males dominate in spring and autumn, the share of females increases in the summer only. All analyzed fish were with maturing gonads at the maturity stage II, except for young-of-the-year (sex unidentified). The studied species were actively feeding in the water area of the Kola Bay, juveniles consumed planktonic organisms, while elder fish fed on juveniles of their own species and other fish. The food spectra of Atlantic cod and saithe was wide; amphipods of the Gammaridae family played a significant role in fish feeding, accounting for up to 90% of the food bolus. Therefore, no shortage of the food resources was observed.

References

1. M. V. Minina, A. V. Mit'ko, Arctic Security, *Business magazine NEFTEGAZ.RU* **9 (81)**, 78–82 (2018)
2. A. M. Vasiliev, Problems and prospects of an integrated approach to the organization of maritime activities in the coastal space of the Murmansk region // *Priorities of Russia* **24 (165),** 2–10 (2012)
3. K. M. Deryugin, Fauna of the Kola Bay and the conditions of its existence, *Zapiski Imparatorskoi Akademii Nauk* **34 (8)** (1915)
4. O. V. Karamushko, et al. *Kola Bay: Oceanography, Biology, Ecosystems, Pollutants,* pp 144–155 (Apatity: KSC RAS, 1997)
5. R. A. Linnikov, Some data on the ichthyofauna of the upper sublittoral of the Kola Bay in summer, *Materials of the XXV anniversary conference of young scientists of the Murmansk Marine Biological Institute* (May 2007) pp 122–128 (Murmansk: MMBI KSC RAS, 2007)
6. O. V. Karamushko, et al., *Ichthyofauna of the Kola Bay: Development and Rational Use of Natural Resources*, pp 249–264 (Moscow: Nauka, 2009)
7. O. V. Bondarev, Ichthyofauna of the littoral and upper sublittoral zones of the Kola Bay in April 2010, *Studies of Arctic ecosystems: Mater. XXXVI conf. young scientists MMBI KSC RAS dedicated to the 40th anniversary of the research vessel "Dalnie Zelentsy"*, pp 22–28 (Murmansk: MMBI KSC RAS, 2018)
8. O. V. Karamushko, O. S. Yunacheva, River flounder of coastal waters of Murman, *Rybnoe Khozyaistvo* **6,** 57–59 (2005)
9. O. Yu. Kudryavtseva, Some aspects of the biology of the European sculpin *Myoxocephalus scorpius* (L.) (Cottidae) in the Kola Bay of the Barents Sea // *Vestnik MGTU* **15** (4) 766–775 (2012)
10. O. Bondarev, Seasonal dynamics of biological indicators of the European flounder *Platichthys flesus* Linnaeus, 1758 in the central part of the Kola Bay in 2018, *KnE Life Sciences: International applied research conference "Biological Resources Development and Environmental Management (BRDEM-2019)"* pp 211–220 (2020)
11. I. F. Pravdin, *Guide to the Study of Fish* (Moscow: Pishchevaya Promyshlennost, 1966)
12. *Methodological manual for the study of nutrition and food relations of fish in natural conditions* (Moscow: Nauka, 1974)
13. *Instructions and guidelines for the collection and processing of biological information in the seas of the European North and North Atlantic* (Moscow: VNIRO, 2004)
14. N. I. Chugunova, *Guidelines for the Study of the Age and Growth of Fish: Method. Textbook on Ichthyology* (Moscow: AN USSR, 1959)
15. E. V. Borutsky, *Methodological Manual for the Study of Nutrition and Food Relations of Fish in Natural Conditions* (Moscow: Nauka, 1974)
16. N. S. Gaevskaya, *Key of the Fauna and Flora of the Northern Seas of the USSR* (Moscow: Sovetskaya Nauka, 1948)
17. K. M. Sokolov, V. G. Rudnev, Ichthyofauna *Coast of East Murman: ecological studies of the areas of implementation of the Shtokman Project* pp 136–149 (Murmansk: PINRO, 2019)
18. O. V. Bondarev, Age-size characteristics of juvenile cod *Gadus morhua* Linnaeus, 1958 and saithe *Pollachius virens* (Linnaeus, 1958) in the coastal zone of the Barents Sea in July 2018, *Proceedings of the Kola Scientific Center. Oceanology* **8,** 12–17 (2020)
19. A. V. Dolgov, Feeding of immature saithe (*Pollachius virens* L.) in the inlets of the western Murman, *VNIRO Proceedings* **144,** 222–235 (2005)
20. N. V. Mironova, Biology and fishing of saithe, *Proceedings of the Murmansk Biological Station,* **3,** 114–129 (1957)
21. H. Armannsson, S. Th. Jonsson, J. D. Neilson, G. Marteinsdottir, Distribution and migration of saithe (*Pollachius virens*) around Iceland inferred from mark-recapture studies, *ICES Journal of Marine Science* **64,** 1006–1016 (2007)

22. T. V. Antepova, 1994 Ichthyofauna and fish feeding conditions in the Kislaya Bay of the Barents Sea, isolated by the dam of the hydroelectric power station, *Hydrological research in the bays and bays of the northern seas of Russi*a, pp 160–162 (Apatity: RAN, 1994)
23. H. A. Sande, A. Persson, S. Mehl, et al., Temperature and age effects on latitudinal growth dynamics of the commercially valuable gadoid Northeast Arctic saithe (*Pollachius virens*), *Fisheries Research* **213,** 94–104 (2019)
24. M. Heino, T. Svåsand, J. T. Nordeide, H. Otterå, Seasonal dynamics of growth and mortality suggest contrasting population structure and ecology for cod, pollack, and saithe in a Norwegian fjord, *ICES Journal of Marine Science* **69,** 537–546 (2012)

Responses of bottom invertebrates to pollution in the Arctic: bioassays with blue mussel *Mylilus edulis L.*

A. V. Gudimov[1,*] and *S. S.* Malavenda[2]

[1] Murmansk Marine Biological Institute of the Russian Academy of Sciences, 17, Vladimirskaya Str., Murmansk, 183010, Russian Federation
[2] Murmansk State Technical University, 13, Sportivnaya Str., Murmansk 183010, Russian Federation

Abstract. Environmental control in the Arctic ecosystems needs to be changed to hi-tech monitoring technologies with IT systems for actual and ecologically effective 'arctic safeguard'. In circumstances of the fragile arctic ecosystems, priority bioassay of possible toxic hazards in natural seawater must be undertaken. It was considered useful to examine toxic responses of some indicator species. Bivalves are well-known indicators of pollution stress arising from the activities of man, such as off-shore drilling and oil transportation, dredging or the release of pollutants. Toxic effects of some drilling fluids on energy balance, behavior and survival of the common mussel *Mytilus edulis* L, were estimated in long-duration bioassays of about 30 days each. The highest sensitivity of behavioral responses to any changes in chemical water composition was found. Mussels of the Barents Sea had significant differences from the control by four behavioral parameters even at 0.01 g/l of the ferrochrome-lignosulfonate drilling fluid, while the significant differences from the control in oxygen consumption and filtration rates were observed only at the 10 g/l. Even small concentration of another standard water-based drilling fluid with barite (of about 0.05 g/l) had a lethal effect on *M. edulis*.

1 Introduction

Environmental impact in the Arctic marine ecosystems is associated with increasing human activity on the shelf and seashore related mainly to exploitation of the oil and gas deposits, transportation and transhipments of mineral raw materials, including oil and condensed gas, with following harbour and terminal operations.

Standard system of environmental monitoring is outdated and failed for safeguard of the aquatic ecosystems, first of all because of its extremely delayed response to any environmental changes and pollution [1, 2]. The solely approach to reliable and advanced environmental control is online biomonitoring (OBM) technology [1, 3] with recently appeared OBM systems [1, 4]. Theoretically as a procedure, online-biomonitoring is raised up from bioassay experiments, especially performed in conditions closed to the natural ones (e.g. in semi-natural conditions).

In circumstances of the fragile arctic ecosystems, priority bioassay of possible toxic hazards in natural seawater must be undertaken.

Effects of drilling activities with accidental discharge of drilling mud in areas of oil and gas production and possible dumping of drill cuttings in the coastal zone of the Arctic Seas are expected among others.

Drilling and dredged material often contains toxic components that are able to pollute the coastal environment and damage the highly productive area of the sea; consequently, fisheries, aquaculture, wildlife habitats and water quality are at risk.

The bioassays of toxic effects on dozens of species are not available. Hence, it was considered useful to examine toxic responses of some indicator species.

Bivalves are well-known indicators of pollution stress and can be found in aquatic habitats ranging from oceanic to freshwater. Some bivalves are often subjected to stresses arising from the activities of man, such as off-shore drilling, dredging or the release of pollutants.

Initially, the toxic effects of some drilling fluids on energy balance, behaviour and survival of the common mussel *Mytilus edulis* L, widespread along the coast of the Barents Sea, have been estimated in our long-duration tests.

2 Materials and methods

The main experiments were carried out at the aquaria laboratory of the Murmansk Marine Biological Institute

* Corresponding author: alexgud@mail.ru

located on the shore of the Barents Sea (Dalnie Zelentsy, Murmansk region, Russia). To be closer to the natural environment and real animal responses the seawater was pumped directly from the Dal'naya Zelentskaya Bay to the main laboratory tank and immediately distributed from there to aquaria.

The most part of bioassay as well as the previous background research of the coastal bivalve's activity in regard to the fluctuations of environmental factors (temperature, salinity, light, seston amount, etc.) in natural sea water flow have been explored in terms of behavioral and physiological responses in blue mussel (*Mytilus edulis*). Then some effects of environmental factors have been proved in the natural conditions of the Barents Sea and some bioassay responses to toxic substances have been verified in stationary conditions [4, 5–6].

Behavioral measurements were made by several parameters on monitoring of shell moving activity (gaping) of mussels with an automatic recorder or SG-valvometer (strain gage or tension sensor).

To measure the oxygen consumption rate with standard oxymeter, animals after the acclimation in aquaria, were placed individually or in pairs depending on their size, in glass respirometer flasks for exposures. The rate of oxygen consumption was determined from the rate of the oxygen tension declining in the respirometers sealed for the exposure time.

Filtration rate in the blue mussels was estimated by the classic indirect method [7], when the filtration rate is calculated from the decrease in particle concentration of experimental medium per unit of time. As a suspended material, a pure culture of the unicellular green alga *Dunaliella salina* was used in our experiments. The algae were cultured according to the method described by Winter [7].

Long-term experiments presented here on biossays of ferrochrome-lignosulfonate (FHLS) drilling fluids were performed in flow-through or stationary conditions for about a month each (28–31 days).

In addition, toxic effects of another standard drilling fluid (with barite, modified clay powder and carboxymethyl cellulose) and an experimental drilling fluid on the survival of *M. edulis* from the Kola Bay were tested in constant conditions [6]. A feature of the experimental drilling fluid ("eco-mud") was a supplemental addition of the ground brown algae *Saccharina latissima* (past *Laminaria saccharina*) to the ferrochrome-lignosulfonate (FHLS-M) drilling fluid.

3 Results and discussion

The highest sensitivity of the behavioral responses in mussels to changes in chemical water composition was found in our bioassays. Mussels of the Barents Sea had significant differences ($P < 0.05$) from the control by gape size-index even at 0.01 g/l of the FHLS drilling fluid (DF), while the significant ($P < 0.05$) differences from the control in oxygen consumption and filtration rates were observed only at 10 g/l concentration of DF (Figures 1–2).

The time course of oxygen consumption and filtration rates in mussels exposed to DF of 10 g/l was typical for any stress response, i.e., it was followed the harsh and jog changes immediately after add of DF.

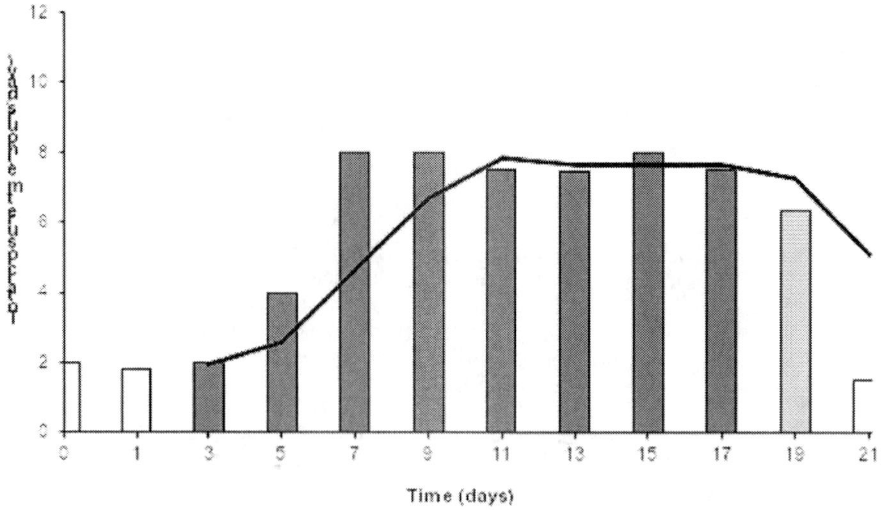

Fig. 1. Daily averaged total closure time in mussels during bioassay with 0.01 g/l concentration of DF (dark bars).

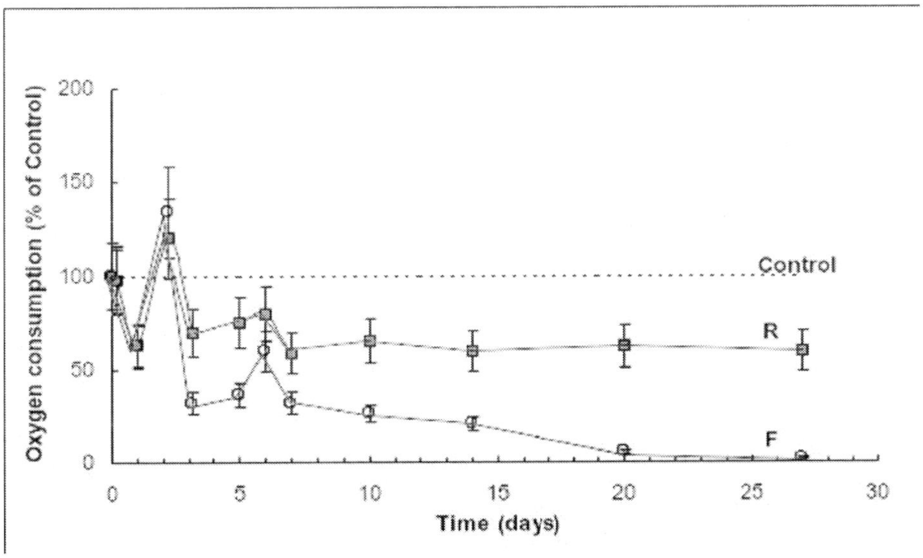

Fig. 2. Oxygen consumption (R) and filtration (F) rates in mussels under 10 g/l concentration of the ferrochrome-lignosulphonate drilling fluid (DF).

Three main phases of a typical acute response to the toxic stress at both concentrations may be distinguished here (Figure 2), including over- or undershoots of rate functions [8]. As for the mussel behaviour it has previously been reported [9, 10] that the immediate response of mussels to environmental stress is closing their shell valves followed by infrequent opening for a short time; either the animal then recovers to normal ventilation or it remains closed and dies. During this initial period of 'shock' when valves are closed tightly, aerobic respiration is reduced or ceases altogether [11].

At the beginning of the exposure in DF an initial decrease and then sudden and abrupt increase in the respiration and filtration rates was observed, stabilizing after a drastic decrease at a lower steady-state level of metabolic rate within 14-16 days of the exposure. The final decreases in the oxygen consumption and filtration rates were 40% and 96% of the control, respectively.

The oxygen uptake and filtration rate in *M. edulis* at 10 g/l and 5 g/l of DF strongly declined during the 28 days of the monitoring period but they were significantly ($P < 0.05$) lower only at 10 g/l of DF than in the control groups. However, no significant difference ($P > 0.05$) existed between the respiration and filtration rates of mussels at 5 g/l concentration of DF and in the control (0 g/l).

On the other hand, the effects of both DF concentrations resulted eventually in the death of 30 and 70% of mussels during a month of deacclimation stage (on the clean seawater flow) after the exposures in 5 and 10 g/l respectively. Hence, despite the sublethal effects obtained in the DF exposures the minimal lethal concentration of DF might be assumed as about 1-5 g/l.

In bioassay with behavioral responses of *M. edulis* the results provided clear evidence that even at 0.01 g/l concentration of DF the behaviour of mussels was significantly affected, indicating that this concentration is high enough to be detected by the animal sensing system.

The pattern of the valve moving activity under DF exposure was abnormal and the adduction/gaping cyclic movements were performed in an abrupt, clear-cut manner. While the average gaping decreased (table 1) the total daily closure time increased significantly ($P < 0.001$) in course of the experiment up to 400% of the average control level (Figure 1).

These daily periods of lasting closures amounted to at least 5 h d^{-1}; the longest duration was obtained within a week from the beginning of the exposure to DF.

After some days of the exposure the total frequency of adductions and quiescences decreased significantly, from 26-30 to 9-11 contractions (partial closing) per day, but then increased slowly within a week of the exposure up to 15-18. This change occurred as a result of shortening the active periods and increasing the number of rest periods, resulting in the decrease of total adduction (frequency of the mussel valves closure contraction) coupled with an increase in the frequency of the alternation of active and rest periods.

The valve movement amplitude decreased twice for the first week of the exposure and then increased later up to the control level. The initial, extremely high value of the daily lasting closures slightly decreased within the exposure but down to a higher level than in the control.

There was a drastic increase of the shell gape size in mussels returned to the clean seawater flow when DF was washing out since the 18th day of the experiment (Figure 1).

https://doi.org/10.1051/bioconf/20225200071

Table 1. Average shell gape size (% of the complete shell closure) in mussels (*Mytilus edulis*) subjected to 0.01 g/l concentration of the ferrochrome lignosulphonate drilling solution (DF) for 2 weeks of the exposure: n.s., not significant at the P < 0.05 level.

Conditions	X	SD	ANOVA (*t*- test)
Control (C)	52.4	13.44	(C - 1DF) n.s.
1st exposure day in DF (1DF)	51.4	15.26	(1DF - 7DF) P < 0.001
7th exposure day (7DF)	71.6	16.20	(C - 7DF) P < 0.001
15th - end exposure day (15DF)	37.8	25.12	(C - 15DF) P < 0.001

All these changes in valve movements are integrated in the average shell gape size of mussels (table 1). Being based on series of the long-term records, we concluded that this main behavioural parameter means the degree of the bivalve organism contact with the media, and it strongly depends on the quality of the environmental conditions.

As generally considered, the toxic effects of FHLS drilling fluids are related to heavy metals of chromium and iron contained in this DF. The maximum permissible concentration (MPC) of ferrochrome lignosulfonate was estimated about 0.002 g/l [12]. The FHLS in seawater may reduce the phytoplankton production, fish embryos survival, and it impedes the development of organs and tissues in fish larvae [13]. On the other hand, at our test [6] concentrations of a modified ferrochrome lignosulfonate (FHLS-M) in the experimental drilling fluid over 5 and even 10 times of MPV did not affect the survival rate of *M. edulis*. Likely the toxic effect of FHLS-M was so mitigated because of the sorbent effect on FHLS-M by the ground algae and clay powder of this "Eco-mud".

Toxic effects of another drilling mud tested in experiments of S. S. Malavenda [6] indicated that despite the previous estimation that barite was a low toxic substance [14], even small concentrations of the standard water-based drilling fluid with barite (of about 0.05 g/l) had lethal effects on *M. edulis*. It can be suggested that only because of the environmental or experimental conditions some aquatic organisms were not affected by barite [14], and the barite concentration of 0.5 g/l did not cause toxic effects on bivalves [15].

In some experiments with low barite concentration the filtration process and gill apparatus were damaged [12, 16]. Besides, the presence of barite enhanced the negative effect of carboxymethyl cellulose, – a component of this standard drilling fluid [12, 15, 16]. In any conditions the most pronounced toxic effect was manifested at chronic pollution [12, 16].

In our experiments carried out on blue mussels [6], the barite content in water from 0.05 g/l to 0.5 g/l, was lower or corresponded to the MPC. In spite of that all mussels died for 9 days of the exposure at any concentration of the standard drilling fluid but were alive in the control.

4 Conclusion

The results of our bioassays revealed that blue mussels possess a sensitive detection system and behavioral mechanism for the avoidance of pollution stress. This mechanism is responsible for both survival and permanent physiological activity of mussels under fluctuated conditions coupled with hazardous levels of contaminant in the environment

A close correlation between the duration of rest periods (valves almost or full closed) and level of contamination has been observed.

Shortening the active periods, increasing the rest time, and then alternating the active and rest periods is the main mechanism of regulation for increasing oxygen demand caused by the pollution stress.

Important to note, that the first sublethal response in blue mussels is the significant change in their shell moving activity and gaping (behavior). Behavioral changes followed by physiological responses such as the rate of oxygen consumption, filtration or cardiac activity with their significant changes at higher concentration of toxic substances.

Starting with behavioral responses the toxic stress ends up with the affected survival indicating the most harmful or lethal conditions.

Priority setting of hazards in water is the main advantage of using bioassay as a means for prevention of environmental impact and ecosystems degradation in the Arctic.

References

1. A. V. Gudimov, The online biological control is a contemporary approach of engineering ecology to environmental safety, *Oil and Gas of the Arctic Shelf* Proc. *4th Inter. Conf.*, 106–109 (2008)
2. A. V. Gudimov, E. P. Komarova, Environmental control: from inertia of the standard biomonitoring to speed of the online one *IOP Conf. Series: Earth and Environmental Science* **625,** 012013 doi: 10.1088/1755-1315/625/1/012013 (2021)

3. A. V. Gudimov, Online biomonitoring system on basis of bioindication *RF Patent* 2437093 (2011)
4. J.-C. Massabuau, A. Gudimov, P. Blanc, Environmental Monitoring of Arctic Waters with Unmanned Bivalve Biosensor Technology: One Year of Background Data Acquisition in the Barents Sea, *Proc, SPE Russian Petroleum Technology Conference Paper SPE* 176681-MS (2015) retrieved from: https://www onepetro org/download/conference-paper/SPE176681MS?id=conferencepaper%2FSPE-176681-MS
5. E. Komarova, A. Gudimov, A. Burdygin Online biomonitoring and immediate bioindication technology: Two years testing in the coastal area of the Barents Sea, *Marine Science and Technology for Sustainable Development 26th Inter. Conf. of Pacific Congress on Marine Science and Technology (PACON-2019)* p 238 2019
6. S. S. Malavenda, A. I. Belukhin, A. O. Bogdanov, A. A. Bannikov The effect of experimental drill fluid on the vital activities of the bivalve mollusk *Mytilus edulis L, IOP Conference Series: Earth and Environmental Science* **539 (1),** 012197 (2020)
7. J. E. Winter, On the influence of food concentration and other factors on filtration efficiency and food utilization in mussels *Arctica islandica* and *Modiolus modiolus, Marine Biology* **4,** 87–135 (1969)
8. O. Kinne, Animals in aquatic environments: crustaceans (Chapter 41), *Adaptation to the Environment: Handbook of physiology* (Section 4 Amer. Physiol. Soc. Washington) pp 669–683 (1964)
9. M. P. Motwani, Experimental and ecological studies on the adaptation of *Mytilus edulis* L to salinity fluctuations, *Proc. National Inst. Sc. India* **21,** 227–246 (1955)
10. S. K. Pierce, Volume regulation and valve movements by marine mussels, *Comp. Biochem. Physiol.* **39A,** 103–117 (1971)
11. B. L. Bayne, The responses of three species of bivalve mollusc to declining oxygen tension at reduced salinity, *Comp. Biochem. Physiol.* **45 A,** 763–806 (1973)
12. M. F. Strachan, P. F. Kingston, A comparative study on the effects of barite, ilmenite and bentonite on four suspension feeding bivalves, *Marine Pollution Bulletin* **64 (10),** 2029–2038 (2012)
13. V. K. Dokholyan, T. P. Akhmedova, The reaction of fish to the presence in the environment of some components of drill fluids, *Proc. I All-USSR Conf. on Fisheries Toxicology* (Riga) **4,** pp 146–147 (1988)
14. C. K. Gratham, J P Sloan, Toxicity study Drill fluid chemicals on aquatic life *Environmental Aspects of Chemical Use in Well-Drilling Operations. Conf. Proc.* Report N EPA-560/1-75-004, (EPA) pp 167–170 (1975)
15. A. A. Movsumov, G. I. Guseinov, The study of the toxicity of chemicals used in drilling and ways to neutralize them, *Gas Industry* **3** (1986)
16. M. F. Strachan, *Studies on the impact of a water-based drilling mud weighting agent (Barite) on some benthic invertebrates*: Doctoral dissertation (Heriot-Watt University, 2010)

Development of safety system in salmon fish farms of the Murmansk region

O. A. Ivanitskaya[1], *N. R.* Kalinina[1], *P. P.* Kravets[2,*], *E. V.* Shoshina[2], and *A. V.* Malaeva[2]

[1]Murmansk Marine Biological Institute of the Russian Academy of Sciences, 45, Shabalin Str., Murmansk, 183053, Russian Federation
[2] Murmansk State Technical University, 13, Sportivnaya Str., Murmansk 183010, Russian Federation

Abstract. The study presents the development of a biosafety system in aquaculture activities on the territory of the Murmansk region, accounting for the existing scientific developments and recommendations. The authors' scientific and practical experience in cultivating the hydrobionts in the conditions of the Polar region is the core of the study. The materials for the development of the regional biosafety program of salmon cages farms were the longstanding epizootic and ichthiopathological studies of the department of fish physiology and diseases, aquaculture objects of the Murmansk Regional Animal Disease Control Center (2013–2019). The most significant risks for salmon net farms of the Murmansk region are: carrying out activities in the zone of risky fish farming; lack (or scarcity) of high-quality nursery material of domestic origin, absence of the center for research and prevention of fish diseases; the absence of the specialized enterprise for aquaculture waste disposal. The developed system of practical safety of aquaculture enterprises in the region will allow managing the risks through certain efforts and events. The principles of biosafety system development in accordance with the environmental requirements and associated conditions for maintaining the epizootic and ichthyopathological safety in the commercial cultivation of fish are considered. Preventive measures to maintain epizootic safety in fish farms have been developed. Suggestions and recommendations are formulated to conduct safe aquaculture activities in the conditions of the Murmansk region. A integrated approach to biosafety through certain risk management activities is aimed at solving the problems of preserving the health of cultivated objects resulting in the maximal feasibility of the enterprise and maintaining the biodiversity of natural communities and environmental sustainability in the region.

1 Introduction

Biosafety is a widely used term in relation to the technology of food production, human habitat and agricultural animals and plants used to manufacture the food products. As for the aquaculture, high quality foods can be obtained only from healthy hydrobionts grown in healthy water environment. Biosafety in aquaculture involves three main components: 1) biosafety of food from hydrobionts produced in aquaculture (food safety); 2) biosafety of cultivated hydrobionts (safety of fish farms); 3) biosafety of the environment (environmental safety) in which aquacultural activity is performed.

Biosafety Program System is one of the methods for reducing the conflicts between aquaculture, the environment and other users of water resources. One of the modern approaches to the solution of biosafety issues is the use of risk analysis [1, 2]. Using an integrated approach, with certain activities, it is possible to solve the problems of preserving the health of cultured hydrobionts, protect the wild populations biodiversity, promote environmental sustainability and ultimately ensure the food safety. An integrated management strategy for commercial, environmental and social risks is aimed at sustainable growth in the Aquaculture sector [3]. The prevention of hydrobiont diseases, the main focus of aquaculture, is a sign of the developing industry, but prevention requiress state support and innovation. Diseases remain an economic and social problem. Understanding the economic consequences of hydrobiont diseases is necessary for the financial assessment of the alternative options and finding potential solutions, which can be provided by the concept of biosafety and preventive measures, and shed light on how to share the limited financial resources, and on the investment opportunities.

In the Murmansk region, as the results of integrated social, economic and environmental research in the northern seas show, aquaculture can be considered as one of the most sustainable types of man-made activities on the inland reservoirs of the Kola Peninsula and in the coastal zone of the White and Barents Seas. The development of aquaculture leads to the creation of new jobs, stabilizes the food situation in the region.

* Corresponding author: ppkravec@mail.ru

The objective of the study is to develop a biosafety system and measures for its implementation in economic entities engaged in the activities of aquaculture on the territory of the Murmansk region, considering the existing scientific developments and recommendations. The main tasks are: to carry out a detailed risk analysis in terms of the occurring possible pathological phenomena and reducing the volume of ichthyomass; develop a number of preventive measures to preserve epizootic safety; devise a technological scheme for arranging the biosafety system in the fish farm. The development of the practical biosafety system of an individual enterprise in this region will allow to manage risks through certain efforts and events.

2 Materials and methods

This study is based on the authors' scientific and practical experience in cultivating the hydrobionts - trout and Atlantic salmon, mussels, laminaria in the conditions of the Murmansk region.

The materials necessary for arranging the biosafety system in fish farms in the North-West region of Russia were analyzed. A survey of legal documentation regulating the activities in aquaculture, aquaculture farms use in their activities, was conducted.

Materials on biotechniques of rainbow trout and Atlantic salmon cultivation in the northern reservoirs, were thoroughly examined taking into account the Norwegian experience and the experience of Murmansk researchers, practitioners dealing with cultivation in the fish farms of various ownership types in the Murmansk region [4, 5, 6].

The basis for the development of the integrated regional program regional program for biosafety on salmon farms were the longlasting materials (2013-2019) on epizootic and ichthyopathological studies of salmon fish grown in the fish farms of the Murmansk region, collected by the employees of the department of fish physiology and diseases, aquaculture facilities of the Murmansk Regional Animal Disease Control Center (MRADCC).

The materials of international agreements on the prevention of the spread of fish diseases in the northern fish-farms and water bodies of Scandinavia and other European countries were analyzed. Instructions and guidelines for assessing the epizootic situation, methods of diagnostics and prevention of invasive and infectious diseases of salmonids in the conditions of the North-Western region were studied [7, 8, 9]. The veterinary regulations of European countries with intensive development of aquaculture have also been examined.

3 Results and discussion

In recent years aquaculture is quite a dynamically developing direction of the Murmansk region's fishery complex activities. A promising direction of aquaculture in the region is the trout culturing in the freshwater sea water and coastal fish farms and the cultivation of Atlantic salmon in marine farms. Currently, the production activities on the territory of the Murmansk region are performed on 32 territorially isolated fish farms of industrial type: fry farms, sea and freshwater cages fish farms, the pilot experimental mussels cultivation farm.

3.1 Developing the safety system is the legal regulation in aquaculture

Aquaculture is in direct conflict with the economic, environmental and social interests of other consumers of the fresh and sea water resources. A more efficient biosafety program is one of the tools for reducing the conflicts. To overcome the conflicts, the most important step is the formation of a rigid legal basis of the coast use, in particular, the definitions of legal rights to water, land, biological resources. In general, key issues include environment and consumers safety, sustainable use of the resources and production development. The rapid expansion of the aquaculture sector has generated a diverse set of international, regional, national and local regulations on veterinary, environmental protection, regulatory functions of various agencies, which reflects the historically established approach to regulation in this area.

Russian aquaculture enterprises rely on their activities are based on the fundamental environmental legislation of the Russian Federation (RF), on the acts regulating the work activities on the preservation of water biological resources and their habitats, and the main documents in the field of veterinary medicine. The legislative framework in the field of aquaculture in the Russian Federation is currently ambiguous and contradictory, there is a permanent process of making changes, and the question of bringing it to the integrated unity is still unresolved.

It is necessary to better coordinate with the legislation at the European and national levels. When developing the safety system, it is necessary to consider the experts' views who works directly with the fish farms and to actualize the existing regulatory framework to the realities of life. It is required to speed up the process of formation and adoption of both regional rules and federal laws.

Despite the development and introduction of regulatory subordinate acts to the fundamental framework in the field of fisheries and preservation of water biological resources, aquaculture, environmental legislation, however the number, complexity and seriousness of the risks of aquaculture activities are not reduced. Therefore, the development of a biosafety system for each economic entity, taking into account considering certain features, first of all, primarily the aquatic object, where the fishing farm is located, the biotechnics of breeding and maintenance of the breeding biotechniques and aquaculture object's maintenance is relevant and justified.

The issue of preserving the Atlantic salmon gene pool in the rivers of the Murmansk region arises in terms of progressive anthropogenic press on fisheries water bodies, including the activities in the aquaculture sector. In this regard, it is important to actualize the system of state control over the habitat of this valuable species of commercial fish in accordance with the current regulatory basis.

At the regional level, problems and trends are determined, including interregional importance, requiring measures taken by the subject of the Russian Federation for the sustainable development of aquaculture in a particular region; programs and recommendations for adopting the measures aimed at solving aquaculture problems of are being developed.

It is especially important for the adjacent regions, where hydraulic connection of the fisheries water objects is observed. In order to plan and implement joint activities, in terms of conservation of the Atlantic salmon habitats and information and experience exchange, interdepartmental agreements are concluded, task teams and commissions working on a regular basis are created. To address the biosafety issues of fish cultivating and processing, it is necessary to improve management in the aquaculture sector.

3.2 Basic environmental risks in salmon fish farms

When developing an integrated aquaculture biosafety program, the analysis and classification of the main environmental risks, occurring in the salmon cages farms of the Murmansk region has been done. The most significant risks in the implementation of commercial fish farming and for the purposes of artificial reproduction are presented in the table.

Table 1. Environmental risks and solutions in the implementation of commodity fish farming in the Murmansk region.

Risks	Solutions	Events in fish farms
Implementation of activities in the risky fish farming zone	Creation of regional scientifically based fish farms, considering the climatic conditions of the Murmansk region, the characteristics of water bodies and aquaculture objects	Implementation of activities in accordance with fish-breeding and biological recommendations or approved instructions, taking into account the adaptation of acceptable methods to polar conditions
Absence (or lack) of high-quality cultivation material of domestic origin	Construction of fish farms on the territory of the Murmansk region. Developing salmon species repair and brood stock. Construction of modern incubatories	The use of high-quality selection cultivation material. Compliance with quarantine measures when importing the cultivation material.
Lack of a center (laboratory) of research and prevention of fish diseases	Creating an interdepartmental center for the research and implementation of methods for fish diseases control and prevention. Development of express methods for diagnosing the diseases. Development of therapeutic feed and vaccines of domestic origin. Immediate informing about monitoring the infectious and especially dangerous fish diseases; outbreaks of new diseases not registered in the region previously	Veterinary control throughout the cycle of cultivation. Compliance with the principles of responsible fish farming. Long-term contracts work with specialized accredited laboratories and conducting production activities under scientific supervision of specialized fishery bodies
Lack of a specialized enterprise for the disposal of aquaculture waste products	Developing a system and the specialized enterprise for the disposal of infected fish and wastes from aquaculture products processing	Instructions for the collection and processing of wastes from aquaculture products processing to ensure epizootic and epidemiological safety on the territory of fish farm

The risks considered for the Murmansk region are generally relevant to other regions of Russia [10].

3.3 Formation of biosafety conditions is the compliance with the environmental principles of safety

The development of commodity fish farming will be successful if the fish farm produces the commodity fish in compliance with the principles of environmental safety and focuses on the preservation of the ecosystem in the location area [11, 12, 13]. Following the environmental principles is necessary to obtain high-quality and safe aquaculture products. Biosafety of salmon fish farms is based on the principles of the so-called precautional approach.

The specification of these principles is reflected in the "biosafety system" of an aquaculture enterprise, which is developed taking into account the characteristics of a particular fish farm, is approved by the enterprise's authority and is obligate for strict execution by the staff. According to the current legislation of the Russian Federation in the design and implementation of works in fishing areas intended for the placement of net modules in a water facility, in fish-water sections and coastal structures owned by the enterprise, measures are also provided for the maximum prevention of the adverse effects of aquaculture facilities on water biological resources, conditions of the habitat and reproduction.

Aquaculture activities are carried out in accordance with the recommendations of the fish-breeding and biological substantiation or approved instructions, considering the adaptation of the methods applied to the northern climatic zone. When implementing the projects related to commodity fish farming in the Murmansk region, it is necessary to consider the following:

1. It is recommended to distribute the objects and the execution of fish-breeding works in places, time and methods that have a minimum unfavorable effect on the aquatic ecosystems.

2. When designing a fish farm in general and, in particular for each fishing area (net complex), scientific studies are conducted to assess the impact of commercial fish farms on the water biological resources of the water body, taking into account other fishery reservoirs included in the hydrological system as a whole.

3. The specific "Plan for a fish farm production activity" contains the section devoted to environmental safety, its essential component being the maximum reduction of anthropogenic load in the area of the aquaculture activities and processing the cultivated production.

4. To ensure the execution of the "Plan for a fish farm production activity" specialized scientific organizations develop a fish-breeding and biological substantiation on the request of the enterprise. It includes the following tasks: determining the hydrology and - hydrochemical conditions of water systems for the net farms purposes; assessing the biodiversity of natural communities in the area of net fish farm's location; evaluation of the environmental consequences of man-made impact on commodity fishing; the proposals of optimal methods of fish-breeding and obtaining high-quality and safe aquaculture products; development of environmental and compensatory events.

Based on the studies conducted and the tasks' solutions proposed, the possible volumes of commercial production are determined, which can later be corrected depending on the degree of influence of the fish farming on the ecological condition of the water used.

In this regard, the obligatory principle of the fish farm is to monitor the breeding conditions of the total volume of the fish cultivated and the pollution level of the water biological resources' habitat. The most effective way to implement such control measures is the work on long-term agreements with specialized accredited laboratories and conducting production activities under scientific support of specialized fishery scientific institutions (Table 1).

The encouraging factor in the development and execution of the biosafety system at aquaculture enterprise is the authority's and manager's understanding of necessity to execute the environmental laws when working with biological objects. Saving the habitat for aquatic bioresources (populations of hydrobionts in the natural environment) and the creation of habitats for salmon species in accordance with their physiological requirements is an essential aspect of the overall biosafety system.

Environmental parameters must be fully integrated into all types of activities in the Arctic zone, by the regional countries accepting the common environmental criteria for the use of natural resources and pollution prevention.

Thus, the implementation of activities in the zone of risky fish farming in the conditions of the Polar region puts the task of creating a unified system in the aquaculture management, which would allow promptly and prospectively, in accordance with the feasibility studies and the objectives of the enterprise's activities, solve the issues from the cultivation of aquaculture objects to the sales in accordance with the market requirements.

3.4 Quality cultivation material of domestic origin is the basis of the fish farm biosafety

Spreading of the diseases dangerous for commercial cultivation facilities and for wild species of valuable associated with the delivery of cultivation materials and fish from abroad and neighboring regions of the Russian Federation cause certain danger. Invasive animals and pathogenic organisms can be carried with living fry and even in the water. International and interregional trade - of live fish trade (larvae, fry, adult) is associated not only with the risk to fish health due to the possibility of transferring pathogens, but also to the risk of breeding material death, for being not adapted to the temperatures of the Kola Peninsula reservoirs during the period of stocking in a fish farm. The physiological non-availability of cultivation material will lead to the death of large amounts of fish, which will cause not only economic losses, but will also lead to negative environmental consequences, primarily in the area of commodity farms [14].

Mandatory measures to verify the safety and health of cultivation material imported to the fish farm are based on the principles and provisions of the current environmental and veterinary regulations of the Russian Federation, as well as harmonized with international standards and requirements of the International Epizootic Bureau (World Animal Health Organization) in the field of aquaculture [15].

There are a few specialists having profound practical experience in the field of commercial cultivation of the Atlantic salmon in Russia, therefore, for the fish farms beginners, simple, affordable and integrated methods of quality control of the acquired cultivation material, harmonized with generally accepted international norms, should be put into practice of the Russian marine fish farming. In Russia, there are no incubators and plants for the production of the Atlantic salmon juvenile for marine commercial farms. In this regard, fish farms are forced to buy fry (smolt) adapted for living in seawater, in Norway, Scotland and other countries with developed industrial fish farming. Considering the longstanding and multifaceted experience of neighboring countries, in particular Norway, attention should be paid to the problems associated with marine fish farming in terms of fish diseases [3, 16]. The scientific studies of Norwegian institutions, both of specialized fisheries and of veterinary, are aimed at developing the sustainable development programs of aquaculture and the national biosafety system in aquaculture.

Thus, one of the first principles of biosafety is the use of high-quality cultivation material.

The system of epizootic safety in the fish farms is arranged through the active implementation of international standards and strengthening the precaution approach "One health for all" based on the association of state and non-state enterprises. It involves strengthening the efforts for the disease's prevention at the level of fish farm, due to responsible fish farming and the expansion of understanding of the necessity to ensure ichthyopathological safety (lack of diseases).

The more intensive the cultivation of salmon and the higher the density of cultivated fish in the cages, the more carefully preventive measures should be carried out, they should become an integral part of all fish-breeding operations [9, 14].

Veterinary observation of aquaculture objects throughout the technological cultivation cycle is the basis for the sustainable development of aquaculture and its safety (Table). Fish of all age groups, including breeders and fish from repair groups, are examined when transferred from one reservoir to another, as well as while test seining in the cultivation season (no less frequency - once a month). If the infectious disease is suspected, fish pathological material from is sent to the veterinary laboratory, simultaneously measures to prevent the spreading of the disease are taken.

3.5 Veterinary and sanitary proposals and recommendations for conducting the activity in salmon fish farming include:

1. Compliance with the requirements of the current environmental and veterinary legislation of the Russian Federation.
2. Considering compulsory, the prevention and control measures to prevent the infection and invasion, as well as precaution measures to prevent the fish loss.
3. Excluding the use of antibiotics and stimulants in cultivation of fish.
4. Using medical drugs and disinfectants for the prevention of diseases and infections caused by parasites only in accordance with the manufacturers' instructions and manuals.
5. Using certified in Russia, standard, granulated salmon feed. Aquafeed is supplied to the farm with the corresponding veterinary accompanying documents.
6. Slaughtering and bleeding of fish in the waters of inlets and bays is prohibited. Slaughtering is done in compliance with sanitary standards and rules, as well as the requirements of environmental legislation in terms of environmental protection against biological and other pollutions.

When implementing the project of cultivating the salmon species, a compulsory condition is the observance of environmental activities and recommendations of state specialized fisheries agencies to preserve water bioresources and habitats. Compliance with the following requirements is supposed:

1. Transportation and movement of salmon fish species should be carried out under the control of veterinary service specialists and in the way to reduce the possibility of pathgens transferring of infectious and invasive diseases and parasites, as well as genetic and other biological interactions. Mechanisms for monitoring the movement of fish should be developed and implemented in the fish farm activities.
2. Implementing the scheme of the escapement catch. This scheme should be coordinated with the veterinary service and the fishery protection bodies in order this catch not to have a negative impact on the stocks of wild salmon.
3. Implementing a specially developed emergency plan in case of an accident, causing great fish escapement.
4. Carrying out the works at all stages of the production process from the delivery of the cultivation material to harvesting from the fish cages, including the transportation of live fish, in accordance with the established veterinary and hygienic standards.
5. Careful applying the appropriate cultivation methods in order to minimize the risk of diseases in the cultivated fish. It involves: ensuring the optimal cultivation densities, careful management, frequent inspection of fish, proper diet and feeding mode, excluding stress, disinfection of equipment used during farm transportation and fish farm maintenance.

6. When clinical signs of diseases appear, it is necessary to inform the state veterinary service immediately. The infected, dead or dying fish should be withdrawn from the cages and disposed, together with production waste, according to the measures agreed on with the veterinary service.
7. Fish farm capacity must correspond to the selected site, and cultivation density should not exceed the level established on the basis of breeding practices. The mode of "fallowing" should be used as a means of reducing the biological burden on the water area and for the prevention of diseases and parasitic invasions outbreaks.

For biologically safe and sustainable aquaculture development on the territory of the Murmansk region the integrated, ecosystem approach should be further improved.

4 Conclusion

The integrated approach to biological safety in fish farms through certain activities is intended to solve problems of preserving the health of cultivated hydrobionts, to protect the biodiversity of wild populations, to promote environmental sustainability and ensure food safety through certain activities. The most significant environmental and epizootological risks for salmon cages farms of the Murmansk region are: carrying out activities in the risky fish farming zone; absence (or lack) of high-quality cultivation material of domestic origin; lack of a specialized center for research and prevention of fish diseases; lack of the system and a specialized enterprise for the disposal of aquaculture wastes. A technological scheme for developing an aquaculture biosafety system allowing to manage the risks through certain efforts and events has been developed.

References

1. *FAO Yearbook. Fishery and Aquaculture Statistics* (Rome, 2014)
2. *FAO Yearbook. Fishery and Aquaculture Statistics* (Rome, 2015)
3. G. L. Taranger, et al., *ICES Journal of Marine Science* **72,** 997–1021 (2015)
4. N. K. Vorobyeva, M. A. Lazareva, L. I. Pestrikova, *Fisheries* **6,** 45–48 (1996)
5. N. K. Vorobyeva, L. I. Pestrikova *Trout Farming in the Arctic* (Murmansk: PINRO, 2011)
6. V. S. Anokhina, N. K. Vorobyeva, L. I. Pestrikova, M. A. Lazareva, *Fisheries* **2,** 40–41 (2000)
7. A. M. Naumova, A. N. Machnev (eds), *Collection of Instructions for Fish Diseases Control* **1** (Moscow: Marketing Department AMB – Agro, 1998)
8. A. M. Naumova, A. N. Machnev (eds), *Collection of Instructions for Fish Diseases Control* **2** (Moscow: Marketing Department AMB – Agro, 1999)
9. R. Rakhkonen, P. Vernistrem, P. Rintamyaki, R. Cannel, *Healthy Fish. Prevention, Diagnostics and Treatment of Diseases* (Helsinki, 2013)
10. L. P. Ryzhkov, T. Yu. Kuchko, *Cage Fish Farming* (Petrozavodsk: PetrSU, 2008)
11. S. B. Cupinsky, *Production Capabilities of Fisheries Water Bodies and Objects of Fish Farming* (Saint-Petersburg: Lan, 2019)
12. R. A. Quiñones, M. Fuentes, R. M. Montes, D. Soto, J. León-Muñoz, *Rev. Aquacult.* **11,** 375–402 (2019)
13. N. Ahmed, S. Thompson, G. M. Turchini, *Food Sec.* **12,** 1253–1267 (2020)
14. N. R. Kalinana, *Vestnik of MSTU* **15,** 517–525 (2012)
15. *FAO Yearbook. Fishery and Aquaculture Statistics* (Rome, 2018)
16. V. K. Zilanov, G. I. Luka, A. V. Zelentsov, *Fish Industry of Norway in the Twentieth Century: From Marine Fisheries to Mariculture* (Moscow: VNIRO, 2008)

https://doi.org/10.1051/bioconf/20225200073

Fucoid seaweeds on the Russian Arctic coast: using traditions and ecological monitoring

V. I. Kapkov[1] and *E. V.* Shoshina[2,*]

[1] Lomonosov Moscow State University, GSP-1, Leninskie Gory, Moscow, 119991, Russian Federation
[2] Murmansk State Technical University, 13, Sportivnaya Str., Murmansk 183010, Russian Federation

Abstract. In recent years, harvest and cultivation of marine biological resources, including commercial seaweeds, has increased significantly. The task of this research was the ecological monitoring of commercial fucoid seaweeds beds in coastal waters of the Barents and White Seas. Data of seasonal supervision at stationary biological station and materials of expedition works on Murmansk coast were analyzed. The species, age, generative and spatial structures of fucoid populations in different biotopes of the littoral were studied. It was found that species composition of phytocenoses and quantitative parameters of commercial fucoids varied significantly in open and closed bays, as well as in algae living in the lower and upper littoral. The average biomass of fucoids in semi-closed inlets varied from 2.8±0.6 to 17.7±2.9 kg/m2, depending on dominant species in community, the structure of substrate and impact of waves. The results indicate that phytocenoses of fucoid algae under the influence of changing environmental factors largely ensure the integrity of littoral ecosystem. For rational harvest and cultivation of commercial fucoids, as well as the restoration of disturbed phytocenoses after the removal of some algae, it is recommended to use the technology of synchronization of harvesting with life cycles of seaweeds.

1 Introduction

Since the fifteen centuries, monks of the Solovetsky monastery have been gathering and using as food the benthic seaweeds, including the brown algae, collected from the coastal zone of the White Sea. In later years, the Monks started extracting iodine, bromine and soda from the collected seaweeds. During the First World War iodine extracted from White Sea seaweeds was utilized. In 1930's, The Algal Institute (Arkhangelsk) developed technologies for extraction of agar from red seaweed and extraction of mannitol and alginate from brown seaweeds. In the White Sea there are commercial quantities of brown algae – two species of Laminaria and three Fucus species plus Ascophyllum nodosum and one red alga species – *Ahnfeltia plicata*. In 1990's, in the Barents Sea, the commercial harvest of *Laminaria saccharina* and *Fucus vesicilosus* was initiated. Currently industrial processing of seaweeds is carried out at Arkhangelsk algal plant which is producing a wide assortment of medical, food, cosmetic and technical production.

Brown, red and green seaweeds play a leading role in formation of littoral biocenoses of the Barents Sea which are characterized by high species biodiversity and biomass of benthic hydrobionts. Due to their long-life cycles fucoid seaweeds are edificators-species in littoral phytocenoses and form the structure of the benthic community. These fucoid seaweeds function as the main primary producers in these communities while also acting as determinants for the rest of the biotic community and in many respects helping to determine the state of coastal ecosystem [1, 2, 3].

In the littoral zone brown seaweeds create multilevel phytocenoses, in which the spatial complexity increases with depth resulting from increasing species diversity due to the numerous accompanying species and epibionts which use the algal thallus as a substratum. Seaweeds are also the important in the food chain as numerous invertebrates are phytophages eating adult plants or, more often juveniles, which plays a role in regulating species diversity and in the formation of spatial structure within the phytocenoses [4]. Benthic seaweeds create the habit environment and serve as a refuge for numerous larvae of invertebrates and fishes.

Coastal seaweeds are known to play an important role in the biological clearing processes, of coastal seawaters from oil and toxic heavy metal pollutants. Seaweed along with others hydrobionts are widely used in reconstruction of damaged biocenoses by construction of artificial reefs for biomediation in coastal ecosystems [5, 6, 7]. Benthic seaweeds have become extremely important trade products for their valuable mineral and organic contents [8, 9]. Development of new or perfection of existing cultivation techniques for seaweed

* Corresponding author: shoshinaev@gmail.com

is becoming increasingly important in the field of ocean aquaculture [10, 11].

In the coastal zone of the Barents Sea brown seaweeds are the dominant species in the littoral zone benthic community. The top littoral horizon is occupied by *Fucus vesiculosus* and *Ascophyllum nodosum*, middle – by *F. distichus*, and *F. serratus* prefers low littoral horizon. Various other seaweeds are cohabitants with the fucoids, however their biomass are on an order of 1-3 times less, than *Fucus sp*. In the low littoral zone the commercial laminarean seaweeds *Saccharina latissima, Laminaria digitata, Alaria esculenta* are grown. In addition here the red alga *Palmaria palmata*, species of genus *Porphyra* and green alga *Ulvaria obscura* are growing which are also used by man. The red alga – agarophyte, *Ahnfeltia plicata* is the dominant species found forming the extensive fields of alga found in the bays of the White Sea.

During the last few decades commercial fucoid seaweeds have received the attention of researchers who have studied their anatomy, morphology, physiology, systematics and biochemistry. Currently further research into the biology and ecology of seaweeds, including their life cycles and growth rate of thallus, the analysis of age and generative structures of populations is needed. In addition, the role of ecological factors in determining the distribution of different biotopes needs further work to determine the role these biotopes play in coastal ecosystem integrity. With increasing use of marine biological resources seaweeds are enter into our life more persistently as food products. Marine aquafarm development in the coastal zones of the Barents and White Seas assures for the fuller utilization of the natural seaweed resources.

Through our research our goal is to facilitate more effective study of our natural seaweed resources. The purpose of our work is hoped to lead to a more rational use of the commercial seaweeds found in the coastal zone of the northern seas. In this research we have monitored the conditions of the fucoid seaweed beds found on the Murmansk coast of the Barents Sea.

2 Materials and methods

Long-term seasonal monitoring of seaweeds data collected at the biological station of Murmansk Marine Biological Institute of the Russian Academy of Sciences (in region of Dalnie Zelentsy) and seaweeds data collected from expeditions on the Murman coast of the Barents Sea and on coast of the Kandalaksha and Onega gulfs of the White Sea were analyzed. Methods of gathering and the materials examined were presented in our publications [1, 7]. The methods used to gather seaweeds data utilized sample plots in typical habit sites of seaweeds in the intertidal zone along transects oriented perpendicular to the coast line. Sampling in fucoid phytocenoses was carried out using a frame of 0,5 x 0,5 m2 (3–5 replicates). Collected seaweeds in each sample were divided into species, then wet weight of each species was measured and summed to determine the total wet weight of seaweeds (B) in communities. The average biomass and the error of average (B ± mB) were calculated. In the population study of fucoids, all plants from samples (n = 3–5) were divided into age groups. Age group determinations were established by the number of dichotomous branches on the thallus (t = d/2). Total number of fucoid plants (Nf), number of plants in each age group (Nt) and wet biomass of each age group (Wt) were determined. In each age group, 5–10 typical thallus was selected, and length (L), weight (W) of plants were measured, then average and standard deviations (σ) were calculated. Statistical processing of the received data was by means of the software package «Microsoft Office Excel».

3 Results and discussion

Fucoids are common and abundant seaweeds in the intertidal zone throughout the coastlines of both the Barents and the White Seas. Fucoids in these areas form complicated perennial phytocenoses. The composition of species comprising the phytocenoses (dominating and accompanying species) show variability depending upon location; open to closed parts of inlets and from upper to low littoral horizons. Over time, the quantitative parameters of phytocenoses and populations of dominating species show a wide corridor of variations.

The high biodiversity found in fucoid phytocenoses of the littoral zone of the Murman coast results many accompanying species of seaweeds, on which the thallus and within the thallus were numerous invertebrates living and clinging to prevent outflow. The spatial structure of phytocenoses is multidimensional. The base level is occupied by the fucoid species as edificators, under their beds and on their thallus litophytes and epiphytes are found growing. The number of seaweed species living in association with the fucoids increase from the upper to the lower portions of the fucoid beds (from 1–7 up to 20–30 species). Similarly, this increasing biodiversity is seen spatially from the closed to the open portions of inlets (from 15–22 up to 20–30 species). For most of the long-lived brown fucoid seaweeds found dominating the research areas, biodiversity was highest in the spring and summer and decreased in the autumn and winter.

The seaweed's biomass was seen to increase with depth from the upper to middle and lower littoral horizons. This being said, even in the upper littoral zone the fucoid's biomass can be quite large. At all sites within the inlet from open to closed portions significant variations in biomass values were seen being dependent upon the conditions of biotope and sampling sites (Table 1).

Greater than 90 % of the phytocenosis biomass was composed of fucoid seaweeds *Ascophyllum nodosum, Fucus vesiculosus, F. distichus, F. serratus*. These species dominate in semi-protected inlets (Table 1) with an average fucoid community biomass of from 2.8±0.6 to 17.7±2.9 kg/m^2. Variables which determined biomass include: the species composition present, the granulometric structure of substratum on which seaweeds were growing and the level of force from the surf. In fucoid phytocenoses the biomass of seaweeds increased since spring to the maximumal values seen in the summer and then decreased during the autumn–winter period. The most significant seasonal changes were observed in association of *A. nodosum* +*F. vesiculosus* in the central part of the inlet with a threefold biomass increase (4–12 kg/m^2).

Table 1. Average biomass variation (kg/m2) in fucoid phytocenoses in inlets of semi-protected and exposed types.

Intertidal	Semi-protected inlet			Exposed inlet
	closed→	middle→	→open part	
upper	2.6	2.3–12.0		1.5–5.1
middle	3.0–17.7	2.8–6.9	2.7–14.4	3.2–6.4
low	0.5–1.1	0.2–0.5	–	9.5

In the protected type of inlets, the maximum average biomass was observed in settlements of *F. vesiculosus* in open portions (3.7 kg/m^2) with a gradual decrease in biomass being seen in closed portions where the minimum values (1.8 kg/m^2) were noted. Analysis of the fucoid population found that as measured by biomass middle aged plants dominated the open and middle portions of inlet, but in the closed areas young individuals dominated due to their high population numbers (Figure 1).

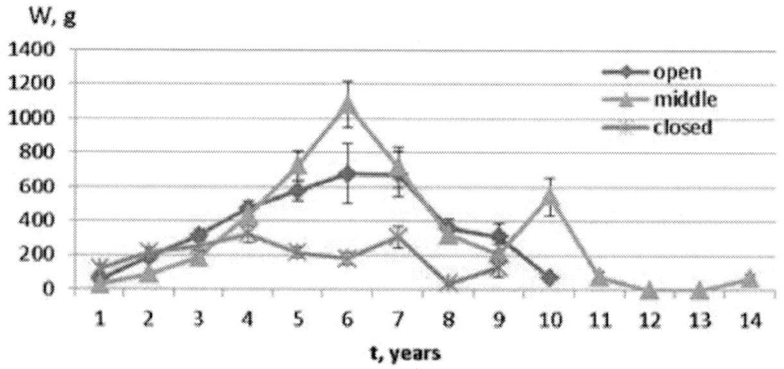

Fig. 1. Total weight of Fucus vesiculosus age groups (Wt, g) in different sites of protected inlet, *n* = 510.

In the littoral zones of the Murmansk coastline, the age structure of fucoid seaweeds was dominated by young plants but all generations from young to old (fertile plants) were present. In the upper littoral horizon seaweeds *A. nodosum* and *F. vesiculosus* were growing in populations with mixed age structure with densities ranging from 750 to 1700 plants/m². While in populations of F. distichus and *F. serratus*, occupying the middle and low portions of the littoral zone, seaweeds were found in groups of similar aged plants in which 4–6 years plants were dominated, with density ranging from 120 to 170 plants/m².

In Figure 2, the age structure of *F. vesiculosus* settlements in the protected (from surf) portions of the inlet is shown. In the near shore locations, the age structure was dominated by young plants with high density (1290 plants/m^2). In the middle part of the inlet, plant ages varied from young to old (up to 14 years) with densities sharply decreasing (809 plants/m^2) as compared to the other parts of the inlet. In the open portions there was settlement with high density (1350 plants/m^2) again, characterized by plants of a wide range of ages. The *Fucus* population age structure in the protected inlet consisted of plants from young to Middle Ages to old plants. The heterogeneous age structure of this population provides evidence that effective management strategies are allowing for renewal of the population.

Fucoid seaweeds are characterized by having long life cycles. In the Barents Sea the maximum age for *F. vesiculosus* plants ranges from 11 to 14 years while longevity for *F. distichus* plants can reach up to 8-9 years, with *F. serratus* living up to 14 years and finally *A. nodosum* has longevity estimates ranging from 19 to 25 years [1].

Fucoids have an unusual type of sexual reproduction for macroalgae. In the diploid phase in life cycle the reproduction bodies act as receptacles where the gametes are formed. Male and female gametes can be formed in the same plant or in different plants, gamete release is synchronized by circadian, lunar and seasonal internal clocks. Species *A. nodosum, F. vesiculosus, F. serratus* – dioecious plants with individuals being either male and

 https://doi.org/10.1051/bioconf/20225200073

female plants; *F. distichus* – a monoecious species. In *Fucus* species the receptacles are formed on top of the branches. These reproductive branches stop growing after reproduction. In *Ascophyllum* receptacles are formed on small stalks on edge of branches. The reproductive receptacles are dumped after reproduction. Following reproduction these shed receptacles can be so numerous, that the coast is covered by them. Fertilization occurs in the water column where the newly formed zygote immediately settles to the bottom and attaches to a substratum. This stage of development is the most critical in life cycle.

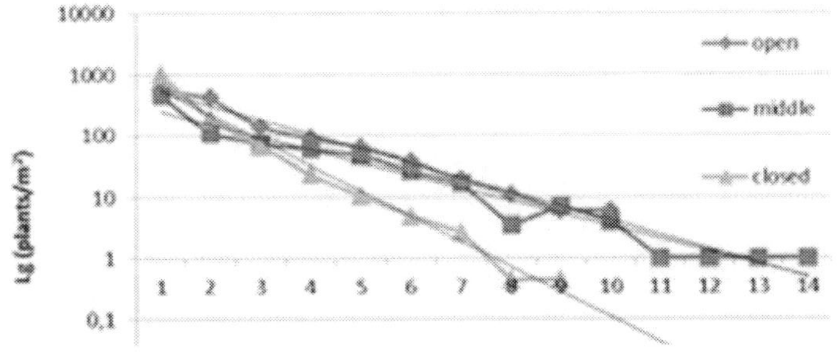

Fig. 2. Age structure of Fucus vesiculosus population in different sides of protected inlet (n = 3450).

Murman fucoid seaweeds start to reproduce in their second to fourth year. This non-reproductive period is considerably longer than is seen in other long-lived seaweeds. Another characteristic feature of fucoids is the long developmental period for receptacle formation which requires from 5 to 12 months. In the White and Barents Seas in *F. vesiculosus, F. distichus, A. nodosum* the receptacles form in the autumn, slowly developing in winter; with the release of numerous gametes in first half of summer of the next year. The quantity of fertile plants in the populations of the different fucoid species depends on ecological factors. The percentage of fertile plants vary from 20-30 % in *F. vesiculosus* and *F. serratus* and up to 50 % in *F. distichus* and *A. nodosum*. The population age group primarily responsible for reproduction is the seaweeds of 4–7 years of age.

In inlet of protected type, the first receptacles in *F. vesiculosus* develops in two-year-old plants. With aging the relative number of fertile plants increases with the majority of adult plants having receptacles. The total percent of fertile plants average from 25-34 %. The ratio of female to male plants found in the central parts of the settlement was estimated as 1:3 female to male while on the edge of the settlement the ratio was found to be approximately 1:1 (Figure 3).

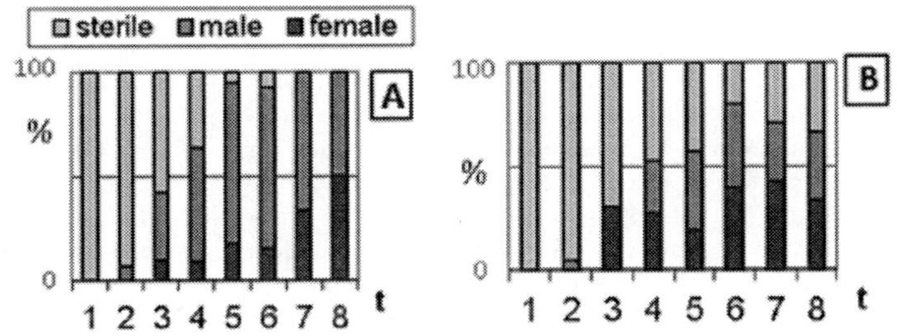

Fig. 3. Generative structure of *Fucus vesi*culosus population in middle (A) and closed (B) sites of protected inlet; t – age, years; n = 426.

In populations of dioecious species in the partially protected regions in dense beds of *A. nodosum* and *F. vesiculosus* the ratio of female to male plants was 1 to 1,5. In populations of *F. serratus* which usually occupies the lower littoral horizon the ratio of female to male plants was 1 to 1. The general it was observed that in stressed habitats the of females to males increased.

Fucoid growth is through the activity of the apical meristem and by dichotomic branching of the thallus, which occurs throughout the entire plant life. It is expected in this case that the growth curve as measured by biomass in a normal population should be exponential in character (Figure 4). The weight of thallus rises almost exponentially up to years 7–10. In later ages the growth of the thallus is not observed. This observation might be due to damage from storm energy by breaking of the thallus as well as by stoppage of branch growth with riped receptacles. However, more often we see that the curve takes the form of an S-curve in more protected areas (Figure 5) where biomass growth data are presented. The shown data were based upon raw thallus weight of plants in protected type inlets. The maximum values of thallus biomass were observed in different age groups in various parts of the inlet of different types.

The first quantitative estimations of commercial seaweed stocks in the White Sea were carried out by employees of the State Oceanographic institute under the leadership of professor Meyer K. I. in 1931. Beginning in the 1950s, the estimation of seaweed stocks was carried out regularly. In recent years mapping of seaweed stocks in the Barents and White Seas was performed through the use of aerial photography, by use of cameras on ships and by underwater observations through the use of diving equipment [12, 13]. Stocks of fucoid seaweeds on the Murmansk coast of the Barents Sea were estimated as 180 thousand tons according to data of PINRO and VNIRO; the regions with the greatest potential for commercial harvest were found in Rybachyi peninsula, Kildin Island, Drozdovskaya and Ivanovskaya Bays [13, 14].

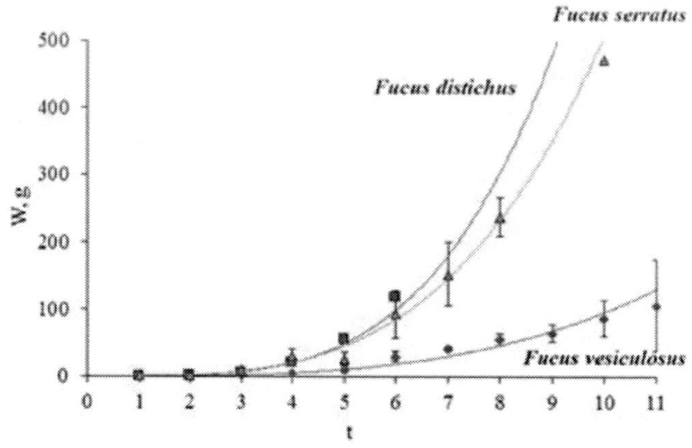

Fig. 4. Weight of thallus (W, g) in ontogenity of *Fucus vesiculosus* (n = 59), *F. disthus* (n = 49), *F. serratus* (n = 40) in semi-protected inlet, the y-axis – t, age, years.

The stock estimation of commercial seaweeds in each location is based on determination of seaweed biomass according to the formula: Q = (B * PC * S)/100 %, where B – an average biomass in a site, kg/m^2; PC – a projective covering (the area occupied with seaweeds), is defined visually, in %; S – the site area, m^2.

Intervention in natural ecological processes which is inevitable with the harvest and trade of seaweeds, can lead to negative consequences, both for the ecosystem and for the environment. Therefore, to minimize the negative ecological consequences of seaweed harvest comprehensive investigations are necessary. The principle of rational wildlife management of stocks of seaweed is based on the estimation of seaweed stocks and on the results of research on the biological communities and specific biology of the commercial seaweed species.

Estimates of seaweed biomass is the primary parameter used for regulation of seaweed harvest. Estimates of seaweed stocks and allocation of quotas for harvest is based on the idea that the algal resources are relatively static. However, seaweed biomass is the result of the habitat conditions under which the seaweeds are growing which can be quite variable, especially on the Arctic coast. The biomass varies considerably with the most important factors being the granular structure of the substrate along with the hydrodynamic conditions (surf conditions, frequency and magnitude of storms, tidal currents, motion of ice, and water outflows). Intensity of water exchange is the most important indicator in determining the makeup of the specific seaweed communities found in the bays and open water zones of the sea, as well morphological and functional structure of dominating species. Seasonal cycle of growth of littoral seaweeds is strongly connected with temperature and photoperiod [1]. The biomass of seaweed has considerable seasonal and inter-annual dynamics, especially in the long-lived fucoid seaweeds, with their highly variability population age structure which is directly dependent upon environmental conditions.

Despite the many advantages which would result from seaweed culture as compared to harvesting from nature, cultivation of seaweeds in the northern Russian seas remains practically undeveloped. There is potential for development of seaweed plantations which could grow seaweeds for raw biomass for direct use or for the processing and extraction of biologically active substances from the seaweeds. Other benefits which could result from the creation of seaweed plantations include use of seaweed as natural biofilters for water treatment in aquafarms when growing invertebrates and fishes or for use in cleaning of sewage dumped at sea. Seaweed as a primary producers can be used as food in aquaculture in the joint cultivation of mussels and

https://doi.org/10.1051/bioconf/20225200073

Laminaria. Fucoid seaweed substrates which float at the sea surface because of their air-filled vesicles can be used for cleaning of oil spills in the sea [6]. Cultivation of *Laminaria* on an experimental trial plantation on the Murmansk coast and on an industrial plantation around Solovetsky islands has shown that seaweed farming can be profitable in the north.

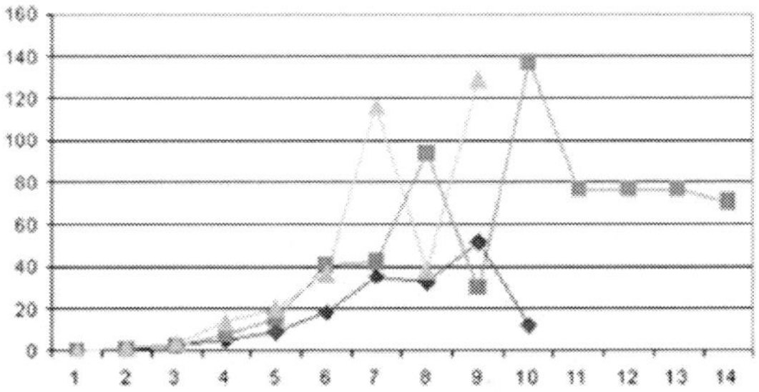

Fig. 5. Weight of thallus (W, g) in ontogenity of *Fucus vesiculosus* (n = 59), *F. disthus* (n = 49), *F. serratus* (n = 40) in semi-protected inlet, the y-axis – t, age, years.

4 Conclusion

The studies of thallus growth dynamics and species, age, and generative structure of fucoid populations can serve as a basis for assessing stocks and justifying quotas for the harvestl of commercial seaweeds from natural phytocenoses and aquaculture plantations in the coastal zone of the Barents and White Seas. The research results allow us to make the recommendations for the rational use of seaweeds and the restoration of phytocenoses disturbed during harvesting, as well as to develop and use a technology for synchronizing the process of harvesting of commercial object with the life cycle of algae, which will provide scientifically based quotas for production and restoration of seaweeds stocks in the littoral zone.

References

1. L. L. Kusnetsov, E. V. Shoshina, *Phytocenosis of the Barents Sea (Physiological and Structural Characteristics)* (Apatity: KSC, 2003)
2. H. Hop, C. Wiencke B. Vogel, N. A. Kovaltchuk, *Bot. Mar.* **55,** 399–414 (2012)
3. Wiencke C, Amsler C D 2012 *Ecol. Studies* **219** 265–291
4. N. Honkanen, V. Jormalainen, *Oecologia* **144,** 196–205 (2005)
5. G. M. Voskoboinikov, D. V. Pugovkin, *Vestnik MSTU* **15,** 716–721 (2012)
6. I. V. Ryzhik, M. V. Makarov, G. M. Voskoboinikov, *Russ. Mar. Biol.* **40,** 131–136 (2014)
7. V. I. Kapkov, E. V. Shoshina, O. A. Belenikina, *Vestnik MSTU* **19,** 286–295 (2016)
8. A. J. Smith, *Appl. Phycol.* **16,** 245–262 (2004)
9. M. Schmid, D. B. Stengel, *J. Phycol.* **51,** 25–36 (2015)
10. J. K. Kim, M. Stekoll, C. Yarish, *Phycol.* **58,** 446–461 (2019)
11. M. Y. Roleda, C. L. Hurd, *Phycol.* **58,** 552–562 (2019)
12. O. A. Pronina, *Fisheries* **5,** 4–7 (2002)
13. T. S. Peltichina *Laminarean Seaweeds of the Barents Sea and Rational Use* (Murmansk: PINRO, 2005)
14. E. I. Blinova, *Seaweeds and Seagrasses of the European part of Russia* (Moscow: VNIRO, 2007)

Marketing research on the study of consumer preferences for sweet dishes

E. E. Kharmich*, *I. E.* Brazhnaia, *O. M.* Kulik, *S. N.* Sudak, and *V. V.* Korchunov

Department of Food Production Technologies, Natural and Technological Institute, Murmansk State Technical University, 13, Sportivnaya Str., Murmansk 183010, Russian Federation

Abstract. The article presents the analysis research results of the confectionery mass market of production and consumption. A comparative characteristic of the confectionery products consumption by residents of Russia and Europe is carried out, the reasons for the differences in consumption are clarified. An analysis of the confectionery market in Russia in 2015–2019, coronavirus impact assessment on it, and a forecast for 2020–2024, according to "Businesstat" data, is presented. Based on the Russian State Policy goals in the field of healthy nutrition, the production priorities of the raw materials food main types that meet modern quality and safety requirements are determined, the introduction development vector of additional regional raw materials is determined, taking into account the of the region population traditions: berry raw materials, algae (kelp). When conducting marketing research, the methods of qualitative and quantitative analysis were used in this work. Marketing research was conducted using the questionnaire method, the sample size was 100 random respondents. The object of the study was consumer preferences when choosing sweet dishes. The survey examined consumer preferences in the category of sweet dishes: whether consumers of Murmansk region buy sweet dishes, where they buy them, what kind of sweet dishes they prefer, what possible additives would be acceptable and preferable for the consumer, what Murmansk-city and the region residents rely on when choosing and buying a sweet dish. Based on marketing research, the sweet dish producing feasibility that meets the potential consumers requirements and preferences was revealed. The consumers interest in this product type enriched with berries of the Kola Peninsula is established.

1 Introduction

The main State policy task in the healthy nutrition field is to develop technologies for functional products, taking into account the traditions of the region population, as well as using local regional raw materials. The Russian Federation population majority nutrition does not comply with the principles of proper and healthy nutrition as a result of the high foods rich in simple carbohydrates consumption. The consequence of this is the various diseases development, that reduce the life quality.

According to the conducted analyses and studies, the northern regions population nutrition is characterized by insufficient consumption of fresh vegetables and fruits, meat, fish and seafood, which leads to a lack of certain vitamins and bio-elements (such as selenium, calcium, and others) [1].

Since it is impossible to exclude the confectionery products and sweet dishes consumption by residents of Russia, as well as northern regions such as Murmansk and the Murmansk region, the development of confectionery products with functional properties enriched with local wild raw materials, with a rich macro-and microelement composition, is a timely and urgent task, based on the wishes of consumers and taking into account their preferences [2].

In the work [3], the authors found that the Russian Federation residents consume confectionery and sweet dishes in smaller quantities, in relation to the consumption of this goods group in some Western countries. At the same time, the bakery products consumption in Russia exceeds this figure in European countries by 49 %. These groups products consumption is influenced by the level of income of the population, traditions, and assortment. Sweet dishes and confectionery products do not belong to the basic necessities, and the demand for this group of goods always increases with the increase in the material well-being of the population [4].

Considering the confectionery market, it can be noted that it is very complex. It is influenced not only by general economic factors of macroeconomics, but also by the annually changing situation on the market of flour, sugar, fat-and-oil and dairy products.

The market supply of this products category also depends on the production and supply of nuts and dried fruits, thickeners and flavorings, preservatives and essences. And at the same time, very often fresh fruits and berries are replaced by analogues, additives identical to natural or artificial variations of fresh berries.

* Corresponding author: sulikova_e@mail.ru

There is also a huge range of confectionery products that are presented on the market. Over the past few years, the number of items has increased dramatically, while the quality has significantly decreased, due to the desire to reduce the cost of production and finished products to generate more profit. This was especially noticeable in relation to the quality of imported confectionery products.

The state policy goals in the healthy nutrition field are to preserve and strengthen the health of the population, to prevent diseases caused by inadequate and unbalanced nutrition. In accordance with this, one of the priority tasks in this area is the domestic production expansion of the food raw materials main types, that meet modern quality and safety requirements, and monitoring the state of population nutrition [5].

Using the possibility of introducing biologically full-fledged, balanced, rich in trace elements, and most importantly, natural and natural additives, such as gelatin, berry raw materials, algae (kelp), will help to reduce the calorie content and make confectionery products more correct in terms of composition, and useful for the consumer [6, 7, 8, 9].

The purpose of this work is to identify the preferences of the main population group of residents of the city of Murmansk and the Murmansk region.

To achieve this goal, the following tasks were formulated:

- to conduct a sociological survey of the Murmansk and the region population consumer preferences in relation to the sweet dish consumption;
- conduct a respondent's reports quantitative analysis of the to identify the sweet dish consumption specifics;
- to determine the possible interest of potential sweet dishes consumers after the developed sweet dish introduction into mass production using raw materials from the North-West region.

2 Materials and methods

To conduct market research, we used the collecting primary data method during a survey of random respondents. To collect primary data, a special questionnaire was developed to study the problem [10].

In this work, the methods of qualitative and quantitative research were used. Marketing research was conducted using the questionnaire method [11], the sample size was 100 random respondents. The object of the study was consumer preferences when choosing sweet dishes.

3 Results and discussion

According to "Businessstat" estimates, in 2015-2019, sales of confectionery products in Russia grew annually. In 2019, 3.6 million tons of confectionery products were sold in the country, which is 11.7% higher than in 2015. The confectionery market growth in 2015–2019 was promoted by a number of factors, including the retail chains development, the confectionery product line expansion, and the strengthening of manufacturers and retailers marketing campaigns [12].

The respondent's survey was conducted in the age group from 18 to 50 years, the purchasing power is different (with low, middle and above-average income).

Initially, to identify a potential and interested audience, respondents were asked the question *Do you buy sweet dishes?* A pie chart of the responses is shown in Figure 1.

Fig. 1. Percentage of respondents' responses to the question: *Do you buy sweet dishes?*

Analyzing the responses received, it turned out that the vast majority of respondents – 80% – answered *positively* and 20% of respondents answered *negatively*. Thus, the relevance of developing a sweet dish is obvious, but more data is needed to understand a more complete picture of consumer preferences.

In the course of studying the priority places for purchasing sweet dishes, respondents were asked the question *Where do you buy sweet dishes?* An analysis of the responses showed that 55% of the respondents surveyed said that they buy sweet dishes in retail stores. At the same time, 18% – in public catering enterprises and various confectionery and bakeries, 5% – in cooking departments and 4% – in cooking stores. Figure 2 shows a pie chart of respondents' responses.

Thus, the maximum number of respondents buy sweet dishes in retail outlets, so the technology recipes development for sweet dishes, adapted to this type of implementation is in demand.

Further, when studying the preferences for the assortment of sweet dishes, the question was asked *What do you prefer to buy sweet dishes?* Figure 3 shows a pie chart of respondent's responses.

Analyzing the results presented in Figure 3, it was found that 34% of respondents prefer to buy desserts based on cream, 29% – sweet gelled dishes, 19 % - just a sweet dessert, at least 18% of respondents choose a dessert based on butter cream.

It can be concluded that when developing new technologies, it is advisable to pay attention to the technology of cream-based desserts, as well as to pay attention to gelatin-based desserts. Desserts based on buttercream are interesting to a small group of consumers, as in their opinion, these products do not meet the requirements of modern healthy nutrition concepts.

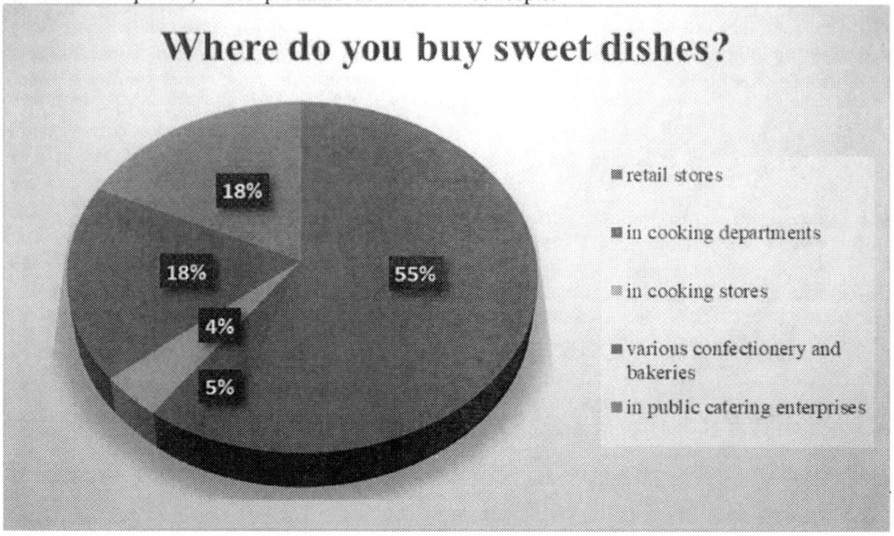

Fig. 2. Percentage of respondents' responses to the question *Where do you buy sweet dishes?*

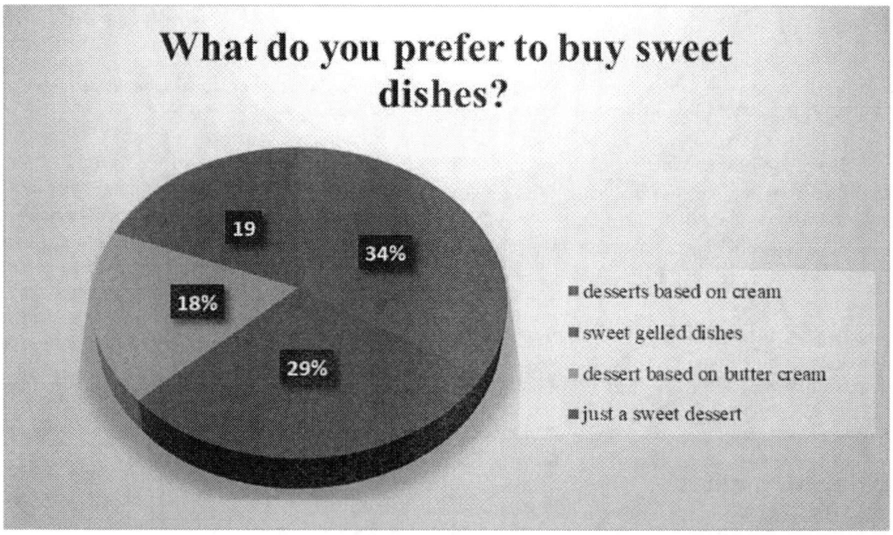

Fig. 3. Percentage of respondents' responses to the question *What sweet dishes do you prefer to buy?*

After analyzing consumer preferences for additional ingredients that are part of desserts and sweet dishes, respondents were asked the question *What do you prefer to buy a sweet dessert with?* Analyzing the answers, we found that 47% of respondents prefer to buy sweet dishes with fresh berries and fruits, 26% – with nuts, 15% – with canned fruits and berries, 6% – with seeds, cereals and dried fruits. Figure 4 presents a respondents' responses pie chart.

Based on the data obtained, it can be concluded that the development of technology for preparing sweet dishes with the use of fresh berries, the richest in vitamin composition, and nuts, rich in vegetable fats, as additional ingredients to improve the taste or as a decoration of dishes, is relevant, since most of the respondents understand the significant advantage of this type of additives in comparison with ingredients that have undergone heat treatment or drying.

To understand a more complete picture of the potential consumers preferences, the respondents were asked the question *What is your first priority when choosing a sweet dish?* The analysis of the respondents' responses showed that preferences were divided equally between the indicators: *price*, *taste*, *quality*, *health benefits and immunity*. A pie chart of respondent's responses is shown in Figure 5.

https://doi.org/10.1051/bioconf/20225200074

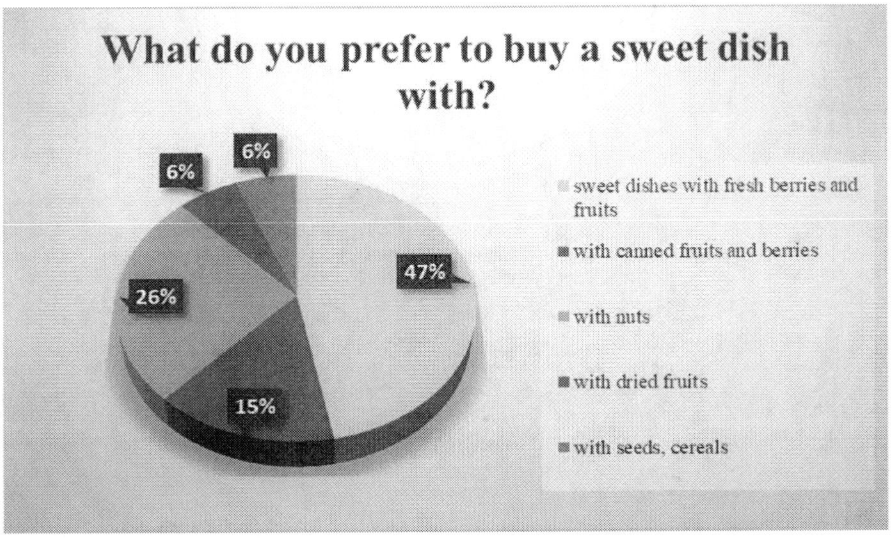

Fig. 4. Percentage of respondents' responses to the question *What do you prefer to buy a sweet dish with?*

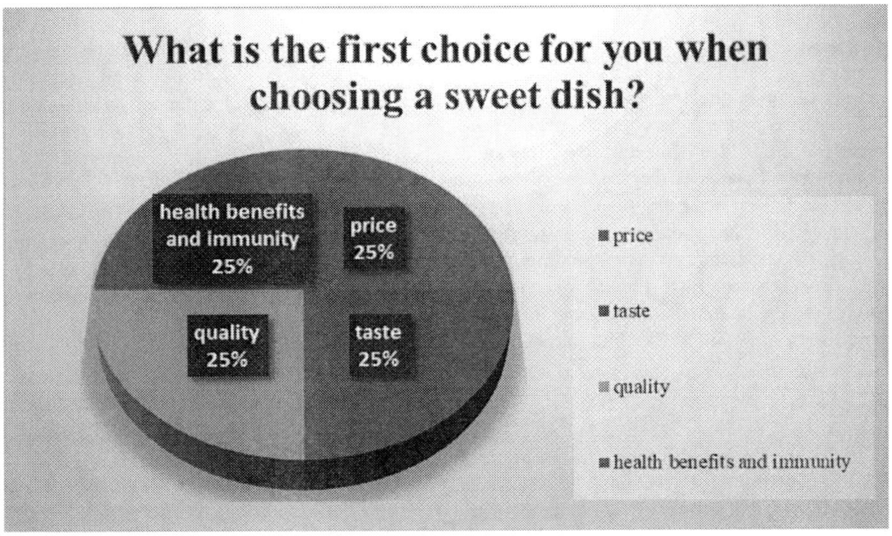

Fig. 5. Percentage of respondents' responses to the question *What is the first choice for you when choosing a sweet dish?*

Based on the data presented in Figure 5, it is revealed that consumers, when choosing a sweet dish, are based not only on monetary costs, but also important aspects are organoleptic indicators, and the impact on the health of the respondents themselves and their family members.

Also, an important aspect for the research and development of the sweet dish preparing technology was the expediency of using rather unconventional regional raw materials in sweet dishes and desserts [13]. To analyze this area, the respondents were asked the question *Would you buy a sweet dish with the addition of berries from the Kola Peninsula?* The analysis of the responses showed that 87% of the respondents responded positively and 13% – negatively. Figure 6 shows a pie chart of respondent's responses.

Based on the data presented in Figure 6, it can be concluded that potential consumers are interested in the appearance of a sweet dish with regional berries on the market, knowing about the microelement value and vitamin benefits of this raw material [14, 15].

Fig. 6. Percentage of respondents' responses to the *question Would you buy a sweet dish with berries from the Kola Peninsula?*

4 Conclusion

Thus, as a result of the conducted research and analysis of the respondent's responses, the confectionery market and the offers for 2020–2024 forecast, the interest of Murmansk and the Murmansk region consumers in purchasing sweet dishes using North-Western region berries was revealed. The value of these marketing research is high, as it gives an understanding of the feasibility of further sweet dish *Cranberry chocolate cream* creature, which includes milk cream, cranberry and cloudberry berries, as well as hazelnut kernels.

References

1. N. I. Didenko, V. A. Davydenko, E. R. Magaril, G. F. Romashkina, D. F. Skripnuk, S. V. Kulik, The nutrition and health status of residents of the Northern Regions of Russia: Outlook of vertical agricultural farms, *Public Health* **18,** doi: 10.3390/ijerph18020414 (2021)
2. V. F. Korolkov, V. V. Bragie Design and manufacture of products based on consumer expectations, *Standards and quality* **11,** 64–65 (2003)
3. A. V. Ryzhakova, O. A. Babina Global confectionery market, *International trade and trade policy* **4 (12),** 59–74 (2017)
4. I. A. Kuznetsov, A. S. Nikolashin, Socio-economic state of the regional confectionery market in the period of import substitution policy, *Economics and Business: theory and practice* **11,** 90–95 (2017)
5. Fundamentals of the state policy of the Russian Federation in the field of healthy nutrition of the population for the period up to 2020 (2010)
6. E. I. Muratova, S. I. Dvoreckii, P. M. Smolikhina, Optimization of the composition of confectionery products, *Vestnik TGTU* **1,** 130–140 (2015)
7. S. Vislouhova, A. Shevchuk, Confectionery products of the new generation, *Science and Innovation* **171,** 30–33 (2017)
8. A. S. Zaharova, L. A. Kozubaeva, E. Y. Egorova, Flour confectionery products with lingonberries, *Polzunov's Bulletin* **4,** 17–20 2019
9. Z. A. Kanarskaya, F. K. Khuzin, A. R. Ivleva, V. M. Gematdinova, Trends in the development of confectionery technology, *Vestnik VGUIT* **3 (69),** 195–204 (2016)
10. E. P. Golubkov, *Marketing Research: Theory, Methodology and Practice* (Moscow: Finpress, 1998)
11. A. G. Bondarenko, *Sociological research*: *Survey Methodology* (Volgograd: VolgGTY, 2006)
12. Analysis of the confectionery market in Russia in 2015–2019, assessment of the impact of coronavirus and forecast for 2020–2024
13. E. G. Turshuk, *Development of Valuable Food Products with The Addition Of Medicinal Natural Raw Materials Of The Far North And Their Commodity Characteristics* (Murmansk: Murmansk State Technical University, 2008)
14. E. E. Kharmich, E. G. Turshuk, Research on the properties of cranberries of the Kola Peninsula for the development of sweet dishes, *Collection of reports of the regional scientific and practical conference "Key factors of development of the*

https://doi.org/10.1051/bioconf/20225200074

Northern territories of Russia" (Murmansk, PINRO named N. M. Knipovich) pp 120–123 (2018)

15. E. E. Kharmich, E. G. Turshuk, Study of the chemical composition and vitamin activity of dishes based on cranberry berries of the Far North, *Collection of reports of the regional scientific and practical conference "Key factors of development of the Northern territories of Russia" (Murmansk, Murmansk PINRO named N. M. Knipovich)* pp 221–233 (2019)

Sanitary and ecological assessment of fishery water-use objects in the Arctic

M. Litvinova*, *E.* Makarevich, *E.* Alekseevskaya, *A.* Garbul, and *E.* Kozhukhova

Microbiology and Biochemistry Department, Murmansk State Technical University, 13, Sportivnaya Str., Murmansk 183010, Russian Federation

Abstract. This work is devoted to the study of sanitary and hygienic monitoring of fishery water use objects in Murmansk region. The paper presents assessment of the ecological and sanitary conditions of fishery reservoirs in the Arctic region of the Kola and Tuloma rivers for the period 2016-2018 based on the use of microbiological, hydrochemical and parasitological indicators, as well as the integral index of surface water pollution. Hydrochemical studies consisted of determining the biochemical oxygen consumption, the mass content of NH4+, NO2-, NO3-, iron, and dissolved O2. The assessment of the river waters sanitary state was carried out by means of microbiological and parasitological methods. The microbiological parameters were used to determine the number of common and thermotolerant coliform bacteria, coliphages, and the presence of intestinal infections pathogens. The water of rivers does not meet the hygienic standards in terms of hydrochemical indicators in seventeen percent and in terms of microbiological indicators in five percent of cases. According to parasitological indicators, hygienic standards meet the requirements. In five points of water sampling, the water of the Tuloma and Kola rivers was evaluated as pure according to the integral hydrochemical index value and belongs to the second category of quality. In general, the high-water quality of Murmansk region rivers was noted. In conclusion, it should be emphasized that systematic monitoring of water supply sources is necessary to obtain an idea of the indicators' variability and to assess the rivers pollution degree.

1 Introduction

One of the most important and relevant environmental issues nowadays is the assessment of the water resources state. More than 15 thousand large and small rivers flow on the Kola Peninsula. Many of these reservoirs are of commercial importance and are included in the fisheries fund.

Natural sources of fresh water experience a constant anthropogenic load, for this reason, it is important to monitor the water quality. Pollution of water bodies occurs directly as a result of the runoff water discharge and indirectly through runoff from the catchment area and groundwater. Hydrochemical and sanitary-microbiological observations are part of monitoring studies of aquatic ecosystems, as they allow us to assess the state of the aquatic environment and its ability to naturally purify itself from most pollutants, including those of anthropogenic origin. All these types of pollution affect the quality of hydrobionts bred and extracted in the reservoir.

The purpose of this work was to assess the ecological and sanitary state of the Tuloma and Kola rivers by means of hydrochemical and microbiological indicators for 2016–2018. When assessing the ecological state of the reservoir, both basic (the content of nitrites, nitrates, biochemical oxygen consumption (BOC_5), the number of common and thermotolerant bacteria, coliphages) and integral (water pollution index) indicators of water quality were used.

Main research objectives:

– to study the spatial and temporal variability and water quality by hydrochemical, microbiological and parasitological indicators of river water masses, as well as to assess the correlation of these indicators;
– to evaluate the ecological state of rivers based on the water pollution index (WPI).

2 Materials and methods

The objects of the study are two rivers of the Kola Peninsula – Kola and Tuloma, as they are located in its densely populated region.

The Tuloma River is one of the major rivers of the Kola Peninsula. The river is 64 km long. The source of the river is located at the outlet of Notozero Lake, it flows into the Kola Bay. There are five villages and two hydroelectric power stations located on the Tuloma; they form reservoirs. Fishing is well developed.

The Kola River (the name originates from the Finno-Ugric word "fish river") is 83 km. The source of the river starts at the Kolozero Lake and flows into the Kola

* Corresponding author: mlit1@rambler.ru

Bay. The largest settlement on the river is the city of Kola [1].

Water sampling was carried out at several points (stations) on both rivers. Sampling points on the Kola River are: station 1 (st. 1) - Molochniy urban-type settlement; station 2 (st. 2) Shonguy settlement; station 3 (st. 3) Magnetity settlement; station 4 (st. 4) Loparskaya station. Sampling points on the Tuloma River: station 5 (st. 5) Tuloma village and station 6 (st.6) the territory of the agricultural enterprise "Tuloma". The sampling points are shown in the diagram (Figure 1).

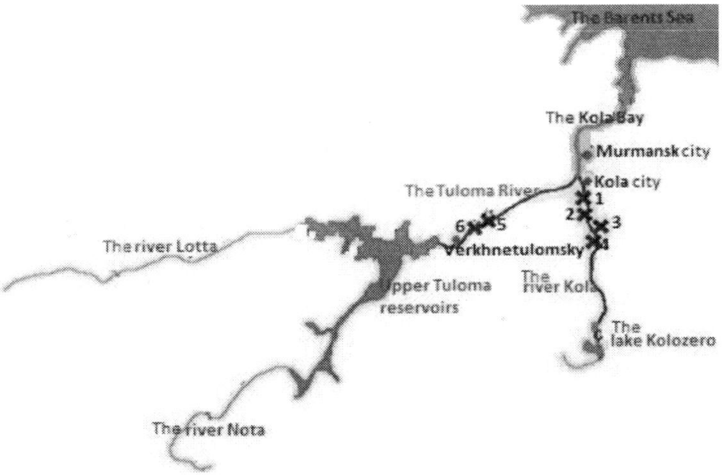

Fig. 1. Sampling scheme on the Tuloma and Kola rivers.

According to microbiological indicators, water samples were taken monthly, according to hydrochemical and parasitological indicators, samples were taken once a quarter from January 2016 to December 2018. From microbiological indicators, river water was evaluated by the number of common, thermotolerant coliform bacteria (CCB and TCB), coliphages, as well as the presence of intestinal infections pathogens (salmonella). Microbiological and parasitological studies were carried out according to the guidelines 4.2.1884-04 [2].

The dissolved O_2, the biochemical oxygen consumption (BOC_5), the amount of nitrites, nitrates, ammonium nitrogen and iron were determined by means of the hygienic standards for chemical parameters.

The results obtained were compared with the standards defined in the sanitary rules and norms 2.1.5.980-00. Statistical data processing was performed using the Statistica 12.0 program. In the correlation analysis, the Spearman rank correlation coefficient was used.

Among other things, the hydrochemical index of water pollution (IWP) was determined in the work. It is the average fraction of the excess of the maximum permissible concentration (MPC) for a strictly limited number of individual components; it is calculated by the formula (1):

$$IWP = \frac{1}{n} \cdot \sum_{i=1}^{n} \frac{C_i}{MPC_i} = \frac{1}{6} \cdot \sum_{i=1}^{6} \frac{C_i}{MPC_{6i}} \quad (1)$$

where n – is the number of indicators used to calculate the index; Ci – is the chemical concentration in water, mg/l; ПДКi (MPC) – maximum permissible concentration of the substance in water, mg/l.

To calculate this index, it is mandatory to use such indicators as dissolved oxygen and BOC5, as well as four other indicators. Such indicators for the Kola and Tuloma rivers were: iron (total), nitrites, nitrates, and ammonium ion [3].

3 Results and discussion

According to the regulatory documentation for open water bodies, parasitic organisms are not allowed in 25 liters. In the period from 2016 to 2018, no helminth eggs, no teniid oncospheres or viable cysts of pathogenic intestinal protozoa were found in any of the selected samples, in addition, the water does not contain intestinal infections pathogens.

Water samples for coliphages determination were carried out monthly. Coliphages are bacterial viruses capable of lysing *E. coli* and forming lysis zones (plaques) on agar [4]. Their amount should not exceed 10 PFU/100 ml. During the period under review, no coliphages were found in the Kola River. In the Tuloma River, coliphages were found at the water intake point near the village of Tuloma in April 2018, their number was 1 PFU in 100 ml. At the water intake point of the agricultural enterprise "Tuloma", coliphages were found in January 2017 (3 PFU in 100 ml) and in August 2018 (4 PFU in 100 ml). This indicator in the fisheries water is within the normal range.

Table 1. Results of microbiological and hydrochemical parameters measurements in water samples of the Kola and Tuloma rivers from 2016 to 2018 (average values).

Sampling station number	Microbiological indicators			Hydrochemical parameters					
	CCB, CFU/100 ml	TCB CFU/100 ml	Coliphages PFU/100 ml	Iron, mg/l	Ammonium nitrogen, mg/l	Nitrates, mg/l	Nitrites, mg/l	Dissolved oxygen, mg/l	BOD_5 mg\l
The Kola River									
st. 1	80.0± 3.2	68 ± 2.7	0	1.63± 0.06	0.25± 0.009	0.51±0.03	0.01± 0.0004	8.32±0.33	1.9±0.076
st. 2	87.2± 3.3	22.1 ±0.8	0	0.19± 0.007	0.24± 0.009	0.78±0.01	0.01± 0.0004	8.77±0.31	1.8±0.070
st. 3	17.6± 0.67	17.6 ±0.79	0	0.17± 0.006	0.19± 0.007	2.06±0.07	0.02± 0.0008	9.11±0.34	1.3±0.06
st. 4	21.5± 0.73	21.5 ±0.67	0	0.37± 0.012	0.20± 0.007	1.79±0.061	0.01± 0.0002	8.68±0.29	1.4±0.05
The Tuloma river									
st. 5	15.6± 0.59	15.6 ±0.47	0.02± 0.001	0.16± 0.006	0.21± 0.008	0.80±0.03	0.01± 0.0001	7.1±0.27	1.6±0.07
st. 6	22.1± 0.82	21.7± 0.803	0.19± 0.007	0.07± 0.002	0.18± 0.006	0.29±0.11	0.01± 0.0003	7.4±0.27	1.4±0.07

CCB is an integral indicator of the fecal contamination degree; it acts as a reservoirs water quality assessment indicator and has indicator properties in relation to intestinal infections caused by pathogens [5]. According to the norms established by Sanpin (Sanitary Consumer Supervision) 2.1.5.980-00, the amount of CCB should not exceed 1000 CFU/100 ml. The results obtained from the study of samples from the Kola and Tuloma Rivers are shown in Figure 2a. In February 2017, in Molochniy village, a violation of the norms for this indicator was recorded, the number of CCB was 1058 CFU/100 ml. Also, the excess of permissible norms was noted in August 2018 in the village of Shongui – 2400 CFU/100 ml. In the Tuloma River, the excess of the norm in the number of CCB was not observed.

It is recommended to determine the TCB at the same time as the CCB to confirm the fecal origin of the contamination [5]. The standard for the amount of TCB is not more than 100 CFU/100 ml. According to the data presented in Figure 2b, in Molochniy village, excess of the norm for this indicator was noted three times: in February 2016 — 758 CFU/100 ml, in April 2018 – 133 CFU/100 ml, in July 2019 – 190 CFU/100 ml. In November 2018, there was an excess of the indicator at the Loparskaya station — 188 CFU/100 ml.

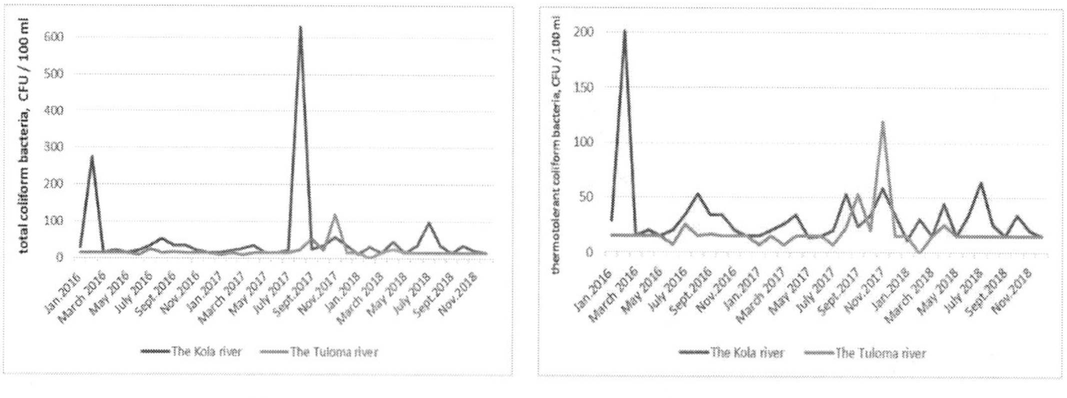

(a) (b)

Fig. 2. Changes in the quantitative content of total coliform bacteria (a) and thermotolerant coliform bacteria (b) in the water of the Kola and Tuloma rivers from year 2016 to 2018 (average values).

According to the standards established in Hygiene standards 2.1.5.1315-03, the iron content should not exceed 0.3 mg/l. It is believed that when this concentration is exceeded, the water gets a glandular taste, and serious health problems begin when the indicator exceeds 2 mg/l. Violation of the MPC of iron in the Kola River was noted at the Loparskaya station in 2016 in the first quarter — 1 mg/l and in the second quarter – 0.6 mg/l. In the first quarter of 2019, the iron content at the Loparskaya station was 0.4 mg/l. In Molochniy village, the maximum permissible concentration of iron was exceeded in 2019: in the second quarter – 0.35 mg/l and in the third – 0.34 mg/l. In Magnetity village, a violation of the norms was recorded in the second quarter of year 2018 – 0.4 mg/l. During the period under review, a violation of the iron MPC was recorded on the Tuloma River in the 3rd quarter of year 2016 with the water intake in Tuloma village – 0.65 mg/l. The soil and rocks of Murmansk region are rich in iron compounds. Therefore, the norms of this indicator are often exceeded.

The presence of ammonium ions in surface waters is mainly associated with the biochemical decomposition of protein substances, urea and amino acids deamination processes. Due to the process of anaerobic regain of nitrates and nitrites, ammonium ions can be formed, in addition, hydrobionts release ammonia in the process of vital activity. During the study period, there were no violations of the ammonium nitrogen content in the Kola and Tuloma rivers.

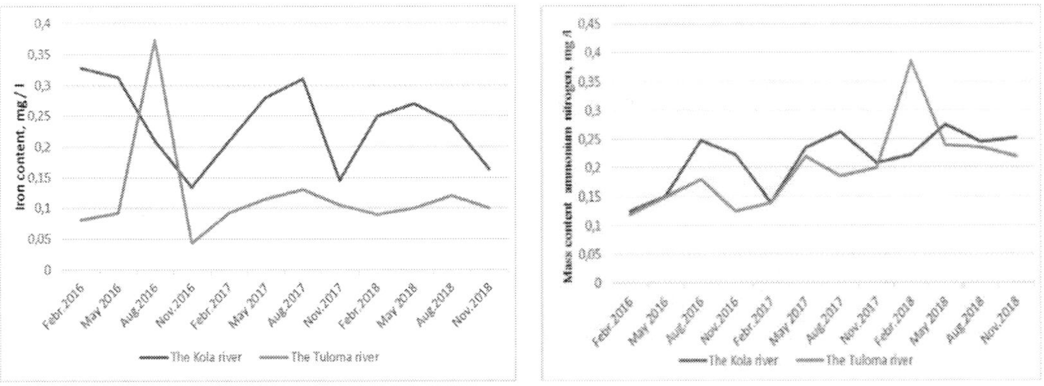

(a) (b)

Fig. 3. Changes in the mass content of iron (a) and ammonium nitrogen (b) in the water of the Kola and Tuloma rivers from 2016 to 2018 (average values).

According to the standards established in Hygiene standards 2.1.5.1315-03, the content of nitrates should not exceed 40 mg/l, and of nitrites should not exceed 3.3 mg/l. The concentrations of these indicators in the water corresponded to the hygienic standards and the water was possible to be used in households and for drinking, according to these indicators.

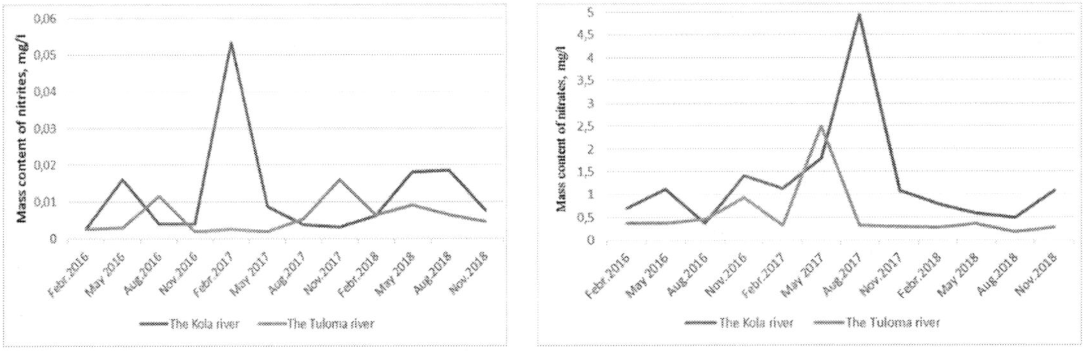

(a) (b)

Figure 4. Changes in the mass content of nitrites (a) and nitrates (b) in the water of the Kola and Tuloma rivers from year 2016 to 2018 (average values).

The oxygen regime determines the chemical and biological state of the water body and affects the vital activity of the reservoir. Oxygen for surface waters is absorbed from the atmosphere, produced as a result of aquatic organisms' photosynthetic activity, as well as obtained with rain and snow waters. The loss of oxygen in the reservoir is determined by the processes of substances oxidation, and hydrobionts respiration. Biochemical oxygen consumption (BOC_5) – the amount of oxygen consumed for aerobic biochemical oxidation influenced by microorganisms and decomposition of unstable organic compounds contained in the tested water. BOC is an important criterion used to determine the level to which the reservoir is polluted with organic substances. It determines the amount of easily oxidizing organic pollutants in the water [6]. The results obtained in terms of dissolved oxygen and BOC_5 meet the requirements set out in the regulatory document Hygiene standards 2.1.5.1315-03 [7].

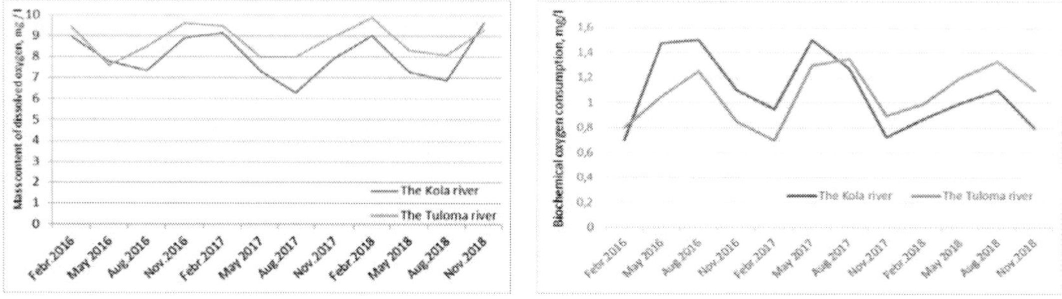

(a) (b)

Fig. 5. Changes in the mass content of dissolved oxygen (a) and BOC_5 (b) in the water of the Kola and Tuloma rivers from years 2016 to 2018 (average values).

The results of studies of the water quality of the Kola and Tuloma rivers on microbiological, hydrochemical and parasitological indicators are presented in Table 2.

Table 2. Comparative analysis of the results obtained.

Name	Number of samples	Number of research	The number of results					
			Microbiological		Hydrochemical		Parasitological	
			Abs.	%	Abs.	%	Abs.	%
The Kola river	144	768	6	4,17	6	12,5	—	—
The Tuloma river	72	384	1	1,39	1	4,17	—	—
Note: the "—" symbol means the absence of results that do not correspond to the regulatory and technical documentation								

It's a common knowledge, the Spearman rank correlation method allows to determine the closeness and direction of the correlation between two features [8]. Let's check the tightness of the data connection in pairs between the available features (BOC_5, concentration of oxygen, iron, nitrites, nitrates, ammonium nitrogen, CCB and TCB). Spearman's rank correlation coefficient conditionally evaluates the closeness of the correlation between the features (Table 3). The correlation analysis of the links between the values of hydrochemical and microbiological parameters in river water showed that a significant direct correlation occurs between the parameters of the CCB and TCB (Table 3). The calculated value of the correlation coefficient turned out to be 0.94 – this conditionally indicates a direct and strong dependence on CCB and TCB content. There is a moderate correlation between the values of dissolved oxygen and ammonium nitrogen, BOC_5, and oxygen concentration. The calculated values of the correlation coefficient were –0.66 and –0.69, respectively. They indicate an inverse and moderate correlation. The feedback is the opposite correlation between the two variables, i.e., as the BOC_5 increases, the amount of dissolved oxygen in the river water decreases.

No significant correlations were found between the remaining concentrations of hydrochemical and microbiological parameters.

The Water Pollution Index indicator is useful for monitoring the dynamics of water quality pollution changes and determining its quality class (Table 4). The calculation Water Pollution Index was carried out according to the formula 1.

Table 3. Values of correlation coefficients between microbiological and hydrochemical parameters in lake waters.

Indicator	Nitrites, mg/l	Nitrates, mg/l	Ammo-nium nitrogen, mg/l	Dissolved oxygen, mg/l	BOC_5, mg/l	Iron, mg/l	CCB, KOE/100 ml
	n = 48	n = 48	n = 48	n = 48	n = 48	n = 48	n = 48
Nitrites, mg/l	–	–0.18	0.22	–0.11	0.36	0.34	–0.02
Nitrates, mg/l	–0.18	–	–0.24	0.14	–0.11	–0.19	0.17
Ammonium nitrogen, mg/l	0.22	–0.37	–	**–0.66**	0.23	0.28	0.33
Dissolved oxygen, mg/l	–0.11	0.14	–0.36	–	**–0.69**	–0.28	0.07
BOC_5, mg/l	0.19	–0.11	0.23	**–0.69**	–	0.38	0.17
Iron, mg/l	0.24	–0.19	0.29	–0.28	0.31	–	0.0004
CCB,	–0.03	0.17	0.48	0.07	0.17	0.0004	–
CFU/100 ml	0.17	–0.01	0.22	0.08	0.17	0.0006	**0.94**

Table 4. Calculated values of WPI of the Kola and Tuloma rivers' waters.

Sampling points	WPI calculated values	Water quality status	Water quality class
Molochniy village	1.12	Moderately polluted	3
Shonguy village	0.32	Pure	2
Magnetity village	0.31	Pure	2
Loparskaya station	0.41	Pure	2
Tuloma village	0.36	Pure	2
'Tuloma' Agricultural enterprise	0.30	Pure	2

4 Conclusion

Thus, as a result of the conducted studies on microbiological indicators for the period (2016-2018) in the Kola River, 4.17 % of the samples taken do not meet the standards. In the Tuloma River, the number of violations is 1.39 %. According to hydrochemical indicators for the period under review, 12.5% of water samples taken in the Kola River do not meet the established standards. In the Tuloma River, the proportion of violations is 4.17% out of the number of samples taken. According to parasitological indicators, the quality of the Kola and Tuloma rivers met the sanitary and epidemiological requirements. The water of the Kola and Tuloma rivers for the three-year period from 2016 to 2018 is characterized by the hydrochemical index of water pollution as clean at the points of water intake near the settlements of Shonguy, Magnetity, Tuloma, Loparskaya station and as moderately polluted at the point of water intake near Molochniy village and belongs to the 2 and 3 classes of water quality, respectively.

References

1. Yu. Elshin, *Resources of Surface Waters in the USSR: Hydrological Studies: Kola Peninsula* (Leningrad: Gidrometeoizdat press, 1969)
2. MUK 4.2.1884-04 2004 Sanitary-microbiological and sanitary-parasitological analysis of surface water bodies guidelines (Moscow: Federal Center for State Sanitary and Epidemiological Supervision of the Ministry of Health of Russia)
3. O. Reshetnyak, *Methods of Assessing the Quality of Surface Waters of the Land: Textbook* (Rostov-on-Don: Southern Federal University Press Publish House, 2017)
4. SanPiN 2.1.5.980–00 2000 Hygienic requirements for surface water protection. Sanitary rules and regulations (Moscow: Federal Center for State Sanitary and Epidemiological Supervision of the Ministry of Health of Russia)
5. O. Andreeva, Studies of drinking water by safety indicators, *Federal Service for Supervision of Consumer Rights Protection and Human Welfare in Omsk region* **6**, 54–58 (2015)
6. Methodological guidelines 4.2.2959-11 2011 Methods of sanitary-microbiological and sanitary-parasitological analysis of coastal waters of the seas in places where the water is used by the population (Moscow: Federal Center for State Sanitary and Epidemiological Supervision of the Ministry of Health of Russia)
7. GN 2.1.5.1315-03 2003 Maximum permissibleconcentration (MPC) of chemicals in

https://doi.org/10.1051/bioconf/20225200075

the water of water bodies for drinking, cultural and domestic water use. (Moscow: Publishing House of Standards)

8. M. Litvinova, et al. Assessment of the state of the open water in water use point of the population the Murmansk region, *KnE Life Sciences* **2020,** 405–412 (2020)

Physical and chemical environmental factors' influence on the proteolytic potential of probiotic microorganisms' individual representatives

E. V. Makarevich*, *E. V.* Kozhukhova, *D. S.* Zaychikova, *M. U.* Litvinova, and *O. G.* Krivenko

Department of Microbiology and Biochemistry, Institute of Nature and Technology, Murmansk State Technical University, 13, Sportivnaya Str., Murmansk 183010, Russian Federation

Abstract. This work is devoted to the study of sanitary and hygienic monitoring of fishery water use objects in Murmansk region. The paper presents assessment of the ecological and sanitary conditions of fishery reservoirs in the Arctic region of the Kola and Tuloma rivers for the period 2016-2018 based on the use of microbiological, hydrochemical and parasitological indicators, as well as the integral index of surface water pollution. Hydrochemical studies consisted of determining the biochemical oxygen consumption, the mass content of NH4+, NO2-, NO3-, iron, and dissolved O2. The assessment of the river waters sanitary state was carried out by means of microbiological and parasitological methods. The microbiological parameters were used to determine the number of common and thermotolerant coliform bacteria, coliphages, and the presence of intestinal infections pathogens. The water of rivers does not meet the hygienic standards in terms of hydrochemical indicators in seventeen percent and in terms of microbiological indicators in five percent of cases. According to parasitological indicators, hygienic standards meet the requirements. In five points of water sampling, the water of the Tuloma and Kola rivers was evaluated as pure according to the integral hydrochemical index value and belongs to the second category of quality. In general, the high-water quality of Murmansk region rivers was noted. In conclusion, it should be emphasized that systematic monitoring of water supply sources is necessary to obtain an idea of the indicators' variability and to assess the rivers pollution degree.

1 Introduction

Protease activity, along with other technologically significant properties of probiotic bacteria, is of considerable interest. First of all, from the point of view of evaluating the potential use of these bacteria in the technology of processing protein-containing raw materials, including hydrobionts. This fact allows not only to reduce the duration of protein-containing raw materials maturation due to the acceleration of physical and biochemical processes, but also to obtain a finished product with functional properties [1,4]. In addition, the resulting free amino acids serve as a source of other nitrogenous compounds that contribute to improving the quality of organoleptic parameters of finished products [2, 3].

The aim of our study was to assess the physical and chemical factors' influence on the proteins hydrolysis intensity of the hydrobionts' muscle tissue sarcoplasmic fraction. Another aim was to study the dynamics of probiotic microorganisms' cultures' proteolytic activity.

The proteolytic potential of the selected research objects (*Lactobacillus acidophilus, Streptococcus thermophilus* and *Propionibacterium freudenreichii*), which have high biological and multifunctional activity, was considered as a possibility of using these bacteria to obtain protein-containing products with increased biological value.

2 Materials and methods

Studying the protease activity of *Streptococcus thermophilus*, *Lactobacillus acidophilus* and *Propionibacterium freudenreichii*, the protein sarcoplasmic fraction of cod (*Gadus morhua*) muscle tissue was used as a protein source for cultivation. In the extract containing the sarcoplasmic fraction, the optical density of the solutions was measured on the SF-2000 spectrophotometer at wavelengths of 260 and 280 nm in three sequences. The concentration of tyrosine formed during the substrate protein components' enzymatic hydrolysis after the addition of a suspension of microorganisms to them was determined spectrophotometrically [5].

During the study, changes in tyrosine levels in culture fluids were determined at different incubation temperatures, pH, and NaCl concentrations after 3, 6, 9, 12, and 24 hours. The amount of tyrosine in the extract was previously measured without exposure or the addition of microorganisms. Further, by comparing the results of bacterial suspensions' enzymatic activity in substrates with different physical and chemical

* Corresponding author: makarevichev@mdtu.edu.ru

characteristics, the influence of incubation conditions and duration on the intensity and direction of metabolic processes was evaluated.

3 Results and discussion

The change in tyrosine concentration occurs as a result of two processes — protein proteolysis, and its assimilation by the bacterial community. The prevalence of the former can be judged by an increase in the concentration of tyrosine in comparison with its initial level. The decrease in the saturation of the culture fluid with tyrosine indicated its withdrawal due to the intensification of assimilation processes. Thus, to assess the enzymatic activity of the culture of living microorganisms, we used the relative change in the level of tyrosine in comparison with its primary values. This method of expressing the results allows us to assess the predominance of processes (dissimilation/assimilation) rather than to measure a specific enzymatic activity [5]. From the point of view of assessing the practically significant properties of bacteria, this methodological approach is informative.

To assess the effect of the fermentation temperature on the change in the tyrosine concentration in the culture fluids, three temperature conditions were selected: 10, 20 and 37°C. These incubation modes correspond to the optimal, minimum and average growth temperatures of the selected cultures of microorganisms. The results of optical density measurements in samples after incubation of *Streptococcus thermophilus*, *Lactobacillus acidophilus* and *Propionibacterium freudenreichii* cultures for 24 hours at three temperatures are shown in Figure 1.

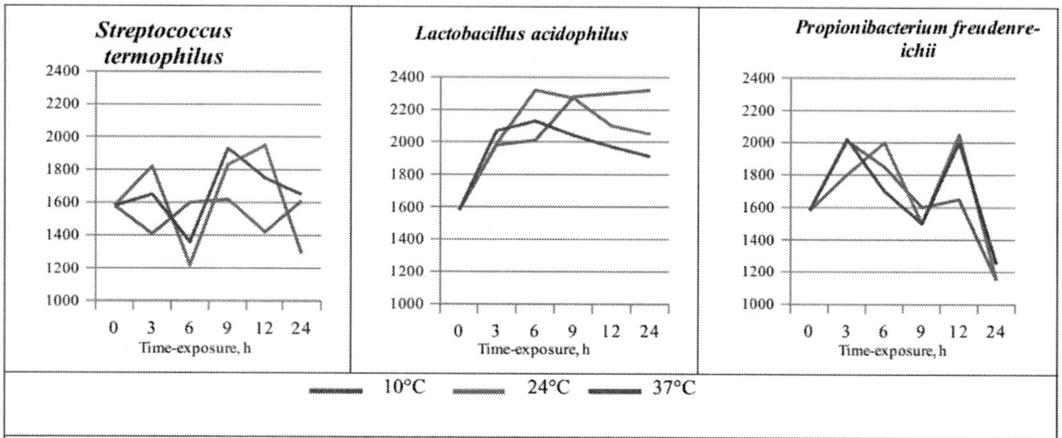

Fig. 1. Dynamics of tyrosine concentration changes in samples, depending on the incubation temperature.

On average, the amount of tyrosine in the substrate containing *Streptococcus thermophilus* culture during incubation ranged from 20–35 %. The predominance of assimilation processes of nitrogen-containing substances was observed at a temperature of 10°C, as evidenced by minimal changes in the tyrosine level. Its level in this case did not exceed the initial value for the entire period of exposure. The minimum marker index was observed at a temperature of 24°C by the 24th hour of the experiment, the concentration of the indicator amino acid decreased by 24%, and the concentration fluctuations ranged from 26 to 35 %.

During the *Lactobacillus acidophilus* cultivation, it was noted that the measured indicator varied in the range of 21–43 %, while the processes of tyrosine release into the environment prevailed over the processes of its consumption. The maximum was recorded by the end of incubation at a temperature of 10°C. The total level of tyrosine increased by 45–52%, depending on the cultivation conditions.

For the *Propionibacterium freudenreichii* culture, the highest levels were observed after 3 and 12 hours (20% higher than the initial concentration). A decrease in the tyrosine concentration was observed by 9 and 24 hours of exposure (25—30 %, depending on temperature conditions).

Temperature conditions can affect the enzymatic reactions, slowing or accelerating them; such a temperature effect is observed in all physico-chemical reactions. It is known that there is a correlation between the rates of enzymatic reactions and temperature [6].

In general, the experiment showed that at a temperature of 10°C, there were less pronounced fluctuations in the tyrosine level over time than at other temperature conditions. This can be attributed to a slowdown in the total enzymatic processes, a decrease in the negative influence of their own metabolites on bacterial community development, and, as a result, the predominance of assimilation processes. With an increase in temperature conditions, significant changes in the concentration of tyrosine in culture fluids were observed due to an increase in the rate of enzymatic multidirectional reactions. A comparative analysis of changes in tyrosine levels in the samples showed that the *Lactobacillus acidophilus* culture had a more pronounced hydrolytic activity against sarcoplasmic fraction proteins at all incubation temperature conditions than other studied objects. *Streptococcus thermophilus* bacteria were characterized by minimal proteolytic activity when incubated at 10°C.

To assess the hydrogen index effect on the metabolic processes of the studied objects, the most significant, from a technological point of view, pH levels were selected – 4.5; 7 and 8. The assessment of the medium hydrogen index influence on the change in the concentration of tyrosine during fermentation is shown in Figures 2–4.

Enzymes, like any protein substances, are sensitive to the hydrogen index of the medium. The pH level determines the ionization of the functional groups in the enzyme's protein molecule. Thus, hydrogen ions affect the active center of the protein molecule [7, 8].

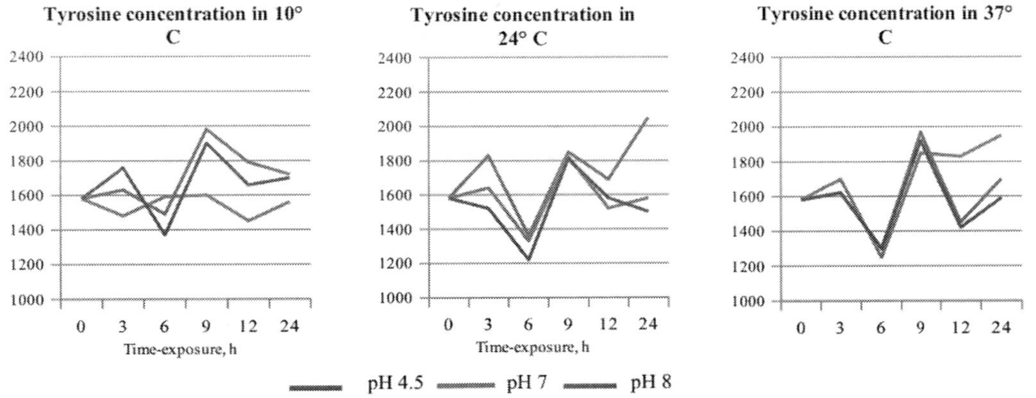

Fig. 2. Dynamics of changes in the tyrosine concentration in samples during incubation of *Streptococcus thermophilus* culture depending on the pH value.

When assessing the pH effect on the enzymatic activity of *Streptococcus thermophilus* culture, it was shown that changes in the tyrosine concentration in the substrate were more significant in a medium with a hydrogen ions concentration, other than neutral. The maximum level of tyrosine was observed in an acidic environment at the 9th hour of sample exposure at temperatures of 10°C and 37°C, at the 24th hour – at 24°C. The concentration in comparison with the initial value increased by 25 and 30%, respectively. In the neutral environment, the predominance of hydrolytic processes reached a maximum after 24 hours of incubation at 37°C. The difference between the maximum and the initial level was 35%.

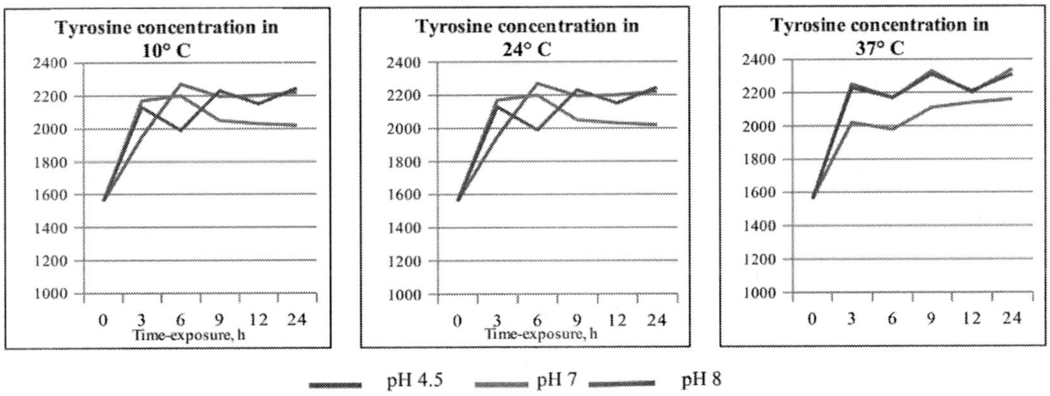

Fig. 3. Dynamics of changes in the tyrosine concentration in samples during incubation of *Lactobacillus acidophilus* culture, depending on the pH value.

The change in the tyrosine concentration during *Lactobacillus acidophilus* cultivation was characterized by an increase in the indicator during the entire incubation period. The concentration increased in the range from 21 to 48.5 %. In acidic and alkaline media, the tyrosine level was 5–10% higher compared to this indicator during cultivation at a pH value of 7. The maximum was at pH 4.5 after 24 hours of fermentation at 37°C (48.5% higher than the initial level).

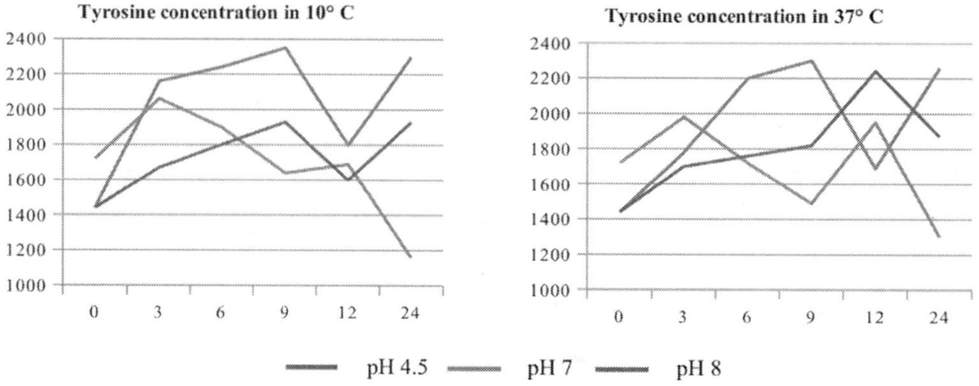

Fig. 4. Dynamics of changes in the tyrosine concentration in samples during *Propionibacterium freudenreichii* culture incubation depending on the pH value.

The tyrosine concentration during *Propionibacterium freudenreichii* cultivation varied in a wider range in an alkaline pH medium. It was noted that in a neutral environment, the difference in tyrosine concentration between the initial and maximum values did not exceed 15%. At pH 4.5, the amount of the marker after a slight increase (no more than 10%) significantly decreased (by 30-50 %).

At different levels of the hydrogen index in the reaction medium, the active center can be ionized weaker or stronger, more or less shielded by neighboring groups. The concentration of hydrogen ions affects the state of the enzyme (its protein part), determining the level of the ratio of anionic and cationic groups, which affects the tertiary structure. During the formation of the tertiary structure, the active enzyme centers are formed. In addition, the medium pH affects the degree of ionization of the reaction products, the substrate, and the enzyme-substrate complex. In general, it can be argued that for all the studied cultures at a pH of 4.5, the predominance of hydrolytic processes was observed, and for assimilation ones, a neutral medium was required.

The evaluation of the effect NaCl has on the metabolic activity of the studied objects was carried out in the culture medium, where salt concentrations were established at the levels of 2%, 4% and 8%. These incubation modes corresponded to the parameters of the expected technological modes of microorganisms' application (Figures 5–7).

NaCl has a direct effect on enzymatic processes. At certain concentrations, salt causes the destruction of the hydrate shell of protein-enzymes, which contributes to a decrease in their activity or a temporary loss of native properties. The functional properties of enzymes are due to the hydration of the protein molecule, in which the proteins are dissolved in water and aqueous solutions. The result is the formation of an electrocharged water layer around the protein, which includes water molecules oriented in relation to the protein molecule in a strictly defined way. The hydrate shell (an electrically charged water layer) prevents the aggregation of protein particles and contributes to the stability of the protein solution. The greater the probability of salting out is, the higher the mass NaCl fraction is [9].

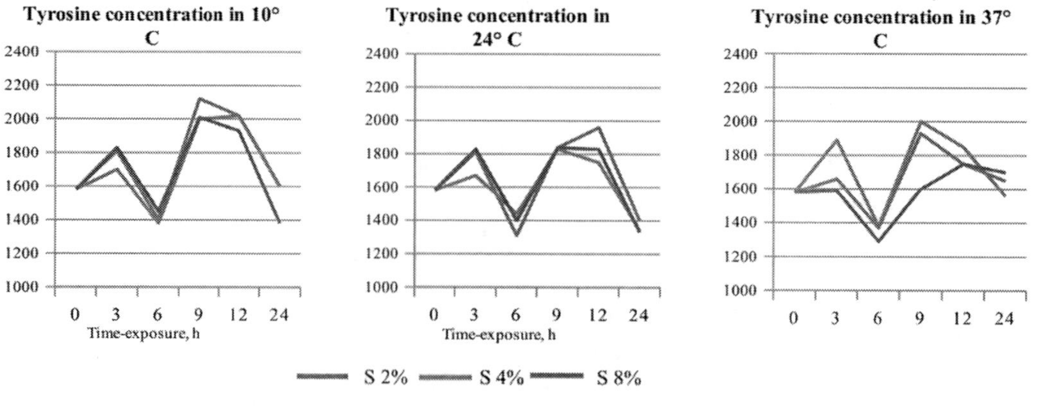

Fig. 5. Dynamics of changes in the tyrosine concentration in samples during incubation of *Streptococcus thermophilus* culture depending on the mass NaCl fraction in the substrate.

 https://doi.org/10.1051/bioconf/20225200076

It was noted that the tyrosine content in the substrate with the NaCl addition in the selected concentrations was less affected by the direction and intensity of the metabolic processes of the studied cultures of microorganisms in comparison with other environmental factors (temperature and pH).

For *Streptococcus thermophilus* cultures, it was noted that the fluctuations in the tyrosine concentration at different NaCl concentrations in the medium had a general trend, but this indicator changed less widely during cultivation in comparison with the control sample (the difference between the initial and minimum did not exceed 20%.

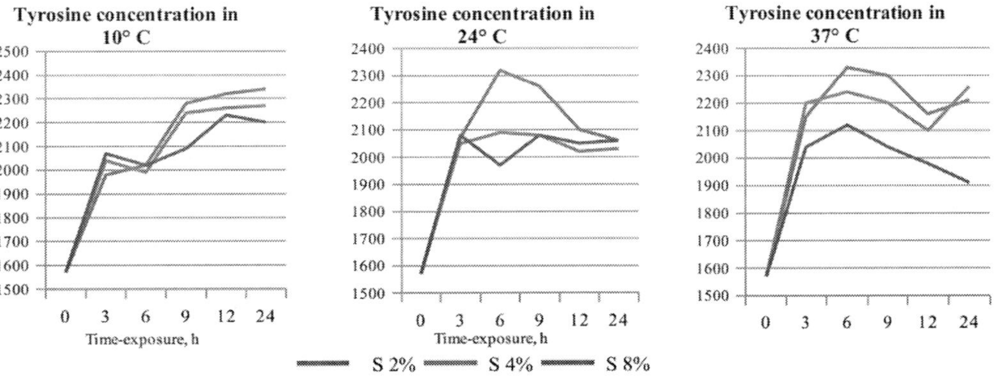

Fig. 6. Dynamics of changes in the tyrosine concentration in samples during incubation of *Lactobacillus acidophilus* culture depending on the mass NaCl fraction of in the substrate.

A comparison of tyrosine levels in *Lactobacillus acidophilus* bacteria cultivation showed that an increase in the mass NaCl fraction in the substrate as a whole led to a decrease in the intensity of multidirectional enzymatic activity of microorganisms.

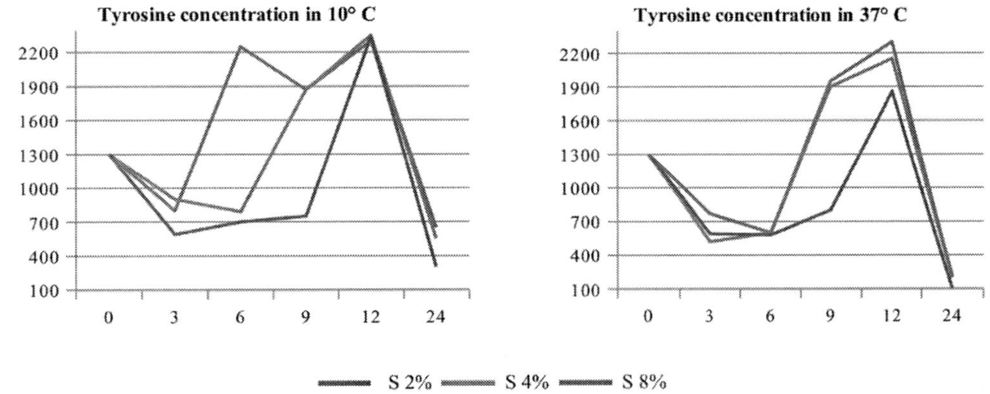

Fig. 7. Dynamics of changes in the tyrosine concentration in samples during incubation of *Propionibacterium freudenreichii* culture depending on the mass NaCl fraction.

In the presence of NaCl, the increase in the tyrosine concentration in the culture fluid of *Propionibacterium freudenreichii* occurred much later than in the control. In a medium with a mass fraction of 2% NaCl, the maximum tyrosine level was observed after 6 hours of cultivation at a temperature of 10 °C, and after 12 hours for substrates with 4 and 8% salt at 37° C. The response of the enzyme systems of the *Propionibacterium freudenreichii* bacteria was expressed in the lengthening of the period of hydrolytic processes predominance onset.

It follows from this that the less intensive course of multidirectional enzymatic processes in the presence of NaCl is determined by the influence of salts of alkaline and alkaline earth metals on both enzymes and individual subcellular structures.

Thus, the experiments carried out in the framework of this work confirm the ability of the selected lactic acid cultures to proteolysis, which allows to consider them as potential probiotic cultures for the production of protein-containing products from aquatic biological resources. The introduction of proteolytic lactic acid bacteria will speed up the physico-chemical and biochemical processes and reduce the duration of protein-containing raw materials maturation.

4 Conclusion

Lactobacillus acidophilus cultures were characterized by the predominance of proteolytic processes, regardless of the physico-chemical conditions of cultivation, for the other studied strains, the alternation of proteolysis and assimilation is inherent.

Streptococcus thermophilus bacteria had minimal proteolytic activity.

It is shown that the incubation temperature was a factor determining the direction of metabolic processes in general. At a temperature of 10° C, less pronounced fluctuations in the level of tyrosine were observed, which indicated a slowdown in the total enzymatic processes and led to the predominance of assimilation processes over hydrolytic ones.

When the pH of the medium changed to the acidic side, the predominance of hydrolytic processes was observed during the cultivation of all the objects of the study, under conditions of neutral hydrogen index of the substrate, assimilation processes prevailed, characterized by a decrease in the level of tyrosine in the samples.

The general reaction of all the studied cultures of microorganisms to an increase in the mass fraction of NaCl in the substrate was to reduce the total intensity of the multidirectional enzymatic activity of microorganisms. *Propionibacterium freudenreichii* cultures were characterized by growth in the time of tyrosine concentration increase in the medium.

References

1. L. V. Kaprelyants, Probiotic properties and biotechnological potential of propionic acid bacteria, *Microbiology and biotechnology* **1**, 6–15 (2017)
2. I. S. Khamagaeva, Study of probiotic properties of the combined starter, *Technique and technology of food production* **1**, 1–5 (2013)
3. K. Piwowarek, *Propionibacterium spp.* – source of propionic acid, vitamin B12, and other metabolites important for the industry, *Applied Microbiology and Biotechnology* **102 (2)**, 515–538 (2018)
4. D. V. Makarevich, On the possibility of using the proteolytic activity of probiotic microorganisms in certain technologies of water bioresources processing, *Intern. Scientific and Pract. Conf. Modern Ecol., Biolog. and Chem. Research, engineering and production technol.* (2018)
5. E. V. Kramarenko, Assessment of Propionibacterium metabolic activity on protein substrate of sarcoplasmic fraction in hydrobionts' muscle tissue, *Intern. Scientific and Pract. Conf. Development of Biological Resources and Rational Use of Natural Resources* pp 230–238 (2020)
6. N. V. Kharchenko, Isolation of bifidobacteria and study of their probiotic properties during long-term storage (2016)
7. Y A. Kirillova, Study of protease activity of *Streptococcus thermophilus* against proteins of the sarcoplasmic fraction *Intern. Scientific and Pract. Conf. Modern Ecol., Biolog. and Chem. Research, engineering and production technol.* (2018)
8. I. S. Khamagaeva, Investigation of acid stress in propionic acid bacteria, *Bulletin of the ESSUTM* **6 (57)**, 5–8 (2015)
9. M. A. Uskova, *Study of the properties of probiotic lactic acid bacteria as biologically active components of food*: PhD dissertation (Research Institute of Nutrition Russian Academy of Medical Sciences, 2010)

Growth rate of Fucales on the Murman Coast, the Barents Sea

S. S. Malavenda[1] * and *S V* Malavenda[2]

[1] Murmansk State Technical University, 13, Sportivnaya Str., Murmansk 183010, Russian Federation
[2] Murmansk Marine Biological Institute of the Russian Academy of Sciences, 17, Vladimirskaya Str., Murmansk, 183010, Russian Federation

Abstract. Fucales are among the key algae on the coast of the Northern Hemisphere and are being actively studied, although minor attention is paid for the age structure, since there are no reliable signs for determining the age of the algae. Direct observations on the thallus growth on the coast evidence on a variation of this trait under different conditions. We assess the rate of linear growth and the number of formed dichotomous branching per year in *Fucus vesiculosus* and Fucus distichus on the Murman Coast of the Barents Sea. These parameters differ in plants sampled from different biotopes. The annual growth of *F. vesiculosus* is two times higher and more dichotomous branches are formed per year in the environment characterized by low water movement intensity and pronounced salinity fluctuations than in plants growing on a weakly protected area of the coast. In the Barents Sea, the branching rate and the annual growth in *F. distichus* and *F. vesiculosus* depend on the intensity of water movement and the salinity regime of a particular biotope. The number of formed branches depends also on the number of already formed dichotomies, i.e., on the physiological age of thallus. It is necessary to modify the equation for calculating the age of these species using the correction factors for salinity, the intensity of water movement, and the number of dichotomies.

1 Introduction

In the Barents Sea, and throughout the Russian Arctic, the largest stock of Fucales is concentrated along the Murman Coast. The biomass of these brown algae may reach up to 20 kg/m2, and the total stock in the Barents Sea are estimated as almost 100 thousand tons of wet weight (Malavenda, 2018). On the one hand, Fucales are commercial species, on the other, it is a key component of coastal ecosystems, supplying a huge amount of organic matter as primary producer, serving as a substrate that increases the bottom surface area several times, and being the bioremediating species. The Fucales communities attract much scientific interest since the very beginning of their studying.

All physiological, population, and partly ecological studies take into account the age of Fucales. It is generally accepted that two dichotomous branches are formed per year in *Fucus vesiculosus* L. and *F. distichus* L. on average, regardless of the characteristics of a particular biotope [2, 3, 4]. However, these data have been obtained for the White Sea populatons, and their application to the Barents Sea populations is still questionable. In addition, the formation of branches and of the air bubbles in F. *vesiculosus* at the coast of the Barents Sea depends on abiotic environmental factors (illumination, littoral level, etc.) [5]. This conclusion has been made on the basis of observations of these plants during the first two years of life. In the population of *F. distichus subsp. evanescens*, inhabiting the Russian Far East, up to six branches can form in the first year of life, then the branching rate and linear growth of plants is significantly reduced [6]. Therefore, this issue may be considered controversial, and so it requires targeted study.

The study aims to search for the effect of the degree of protection of the biotope and its salinity regime on the rate of formation of dichotomous branching and on the linear growth in the populations of *Fucus vesiculosus* and *F. distichus* inhabiting the Murman Coast of the Barents Sea. We have tested the hypotheses if two branches are formed per year and if the growth rate may be considered constant.

2 Materials and methods

2.1 Study area

The growth rate of Fucales was studied mainly in vicinity of the Dalnie Zelentsy Biological Station of the Murmansk Marine Biological Institute, in particular, in Yarnyshnaya, Zelenetskaya, and Shelpinskaya bays (figure 1). This area is the most accessible as one of the pristine biotopes of the Murman Coast of the Barents Sea and is the most studied. The studies have been also performed in the Pechenga Bay of the western Murman Coast to check if the obtained data may be used for extrapolation to the entire Murman Coast (figure 1). The study has been carried out from 2008 to 2014.

* Corresponding author: msergmstu@yandex.ru

 https://doi.org/10.1051/bioconf/20225200077

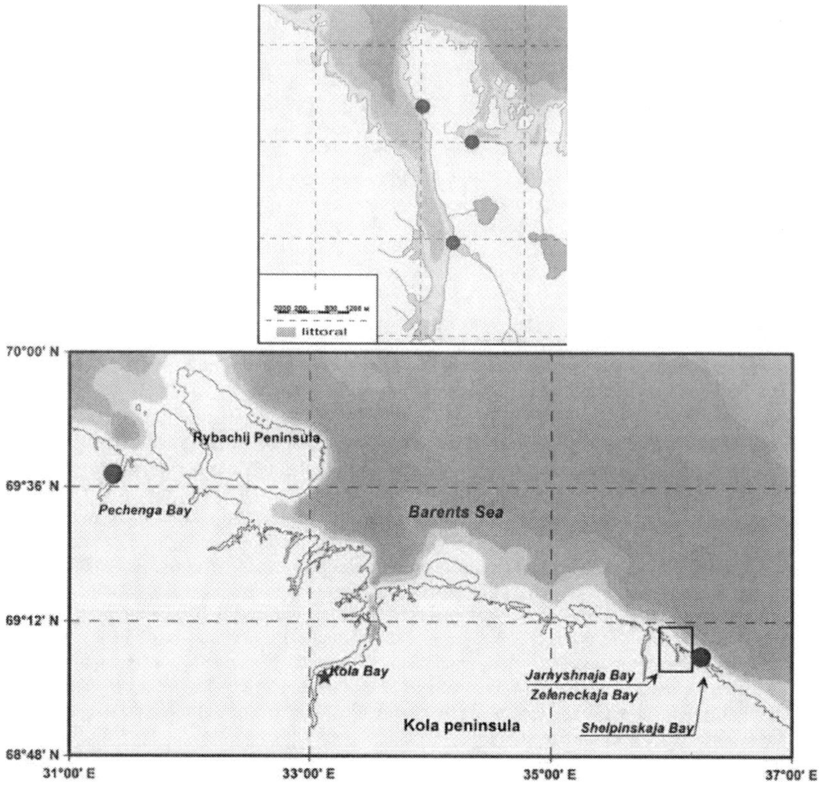

Fig. 1. Study area.

2.2 Hydrological observations

The intensity of water movement in algae thickets was estimated by the method of plaster balls [7, 8]. At each test site, at least twenty 30-g plaster balls with a diameter of 40 mm were exposed during the day at syzygy and quadrature tides. The water salinity was measured using a portable refractometer with an accuracy of 1‰. At each test site, several salinity measurements were carried out during syzygy and quadrature tides (both at high tide and low tide) in summer.

2.3 Labeling

Adult *Fucus vesiculosus* and *F. distichus* were labeled in the littoral zone in the first decade of August each study year. The label was a 2 × 2 cm cardboard with a number, pasted over with adhesive tape and tied with a cotton thread to the base of the thallus, more precisely to the first cylindrical elements. This method of labeling was tested by us earlier and provided a greater safety of the material in comparison with other methods. Labels were placed in the littoral areas with known values of the intensity of water movement and salinity fluctuations (table 1). In order to take into account, the effect of illumination, plants were labeled at the same levels of the littoral zone: *F. distichus*, at –0.6 m, *F. vesiculosus*, at 1.3 m. There were 3 to 10 dichotomous branches on the main branch of plant, i.e., these specimens were mature and able to reproduce; since the number of apexes was less than the number of receptacles in such algae, their growth potential was still high [9]. The length of the main plant axis and the number of branches were determined, and the apical parts of the plant were additionally labeled to determine their growth rate. In total, about 1 000 plants were tagged.

Starting from 2009, the labeled thalli were sampled and measured each year in the first decade of August. However, in all studied biotopes, not all of the noted thalli were found: *F. distichus* was found at all experimental sites, but *F. vesiculosus* was found only in the littoral zone of the Yarnishnaya Bay. In total, over the entire period of the experiment, after a year of life, we have found 399 labeled thalli. High rate of loss of materal due to breakage of thallus parts is a serious methodological obstacle to the widespread use of this method for determining the growth rate and the branching rate.

2.4 Statistics

The average number of formed branches and the average length increase were estimated for *F. vesiculosus* and *F. distichus* at each site. Variation of the trait was assessed by standard deviation. The role of physiological age and the influence of hydrological parameters were assessed using one-way ANOVA for each factor separately using PAST 3.0 software package. The calculations were performed in Excel (primary data) and in the PAST 3.0 software package (analysis of variance) [10, 11].

3 Results and Discussion

3.1 Hydrological regime

During the observation period, it was possible to identify the main features of the hydrological regime of the study sites. The data obtained formed the basis for factor analysis of the growth rate of Fucales (Table 1).

Table 1. Intensity of the water movement (IWM) and water salinity in summer at the study sites.

Study site	IWM, 10^{-3} mgCaSO$_4$/g*h	Salinity, ‰	
		High tide	Low tide
Bobrovaya Inlet, Yarnyshnaya Bay	7	30	5
Devkina Zavod' Inlet, Pechenga Bay	8	34	25
Oskara Inlet, Zelenetskaya Bay	10	34	25
Shel'pinskaya Bay (central part)	14	34	30
Cape Krasnaya Skala, Yarnyshnaya Bay	20	34	30

The hydrological regime at the study sites did not differ from that observed in 2004–2007 [12]. According to literature data, there was a slight increase in water temperature in the coastal regions of the Kola Peninsula [13]. In 1983–1984, the water temperature in the 0–15-m layer in January could drop below –1.5°C, in June-July, it was stably below 7°C and reached 9°C only at the end of August–September. In 2002–2012, the average surface water temperature was 3.8°C, in July–September, it exceeds 7°C in average, while in winter it did not fall below 0.7°C [13].

It is important that salinity and intensity of water movement correlate in the study area. A large stream flows to the apex of the Yarnyshnaya Bay, and a river flows to the Pechenga Bay, which causes a decrease in salinity from the mouth to the apex. The Zelenetskaya Bay is distinguished by its wide mouth, blocked off by a group of islands.

3.2 Branching

Our observations evidence that from three to four branches of the thallus are formed per year on average, but this parameter varies. There is a relationship between the number of newly formed dichotomies and the number of already existing dichotomies at the time of labeling (Figure 2). The maximum growth rate has been observed for *F. distichus* thalli carrying six already formed dichotomies: in average, 4 branches per year were formed; the plants with 3–5 existing branches formed somewhat less (3, less often 4) branches; the oldest thalli with 9–10 branches formed two new ones.

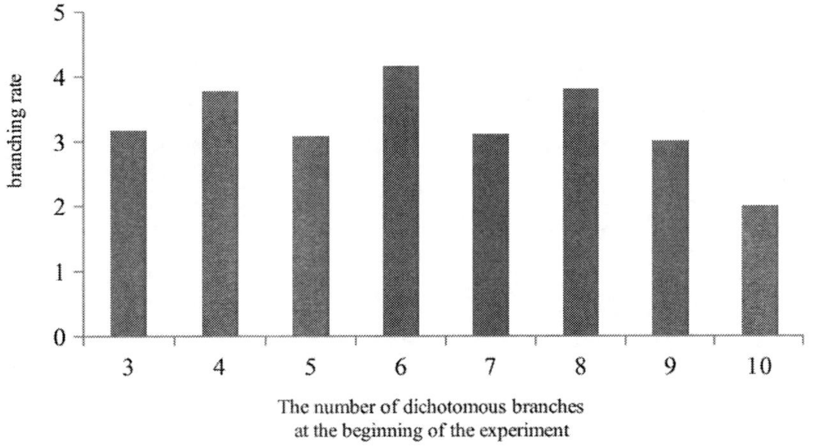

Fig. 2. Average number of branches formed per year in *F. distichus* in regard to the age (total sample).

The branching rates differ in plants from different biotopes (Figure 3). *F. distichus* forms 4–5 branches per year in the close bays, but only 2–3, in a biotope with high intensity of water movement; in *F. vesiculosus*, from 3 to 6 branches formed in the well-protected part of the coast, and from 1 to 4, in the weakly protected areas (Figure 3). Therefore, on average, twice as many dichotomies were formed in thalli of both species in the well-protected area of the littoral zone comparing to the weakly protected area, but the variation of this parameter was higher in *F. vesiculosus*.

Analysis of variance evidenced on a high dependence of the number of formed branches on the intensity of water movement and salinity (Table 2). Water salinity had a greater effect on the number of branches than the intensity of water movement did.

Table 2. ANOVA results.

The intensity of water movement (IWM)					
	Sum of squares	*df*	Mean square	*F*	*p*(same)
Between groups	361.79	39	9.27	6.33	3.101E-17
Total	658.99	199			
Omega²	**0.44**				
Levene's test for homogeneity of variance, based on means: *p*(same) = 1.127E-08					
Based on medians: *p* (same) = 0.05053					
Salinity					
	Sum of squares	*df*	Mean square	*F*	*p*(same)
Between groups	2968.37	39	76.11	44.88	1.057E-66
Total	658.99	199			
Omega²	**0.88**				
Levene's test for homogeneity of variance, based on means: *p*(same) = 1.088E-05					
Based on medians: *p*(same) = 0.5452					

Previously, it has been experimentally proven that constant fluctuations in salinity increased the rate of photosynthesis and enhanced the linear growth of Fucales in the Barents Sea. Periodic decrease in salinity to threshold values (5 ‰) contributes to an increase in the overall productivity of Fucales [14]. Such adaptations are key for Fucales. The intensity of water movement does not affect the salinity tolerance range of *F. vesiculosus* and *F. distichus*, but increases the time of survival in unfavorable conditions and contributes to the development of adaptive changes [15].

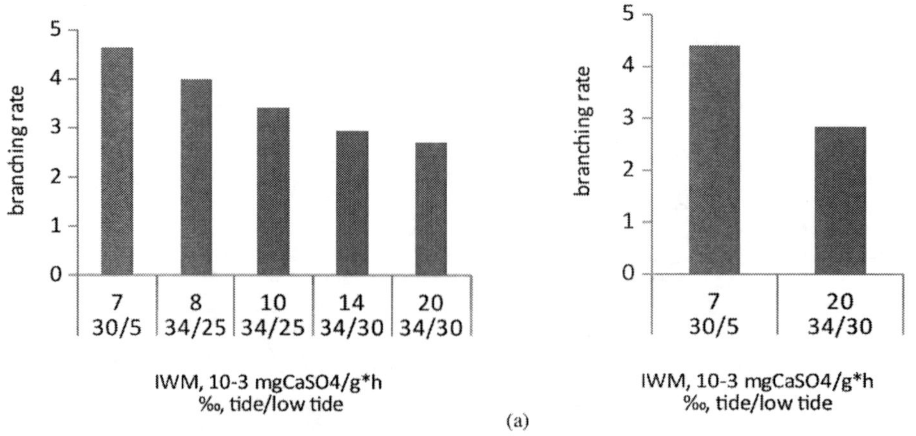

(a) (b)

Fig. 3. Branching rate, levels per year, in *F. distichus* (a) and *F. vesiculosus* (b) in the gradient of intensity of the water movement (IWM) and salinity (‰).

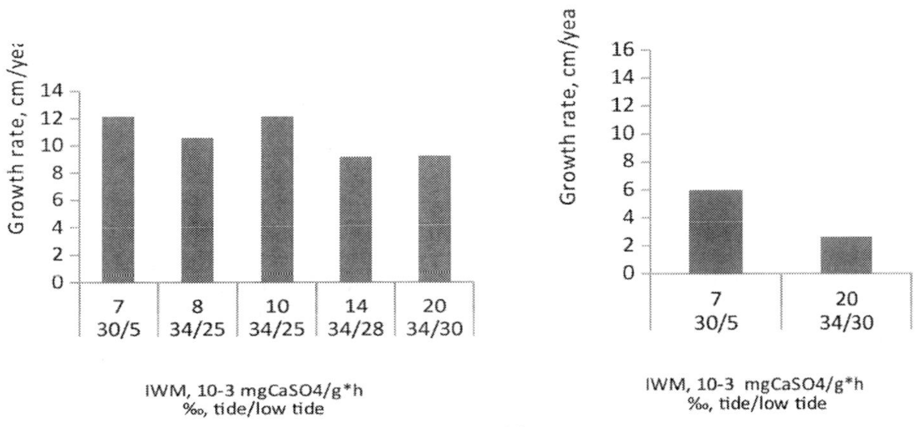

Fig. 4. Linear growth rate (cm/year) in *F. distichus* (a) and *F. vesiculosus* (b) in the gradient of intensity of the water movement (IWM) and salinity (‰).

3.3 Linear growth rate

In the plants from the weakly protected part of the bay, the average length of the element was on average twice as long comparing to the plants from the closed inlets (Figure 4). The annual increase in the thallus length of *F. distichus* in the Bobrovaya Inlet exceeds that in plants inhabiting the area nearby the Cape Krasnaya Skala. The length of one element and the number of elements seem to balance each other, and larger, branched thalli are formed. The significance and mechanism of this adaptation in *F. vesiculosus* have been described earlier [16]. The length of the elements of whole thalli of *F. distichus* also varies in different orders of branching and may vary significantly under different conditions of water exchange [17]. The average length of the thallus elements of *F. distichus* in the weakly protected area decreases from the basal to the apical part of the thallus, i.e., from parts that perform a supporting function to the parts that perform a photosynthetic function. In the protected inlets, the length of the elements is somewhat higher in the apical part of the plant, but it is practically the same in the basal and middle parts. This pattern is due to the increased relationship of particular thallus elements under conditions of reduced water exchange; s, called determinism.

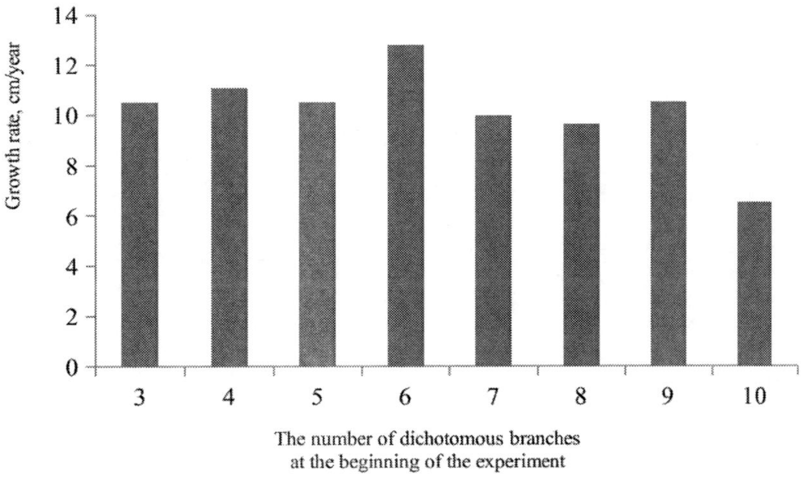

Fig. 5. Linear growth rate in *F. distichus* in regard to the age (total sample).

It was believed that the rate of linear growth in Fucales weakly depended on the number of branching and averaged 5 cm per year [3, 4]. However, the results of our experiment evidenced that the annual growth was from 7 to 13 cm per year, although this indicator was highly variable (Figure 5), depending on the biotope conditions as described above. As well as the branching rate, the maximum growth rate was observed in *F.*

distichus plants carrying six branches, then it decreased with age and reached a minimum in plants with ten dichotomies (7 cm per year).

The dichotomic branching rate in *F. distichus* depends on the age of the thallus; these results are in good agreement with the concept of the physiologically optimal age of this species. It has been found that the highest rate of photosynthesis is a characteristic of thalli with 8–9 branching; in addition, the loss of parts of thalli begins to prevail over their growth as the alga reaches this age [17].

Our data evidence that *F. distichus* and *F. vesiculosus* grow intensively during the first two years of life, when they form two or three dichotomies on average; in the third year, they are the peak of their growth rate, reaching maximum biomass, when 4 branches are formed (in total, there are 9 to 10 branches are formed on average). In the fourth year of life, both growth rate and branching rate slow down, as well as photosynthesis does, about two branches are formed, very few thalli reach the fifth year of life, and aging processes prevail. These simple calculations force us to look completely differently at the age structure of the population and the lifespan of Fucales.

The annual growth of *F. vesiculosus* is two times higher and more dichotomous branches are formed per year in the environment characterized by low water movement intensity and pronounced salinity fluctuations than in plants growing on a weakly protected area of the coast.

4 Conclusion

Therefore, in the Barents Sea, the branching rate and the annual growth in *Fucus distichus* and *F. vesiculosus* depends on the intensity of water movement and the salinity regime of a particular biotope. It is necessary to modify the equation for calculating the age of these species using the correction factors for salinity, the intensity of water movement, and the number of dichotomies.

Acknowledgments

The studies were reported according to the research plan of the Laboratory of Algology (MMBI, Kola Science Center RAS), on the topic Mechanisms of Adaptation, Regulation of Growth, Reproduction and Rational Use of Macroalgae of the Arctic Seas (State Task no. AAAA-A17-117052310082-8)

References

1. S. V. Malavenda, Stocks of Fucales in the Barents Sea: The level of knowledge and new data, *Materials of the international scientific-practical conference "Modern ecological, biological and chemical research, technology and production technology"* (Murmansk, April 25), pp 43–50 (2018)
2. V. V. Kuznetsov, *The White Sea and the Biological Characteristics of its Flora And Fauna* (Moscow: Publishing house of the Academy of Sciences of the USSR, 1960)
3. O. N. Maksimova, Some seasonal features of development and determination of the age of Fucales of the White Sea in book: *Bottom flora and production of the margin seas of the USSR* pp 73–78 (Moscow: Nauka, 1980)
4. V. B. Vozzhinskaya, *Bottom Macrophytes of the White Sea* (Moscow: Nauka, 1986)
5. Z. P. Tikhovskaya, The life cycles of Fucus vesiculosus on the coast of the Eastern Murman, *Proceedings of Murmansk Biological Station* **2,** 93–107 (1955)
6. A. N. Kashutin, A. V. Klimova, Dynamics of growth of Fucus distichus subsp. evanescens (Phaeophyceae, Fucales) in the Avacha Bay in 2017, *LBC* **20.1,** 77 (2018)
7. B. J. Muus, A field method for measuring "exposure" by means of plaster balls, *Sarsia* **34,** 61–68 (1968)
8. A. V. Moshchenko, Study of water movement in hydrobiological research of the bottom boundary layer using gypsum-plaster balls, *Russian Journal of Marine Biology* **26 (6),** 457–464 (2000)
9. Z. P. Tikhovskaya, Primary productivity of Fucales in the bays of the Eastern Murman, *Proceedings of Murmansk Biological Station* **1,** 164–189 (1948)
10. E. C. Pielou, *Mathematical Ecology* (New York: John Wiley & Sons, 1977)
11. Ø. Hammer, D. A. T. Harper, P. D. Ryan, PAST: Paleontological statistics software package for education and data analysis, *Palaeontologia Electronica* **4 (1)**, 9 (2001) Retrieved from: http://palaeo-electronica.org/2001_1/past/issue1_01.htm
12. S. V. Malavenda, G. M. Voskoboinikov, G. G. Matishov, The role of salinity and intensity of water flow in the formation of the population structure of Fucus vesiculosus L. (Phaetophyta) in the Barents Sea, *Biological Sciences* **413 (1)**, 137–139 (2007)
13. G.V. Il'in, D. V. Moiseev, D. V. Shirokolobov, A. A. Deryabin, L. G. Pavlova, Long-term dynamics of the hydrological conditions of Zelenetskaya Bay, East Murman, *Vestnik MSTU* **19,** 268–277 (2016)
14. S. V. Malavenda, Resistance of Fucales of the Barents Sea to variable salinity, *Botanical Journal* **96 (3)** 342–349 (2011)

https://doi.org/10.1051/bioconf/20225200077

15. S. V. Malavenda, S. S. Malavenda, Morphological adaptations of Fucus distichus L. (Phaeophyta) to salinity and water movement intensity, *Complex hydrobiological databases: Resources, technologies and use. Adaptation of aquatic organisms*, pp 195–198 (2005)
16. K. M. Khailov, V. P. Parchevsky, *Hierarchical Regulation of the Structure and Function of Marine Plants* (Kiev: Naukova Dumka, 1983)
17. S. S. Malavenda, *Morphophysiological features of the brown alga Fucus distichus L. in the ecosystems of the Barents Sea*: PhD thesis (Moscow: Lomonosov Moscow State University, 2009)

Technical and economic indicators calculation for rainbow trout juveniles growing nursery in Murmansk region

P. G. Priymak[1,*], *P. P.* Kravets[1], *D. S.* Boroukhin[2], and *A. I.* Lyutkevich[1]

[1] Murmansk State Technical University, 13, Sportivnaya Str., Murmansk 183010, Russian Federation
[2] Cherepovets State University, 5, Lunacharskogo Ave., Cherepovets, 162600, Russian Federation

Abstract. The paper describes characteristic features of rainbow trout fish seed material market in Murmansk region and in Russia, gives the description of the main elements of the organizational and production plans of the farm enterprise, presents the fish-breeding calculation of the proposed farm as well as the main technical and economic indicators. The urgency of creating a nursery for growing rainbow trout juveniles in Murmansk region is justified.

1 Introduction

Industrial aquaculture is of particular importance in the northern regions of our country, which have significant reserves of water and energy resources on their territories. Important prerequisites for this are the vulnerability of the ecosystems in these regions and the intensive use of some fish resources, leading to their depletion and the prospect of long-term recovery. Ensuring the food security of the northern regions of the Russian Federation and the country as a whole is based on the complexly integrated economic relations of the Russian Federation subjects with each other and with neighboring states.

The rainbow trout, *Oncorhynchus mykiss*, is one of the most common aquaculture objects. According to the FAO, in 2018, more than 35 thousand tons of rainbow trout were grown in the Russian Federation, which accounted for 12% of the rainbow trout production in Europe. The popularity of this type of trout as an object of aquaculture in Russia is due to the well-developed biotechnology of cultivation, consistently high demand from the buyer and other reasons. These prerequisites create a demand for fish planting material for part-time farms engaged in trout rearing.

Trout farms of Murmansk region mainly buy fish seed material (FSM) in other regions of the Russian Federation. The organization of the farm for the young fish breeding (fish hatchery) on Murmansk region territory is promising, as it will eliminate long-distance transportation and offer commercial farms young rainbow trout at affordable prices.

According to preliminary data from Rosrybolovstvo (Russian Fishery Ministry), in the first quarter of year 2020, the production of fish seed material in the Russian Federation increased by 29.7% and amounted to about 9.2 thousand tons.

The work is relevant due to the need to create an economy that can provide commercial trout producers with seed material at an affordable price on Murmansk region territory.

Objective is to calculate the main technical and economic indicators of a nursery establishment for rainbow trout juveniles breeding and to assess the relevance of this project implementation in Murmansk region. The research materials can be used in a real fish farm projecting.

2 Rainbow trout Oncorhynchus mykiss as an object of cultivation

Rainbow trout is native to the Pacific coast of North America. It is the most eurythermal among salmonids, characterized by a maximum latitudinal distribution.

The migratory form of *O. mykiss* spends the early part of its life in the river, growing and feeding in the sea or in the ocean, after which it returns to rivers and streams to spawn. The living form spends its entire life in fresh water [1]. Anadromous trout are known for their rapid growth and can reach a weight of 7–10 kg in three years, while lake trout can gain up to 4.5 kg in the same period.

2.1 Current state of trout farming

In Russia, in recent years, there has been a rapid growth in trout production, which is currently estimated at 33 thousand tons. Trout farming is carried out in Leningrad Region, Karelia, Altai Territory, Krasnoyarsk Territory, Krasnodar Territory (Adler trout farm), Stavropol Territory (Kislovodsk trout farm) and in other areas. Most trout farms have a small capacity of up to 10 tons of marketable products [2].

* Corresponding author: priymakpg@mstu.edu.ru

In the Russian Federation, there are breeding trout farms located in the south of the country, in particular, the Federal State Unitary Enterprise "Trout Breeding Plant 'Adler'" (Krasnodar Territory). Caviar is produced from September to March and is supplied to the farms of the Russian Federation and the CIS countries. "Adler trout" can be found in Irkutsk region, Krasnodar Territory, the Urals, Karelia, Kazakhstan and Belarus [3].

"'Forelevy'Agricultural Breeding Plant" JSC has been operating in Kislovodsk since 1935. In addition to caviar, the 'Forelevy' plant sells seed material, its cultivation volume is 800 thousand fingerlings and about 100 tons of commercial trout per year.

In general, according to the FAS, more than 30 economic entities in the Russian Federation produce and sell fish seed material of various breeds of rainbow trout. The leaders in the market of trout seed material are: "Trout Breeding Plant 'Adler'", Federal Selection and Genetic Center of Fish Farming "Ropsha", "'Forelevy' Agricultural breeding plant" JSC, "Virta" JSC, "Chistye Prudy" LLC. Fish farms "Lapland" LLC and "Primorskoe" LLC, located in Leningrad region also play an important role for Murmansk region, due to economic and geographical reasons.

2.2 Trout farming in Murmansk region

Trout farming in the Kola North has been going on for more than 30 years. Rainbow trout are imported to the north-western region from farms in the central and southern parts of Russia, as well as from Finland. The first cage farms were created in the warm waters of nuclear power plants, thermal power plants and in natural fresh water bodies in the 60–70 years of the last century. Their turnover was small and ranged from 5 to 10 tons.

In Murmansk region, the development of trout farming was due to the launch of the Kola nuclear power plant, which discharges heated water into Imandra Lake.

In the mid-80s of the twentieth century, the possibility of keeping trout in the coastal zone of the northern seas was proved [4,5]. Currently, this industry is developing quite intensively in the northern region, and commercial trout farms are being created in the coastal zone of the Barents and White Seas.

Fish production was constrained by the need for annual seed material import from other regions. Thanks to the creation of a brood trout herd with a capacity of 4 million spawn at the Imandra fish hatchery, 350 tons of trout were raised in 1991 [4]. In the early 90s, a new plant for the seed material production – the Arctic Salmon plant – was created in Murmansk region. The plant is located on the territory of Verkhnetulomskaya HPP, it contains a breeding and repair herd, incubates caviar and raises young trout. The farm uses warm water from the cooling system of HPP generators.

Other closest suppliers in the Russian Federation are located in Karelia, where several companies grow rainbow trout for sale: "Virta" JSC (part of "Kala-Ranta" JSC group), "Karelproducts" LLC, "Kala ya maryapoyat" LLC, "Kintizma" LLC, "Yanisjarvi" LLC. In St. Petersburg and Leningrad region, trout fry is supplied by "Volna" LLC, "Rybstandart" LLC, Federal Selection and Genetic Center of Fish Farming "Ropsha", "Lapland" LLC and "Primorskoe" LLC. In Novgorod region, "Yazhelbitsky Fish Farm" LLC, "RIF" LLC grow trout seed material for sale.

Fish seed material is divided into two main categories:

1. Fertilized spawn at the stage of eye pigmentation
2. Young fish divided into the following categories depending on the age: fry (1–100 grams), fingerlings (50–300 grams), yearling (100–700 grams).

The average prices for young rainbow trout (females) from Russian producers depend on the weight and are presented in Table 1. Prices are taken from open sources, the values are indicated with VAT and without delivery, and are relevant at the beginning of 2020.

Table 1. Average prices for young rainbow trout (female) from Russian producers (including VAT).

Fish weight, g	Price per piece, rub.	Price per kg, rub.
1	10	10000
5	18	3600
10	23	2300
20	30	1500
50	47.5	950
70	51	730
100	77	770

In Murmansk region, commodity farms are mainly stocked in late spring – early summer and in the autumn before the ice period. The greatest demand is for fry weighing 5 grams. Commercial farms prefer fish-planting material, the batches of which are 95% or more are represented by females, who spend less energy on spawning changes in the body, and have better growth rates. In addition, breeding females provides an additional opportunity to obtain spawn.

It is less advisable to order the delivery of juveniles from regions remote from Murmansk region, since it is recommended to deliver fry and fingerlings within a day because it is necessary to control the water temperature and its oxygen saturation. This fact determines the importance of having some own fry farms in Murmansk region, even if they work on imported fertilized spawn. If the rules of transportation are followed, the caviar at the eye stage can tolerate the road for up to five days.

Prices for fertilized rainbow trout spawn for the final buyer will depend on the purchased batch volume, transportation containers and the transportation distance. Some idea of the cost of fertilized spawn can be made according to Table 2, prices are relevant at the beginning of 2020.

Table 2. Prices for fertilized rainbow trout spawn at the eye-stage from Russian producers.

Batch size	Price per piece, rub
"FORELEVY" CLOSED JOINT-STOCK COMPANY, AGRICULTURAL BREEDING PLANT	
The cost of delivery – from 48 rubles/km and up to 63 rubles/km, depending on the volume of containers transported. The calculation is for the full there and back trip	
Up to 50 thousand items	1.9
From 50 thousand items	1.6
'Adler' Breeding plant JSC	
Prices including delivery to the airport, nearest to the buyer (Russia)	
Monoecious spawn (females)	
Up to 50 thousand items	1.8
From 50 thousand items to 100 thousand items	1.6
From 100 thousand items to 500 thousand items	1.5
From 500 thousand items	1.2
Double-gender spawn	
Up to 50 thousand items	1.6
From 50 thousand items to 100 thousand items	1.5
From 100 thousand items to 500 thousand items	1.3
From 500 thousand items	1.2
'Lapland' Group of Companies: "Lapland" LLC and "Primorskoe" LLC Regardless of the volume of the batch, excluding delivery	2

In Murmansk region, there are currently 13 commercial trout farming enterprises, two of which are focused on commercial mariculture and purchase imported seed material, while the volume of commercial trout produced is steadily growing mainly due to large firms. Thus, there is practically no competition in the market of rainbow trout seed material in Murmansk region.

2.3 Biological and fish-breeding characteristics of rainbow trout

The average life expectancy of rainbow trout is 6-8 years, the maximum is about 11–12 years, and sexual maturity is reached in females by 2–4 years, in males–by 1–3 years.

There are seven stages of rainbow trout embryogenesis: from the blastodisk formation and crushing to the eyes' pigmentation and, then, hatching [6]. The duration of the embryogenesis stages depends on the temperature. In the range of 6–12^{0}C, the period between insemination and hatching can take from 26 to 50 days [7].

The postembryonic development of trout includes the stages of pre-larvae, larvae afloat, fry, fingerlings and sexually mature fish. The duration of development of different age groups depends not only on the water temperature and its oxygen content, but also on the quality and quantity of feed consumed. The fry development from the larva afloat takes 1.5–3 months. The development of fingerlings from fry takes 3–4.5 months. The development of commercial fish from a fingerling takes 4–6.5 months.

When breeding different age groups of rainbow trout, it is necessary to ensure optimal or close to optimal growing conditions for them. The optimal pH range for embryos and fry is 6.5–8.0; in older age groups, both the optimal (6–8.5) and acceptable (4–11) pH ranges are wider [8]. The water temperature ranges for caviar and fry are optimal between 6–12°C, acceptable from 4 to 16°C; for older age groups of trout, the optimal temperature range is 7–20°C, acceptable is 0–25°C.

The appetite of rainbow trout increases when the temperature rises in the optimal range of 7–18 °C. If the water temperature is too low or too high, the fish stops eating [8]. The best increase due to the feed consumed and the optimal efficiency of feed use are observed in the range from 13°C to 15°C [9].

During the spawn incubation and at the first stages of fry ontogenesis, the optimal and acceptable concentrations of O_2 are 5–6 mg/l. In older age groups, the minimum acceptable oxygen content in water can be about 4–5 mg/l. It is important to know that during and after feeding, the oxygen consumption of fish increases significantly [8].

The required amount of water supply depends on the age and number of developing fish in the fish tank, which is determined by the O_2 content in the supplied water. Thus, the actual volume of water required to contain the same number of developing embryos, fry, and fish will vary. At low water temperatures, the volume of water supplied may be less, at higher temperatures – more. It is obvious that the best opportunities to control the environmental parameters at different young trout development stages are provided by the technology of growing in recirculating aquaculture systems.

These calculations are indicative in nature and are intended only to show the general structure of technical and economic indicators in the business planning. The legal support of the project and the time spent on it have not been considered. The calculations are carried out for the creation of a fish hatchery, which provides three production cycles per year. In one cycle, the nursery will produce ~ 330 thousand pieces or 1.65 tons of fry with a weight of 5 g. Thus, the capacity of the projected nursery is 1 million pieces or 5 tons of fish planting material per year. In the end, such a number of juveniles can ensure the production of up to 1.5 thousand tons of commercial trout weighing 1.5 kg.

3 Designing a trout nursery

Table 3. Main characteristics of the project.

The project aim	Creation of an industrial fish hatchery for the cultivation of rainbow trout juveniles in a recirculating aquaculture system
Organizational and legal form of the project initiator	Common household farming
Project location	Murmansk region, Murmansk city
Main types of economic activity under All-Russian classifier of economic activity types	03.2 "Fish farming" 03.22 "Freshwater fish farming" 03.22.1 "Industrial freshwater fish farming"
Planned number of jobs	6 people (including the founder)
Investments	13 mln. rubles

The planned nursery for rearing young rainbow trout *Oncorhynchus mykiss* is a type of cold-water (trout) industrial nurseries.

The nursery will be organized by the principle of closed water supply. Recirculation is based on mechanical filtration, biological purification, aeration and degassing. Taking into account the modular nature of the recirculating aquaculture system, it is possible to add modules for UV disinfection, heat exchange [10], ozonation, and water treatment at the stage of water intake.

The use of a water recirculation system allows to save water and energy resources. For the normal juveniles' development in the direct-flow water supply system, it is necessary to ensure water exchange every 15 minutes at least and the temperature of at least 6 °C. In the conditions of Murmansk region, this will require high costs for water heating. In general cases, the recirculating aquaculture system requires a daily supply of fresh water in the amount of up to 10 % of the total volume of water in the pools.

The use of the recirculating aquaculture system provides the following advantages:

- no need to place the farm near the reservoir;
- lower volumes of water used and discharged;
- the ability to provide stable conditions, minimizing the impact of external factors, including pathogenic ones;

The disadvantages include the general high cost of operation and high requirements for the personnel qualification.

Monitoring of production parameters allows to grow trout throughout the year. The advantage in the growth rate of fish makes it possible to conduct more production cycles per year, get more profit and compensate for the higher initial investment in the construction of a closed water supply farm compared to a flow-through farm.

3.1 Elements of the organizational plan

The fish hatchery will be registered as a common household farm, the head of which is an individual entrepreneur (IE). The simplified tax system, with a tax rate of 6% of income, will be used. Such farms can receive land acquisition benefits, land lease benefits, and participate in a competition for a business development grant through federal or regional programs to support domestic agricultural producers. The project budget includes an amount of 100,000 rubles for all the costs connected with registering the company, obtaining all the necessary permits and approvals.

To organize a fish hatchery, one will need to rent a plot or a room with an area of 1,000 m^2 with connected utilities: electricity and/or gas supply, water and sewerage, access roads. Staff with appropriate qualifications and work experience will be required to carry out work in the nursery (Table 4).

3.2 Elements of the production plan

The main source of water supply for the fish hatchery will be an artesian well; 250 000 rubles are budgeted for its drilling and a water intake node construction. With such a water supply of the recirculating aquaculture system, it will be necessary to install modules for aeration and de-ironization of artesian water in the water treatment system. Water well licensing is carried out for 4 years, repeated for 25 years. Approximately 250 000 rubles are included in the project budget for licensing.

Table 4. Fishery staff.

Position	Number	Shift	Salary, rub
Leading fish breeder	1	from 9 to 17 o' clock, 5/2 schedule	60 000
Fish breeder-worker	4	12 hours day time, 12 hours night time, 2/3 schedule	28 500

3.2.1. Construction of insulated hangar room

In order to create a fish hatchery with the use of such a technology, this project involves the construction of an insulated hangar - type room, an arched or a straight-walled one, on the leased site. The hangar will have an effective area of 700 m^2, accommodating the main fish-breeding and hatching shops, a zone for technological equipment, a household area, a laboratory and a "cold" annex for storing feed. The hangar walls shall be made of sandwich panels, overlappings that allow it to maintain the necessary positive temperature of at least 15–18 °C.

3.2.2. Equipment purchases and installation

Equipment for the fish hatchery can be purchased from one of the companies that specialize in closed water supply systems: these companies also install equipment and offer the engineering services for the technical projects development, recirculating aquaculture systems' launch and configuration.

The recirculation system consists of fish-breeding pools, incubators, water treatment section, pumps, pipelines and trays for water supply and discharge (Figure 1).

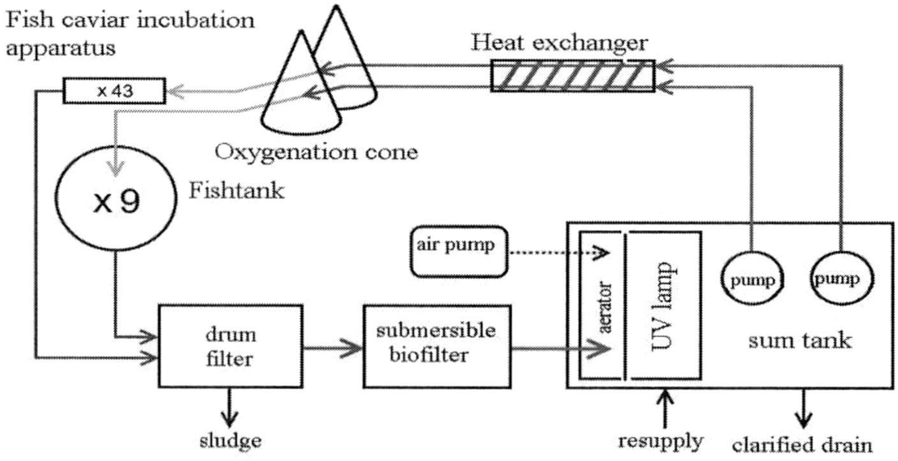

Fig. 1. Closed water supply system.

The project budget includes 400 000 rubles for the equipment installation; 500 000 rubles for the technical project (according to 'FISH-AGRO' LLC at the beginning of 2020).

3.2.3. Production assurance

The production cycle of growing fry in the projected nursery begins with the purchase of caviar at the stage of eye pigmentation. To start one production cycle of the nursery, it is necessary to purchase 500 thousand spawns at this stage.

The supplier of the spawn was selected to be the breeding trout breeding plant "Adler", the leader in the Russian market of rainbow trout, which offers a choice of seed material of various breeds. The contract for the spawn supply is concluded 3 months before incubation.

To those purchasing a batch of 500 thousand pieces, this plant offers spawn at a price of 1 200 rubles for 1 thousand pieces of same-sex (only female) spawn at the stage of eye pigmentation. The cost of 500 thousand spawn will be 600 thousand rubles.

The nursery will use starter feeds from the 'Biomar' manufacturer, the INICIO line for different age groups.

3.2.4. Production cycle

In the projected fish hatchery, it is planned to grow 330 thousand pieces of fry weighing 5 g in one production cycle, including: spawn purchase (3 months before the start of incubation), spawn pre-incubation (10–15 days), obtaining larvae weighing 0.5 g (about 30 days), obtaining fry weighing 5 grams (50–60 days), sale.

The nursery will carry out three cycles per year. The first cycle begins with the laying of spawn in February and ends with the implementation of the 5-gram fry at the end of May. The second cycle lasts from the second half of April to mid-August. The third cycle begins in July, the fry is sold in the second half of October.

From November to February, the nursery does not contain fish, repairs, adjustment and cleaning of equipment are carried out.

4 Fish farming calculations

4.1 Calculation of the number of different age groups of juveniles grown on the farm

The trout nursery under consideration is designed to carry out three production cycles per year (capacity – 5 tons per year). When purchasing a batch of 500 thousand spawn, we predict the following waste: up to the hatching stage – 10 %, up to an average weight of 0.5 g – 15 %, up to a weight of 5 g – 20 % [7]. In one cycle, the farm grows 330 thousand fries with a weight of 5 g.

4.2 Calculation of the area, volume and number of fish tanks

The spawn at the eye stage in the amount of 500 thousand are planned to be placed in the incubation devices of the IL-4 "Trout" tray type. The useful area of the incubator is 0.84 m^2, places up to 42 thousand spawns. After hatching, the larvae are kept in pallets filled with water up to a level of 15 cm, until they reach an average weight of 0.5 g and rise to float. The number of devices is calculated in accordance with the requirements for the placement of pre-plates (10 thousand pieces/m^2). Taking into account the spawn waste, 40 m^2 of free space will be required for 400 thousand pre-larvae. Taking into account the area of the pallet (0.93 m^2), it is easy to calculate that 43 incubation devices will be needed for the maintenance of larvae. One incubation tray of spawn will contain 2900–3000 eggs. The larvae, after being raised on the float, are transplanted into round pools with the following parameters d = 2.8 m; h = 1.3 m; V_{aqua} = 6.2 m^3, where they will be kept until an average weight of 5 g is reached. Based on the planting rate of 30 kg/m^3, the required volume of pools for the content of 330 thousand fry with a weight of 5 g is calculated the following way:

$$\sum V_{tank} = (330\,000 \cdot 5 : 1\,000) : 30 = 55 \text{ m}^3$$

Let's calculate the required number of fish-breeding pools:

$$N_{tank} = 55 : 6{,}2 = 9 \text{ pools}$$

4.3 Calculation of feed consumption

The daily norms of juveniles feeding are taken from the manufacturer's website, the values are recommended in order to achieve the lowest feed coefficient, which is especially important in the RAS to reduce the amount of waste and the load on the biofilter.

Feeding of the larvae begins on average 20 days after their hatching at a temperature of 10°C. When switching to external nutrition (about 10 days), the larvae are fed with starter feed with granules of the smallest size – 0.2 mm. Then you can transfer the larvae to the starter feed with granules of 0.8 mm and 1.1 mm. Table 5 shows the daily feeding rates of juveniles in accordance with the manufacturer's recommendations. Trout feeding rates depend on the body weight of the fish and on the water temperature.

Table 5. The amount of feed a trout nursery needs at different stages of the cycle

Mass, g	Period, 10°C, 24 h	Day norm % from the mass	Feed need		Cost, rub/kg
			kg/day	per growing period, kg	
0.3–0,5	~ 10	2.76	4.5	45	560
0.5–1,5	~ 15	2.41	9.3	140	360
1.5–5	~ 45	2.11	22.3	1 000	360

The cost of feed, necessary for the production cycle – 435 600 rubles; for one year – 1 306 800 rubles.

5 Technical and economic indicators of the fish hatchery

The cost of one fry (5 gram) is equal to the cost of its production. Dividing the cost of production by the number of fry obtained for the year, we get the cost of one fry weighing 5 g, which is equal to 5.5 rubles. The selling price will be set equal to the average price in Russia for fry weighing 5 g – 18 rubles/one.

Table 7 shows how the price of products is formed for the consumer.

The economic efficiency indicators of the projected fish hatchery are shown in Table 8.

Table 6. The main technical and economic indicators of the projected fish hatchery.

Indicator	Measurement unit	Value

1. Production output in kind (per year)	pieces	1 000 000
	kg	5 000
2. The cost of products per piece/ per kg (with VAT 20%)	Rub.	18/3 600
3. Production cost per piece/ kg	Rub.	5.5/1 100
4. Number of employees (including the founder)	people	6
5. Staff Salary Fund	Rub.	171 000/2 052 000
6. Accruals for labor remuneration (contributions to the FIU, Social Insurance Fund, Federal Compulsory Health Insurance Fund, the total amount of contributions - 30%) month/year	Rub.	51 300/615 600
7. Amount of initial investment	Rub.	13 000 000
8. Revenue (per year)	Rub.	18 000 000
9. Costs (per year)	Rub.	11 167 600
10. Income (per year)	Rub.	6 832 400
11. Taxes (Simplified tax system 6%, per year)	Rub.	1 080 000
12. Net profit (per year)	Rub.	5 752 400
13. Payback period	months	28

Table 7. Price structure for nursery products.

Indicator	Measurement unit	Value
Unit cost/kg of production	Rub.	5.5/1100
Price per unit/kg of products without VAT	Rub.	15/3000
Unit price/kg of products for the consumer (with VAT 20%)	Rub.	18/3600
Extra charge	%	63
Profit from 1 unit of production	Rub.	9.5

Table 8. Indicators of economic efficiency of the projected fish hatchery.

Indicator	Measurement unit	Value
Project profitability ratio	%	88.50
Return on sales	%	31.96
Profitability of production	%	62.04
Costs per 1 rub of products sold	Rub.	0.31
Average annual cost of fixed assets	Rub.	12 000 000
Return on funds	Rub.	1.5
Capital intensity	Rub.	0.67

6 Conclusion

1. The profitability ratio of the project with the initial indicators laid down in this work is quite high (88.50%). However, this amount will be significantly adjusted when preparing a specific project, taking into account the full structure of its costs.
2. Return on sales. According to Rosstat (Russian Statistics Department), for the "Fishing and fish farming" activity in 2018, the return on sales was 31.96%. Thus, the activities of the projected fish hatchery will bring profit, which is typical for the fishing industry of the last decade.
3. The profitability of production (62.04%) can be estimated as high. It's necessary to note that this value will also change when production costs are fully accounted for.
4. The calculation of the fund return indicator showed that the unit cost of fixed assets of the fish hatchery accounts for 1.5 rubles of revenue, which shows the economic efficiency of this project.
5. The calculation of the capital intensity index showed that 1 ruble of the revenue received by the nursery accounts for 0.67 rubles of the company's fixed assets. This indicator is normal for a fish-breeding enterprise, since fish farming is a fund-intensive industry.
6. The foundation of such a farm will require large investments. However, the calculated technical and economic indicators of the nursery for the cultivation of rainbow trout juveniles indicate the feasibility of implementing such a project in Murmansk region. The nursery, designed to produce about 1 million fry weighing 5 g for three production cycles per year, is able to generate high profits. The payback period for the initial investment can be 28 months. The creation and operation of a new economic entity in

Murmansk region will help increase tax revenues to the regional budget system, increase and accelerate cash flows between economic entities in the sphere of aquaculture.

References

1. J. Guziur, K. Goryczko, *History of Rainbow Trout Farming: The Quality of Rainbow Trout (Oncorhynchus Mykiss, Walbaum 1792) From Technologies Applied on Poland* pp 11–13 (Olsztyn: University of Warmia and Mazury, 2013)
2. A. E. Arinzhanov, E. E. Serezhina, *The Current State of Trout Farming in Russia* (Orenburg: Orenburg State University, 2017)
3. V. M. Golod, *Russian Fish Breeds* (Moscow: "Mir" Publishing house, 2010)
4. N. K. Vorob'eva, Commodity trout breeding in the Arctic, *Problems of commodity cultivation of salmon fish in Russia: Collection of reports All–Russian meeting* pp 42–45 (Murmansk: PINRO Publishing House, 1995)
5. A. V. Altov, *Specifics of biotechnics of the rainbow trout cage cultivation in the coastal waters of the White Sea*: Abstract of PhD dissertation (Petrozavodsk, 2002)
6. A. T. Gasanova, Rainbow trout embryonic development stages (*Salmo gairdneri Richardson*, 1836), Bulletin of the Azerbaijan Medical University **1 (68),** pp 49–53 (2013)
7. N. V. Barulin, et al., *Recommendations for the Cultivation of Fish Planting Material of Rainbow Trout in Fish-Breeding Industrial Complexes (With Temporary Standards)* (Gorki: Buryat State Agricultural Academy, 2016)
8. A. Voynarovich, D. Hoichi, T. Mot-Poulsen Small-scale breeding of rainbow trout, *FAO Technical Document on Fisheries and Aquaculture* **561** (Rome: FAO, 2014)
9. B. Molony, Environmental requirements and tolerances of rainbow trout (*Oncorhynchus mykiss*) and brown trout (*Salmo trutta*) with special reference to Western Australia: A review, *Fisheries Research Report* **130** (Perth, Australia: Fisheries Research Division, 2011)
10. Ya. Brynballe, *Guide to Aquaculture in Closed Water Supply Management. Introduction to New Ecological and Highly Productive Closed Fish Farming Systems* (Copenhagen, 2010)

Details of artificial reproduction of Atlantic salmon (Semga) at the Umba fish hatchery

M. Semenikhina[1], *E.* Minchenok[2,*], and *T.* Kovaleva[2]

[1] Murmansk Branch of the FSFI «Glavrybvod», 7, Comintern Str., Murmansk, 183038, Russian Federation
[2] Murmansk State Technical University, 13, Sportivnaya Str., Murmansk, 183010, Russian Federation

Abstract. The world's demand for fish products for food and industrial purposes has already significantly exceeded the reproduction capacity of wild fish. Further increasing exploitation of natural fish populations leads to a sharp reduction in their numbers and degradation of population biodiversity. The only way to solve the problem is to develop artificial reproduction of fish. This applies primarily to salmon, which are the most valuable and popular types of fish. In the Murmansk region, there are three fish hatcheries that carry out artificial reproduction of Atlantic salmon. Factory reproduction at the Umba Fish Hatchery (URZ) is currently an important and necessary process, since the plant is the only fish hatchery for reproductive purposes in the Tersk district. The article reveals the peculiarities of reproduction of Atlantic salmon (salmon) in the URZ. A brief description of the Umba River is given. The main fish-breeding activities are covered: catching and keeping of producers, collection of sexual products and fertilization, incubation of eggs, keeping of larvae and their transfer to external nutrition, factory tagging of juveniles, release of juveniles in the Umba River and its tributaries. The peculiarity of the Umba plant in contrast to the other two fish hatcheries of the Kola Peninsula is shown. The question of the necessity of the existence of the Umba fish hatchery and the prospects for its development is highlighted. The reasons for the decline in the number of salmon in Umba are listed. Recommendations for restoring the abundance and preserving the intraspecific biodiversity of the Umba River salmon are provided.

1 Introduction

Fish products are known to be highly delicious and dietary. They are an essential source of animal proteins. Fish also has therapeutic and disease-prevention properties. It contains all essential compounds necessary for a human, including amino acids, unsaturated fatty acids that inhibit development of atherosclerosis, vitamins and trace elements. In terms of vitamins, except for vitamin C, fish surpasses fruit and vegetables. Thus, the importance of fish is not limited to its nutritional value only. That is the reason why the need for fish products will increase even if consumption of meat and dairy products in Russia increases at some point of time [1].

As the demand for fish products goes up, so does the pressure on natural fish populations, which leads to their reduction and the depletion of populational biological diversity. Artificial reproduction of aquatic bioresources provides an opportunity to solve the problem described above.

Salmon fish and red caviar extracted from it are highly valued and considered a delicacy. Restrictions imposed upon catching of salmon in seas and rivers and constantly growing demand necessitated artificial breeding of salmon which became a very profitable line of business.

In the Murmansk region three fish-farming factories – Umba, Kandalaksha, Knyazheguba – are engaged in young salmon breeding for reproduction purposes.

The Umba fish hatchery (UFH) is a subdivision of FGU "Murmanrybvod". UFH mostly deals with artificial reproduction of Atlantic salmon of the Umba River of the White Sea basin.

The Umba River is one of the five most productive rivers of the Kola Peninsula. The length of the river is 125 km. The catchment area is 6248.5 km^2. There are 599 rivers of different sizes in the basin. The total water area of all the lakes located within the Umba River basin is 807.3 km^2. The lakes quotient is 12.9%. Potentially, the number of smolts living in the Umba River can be up to 700 000 specimens. The potential number of spawners, at 5% come-back rate, is 35 000 [2].

The first phase of facility evolution (1932–1959) was characterized by a sharp decrease in the number of salmon fished in specific years, increased quantities of timber rafted down spawning rivers of the White Sea basin, considerable immigration of new people, construction of river infrastructure.

In 1932, fish hatcheries were built on the Varzuga and Umba rivers, the technology was based upon "salmon eggs and larvae field incubation method" proposed for use by I. L. Zhukovsky.

Today, the UFH, despite its more than 89-year history, continues to deal with artificial reproduction of

* Corresponding author: minchenok.elena@yandex.ru

the Umba River population of Atlantic salmon (semga).

2 Materials and methods

Fish breeding methodology utilized at the fish hatchery is based upon "Atlantic salmon breeding procedure" [3] that has been originally customized to suit physical and climatic conditions of the Umba River.

Fish breeding activities at the UFH can be divided into the following phases: catching and holding of spawners, collection of eggs and sperm and fertilization, incubation of eggs, holding of larvae and their switch to external nutrition model, followed by summer and winter rearing until they become yearlings, release of juveniles into the Umba River and its tributaries.

New biotechnical requirements that regulate breeding practices of young (larvae) of Atlantic salmon (semga), that were reared at fish farms of the Murmansk region, became effective January 30, 2015 [6].

Every year, the Umba fish hatchery has been ensuring successful compliance with all biological requirements that regulate salmon farming.

Spawners are caught for fish-breeding purposes at the following fish-counting fences (FCF): "Rybovodny zavod" (June-October) and "Maly Krivets" (June-August). Every year, FGU "Murmanrybvod" obtains a license allowing it to catch Atlantic salmon spawners at the Umba River. Within the period from September 1 through September 20, Atlantic salmon (semga) spawners are transported from fish counting fence "Maly Krivets" in a "wooden fish-transportation-boat" to fish hatchery where they are subsequently held in plastic cages.

As required by "Atlantic salmon breeding procedure", only those Umba salmon spawners that are in good health and have specified average size-and-weight characteristics are allowed to be put into cages for long-term holding of Atlantic salmon [3].

Spawners are accounted for on an yearly basis – Table 1.

Table 1. Atlantic salmon spawners accountability [4].

	Year 2017	Year 2018	Year 2019	Year 2020
Total spawners caught (each)	1147	360	98	13
Including females (each)	52	198	61	6
Females used (each)	46	198	Within the 2019–2020 period, due to disease of unknown origin that has affected spawners in the Umba River of the White Sea basin, no eggs and sperm were collected and no eggs were put into incubators.	

When water temperature drops to 7-8° and spawning period is about to come, females and males are moved into separate holding areas. Once that happens, all the spawners get tested for maturity individually every 4 or 5 days.

In order to preserve the genetically established structure of Atlantic salmon populations, eggs should be collected from those spawners that represent all spawning periods and caught throughout the entire spawning period. Due to compliance with research recommendations of FGUP "PINRO", the Umba fish hatchery has been a success in caging autumn race Atlantic salmon spawners since 2008.

A sign of males' maturity is the emergence of a drop of roe when light pressure is applied near its anus.

The maturity degree of a female fish is assessed by lifting a female fish by its tail to see how soft and sagged its abdominal cavity walls are at the rear part of the body. This condition occurs because part of mature eggs, that have fallen out of the ovary, have moved to the front part of body cavity. As a rule, males mature earlier and at faster rate than females.

Salmon specimens that participated in autumn and summer runs are kept in different cages at the Umba fish hatchery.

Eggs and sperm are collected in October-November at the hatchery by straining them off. Atlantic salmon spawners are released back into the Umba River once their eggs and sperm have been collected and fish has gone through recuperation process for one full day (24 hours).

Eggs and sperm are only collected from those spawners that are in good health and whose eggs and sperm are mature (fluid).

Fig. 1. Eggs fertilization.

https://doi.org/10.1051/bioconf/20225200079

In order to reduce influence of individual males, each batch of eggs must be fertilized using sperm pre-mix collected from two or three males. Pre-mixing must be done right before the use of pre-mix. Pre-mix can be used only after it has been tested for quality (Figure 1) [1].

Far-east-type troughs with Zhukovsky machines installed inside are utilized for incubation at the Umba fish hatchery.

Total amount of eggs obtained: in 2017 – 319.4 thousand eggs, in 2018 – 1721 thousand eggs. Eggs incubation periods: 2017 – 218 days, 2018 – 223 days. Survival rate during incubation period: 2017 – 110.824 (98%), 2018 – 431.52495 (95%) [4]. The number of eggs placed for incubation and transferred to other fish hatcheries in 2018 is shown in Figure 2.

Holding and rearing of larvae, summer and winter holding of Atlantic salmon juveniles.

Young salmon fish were fed with pelleted fish food: Biomar (Denmark); EWOS Micro (Great Britain). Food consumption and average water temperature by month at the URZ are shown in Figure 3.

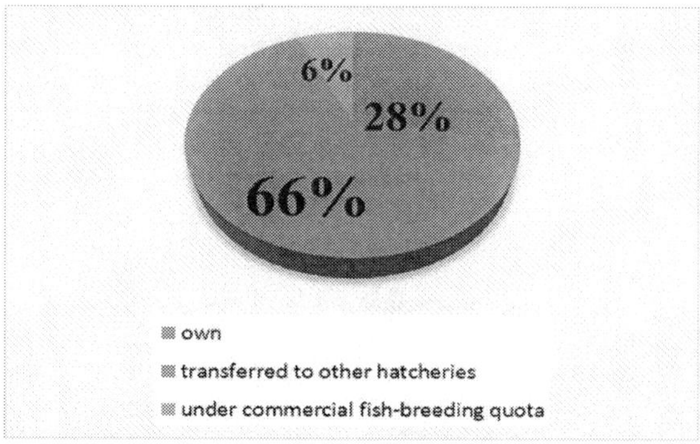

Fig. 2. Number of eggs placed for incubation.

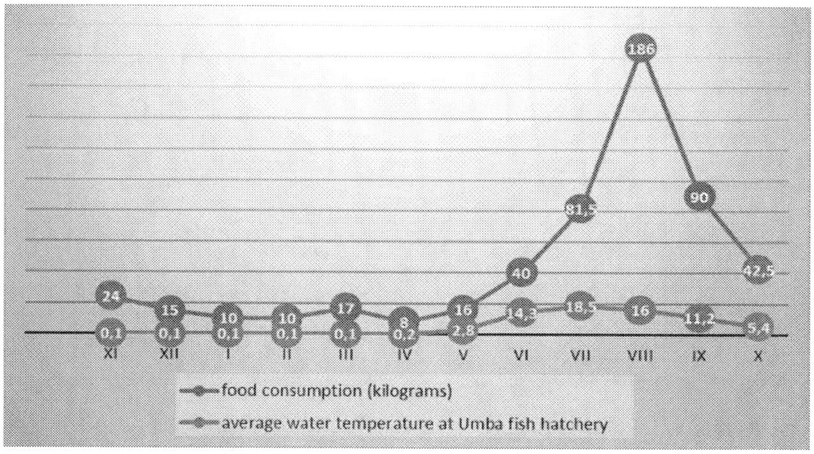

Fig. 3. Fish food consumption rates and average water temperature at UFH.

Based upon food tables, food consumption rates go up as water temperature goes up.

In order to be able to determine how many fish have come back to hatchery, juveniles to be released are tagged at the hatchery by amputating their fat fin (Figure 4.).

Fig. 4. Tagging of juveniles.

3 Results and discussion

Since January 03 through February 24 2020, 202 842 thousand fish that had been born that year were tagged; the average weight of each specimen was 1.5 grams. Under a government contract, 187 thousand of young salmon have been tagged; 15 842 thousand young salmons have been tagged above that number [4].

The term "release of juveniles" is interpreted as "purposeful introduction of Atlantic salmon into its natural habitat at any phase of its life cycle in order to increase, improve, replenish, restore or breed the species"[5]. During release of juveniles held in the hatchery, juveniles were accounted for by total accounting method.

In 2020, the number of fish returned to hatchery were accounted for at the "Maly Krivets" fish counting fence within the period since 22 July 2020 through 07 August 2020 with river completely fenced up (Table 2). Fishing mode: daily accounting and caging of Atlantic salmon (Semga) spawners for fish-breeding purposes [4].

Fish counting fence "Rybovodny zavod" has been in the shut-down status in 2020.

Data on how many eggs and juveniles have perished during rearing process are presented in Table 3 [4].

Table 2. Number of salmon entered "Maly Krivets" fish counting fence in 2020 [4].

Period:	Total:	Fish hatchery's	
		Specimen	% of total quantity
22 July 20 – 31 July 20	275	2	0.7
01 Aug 20 – 07 Aug 20	97	2	2.1
Grand total:	372	4	1.1

Table 3. Salmon eggs and juveniles perished during rearing, by periods (%).

Age	Period	Requirement	Umba fish hatchery
eggs	01 November 2018 – 01 June 2019 (generation year 2018)	10	5.4
0+	01 June 2019 – 01 November 2019 (generation year 2018)	53	28
	01 November 2018 – 01 June 2019 (generation year 2017) 01 November 2019 – 01 June 2020 (Generation year 2018)	20 20	5.6 5.2

All fish-breeding activities associated with release of young salmon are subject to coordination with experts of the Polar branch of FGBNU "VNIRO" ("PINRO" named after N.M. Knipovich) and the Murmansk branch of "Glavrybvod". Release of fish was made in April - August 2020.

Atlantic salmon generation-2018 release plan is 100% fulfilled: 187.0 thousand yearlings have been released; average weight of each specimen is 1.5 grams. All juveniles were placed in locations recommended by the Polar branch of FGBNU "VNIRO" ("PINRO" named after N.M. Knipovich) [4]. Fish release data is presented in Table 4.

Table 4. Release (transfer) of fish in 2020 [4].

Item #	Name of species	Fish age	Number of juveniles		Average mass, kg	Released (transferred) within	Release location or entity-transferee
			Total, thousand each	Including down-streams, thousand each			
1	2	3	4	5	6	7	8
1	Atlantic salmon (Semga), generation year 2018	1·	187,00	–	1.5	01 Apr – 02 Apr 2020	Umba river, Kanozero rapid – ***102*** thousand specimens, Karelskiy rapid – ***85*** thousand specimens
2	Atlantic salmon (Semga), generation year 2018	1·	26,151	–	1.5	28 Apr – 17 Aug 20	Vyala tributary (distance from dam to mouth) – ***2 650*** thousand specimens; ***3 036*** thousand specimens; ***4 000*** thousand specimen; ***1 000*** thousand specimens; Nyizma channel, Kryvoy porog – ***9 362*** thousand specimens; ***2 428*** thousand specimens; ***3 493*** thousand specimens; ***0 182*** thousand specimens.

In summer, research is conducted in young fish habitats to see how well they have managed to get accustomed to natural conditions. Based on research results, adjustments are made associated with release of juveniles in subsequent years.

Spawners that entered the trap of the "Maly Krivets" fish counting fence had the signs of being affected by ulcerative dermal necrosis disease, that is why only a little number of them have been caged. On August 11, 2020, 13 spawners were transported to a cage installed at the Umba fish hatchery so that possible scenarios of pathological process could be monitored. Out of these, 3 specimens were transferred to the aquaculture, fish physiology and disease sector laboratory "Murmansk region's animal disease control station". The remaining spawners perished over the time period August 28, 2020 – September 25, 2020 and were disposed of at the cremation facility at Knyazhegubsky fish hatchery [4].

4 Conclusion

The difference between UFH and other FGU "Murmanrybvod" fish hatcheries is that the former is located right on the Umba River. This provides an opportunity to release juveniles right at the location where eggs have been collected, thus the need to transport the fry is obviated.

Autumn-race spawners are caged and held at the hatchery.

In the future, UFH can also handle reproduction of other valuable fish species (whitefish, kumza, trout).

The main reason why the number of salmon in the Umba River has decreased are the aftermath of timber downstream rafting and illegal fishing (poaching). At present, there are as few as two pay-roll water-bailiffs providing fish conservation activities out in the field. In 2020, the captured producers died as a result of the disease ulcerative dermal necrosis.

In order to replenish the number of salmon quantities and preserve intra-species biological diversity of the Umba River, the following needs to be provided: artificial reproduction of aquatic biological resources; reclamation of spawning-and-rearing areas; technical (retrieval of sunken wood) and biological (catching of predator fish) reclamation; fish-preservation activities.

References

1. G. G. Serpunin, *Biological Bases of Fish Farming* (Moscow, 2009)
2. O. G. Kuzmin, Biological characteristics Salmon of the South-Eastern part of the Cola Peninsula, *Problems of studying, rational use and protection of natural resources of the White Sea,* p 92 (Leningrad, 1987)
3. N. I. Yandovskaya, *Instructions for Breeding Atlantic Salmon,* p 95 (Leningrad, 1979)
4. Report on fish-breeding works at the Umba fish-breeding plant for 2017–2020
5. N. I. Chirkov, *The Williamsburg Resolution,* p 33 (Murmansk)
6. On the approval of the Methodology for calculating the volume of production (catch) of aquatic biological resources necessary to ensure the conservation of aquatic biological resources and ensure the activities of fish farms, when fishing for the purposes of aquaculture (fish farming): Order of the Ministry of Agriculture of the Russian Federation of 30.01.2015 No. 25-FZ: ed. from 25.08.2015 (2015) Registered in the Ministry of Justice of Russia 20.02.2015 Np 36147 Retrieved from: http://www.consultant.ru/document/cons_doc_LAW_176258/

The role played by the Umba fish hatchery in artificial reproduction and conservation of water biological resources in the Murmansk region

M. Semenikhina[1], *E.* Minchenok[2,*], and *T.* Kovaleva[2]

[1] Murmansk Branch of the FSFI "Glavrybvod", 7, Comintern St., Murmansk, 183038, Russian Federation
[2] Murmansk State Technical University, 13, Sportivnaya St., Murmansk, 183010, Russian Federation

Abstract. The article provides information on the need for artificial reproduction of water biological resources. It is shown that the conservation of biological diversity of fish, maintaining stocks of valuable fish species, preserving the natural habitat is impossible without effective measures in the field of artificial reproduction. A brief description of the activities of the Murmansk branch «Glavrybvod» is given. The greatest attention is paid to the oldest enterprise in the Murmansk region – the Umba fish hatchery. The historical information about the activity of this enterprise is given. An assessment of the current state of the Atlantic salmon stocks in the Umba River basin is given. The main problems of the fish-breeding enterprise are identified and analyzed: a high level of illegal fishing, the consequences of molten timber rafting and the influence of the activities of enterprises of the mining complex. Measures aimed at solving these problems are proposed. It has been proved that thanks to the fish breeders of the Umba fish hatchery, the stocks of the Atlantic salmon population of the Umba River are replenished. It is shown that a promising direction of the plant's activity is the cultivation of other valuable fish species (whitefish, pink salmon, brown trout), which will significantly contribute to the replenishment of valuable fish stocks in the water bodies of the Murmansk region. The role and importance of artificial reproduction and conservation of water biological resources will increase more and more as the anthropogenic impact on the environment intensifies. This type of activity will be the only way to preserve the commercial value of valuable species of fish and other biological resources, which makes artificial reproduction artificial reproduction one of the highest priorities in development of fishery system of the Russian Federation.

1 Introduction

Humanity's demand for fish food products in the contemporary world significantly exceeds potential capacity of natural fish reproduction. Further intensification of exploitation of natural populations of fish can cause a sharp decrease in their number, disruption of natural habitats, reduction of water bodies' productivity, and loss of biological diversity. Solution of these problems is being implemented through artificial reproduction of fish.

Artificial reproduction of fish includes work associated with obtaining offspring and rearing of young water biological resources at fish-farms which are subsequently released into natural water bodies and reservoirs.

Artificial reproduction of fish in the Russian Federation has the following goals:

- preservation of fish biological diversity;
- maintenance of valuable fish species' stocks taken away by fishing;
- improvement of species composition of fish-fauna that lives in fishery water bodies.

2 Materials and Methods

2.1 Description of the activity of the Murmansk branch "Glavrybvod"

The Murmansk branch of FGBU "Glavrybvod" (hereinafter referred to as the Branch) is the only entity on the Kola Peninsula involved in artificial reproduction and conservation of water biological resources business. The Branch has three fish hatcheries: Knyazhegubskiy fish hatchery, Kandalakshsky experimental salmon fish hatchery and Umbsky (Umba) fish hatchery. These fish hatcheries have been primarily dealing with reproduction of Atlantic salmon (Semga), in particular, rearing of young fish until they become yearlings (one year old) and then releasing them into natural water bodies of the Murmansk region. The work described above is performed by fish-farms under government contracts awarded every year [1].

Atlantic salmon (Semga) has been traditionally reproduced at fish hatcheries of Murmansk region as well as pink salmon and trout. Since 2017, the Murmansk region's fish farmers have been in business of rearing whitefish for reproduction purposes. The plan is to bring in coho salmon from Arkhangelsk region to the

* Corresponding author: minchenok.elena@yandex.ru

Umba fish hatchery for purposes of biological processes optimization.

The subject of research completed in this article is to review the role played by the Umba fish hatchery in artificial reproduction and conservation of water biological resources of the Murmansk region.

The Umba fish hatchery is involved in the following activities:

- artificial reproduction of water biological resources;
- implement measures to conserve water biological resources and their habitat;
- improve water bodies for fish-farming purposes;
- create, maintain, operate and account for brood stock, including spawners;
- produce (catch) water biological resources for aquaculture (fish farming) purposes;
- maintain and operate water works and structures;
- provide security for biological resources and property controlled by the hatchery.

2.2 Historical background

The Umba fish hatchery (UFH) was commissioned in 1932. It is located at the mouth of the Umba River that flows into the Kandalaksha bay of the White Sea (Figure 1).

Fig. 1. Umba fish hatchery. 1949 г. (courtesy of "Murmanrybvod").

Umba fish hatchery has been mostly involved in artificial reproduction of Atlantic salmon (Semga) of the Umba River of the White Sea basin.

The first phase of facility operations in 1932–1959 is characterized by a sharp decrease in the amount of salmon fished in specific years, increased quantities of timber rafted down spawning rivers of the White Sea basin, considerable immigration of new population, construction of river infrastructure [2].

During these years, the majority of eminent Russian fish-breeding researchers believed that natural reproduction of fish was inefficient, especially at the embryonic phase. This circumstance has determined specific organizational forms of fish breeding.

In 1932, fish hatcheries have been built on the Varzuga and Umba Rivers. I. L. Zhukovsky's field method of salmon eggs and larvae incubation has been practiced for fish breeding. The production process consisted of catching and holding of spawners, getting the eggs fertilized, incubation of eggs in the river, short-term holding of larvae and releasing them with yolk sac undissolved into water body. During ice formation and ice drift periods, the eggs were incubated in a humid environment inside specialized incubating iceboxes.

At the hatchery, egg production, incubation, and larvae holding areas were located at great distances from each other – 5 to 50 km. This added challenges to fish breeding which was by itself very labor-intensive.

The production capacity of the UFH, based on its ability to incubate eggs at the initial phase of development, was determined to be 3 million fish larvae.

In 1940s it was generally firmly believed that salmon-farming in the Polar region, the way it was practiced back then, was, for no sufficient reason, opposed to natural reproduction and was implemented, in a number of cases, not to supplement it, but to damage it. Efforts made by fish-farming science at that time were concentrated upon in-depth study of salmon biology and conditions that drove the state of its stock as well as construction of foundation for intensive fish breeding technology.

Within the period from March 7 to April 30, 1949, the fish hatchery management has built several natural nurseries by constructing three log dams in the tributary of Umba River between the «Upper intake» islands thus creating 800 m^2 of total water surface area. Out of this number, 300 m^2 of water surface was dedicated to natural nurseries for holding fry; and the main pool –500 m^2 – was used for salmon spawners.

Due to challenges faced throughout the renovation of national economics after the Great Patriotic War, it took around 10 years for the real opportunity to emerge to start doing fish farming in the Murmansk region.

Drastic changes in salmon hatchery reproduction methods and technology in the Murmansk region have been driven by the fact that a number of activities associated with acclimatization of far-eastern salmon (pink and chum salmon) in the White Sea and the Barents Sea that commenced in 1959, have required appropriate production facilities to be constructed.

Within the period from 1959 through 1961, the Umba fish hatchery was redesigned to match Sakhalin fish hatcheries. In 1958 an incubating hatchery was constructed having 290 m^2 of usable area as well as auxiliary premises. In 1961 another incubating hatchery, having 510 m^2 of usable area, was constructed as well as channel cage having 2000 m^2 of water surface area.

The design capacity of UFH in 1999 was 70 thousand pieces of downstream fry of salmon (three-year-old fish). Since 2005, pursuant to decision made by the Inter-industry ichthyologic commission, the Umba fish hatchery has been producing 140 thousand yearlings of

Atlantic salmon (Semga). Since 2006, after Taybolsky fish hatchery has been shut down, the number of fish to be provided under government contract went up to 187,0 thousand.

At present, the fish hatchery continues to artificially reproduce the Umba River population of Atlantic salmon (Semga). The Far East salmon (pink salmon) is being dealt with as well. Since 2019, the hatchery has been engaged in breeding the whitefish (freshwater live form). Whitefish eggs, at the "eyelet" phase, are delivered to the hatchery from Leningrad region.

3 Results and Discussion

3.1 The status of Atlantic salmon (semga) stock in the Umba River basin

The Umba River basin is located in the southwest of the Kola Peninsula and is characterized by high presence of lakes (12.9 %), marsh-lands and wood-lands. The river flows out of Umbozero and flows into the Kandalaksha bay of the White Sea. The river length is 125 km. The catchment area is 6.25 thousand km^2 (Figure 2).

Based on the nature of its hydrological regime, the Umba River is a lake-river type of water body. Half of the total length of the lake-river system are channel lakes, the total area of which is 807.3 km^2. Quite large lakes are located at sources and channels of major salmon-inhabited tributaries of Umba – the rivers of Vyala and Muna. The fact that the flow of rivers is controlled by lakes affects their temperature and water levels resulting in belated water warming-up in spring and prolonged cooling-down in autumn [3].

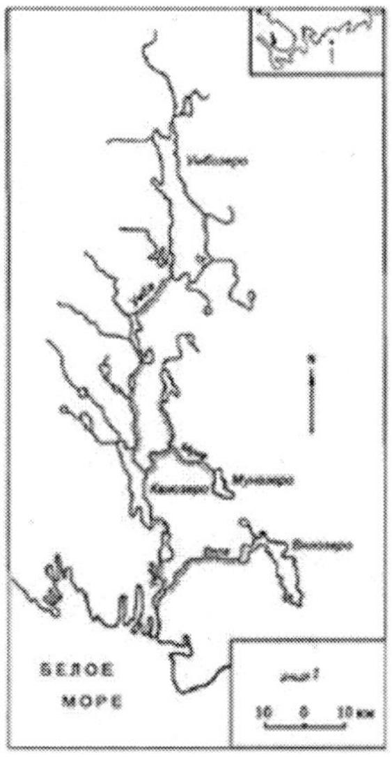

Fig. 2. Umba River basin map. 1 – Atlantic salmon spawning-and-rearing areas [3].

The total area of the Umba River spawning-and-earing areas (SRF) is 450 hectares, pestle fish density varies from 6 to 140 specimens/100 m^2.

Based on calculations that were made, production of smolts in the Umba River can be about 700 thousand specimens, and potential number of spawners, if the return rate is at 5%, is around 35 thousand spawners. The salmon spawning run lasts from May through October. Based upon many years' statistics, most of the spawners enter since middle of July through the end of September (Figure 3).

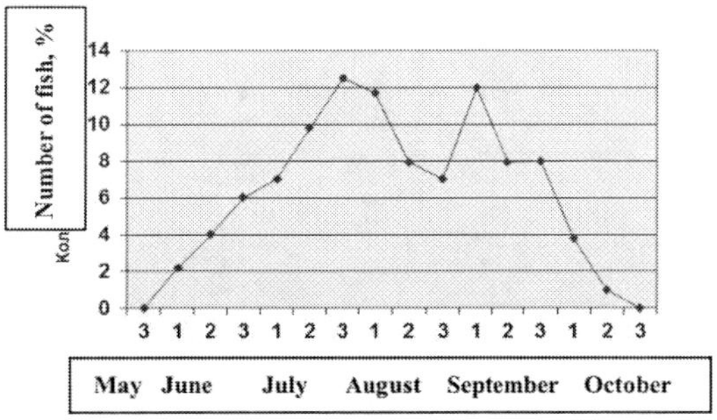

Fig. 3. Atlantic salmon run statistics in the Umba River based upon multiple year/average data [3].

The total spawning-and-rearing stock of the Umba River, including 1-st and 2-nd order tributaries – Vyala, Muna, Inga, Lyamuksa and channels Kitsa, Rodvinga, Nizma, Bolshoi Krivets is 4 477 800 m^2, out of which 1 909 700 m^2 are spawning areas and 2 568 100 m^2 are rearing areas [3].

The density of young salmon distribution is dependent upon water temperature, current, presence of predators and many other factors. The highest density – 45 fish/m^2 has been observed up in the mid- stream of the Umba River as well as in a tributary – the Vyala River – 40 fish /m2.

Based upon various assessments made, the potential number of salmon spawners in the Umba River can be anywhere from 34.6 to 40.5–80.9 thousand. However, the value of conservation limit, i.e., the minimum level of fish stock that provides the maximum sustainable catch during fishing, is estimated at 6 270 specimens [3, 4].

The salmon run dynamics in the Umba River in 2009–2020 as recorded by "Maly Krivets" fish accounting fence and "Rybovodny Zavod" fish accounting fence is presented in Figure 4. In 2018, as recorded by «Maly Krivets» fish accounting fence, the number of salmon that came in was 3208 fish which is much more than in the years before. This is probably due to increased protection of fish migration routes that was available that year.

Currently, the number of Atlantic salmon (Semga) in the Umba River is in depressive status, which is driven by anthropogenic factors: high level of illegal fishing, consequences of timber rafting as well as mining businesses' operations [5, 6].

The number of Atlantic salmon fished illegally in the Umba River in the second half of the 20th century has been at 25–26% of the annual number of spawning migrants. At the beginning of the new century it increased sharply, and now illegal fishing quantities are at 67–73% of the fall biological group of salmon. As a matter of fact, the critical level fish perished through fishing and below which the fish population becomes extinct, varies between 82–85% [4].

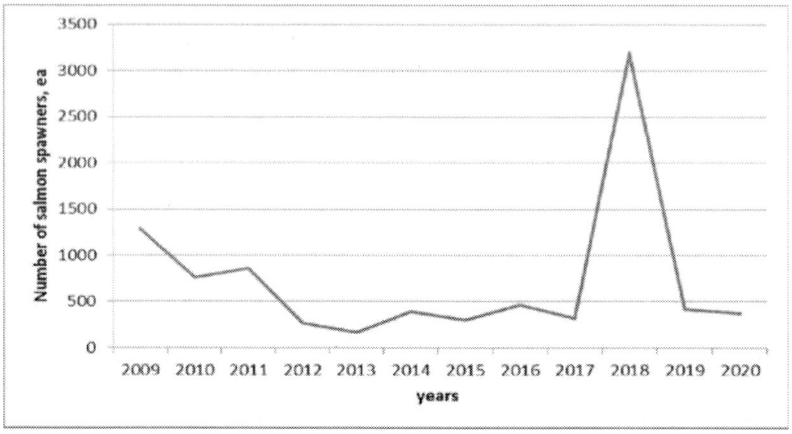

Figure. 4. Number of spawners entering the Umba River in 2009–2020.

For the Umba River, based upon high percentage of lakes present within its basin, pressure exerted by predatory fish upon other fish is believed to be widespread. At the same time, the quantities of biological reclamation have decreased noticeably over recent years.

Timber rafting downh the Umba River and its major 1-st and 2-nd order tributaries - the Muna, Inga, Vyala and Lyamuksa – has been on-going for a long time and was completely terminated in 1993. The timber rafting has changed hydrological conditions of the Umba River and worsened the condition of spawning-and-rearing areas. Consequences of timber rafting are felt to this day [5–7].

In recent years (2019–2020), the Umba River has been facing another problem - salmon spawners infected with Ulcerative Dermal Necrosis (UDN). This is a fish skin disease that causes skin ulcerations and is most commonly seen in Atlantic salmon (semga) and sea trout (kumja) every time they come back to fresh water for spawning. Salmon having this disease cannot be used for fish-farming and should be disposed of (cremated). For this particular reason, no work was done at the fish hatchery associated with selection of germplasm and placement of salmon eggs for incubation in 2019–2020.

3.2 The role played by the Umba fish hatchery in artificial reproduction and conservation of Atlantic salmon (Semga) stocks

Since the mid-1980s, there have been negative changes in the population characteristics of the Umba River salmon that were caused by aftermath of timber rafting and over-fishing. There are no smolts aged over 5 and over 6 anymore. The annual maximum and minimum numbers of spawning migrants now differs by 18 times [6].

The Umba fish hatchery is the only hatchery doing artificial reproduction of Atlantic salmon on the White Sea coast. The hatchery is located right on the Umba River where salmon spawners are penned for reproduction. Young salmon is released into this river basin without having to be transported over a long distance. It has a positive effect on fish survival rate and further adaptation in the natural environment.

There are many rivers where salmon has become extinct or salmon numbers are depressed. Construction of salmon hatcheries at such rivers is well justified, and ideally, it should be supported by restoration of natural spawning areas so that over time, due to contribution made by fish hatcheries, wild salmon herds recover their numbers.

Construction of salmon hatcheries for the purpose of re-acclimatization is also well justified. There are rivers where salmon used to live, but due to certain natural reasons it has disappeared. This also includes rivers within the habitat that have no conditions to support natural spawning. That is the reason why Umba is the river where artificial restoration of natural salmon population is necessary and well justified.

The UFH successfully complies with all requirements associated with holding of spawners, holding of eggs, holding and rearing of larvae, rearing and wintering of fish born the current year. The amounts of waste generated at each phase are minimal (Table 1). Based upon the available data (Table 1), the actual survival rate of salmon throughout all the phases of the process exceeds requirements.

To assess UFH operational efficiency, young fish to be released goes through tagging every year. The fat fin amputation method is used to tag young fish. Fish hatchery efficiency is assessed by the number of tagged fish entering "Maly Krivets" fish accounting fence. The number of salmon fish that have entered "Maly Krivets" fish accounting fence is presented in Table 2.

Table 1. Compliance with major biological requirements associated with Atlantic salmon (semga) farming throughout all the phases of fishing breeding process at the Umba fish hatchery (based upon inventory results).

Item	Parameter	UOM	Required value	Generation	
				2017	2018
1	Holding of spawners	%	**75**	100	96.4
2	Average fertility	thous. ea.	**5.6**	6.9	8,6
3	Holding of eggs	%	**90**	98	97
4	Holding of larvae	%	**93**	97	97
5	Rearing of larvae	%	**76**	86	90
6	Rearing of fish born the current year	%	**66.5**	72*	81*
7	Average mass of fish born the current year	g.	**0.8–1.0**	2.0	1.5
8	Wintering of fish born the current year	%	**80**	98	90
9	Rearing of two-year-olds	%	**80**	90*	57*
10	Average mass of two-year-olds	g	**6.0–9.0**	13.2	12.1
11	Wintering of two-year-olds	%	**94**	67	

Table 2. Number of salmon that entered "Maly Krivets" fish accounting fence in 2017–2020.

Date:	Total:	Fish hatchery	
		Each	% of total number
19 July, 17 – 31 July, 17	168	2	1.2
01 Aug, 17 – 10 Aug, 17	89	3	3.4
11 Aug, 17 – 26 Aug, 17	61	0	0
Total for 2017:	**318**	**5**	**1.6**
18 June, 18 – 30 June, 18	104	0	0
1 July, 18 – 10 July, 18	134	3	2.2
11 July, 18 – 20 July, 18	1467	10	0.7
21 July, 18 – 31 July, 18	1098	4	0.4
1 Aug, 18 – 10 Aug, 18	405	1	0.2
Total for 2018:	**3208**	**18**	**3.5**
16 June, 19 – 30 June, 19	65	0	0
1 July 19 – 10 July, 19	95	4	2.5
11 July, 19 – 20 July, 19	180	0	0
21 July, 19 – 1 August, 19	81	2	0.6
Total for 2019:	**421**	**6**	**3.1**
22 July, 20 – 31 July, 20	275	2	0.7

1 Aug, 20 – 7 Aug, 20	97	2	2.1
Total for 2020:	**372**	**4**	**1.1**

4 Conclusion

Thus, the number of fish that came back to fish hatchery over period under study ranged from 1.1% (2020) to 3.5% (2018). In recent years, fish return rates have decreased markedly. For comparison, within 2000-2010, the number of hatchery fish that came back was steadily at 7%. 5% of fish that came back is a good percentage (close to natural values). Therefore, it is necessary to take measures to increase percentage of hatchery fish coming back.

Due to significant level of unauthorized fishing (67-73%) and pollution caused by timber rafting, the number of salmon spawners in the Umba River has been decreasing. At present, most of the Atlantic salmon spawners in the lower and upper reaches of the river are being fished illegally. Therefore, urgent measures are required to protect the Umba River salmon population from complete extinction as well as measures to replenish its inventories.

Measures aimed at improving operational efficiency of the Umba fish hatchery:

- resume biological reclamation at locations where young fish is released and reach areas of Umba River and its tributaries (downstream migration of fish);
- propose to do experiments associated with changing biological methods of rearing juveniles in hatcheries, in particular, juveniles can be released via adaptation pond when they turn three years of age;
- continue cleaning the river of timber rafting debris;
- at the «Maly Krivets» fish accounting fence, put all the salmon (if fish comes in small quantities) into penns and release right before the spawning time. This will prevent poachers from fishing them at fish wintering locations (Middle Umba of Kanozero Lake);
- augment presence of fish water bailiffs at the Umba River;
- conduct joint operations involving local law enforcement agencies and executive authorities. This will expose poaching, corruption bonds with poachers and users of water biological resources.

The measures above should be supported by development and implementation of social-and-economic programs, development of recreational fishing and development of fishing tourism infrastructure.

Even though problems were found, it should be noted that due to efforts made by Umba fish hatchery fish-farmers, the Umba River Atlantic salmon population is growing. Additionally, now that fish hatchery has gone through renovation, other fish species (whitefish, humpback salmon, kumzhi) can be bred there. This would significantly contribute to replenishment of valuable fish species in water bodies of the Murmansk region [9].

Thus, the role and importance of artificial reproduction and conservation of water biological resources will increase as environment anthropogenic impact grows larger. This type of activity will be the only way to preserve the commercial value of valuable fish species and other biological resources. This, even at present time, makes artificial reproduction one of the highest priorities in development of fishery system of the Russian Federation.

References

1. Report on the state and protection of the environment of the Murmansk region in 2019, retrieved from: https://mpr.gov-murman.ru/activities/napravleniya/okhrana-okruzhayushchey-sredy/00.condition/index.php
2. V. V. Kostyunichev, et al., *Proceedings of VNIRO* **153,** pp 26–41 (2015)
3. V. G. Martynov, *Atlantic Salmon (Salmo Salar L.) in the North of Russia* (Yekaterinburg: Ural Branch of the Russian Academy of Sciences, 2007)
4. A. V. Zubchenko, et al., *Reproduction Characteristics of Atlantic Salmon (Salmo Salar L.) in the Umba River (Kola Peninsula)* (Petrozavodsk, 2007)
5. A. E. Veselov, et al., *Fish Industry*, 81–84 2006
6. M. Yu. Alekseev, et al., *Fishing Industry Agenda* **7-2 (26),** 318–325 (2006)
7. M. Yu. Alekseev, et al., *Proceedings of the II All-Russian scientific conference with international participation "Fishery reservoirs of Russia: fundamental and applied research,* pp 20–26 (2018)
8. Z. M. Sergieva, et al., *Proceedings of VNIRO* **153,** 3–25 (2015)

Assessment of vegetable and animal products in the Murmansk region within the implementation of the federal project “Public Health Promotion”

I. V. Uskova*, *O. G.* Krivenko, *I. N.* Rogacheva, *V. A.* Poteshkina, and *L. A.* Mishanina

Department of Microbiology and Biochemistry, Institute of Nature and Technology, Murmansk State Technical University, 13, Sportivnaya Str., Murmansk 183010, Russian Federation

Abstract. Monitoring studies of the quality of food products of plant and animal origin, including vegetable oils, some confectionery products and dietary supplements were carried out in the Murmansk region as a part of the implementing the federal project “Public Health Promotion” of the National Project “Demography”. Within the framework of monitoring studies, data allowing to evaluate the quality of food products in circulation on the territory of the Murmansk region were obtained, the share of products, up to 12% of the volume of the studied range not meeting the established requirements in safety quality indicators is revealed. The highest percentage of food products that do not conform to the technical regulation requirements is defined in the category of animal products and made up 7.8% of the total amount the of studied samples, the second place (2.5%) is given to the products of plant origin, other products – 1.8% complete the list. Toxic elements, pesticides, antibiotics, preservatives, mycotoxins, nitrates, dry low-fat milk residue, genetically modified enzymes (microbial transglutaminase) and sanitary-indicative microorganisms were determined among the safety indicators. Products of animal origin like eggs, milk, yogurt, kefir, sour cream, cake cheese, cheeses, condensed milk, butter, bockwurst sausages, sausages, were not safe for the consumer as a result of violating the production technology, including storage conditions failure, the use of low-quality raw material, or product falsification. The assessment of apples, cucumber, tomato, sunflower and olive oil, cookies and dietary supplements samples showed unsatisfactory results in the residual content of pesticides, nitrates, toxic elements, antibiotics and peroxide.

1 Introduction

State laboratory monitoring of prohibited and harmful substances traces in the raw materials and products of animal and vegetable origin is the main complex of measures capable of ensuring the safety of food products in order to protect people’s life and health [1, 2].

In August 2012, Russia joined the World Trade Organization (WTO), thus our products must meet the international standards and be competitive. With this goal, on the territory of the Russian Federation, within the framework of the national program, the veterinary services carry out laboratory monitoring of residual traces of prohibited and hazardous substances, both in the feed and in the organisms of farm animals, as well as in the finished products of animal origin [3].

When conducting socio-hygienic monitoring, one of the main tasks of the Office of the Federal Service for Supervision of Consumer Rights Protection and Human Welfare is collecting reliable and objective information about the quality and safety of food products manufactured by the Russian enterprises, imported foods and manufactured products for free circulation on the territory of the Russian Federation and already found in the circulation, as well as the data concerning nutrition and state of population health in terms of the conditions and quality of foodstuffs [3].

2 Materials and Methods

The studies determined general safety indicators regulated for food products taking into account the peculiarities of their component composition and processing technology.

The method of highly efficient liquid chromatography with various detectors was used to reveal the mycotoxins content. Pesticides were determined by gas chromatography. The content of toxic elements was determined by the spectral analysis method (atomic absorption spectroscopy). To identify genetically modified organisms and microorganisms typing, the method of polymerase chain reaction (PCR) was used. Accounting and identification of sanitary microorganisms were performed according to the generally accepted methods of microbiology.

The results were evaluated in accordance with the regulatory technical documentation, which acts towards the specific food category.

According to the technical regulations of the Customs Union "Requirements for the safety of food

* Corresponding author: uskovaiv@mstu.edu.ru

additives, flavors and processing aids" (TR CU 029/2012), the presence of food additives was assessed, as well as their quantitative content [4].

3 Results and Discussion

In 2020 489 samples of food products in circulation on the territory of the Murmansk region were investigated to assess the quality of foods, of them 140 samples (28.6%) were the products of plant origin, 279 samples (57.1%), – of animal origin and 70 (14.3%) – other products, including dietary supplements (Table 1).

Table 1. The amount of food products in circulation on the territory of the Murmansk region, taken for quality assessment in 2020.

Category of Food Product	Total number studied (number of tests, items.)	Total number studied (number of tests, %)
Foods of plant origin	140	28.6
Foods of animal origin	279	57.1
Other products	70	14.3
Total	489	100

Of the 489 samples studied, 58 were found as not corresponding to safety indicators, established values of regulatory and technical documentation (RTD), which is almost 12% of the total number (Table 2).

The largest percentage of samples not meeting the requirements of the regulatory technical documentation has been identified in animal products – 7.8% of the total number of the samples studied. The second place (2.5%) as unsafe products were taken by products of plant origin, and the category "Other products" count for 1.8% (Table 2).

Table 2. The amount of food products in circulation on the territory of the Murmansk region studied for safety assessment in 2020.

Category of Food Product	Number of tests, conforming to RTD		Number of tests, not conforming to RTD	
	Abs. number	%	Abs. number	%
Foods of plant origin	128	26.2	12	2.5
Foods of animal origin	242	49.5	38	7.8
Other products	61	12.4	9	1.8

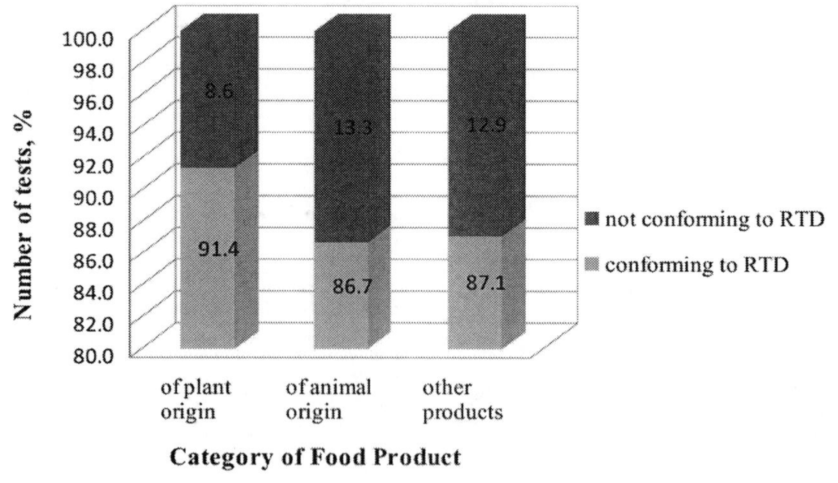

Fig. 1. The share of low-quality samples of food products of various categories in circulation on the territory of the Murmansk region in 2020.

Foods from the category "Products of animal origin" turned out to be the most unfavorable in terms of safety for a consumer. It should be noted that, in this category a broader range of products compared to two other groups (Figure 1, Table 3) was assessed. Of the 279 samples 38 (13.6%) do not correspond to the safety indicators of established RTD values, or 11 positions from 22 taken for tests: eggs, milk, dairy products and meat processing foods (Figure 1, Table 3).

Table 3. The number of food products of animal origin, in circulation on the territory of the Murmansk region, taken for safety assessment in 2020.

No	Foodstuffs	Number of tested samples	Conform to RTD	Do not conform to RTD
1.	Cottage cheese	15	15	–
2.	Ice cream	10	10	–
3.	Boiled sausages	10	10	–
4.	Ft included short bratwursts	4	4	–
5.	Semi-finished products	20	20	–
6.	Canned meat	10	10	–
7.	National meat products	10	10	–
8.	Oceanic chilled and frozen fish	18	18	–
9.	Fish preserves	7	7	–
10.	Canned fish	8	8	–
11.	Curd	10	10	–
12.	Eggs	15	13	2
13.	Milk	20	19	1
14.	Yogurt	20	13	7
15.	Kefir	10	0	10
16.	Sour cream	20	19	1
17.	Cake cheese	10	9	1
18.	Cheese	15	13	2
19.	Condensed milk	15	14	1
20.	Butter	15	9	6
21.	Bbockwurst sausages	8	3	5
22.	Sausages	9	7	2
	Total	279	241	38
	%	100	86.4	13.6

Among the safety indicators, toxic elements, pesticides, antibiotics, preservatives, mycotoxins (aflatoxin), nitrates, DLFMR (dry low-fat milk residue), genetically modified enzymes (microbial transglutaminase) and sanitary-indicatory microorganisms were identified (Table 4).

All chemical and biological substances that may be dangerous for functioning of a macroorganism, either enter the finished products in technological processing, including the stages of processing the raw materials, packaging, storage and selling of goods, or accumulate in a food chain, on the one hand, and in the environment - on the another.

Therefore, from this point of view, food chemicals can be conventionally divided into the substance's characteristic or specific for a product, substances introduced into the food in order to achieve technological effect and contaminants.

One of the main problems of the food industry is to increase the shelf life of foods preserving and preferably enhancing the quality of products. For this purpose, various possibilities associated with inhibiting the chain mechanism of organic substances oxidation, in particular fat, by adding various food antioxidants to the product are utilized.

Considering that the toxicity of antioxidants depends on their concentration, any complications are not excluded when using antioxidants in the production of food products, thus the stringent control is needed.

No less danger to the consumer is contaminants, both of natural and man-made origin, up to 80% of the total amount of foreign chemicals entering with food.

Contaminants and nutritional supplements, like micotoxins and genetically modified enzymes, were not detected in any of the tested food categories.

Among the sanitary-indicatory microorganisms' bacteria from the Escherichia coli genera were found in cake cheeses in one of the tests, 12 samples of fermented milk products were revealed to have the excessive amount of mold and yeast fungi (sour cream, kefir, yogurt). Most likely, this indicates the unfavorable sanitary and hygienic condition of manufacturing or violating the shelf-life of these products.

The three products: condensed milk (1 sample), kefir (1 sample) and yogurt (7 samples) do not meet the standards and technical documentation in terms of the quality of the lactic dairy products DLFMR (dry low-fat milk residue), reflecting the natural origin and full value of the raw materials. The low rate of DLFMR in these products may indicate the use of poor-quality raw materials or falsification of the product, that is, the replacement of the product's part with some additives (for example, starch and water).

Determining the presence of toxic elements in foods is one of the most commonly monitored indicators (Hg, Cd, As, Pb, Cu, Fe). Toxic elements were found in all types of raw materials, largely in the final foods. In the raw materials of animal origin, the accumulation of toxic elements is possible when animals are fed on contaminated feed. In finished food products, toxic substances or, as they are called, impurities of endogenous origin are formed as a result of various processing methods (enzyme, heat, ionizing irradiation) [5].

Table 4. Foods of animal origin circulating on the territory of the Murmansk region in 2020 and tested for safety indicators required by the standards and technical documentation.

Foodstuffs	Safety indicator,

	% from the total number of animal products samples						
	Sanitary-indicative m/o		DLFMR	TE	Antibiotics	Preservation agents (acids)	
	conditionally pathogenic (including pathogenic)	sanitary-hygienic indicators				sorbic	benzoic
Eggs				0.7			
Milk				0.35			
Yogurt		0.35	2.5				
Kefir		3.6	0.35				
Sour cream		0.35					
Cake cheese	0.35						
Cheese				0.35	0.35		
Condensed milk			0.35				
Butter				2.15			
Bockwurst sausages						0.35	0.35
						1.1	
Sausages						0.35	0.35

The excess of the maximum permissible concentration of toxic elements was found in some samples of the foods of animal origin. The "leader" in the toxic elements content was butter (ferrum was found in 5 of the 15 samples, in 1 sample copper was detected), the excess of the maximum permissible concentration of arsenic and cadmium (2 samples) was revealed in eggs, in milk – lead (1 sample), in cheese – mercury (1 sample).

Moreover, the residual content of antibiotics (1 sample) was found in cheese. Antibiotics are not used in the cheese-making technology; therefore, the medical drugs can get there only with the dairy raw materials. In the process of cow breeding, a lot of drugs, including antibiotics are used for medical and preventive measures. Also, manufacturers of animal feed increase the shelf-life of their production by adding antibiotics. In turn, the feed enriched with antibiotics stimulates the animal growth.

In the process of cheese ripening, antibiotics present in the dairy raw materials can suppress the enzymatic activity of the lactic acid and aroma-forming microbiota, which cause the taste-aromatic bouquet of the finished product.

Moreover, the risk of eating the foods with antibiotic components which enter the environment passing through the consumer's organism leads to the formation and propagation of antibiotic-resistant microorganisms and genes, as well as increases the costs of subsequent medical and preventive measures and causes long-term bacteria carrying [6].

In bockwurst sausages and sausages, a high content of preservatives (sorbic and benzoic acid) was found. When analyzing the sausages, 2 negative samples from the 9 tested ones were obtained. The analysis of bockwursts showed five samples of non-compliance, and 3 samples were found to fail in both preservatives simultaneously. Sorbic and benzoic acids are preservatives, their action is essential, mainly against the microscopic fungi, including aflatoxin-forming. Dosage of preservatives depends on the initial degree of bacterial dissemination, from the functional properties of the product, processing and storage technology [7]. However, the rigid manufacturers' competition in the food industry market leads to an increase in the number of preservative additives used [8].

It should be noted that all fish products samples were found to be safe for the consumer.

The residents of Murmansk and the Murmansk region, can eat such foods as boiled sausages, meat semi-finished products and canned food, national meat products, fat included short bratwursts, as well as cottage cheese, curds and ice cream without fear for their health.

In terms of safety indicators, 90% of samples of the plant origin products and the ones referred to the category "other foods" met the requirements specified for the products (Table 5).

When analyzing some samples of apples, cucumbers, tomatoes, sunflower and olive oils, cookies and dietary supplements unsatisfactory results were obtained, constituting 10% of the total number of the tested products of the corresponding categories (Table 5).

The content of pesticides unstated in RTD exceeded the maximum permissible level in fresh vegetables and fruits (5.3% of samples in the category of products of plant origin and other products) (Table 6).

In order to preserve and increase the crop, the use of pesticides in agriculture is quite justified. At the same time, these substances negatively affect the environment, violating the balance in the environmental system of flora and fauna. Their presence in agricultural raw materials and food is a significant danger to public health.

The constant admission of low-toxic pesticides to the body, even in small quantities, can lead to poisoning, decrease of immunity, development of allergic responses of different severity [2].

Table 5. The number of food product samples of plant origin and other products assessed for safety indicators in circulation on the territory of the Murmansk region in 2020.

No	Foodstuffs	Number of tested samples	Conform to RTD	Do not conform to RTD
1.	Bakery products	40	40	–

2.	Fortified grains	15	15	–
3.	Pasta	15	15	–
4.	Gluten-free foods	15	15	–
5.	Juice, fruit nectar	15	15	–
6.	Canned fruit and berries	15	15	–
7.	Apples	10	8	2
8.	Cucumbers	8	4	4
9.	Tomatoes	7	1	6
10.	Iodinated salt	10	10	–
11.	Confectionery	10	10	–
12.	Sunflower oil	10	9	1
13.	Olive oil	10	6	4
14.	Cookies	10	9	1
15.	Dietary supplements	20	17	3
	Total	210	189	21
	%	100	90	10

When analyzing cucumbers from the 1st sample, besides pesticides, an excessive concentration of nitrates was found. The presence of nitric acid salts (nitrates) in cucumbers is a common phenomenon, especially in the early ripening varieties. Nitrates enter the plants not only from fertilizers, but also from the soil. Not harmful to the fruits themselves, nitrates (already in the form of nitrites) harm a human organism [2], being hematoxic, they violate the properties of hemoglobin. As a result, tissue hypoxia develops, which is accompanied by weakness, dizziness, headache, shortness of breath, heartbeat.

Excess of the permissible values of cadmium and iron was noted in single samples of cookies and sunflower oil, respectively (Table 6). In 4 samples of olive oil, excess of the maximum permissible level of arsenic, lead, copper and a peroxide value were revealed. In one of the samples, non-compliance in the content of cadmium and lead was defined in the assessment of dietary supplement samples. In 4 olive oil samples and 2 dietary supplement samples, the discrepancy in peroxide value reflects the degree of oxidative spoilage of fat. In oils and dietary supplements with high peroxide number, partial or complete loss of biological activity occurs. Ultimate oxidation products are toxic and can cause a carcinogenic effect [2].

Table 6. Food products of plant origin and other products assessed for the safety indicators from regulatory technical documentation requirements, in circulation on the territory of the Murmansk region in 2020.

Foodstuffs	Safety indicator, % from the total number of plant products and other products samples						
	Pecticides	Nitrates	Cd	As	Pb	Cu	Fe	Peroxide value
Apples	0.95	–	–	–	–	–	–	–
Cucumbers	1.4	0.475	–	–	–	–	–	–
Tomatoes	2.9	–	–	–	–	–	–	–
Sunflower oil	–	–	–	–	–	–	0.475	–
Olive oil	–	–	–	0.475	0.475	0.475	–	0.475
Cookies	–	–	0.475	–	–	–	–	–
Dietary supplements	–	–	0.475	–	0.475	–	–	0.475

4 Conclusion

As a result of the study the range of products in circulation on the territory of the Murmansk region, capable of providing the body's needs with essential nutrients, most of meat and milk processing products samples, as well as chicken, egg and milk do not meet the stated safety indicators. This reduction in the quality of the products is mostly associated with violations in production technology, including falsification of raw materials and the conditions of selling the finished product.

Fresh fruits and vegetables (apples, tomatoes and cucumbers) were not of good quality for the presence of residual content of pesticides (apples, tomatoes and cucumbers).

Vegetable oils (in particular olive oil) and dietary supplies are featured with high content of toxic elements and peroxide value, which is also an indicator of violation in processing and storage conditions of the product.

References

1. MR 2.3.7.0168-20. Assessment of the quality of food products and evaluating the access of the population to domestic food products promoting the elimination of macro and micronutrients

deficiency, retrieved from: http://www.consultant.ru/document/cons_doc_LAW_359586/
2. G. G. Onishchenko, et al., Health risk analysis in the strategy of state social and economical development monography, *Perm National Research Polytechnic University* p 737 (2014)
3. About the state veterinary laboratory monitoring of safety and reinforced laboratory control of food raw materials of animal origin in the Russian Federation, retrieved from: http://www.consultant.ru/document/cons_doc_LAW_137013/
4. On the organization and conduct of monitoring of quality, food safety and public health: Government Resolution Ros. Federation No 883, retrieved from: https://base.garant.ru/12121276/
5. I. A. Rogov, et al., *Food Biotechnology. Basics of Food Biotechnology* vol 1 (Moscow: Colossus, 2013)
6. L. V. Shulgina, Antibiotics in Aquaculture Facilities and Their Ecological Significance, *Tinro News* **181,** 217 (2015)
7. M. F. Tsukanov, et al., *Methodology Human Nutrition* (St. Petersburg: Publishing House of SPbGEU, 2015)
8. N. V. Zaitseva, P. Z. Shur, Actual issues of methodological support for public health risk assessment while ensuring the safety of products: World Foreign Experience and Practice of the Customs, Union *Risk analysis of health* **4,** 4–16 (2013)

Role of agriculture in climate change and adaptability: Use of innovative technologies in the Russian agricultural sector for climate conservation and protection

I. V. Kryukova[1*], *A. A.* Lagun[1], *S. G.* Golubeva[1], and *E. N.* Yakovleva[2]

[1] Vologda State Dairy Farming Academy named after N.V. Vereshchagin, 2 Schmidt St., Molochnoe village, Vologda, 160555, Russian Federation
[2] Russian Presidential Academy of Natural Economy and Public Administration, 71 Leningradskaya, Str., Vologda Oblast, Vologda, 160017, Russian Federation

Abstract. Global climate change poses two major challenges to the agricultural sector of the economy. First, it is necessary to assess the value and extent of its impact on atmospheric pollution by greenhouse gases, and, based on this, to determine priorities and the emission reduction strategy. The second concerns the adaptation of agriculture to threats and the use of causes from global climate change. Therefore, the purpose of the study is to disclose these two aspects. First, the national contribution of agriculture to greenhouse gas pollution was studied using the example of the Russian Federation according to the author's methodology. Then the directions of the state adaptation policy towards the studied sector in the light of climate change are proposed. The results of the study can be useful in developing the adaptation strategy, increasing the flexibility of the agricultural sector of different countries to climate change.

1 Introduction

The interaction of agriculture with climate risk factors is quite complex and requires careful study. On the one hand, this sector makes a significant contribution to air pollution with greenhouse gases: nitric acid emissions account for 90% of the world's emissions, methane – 70% and carbon dioxide – 20% of all emissions [1]. Therefore, there is a need for innovations to reduce the technogenic burden of the agricultural sector on the environment without increasing average costs and prices for agricultural products. On the other hand, the natural interest is caused by an increase in the flexibility of agriculture response to changing climatic conditions. This increases the relevance of developing and implementing an effective climate change adaptation mechanism. Many works of foreign and domestic authors are devoted to the study of these problems. Since traditional agricultural methods are associated with greenhouse gas emissions [1], it becomes necessary to look for opportunities to reduce the negative impact of this sector on the climate [2]. In turn, the efforts to reduce greenhouse gas emissions also have a negative impact on the capacity of agricultural markets and their pricing. In this regard, the threats to ensuring food security of the world community as a whole and the problems of hunger in certain poor regions of Africa and South Asia are increasing [3]. The issue of methodical provision of climate risk assessment for the agricultural sector remains unresolved [4, 5, 6]. It is critical to address this issue since regional climate change forecasts and modeling of the associated future change in socio-economic systems are needed to regulate, adapt and develop the climate risk management strategies [4]. In addition, global warming is causing changes in land use and increases competition for land and water resources. The existence of a number of consequences of climate change seems quite positive, for example, an increase in the growing season and the involvement of previously unused northern territories in agricultural circulation [7], the development of agriculture due to an increase in demand for biofuels as a low-carbon energy resource [2], and so on. Therefore, the development of adaptation strategies and measures for agriculture is one of the most debatable issues [8, 9]. Particular attention is paid to the study of ways to intensify the adaptation of the agricultural industry to climate change, since the incentives for such adaptation by private enterprises are underestimated [7]. A reactive approach prevails in agriculture, when farmers make short-term management decisions in response to climate challenges [10]. At the same time, it is the agricultural sector that requires the development and application of proactive strategic adaptation policies, since long-term climate risks may be much greater than the current ones [6]. An attempt to solve methodological problems at the national level was made earlier by the authors [11]. Within the framework of this study we will test the proposed methodology to determine the role of agriculture in the anthropogenic component of climate change using the example of the Russian Federation and propose recommendations for the development of appropriate adaptation policies.

* Corresponding author: iri917717@yandex.ru

https://doi.org/10.1051/bioconf/20225200082

Let us study the intersectoral structure of greenhouse gas emissions in the Russian Federation to identify the position of the industry in the Russian economy in terms of natural and climatic impact. The energy sector dominates in this structure (according to Rosgidromet), the share of which in the total emission was 78.7% in 2019, in the second place – industry (11.2%), in the third – agriculture (5.4%). This state of affairs is also characteristic of other countries. For example, in Turkey in 2016 the energy sector took first place with a share of 72%, followed by industry (12.6%) and agriculture (11.4%) [1]. In 2019, greenhouse gas emissions from the agricultural sector of the Russian Federation amounted to 114.2 million tons of CO_2-eq., which corresponds to 46.1% of the 1990 level (247.5 million tons of CO_2-eq.). The reasons for this reduction are a decrease in the number of farm animals (for example, during this period the number of livestock decreased by almost 2.4 times and poultry – by 17.2%), a decrease in sown areas (decreased by 30.6%) and the norms for introduced mineral fertilizers (application of mineral nitrogen fertilizers decreased by 59.4%). Soils (mainly from nitrogen fertilizers) and internal fermentation (CH_4) account for the largest share of CO_2-eq. agricultural emissions (Figure 1).

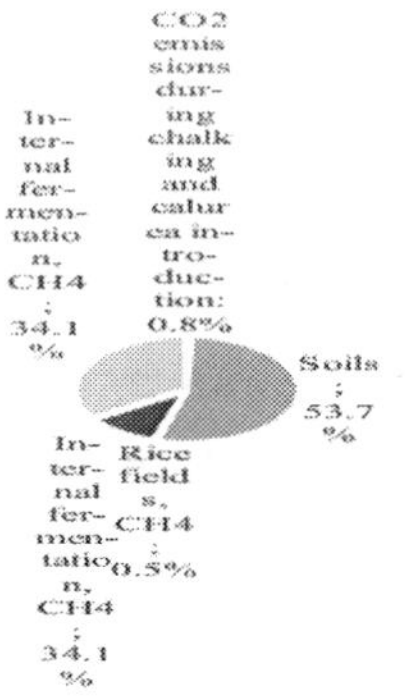

Fig. 1. Structure of greenhouse gas emissions in Russian agriculture by sources (thousand tons of CO_2-eq.) in 2019.

2 Materials and methods

The previously conducted comparative assessment of the Russian regions using the author's methodology [11] in the coordinates "climate intensity-energy intensity" showed that the majority of the Russian regions in both indicators is around the median [12]. The exception is energy intensity – Vologda Region (78.5% of the turnover – enterprises in the manufacturing industries, mainly ferrous metallurgy and chemistry), Kemerovo Region (72% of turnover – mining and manufacturing), Lipetsk Region (69.5 – enterprises of manufacturing industries, mainly ferrous metallurgy) and the Republic of Khakassia (44% of turnover – manufacturing enterprises), and in terms of climate intensity – Nenets Autonomous Okrug (70% of the turnover – mining), Yamalo-Nenets Autonomous Okrug (66% of the turnover – mining), Khanty-Mansi Autonomous Okrug (73% of the turnover – mining), Krasnoyarsk Territory (59% of the turnover – mining and manufacturing) and the Republic of Komi (42% of the turnover – mining). It seems clear that the high energy intensity of the regional economy is formed by enterprises in the manufacturing industries, mainly ferrous metallurgy, and significant climate intensity is formed by mining.

Figure 2 shows the TOP-15 in the ranking of the Russian regions by the volume of agricultural products produced in 2019. According to the figure, the presence of a developed agricultural sector in the region's economy does not have a decisive impact on its position in the climate risk matrix [12]: agricultural territories are found in almost all groups of the matrix. Thus, the leader of the agricultural industry – Krasnodarsky Krai has high climate intensity and low energy intensity. Eight subjects from TOP-15 have low climate intensity and high or medium energy intensity.

The application of the author's methodology for the comparative characterization of sectors [13] revealed that in terms of the combination of two indicators, the highest natural and climatic risks are inherent in the energy sector (due to the predominant use of traditional energy sources) and the mining industry (mainly due to the high energy intensity of production). Waste sector holds the second place and Agriculture – the third. All three types of economic activity fell into the group with the highest level of climate intensity. Mining turned out to be the most energy-intensive activity. Transport and Manufacturing sectors have average climate risks. No sector has low risks.

The subject of this study is agriculture as one of the climate-intensive types of economic activity and a solution to the problem of developing the adaptation policies due to the impact on climate.

The purpose of the study is to analyse the national contribution of agriculture to greenhouse gas pollution using the example of the Russian Federation via the author's methodology and to propose directions for the state adaptation policy towards the studied sector in the light of climate change.

A set of methods of scientific knowledge were used in the study, including analysis, synthesis, induction, deduction, abstractions, tabular method, etc.

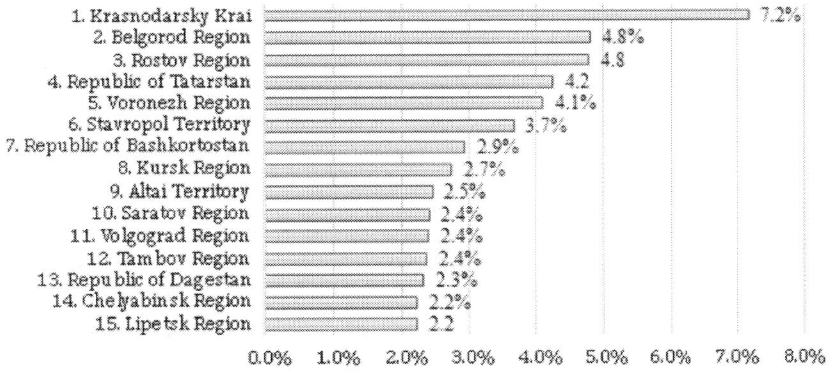

Fig. 2. Relative share of agricultural products of the regions from the total agricultural products in the country (in farms of all categories in actual prices), %.

3 Results and discussion

The study showed the average impact of the agricultural sector of the Russian Federation on the level of climate threats. The application of this method is of scientific and practical interest not only for inter-industry, but also for intra-industry analysis, which requires the statistical accounting of greenhouse gas emissions and energy consumption in the context of industries and sub-sectors of the country's national economy.

Since agriculture, in either absolute or relative terms, clearly does not belong to the leaders of the negative impact on climate, it is possible to recommend that the national climate policy focus on the adaptation of this sector to climate threats. This is especially true because global climate change is causing a whole set of new problems in agricultural production that can be combined into three groups.

1) The climate factor is decisive in ensuring agricultural productivity, and the increase in climate risks and their negative consequences (expansion of pest ranges, the spread of bacterial and viral diseases, etc.) increase threats to the country's food security. Therefore, there are conceptual and methodological problems related to the need to take into account, assess the likelihood of occurrence and scale of the spread of the consequences of natural and climatic risks for agriculture, and their management tools. Theoretical and applied solutions are also needed to adapt the industry to this risk group.

2) Global warming leads to a longer growing season in the northern regions, creating conditions for an increase in the area of farmland, the development of which requires the attraction of very scarce investment resources. The enrichment of the atmosphere with CO_2 creates conditions for the increased yield of grain and feed crops. However, the positive effects of climate change on agriculture may be less significant than the negative ones (Table 1).

Since increased climate risks lead to both additional threats and chances for agricultural development (Table 1), the adaptation measures are needed to reduce the former and use the latter.

3) There is a need for a mechanism to stimulate the reduction of anthropogenic greenhouse gas emissions from the agricultural sector, as well as technologies and methods for such a reduction. However, if the state imposes stringent requirements to reduce greenhouse gas emissions, this could have a more significant negative impact on the country's food security than direct losses from climate change [3].

All three of the above groups of problems are most effectively solved through the activation of innovative processes. Innovation plays a decisive role in intensifying economic development with qualitative improvement in farming. Innovative activity in agriculture has significant specifics in connection with the peculiarities of the industry itself: low profitability, competitiveness, and as a result, the investment attractiveness of the industry, high land capacity, long production cycle, stationary and seasonal nature of production, high dependence on climatic conditions, low rates of scientific and technical innovations in comparison with other industries, etc. Strengthening the innovative development of the agricultural sector is possible only by increasing the volume and list of measures of state support for this process.

The interests of the state in this process lie in the plane of ensuring food security, developing import substitution, increasing the competitiveness of the industry, developing rural areas, and, as a result, solving such social problems as improving the quality of life of the population, reducing poverty, differentiating incomes and others. In our opinion, the state policy on the adaptation of the agricultural sector to climate change should be focused on the innovative development of the industry using the following support measures:

- distribution of state resources (orders, grants, loans) between priority R&D areas;
- stimulation of innovative cooperation between agricultural, educational and scientific

enterprises;
- support for innovative entrepreneurship in the agricultural sector;
- application of tax, depreciation, antimonopoly, patent, foreign trade policy instruments through differential subsidies and the provision of benefits that create conditions for accelerating scientific and technological progress in the field of adaptation of agriculture to climate change;
- stimulation of public-private partnership in the considered sphere;
- regional support for the development of innovations, including digital technologies.

Table 1. Impacts of climate change on agricultural sectors and their adaptation to climate risks.

Types of risks	Climate change impacts		Risk adaptation pathways
	negative	positive	
Increase in soil erosion, increase in groundwater level, flooding of areas, floods, droughts, fires.	Expansion of the habitats of some weeds, parasites, insects – pests, zones of viral and bacterial diseases of animals and plants. Soil erosion due to increased wind loads, landslides, etc. Deterioration of agricultural conditions due to an increase in the level of groundwater, underfloods, flooding, floods, desertification of territories. Reduction of pasture areas.	Expansion of possibilities of growing forage, grain and vegetable crops in the northern regions due to an increase in the growing season. Increase in the yield of grain and forage crops due to the enrichment of atmospheric air with CO_2.	Selection of crops and zoned varieties, measures to preserve moisture, irrigation, application of fertilizers, application of new technologies, improvement of material and technical base for full implementation of agricultural technology of cultivation, diversification of agricultural production, creation of food stocks, insurance against hazardous hydrometeorological phenomena, prevention of negative human interference in natural processes (drainage, plowing of floodplains, errors in the construction of irrigation systems, etc.), enhancement of the competitiveness of the industry and its investment attractiveness.

4 Conclusion

The study revealed the following:
- first, the agricultural sector contributes several times lower to the anthropogenic component of the greenhouse effect than energy, and lower, but comparable, than the manufacturing industry;
- second, the relationship between the regional contribution to the greenhouse effect and the level of agricultural development was not identified: the agricultural sector is developed in territorial entities with both high and low climatic impact;
- third, the agricultural sector belongs to the types of economic activities with an average level of risk of impact on the climate;
- fourth, since agriculture is not a priority in the management impact of reducing greenhouse gas emissions, it is recommended that key attention be paid to developing a government mechanism for adapting the industry to climate change. The best direction of adaptation, according to the authors, is innovative development. The study gives general recommendations on the state stimulation of the introduction of innovations at agricultural enterprises.

References

1. M. Çetin, S. Saygın, H. Demir, The impact of agriculture on environmental pollution, *Turkish Economy For A Cointegration and Causality Analysis* **17 (3),** 329 (2020)
2. K. M. Findlater, T. Satterfield, M. Kandlikar, S. D. Donner, Integration anxiety: the cognitive isolation of climate change. *Global Environmental Change* **50,** 178 (2018)
3. T. Hasegawa, et al., Risk of increased food insecurity under stringent global climate change mitigation policy, *Nature Climate Change* **8 (8),** 699 (2018)
4. W. J. W. Botzen, J. C. J. M. Van Den Bergh, Managing natural disaster risks in a changing climate, *Environmental Hazards* **8 (3),** 209 (2009)
5. F. Ewert, et al. Crop modelling for integrated assessment of risk to food production from climate change, *Environmental Modelling & Software* **72** (2015)
6. I. Brown, Assessing climate change risks to the natural environment to facilitate cross-sectoral adaptation policy, *Philosophical Transactions: Mathematical, Physical and Engineering Sciences (Series A)* **376 (2121)** (2018)
7. N. N. Yashalova, T. K. Molchanova, D. A. Ruban Prerequisites for effective management of climate risks in Russian agriculture: investment and innovative aspect, *Bulletin of UrFU. Economics and Management* **18 (5),** 637 (2019)
8. R. P. Motha, M. V. K. Sivakumar, Managing weather and climate risks in agriculture summary and recommendations, *Managing Weather and Climate Risks in Agriculture* **477** (2007)
9. R. E. Schattman, G. Roesch-Mcnally, S. Wiener, M. T. Niles, D. Y. Hollinger, Farm service agency employee intentions to use weather and climate data in professional services, *Renewable Agriculture and Food Systems* **33 (3),** 212 (2008)
10. M. T. Niles, M. Lubell, V. R. Haden, Perceptions and responses to climate policy risks among California farmers, *Global Environmental*

Change **23 (6),** 1752 (2013)
11. E. N. Yakovleva, N. N. Yashalova, D. A. Ruban, V. S. Vasiltsov, Methodological approaches to natural and climatic risks for the sustainable development of the state, *Scientific notes of the Russian State Hydrometeorological University* **52,** 120 (2018)
12. V. S. Vasiltsov, N. N. Yashalova, E. N. Yakovleva, A. V. Kharlamov, National climate policy: Conceptual frameworks and adaptation challenges, *Economy of the Region* **17 (4),** 1123
13. E. N. Yakovleva, N. N. Yashalova, V. S. Vasiltsov, Analysis of natural and climatic risks of economic activities in the Russian Federation, *Scientific notes of the Russian State Humanitarian University* **61,** 421 (2020)

The prospects of AI and IoT technology development effect on the carbon market regulation

Dmitry Mikhailov[1], *Ilman* Shazhaev[2], *Vera* Chumanskaya[3], *Victor* Abramov[4], *Zafar* Iskandarov[5], and *Valery* Grempel[6]

[1] China Branch of BRICS Institute of Future Networks, Shenzhen, China
[2] Shanghai Jiao Tong University, Shanghai, China
[3] Primakov National Research Institute of World Economy and International Relations of the Russian Academy of Sciences, 23, Profsoyuznaya St., Moscow, 117997
[4] National Research Nuclear University MEPhI, 31, Kashirskoe shosse, Moscow, 115409, Russian Federation
[5] Tashkent State Agrarian University, 2a, University St., Tashkent, 100140, Uzbekistan
[6] Moscow State Institute of International Relations (MGIMO), 76, Vernadsky Prospekt, Moscow, 119454, Russian Federation

Abstract. The problems of regulating the carbon quota market, which is formed both at the state level of individual countries and at the supranational level within the framework of international associations, for example, such as the Conference of the Parties to the UN Framework Convention on Climate Change, are considered. At the moment, the issue of rational regulation of measures to reduce the negative effects of climate change while maintaining the pace of economic development is extremely urgent. Artificial intelligence and the Internet of Things technologies play an important role in solving this issue, which allow optimising production processes and achieving zero emissions in a growing economy. The authors see the prospect of further research in the search for specific measures to regulate the carbon market, in creating favourable economic, social and political conditions for their implementation, as well as in maintaining and stimulating the exchange of experience in this area at the international level.

1 Introduction

The global COVID-19 pandemic has led to the realisation of how important it is for humanity to be prepared for impending challenges and to search for sustainable development approaches in advance. The UN believes that the main goal of the concept of sustainability, in fact, is to ensure an appropriate balance between society, the economy and the environment in terms of the stability of the planet and the reliability of life-supporting ecosystems [1]. Modern theories of sustainable development seek to prioritise and integrate social, environmental and economic models in solving human problems in such a way that it constantly benefits people [2].

Climate change is an existential threat that humanity is facing today. Following the results of the 26th Conference of the Parties to the UN Framework Convention on Climate Change (hereinafter - COP26) held in November 2021, the participants decided to reduce emissions by 2030 and submit updated plans by the end of 2022. The most important outcome of the COP26 Conference was the harmonisation of the rules of international trade in carbon units, which involves the introduction of a single set of rules and a gradual reduction of quotas. The implementation of COP26 solutions is important for the image of the country, and also opens up opportunities for attracting investments to the country and creating new jobs through the implementation of the carbon trading program.

Rapidly developing artificial intelligence (AI) technologies are already being used in various sectors of the economy to develop and implement environmental impact management methods. Examples of such practices are: precision farming and environmental monitoring. And it has a huge potential to reduce greenhouse gas emissions without reducing the pace of economic growth.

The Internet of Things (IoT) is a digital control system, a network of objects connected via the Internet by sensors, software and other technologies that exchange data with each element of the network. Open access to the introduction of the Internet of Things around the world can provide a truly waste-free economy. Thus, it is possible to establish industries where related production processes are linked so that there is no waste and emissions.

The purpose of this study is to identify promising areas of carbon market regulation to achieve sustainable development goals using AI and IoT technologies.

2 Materials and methods

In the course of numerous discussions of climate change problems by the expert community, scientists mainly oppose temperature increases due to anthropogenic gases (primarily CO2), the concentration of which has increased by almost 50% since the beginning of the industrial revolution [3].

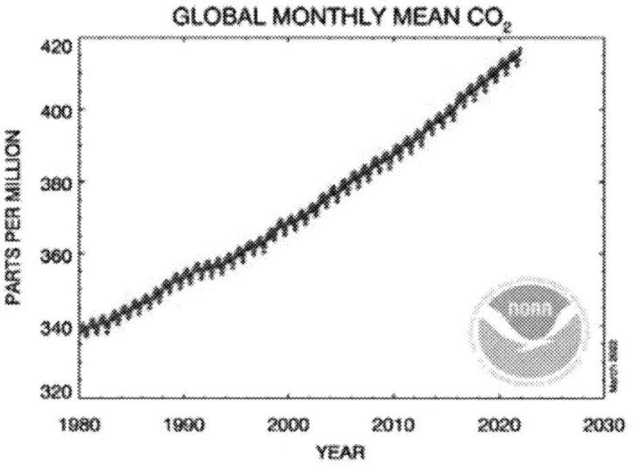

Fig. 1. The average annual global concentration of released hydrocarbons from 1980 to 2020 according to the Mauna Loa Observatory [4].

Figure 1 shows a graph of the growth of CO2 concentration in the atmosphere from 1960 to 2022, where on the ordinate axis is the time measured in years, and on the abscissa axis is the concentration of CO2 in the atmosphere in millionths (ppm). The graph clearly shows an increase in the concentration of CO2 in the atmosphere: from 1980 to 2020, the concentration of carbon in the atmosphere increased from 340 ppm to 420 ppm, that is, by 80 ppm.

Back in 1997, the Kyoto Protocol stated that there are six anthropogenic greenhouse gases whose emissions need to be reduced: carbon dioxide (CO2), methane (CH4), nitrous oxide (N2O), hydrofluorocarbons, perfluorocarbons and sulphur hexafluoride (SF6). In this sense, CO2 is one of the first anthropogenic greenhouse gases, which accounts for approximately 76% of the total amount of greenhouse gases considered by the Kyoto Protocol [5]. Data-t-on anthropogenic greenhouse gas emissions are presented in Fig. 2 in the form of four graphs.

https://doi.org/10.1051/bioconf/20225200083

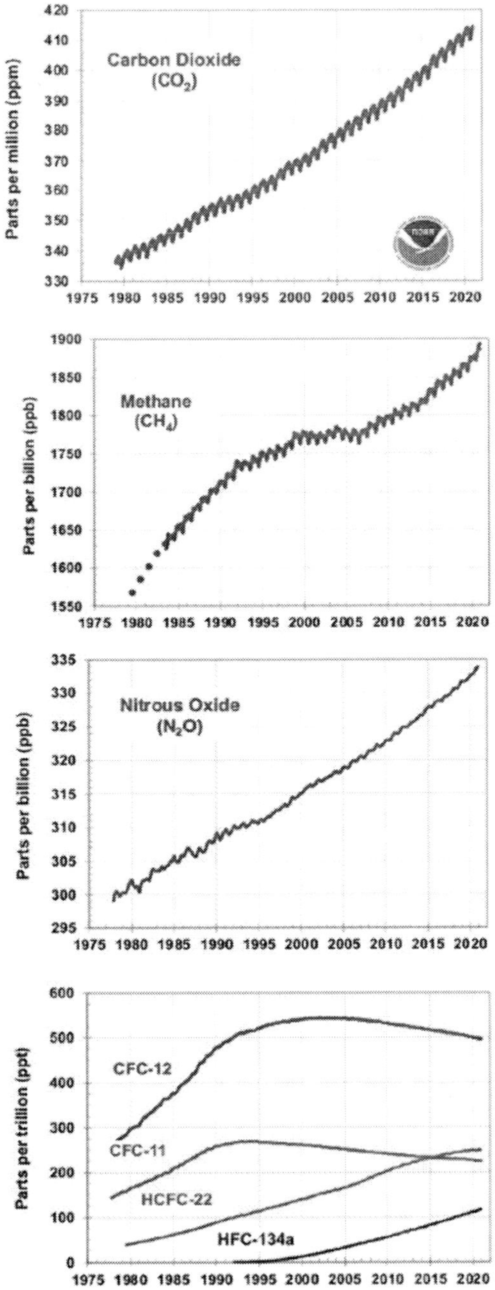

Fig. 2. Anthropogenic greenhouse gas emissions from 1980 to 2020 according to the Mauna Loa Observatory [4].

Meanwhile, the Kyoto Protocol has established a number of innovative cooperation mechanisms to reduce the costs of mitigating greenhouse gas emissions, namely: mechanisms for international emissions trading, joint implementation and clean development. These three mechanisms include carbon emissions trading between countries. Since the carbon trading mechanism has become an important means of effectively reducing energy consumption, large-scale trading platforms have emerged at the international level, such as the European Union Emissions Trading System (EU-ETS) and the Regional Greenhouse Gas Initiative.

Currently, two points of view on the impact of environmental legislation on pollution reduction are discussed in the literature. The first point of view is that environmental regulation harms the economic interests of enterprises, causing pollution control costs, and does not contribute to improving the efficiency of carbon emissions [6]. The second one is based on the "Porter hypothesis", according to which environmental regulation can force enterprises to carry out innovative activities, increase their productivity in order to achieve an innovative compensatory effect [7], eliminate or modernise lagging production capacities, optimise resource allocation using AI and IoT technologies, and thus increase the effectiveness of measures to reduce carbon emissions. The authors confirmed that modernization of the industrial structure and "green" technological innovations are two effective channels through which environmental regulation contributes to the "green" transformation of the economy [8]. From the point of view of enterprises, when the trading price of the carbon emission right in the market is higher than the marginal cost of pollution control of the enterprise, it makes sense for enterprises to take the initiative to improve the efficiency of carbon emissions control, based on the goal of maximising profits. At the same time, it is possible to effectively use the carbon emissions quota to prevent additional costs and sell the excess carbon emissions quota, increasing emission reductions to generate additional income [9].

After 6 years of discussions within the framework of COP26, a concrete plan for working with carbon credits in international markets was adopted. Carbon credit is all types of reductions in greenhouse gas emissions as a result of project activities, including emission reduction units (quotas) and units of certified emission reductions. A carbon (carbon) loan is a permit that allows the company that holds it to emit a certain amount of carbon dioxide or other greenhouse gases. One credit permits the release of a mass equal to one ton of carbon dioxide. The ultimate goal of carbon credits is to reduce greenhouse gas emissions into the atmosphere.

Units of measurement:

1 carbon credit = 1 ton of CO2 (carbon dioxide) emissions;

1 quota = 1 ton of permitted CO2 emissions.

The mandatory carbon market is controlled by the UN Framework Convention and the Kyoto Protocol. The principle of functioning of the mandatory carbon market is that polluting companies buy loans, which are a kind of permit for the emission of a specific amount of CO2. Such loans enable companies to continue to emit carbon dioxide up to a certain limit. The emission limit is reduced with a certain frequency. Thus, companies can make changes to their production plans, a plan to optimise and modernise production in order to reduce carbon dioxide emissions. In addition, the company can sell unused loans to another company that needs them. Thus, production companies have incentives to reduce greenhouse gas emissions: firstly, the company may be fined for exceeding the emission limit, and secondly, the company may receive additional profit from the sale of unused loans that appear as a result of optimization and modernization of production.

Countries that have difficulties with reducing greenhouse gas emissions can buy quotas from those countries that have exceeded their goals for cleaning the atmosphere. At the same time, various organisations, whether private companies or state institutions, can invest in projects that reduce emissions.

When working in the UN-controlled market, each State must confirm to the supervisory authority that the emission reduction project directly contributes to sustainable development and can be taken into account in the contribution determined at the national level. Currently, carbon units in the country should be recorded by an authorised and accredited organisation in international bodies in special electronic accounting registers. The results of each climate project must be verified in order for the carbon units of such a project to be reflected in the register. Further, carbon units can be put into circulation both in exchange and over-the-counter segments.

There are several carbon unit trading systems in the world today:

 https://doi.org/10.1051/bioconf/20225200083

• American Carbon Registry (ACR);

• Climate Action Reserve (CAR);

• Gold Standard (Gold Standard).

Technical trading conditions are provided by exchanges, in particular the Carbon Trade Exchange in London and Sydney, the AirCarbon exchange in Singapore.

3 Results and discussion

The creation of a state inventory system based on AI and IoT will allow the implementation of a mechanism involving the involvement of all sectors of the economy of large enterprises in the fulfilment of obligations and, accordingly, the inclusion in the inventory of all existing sources of emissions in the country [10]. It should be noted the important role of state regulation of innovative activity of organisations in order to reduce greenhouse gases [11].

Within the framework of the proposed mechanism, the State creates an aggregated limit of greenhouse gas emissions, which represents the maximum allowed amount of emissions from pollution sources (enterprises) participating in this system for a certain period of time. Then the State issues a permit (quota) for each unit of emissions (usually 1 ton). The emission limit is determined by the sum of all existing quotas. This is followed by the process of allocating emission quotas between enterprises. Further, market mechanisms come into effect, where the functions of an observer and controller are performed by a responsible body determined by the state. The end of each reporting period is the reference point for checking the compliance of the actual emissions of each enterprise with the quotas received or purchased by it. If there are not enough existing quotas, the state establishes penalties for violation of obligations. At the same time, the size and conditions of punishments should be formed so that compliance with obligations becomes the most favourable and profitable way of doing business. In other words, the fines should be significantly higher than the quota price.

According to this principle, a monetary value is set for each greenhouse gas emission quota, which is equal to the alternative cost of reducing emissions. That is, in the case when the price of the quota exceeds the costs of the company to reduce emissions, the company can sell its quotas. However, if the costs of reducing emissions exceed the cost of the quota, then it is more profitable for such enterprise to acquire more quotas without reducing the level of emissions.

The transition to zero emissions requires significant changes at the public and industrial level, and governments, as well as corporations, are increasingly turning to technological innovations to achieve zero emissions goals [12]. Digital technologies such as AI and IoT make it possible to find sustainable solutions to many seemingly intractable social problems related to climate change [13]. The WEF report shows how digital technologies can help automate and significantly improve the efficiency of industrial, manufacturing and agricultural processes, and how artificial intelligence-based systems can contribute to a 4% reduction in global emissions by 2030 [14].

Digital transformation has accelerated in various government and business sectors, and it is necessary to continue to use the potential of information systems to solve key social, economic and environmental problems. With the support of well-designed information systems, researchers, economic actors, politicians and civil society can join forces to build a more sustainable future.

An important sign of digital transformation is the change not only of the models of the implementation of state functions, but also the rethinking of their composition and essence in the light of new technological opportunities. For example, the introduction of the Internet of Things in the field of public and municipal administration [15], artificial intelligence and big data technology is promising. As a result, all organisations will need to create a management accounting system that would be the basis for the transition to a culture of data-based decision-making [16].

Currently, the world is dominated by sequential final production processes, rather than interconnected value chains between industries. Gradually, with the help of IoT, powerful technologies are being created that can compensate for the losses from each process and exchange materials between related companies. Both materials and energy consumption can be separated to dramatically reduce CO_2 emissions. Thus, IoT will allow creating closed-loop economy technologies.

4 Conclusion

For several decades, the search has been conducted for effective ways to reduce greenhouse gas emissions in order to achieve sustainable development goals. The work on the creation of mechanisms for controlling greenhouse gas emissions has led to the formation of voluntary and mandatory carbon quota markets, which continue to develop both at the national levels of individual States and at the international level. In order to achieve the goals set within the framework of the COP26 to reduce greenhouse gas emissions, it is necessary to support and stimulate the development of the existing carbon trading infrastructure to increase accessibility for as many enterprises as possible, as a result of which greenhouse gas emissions occur.

References

1. *United Nations Statistics Division UNSD*, retrieved from: https://unstats-undesa.opendata.arcgis.com/ (2022)
2. I. Trusina, E. Jermolajeva, The scientific discourse on the concept of sustainable development, *East. Jour. of Eur. Stud.* **12** (2021)
3. V. Tetelmin, Physical foundations of anthropogenic climate change, *Gidrotechnica* **1,** 58 (2020)
4. Global Monitoring Laboratory. *Trends in Atmospheric Carbon Dioxide. Global Monthly Mean CO2*, retrieved from: https://gml.noaa.gov/ccgg/trends/global.html (2022)
5. E. Kissel, S. Eric, et al., *Climate Change 2014: Impacts, adaptation and vulnerability* (2014)
6. D. Zheng, M. J. Shi, Clean Multiple environmental policies and pollution haven hypothesis, *Evidence from China's polluting industries* **141**, 295–304 (2017)
7. M. E. Porter, C. Linde, J. Econ. Perspect. *Toward a New Conception of the Environment-Competitiveness Relationship* **9,** 97–118 (1995)
8. K. Du, Y. Cheng, X. Yao, Environmental regulation, green technology innovation, and industrial structure upgrading: the road to the green transformation of chinese cities *Energ. Econ.* (2021)
9. M. H. Shakil, N. Mahmood, M. Tasnia, Z. H. Munim, *Do environmental, Social and Governance Performance Affect the Financial Performance of banks? A Cross-Country Study of Emerging Market banks* **30,** 1331-1344 (2019)
10. V. I. Abramov, A. V. Borzov, K. Yu. Semenkov, Theoretical and methodological analysis of digital maturity models for Russian companies, *News of high. educ. inst.* **4,** 42-51 (2021)
11. V. I. Abramov, V. V. Asaul, S. G. Vagin, K. S. Nikolaeva, State regulation of innovative activity of organizations, *Bulletin of Samara State University of Economics* **11,** 91–96 (2014)
12. J. Miller, *Climate change solutions: The role of technology House of Commons Library*, retrieved from: https://commonslibrary.parliament.uk/climate-change-solutions-the-role-of-technology (2022)
13. G. George, R. K. Merrill, Digital sustainability and entrepreneurship: How digital innovations are helping tackle climate change and sustainable development *S. J., Entrep. Theory and Pract.* **45,** 999–1027 (2021)
14. WEF 2021, retrieved from: https://www.weforum.org/agenda/2021/10/as-the-world-gathers-for-cop26-here-s-how-leaders-can-dispel-esg-confusion-ac418a72bc (2022)
15. V. I. Abramov, V. D. Andreev, The possibilities of the Internet of Things for the digital transformation of public and municipal administration in Russia, *Collect. of art. of the V All-Rus. Scien. and Pract. Conf.* pp 3–11 (2021)
16. V. I. Abramov, D. M. Mikhailov, A. D. Stolyarov, The role and features of management accounting in the implementation of the digital transformation of the company, *Econ. and Manag.: probl., solut.* **4,** 165–168 (2021)

Impact of mineral fertilizers on winter wheat quality using No-Till technology

Yaroslav Ilchenko*, *Olga* Buryukova, *Vitaly* Ilyushechkin, and *Vladimir* Erin

Southern Federal University, Stachki Ave, 194/1, Rostov-on-Don, 344090, Russian Federation

Abstract. The paper presents the study of the influence of ammophos, potassium magnesia and ammonium nitrate on the content of protein, gluten and grain unit of winter wheat using No-Till technology in the edaphoclimatic conditions of the southern zone of Rostov Region. It was shown that the application of mineral fertilizers to the surface soil layer during winter wheat cultivation using No-Till technology improves the grain quality. The highest grain quality parameters (protein content – 13.4% and gluten content – 25.3%, grain unit – 780.0 g/l) were obtained when ammophos and potassium magnesia were jointly added prior to sowing at a depth of 10 cm with double dressing of ammonium nitrate in the tillering and booting phases. High direct dependence of protein content in winter wheat grain on the accumulation of nitrate nitrogen in soil in the booting phase was revealed in all test layers: 0–10 (r = 0.85), 10–20 (r = 0.72), 0–30 cm (r = 0.76). The gluten content in the grain and its grain unit significantly depended on the mobility of phosphorus in the surface soil layer – 0–10 cm (r = 0.66 and 0.52, respectively). Along with the quality improvement of winter wheat grain, the rational use of mineral fertilizers in the No-Till system increases its yield. There is a direct positive relationship between grain protein content and winter wheat yield (r = 0.93).

* Corresponding author: 10yaroslav@mail.ru

1 Introduction

Russian grain is a strategic export product and positions Russia in the international arena as one of the guarantors of global food security [17]. Grain production is the basis of crop production and the entire agricultural sector in the vast majority of agricultural zones of the world [1]. Currently, Rostov Region is the leader among other regions of the country in gross grain yield, and the largest producer of winter wheat. However, it is impossible to increase yield and improve the quality of winter wheat in conditions of intensive farming without the use of resource-saving technologies for its cultivation, including No-Till technology. One of the elements of these technologies is the rational use of mineral fertilizers, which is the main factor that increases the yield and improves the quality of crops while maintaining soil fertility [10, 18, 20]. The problem of optimizing the mineral nutrition of crops with the introduction of resource-saving technologies is relevant and at the same time the least studied in the edaphoclimatic conditions of the European part of southern Russia.

2 Materials and Methods

The studies were conducted from 2016 to 2018 within field tests [9] in the territory of Kirov ZAO, Peschanokopsky District, located in the southern natural and climatic zone of Rostov Region.

The climate pattern of this zone includes scarcity of precipitation, hot summer, low relative humidity in summer. Annual precipitation amounts to 410-460 mm, hydrothermal index = 0.7–0.76, average annual temperature – 8.7–9.5°C [15]. The agrometeorological conditions were different over the years of study, which made it possible to objectively assess the impact of mineral fertilizers on the quality of winter wheat.

Soil – ordinary calcareous medium-phase heavy loamy chernozem. According to the World Research Base, the studied soils belong to Calcic Chernozems [23]. Ordinary calcareous chernozems are among the best arable soils in the region, favorable for the cultivation of all field crops. The agrochemical characteristics of the arable soil layer of the test site are as follows: content of nitrate nitrogen and mobile sulfur – low, content of mobile phosphorus and magnesium – average; exchange potassium – high.

Test crop – winter wheat (Triticum aestivum L.), Grom variety. This variety is characterized by high winter hardiness, good drought resistance, therefore, it is widely cultivated not only in the Central Chernozem Region, but also in the North Caucasus and Lower Volga regions characterized by unstable and insufficient moisture. Predecessor: flax Linum L.

Experimental design: 1. Control (without fertilizers); 2. N12P52 in the sowing phase + N30 in the tillering phase + N70 in the booting phase; 3. K32Mg12S20 in the sowing phase + N30 in the tillering phase + N70 in the booting phase; 4. N12P52 + K32Mg12S20 in the sowing phase + N30 in the tillering phase + N70 in the booting phase; 5. N12P52 + K32Mg12S20 in the sowing phase to a depth of 10 cm + N30 in the tillering phase + N70 in the booting phase.

The following fertilizers were used: ammophos (N12P52), potassium magnesia (K32Mg12S20), ammonium nitrate (N34). The test was repeated four times. The total area of the plot – 110 m^2. MTZ 1523 tractor and Semeato TDNG 420 seeder (Brazil) were used for sowing. Seed application rate – 5 million pieces of viable seeds per 1 hectare, seeding depth – 4 cm. The yield was taken into account by the ACROS 585 combine to achieve the full ripeness of winter wheat grain. Sampling of winter wheat grain and their preparation for analysis were performed using generally accepted methods [13]. The gluten content in the grain was determined according to GOST 13586.1-68 [3], protein – according to GOST 10846-91 [5], grain unit – according to GOST 10840-64 [4]. The main quality indicators were evaluated in accordance with GOST 9353-2016 [6].

Correlation and dispersion analyses were performed using STATISTICA 13.

3 Results

The protein content of winter wheat with the application of mineral fertilizers varied in 2016 – from 11.5 to 14.1%; in 2017 – from 13.5 to 13.7%; in 2018 – from 12.0 to 12.4%. The gluten content changed in 2016 – from 14.7 to 21.1%; in 2017 – from 28.9 to 29.5%, in 2019 – from 24.6 to 25.5%. The grain unit according to the test variants varied in 2016 – from 691.3 to 742.8 g/l; in 2017 – from 795.5 to 811.8 g/l; in 2018 – from 788.5 to 800.0 g/l (Figures 1, 3, 5).

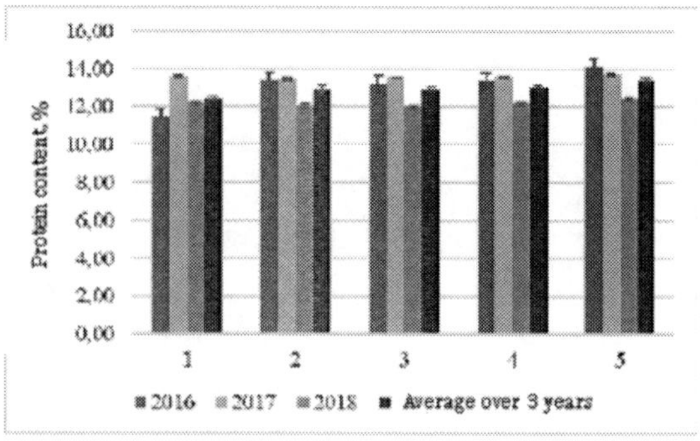

Fig. 1. Protein content of winter wheat with the application of fertilizers in No-Till system, %.

* Corresponding author: 10yaroslav@mail.ru

On average, over the years of the study the content of protein, gluten and grain unit of winter wheat in the control variant made 12.4%, 23.1%, 760.7 g/l, respectively. The application of mineral fertilizers improves the quality of grain. The presowing application of ammophos with double dressing of ammonium nitrate increases the protein content to 13.0%, which is 4.0% more in relation to the control sample. The addition of potassium magnesia with ammonium nitrate also increases the protein content by 4.0% compared to the control sample (Figure 2).

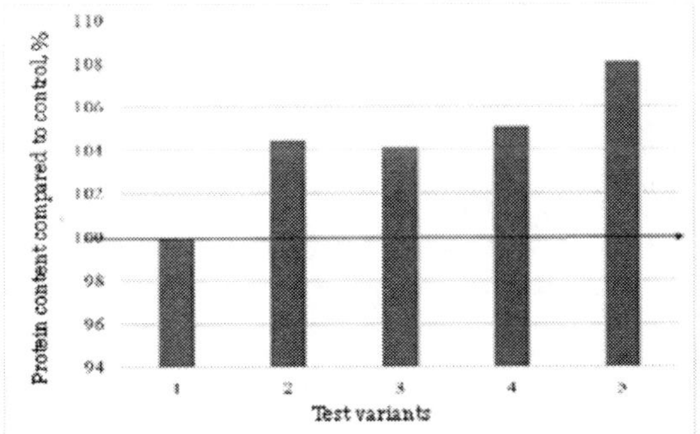

Fig. 2. Change in protein content of winter wheat with the application of mineral fertilizers in relation to control sample, average over 3 year.

The protein content in the grain with the combined use of ammophos and potassium magnesia is almost the same as with the introduction of ammophos alone.

The maximum protein content (13.4%) was revealed in variant N12P52 + K32Mg12S20 (to a depth of 10 cm) + N30 in the tillering phase + N70 in the booting phase. The increase in protein content was 8.0% relative to the control.

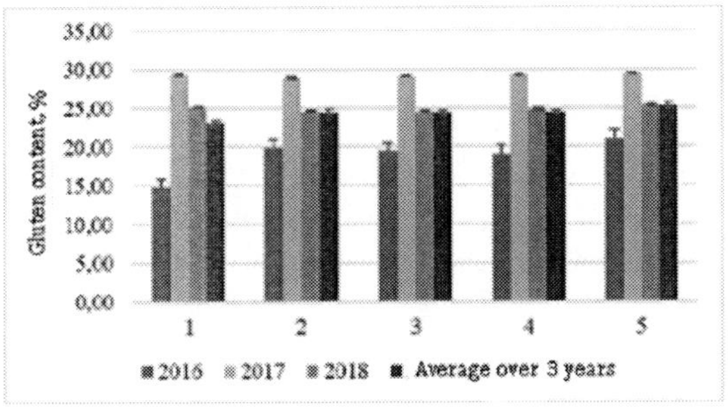

Fig. 3. Gluten content of winter wheat with fertilizer treatment within the No-Till system.

Fig. 4. Change in gluten content of winter wheat with the application of mineral fertilizers in relation to control sample, average over 3 years.

A similar pattern was established according to the effect of fertilizers on gluten content of winter wheat (Figure 3). Over three years of studies, the increase in gluten both with the separate application of ammophos and potassium magnesia, and when used together to the sowing depth was 6.0% compared with the control sample. The introduction of ammophos and potassium magnesia to a depth of 10 cm against the background of double dressing with ammonium nitrate increases the gluten content to the highest value – 25.3., which is 10.0% more than in the control version (Figure 4).

The analysis of variance showed that the established increase in protein and gluten content of winter wheat with the application of the studied mineral fertilizers is significant and reliable (Figure 5).

The grain unit is less than the content of protein and gluten, and changes when mineral fertilizers are applied. On average, there is a slight increase in the fertilizer variants – 2.0% to 3.0% compared to the control sample.

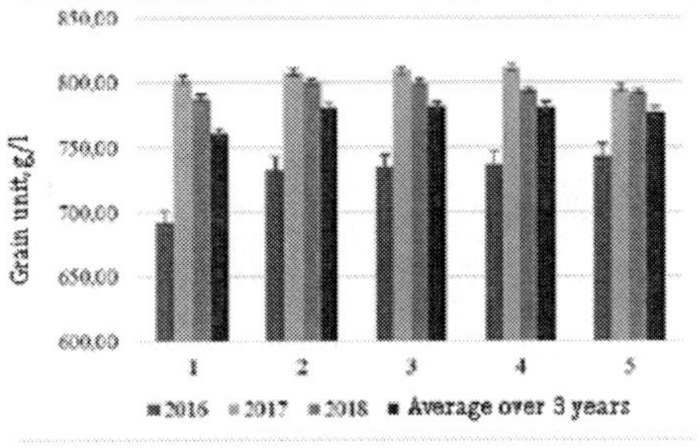

Fig. 5. Winter wheat grain unit with the application of fertilizers within No-Till system, g/l.

https://doi.org/10.1051/bioconf/20225200084

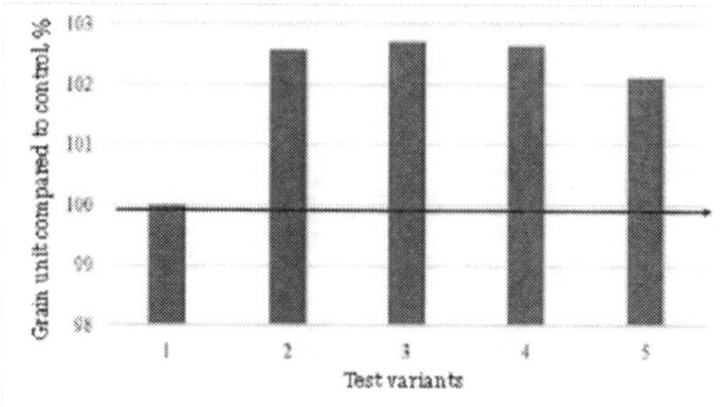

Fig. 6. Change in the grain unit of winter wheat with the application of mineral fertilizers in relation to control sample, average over 3 years.

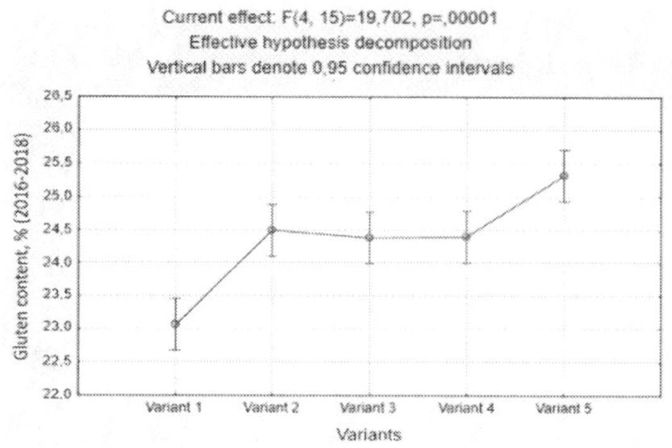

(b)

Fig. 7. Influence of mineral fertilizers on the content of protein (a) and gluten (b) in winter wheat grain

4 Discussion

The term "wheat quality" includes more than two dozen features that characterize the composition of grain, as well as the baking properties of flour. Among them the protein content serves a determining indicator of the quality of winter wheat grain. This indicator is an inherited feature and depends to a greater extent on the variety, but under the influence of agrotechnical factors its fluctuations may reach 8.0% or more [16]. Fertilizers are the main factor in regulating the nutritional regime of the soil, thus improving its biological and physical properties, which ensures the production of high-quality and stable crops [7, 10, 12]. The correlation analysis revealed a significant dependence of winter wheat grain quality on effective fertility of ordinary calcareous chernozem (Figure 5).

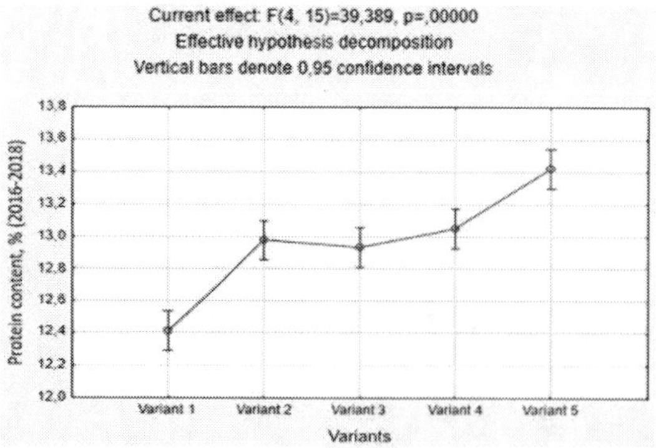

(a)

The direct dependence of the protein content of winter wheat on the amount of nitrate nitrogen in the soil in the booting phase was established. The high correlation coefficient between the above parameters is typical for all the studied soil layers: 0–10 (r = 0.85), 10–20 (r = 0.72), 0–30 cm (r = 0.76). The gluten content and the grain unit of winter wheat significantly depended on the mobility of phosphorus in the surface soil layer – 0–10 cm (r = 0.66 and 0.52, respectively).

The findings are supported by research data stating that nitrogen fertilizers play a major role in improving winter wheat quality in the arid zone of the North Caucasus. The application of nitrogen at a dose of N30 in the tillering and booting phases increases the protein content in the grain by 2–3%, crude gluten – by 4–5%.

According to [16], the protein content should be at the level of 11.0–17.0%. If the protein content increases more than 17.0–19.0% or decreases by less than 11.0%, bread quality deteriorates [19, 22].

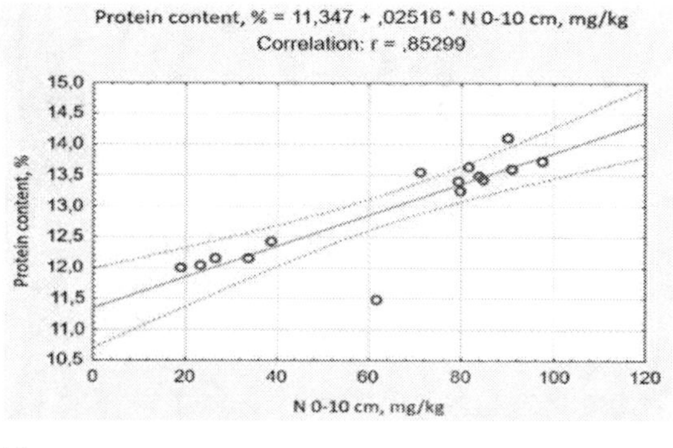

(a)

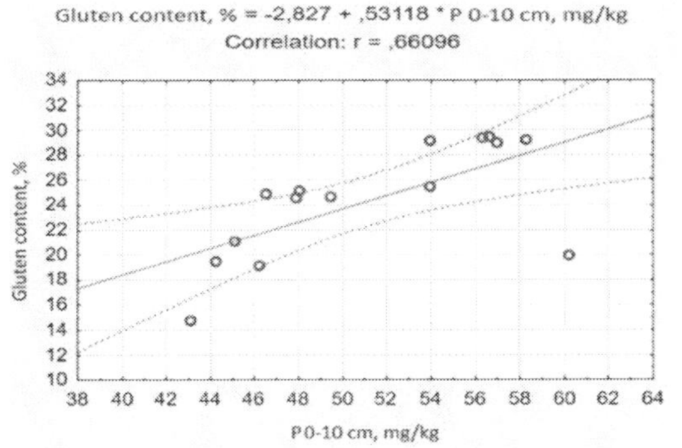

(b)

Fig. 8. Dependence of protein (a) and gluten (b) content in grain on soil fertility indicators.

In our experiments (on average over three years) the content of protein and gluten in the grain was not high (<13.5% and <28.0%, respectively), which is probably associated with precipitation during the grain-filling period. According to these indicators, grain belongs to the third class in accordance with the requirements of GOST 9353-2016.

According to [8, 22], edaphoclimatic conditions significantly affect the protein content of winter wheat grain. Temperature and moisture from heading to yellowing periods have the greatest influence on gluten content [11]. According to the obtained data, in 2017 grain filling was observed in hot, dry weather, so the content of protein and gluten in the grain in all test variants was higher than in 2016 and 2018.

The impact of mineral fertilizers on winter wheat grain quality was found to depend on weather conditions during the years of study. In 2016 the content of protein, gluten and grain unit at the control sample was less, and the efficiency of mineral fertilizers was higher compared to 2017 and 2018. A similar dependence was revealed on the yield of winter wheat. At the same time, high positive relationship ($r = 0.93$) was established between the protein content in winter wheat grain and its yield (Figure 9).

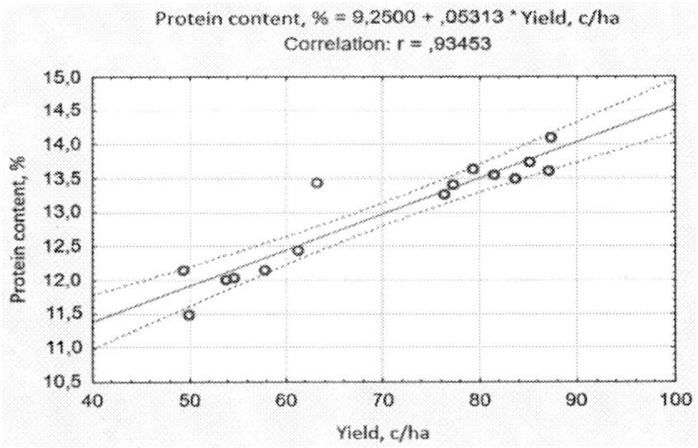

Fig. 9. Dependence of protein content in grain on winter wheat yield.

The protein content of wheat grain is significantly correlated with the accumulation of nitrogen in plants. In order to obtain stable high-quality winter wheat yields in the studied area, the nitrogen content in the top of plants in the booting phase should be at least 3.0% [2]. Protein content in winter wheat grain depends not only on the degree of nitrogen deficiency, but also on its correlation with manganese, phosphorus and iron. Earlier it was proved [21] that the use of No-Till technology stabilizes the nitrogen state of the soil of winter wheat agrocenoses. Therefore, with an average supply of soil with phosphorus and manganese against the background of double dressing with ammonium nitrate in the first half of the vegetation, the intensity of nitrogen absorption by plants was quite high. This contributed not only to the increase in protein content of winter wheat grain, but also in its yield. The obtained data are confirmed by studies conducted in various edaphoclimatic conditions [14]. It was shown that in a certain range of doses of nitrogen fertilizers, the correlation coefficients of grain yield – protein content, are usually significantly positive. The increase of the dose of N-fertilizer above a certain value can lead to the reduction in yields.

5 Conclusion

In conditions of insufficient humidification, the rational use of mineral fertilizers in ordinary calcareous chernozems when cultivating winter wheat using zero technology increases the quality of winter wheat For the studied variety of winter wheat (Grom), the most effective joint application of ammophos (N12P52) and potassium magnesia (K32Mg12S20) before sowing to a depth of 10 cm with double dressing with ammonium nitrate in the tillering (N30) and booting (N70) phases. A significant effect of potassium magnesia on the content of protein, gluten and grain unit was revealed, which indicates the need to use mineral fertilizers containing sulfur and magnesium when cultivating winter wheat. The grain obtained in the test sample corresponds to class 3. To improve the quality of winter wheat grain calcareous soils, it is necessary to optimize the mineral nutrition of the culture not only with macro-, but also with microelements.

Acknowledgments

The research was financially supported by the "Priority 2030" program of the Ministry of Science and Education of the Russian Federation, project No. SP02/S4_0708 Priority_01/SP02/S4_0706 Priority_01 and by the project of the Ministry of Science and Higher Education of Russia on the Young Scientist Laboratory within the framework of the Interregional Scientific and Educational Center of the South of Russia (No. LabNOTs-21-01AB, FENW-2021-0014).

References

1. A. V. Alabushev, A. A Sukharev., A. S. Popov, S. I. Kambulov, A. Ya. Logvinov, Changing the productivity of crops under the influence of the same type of basic tillage methods, *Agriculture* **8,** 25–28 (2015)
2. O. A. Biryukova, I. I. Yelnikov, V. S. Kryshchenko, *Operational Diagnostics of Plant Nutrition* (Rostov-on-Don: Publishing House of SFU, 2010)
3. GOST 10840-64 Grain. Methods of determining the grain unit (Moscow: Standardinform, 2009)
4. GOST 10846-91 Grain and products of its processing. Method of determining the protein (Moscow: Standardinform, 2009)
5. GOST 13586.1-68 Grain. Methods for determining the quantity and quality of wheat gluten (Moscow: Standardinform, 2009)
6. GOST 9353-2016 Wheat. Specifications (Moscow: Standardinform, 2019).
7. Y. V. Gubanov, N. N. Ivanov, *Winter Wheat* (Moscow, 1983)
8. Yu. A. Gulyanov, M. S. Karpov, Influence of the autumn-summer regime of nitrogen nutrition of winter wheat on the grain yield in the chernozem regions of the Southern Urals, *Izvestia of Orenburg State Agrarian University* **1 (57),** 23–26 (2016)
9. B. A. Dospekhov, *Field Methodology (with the Basics of Statistical Processing of Research Results)* (Moscow: Kolos, 1985)
10. Ya. I. Ilchenko, O. A. Biryukova, A. M. Medvedev, Winter wheat yield using No-Till technology in chernozem regions of the Lower Don, *Collection of scientific works on the materials of the International Scientific Environmental Conference "Agrarian Landscapes, their Sustainability and Development Features"* (March 24-26, 2020, Krasnodar), pp 490–493 (2020)
11. A. R. Konstantinov, *Weather, Soil and Winter Wheat Yield* (Leningrad: Hydrometeoizdat, 1978)
12. N. S. Matyuk, A. I. Belenkov, M. A. Mazirov, V. D. Polin, A. A. Rassadin, E. D. Abrashkina, *Ecological Agriculture with the Basics of Soil Science and Agrochemistry* (Moscow: Russian State Agrarian University – Moscow Timiryazev Agricultural Academy, 2011)
13. V. G. Mineev, *Workshop on Agrochemistry* (Moscow: Moscow State University, 2001)
14. V. F. Fedorenko, A. A. Zavalina, N. Z. Milashchenko, *Scientific Foundations for the Production of High-Quality Wheat Grain: Scientific Edition* (Moscow: Rosinformagrotech, 2018)
15. Yu. P. Khrustalev, V. N. Vasilenko, I. V. Svisyuk, V. D. Panov, Yu. A. Larionov, *Climate and Agroclimatic Resources of Rostov Region* (Rostov-on-Don, 2002)
16. A. Kh. Sheudzhen, T. N. Bondareva, L. M. Onishchenko, *Nutrition and Fertilization of Grain, Cereal and Leguminous Crops,* pp 31–78 (Krasnodar: KubSAU, 2012)
17. D. Shpaar, et al. *Cereals (Cultivation, Harvesting, Refinement and Use)* (Moscow: Publishing house LLC DLV AGRODELO, 2008)
18. A. Bruce, E. Mark, L. Xinqiang, J. Groenigen, N. Gestel, J. Six, T. Rodney, When does No-Till yield more? A global meta-analysis, *Field Crops Research*, 156–168 (2015)
19. L. Hajasa, K. Scherfb, K. Toroka, Z. Bugyia, E. Schalla, R. Pomsc, P. Koehlerb, S. Tomoskozi Variation in protein composition among wheat (Triticum aestivum L.) cultivars to identify cultivars suitable as reference material for wheat gluten analysis, *Food Chemistry* **267,** 387–394 (2018)
20. F. Lessiter, 29 reasons why many growers are harvesting higher No-Till yields in their fields than some university scientists find in research plots, *No-Till Farmer* **44(2),** 8 (2015)
21. A. Medvedeva, O. Biryukova, Y. Ilchenko, T. Minkina, A. Kucherenko, T. Bauer, S. Mandzhieva, M. Mazarji, Nitrogen state of haplic chernozem of the European part of Southern Russia in implementation of resource-saving technologies, *Journal of the Science of Food and Agriculture* doi: 10.1002/jsfa.10852 (2020)
22. J. Tomic, A. Torbica, L. Popovic, N. Hristov, B. Nikolovski, Wheat breadmaking properties in dependance on wheat enzymes status and climate conditions, *Food Chemistry* **199,** 565–572 (2016)
23. World reference base for soil resources, International soil classification system for naming soils and creating legends for soil maps, *World Soil Resources Reports* **106** (Rome: FAO, 2014)

Potential of corn hybrids of universal use

Marina Yurievna Mikhailova*, *Rogat Vagizovic* Minikaev, *Fannur Shaukatovich* Faskhutdinov, and *Liliya Talgatovna* Vafina

Kazan State Agrarian University, 65, Karl Marx St., Kazan, Republic of Tatarstan, 420015, Russian Federation

Abstract. This article presents the results of studies of corn hybrids of a universal direction. Their responsiveness to the application of mineral fertilizers, incrustation of seeds with biological preparations, and foliar fertilizing with zinc-containing preparations is shown. NK background at 60 t/ha + foliar application of Zn ensures maximum grain yield and better disclosure of the grain potential of hybrids of universal use (61.3 c/ha for the Nur hybrid, 67.41 c/ha for the Bilyar-160, 77, 39 c/ha for the Voronezhsky-279 hybrid). Early ripe hybrids in the conditions of the Fore- Kama region of the Republic of Tatarstan form stable grain yields due to lower pre-harvest moisture. Mid-early hybrids of universal use respond well to the application of mineral fertilizers and foliar feeding with zinc-containing preparations, providing high grain yields.

* Corresponding author: marisha.m.u@mail.ru

1 Introduction

Corn has a huge potential to produce high yields of grain. Under production conditions, the gross harvest of corn grain is 1.5 times higher than winter wheat. Also, in comparison with other crops of the grain wedge, corn performs better in the conditions of more frequent spring-summer droughts [1]. It is the scientific approach when choosing a hybrid, precisely adjusted dates and seeding rates that minimizes the negative impact of dry periods of vegetation [2]. Therefore, a hybrid selected taking into account the soil and climatic conditions of the region, with the full disclosure of the genetic potential, is able to provide an increase in yield up to 40% [3].

Mineral fertilizers play an important role in the life of corn. The provision of mineral nutrition activates the growth of corn supporting roots, which increases stress resistance [4]. The dimensions of the leaf surface depend on nitrogen nutrition [5]. The introduction of macronutrients increases the photosynthetic potential of maize leaves [6]. And most importantly, changing the conditions of mineral nutrition increases the yield of corn, and single doses of mineral fertilizers provide a net income, the absolute value of which increases with the doses of their application [7].

Foliar fertilizing with microelements in crops of any agricultural crops has been popularized on a global scale [8, 9, 10]. On corn, foliar feeding is carried out starting from the 3-leaf phase. The positive effect is especially noticeable from zinc-containing preparations, since corn is a sensitive crop in relation to zinc, which, being part of enzymes, ensures the normal course of corn respiration processes [11].

The purpose of the research is to show the grain potential of corn hybrids of universal use when cultivated for fodder purposes.

2 Materials and methods

The studies were carried out in 2019-2021. on typical gray forest soils with a humus content of 3.8% according to Tyurin, mobile phosphorus of 288 mg/kg and exchangeable potassium of 153 mg/kg of soil (according to Kirsanov). The reaction of the soil environment was close to neutral (pH 6.6).

Sowing was carried out on May 22 in 2019, on May 15 in 2020 and on May 14 in 2021 with the Vesna-8 (Favorit) seeder with a row spacing of 70 cm and a seeding rate of 85 thousand pieces/ha of germinating seeds. The calculated norms of fertilizers were applied under pre-sowing cultivation in a scattered way (ammonium nitrate + potassium sulphate). To obtain the planned yield of 50 t/ha of green mass of corn, 157 kg of AI/ha of nitrogen and 267 kg of AI/ha of potassium were introduced. For the planned yield of 60 t/ha, 207 kg a.i./ha of nitrogen and 357 kg a.i./ha of potassium were applied. Before sowing, the seeds of individual variants were treated with the biopreparation GSN-2004 (0.5 l/t). This is an immunostimulant, an anti-stress preparation from enzymes of natural origin, which includes amino acids, humic acids. For foliar feeding, the drug Batr Zn (1 l/ha) was used in the phase of 5-6 leaves. Batr Zn is a microfertilizer with organic acids (succinic, citric, ascorbic) and 5% zinc in chelated form. Two early ripe hybrids Nur (FAO 150) and Bilyar-160 (FAO 160) and a mid-early hybrid Voronezhsky-279 (FAO 290) were sown. The studied hybrids, according to the recommendations of the authors, belong to hybrids of universal use. Two-factor field experience was laid. The first factor is hybrids. The second factor is nutrient backgrounds: 1) without fertilizers (control), 2) NK for 50 t/ha, 3) NK for 50 t/ha + foliar feeding, 4) NK for 60 t/ha, 5) NK for 60 t / ha + foliar application, 6) GSN-2004 incrustation, 7) GSN-2004 incrustation + foliar application.

During the research, the following methods were used:

- the yield in field experiments was taken into account on test plots (14.3 m) and at the same time the moisture content of corn grain was determined using a moisture meter. The recalculation of the yield of the studied options was carried out according to the basic moisture content of corn grain of 15%;
- height of the plant was measured in three accounting phases of corn development in two non-adjacent repetitions in five equidistant places of the plot;
- biometric analysis of corn cobs included measuring the length of the cob, the length of the unfinished part of the cob, the number of rows of grains in the cob, the number of grains in a row, the grain content, the weight of the cob, the weight of the grain from the cob, the yield of grain from the cob, the weight of 1000 seeds;
- statistical processing of the results of the experiment was carried out by the method of dispersion analysis.

3 Results and discussion

When conducting research to study the effect of nutritional backgrounds on the nutritional quality of corn hybrids of universal use, the grain potential of these hybrids was also taken into account. The analysis was done on average over three years of research. When growing corn for grain, it is important to form a vigorous plant with a strong, lodging-resistant stem in order to be able to keep the cob in time for the right harvest. The studied hybrids during the growing season form quite powerful plants (Table 1). Already by the phase of 7-8 leaves, the positive effect of nutrition backgrounds on the linear growth of plants became noticeable. If in the control plant height was from 76.5 to 79.3 cm, then the application of only mineral fertilizers increased the plants by 8-14% and 14-25% (by 8.3 cm and 15.3 cm).

* Corresponding author: marisha.m.u@mail.ru

Carrying out only incrustation with GSN-2004 increases the growth of hybrids in height by 15 percent. Adding additional foliar feeding to each of the nutritional backgrounds increases the plant height by 4 cm (NK at 50 t/ha + foliar feeding), by 1 cm (NK at 60 t/ha + foliar feeding) and by 0.7 cm (encrustation GSN-2004 + foliar feeding).

The maximum height of plants in the heading phase was on the NK variant at 60 t/ha + foliar feeding for all the studied hybrids. The Nur hybrid had the highest value of 175.4 cm.

In the phase of milky ripeness, the grains of the plant reached the maximum values of linear growth in height. The mid-early hybrid Voronezhsky-279 in the control reached 194.5 cm. This is the highest value in this variant. The application of nitrogen-potassium fertilizers against the calculated background of NK by 50 t/ha increased the height of plants by 5-10%, compared with the control. And against the background of NK by 60 t/ha by 16-30%. During incrustation, the plant height was 171.4 cm for the Nur hybrid, 177.0 cm for the Bilyar-160 hybrid, and 213.4 cm for the Voronezh-279 hybrid. Carrying out additional foliar feeding in the phase of 5-6 leaves with a zinc-containing preparation provided an increase in plant height by 2.3% against the background of NK by 50 t/ha + foliar feeding, by 3.4% against the background of NK by 60 t/ha + foliar feeding, and by 2.8% on the GSN-2004 encrustation + foliar application. On all backgrounds of nutrition, the Voronezhsky-279 hybrid formed taller plants, in comparison with early-ripening hybrids.

Table 1. Average plant height for 2019–2021, cm.

Food backgrounds	Hybrids	Development phase		
		7-8 leaves	Balayage	Milky ripeness
Without fertilizer (control)	Bilyar -160	79.3	143.1	157.2
	Voronezhskiy -279	76.5	127.4	194.5
	Nur	76.6	143.1	152.4
NK per 50 t/ha	Bilyar-160	87.4	146.0	170.5
	Voronezhskiy -279	82.7	140.0	203.6
	Nur	87.2	154.5	167.3
NK per 50 t/ha + foliar feeding	Bilyar -160	90.6	152.1	173.9
	Voronezhskiy -279	86.8	156.7	206.4
	Nur	91.9	161.9	173.2
NK per 60 t/ha	Bilyar -160	94.9	160.6	184.0
	Voronezhskiy -279	87.3	168.4	225.7
	Nur	96.1	172.3	197.1
NK per 60 t/ha + foliar feeding	Bilyar -160	96.8	164.8	189.5
	Voronezhskiy -279	88.5	173.6	233.4
	Nur	96.2	175.4	204.4
Inlay GSN-2004	Bilyar -160	93.9	148.8	177.0
	Voronezhskiy -279	85.6	147.8	213.4
	Nur	87.8	160.6	171.4
Inlay GSN-2004 + foliar feeding	Bilyar -160	94.3	152.6	181.1
	Voronezhskiy -279	86.6	150.0	219.7
	Nur	88.6	168.6	176.7

The combined use of incrustation and foliar application produces plants similar to the NK variant at 50 t/ha + foliar application.

The biometric indicators of corn had a close relationship with nutritional backgrounds. Some indicators depended on the characteristics of the hybrid (Tables 2, 3). The length of the cob strongly correlated with nutritional conditions. If in the control the length of the cobs of the studied hybrids was from 15.6 to 18.0 cm, then when the calculated norms of mineral fertilizers were applied, the length of the cobs became 17.5 to 22.0 cm.

Encrustation GSN-2004 + foliar feeding Batr Zn 19.1 cm, in Voronezh-279 and Nur hybrids against the background of NK at 60 t/ha + foliar feeding of 23.0 cm and 19.0 cm.

The length of the unfulfilled part of the cob did not depend on the food backgrounds. This value is more related to climatic conditions and characteristics of the hybrid. The number of rows of grains on the cob is also a feature of the hybrid. In the studied hybrids, this value ranged from 13.4 to 15.5 pcs in the Bilyar-160 hybrid, from 14.0 to 15.7 pcs in the hybrid Voronezhsky-279 and from 12.7 to 14.2 pcs in the Nur hybrid.

Table 2. Biometric indicators of corn hybrids.

Food backgrounds	Hybrids	Cob length, cm	The length of the unfinished part of the cob, cm	Number of rows of grains on the cob, pcs.	Number of grains in a row, pcs.	Cob grain size, pcs.
Without fertilizer (control)	Bilyar -160	15.6	0.1	14.8	28.0	414.4
	Voronezhskiy -279	18.0	2.7	14.0	30.8	431.2
	Nur	15.7	1.1	13.4	32.6	436.8
NK per 50 t/ha	Bilyar-160	18.5	1.7	13.4	34.7	465.0
	Voronezhskiy -279	18.6	2.9	14.6	31.4	458.4
	Nur	15.5	0.5	12.7	30.8	391.2
NK per 50 t/ha + foliar feeding	Bilyar -160	17.9	1.5	14.6	32.8	478.9
	Voronezhskiy -279	19.1	2.8	14.4	33.3	479.5

	Nur	16.1	0.9	13.7	33.6	460.3
NK per 60 t/ha	Bilyar -160	17.5	1.4	14.8	32.0	473.6
	Voronezhskiy -279	22.0	2.8	15.4	37.8	582.1
	Nur	18.9	1.3	13.4	34.8	466.3
NK per 60 t/ha + foliar feeding	Bilyar -160	18.7	0.5	15.2	36.8	559.4
	Voronezhskiy -279	23.0	2.7	15.4	39.4	606.8
	Nur	19.0	1.5	13.6	35.0	476.0
Inlay GSN-2004	Bilyar -160	18.0	0.5	15.1	35.6	537.6
	Voronezhskiy -279	20.9	2.9	15.6	37.8	589.7
	Nur	16.0	0.9	13.8	33.7	465.1
Inlay GSN-2004 + foliar feeding	Bilyar -160	19.1	0.5	15.5	36.3	562.7
	Voronezhskiy -279	21.6	2.7	15.7	38.0	596.6
	Nur	16.3	1.0	14.2	34.1	484.2

Table 3. Biometric indicators of corn hybrids.

Food backgrounds	Hybrids	Cob mass, gr.	Mass of grain from the cob, gr.	Grain output from the cob, %	Weight of 1000 grains, gr.	Grain moisture content, %	Productivity, c/ha
Without fertilizer (control)	Bilyar -160	129.8	99.2	76.4	294	23.0	41.36
	Voronezhskiy -279	131.2	121.8	92.8	311	26.4	51.17
	Nur	111.1	91.6	82.4	270	21.5	45.74
NK per 50 t/ha	Bilyar-160	146.6	111.2	75.9	282	20.8	58.69
	Voronezhskiy -279	169.3	125.4	74.1	321	28.4	61.70
	Nur	120.9	94.9	78.5	320	23.1	47.53
NK per 50 t/ha + foliar feeding	Bilyar -160	147.1	113.5	77.2	288	20.5	59.37
	Voronezhskiy -279	174.1	128.5	73.8	334	29.3	67.25
	Nur	126.7	98.6	77.8	341.2	23.4	50.70
NK per 60 t/ha	Bilyar -160	148.1	114.6	77.4	312	20.3	61.18
	Voronezhskiy -279	204.1	155.7	76.3	313	23.8	74.73
	Nur	180.0	145.8	81.0	363	21.7	59.26
NK per 60 t/ha + foliar feeding	Bilyar -160	155.9	123.8	79.4	276.0	19.5	67.41
	Voronezhskiy -279	229.7	169.3	73.7	332	22.6	77.39
	Nur	182.5	147.6	80.9	365	21.5	61.30
Inlay GSN-2004	Bilyar -160	146.7	112.5	76.7	285	22.9	58.14
	Voronezhskiy -279	196.0	146.1	74.5	305	22.7	58.03
	Nur	131.0	110.3	84.2	281.0	19.9	57.87
Inlay GSN-2004 + foliar feeding	Bilyar -160	152.4	114.8	75.3	288	22.5	59.43
	Voronezhskiy -279	201.4	151.8	75.4	318.0	22.5	59.68
	Nur	134.6	124.7	92.6	301	20.0	59.80

Improving nutritional conditions increased the number of formed grains in a row. If in the control this value was 30.5 pcs on average for the studied hybrids, then on the NK variant per 50 t/ha, the value became 32.3 pcs, NK per 50 t/ha + foliar feeding - 33.2 pcs, NK per 60 t/ha of 34.9 pcs, NK per 60 t/ha + foliar application of 37.1 pcs. It can be said that with each subsequent increase in the rate of applied fertilizers, the number of formed grains in the cob increases. Without mineral fertilizers, the number of grains in a row was 35.7 pcs with inlay and 36.1 pcs when encrusting and foliar feeding. This means that biological preparations and zinc-containing preparations also increase the number of grains in a row. Among the hybrids studied, the Voronezhsky-279 hybrid was characterized by the largest number of grains in the row. On the NK variant at 60 t/ha + foliar application, the cultivation of this hybrid makes it possible to obtain 39.4 pcs grains in a row.

The more rows of grains and grains in a row, the higher the grain content of the cob. The studied hybrids were characterized by good cobs grain content. On the fertilized variants, the grain content was also better. The cobs of the medium-early hybrid Voronezhsky-279 turned out to be the most grainy - 606.8 gr. against the background of NK at 60 t/ha + foliar application.

Although, when only incrustation was carried out, the graininess of ears of the Voronezhsky-279 hybrid was only 17.1 g lower, and when incrustation and foliar feeding were lower, by 10.2 g. The graininess of the cobs of the studied hybrids on the variants with incrustation and foliar feeding is at the level of backgrounds with mineral fertilizers, and in the Bilyar-160 hybrid, it is even higher.

Despite the fact that the mass of the cob is a value that depends on the characteristics of the hybrid, it is strongly influenced by nutritional conditions, namely the calculated norms of mineral fertilizers. If in the control the mass of ears of the Bilyar-160 hybrid was 129.8 g, Voronezhsky-279 of 131.2 g, and Nur of 111, g, then the introduction of the calculated norms of NK by 50 t/ha increased these values by 16, eight; 38.1 and 9.8 gr. This is a significant increase of 10-30%. And the addition of foliar feeding with a zinc-containing preparation in the phase of 5-6 leaves additionally gives an additional 3-5%

increase in the weight of the cob. An increase in fertilizer rates to the calculated NK by 60 t/ha also increases the weight of the cob. When carrying out the incrustation with a biological product, the weight of the cob increased by 16.9-64.8 g, and in the variant GSN-2004 encrustation + foliar feeding by 17-50% g. The most responsive one to biopreparations and microfertilizers was the hybrid Voronezh-279.
The yield of grain from the cob was quite high. For all options, it did not fall below 73.7%. Grain yield did not depend on the methods studied.

The weight of 1000 grains correlated with the studied hybrids and nutrition backgrounds. The largest weight of 1000 grains was in the Bilyar-160 hybrid on the NK variant at 60 t/ha of 312 g. In the Voronezhsky-279 hybrid, the highest value of the weight of 1000 grains was 334 g against the calculated background of NK at 50 t/ha + foliar feeding, and in the Nur hybrid on the NK variant at 60 t/ha + foliar feeding 365 g. This is the highest value among the studied hybrids. Smallest value mass 1000 grains in the Voronezhsky-279 hybrid was 305 gr. in the variant with seed incrustation with a biological product, and in the Nur hybrid of 270 gr under control.

Despite the fact that early-ripening hybrids are inferior to the medium-early hybrid in a number of indicators, early-ripening ones are characterized by low pre-harvest grain moisture.

The biological yield in the control was 41.36 c/ha for the Bilyar-160 hybrid, 45.74 c/ha for the Nur hybrid, and 51.17 c/ha for the Voronezhsky-279 hybrid in the control. Let us pay attention to the potential of mid-early corn hybrids in the conditions of the Fore-Kama region of the Republic of Tatarstan. Against the calculated background of the NK variant at 50 t/ha, the grain yield increased by 1.8 c/ha (Nur), 10.5 c/ha (Voronezhsky-279), 17.3 c/ha (Bilyar-160). An increase in the application rate of mineral fertilizers makes it possible to obtain 59.26 centners/ha of grain in the Nur hybrid, 61.18 centners/ha in the Bilyar-160 hybrid and 74.73 centners/ha in the Voronezh-279 hybrid. When macroelements are introduced and foliar feeding is carried out, the yield additionally increases by 7-9% and 4-10%, compared with the options where only mineral fertilizers were applied. In the variant with incrustation with biological preparations, the yield was 57-58 centners/ha, which is higher than the control by 6-16 centners/ha. And when carrying out encrustation and foliar feeding, the yield level for the studied hybrids leveled up to 59.6 c/ha. As a result, the NK background at 60 t/ha + foliar application of Batr Zn provides the maximum grain yield and better disclosure of the grain potential of hybrids of universal use (61.3 c/ha for the Nur hybrid, 67.41 c/ha for the Bilyar-160 hybrid, 77.39 c/ha for the Voronezhsky-279 hybrid).

4 Conclusion

Hybrids of universal use have a fodder potential and are capable of producing high grain yields under equal conditions. Early ripe hybrids are characterized by lower pre-harvest grain moisture, which ensures stable yields. Mid-early hybrids respond well to the application of both mineral fertilizers and the use of biological products in the form of seed encrustation before sowing and foliar fertilization with microelements. As a result, the NK background at 60 t/ha + foliar application of Batr Zn (for green mass) provides the maximum grain yield and better disclosure of the grain potential of hybrids of universal use (61.3 dt/ha for the Nur hybrid, 67.41 dt/ha in the Bilyar-160 hybrid, 77.39 c/ha in the Voronezhsky-279 hybrid).

References

1. V. M. Bagryantseva, V. V. Bukharev, S. V. Mikitin, The value of early sowing of corn in the zone of unstable moisture in the Stavropol Territory, *Corn and Sorghum* **1,** 3–8 (2015)
2. A. P. Potapov, The yield of corn hybrids depending on the timing and density of sowing, *Modernization of agricultural technologies in adaptive - landscape agriculture of the Central Chernozem region: a collection of scientific reports of the All-Russian scientific and practical conference* (June 18–19, 2014, Kamennaya Step) pp 96–100 (Voronezh: Publishing House "Istoki", 2014)
3. M. U. Mikhailova, P. I. Talanov, Cultivation of corn hybrids on the expected nutritional background in the Volga region of the Republic of Tatarstan, *IOP Conference Series: Earth and Environmental Science* **341,** 012008 doi: 10.1088/1755-1315/341/1/012008 (2019)
4. Zh. O. Kanukova, M. V. Kashukoev, V. Kh. Kalova, Zh. O. Kanukova, Correlation analysis of indicators of corn hybrids using mineral fertilizers in the mountainous zone of the Kabardino-Balkarian Republic, *Agro-food policy of Russia* **2 (38),** 71–74 (2015)
5. A. I. Nevzorov, M. A. Nevzorov, Effect of various doses and methods of applying mineral fertilizers on the yield of silage corn, *Science and Education* **4 (2)** (2021)
6. M. Mikhailova, I. P. Talanov, The effect of nutritional backgrounds on the formation of leaf surface and yield and green mass of corn, *BIO Web of Conferences* **17,** 00074 (2020)
7. E. Elmukhanova, Zh. B. Bakenova, G. R. Tastanbekova, The effectiveness of the use of mineral fertilizers on the yield of corn, *Soil Science and Agrochemistry* **1,** 67–72 (2016)
8. R. M. Sabirova, R. S. Shakirov, Z. M. Bikmukhametov, Bioplant flora - fertilizer of a new generation, *Bulletin of Kazan State Agrarian University* **14-2 (53),** 37–42 doi: 10.12737/article_5d3e15f17c3223.64554857 (2019)

9. V. A. Kolesar, G. F. Sharipova, D. Safina, R. I. Safin, Use of foliar fertilizers on soybeans in the Republic of Tatarstan, *BIO Web of Conferences* **17,** 00069 (2020)
10. R. Minikayev, L. Gaffarova, The effect of bacterial preparations on the growth, development and quality indicators of sugar beet yield, *BIO Web of Conferences* **17,** 00250, doi: 10.1051/bioconf/20201700250 (2020)
11. A. V. Tyurin, A. L. Toygildin, M. I. Podsevalov, Efficiency of methods of cultivation of corn for grain in the conditions of the Middle Volga regionб *Bulletin of Ulyanovsk State Agricultural Academy* **3 (55),** 55–65 doi: 10.18286/1816-4501-2021-3-55-62 (2021)

Study of vortex pneumatic sprayer for liquid disinfection

Boris Ivanov*, *Bulat* Ziganschin, *Andrei* Dmitriev, *Vladimir* Medvedev, and *Ilnur* Gayfullin

Kazan State Agrarian University, 65, K. Marx Str., Kazan, 420015, Republic of Tatarstan, Russian Federation

Abstract. The production of high-quality livestock products can be achieved by upgrading livestock breeds and complying with zootechnical requirements of ensuring sanitary well-being in animal houses. Different ways are applied to gain the proper clean condition of livestock houses and equipment, disinfection preparations for working surfaces. Besides, non-compliance with the requirements for operating devices and standards for disinfectant consumption leads to increased antimicrobial resistance of microorganisms [1]. The efficiency of disinfection depends mainly on the technical capabilities of the units, mainly on the spray devices. Many researchers solve complex problems by optimizing spray devices' design and technological parameters to improve their spray quality [2]. The authors developed a vortex pneumatic sprayer for liquids for the disinfection process. The dispersion composition of disinfectant liquid aerosols was obtained in different operating regimes of the proposed sprayer design. Based on experimental data, dependencies of the influence working pressures of disinfectant liquid and compressed air in their supply channels on the dispersion were obtained. The flow rates of the vortex pneumatic sprayer during its operation in various regimes were determined. Dependencies of disinfectant liquid flow rate through the sprayer were constructed at different pressure ratios in supply channels of working fluids (disinfectant liquid and compressed air) and the angle between them. Optimal design parameters of the vortex pneumatic sprayer for aerosol disinfection of livestock houses and equipment were determined.

1 Introduction

The techno-economic efficiency of the disinfection process of livestock houses and equipment depends mainly on the design characteristics of the sprayers. The following parameters determine the sprayer performance: the spray dispersion, coating degree of the surface to be treated, the uniform distribution of the disinfectant [1; 2]. Despite the diversity of sprayer designs, there is a pressing issue to reduce the cost of disinfectant liquids and increase the dispersion efficiency, the consumption of the working fluid, and its losses [3; 4]. It is known that the fine droplets improve the penetrating capacity of the aerosols [5]. Fine dispersion is achieved by increasing pressure in the supply line for the hydraulic sprayers and requires high energy costs. Therefore, pneumatic sprayers are the most promising devices for the dispersion of disinfectant liquids. The dispersion is changed by controlling the pressure in the supply line, and the flow rate varies slightly [6].

Centrifugal (vortex) sprayers make it possible to carry out a high-quality disinfection process with lower pressure in the supply line. Moreover, they have a larger diameter of the outlet openings (nozzles) than other types of sprayers, leading to lower requirements for cleaning feed liquid and an increased service life associated with the nozzle wear and uniform covering treated surface [7]. When vortex pneumatic sprayers disperse fluids, the jet exits the nozzle in the form of a swirling flow of aerosols, forming an annular trace on the working surface [8; 9].

The work aims to investigate the influence of the design of the developed vortex pneumatic sprayer on the characteristics of the disinfectant liquid at different operating regimes.

2 Materials and methods

In the previous study [10], the most important factors influencing the dispersion of the spray liquid have been identified:

1. Spray design parameters.
2. Working pressure of the spray liquid (P_{liq} , MPa).
3. Working pressure of the compressed air (P_{air} , MPa).
4. Angle between input channels (α, degrees).

The vortex pneumatic sprayer, as shown in Figure 1, was manufactured to determine the influence of the factors mentioned above on the dispersion of the disinfectant liquid. The sprayer includes body 1 with a cylindrical vortex chamber 2 and two channels arranged at an angle α to each other. A radial channel 3 is used to supply disinfectant liquid, tangent channel 4 – compressed air. A conical cover 5 forms a conical vortex chamber 6 with a conical angle of 120 degrees. An output hole 7 for aerosols is arranged on the top corner of the conical vortex chamber.

* Corresponding author: littab@mail.ru

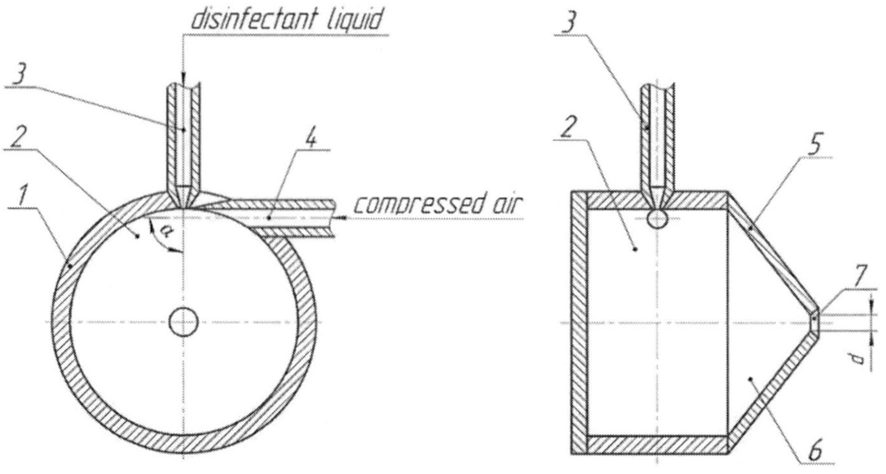

Fig. 1. Vortex pneumatic sprayer for disinfectant liquid: 1 – body; 2 – cylindrical vortex chamber; 3 – disinfectant liquid channel; 4 – compressed air channel; 5 – conical cover; 6 – conical vortex chamber; 7 – hole; α – angle between the input channels; d – diameter of the output hole.

In the proposed design of the sprayer, the liquid is dispersed in two stages: first utilizing a compressed air jet and then increasing the rotation speed and centrifugal forces in the conical chamber [11].

3 Results and discussion

Experimental studies have been carried out to determine optimum design parameters and dispersion composition of aerosols for different operating regimes of the vortex pneumatic sprayer [12]. The droplet size of aerosols D (Table 1) is determined at different pressure ratios in the pneumatic and liquid channels $\frac{P_{air}}{P_{liq}} = \mu$ and by changing the angles between the input channels α.

Based on the averaged size values of aerosols D_{sr}, we constructed the dependencies (Fig. 2) against different pressure ratios in the pneumatic and liquid channels μ for different angles between the input channels α.

Table 1. Experimental sizes of aerosol droplets D at different values of μ and α.

$\frac{P_{air}}{P_{liq}} = \mu$	D, μm, at $\alpha = 45°$	D, μm, at $\alpha = 60°$	D, μm, at $\alpha = 90°$
0.5	86	78	69
1.0	71	64	54
1.5	57	50	40
2.0	44	36	26
2.5	35	26	15
3.0	28	20	8

Figure 2 shows that when the pressure ratio μ increases, the droplet size of the aerosols decreases. It is caused by increasing the efficiency of pneumatic dispersion of disinfectant liquid due to rising the pressure of compressed air in the pneumatic channel [13].

The flow rate of disinfectant liquid Q (Table 2) was determined at different pressure ratios in the pneumatic and liquid channels $\frac{P_{air}}{P_{liq}} = \mu$ and at angles between the input channels α during experiments.

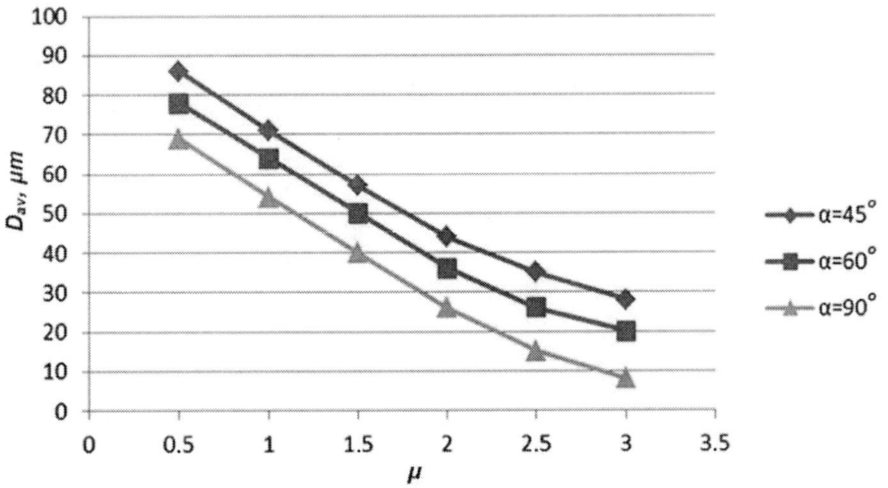

Fig. 2. Dependence of the average droplet size of aerosols against different pressure ratios in the pneumatic and liquid channels μ under changing the angles between the input channels α

Table 2. Experimental values of flow rate Q for disinfectant liquid at different values of μ and α.

$\frac{P_{air}}{P_{liq}} = \mu$	Q, ml/min, at $\alpha = 45°$	Q, ml/min, at $\alpha = 60°$	Q, ml/min, at $\alpha = 90°$
0.5	0.32	0.32	0.29
1.0	0.31	0.30	0.27
1.2	0.29	0.27	0.24
1.4	0.26	0.24	0.21
1.6	0.24	0.21	0.17
1.8	0.21	0.17	0.09
2.0	0.17	0.12	0.05

Figure 3 represents the dependencies of the obtained flow rates of disinfectant liquid Q from different pressure ratios in the pneumatic and liquid channels μ and angles between the input channels α.

Figure 3 shows that as the pressure ratio μ increases, the flow rate of disinfectant liquid decreases. It has to do with the partial closing of the channel for supplying disinfectant liquid by increasing the pressure of compressed air in the pneumatic channel [5; 14; 15].

The results showed that the optimum design parameters of the sprayers are the following: the angle between the pneumatic and liquid input channels α is 90°; the diameter of the output hole d is 2.0 mm.

The developed design of the vortex centrifugal sprayer provides controlling the dispersion of the spray at a constant flow rate, uniform covering the surface to be treated, increasing the disinfection efficiency.

4 Conclusion

A new design of a vortex pneumatic sprayer was developed, which ensures obtaining monodispersed aerosols and reliable experimental data affecting the dispersion quality of disinfectant liquids.

The design and technological dependencies of the vortex pneumatic sprayer were determined, which form aerosols with defined dispersion characteristics under different operating regimes.

The graphical dependencies for the droplet size of the aerosol and the flow rate of disinfectant liquid from the technological parameters of the vortex pneumatic sprayer are constructed, providing the operating regimes with the required dispersion.

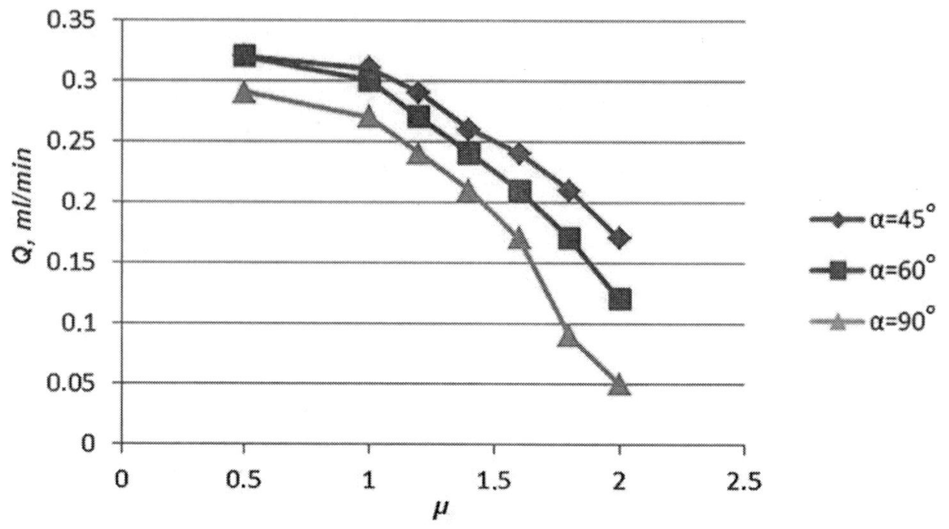

Fig. 3. Dependence of disinfectant flow rate Q through the sprayer against different pressure ratios in the pneumatic and liquid channels μ under changing angles between the input channels α.

References

1. B. L. Ivanov, B. G. Ziganshin, A. V. Dmitriev, G. V. Pikmullin, A. A. Mustafin, Droplet size of virocide disinfectant liquid from vortex injector sprayer under different operating conditions, *Engineering for rural development: 20th International Scientific Conference* pp 564–571 doi: 10.22616/ERDev.2021.20.TF122 (Jelgava: Latvia University of Life Sciences and Technologies, 2021)
2. I. Galiev, C. Khafizov, R. Khusainov, E. Galimov Optimization of tractor operation level, *Engineering for Rural Development. 20. Ser. «20th International Scientific Conference Engineering for Rural Development, ERD 2021 – Proceedings»* pp 38–42 (2021)
3. M. Yu. Kostenko, R. V. Beznosyuk, I. N. Goryachkina, P.A. Kostychev, Study of the operation of a hot fog generator during the processing of stalks, *Bulletin of the Ryazan State Agrotechnological University* **4 (44),** 87–92 (2019)
4. G. K. Rembalovich, A. I. Ryazantsev, M. Yu. Kostenko, Study of the trajectories of the movement of drops of a sprinkling machine, *Bulletin of the Ryazan State Agrotechnological University. P. A. Kostychev* **4 (40),** 138–142 (2018)
5. J. C. Beck, A. P. Watkins, On the development of a spray model based on drop-size moments, *Proc. R. Soc.* A, **459 (2034),** 1365–1394 (2003)
6. B. L. Ivanov, B. G. Ziganshin, R.F. Sharafeev, I. R. Sagbiev, The theory of liquid atomization by injectors, *Bulletin of Kazan State Agrarian University* **14-2 (53),** 95–99 (2019)
7. L. M. Nuriev, S. M. Yakhin, I. I. Aliakberov, I. G. Galiev, S. A. Sinitsky, Kinematics and parameters for spiral-helical machinery unit used for secondary tillage, *IOP Conference Series: Earth and Environmental Science* **488 (1),** 012051 (2020)
8. V. V. Kharkov, Mathematical modelling of thermolabile solutions concentration in vortex chamber, *J. Phys. Conf. Ser.* **980 (1),** 012006 (2018)
9. P. L. Lekomtsev, A. V. Savushkin, E. V. Dresvyannikova, A. M. Niyazov, Study of aerosol charging in electro-aerosol generator, *Journal of Applied Engineering Sciences* **7-2 (20),** 69–77 (2017)
10. B. L. Ivanov, B. G. Ziganshin, A. I. Rudakov, M. A. Lushnov, Evaluation of the distribution of drops of disinfectant liquid on the treated surface, *Bulletin of Kazan State Agrarian University* **14-3 (54),** 103–107 (2019)
11. P. S. Kuts, V. K. Samsonyuk, P. V. Akulich, Investigation of the hydrodynamics of vortex pneumatic sprayers, *Journal of Engineering Physics and Thermophysics* **74 (4)** 978–981 (2001)
12. D. M. Balakhanov, Metrological support for measurements of the disperse parameters of aerosols, suspensions, and powdered materials. *Al'Manac of Modern Metrology* **1 (6),** 87–97 (2016)
13. B. H. Akhalaia, Y. K. Shogenov, The turbulent airstream influence on the seeding quality, *Bulletin of the Russian Agricultural Science* **1,** 54–57 (2018)
14. N. Barai, N. Mandal, Breakup versus coalescence of closely packed fluid drops in simple shear

flows, *International Journal of Multiphase Flow* **111,** 1–15 (2019)
15. I. N. Goryachkina, O. A. Teterina, M. Yu. Kostenko, Installation for applying aerosol humates in the flow of agricultural products, *Bulletin of the RESKH* **4 (29),** 124–128 (2017)

AUTHOR INDEX